The Ultimate Digital Study Tool

Self-Test

A comprehensive quizzing section lets you test your ability to identify anatomical structures in a simulated practical exam. Additional quizzes provide review of structure functions and features.

Diagnostics displayed throughout the test track your progress.

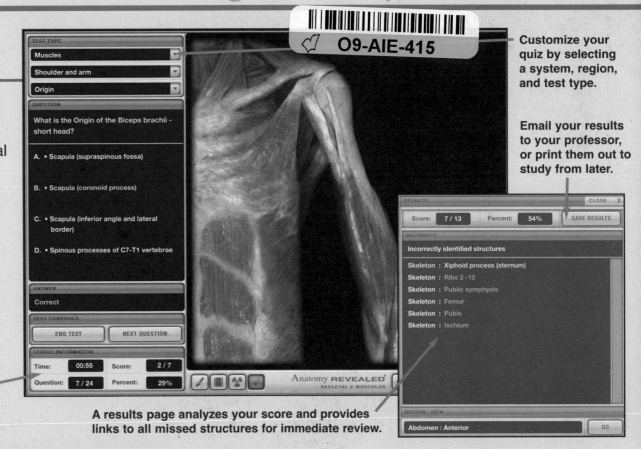

TEST TYPE
Muscles
Shoulder and arm
Origin

QUESTION
What is the Origin of the Biceps brachii - short head?

A. • Scapula (supraspinous fossa)

B. • Scapula (coronoid process)

C. • Scapula (inferior angle and lateral border)

D. • Spinous processes of C7-T1 vertebrae

ANSWER
Correct

TEST CONTROLS
| END TEST | NEXT QUESTION |

STATUS INFORMATION
| Time: | 00:55 | Score: | 2 / 7 |
| Question: | 7 / 24 | Percent: | 29% |

Anatomy REVEALED
SKELETAL & MUSCULAR

O9-AIE-415

Customize your quiz by selecting a system, region, and test type.

Email your results to your professor, or print them out to study from later.

RESULTS — CLOSE X
| Score: | 7 / 13 | Percent: | 54% | SAVE RESULTS |

INCORRECT
Incorrectly identified structures

Skeleton : Xiphoid process (sternum)
Skeleton : Ribs 2 -12
Skeleton : Public symphysis
Skeleton : Femur
Skeleton : Pubis
Skeleton : Ischium

REGION - VIEW
| Abdomen : Anterior | GO |

A results page analyzes your score and provides links to all missed structures for immediate review.

Animation

Compelling animations help clarify anatomical relationships or explain difficult physiological concepts.

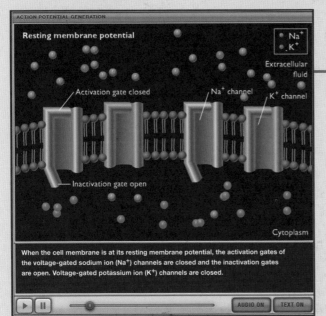

ACTION POTENTIAL GENERATION

Resting membrane potential

• Na$^+$
• K$^+$

Extracellular fluid

Activation gate closed

Na$^+$ channel

K$^+$ channel

Inactivation gate open

Cytoplasm

When the cell membrane is at its resting membrane potential, the activation gates of the voltage-gated sodium ion (Na$^+$) channels are closed and the inactivation gates are open. Voltage-gated potassium ion (K$^+$) channels are closed.

AUDIO ON   TEXT ON

Pin and label important structures.

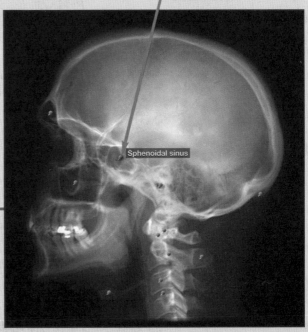

Sphenoidal sinus

Imaging

Labeled X-ray, MRI, and CT images help you learn to recognize key anatomical structures as seen through various medical imaging techniques.

Mader's

UNDERSTANDING

HUMAN ANATOMY & PHYSIOLOGY

SIXTH EDITION

Susannah Nelson Longenbaker

Columbus State Community College, Columbus, OH

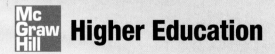

Boston   Burr Ridge, IL   Dubuque, IA   New York   San Francisco   St. Louis
Bangkok   Bogotá   Caracas   Kuala Lumpur   Lisbon   London   Madrid   Mexico City
Milan   Montreal   New Delhi   Santiago   Seoul   Singapore   Sydney   Taipei   Toronto

Higher Education

MADER'S UNDERSTANDING HUMAN ANATOMY & PHYSIOLOGY, SIXTH EDITION

Published by McGraw-Hill, a business unit of The McGraw-Hill Companies, Inc., 1221 Avenue of the Americas, New York, NY 10020. Copyright © 2008 by The McGraw-Hill Companies, Inc. All rights reserved. No part of this publication may be reproduced or distributed in any form or by any means, or stored in a database or retrieval system, without the prior written consent of The McGraw-Hill Companies, Inc., including, but not limited to, in any network or other electronic storage or transmission, or broadcast for distance learning.

Some ancillaries, including electronic and print components, may not be available to customers outside the United States.

⊕ This book is printed on recycled, acid-free paper containing 10% postconsumer waste.

1 2 3 4 5 6 7 8 9 0 DOW/DOW 0 9 8 7 6

ISBN 978–0–07–294583–6
MHID 0–07–294583–4

Publisher: *Michelle Watnick*
Senior Developmental Editor: *Kathleen R. Loewenberg*
Marketing Manager: *Lynn M. Breithaupt*
Project Manager: *April R. Southwood*
Lead Production Supervisor: *Sandy Ludovissy*
Senior Media Project Manager: *Tammy Juran*
Lead Media Producer: *John J. Theobald*
Senior Coordinator of Freelance Design: *Michelle D. Whitaker*
Cover/Interior Designer: *Elise Lansdon*
(USE) Cover Image: *©Mike Powell / Stone+*
Senior Photo Research Coordinator: *Lori Hancock*
Photo Research: *Connie Mueller*
Compositor: *Techbooks*
Typeface: 10/12 *Giovanni Book*
Printer: *R. R. Donnelley Willard, OH*

The credits section for this book begins on page 470 and is considered an extension of the copyright page.

Library of Congress Cataloging-in-Publication Data

Longenbaker, Susannah Nelson.
  Mader's understanding human anatomy & physiology / Susannah Nelson Longenbaker.–6th ed.
    p. cm.
  Rev. ed. of: Understanding human anatomy & physiology / Sylvia S. Mader. 5th ed. c2005.
  Includes index.
  ISBN 978–0–07–294583–6 — ISBN 0–07–294583–4 (hard copy : alk. paper)
    1. Human physiology. 2. Human anatomy. I. Mader, Sylvia S. Understanding human anatomy & physiology. II. Title. III. Title: Mader's understanding human anatomy and physiology. IV. Title: Understanding human anatomy & physiology.
QP34.5.M353 2008
612–dc22
                                                                    2006036985

www.mhhe.com

BRIEF CONTENTS

CONTENTS

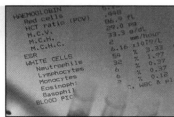
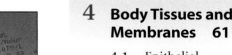
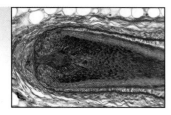
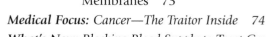

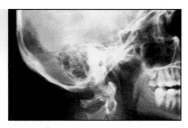
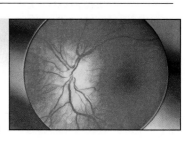
PART 3
Integration and Coordination

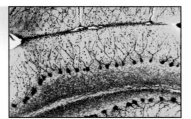

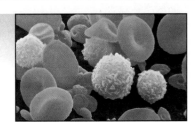

PART 4
Maintenance of the Body

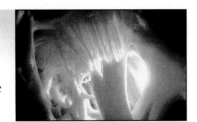

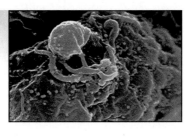
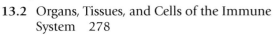

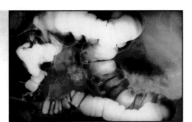
PART 5
Reproduction and Development

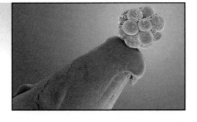
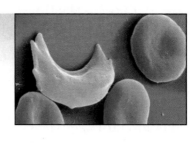

About the Authors

Susannah Nelson Longenbaker After acquiring a B.S. in Biology from St. Mary's College (Notre Dame, IN), and an M.S. in Physiology from Ohio State University, Sue Longenbaker began her educational career at Columbus State Community College, teaching biology and human anatomy and physiology. During her tenure there, she accepted two promotions, and co-authored and received several grants. She was the recipient of both the Columbus State Community College Distinguished Teaching Award and *Ohio Magazine's* Excellence in Education Award, all while authoring four different lab manuals and serving as a contributing author on the fifth edition of *Understanding Human Anatomy and Physiology*. Sue continues to teach at the college and loves every minute of it. She feels privileged to pass along the gift of learning that she received from many excellent teachers.

Sue is honored to be named the author to continue Dr. Mader's work on this textbook and looks forward to hearing suggestions or comments from both instructors and students alike. Feel free to contact her at the following address:

Sue Longenbaker
Department of Biological/Physical Science
Columbus State Community College
Columbus, OH 43215
(614) 287-2430

To Bill, and to my parents and family, for their love and support.

Dr. Sylvia Mader In her 20-year career with McGraw-Hill, Dr. Sylvia Mader has written an impressive collection of textbooks. Aside from five editions of *Understanding Human Anatomy and Physiology*, she is the author of *Biology*, ninth edition; *Human Biology*, tenth edition, and *Inquiry into Life*, twelfth edition, through which she has successfully helped innumerable students learn biology as well as human anatomy and physiology.

Dr. Mader's interest in anatomy and physiology began when she took courses at the Medical School of St. Andrews University, in Scotland, during her junior year abroad. As a fledgling faculty member, she was called upon to teach a variety of courses, among them human anatomy and physiology. As a textbook writer she discovered that the teaching and learning techniques she so successfully used in the classroom were also appropriate for her biology texts and then later for her anatomy and physiology text. Dr. Mader's direct writing style and carefully constructed pedagogy combine to provide students with unmatched opportunities to master the basics of biology and anatomy and physiology.

PREFACE

Three years ago, I accepted a challenge to walk in some pretty big footsteps when I was asked to revise the sixth edition of Dr. Sylvia Mader's *Understanding Human Anatomy and Physiology*. I was thrilled and honored by the invitation, but also just a bit intimidated by the prospect of what lay ahead. I soon realized, however, that Sylvia and I share the same vision for this book. We both want the book to appeal to a wide audience, from students studying the traditional allied health fields to non-majors who are interested in the function of their bodies. We both recognize the importance of the text being clear, direct, and user-friendly, with objectives that are achievable by students with no previous training in anatomy and physiology. Finally, understanding the value of "pictures are worth a thousand words," we both insist on excellent illustrations.

I believe that this sixth edition successfully demonstrates our shared vision. I think you will see that the style of the text is still the crisp, concise writing you've come to expect from Sylvia, and it remains a book that was originally written for the one-semester course. Each chapter is enriched with clinical applications to engage student interest and medical terminology to increase student confidence. Throughout the text, the new "Content Check-Ups" allow students to test their comprehension before moving on to a new section in the chapter.

I've drawn from over 20 years of experience as a teaching professional to add classroom-tested features to this edition. Illustrations using simple analogies (like the "box with a lining" for serous membranes in Chapter 1) work in my classroom because of their relevance to everyday objects or examples. My students love to talk about anatomy, physiology, and pathophysiology examples they've seen in the media or come across on the job. The new "Focus on Forensics" readings are simplified anatomical and physiologic explanations for real-life criminal investigations, and were written with these students in mind. The popular "Medical Focus" and "What's New" features have been updated to reflect advancements in technology and clinical practice. All of the important homeostasis sections in each chapter now include new art, rich in necessary detail and realism.

I've been a Mader fan through many editions of several different textbooks. Like me, Sylvia was a classroom instructor for a number of years before she wrote her first college textbook. We are both passionate about educating students. It's been a delight to work with Sylvia and I look forward to future successful collaborations. I know that this text will help you, the instructor, to engage and excite your students about the fascinating study of the human body. It works for me, and it will work for you too!

Acknowledgements

I have a fabulous team at McGraw-Hill, and each deserves recognition and thanks for their hard work: Publisher Michelle Watnick, Sponsoring Editor Jim Connely, Marketing Manager Lynn Breithaupt, Project Manager April Southwood, Photo Research Coordinator Lori Hancock, Design Coordinator Michelle Whitaker, Production Supervisor Sandy Ludovissy, Media Producer Jake Theobald, Media Project Manager Tammy Juran, and Project Coordinator Tracy Konrady. Finally, a special thanks to my Developmental Editor, Kathy Loewenberg, for her kindness, encouragement, and never-failing good humor.

Reviewers

The sixth edition of *Understanding Anatomy and Physiology* would not have been possible without the valuable input from instructors who have provided comments and suggestions in their reviews. I would like to acknowledge the following individuals and thank all of them for contributing their time and talents. The observations and detailed recommendations they shared with me have served as a guideline for improvements to chapter content and illustrations. I deeply appreciate their help. My deepest thanks to Lynn Atkins and Deb McCool. Their careful reading helped to correct errors and improve the text. As an author, it is very comforting to know you have such skilled and talented scientists reviewing the content. Ladies, thank you both.

Susannah Longenbaker

Shyla Akkaraju
Bronx Community College of CUNY

Lynn Atkins
University of Maine

Sharon Rohr Barnewall
Columbus State Community College

Wendy D. Bircher
San Juan College

Patricia Bostwick
Florence-Darlington Technical College

Barbara M. Burckart
Columbia College, Chicago

Pamela A. Cole
Shelton State Community College

Amy L. Cooper
San Juan College

Jessica Elliott
Holmes Community College— MS Grenada Center

Bridget A. Falkenstein
Sierra College

Ellen E. Faszewski
Wheelock College

Ralph E. Ferges
Palomar College

Michael J. Harman
North Harris College

Duane P. Hill
Newbury College

Nathan R. Hoverkamp
Central Oregon Community College

Gary W. Hunt
Tulsa Community College

Edward W. Johnson
Central Oregon Community College

Thomas Jordan
Pima Community College

Jason LaPres
North Harris College

Ken Malachowsky
*Florence-Darlington
Technical College*

Sara L. Parr
*Madison Area Technical
College*

D. B. Pelletier
*Green River Community
College*

David Quadagno
Florida State University

Stephanie A. Shumate-Vance
Jefferson Community College

Mark L. Wygoda
McNeese State University

Tell Me About This Book

Understanding Anatomy and Physiology features the **perfect writing style for the one-semester course**. The flow of the text is logical and accessible without being overly "chatty" and consistently makes use of relevant examples and analogies.

"Teaching and learning work well when the book and the material taught in the course go together like a hand and a glove. That's the way this textbook works for our Biology 110 course—it's a well-written and well-organized textbook that suits the needs of our course and students. Its pictures and diagrams complement the concepts and it's a good resource for our students to take with them to higher-level courses. Students have mentioned that they have kept this book when they go on to Anatomy and Physiology I and II because it discusses concepts simply."

—*Patty Bostwick, Florence Darlington Technical College*

Easy-to-understand art covers what's important but leaves out unnecessary, confusing detail. Good examples of this are the homeostasis illustrations; instead of showing lots of various colored arrows and boxes with explanations, these simple visual pieces get the message across beautifully. Another example is stepped-out art, which shows key stages of an illustration identified by numbered circles. This type of explanation builds comprehension sequentially. (New with this edition—numbered circles are now in bright yellow to better highlight the steps.)

Built-in study aids such as the new "Content Check-Up" segments allow students to test themselves over major

sections of text before moving on to new material. When students add these review questions all together, they have a practice quiz over the entire chapter. Key terms are divided into basic and clinical terms and include page references. Two levels of additional questions, along with exercises that reinforce medical terminology, are also included with every chapter.

Unsurpassed clinical coverage is evident all through this text. "What's New," "Medical Focus," and new "Focus on Forensics" readings relate the very latest research and/or developments in applied aspects of anatomy and physiology to important concepts in the text. New "Focus on Forensics" readings engage students in problem-solving scenarios that challenge them to use, and expand upon, their recently acquired knowledge. Examples of new topics in the "What's New" and "Medical Focus" boxes include "Body Art—Buyers Beware," "Spinal Cord Injuries," "To Botox or Not to Botox," and "Avian Flu, Potential Pandemic?"

"*Understanding Anatomy and Physiology* is excellent and very well-suited for students taking an introductory course in human anatomy and physiology. Not only does it provide a clear and easy-to-read primer for the non-major, but it is also visually engaging with useful, professionally prepared graphics. Issues of clinical significance are included to illustrate key pathologies and to create connections for the reader to everyday issues of health and disease. The 'What's New' readings feature some of the latest advances in medical technology."

—*Duane P. Hill, Newbury College*

What's New in this Edition?

Understanding Human Anatomy and Physiology has undergone **a thorough revision**. Every chapter has been improved to include new material, pedagogy, examples, and data based on the latest research available and valued reviewer feedback.

Content Check-Ups are small, nonobtrusive review quizzes that have been added after major sections of the text to help students test their comprehension as they are reading. These questions are also excellent study tools for chapter tests.

Focus on Forensics readings are designed to increase students' curiosity and interest through the perspective of forensics. These new boxes are rich in discoveries and methods of solving problems with knowledge of anatomy and physiology. Just a few of the topics include

"DNA Fingerprinting," "Retinal Hemorrhage in Shaken Baby Syndrome," "Urinalysis," and "Skeletal Remains."

New material relevant to various careers in health-related disciplines has been added to this edition. Careful thought was given to choosing content that students pursuing careers in massage therapy, emergency medical technology, forensic nursing, respiratory therapy, X-ray technology, etc., could identify with and use in their respective disciplines.

Updates to the art include new chapter-opening photos, better scanning electron micrographs, all new systems illustrations, improved stepped-out art, and changes to line art and photos based on recent data and/or reviewer input.

The sixth edition sports an **entirely new design.** Fresh, bright openings to every chapter, and newly designed tables and interior colors give the sixth edition a modern, contemporary feel.

Teaching and Learning Supplements

McGraw-Hill offers various tools and technology products to support the sixth edition of *Understanding Anatomy and Physiology.* Students can order supplemental study materials by contacting their campus bookstore. Instructors can obtain teaching aids by calling the McGraw-Hill Customer Service Department at 1-800-338-3987, visiting our A&P catalog at www.mhhe.com/ap, or contacting their local McGraw-Hill sales representative.

Text Website

The ARIS website that accompanies this textbook includes tutorials, animations, practice quizzes, helpful Internet links, and more—a whole semester's worth of study help for students. Instructors will find a complete electronic homework and course management system where, with just a few clicks of the mouse, they can create and share course materials and assignments with colleagues. Instructors can also edit questions, import their own content, and create announcements and/or due dates for assignments. ARIS offers automatic grading and reporting of easy-to-assign homework, quizzes, and testing.

Check out www.aris.mhhe.com, select your subject and textbook, and start benefiting today!

New Online Presentation Center for Instructors

Build instructional materials where-ever, when-ever, and how-ever you want! The Presentation Center, part of the ARIS Website, is a digital library containing assets such as photos, artwork, animations, PowerPoints, and other media resources that can be used to create customized lectures, visually enhance tests and quizzes, and design compelling course websites or attractive printed support materials.

Nothing could be easier!

Accessed from your textbook's ARIS website, the Presentation Center's dynamic search engine allows instructors to explore by discipline, course, textbook chapter, asset type, or keyword. Simply browse, select, and download the files needed to build engaging course presentations. All assets are copyrighted by McGraw-Hill Higher Education but can be used by instructors for classroom purposes. The visual resources included in this collection are located easily within each chapter:

- **Art**—Full-color digital files of all illustrations in the book can be readily incorporated into lecture presentations, exams, or custom-made classroom materials. In addition, all files are pre-inserted into blank PowerPoint slides for ease of lecture preparation.
- **Photos**—The photo collection contains digital files of instructionally significant photographs from the text that can be reproduced for multiple classroom uses.
- **Tables**—Every table that appears in the text has been saved in electronic form for use in classroom presentations and/or quizzes.

In addition to the content found within each chapter, the Presentation Center contains the following multimedia instructional materials:

- **Active Art**—Active Art consists of art files that have been converted to a format that allows the artwork to be edited inside of PowerPoint. Each piece can be broken down to its core elements, grouped or ungrouped, and edited to create customized illustrations.
- **Animations**—Numerous full-color animations illustrating physiological processes are also provided. Harness the visual impact of processes in motion by importing these files into classroom presentations or online course materials.
- **Lecture Outlines**—Specially prepared custom outlines for each chapter are offered in easy-to-use PowerPoint slides.

Instructor's Testing and Resource CD

This cross-platform CD-ROM provides a wealth of resources for the instructor. One of the supplements featured on this CD is EZ Test, a flexible and easy-to-use electronic testing program. This program allows instructors to create tests from book-specific items and accommodates a wide range of question types, including the option for instructors to add their own questions. Multiple versions of the test can be created and any test can be exported for use with course management systems such as WebCT, BlackBoard, or PageOut. The *Instructor's Manual* is also included with this CD.

Anatomy & Physiology | REVEALED® Student Tutorial

 AP|REVEALED® is a unique multimedia study aid designed to help you learn and review human anatomy using digital cadaver specimens. Dissections, animations, imaging, and self-tests all work together as an exceptional tool for the study of structure and function.

The AP|REVEALED® CD series includes:

Volume 1—Skeletal and Muscular Systems

Volume 2—Nervous System

Volume 3—Cardiovascular, Respiratory, and Lymphatic Systems

Volume 4—Digestive, Urinary, Reproductive, and Endocrine Systems

A new online version of APR includes the Integumentary System and expanded histology content.

Visit www.mhhe.com/aprevealed for more information.

Virtual Anatomy Dissection Review (available online or as a CD-ROM)

- This multimedia program contains high-quality cat dissection photographs that are correlated to illustrations and photos of human structures. The format makes it easy to identify and review cat anatomy and also relate the cat specimen to corresponding human structures.

Physiology Interactive Lab Simulations (Ph.I.L.S.)

This unique study tool contains 26 lab simulations that allow students to perform experiments without using expensive lab equipment or live animals. The easy-to-use interface offers students the flexibility to change the parameters of every lab experiment, with no limit to the number of times they can repeat experiments or modify variables. The power to manipulate each experiment reinforces key physiology concepts by helping students to view outcomes, make predictions, and draw conclusions.

"MediaPhys" Tutorial

This physiology study aid offers detailed explanations, high-quality illustrations, and animations to provide a thorough introduction to the world of physiology. MediaPhys is filled with interactive activities and quizzes to help reinforce physiology concepts that are often difficult to understand.

e-Instruction with CPS

The Classroom Performance System (CPS) is an interactive system that allows the instructor to administer in-class questions electronically. Students answer questions via handheld remote control keypads (clickers), and their individual responses are logged into a gradebook. Aggregated responses can be displayed in graphical form. Using this immediate feedback, the instructor can quickly determine if students understand the lecture topic or if more clarification is needed. CPS promotes student participation, class productivity, and individual student confidence and accountability. Specially designed questions for e-Instruction to accompany *Understanding Anatomy and Physiology* are provided through the book's ARIS website.

Transparencies

The set of transparency acetates that accompanies this text includes 200 full-color images identified by the author as the most useful figures to incorporate into lecture presentations.

Course Delivery Systems

In addition to McGraw-Hill's ARIS course management options, instructors can also design and control their course content with help from our partners WebCT, Blackboard, TopClass, and eCollege. Course cartridges containing website content, online testing, and powerful student tracking features are readily available for use within these platforms.

Understanding is the Key ...

What Sets This Book Apart?

- Students develop an understanding of basic concepts, laying the foundation for a working knowledge of structure and function, via an *accessible writing style*.

- Self-confidence increases as students master important medical terminology with *simply designed art and built-in study aids*.

- *Clinical applications* build on core principles and introduce students to anatomy and physiology put into practice.

Written with an Understanding of the One-Semester Course

The authors' writing style is more important than any other component of a textbook and must be appropriate for the level of the reader. *Understanding Human Anatomy and Physiology* is known for its engaging writing style and has always been written and designed for the one-semester course, not adapted from a two-semester textbook. Paragraph introductions, explanations, comparisons, and relevant, everyday examples are used with these students in mind.

How Hormones Function

Along with fundamental differences in structure, peptide and steroid hormones also function differently. Most peptide hormones bind to a receptor protein in the plasma membrane and activate a "second messenger" system (Fig. 10.2). The "second messenger" causes the cellular changes for which the hormone is credited. As an analogy, suppose you are the person in charge of a crew assigned to redecorate a room. As such, you stand outside the room and direct the workers inside the room. The workers clean, paint, apply wallpaper, etc. Like the "boss" in this analogy, the peptide hormone stays outside the cell and directs activities within. The peptide hormone, or "first messenger," activates a "second messenger"—the crew workers inside the cell. Common second messengers found in many body cells include **cyclic AMP** (made from ATP, and abbreviated cAMP) and calcium.

Ample use of analogies and examples connect to everyday life and help students relate to new concepts and ideas.

Content CHECK-UP!

1. The term for the expanded portions at the ends of a long bone is:
   a. diaphysis.
   c. periosteum.
   b. epiphysis.
   d. articular cartilage.
2. Which type of bone cell breaks down bone and deposits calcium into the blood?
   a. osteoblast
   c. osteoprogenitor
   b. osteocyte
   d. osteoclast
3. The term for a rounded opening through a bone is:
   a. foramen.
   c. trochanter.
   b. tuberosity.
   d. condyle.

New Content Check-Ups test students' comprehension as they read through the chapter. These mini-quizzes follow major sections of text, and are an excellent study tool.

Macro-to-micro presentation helps students make the connection between gross anatomy and microscopic anatomy.

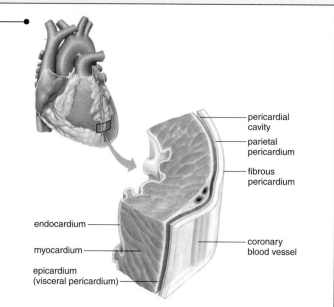

pericardial cavity

parietal pericardium

fibrous pericardium

coronary blood vessel

endocardium

myocardium

epicardium (visceral pericardium)

Figure 12.2 The coverings and wall of the heart. The heart wall has three layers, from deep to superficial: endocardium, myocardium, and epicardium.

Easy-to-Understand Art

The art for *Understanding Human Anatomy and Physiology* covers what's important, leaving out unnecessary detail. Clear, expertly drawn illustrations are paired perfectly with the surrounding text, enhancing the student's overall understanding of the material.

Stepped-out art in clearly marked, consecutive steps develops student comprehension.

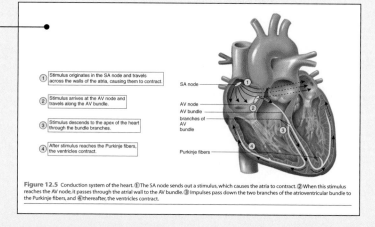

① Stimulus originates in the SA node and travels across the walls of the atria, causing them to contract.

② Stimulus arrives at the AV node and travels along the AV bundle.

③ Stimulus descends to the apex of the heart through the bundle branches.

④ After stimulus reaches the Purkinje fibers, the ventricles contract.

SA node

AV node

AV bundle

branches of AV bundle

Purkinje fibers

Figure 12.5 Conduction system of the heart. ①The SA node sends out a stimulus, which causes the atria to contract. ②When this stimulus reaches the AV node, it passes through the atrial wall to the AV bundle. ③Impulses pass down the two branches of the atrioventricular bundle to the Purkinje fibers, and ④thereafter, the ventricles contract.

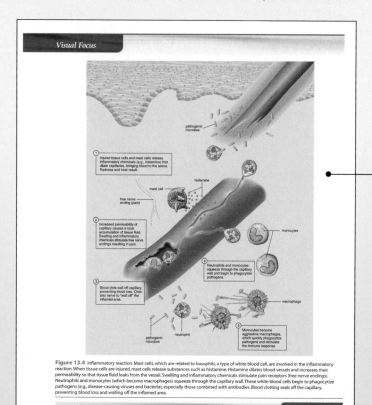

Visual Focus

pathogenic microbes

① Injured tissue cells and mast cells release inflammatory chemicals (e.g., histamine) that dilate capillaries, bringing blood to the scene. Redness and heat result.

histamine

mast cell

free nerve ending (pain)

② Increased permeability of capillary causes a local accumulation of tissue fluid. Swelling and inflammatory chemicals stimulate free nerve endings resulting in pain.

③ Blood clots wall off capillary, preventing blood loss. Clots also serve to "wall off" the inflamed area.

④ Neutrophils and monocytes squeeze through the capillary wall and begin to phagocytize pathogens.

monocytes

macrophage

pathogenic microbes

neutrophil

⑤ Monocytes become aggressive macrophages, which quickly phagocytize pathogens and stimulate the immune response.

Figure 13.4 Inflammatory reaction. Mast cells, which are related to basophils, a type of white blood cell, are involved in the inflammatory reaction. When tissue cells are injured, mast cells release substances such as histamine. Histamine dilates blood vessels and increases their permeability so that tissue fluid leaks from the vessel. Swelling and inflammatory chemicals stimulate pain receptors (free nerve endings). Neutrophils and monocytes (which become macrophages) squeeze through the capillary wall. These white blood cells begin to phagocytize pathogens (e.g., disease-causing viruses and bacteria), especially those combined with antibodies. Blood clotting seals off the capillary, preventing blood loss and walling off the inflamed area.

Visual Focus figures spotlight illustrations that warrant closer inspection.

Photomicrographs shown side-by-side with line art offer two different perspectives to students.

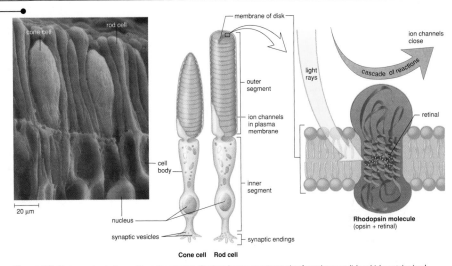

Figure 9.9 Photoreceptors in the eye. The outer segment of rods and cones contains stacks of membranous disks, which contain visual pigments. In rods, the membrane of each disk contains rhodopsin, a complex molecule containing the protein opsin and the pigment retinal. When retinal absorbs light energy, it splits, releasing opsin, which sets in motion a cascade of reactions that cause ion channels in the plasma membrane to close. Thereafter, nerve impulses go to the brain.

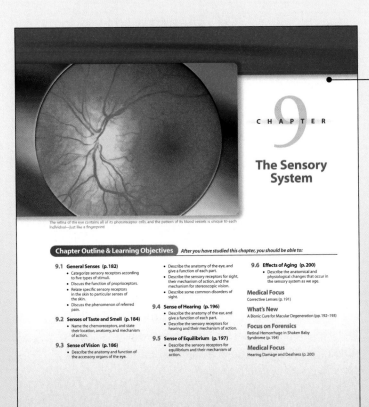

Dynamic photos give students a closer look inside the wonders of the human body through the technology of scanning electron micrographs.

C H A P T E R
9

The Sensory System

The retina of the eye contains all of its photoreceptor cells, and the pattern of its blood vessels is unique to each individual—just like a fingerprint!

Chapter Outline & Learning Objectives *After you have studied this chapter, you should be able to:*

9.1 General Senses (p. 182)
- Categorize sensory receptors according to five types of stimuli.
- Discuss the function of proprioceptors.
- Relate specific sensory receptors in the skin to particular senses of the skin.
- Discuss the phenomenon of referred pain.

9.2 Senses of Taste and Smell (p. 184)
- Name the chemoreceptors, and state their location, anatomy, and mechanism of action.

9.3 Sense of Vision (p. 186)
- Describe the anatomy and function of the accessory organs of the eye.

- Describe the anatomy of the eye, and give a function of each part.
- Describe the sensory receptors for sight, their mechanism of action, and the mechanism for stereoscopic vision.
- Describe some common disorders of sight.

9.4 Sense of Hearing (p. 196)
- Describe the anatomy of the ear, and give a function of each part.
- Describe the sensory receptors for hearing and their mechanism of action.

9.5 Sense of Equilibrium (p. 197)
- Describe the sensory receptors for equilibrium and their mechanism of action.

9.6 Effects of Aging (p. 200)
- Describe the anatomical and physiological changes that occur in the sensory system as we age.

Medical Focus
Corrective Lenses (p. 191)

What's New
A Bionic Cure for Macular Degeneration (pp. 192–193)

Focus on Forensics
Retinal Hemorrhage in Shaken Baby Syndrome (p. 194)

Medical Focus
Hearing Damage and Deafness (p. 200)

In addition to selected new terms, a summary, and two types of study questions, the end-of chapter material includes **Medical Terminology Exercises** that help reinforce the pronunciation and meaning of new medical terms.

MEDICAL TERMINOLOGY EXERCISE

Consult Appendix B for help in pronouncing and analyzing the meaning of the terms that follow.

1. orchidopexy (or″ki-do-pek'se)
2. transurethral resection of prostate (TURP) (trans″yū-re'thral re-sek'shun ov pros'tăt)
3. gonadotropic (go″nad-o-tröp'ik)

4. contraceptive (kon″truh-sep'tiv)
5. gynecomastia (jin″ě-ko-mas'te-uh)
6. hysterosalpingo-oophorectomy (his″ter-o-sal-ping'go-o″ahf-or-ek'to-me)
7. colporrhaphy (kol-por'uh-fe)
8. menometrorrhagia (men'o-me-tro-ra-je-uh)
9. multipara (mul-tip'uh-ruh)

10. balanitis (bal″uh-ni'tis)
11. seminoma (sem'ĭ-no'muh)
12. genitourinary (jen'ĭ-to-yū'rĭ-năr'e)
13. prostatic hypertrophy (pros-tat'ik hy'per-tro'fe)
14. azoospermia (ă-zo'o-sper'me-uh)

WEBSITE LINK

Visit this text's website at http://www.mhhe.com/maderap6 for additional quizzes, interactive learning exercises, and other study tools.

Atlas figures of the human body give students an additional anatomical study aid.

Clinical Connections Foster Understanding

The authors of *Understanding Human Anatomy and Physiology* recognize that students need to see applications of the important concepts and terminology presented in lectures. Years of teaching experience, combined with countless hours of research, have given Susannah Longenbaker a rich supply of examples, articles, and data to demonstrate the hands-on side of anatomy and physiology. Everyday scenarios, cutting-edge forensic methods, just-released discoveries, and new medical procedures all combine to present clinical connections that are unsurpassed among one-semester A & P textbooks.

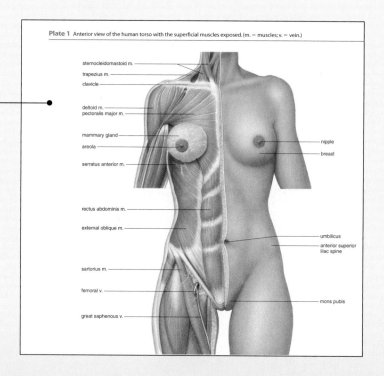

Plate 1 Anterior view of the human torso with the superficial muscles exposed. (m. = muscles; v. = vein.)

sternocleidomastoid m.
trapezius m.
clavicle
deltoid m.
pectoralis major m.
mammary gland
areola
serratus anterior m.
rectus abdominis m.
external oblique m.
sartorius m.
femoral v.
great saphenous v.
nipple
breast
umbilicus
anterior superior iliac spine
mons pubis

FOCUS on FORENSICS

Urinalysis

Since ancient times, urinalysis has been used to diagnose disease. As early as 600 B.C., Hindu physicians in India noted that the urine of a diabetic was sweet to the taste. In diabetes mellitus, blood glucose is abnormally high, either because insulin-secreting cells have been destroyed or because cell receptors don't respond to the insulin that is present. Thus, the filtrate level of glucose is extremely high, and the proximal convoluted tubule can't reabsorb it all. Tasting urine to diagnose diabetes persisted through the 1800s (thankfully, modern techniques have made it obsolete). Similarly, the great Greek physician Hippocrates studied and wrote about urinalysis, producing perhaps the first written work about renal failure. Hippocrates noted that shaking the urine of renal failure patients produced frothy bubbles at the surface of the sample. Today, we know that the froth is a sign of *proteinuria*, or protein in the urine. Proteinuria indicates that the glomerulus is more permeable than normal, and is an early indicator of chronic renal failure.

Today, the use of urinalysis has expanded beyond medical applications to include forensic diagnosis of drug abuse. Screening for illegal drug use is now mandated by federal and state agencies as a condition of employment, and most private employers now require it as well. Urinalysis can be court ordered if drug abuse is involved in the commission of a crime. The National Collegiate Athletic Association requires all student athletes to undergo testing, as do many high school athletic programs.

Urinalysis is not used to screen for the drugs themselves, but for drug metabolites—the breakdown products of drugs that are consumed or injected. Once in the body, drugs are metabolized by the liver and filtered by the kidney. Thus, metabolites will be present in the urine of a drug abuser. Two types of techniques can be used for metabolite detection. The first, a screening exam, involves

a test strip placed into freshly voided urine. The test strip contains monoclonal antibodies (see pages 292 and 294) specific for metabolites of street drugs. Strips can test for 12 or more different drugs at once, but most screen for 5 commonly abused drugs: marijuana, amphetamine, PCP, cocaine, and opiates such as heroin. Urine strip testing will give results within minutes, but certain legal, over-the-counter medications can give false-positive results. If a sample tests positive, it can be immediately sent for a sophisticated chemical analysis such as gas chromatography. Tracking long-term drug use may require using hair samples as well as urine samples because drug metabolites can be incorporated into the hair.

The urine specimen must be properly collected to avoid possible tampering by the individual being tested. Specimens can be altered by simple dilution with sink or toilet water, or by contaminating the sample with any number of additives. Bleach, drain cleaner, soft drinks, etc., can be used, as well as products specifically sold for the express purpose of "helping to beat a drug screen." Drug abusers may also attempt to substitute the urine of a "clean" individual for their own. Tampering may be prevented by simply requiring the presence of a witness at all times while the sample is being collected and stored. In addition, proper documentation must accompany any urine sample. Chain-of-custody forms record each step of the handling of a specimen, from collection to disposal. This provides proof of everything that happens to the specimen, and prevents the possibility of the specimen from being rejected as evidence in court proceedings.

The goal of any forensic urinalysis testing program should go beyond mere *detection* of illegal drug use. The more important aim of screening should be *intervention*. With proper treatment and counseling, drug abusers can overcome addiction and lead healthier and more productive lives.

New Focus on Forensics readings combine modern intrigue with a demonstrated need for understanding how the human body works. Captivating scenarios grab students' interest and detailed explanations illustrate how knowledge of anatomy and physiology is put to use in a practical setting.

Students planning to enter into any health-related career will appreciate the **Medical Focus** readings and their up-to-date, interesting topics.

MEDICAL FOCUS

Bariatric Surgery for Obesity

It's a final measure sought by increasing numbers of people: **bariatric surgery**, or surgical intervention with the specific goal of causing drastic weight loss. For many overweight people, years of dieting haven't worked. Name a diet plan, and they have tried it—sure, they lose weight only to regain it. Obesity is approaching epidemic levels in the United States, with more than 6 million adults classified as obese. In addition, pediatricians are especially concerned with the dramatic rise in childhood and teenage obesity.

Obesity is defined as a body mass index (BMI) greater than 35,[1] or a weight that is 41% or higher than the ideal weight for one's height. It's a physical and emotional challenge for the overweight patient. Obesity is a primary risk factor for hypertension, type II diabetes mellitus, atherosclerosis, stroke, coronary artery disease, and early heart attack. It has been linked to increased risk of breast, ovarian, uterine, and prostate cancer. Obese individuals suffer disrespect and ridicule from society and discrimination on the job. After years of struggling with the problem, many are willing to undergo surgery as a last-chance option.

However, reputable programs offering bariatric surgery have strict requirements for patients. To be admitted as a surgical candidate, the patient must be morbidly obese (BMI greater than 40; or greater than 100 pounds over ideal weight). Patients cannot have respiratory or cardiac problems that might complicate surgery. Most important, the patient must understand the risks of surgery as well as its benefits. Patients must also understand that lifestyle changes will be necessary even after successful surgery and recovery. Psychological and nutritional counseling is usually required before surgery to prepare the patient for new eating habits and ways of thinking about food. Follow-up counseling tracks the patient's progress in adapting.

The two most commonly used interventions are laparoscopic banding and the Roux-en-Y gastric bypass (Fig. 15D). A more conservative approach, laparoscopic banding, requires making a series of small incisions around the stomach and using an instrument called a laparoscope to illuminate structures in the abdomen. A band is placed around the stomach. Once tightened like a belt, the stomach is divided into a smaller upper chamber, which receives food, and a lower chamber that remains connected to the duodenum. The belt can later be tightened further, or removed if necessary. In Roux-en-Y gastric bypass, the top section of the stomach is cut free and stapled shut to make a pouch about the size of an egg. The small intestine is cut free between the duodenum and jejunum, and the jejunum is sewn to the end of the stomach pouch. Finally, the duodenal segment is sewn back to the jejunum (forming the Y-shaped branch for which the procedure is named). The lower stomach and duodenum remain healthy and continue to secrete digestive enzymes, but never receive food. Regardless of the approach used, the person will only be able to eat small amounts of food, but should feel full due to the small size of the stomach after surgery.

It's important to note that bariatric surgery comes with extremely serious potential complications. Postoperative bleeding and infection are risks of any surgical procedure. Blood clots in the legs can form during hospitalization, causing pulmonary embolism or stroke. After gastric bypass, staple lines in the stomach can leak. In rare cases, the connection between the stomach pouch and jejunum narrows, requiring additional corrective surgery. Worst of all, a small percentage of patients die during the surgery itself. After surgery, vitamin and mineral deficiencies are possible.

Further, bariatric surgery offers no guarantees of permanent weight loss—patients can in fact regain any weight that is lost, even with a drastically smaller stomach. However, with proper nutrition and behavioral changes, bariatric surgery can result in dramatic weight loss and improvement to health.

[1]To calculate BMI, use the following formula: $\dfrac{\text{weight in pounds}}{\text{(height in inches)} \times \text{(height in inches)}} \times 703.$
BMI is not always accurate in determining obesity. Other factors such as percent body fat may also need to be used.

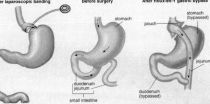

After laparoscopic banding   **Before surgery**   **After Roux-en-Y gastric bypass**

stomach

stomach (bypassed)

pouch

jejunum

duodenum
jejunum

duodenum (bypassed)

small intestine

Figure 15D (*far left figure*) A laparascopic band is placed around the stomach and then tightened to decrease stomach size. (*center*) The normal movement of food prior to surgery. (*right*) The flow of food after Roux-en Y gastric bypass surgery.

What's New

Swallowing a Camera?

"I really just don't even want to eat anymore. My gut just hurts so badly, and then I can't even control the diarrhea. I moved out of the dorm last year because I was so embarrassed to always be in the bathroom." The patient, a 20-year-old college student, is pale and underweight. A complete blood count reveals that she is anemic and her white blood cell count is slightly elevated. She identifies the lower right abdominal quadrant as the source of her pain. The pain has persisted, off and on, for more than six months.

Actual size

INSIDE THE *Pillcam*

1. **Optical dome**
2. **Lens holder**
3. **Lens**
4. **Illuminating LEDs (Light Emitting Diode)**
5. **CMOS (Complementary Metal Oxide Semiconductor) imager**
6. **Battery**
7. **ASIC (Application Specific Integrated Circuit) transmitter**
8. **Antenna**

Figure 15B The camera pill is about the size of a large vitamin. A belt pack receives video images, which are later viewed by a physician.

To diagnose troenterologists have previously and barium X r with a video cam upper GI tract: th Colonoscopy is tube is inserted

---

What's New

Avian Flu—Potential Pandemic?

"Over there, over there, Send the word, send the word over there—
That the Yanks are coming, The Yanks are coming . . . And we
won't come back till it's over, Over there!"
—"Over There," George M. Cohan

In April 1918, American soldiers who proudly marched off to World War I battlefields "over there" (referring to Europe, in the words of the popular patriotic ballad of the time) took an unwanted hitchhiker with them. A particularly virulent strain of influenza A, nicknamed the "Spanish flu" because of the huge numbers of victims in that country, was carried to Europe. It quickly became a worldwide epidemic, or **pandemic**, causing an estimated 50 million deaths. Spanish flu caused twice as many deaths among the soldiers as those that resulted from combat. Unlike previous strains of influenza, this virus was capable of striking down young, otherwise healthy people as well as the elderly and infirm. This particularly savage virus is once again a concern because of characteristics it shares with a modern-day virus: the avian influenza, or "bird flu" virus.

Influenza viruses, like all viruses, have two basic parts. The core of the virus is genetic material, which can be either DNA or RNA. Influenza A viruses are RNA viruses. Their genetic material is covered by a protective protein coat called a **capsid**. An additional third piece, called the lipid envelope, contains protein spikes that allow the virus to attach only to its own specific kind of host cell. The two types of protein spikes, abbreviated "H" (for hemagglutinin) and "N" (for neuraminidase) allow each type of influenza virus to be categorized. There are 16 known types of H proteins, and 9 known types of N proteins. Thus, many different combinations are possible, meaning there are many different ways for viruses to attach to different kinds of host cells. Each different combination of H and N represents a new form of flu virus. Spanish flu virus is denoted "H1N1," and bird flu, avian influenza A, is denoted "H5N1." The ever-changing nature of influenza viruses explains why each year a new—and different—flu shot is needed.

Avian flu is highly contagious among birds and causes the death of entire flocks of wild and domesticated birds in a very short period. Yet, if avian influenza, and other animal viruses, were limited to infecting birds and/or other animals, avian and other animal influenzas would be largely just a veterinary problem. However, the genetic material of influenza virus is well known for its high mutation rate. Some of these mutations affect the structure of the protein spikes, so that a virus that previously could only infect a particular animal species can "jump species" to infect humans. In the case of

H5N1, the jump has already occurred. More than 130 cases of H5N1 infection in humans were reported between January 2004 through January 2006 in Southeast Asia and China. Most infections occurred when the victims came into contact with infected birds, but a troubling few seem to have been due to human-to-human transmission. Scientists studying H1N1, the deadly Spanish flu virus of 1918, have noted that it probably also originated as a "bird flu" virus that mutated to become transmissible among human beings.

Up until now, H5N1 viruses have not usually infected human beings, so the recent human infection cases have public health officials especially concerned. Because the human population has not had ongoing exposure to this virus, little or no naturally immunity is thought to exist among humans. During recent outbreaks, avian flu has killed more than half of those infected with the virus, and most of its victims were previously healthy children and young adults—a fact eerily reminiscent of the 1918 Spanish flu pandemic.

Researchers and public health officials are trying to be prepared, just in case the virus achieves widespread human infection. Fortunately, thanks to the advance of modern technology, we will have a few more weapons in our antiviral arsenal than our grandparents or great-grandparents had back in 1918. Vaccinations, drugs that prevent viral replication, antibotics for secondary infections such as pneumonia, as well as communication tools like radio, television, and the Internet—all will hopefully create a different outcome if a pandemic threatens the world again.

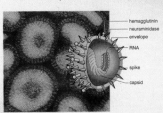

hemagglutinin
neuraminidase
envelope
RNA
spike
capsid

Figure 13C Influenza virus is an RNA virus, with protein spikes on its outer envelope that make it specific to one kind of host cell.

What's New boxes enlighten and educate readers on new trends, data, discoveries, etc., that are relevant to anatomy and physiology.

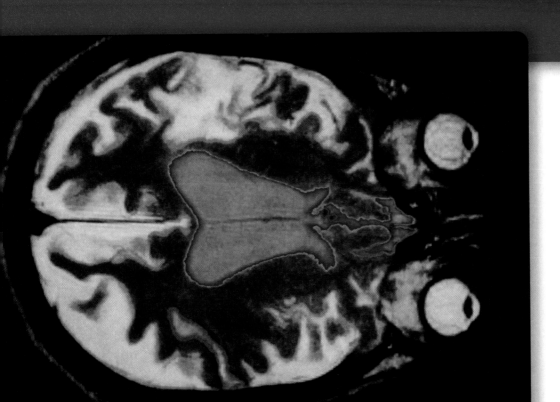

PART I

CHAPTER

1

Organization of the Body

Magnetic resonance imaging of the head in a transverse section. MRI allows clinicians to view soft tissues such as the brain and eyes, and to locate abnormalities like blood clots or tumors.

Chapter Outline & Learning Objectives
After you have studied this chapter, you should be able to:

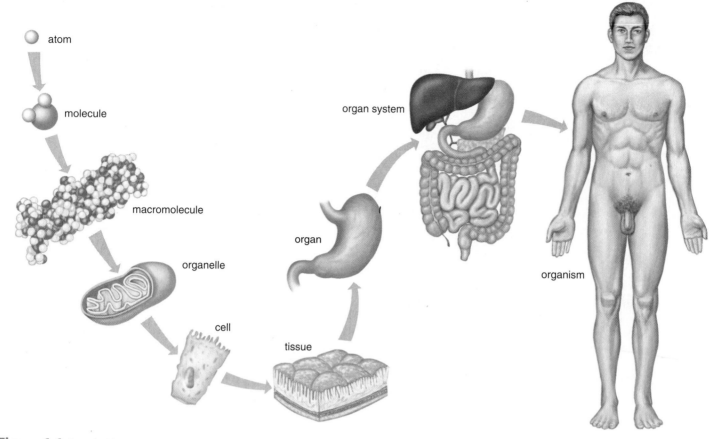

atom

molecule

macromolecule

organelle

cell

tissue

organ

organ system

organism

Figure 1.1 Levels of organization of the human body. Each level is more complex than the previous level.

1.1 The Human Body

Anatomy and physiology is the study of the human body. **Anatomy** is concerned with the structure of a part. For example, the stomach is a J-shaped, pouchlike organ (Fig. 1.1). The stomach wall has thick folds, which disappear as the stomach expands to increase its capacity. **Physiology** is concerned with the function of a part. For example, the stomach temporarily stores food, secretes digestive juices, and passes on partially digested food to the small intestine.

Anatomy and physiology are closely connected in that the structure of an organ suits its function. For example, the stomach's pouchlike shape and ability to expand are suitable to its function of storing food. In addition, the microscopic structure of the stomach wall is suitable to its secretion of digestive juices, as we shall see in Chapter 15.

Organization of Body Parts

The structure of the body can be studied at different *levels of organization* (Fig. 1.1). First, all substances, including body parts, are composed of chemicals made up of submicroscopic particles called **atoms.** Atoms join to form **molecules,** which can in turn join to form **macromolecules.** For example, molecules called amino acids join to form macromolecules called proteins. Different proteins make up the bulk of our muscles.

Macromolecules compose the cellular **organelles,** which are found within all cells. Organelles are tiny structures that perform cellular functions. For example, the organelle called the nucleus is especially concerned with cell reproduction; another organelle, called the mitochondrion, supplies the cell with energy. **Cells** are the basic units of living things.

Tissues are the next level of organization. A **tissue** is composed of similar types of cells and performs a specific function. An **organ** is composed of several types of tissues and performs a particular function within an **organ system.** For example, the stomach is an organ that is a part of the digestive system. It has a specific role in this system, whose overall function is to supply the body with the nutrients needed for growth and repair. The other systems of the body (see page 11) also have specific functions.

All of the body systems together make up the **organism**— for example, a human being. Human beings are complex animals, but this complexity can be broken down and studied at ever simpler levels. Each simpler level is organized and constructed in a particular way.

1.2 Anatomical Terms

Certain terms are used to describe the location of body parts, regions of the body, and imaginary planes by which the body can be sectioned. You should become familiar with these terms

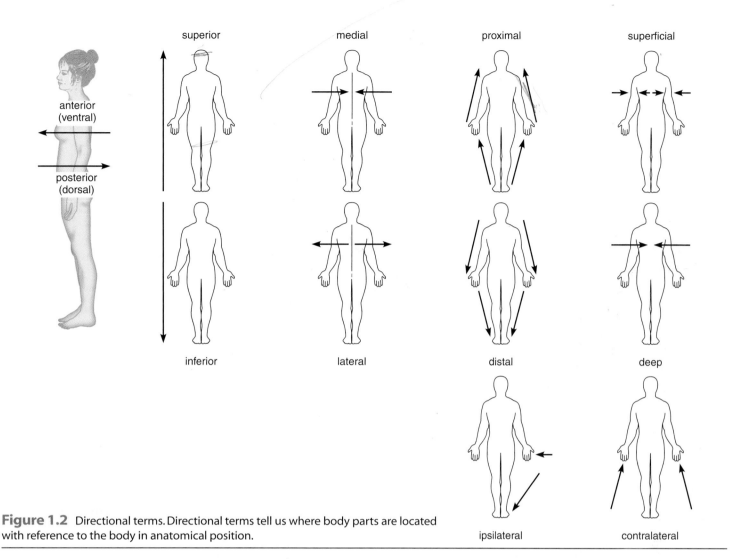

Figure 1.2 Directional terms. Directional terms tell us where body parts are located with reference to the body in anatomical position.

before your study of anatomy and physiology begins. Anatomical terms are useful only if everyone has in mind the same position of the body and is using the same reference points. Therefore, we will assume that the body is in the *anatomical position:* standing erect, with face forward, arms at the sides, and palms and toes directed forward, as illustrated in Figure 1.1.

Directional Terms

Directional terms are used to describe the location of one body part in relation to another (Fig. 1.2):

Anterior (ventral)—a body part is located toward the front; the windpipe (trachea) is anterior to the esophagus.

Posterior (dorsal)—a body part is located toward the back; the heart is posterior to the sternum (breastbone).

Superior—a body part is located above another part, or toward the head; the face is superior to the neck.

Inferior—a body part is below another part, or toward the feet; the navel is inferior to the chin.

Medial—a body part is nearer than another part to an imaginary midline of the body; the bridge of the nose is medial to the eyes.

Lateral—a body part is farther away from the midline; the eyes are lateral to the nose.

Proximal—a body part is closer to the point of attachment or closer to the trunk; the elbow is proximal to the hand.

Distal—a body part is farther from the point of attachment or farther from the trunk or torso; the hand is distal to the elbow.

Superficial (external)—a body part is located near the surface; the skin is superficial to the muscles.

Deep (internal)—the body part is located away from the surface; the intestines are deep to the spine.

Central—a body part is situated at the center of the body or an organ; the central nervous system is located along the main axis of the body.

Peripheral—a body part is situated away from the center of the body or an organ; the peripheral nervous system is located outside the central nervous system.

Ipsilateral—a body part is on the same side of the body as another body part; the right hand is ipsilateral to the right foot.

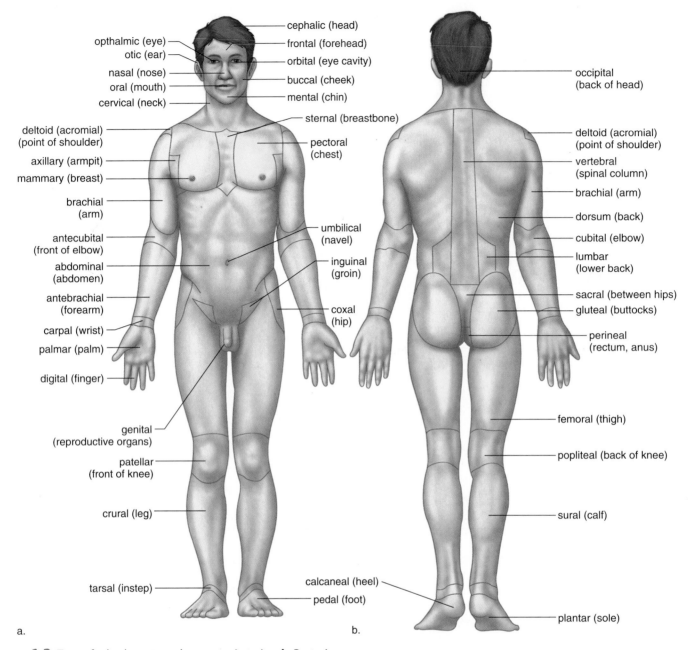

Figure 1.3 Terms for body parts and areas. **a.** Anterior. **b.** Posterior.

Contralateral—a body part is on the opposite side of the body from another body part; the right hand is contralateral to the left hand.

Regions of the Body

The human body can be divided into axial and appendicular portions. The **axial portion** includes the head, neck, and trunk. The trunk can be divided into the thorax, abdomen, and pelvis. The pelvis is that part of the trunk associated with the hips. The **appendicular portion** of the human body includes the limbs—that is, the upper limbs and the lower limbs.

The human body is further divided as shown in Figure 1.3. The labels in Figure 1.3 do not include the word "region." It is

understood that you will supply the word region in each case. The scientific name for each region is followed by the common name for that region. For example, the cephalic region is commonly called the head.

Notice that the upper arm includes among other parts the brachial region (arm) and the antebrachial region (forearm), and the lower limb includes among other parts the femoral region (thigh) and the crural region (leg). In other words, contrary to common usage, the terms arm and leg refer to only a part of the upper limb and lower limb, respectively.

Most likely, it will take practice to learn the terms in Figure 1.3. One way to practice might be to point to various

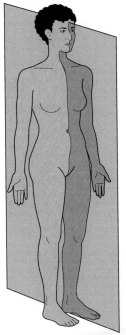

a. Midsagittal (median) plane

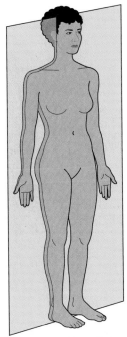

b. Frontal (coronal) plane

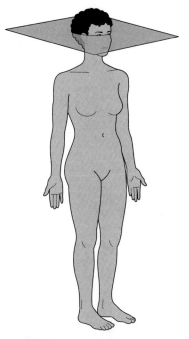

c. Transverse (horizontal) plane

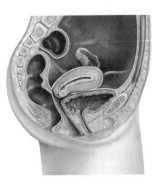

d. Midsagittal section of
pelvic cavity

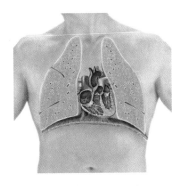

e. Frontal section of
thoracic cavity

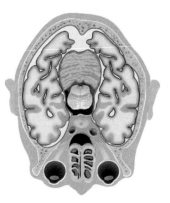

f. Transverse section of
head at eye level

Figure 1.4 Body planes and sections. The planes shown in **(a)**, **(b)**, and **(c)** are typically used as sites for sectioning the body as shown in **(d)**, **(e)**, and **(f)**.

regions of your own body and see if you can give the scientific name for that region. Check your answer against the figure.

Planes and Sections of the Body

To observe the structure of an internal body part, it is customary to section (cut) the body along a plane. A plane is an imaginary flat surface passing through the body. The body is customarily sectioned along the following planes (Fig. 1.4):

A **sagittal** (median) **plane** extends lengthwise and divides the body into right and left portions. A midsagittal plane passes exactly through the midline of the body. The pelvic organs are often shown in midsagittal section (Fig. 1.4d). Sagittal cuts that are not along the midline are called parasagittal sections.

A **frontal** (coronal) **plane** also extends lengthwise, but it is perpendicular to a sagittal plane and divides the body or an organ into anterior and posterior portions. The thoracic organs are often illustrated in frontal section (Fig. 1.4e).

A **transverse** (horizontal) **plane** is perpendicular to the body's long axis and therefore divides the body horizontally to produce a cross section. A transverse cut divides the body or an organ into superior and inferior portions. Figure 1.4f is a transverse section of the head at the level of the eyes.

The terms *longitudinal section* and *cross section* are often applied to body parts that have been removed and cut either lengthwise or straight across, respectively.

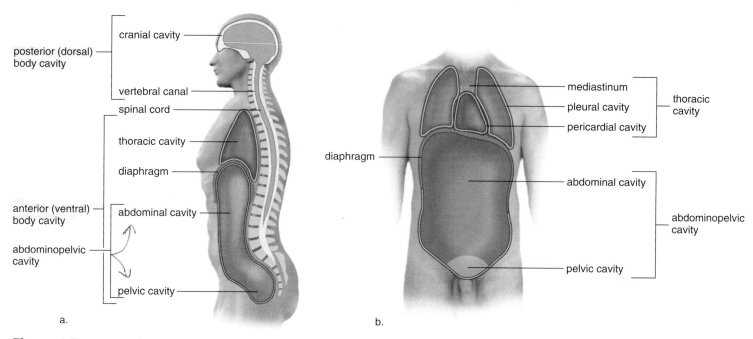

Figure 1.5 The two major body cavities and their subdivisions. **a.** Left lateral view **b.** Frontal view.

1.3 Body Cavities and Membranes

During embryonic development, the body is first divided into two internal cavities: the posterior (dorsal) body cavity and the anterior (ventral) body cavity. Each of these major cavities is then subdivided into smaller cavities. The cavities, as well as the organs in the cavities (called the **viscera**), are lined by membranes.

Posterior (Dorsal) Body Cavity

The posterior body cavity is subdivided into two parts: (1) The **cranial cavity**, enclosed by the bony cranium, contains the brain. (2) The **vertebral canal**, enclosed by vertebrae, contains the spinal cord (Fig. 1.5a).

The posterior body cavity is lined by three membranous layers called the **meninges**. The innermost of the meninges is tightly bound to the surface of the brain and the spinal cord. The space between this layer and the next layer is filled with cerebrospinal fluid. Cerebrospinal fluid supports and nourishes the brain and the spinal cord, and enables their cells to transmit electrical signals. The most superficial of the three meninges lies directly under the bone of the skull and vertebral column.

Anterior (Ventral) Body Cavity

The large anterior body cavity is subdivided into the superior **thoracic cavity** and the inferior **abdominopelvic cavity** (Fig. 1.5a). A muscular partition called the **diaphragm** separates the two cavities. Membranes that line these cavities are called **serous membranes** because they secrete a fluid that is similar to blood **serum**. Serum is the fluid that remains if all of the clotting proteins are removed from the blood. Serous fluid between the smooth serous membranes reduces friction when the viscera rub against each other or against the body wall.

To understand the relationship among serous membranes, the outer body wall, and an organ, consider the following example. Imagine lining a box (the outer body wall) with a plastic bag (the **parietal serous membrane**) (Fig. 1.6). Next, picture the organ placed in a separate zipper style plastic bag (the **visceral serous membrane**) that is zipped shut. The zipper bag is placed inside the plastic-lined cardboard box. Finally, pour water between the two plastic bags—this represents **serous fluid**. Likewise, in the body, organs of the anterior body cavity are enclosed in a visceral serous membrane and sit in cavities that are lined by parietal serous membrane. Inflammation of the serous membrane, or infection of the serous fluid in the body cavities causes serious and potentially fatal illness (see Medical Focus, p. 9).

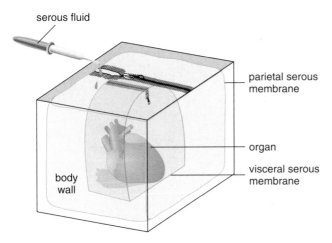

serous fluid

parietal serous membrane

organ

visceral serous membrane

body wall

Figure 1.6 The body cavities (represented by a box) have an inner lining, the parietal serous membrane. Organs are covered by visceral serous membrane. Serous fluid fills the space between membranes.

Thoracic Cavity

The thoracic cavity is enclosed by the rib cage and has three portions: the left, right, and medial portions. The medial portion, called the **mediastinum,** contains the heart, thymus gland, trachea, esophagus, and other structures (Fig. 1.5*b*).

The right and left portions of the thoracic cavity contain the lungs. The lung tissue is covered by a serous membrane—the **visceral pleura.** The **parietal pleura** lines the thoracic cavity. In between these two **pleurae** is the pleural cavity, which contains a small amount of pleural fluid. Similarly, in the medial thoracic cavity, the heart is covered by the **visceral pericardium.** The visceral pericardium contributes to the outermost connective tissue layer of the heart. Forming a tough connective tissue sac around the heart is the **parietal pericardium,** which creates the **pericardial cavity.** The heart,

inside of its visceral pericardial sac, is separated from the outer parietal pericardium by a small amount of pericardial fluid.

Abdominopelvic Cavity

The abdominopelvic cavity has two portions: the superior **abdominal cavity** and the inferior **pelvic cavity.** The stomach, liver, spleen, gallbladder, and most of the small and large intestines are in the abdominal cavity. The pelvic cavity contains the rectum, the urinary bladder, the internal reproductive organs, and the rest of the large intestine. Males have an external extension of the abdominal wall, called the **scrotum,** where the testes are located.

Many of the organs of the abdominopelvic cavity are covered by the **visceral peritoneum,** whereas the wall of the abdominal cavity is lined with the **parietal peritoneum.** Peritoneal fluid fills the cavity between the visceral and parietal peritoneum. Table 1.1 summarizes our discussion of body cavities and membranes.

It is important that clinicians use the same terminology to reference various regions of the abdominopelvic cavity. Either of two systems can be used. The first uses nine regions (imagine a "tic-tac-toe" grid, with the umbilicus in the center square). The upper regions are right hypochondriac, epigastric, and left hypochondriac. The center regions are right lumbar, umbilical, and left lumbar. The lower regions are right inguinal (iliac), pubic, and left inguinal (iliac). Note that the terms used are those for each body area, as illustrated in Figure 1.3.

Alternatively, the abdominopelvic cavity can be divided into four quadrants by running a horizontal plane across the median plane at the point of the navel (Fig. 1.7*a*). Physicians commonly use these quadrants to identify the locations of patient's symptoms. The four quadrants are: (1) right upper quadrant, (2) left upper quadrant, (3) right lower quadrant, and (4) left lower quadrant.

TABLE 1.1 Body Cavities and Membranes			
Name of Cavity	**Contents of Cavity**	**Membranes**	
POSTERIOR BODY CAVITY			
Cranial cavity	Brain	Meninges	
Vertebral canal	Spinal cord	Meninges	
ANTERIOR BODY CAVITY			
Thoracic Cavity		*Parietal Membrane*	*Visceral Membrane*
Pleural cavity	Lungs	Parietal pleura	Visceral pleura
Pericardial cavity	Heart	Parietal pericardium	Visceral pericardium (epicardium)
Abdominopelvic Cavity			
Abdominal cavity	Digestive organs, liver, kidneys	Parietal peritoneum	Visceral peritoneum
Pelvic cavity	Reproductive organs, urinary bladder, rectum	Parietal peritoneum	Visceral peritoneum

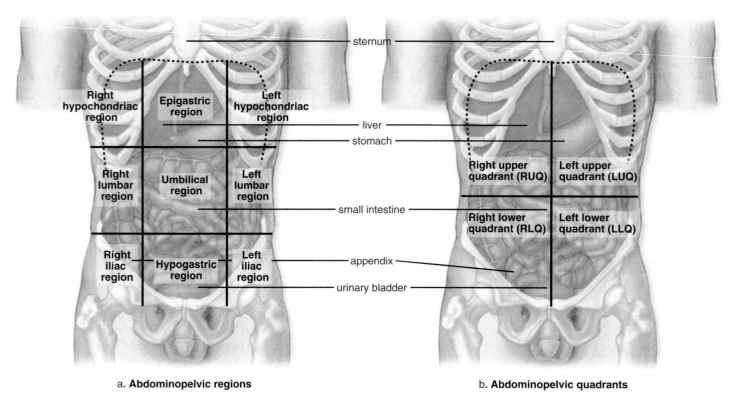

a. **Abdominopelvic regions**

b. **Abdominopelvic quadrants**

Figure 1.7 The abdominopelvic cavity can be subdivided into **a.** nine regions or **b.** four quadrants.

Figure 1.7*b* compares the two methods of referencing the abdominopelvic region and shows the organs within each region.

Content **CHECK-UP!**

4. Match each of the serous membranes to its function.

parietal pleura _____   a. covers the heart

visceral pericardium _____   b. covers walls of right and
                                   left portions of thoracic cavity

visceral peritoneum _____   c. covers the abdominal
                                   organs

5. Pleurisy refers to infection or inflammation of the pleurae. The most logical symptom of pleurisy is _____.

1.4   Organ Systems

The organs of the body work together in systems. The reference figures in Appendix A can serve as an aid to learning the 11 organ systems and their placement. The type of illustration that will be used at the end of each of the organ system chapters is introduced on page 14. In this chapter, the illustration demonstrates the general functions of the body's organ systems. The corresponding illustrations in the organ system chapters will show how a particular organ system interacts with all the other systems. In this text, the organ systems of the body have been divided into four categories, as discussed next.

Support, Movement, and Protection

The **integumentary system,** discussed in Chapter 5, includes the skin and accessory organs, such as the hair, nails, sweat glands, and sebaceous glands. The skin protects underlying tissues, prevents infection and water loss, helps regulate body temperature, contains sense organs, and even synthesizes certain chemicals that affect the rest of the body.

The **skeletal system** and the **muscular system** give the body support and the ability to move. The skeletal system, discussed in Chapter 6, consists of the bones of the skeleton and associated cartilage, as well as the ligaments that bind these structures together. The skeleton protects body parts. For example, the skull forms a protective encasement for the brain, as does the rib cage for the heart and lungs. Some bones produce blood cells, and all bones are a storage area for calcium and phosphorus salts. The skeleton as a whole serves as a place of attachment for the muscles.

Contraction of *skeletal muscles,* discussed in Chapter 7, accounts for our ability to move voluntarily and to respond to outside stimuli. These muscles also maintain posture and are responsible for the production of body heat. *Cardiac muscle*

Meningitis and Serositis

The anterior and posterior body cavities are enclosed areas that are protected by bone, muscle, connective tissues, and skin. Inflammation of the membranes lining these cavities is a fairly rare, but serious, illness. If body defenses are overcome by bacteria, viruses, or other microbes, the result is a serious, potentially fatal infection and inflammation of the meninges (meningitis) or the serous membranes (**serositis**). Pleurisy, pericarditis, and peritonitis are all forms of serositis.

Meningitis is the term for inflammation of the meninges—linings of the posterior body cavity that cover the brain and spinal cord. The most dangerous form is caused by bacteria that commonly inhabit the nose. In the bacterial meningitis patient, a previous viral infection (which may be a simple common cold) allows these bacteria to enter the bloodstream and infect the meninges. Symptoms of bacterial meningitis include a severe headache and stiff neck, sensitivity to light, fever, weakness, and fatigue. Even with aggressive antibiotic treatment, bacterial meningitis is fatal in 25% of adults. The best treatment is prevention by immunization—especially important for young college students living in the close quarters of a college dorm.

Pleurisy is an inflammation of the pleurae—linings of the thoracic cavity that also cover the lungs. It is often caused by a cold virus, although it can signal the presence of more serious infections or even lung cancer. Its symptoms include chest pain that worsens with deep breathing, and *pleural friction rub*—a rough, grating sound in the chest that can be heard with a stethoscope placed over the painful area. Treatment for pleurisy depends on its cause; most often, pleurisy that results from a common cold requires only pain medication such as aspirin or ibuprofen. Treatment for bacterial infection requires antibiotics.

Pericarditis affects the linings surrounding the heart. Like meningitis, it often results from previous infections and can be extremely dangerous. It is a common complication in drug abusers who use dirty needles for injections. Symptoms include severe chest pain (which may be mistaken for a heart attack), fever, and weakness. Physicians can hear *pericardial friction rub* by placing a stethoscope over the patient's heart. Fluid accumulation inside the pericardial sac surrounding the heart may interfere with blood flow to and from the heart. Bacterial pericarditis is treated with antibiotics, pain medications, and drugs that reduce swelling.

Peritonitis affects the lining of the abdominopelvic cavity. It usually results from bacterial infection; a common cause of infection is a ruptured appendix from appendicitis. Severe pain, fever, elevated white blood cell counts, and tenderness are common symptoms. Aggressive treatment with antibiotics is necessary to prevent bacteria from invading the blood.

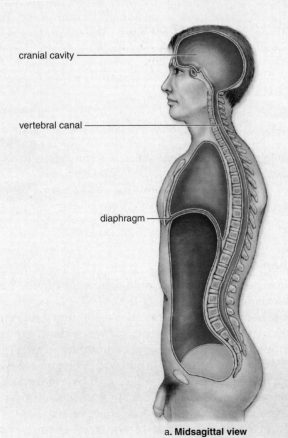

cranial cavity

vertebral canal

diaphragm

a. **Midsagittal view**

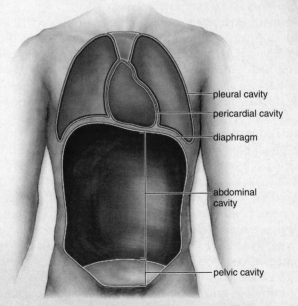

pleural cavity

pericardial cavity

diaphragm

abdominal cavity

pelvic cavity

b. **Coronal (frontal) view**

Figure 1A **a.** Meningitis is infection or inflammation of the linings of the cranial cavity and vertebral canal. **b.** Serositis is infection or inflammation of the ventral body cavities. Pleurisy affects the pleural cavities, pericarditis affects the pericardial cavity, and peritonitis affects the abdominopelvic cavities.

and *smooth muscle* are called involuntary muscles because they contract automatically. Cardiac muscle makes up the heart, and smooth muscle is found within the walls of internal organs and blood vessels.

Integration and Coordination

The **nervous system,** discussed in Chapter 8, consists of the brain, spinal cord, and associated nerves. The nerves conduct sensory nerve signals to the brain and spinal cord. They also conduct nerve impulses from the brain and spinal cord to the muscles, glands, and organs.

The sense organs (discussed in Chapter 9) provide us with information about our internal and external environment. The brain and spinal cord then process this information, and the individual responds to environmental stimuli using organs, glands, and/or muscles.

The **endocrine system,** discussed in Chapter 10, consists of the hormonal glands, which secrete chemicals that serve as messengers between body parts. Both the nervous and endocrine systems help maintain a relatively constant internal environment by coordinating and regulating the functions of the body's other systems. The nervous system acts quickly but has a short-lived effect; the endocrine system acts more slowly but has a more sustained effect on body parts. The endocrine system also helps maintain the proper functioning of the male and female reproductive organs.

Maintenance of the Body

The internal environment of the body is the blood within the blood vessels and the tissue fluid that surrounds the cells. Five systems add substances to and/or remove substances from the blood and tissue fluid: the cardiovascular, lymphatic, respiratory, digestive, and urinary systems.

The **cardiovascular system,** discussed in Chapter 12, consists of the heart and the blood vessels that carry blood through the body. Blood transports nutrients and oxygen to the cells, and removes waste molecules to be excreted from the body. Blood also contains cells produced by the **lymphatic system,** discussed in Chapter 13. The lymphatic system protects the body from disease.

The **respiratory system,** discussed in Chapter 14, consists of the lungs and the tubes that take air to and from the lungs. The respiratory system brings oxygen into the lungs and takes carbon dioxide out of the lungs.

The **digestive system** (see Fig. 1.1), discussed in Chapter 15, consists of the mouth, esophagus, stomach, small intestine, and large intestine (colon), along with the accessory organs: teeth, tongue, salivary glands, liver, gallbladder, and pancreas. This system receives food and digests it into nutrient molecules, which can enter the cells of the body.

The **urinary system,** discussed in Chapter 16, contains the kidneys and the urinary bladder. This system rids the body of nitrogenous wastes and helps regulate the fluid level and chemical content of the blood.

Reproduction and Development

The male and female **reproductive system,** discussed in Chapter 17, contain different organs. The *male reproductive system* consists of the testes, other glands, and various ducts that conduct semen to and through the penis. The *female reproductive system* consists of the ovaries, uterine tubes, uterus, vagina, and external genitalia. Both systems produce sex cells, but in addition, the female system receives the sex cells of the male and also nourishes and protects the fetus until the time of birth. Development before birth and the process of birth are discussed in Chapter 18.

1.5 Homeostasis

Homeostasis is the relative constancy of the body's internal environment. Because of homeostasis, even though external conditions may change dramatically, internal conditions stay within a narrow range. For example, regardless of how cold or hot one's environment gets, the temperature of the body stays around 37°C (97° to 99°F). No matter how acidic your meal, the pH of your blood is usually about 7.4, and even if you eat a candy bar, the amount of sugar in your blood is just about 0.1%.

It is important to realize that internal conditions are not absolutely constant; they tend to fluctuate above and below a particular value. Therefore, the internal state of the body is often described as one of *dynamic* equilibrium. If internal conditions change to any great degree, illness results. This makes the study of homeostatic mechanisms medically important.

Negative Feedback

Negative feedback is the primary homeostatic mechanism that keeps a variable close to a particular value, or set point. A homeostatic mechanism has three components: a sensor, a control center, and an effector (Fig. 1.8*a*). The sensor detects a change in the internal environment; the control center activates the effector; the effector reverses the initial change and brings conditions back to normal again. Now, the sensor is no longer activated.

Mechanical Example

A home heating system illustrates how a negative feedback mechanism works (Fig. 1.8*b*). You set the thermostat at, say, 68°F. This is the set point. The thermostat contains a thermometer, a sensor that detects when the room temperature falls below the set point. The thermostat is also the control center; it turns the furnace on. The furnace plays the role of the effector. The heat given off by the furnace raises the temperature of the room to 70°F. Now, the furnace turns off because the sensor is no longer activated.

Notice that a negative feedback mechanism prevents change in the same direction; the room does not get warmer and warmer because warmth inactivates the system.

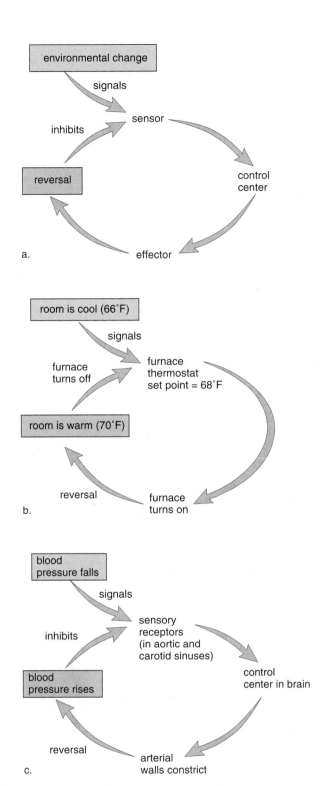

Figure 1.8 Negative feedback. In each example, a sensor detects an internal environmental change and signals a control center. The center activates an effector, which reverses this change. **a.** The general pattern. **b.** A mechanical example. **c.** A human example.

Human Example: Regulation of Blood Pressure

Negative feedback mechanisms in the body function similarly to the mechanical model. For example, when blood pressure

falls, sensory receptors signal a control center in the brain (Fig. 1.8c). This center sends out nerve impulses to the arterial walls so that they constrict. Once the blood pressure rises, the system is inactivated.

Human Example: Regulation of Body Temperature

The thermostat for body temperature is located in a part of the brain called the hypothalamus. When the body temperature falls below normal, the control center directs (via nerve impulses) the blood vessels of the skin to constrict (Fig. 1.9). This conserves heat. If body temperature falls even lower, the control center sends nerve impulses to the skeletal muscles, and shivering occurs. Shivering generates heat, and gradually body temperature rises to 37°C. When the temperature rises to normal, the regulatory center is inactivated.

When the body temperature is higher than normal, the control center directs the blood vessels of the skin to dilate. This allows more blood to flow near the surface of the body, where heat can be lost to the environment. In addition, the nervous system activates the sweat glands, and the evaporation of sweat helps lower body temperature. Gradually, body temperature decreases to 37°C.

Positive Feedback

Like negative feedback, in a **positive feedback** mechanism a sensor will trigger the control center. The control center in turn activates the effector. However, the effector in this type of feedback mechanism produces a response that continues to stimulate the sensor. Thus, positive feedback brings about an even greater change in the same direction. A positive feedback mechanism can be harmful, as when a fever causes metabolic changes that push the fever still higher. Death occurs at a body temperature of 45°C because cellular proteins denature at this temperature and metabolism stops.

Still, positive feedback loops such as those involved in blood clotting, the stomach's digestion of protein, and child-birth assist the body in completing a process that has a definite cutoff point.

Consider that when a woman is giving birth, the head of the baby begins to press against the cervix, stimulating sensory receptors there. When nerve impulses reach the brain (control center), the brain causes the pituitary gland (effector) to secrete the hormone oxytocin. Oxytocin travels in the blood and causes the uterus to contract. As labor continues, the cervix is ever more stimulated, and uterine contractions become ever stronger until birth occurs (Fig. 1.10).

Homeostasis and Body Systems

The internal environment of the body consists of blood and tissue fluid. Tissue fluid bathes all the cells of the body. Oxygen and nutrients move from blood to tissue fluid, and wastes move from tissue fluid into the blood (Fig. 1.11). Tissue fluid remains constant only as long as blood composition remains constant. All systems of the body contribute

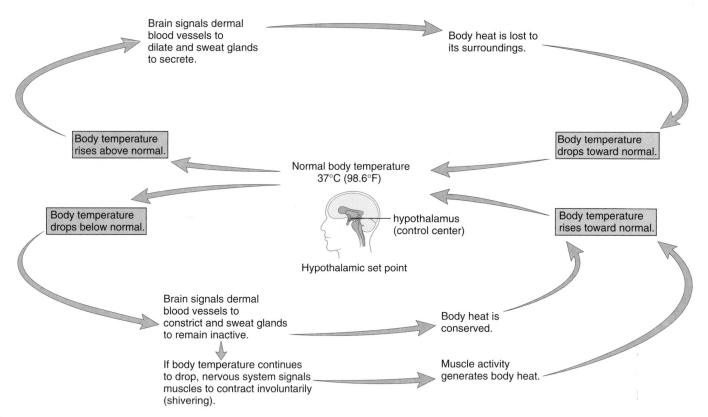

Figure 1.9 Homeostasis and body temperature regulation. Negative feedback mechanisms control body temperature so that it remains relatively stable at 37°C. These mechanisms return the temperature to normal when it fluctuates above and below this set point.

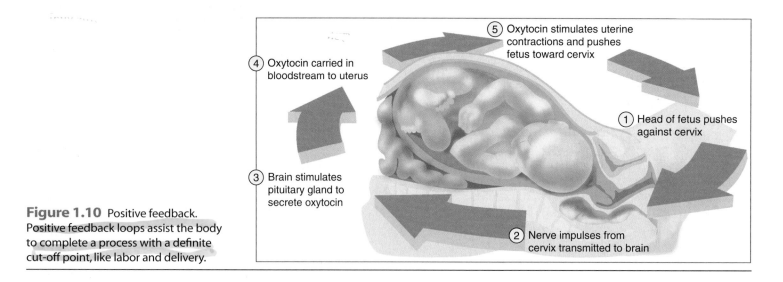

Figure 1.10 Positive feedback. Positive feedback loops assist the body to complete a process with a definite cut-off point, like labor and delivery.

toward maintaining homeostasis and, therefore, a relatively constant internal environment.

As you use the text, you will be introduced to each of the organ systems. Remember the job of body systems in maintaining homeostasis as you study each:

Integumentary System (skin)—*Support and protection* of delicate internal structures.

Skeletal System—*Movement.* In addition, the skeleton stores minerals and produces the blood cells.

Muscular System—*Movement.* Skeletal muscles and bones work together for movement. Smooth muscle inside internal structures moves substances inside a tube, as when a meal is moved through the digestive tract. Muscle activity also generates heat, thus helping to maintain body temperature.

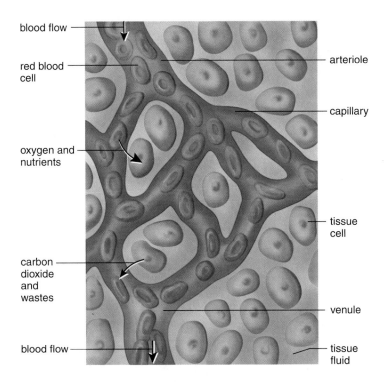

blood flow

red blood cell

oxygen and nutrients

carbon dioxide and wastes

blood flow

arteriole

capillary

tissue cell

venule

tissue fluid

Figure 1.11 Regulation of tissue fluid composition. Cells are surrounded by tissue fluid (blue), which is continually refreshed because oxygen and nutrient molecules constantly exit the bloodstream, and carbon dioxide and waste molecules continually enter the bloodstream.

Nervous and Sensory Systems—*Control.* We have already seen that in negative and positive feedback mechanisms, sensory receptors send nerve impulses to control centers in the brain, which then activate effectors: muscles, glands, or organs. The nervous system can cause rapid change, if needed, to maintain homeostasis.

Endocrine System—*Control.* Endocrine glands secrete hormones that bring about a slower, more lasting change that keeps the internal environment relatively stable.

Blood and Cardiovascular System—*Transportation and defense.* Red blood cells and blood plasma (the liquid fraction of blood) transport oxygen, carbon dioxide, nutrients, and wastes. Platelets in blood participate in the clotting process, preventing excess blood loss. White blood cells defend against infection. The cardiovascular system conducts blood to and away from capillaries, where exchange occurs. The heart pumps the blood and thereby keeps it moving toward the capillaries.

Lymphatic System—*Transportation and defense.* The lymphatic system assists the cardiovascular system. Lymphatic capillaries collect excess tissue fluid, which is returned via lymphatic vessels to the cardiovascular veins. Lymph nodes help to purify lymph and keep it free of pathogens. This action is assisted by the white blood cells that are housed within lymph nodes.

Respiratory System—*Gas exchange.* The respiratory system adds oxygen to and removes carbon dioxide from the blood. It also plays a role in regulating acid-base balance in blood and tissue fluid. Removal of CO_2 helps to eliminate acid and prevent acidosis.

Digestive System—*Nourishment and waste removal.* The digestive system takes in and digests food, providing nutrient molecules to replace the nutrients that are constantly being used by the body cells. Substances that cannot be digested are eliminated. The liver, an accessory digestive organ, also manufactures urea, a waste product of protein digestion. The liver also removes toxic chemicals such as alcohol and other drugs. Additionally, the liver regulates blood glucose (sugar). As glucose enters the blood after a meal, any excess is removed by the liver and stored as glycogen. Later, the glycogen can be broken down to replace the glucose used by the body cells. In this way, the glucose composition of blood remains constant.

Urinary System—*Waste removal.* Urea and other metabolic waste molecules are excreted by the kidneys, which are a part of the urinary system. Urine formation by the kidneys is extremely critical to the body, not only because it rids the body of unwanted substances but also because urine formation offers an opportunity to carefully regulate blood volume, salt balance, and acid-base balance.

Reproductive System—*Survival of the species.* Although individuals can survive and thrive without reproducing, the human species cannot continue without this vital system.

The contributions of each of the body's systems are summarized in the Human Systems Work Together illustration on page 14.

Disease

Disease occurs when homeostasis fails and the body (or part of the body) no longer functions properly. The effects may be limited or widespread. A **local disease** is more or less restricted to a specific part of the body. On the other hand, a **systemic disease** affects the entire body or involves several organ systems. For example, streptococcal tonsillitis, or "strep throat," is a local disease. If not effectively treated with antibiotics, strep throat can progress to become a dangerous systemic disease—rheumatic fever. Diseases may also be classified on the basis of their severity and duration. **Acute diseases** occur suddenly and generally last a short time. **Chronic diseases** tend to be less severe, develop slowly, and are long term. Acute bronchitis is a bacterial infection of lung airways, and a frequent complication of the common cold; chronic bronchitis is typically caused by years of smoking.

The medical profession has many ways of diagnosing disease including imaging internal body parts (see Medical Focus, page 16).

Human Systems Work Together

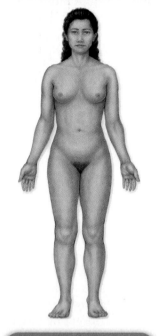

Inegumentary System

External support and protection of body; helps maintain body temperature.

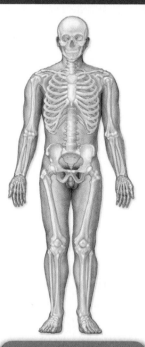

Skeletel System

Internal support and protection: body movement; production of blood cells.

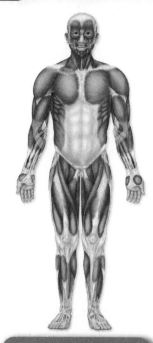

Muscular System

Body movement; production of heat that maintains body temperature.

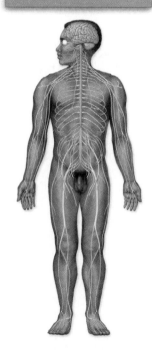

Nervous System

Regulatory centers for control of all body systems; learning and memory.

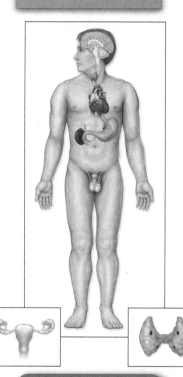

Endocrine System

Secretion of hormones for chemical regulation of all body systems.

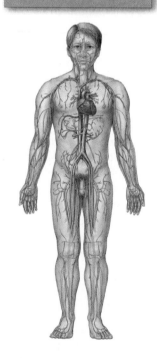

Blood/Cardiovascular System

Transport of nutrients to body cells and transport of wastes away from cells

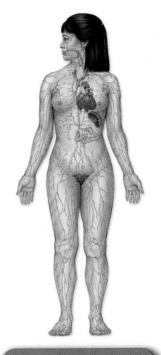

Lymphatic System/Immunity

Drainage of tissue fluid; purifies tissue fluid and keeps it free of pathogens.

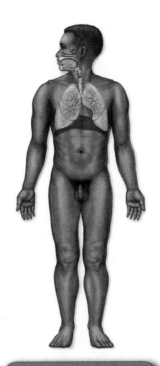

Respiratory System

Rids the blood of carbon dioxide and supplies the blood with oxygen; helps maintain the pH of the blood.

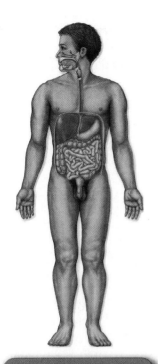

Digestive System

Breakdown of food and absorption of nutrients into blood.

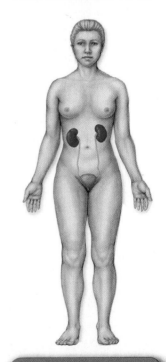

Urinary System

Maintenance of volume and chemical composition of blood.

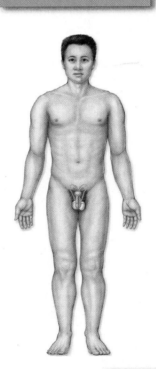

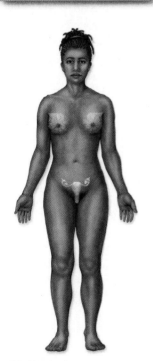

Reproductive System

Production of sperm and egg; transfer of sperm to female system where development occurs.

Imaging the Body

Imaging the body for diagnosis of disease has certainly changed since Wilhelm Roentgen's accidental invention of the X ray in 1895. Many new techniques allow clinicians to visualize internal structures with great accuracy. The least expensive and most widely used imaging technique remains the X ray. X rays are produced when high-speed electrons strike a heavy metal. Dense structures (bone, teeth, cancer tumors) absorb X rays well and show up as light areas on photographic film. Soft tissues absorb X rays to a lesser extent and show up as dark areas. Injecting opaque dye into blood vessels allows X-ray imaging of blood vessels, as in a coronary angiogram (an X ray of the arteries supplying the heart). X-ray imaging of the digestive tract is possible on patients who have first consumed opaque dye solutions.

A bone density scan is another specialized type of X ray. The patient is first injected with a harmless radioactive tracer, which is rapidly taken up by bone. Subsequent X rays can reveal areas of increased bone metabolism (as in a cancerous tumor) or decreased metabolism (as in osteoporosis). A similar technique uses radioactive iodine to study the thyroid gland; the thyroid is the only tissue to use the element iodine (Fig. 1B, 1C).

During computed tomography, or CT scan (previously called computerized axial tomography), X rays are sent through the body at various angles, and a computer uses the X-ray information to form a series of cross sections (Fig. 1D). CT scanning has reduced the need for exploratory surgery and can guide the surgeon in visualizing complex body structures during surgical procedures.

PET (positron emission tomography) is a variation on CT scanning (Fig. 1D). Radioactively labeled sugar is injected into the body; metabolically active tissues tend to take up this substance and then emit gamma rays. A computer uses the gamma ray information to again generate cross-sectional images of the body, but this time the image indicates metabolic activity, not structure. PET scanning is used to diagnose brain disorders, such as a brain tumor, Alzheimer disease, epilepsy, or stroke (Fig. 1D).

During MRI (magnetic resonance imaging), the patient lies in a massive, hollow, cylindrical magnet and is exposed to short bursts of a powerful magnetic field. This causes the protons in the nuclei of hydrogen atoms to align. When the protons are exposed to strong radio waves, they move out of alignment and produce signals. A computer changes these signals into an image (see p. 1). Tissues with many hydrogen atoms (such as fat) show up as bright areas, while tissues with few hydrogen atoms (such as bone) appear black. This is the opposite of an X ray, which is why MRI is more useful than an X ray for imaging soft tissues. However, many people cannot undergo MRI because the magnetic field can actually pull a metal object (like a tooth filling or an artificial hip) out of the body!

The least-expensive method of creating tissue images is sonography, or ultrasound. High-frequency soundwaves are transmitted into tissues. Organs and tissues reflect the soundwaves; a receiver creates an image using the reflected waves. Use of sonography avoids any radiation exposure to the patient. The

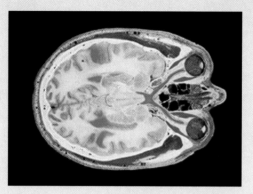

Figure 1B CT (computed tomography) of the brain. This transverse section is at eye level.

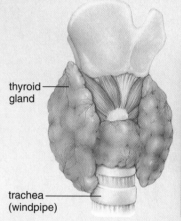
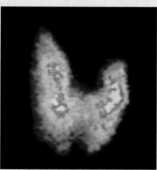

thyroid gland

trachea (windpipe)

a. **Drawing of thyroid**   b. **Scan of thyroid**

Figure 1C Use of radiation to aid a diagnosis. After the administration of radioactive iodine, a scan of the thyroid reveals pathology. The missing portion of the gland is cancerous and therefore failed to take up the iodine.

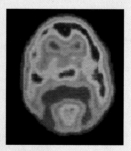

a. **Patient entering PET scanner**   b. **PET scan**

Figure 1D Use of radiation to study the brain. After the administration of radioactively labeled glucose, a PET scan reveals which portions of the brain are most active.

rare adverse reaction to dyes or radioactive compounds can also be avoided. Ultrasound is safe for imaging the fetus in a pregnant woman and can show an amazing amount of detail.

6. The two organ systems responsible for controlling other body systems are the _____ and _____ systems.

7. The two organ systems that defend the body against infection are:

a. cardiovascular and urinary

b. cardiovascular and respiratory

c. cardiovascular and skeletal

d. cardiovascular and lymphatic

8. A person with osteoarthritis, a degenerative disorder that causes stiff, painful joints, developed this condition at age 60 after decades of "wear and tear." Osteoarthritis can best be described as:

a. acute

b. chronic

SELECTED NEW TERMS

Basic Key Terms

abdominal cavity (ab-dom′ĭ-nal kav′ĭ-te), p. 7

abdominopelvic cavity (ab-dom″ĭ-no-pel′vik kav′ĭ-te), p. 6

anatomy (uh-nat′o-me), p. 2

anterior (ăn-tîr′ē-ər), p. 3

appendicular portion (ăp′ən-dĭk′yə-lər pôr′shən), p. 4

atoms (ăt′əms), p. 2

axial portion (ăk′sē-əl pôr′shən), p. 4

cells (sĕls), p. 2

central (cĕn′trəl), p. 3

contralateral (kŏn′trə-lăt′ər-əl), p. 4

cranial cavity (kra′ne-al kav′ĭ-te), p. 6

deep (dēp), p. 3

diaphragm (dĭ′ə-frăm′), p. 6

distal (dis′tal), p. 3

frontal plane (frŭn′tl plăn), p. 5

homeostasis (ho″me-o-sta′sis), p. 10

inferior (ĭn-fĭr′ē-ər), p. 3

ipsilateral (ĭp′sə-lăt′ər-əl), p. 3

lateral (lat′er-al), p. 3

macromolecules (măk′rŏ-mŏl′ĭ-kyoōl′s), p. 2

medial (me′de-al), p. 3

mediastinum (me″de-uh-sti′num), p. 7

molecules (mŏl′ĭ-kyoōl′s), p. 2

negative feedback (neg′uh-tiv fēd′bak), p. 10

organ (ôr′gən), p. 2

organelles (ôr′gə-nĕls), p. 2

organism (ôr′gə-nĭz′əm), p. 2

organ system (ôr′gən sĭs′təm), p. 2

parietal pericardium (per″ĭ-kar′de-um), p. 7

parietal peritoneum (pə-rī′ĭ-təl pĕr′ĭ-tə-nē′əm), p. 7

parietal pleurae (pə-rī′ĭ-təl ploŏr′ə), p. 7

parietal serous membrane (pə-rī′ĭ-təl sîr′əs mĕm′brăn′), p. 6

pelvic cavity (pel′vik kav′ĭ-te), p. 7

pericardial cavity (pĕr′ĭ-kăr′dē-əl kăv′ĭ-tē), p. 7

peripheral (pə-rĭf′ər-əl), p. 3

peritoneum (per″ĭ-to-ne′um), p. 7

physiology (fĭz″e-ol′o-je), p. 2

positive feedback (poz′ĭ-tiv fēd′bak), p. 11

posterior (pŏ-stîr′ē-ər), p. 3

proximal (prok′sĭ-mal), p. 3

sagittal plane (saj′ĭ-tal plăn), p. 5

serous fluid (sîr′əs floō′ĭd), p. 6

serous membrane (sēr′us mem′brăn), p. 6

serum (sîr′əm), p. 6

superficial (soō′par-fĭsh′əl), p. 3

superior (soō-pîr′ē-ər), p. 3

thoracic cavity (tho-ras′ik kav′ĭ-te), p. 6

tissue (tĭsh′oō), p. 2

transverse plane (trans-vers′ plăn), p. 5

vertebral canal (vŭr′tə-brəl kə-năl), p. 6

viscera, visceral pleura (vis′-er-uh, vis′-er-ŭl), p. 6

visceral pericardium (vĭs′ər-əl pĕr′ĭ-kär′dē-əm), p. 7

visceral peritoneum (vĭs′ər-əl pĕr′ĭ-tə-nē′əm), p. 7

visceral pleura (vĭs′ər-əl ploŏr′a), p. 7

visceral serous membrane (vĭs′ər-əl sîr′əs), p. 6

Clinical Key Terms

acute (uh-kyoot′) disease, p. 13

chronic (kron′-ik) disease, p. 13

disease (di-zez′), p. 13

local disease (lō′kəl dĭ-zēz′), p. 13

meningitis (men″in-ji′tis), p. 9

pericarditis (per″i-kar-di′tis), p. 9

peritonitis (per″i-to-ni′tis), p. 9

pleurisy (ploo′ri-see), p. 9

serositis (se″ro-si′tis), p. 9

systemic disease (sis-tem′ik di-zez′), p. 13

SUMMARY

1.1 The Human Body

A. Anatomy is the study of the structure of body parts, and physiology is the study of the function of these parts. Structure is suited to the function of a part.

B. The body has levels of organization that progress from atoms to molecules, macromolecules, organelles, cells, tissues, organs, organ systems, and finally, the organism.

1.2 Anatomical Terms

Various terms are used to describe the location of body organs when the body is in the anatomical position (standing erect, with face forward, arms at the sides, and palms and toes directed forward).

A. The terms *anterior/posterior*, *superior/inferior*, *medial/lateral*, *proximal/distal*, *superficial/deep*, *central/peripheral*, and *contralateral/ipsilateral* describe the relative positions of body parts.

B. The body can be divided into axial and appendicular portions, each of which can be further subdivided into specific regions.

For example, *brachial* refers to the arm, and *pedal* refers to the foot.

C. The body or its parts may be sectioned (cut) along certain planes. A median or midsagittal (vertical) cut divides the body into equal right and left portions. A frontal (coronal) cut divides the body into anterior and posterior parts. A transverse (horizontal) cut is a cross section.

1.3 Body Cavities and Membranes

The human body has two major cavities: the posterior (dorsal) body cavity and the anterior (ventral) body cavity. Each is subdivided into smaller cavities, within which specific viscera are located. Specific serous membranes line body cavities and adhere to the organs within these cavities.

1.4 Organ Systems

The body has a number of organ systems. These systems have been characterized as follows:

A. Support, movement, and protection. The integumentary system, which includes the skin, not only protects the body, but also has other functions. The skeletal system contains the bones, and the muscular system contains the three types of muscles. The primary function of the skeletal and muscular systems is support and movement, but they have other functions as well.

B. Integration and coordination. The nervous system contains the brain, spinal cord, and nerves. Because the nervous system communicates with both the sense organs and the muscles, it allows us to respond to outside stimuli. The endocrine system consists of the hormonal glands. The nervous and endocrine systems coordinate and regulate the activities of the body's other systems.

C. Maintenance of the body. The cardiovascular system (heart and vessels), lymphatic system (lymphatic vessels and nodes, spleen, and thymus), respiratory system (lungs and conducting tubes), digestive system (mouth, esophagus, stomach, small and large intestines, and associated organs), and urinary system (kidneys and bladder) all perform specific processing and transporting functions to maintain the normal conditions of the body.

D. Reproduction and development. The reproductive system in males (testes, other glands, ducts, and penis) and in females (ovaries, uterine tubes, uterus, vagina, and external genitalia) carries out those functions that give humans the ability to reproduce.

1.5 Homeostasis

Homeostasis is the relative constancy of the body's internal environment, which is composed of blood and the tissue fluid that bathes the cells.

A. Negative feedback mechanisms help maintain homeostasis. Positive feedback occurs in processes with a definite cut-off point.

B. All of the body's organ systems contribute to homeostasis. Some, including the respiratory, digestive, and urinary systems, remove and/or add substances to blood.

C. The nervous and endocrine systems regulate the activities of other systems. Negative feedback is a self-regulatory mechanism by which systems and conditions of the body are controlled.

STUDY QUESTIONS

1. Distinguish between the study of anatomy and the study of physiology. (p. 2)
2. Give an example that shows the relationship between the structure and function of body parts. (p. 2)
3. List the levels of organization within the human body in reference to a specific organ. (p. 2)
4. What purpose is served by directional terms as long as the body is in anatomical position? (p. 3)
5. Distinguish between the axial and appendicular portions of the body. State at least two anatomical terms that pertain to the head, thorax, abdomen, and limbs. (p. 4)
6. Distinguish between a midsagittal section, a transverse section, and a coronal section. (p. 5)
7. Distinguish between the posterior and anterior body cavities, and name two smaller cavities that occur within each. (pp. 6–7)
8. Name the four quadrants and the nine regions of the abdominopelvic cavity. (p. 7)
9. Name the major organ systems, and describe the general functions of each. (p. 8)
10. List the major organs found within each organ system. (p. 10)
11. Define homeostasis, and give examples of negative feedback and positive feedback mechanisms. (pp. 10–11)
12. Discuss the contribution of each body system to homeostasis. (p. 10)

LEARNING OBJECTIVE QUESTIONS

I. Match the terms in the key to the relationships listed in questions 1–5.

Key:
   a. anterior
   b. posterior
   c. superior
   d. inferior
   e. medial
   f. lateral
   g. proximal
   h. distal

1. the esophagus in relation to the stomach
2. the ears in relation to the nose
3. the shoulder in relation to the hand
4. the intestines in relation to the vertebrae
5. the rectum in relation to the mouth

II. Match the terms in the key to the body regions listed in questions 6–12.

Key:
   a. oral
   b. occipital

c. gluteal
d. carpal
e. palmar
f. cervical
g. axillary
6. buttocks
7. palm
8. back of head
9. mouth
10. wrist
11. armpit
12. neck

III. **Match the terms in the key to the organs listed in questions 13–18.**

Key:

    a. cranial cavity
    b. vertebral canal
    c. thoracic cavity

d. abdominal cavity
e. pelvic cavity
13. stomach
14. heart
15. urinary bladder
16. brain
17. liver
18. spinal cord

IV. **Match the organ systems in the key to the organs listed in questions 19–25.**

Key:

    a. digestive system
    b. urinary system
    c. respiratory system
    d. cardiovascular system
    e. reproductive system
    f. nervous system
    g. endocrine system

19. thyroid gland
20. lungs
21. heart
22. ovaries
23. brain
24. stomach
25. kidneys

V. **Fill in the blanks.**

26. A(n) _____ is composed of several types of tissues and performs a particular function.
27. The imaginary plane that passes through the midline of the body is called the _____ plane.
28. All the organ systems of the body together function to maintain _____, a relative constancy of the internal environment.

MEDICAL TERMINOLOGY EXERCISE

Consult Appendix B for help in pronouncing, analyzing, and filling in the blanks to give a brief meaning to the terms that follow.

1. Suprapubic (su"pruh-pyū′bik) means _____ the pubis.
2. Infraorbital (in"fruh-or′bĭ-tal) means _____ the eye orbit.
3. Gastrectomy (gas-trek′to-me) means excision of the _____.
4. Celiotomy (se"le-ot′o-me) means incision (cut into) of the _____.
5. Macrocephalus (mak"ro-sef′uh-lus) means large _____.
6. Transthoracic (trans"tho-ras′ik) means across the _____.

7. Bilateral (bi-lat′er-al) means two or both _____.
8. Ophthalmoscope (of-thal′mo-skōp) is an instrument to view inside the _____.
9. Dorsalgia (dor-sal′je-uh) means pain in the _____.
10. Endocrinology (en"do-krĭ-nol′o-je) is the _____ of the endocrine system.
11. The pectoralis (pek-to-ral′is) muscle can be found on the _____.
    a. chest b. head c. buttocks
    d. thigh

12. The sacral (sa′krul) nerves are located in the _____.
    a. lower back b. neck c. upper back
    d. head
13. Hematuria (he-muh-tu′re-uh) means _____ in the urine.
14. Nephritis (nef-ri′tis) is _____ of the _____.
    a. lungs b. heart c. liver d. kidneys
15. Tachypnea (tak-ip-ne′uh) is a breathing rate that is _____.
    a. faster than normal b. slower than normal

WEBSITE LINK

Visit this text's website at <u>http://www.mhhe.com/maderap6</u> **for additional quizzes, interactive learning exercises, and other study tools.**

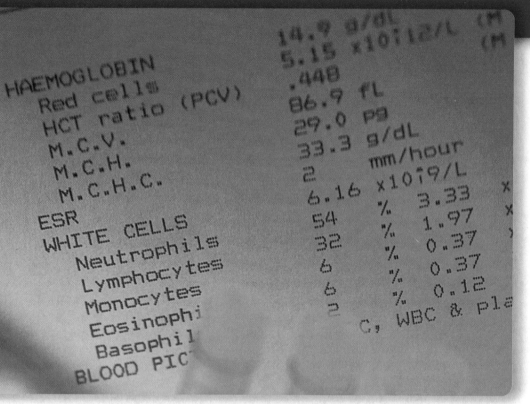

HAEMOGLOBIN 14.9 g/dL (M
  Red cells 5.15 x10↑12/L (M
  HCT ratio (PCV) .448
  M.C.V. 86.9 fL
  M.C.H. 29.0 pg
  M.C.H.C. 33.3 g/dL
ESR 2 mm/hour
WHITE CELLS 6.16 x10↑9/L
  Neutrophils 54 % 3.33 x
  Lymphocytes 32 % 1.97 x
  Monocytes 6 % 0.37
  Eosinophi 6 % 0.37
  Basophil 2 % 0.12
BLOOD PIC C, WBC & pla

Studying the chemistry of blood, including the concentrations of cells and dissolved solutes, is an important first step in diagnosing a disease.

CHAPTER

2

Chemistry of Life

Chapter Outline & Learning Objectives

After you have studied this chapter, you should be able to:

2.1 Basic Chemistry (p. 21)
- Describe how an atom is organized, and tell why atoms interact.
- Define radioactive isotope, and describe how they can be used in the diagnosis and treatment of disease.
- Distinguish between an ionic bond and a covalent bond.

2.2 Water, Acids, and Bases (p. 24)
- Describe the characteristics of water and three functions of water in the human body.
- Explain the difference between an acid and a base with examples.
- Use and understand the pH scale.

2.3 Molecules of Life (p. 26)
- List the four classes of macromolecules in cells, and distinguish between a dehydration reaction and a hydrolysis reaction.
- Name the individual subunits that comprise carbohydrates, lipids, proteins, and nucleic acids.

2.4 Carbohydrates (p. 26)
- Give some examples of different types of carbohydrates and their specific functions in cells.

2.5 Lipids (p. 28)
- Describe the composition of a neutral fat, and give examples of how lipids function in the body.

2.6 Proteins (p. 31)
- State the major functions of proteins, and tell how globular proteins are organized.

2.7 Nucleic Acids (p. 34)
- Describe the structure and function of DNA and RNA in cells.
- Explain the importance of ATP in the body.

Medical Focus
Nutrition Labels (p. 28)
Prions: Malicious Proteins? (p. 33)

2.1 Basic Chemistry

Matter is anything that takes up space and has weight; it can be a solid, a liquid, or a gas. Therefore, not only are we humans matter, but so are the water we drink and the air we breathe.

Elements and Atoms

All matter is composed of basic substances called **elements.** It's quite remarkable that there are only 92 naturally occurring elements. It is even more surprising that over 90% of the human body is composed of just four elements: carbon, nitrogen, oxygen, and hydrogen.

Every element has a name and a symbol; for example, carbon has been assigned the atomic symbol C (Fig. 2.1*a*). Some of the symbols we use for elements are derived from Latin. For example, the symbol for sodium is Na because the Latin word *natrium* means sodium.

Elements are composed of tiny particles called atoms. The same name is given to both an element and its atoms.

Atoms

An **atom** is the smallest unit of an element that still retains the chemical and physical properties of the element. Although it is possible to split an atom by physical means, an atom is the smallest unit to enter into chemical reactions. For our purposes, it is satisfactory to think of each atom as having a central nucleus and pathways about the nucleus called *shells*. The subatomic particles called **protons** and **neutrons** are located in the nucleus, and **electrons** orbit about the nucleus in the shells (Fig. 2.1*b*). Most of an atom is empty space. If we could draw an atom the size of a football stadium, the nucleus would be like a gumball in the center of the field, and the electrons would be tiny specks whirling about in the upper stands.

Protons carry a positive (+) charge, and electrons have a negative (−) charge. The atomic number of an atom tells you how many protons, and therefore how many electrons, an atom has when it is electrically neutral. For example, the atomic number of carbon is six; therefore, when carbon is neutral, it has six protons and six electrons. How many electrons are in each shell of an atom? The inner shell is the lowest energy level and can hold only two electrons; after that, each shell, for the atoms noted in Figure 2.1*a*, can hold up to eight electrons. Using this information, we can calculate that carbon has two shells and that the outer shell has four electrons.

The number of electrons in the outer shell determines the chemical properties of an atom, including how readily it enters into chemical reactions. As we shall see, an atom is most stable when the outer shell has eight electrons. (Hydrogen, with only one shell, is an exception to this statement. Atoms with only one shell are stable when this shell contains two electrons.)

The subatomic particles are so light that their weight is indicated by special designations called *atomic mass units*. Protons and neutrons each have a weight of one atomic mass unit, and electrons have almost no mass. Therefore, the mass number of an atom generally tells you the number of protons plus the

Common Elements in Living Things				
Element	Atomic Symbol	Atomic Number	Atomic Weight	Comment
hydrogen	H	1	1	These elements make up most biological molecules.
carbon	C	6	12	
nitrogen	N	7	14	
oxygen	O	8	16	
phosphorus	P	15	31	
sulfur	S	16	32	
sodium	Na	11	23	These elements occur mainly as dissolved salts.
magnesium	Mg	12	24	
chlorine	Cl	17	35	
potassium	K	19	39	
calcium	Ca	20	40	

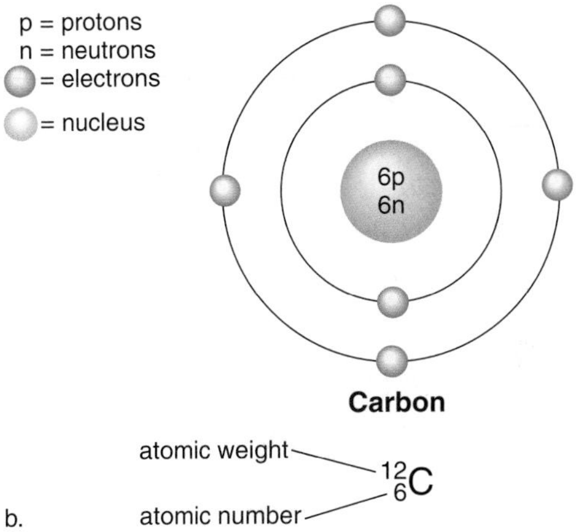

a.

p = protons
n = neutrons
= electrons
= nucleus

Carbon

atomic weight — $^{12}_{6}C$
atomic number

b.

Figure 2.1 Elements and atoms. **a.** The atomic symbol, number, and weight are given for common elements in the body. **b.** The structure of carbon shows that an atom contains the subatomic particles called protons (p) and neutrons (n) in the nucleus (colored pink) and electrons (colored blue) in shells about the nucleus.

number of neutrons. How could you calculate that carbon (C) has six neutrons? Carbon's mass number is 12, and you know from its atomic number that it has six protons. Therefore, carbon has six neutrons (Fig. 2.1*b*).

Also, as shown in Figure 2.1*b*, the atomic number of an atom is often written as a subscript to the lower left of the atomic symbol. The mass number is often written as a superscript to the upper left of the atomic symbol.

Isotopes

Isotopes of the same type of atom differ in the number of neutrons and therefore in the weight. For example, the element carbon has three common isotopes:

$$^{12}_{6}C \qquad ^{13}_{6}C \qquad ^{14}_{6}C*$$
$$*radioactive$$

Carbon 12 has six neutrons, carbon 13 has seven neutrons, and carbon 14 has eight neutrons. (Note that the mass number minus the atomic number equals the number of neutrons.) Unlike the other two isotopes of carbon, carbon 14 is unstable and breaks down over time. As carbon 14 decays, it releases various types of energy in the form of rays and subatomic particles, and therefore it is a **radioactive isotope.** The radiation given off by radioactive isotopes can be detected in various ways. You may be familiar with the use of a Geiger counter to detect radiation.

Low Levels of Radiation

The importance of chemistry to biology and medicine is nowhere more evident than in the many uses of radioactive isotopes. A radioactive isotope behaves the same as do the stable isotopes of an element. This means that you can put a small amount of radioactive isotope in a sample, and it becomes a **tracer** by which to detect molecular changes (see the Medical Focus, on page 16 for a description of the uses of radioactive tracers).

High Levels of Radiation

Radioactive substances in the environment can harm cells, damage DNA, and cause cancer. The release of radioactive particles following a nuclear power plant accident can have far-reaching and long-lasting effects on human health. The harmful effects of radiation can also be put to good use, however. Radiation from radioactive isotopes has been used for many years to sterilize medical and dental products. Now the possibility exists

that it can be used to sterilize the U.S. mail to free it of possible pathogens, such as anthrax spores.

The ability of radiation to kill cells is often applied to cancer cells. Radioisotopes can be introduced into the body in a way that allows radiation to destroy only the cancerous cells, with little risk to the rest of the body.

Molecules and Compounds

Atoms often bond with each other to form a chemical unit called a **molecule.** A molecule can contain atoms of the same kind, as when an oxygen atom joins with another oxygen atom to form oxygen gas. Or the atoms can be different, as when an oxygen atom joins with two hydrogen atoms to form water. When the atoms are different, a **compound** results.

Two types of bonds join atoms: the **ionic bond** and the **covalent bond.** Ionic bonds are created by electrical attraction between **ions**—electrically charged atoms. Covalent bonds are created by the sharing of electrons between atoms.

Ionic Bonds

Recall that atoms with more than one shell are most stable when the outer shell contains eight electrons. Sometimes during a reaction, atoms give up or take on an electron(s) in order to achieve a stable outer shell.

Figure 2.2 depicts a reaction between a sodium (Na) atom and a chlorine (Cl) atom. Sodium, with one electron in

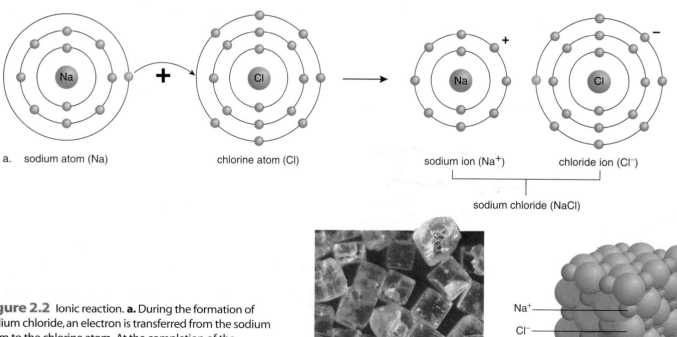

a. sodium atom (Na)   chlorine atom (Cl)

sodium ion (Na$^+$)   chloride ion (Cl$^-$)

sodium chloride (NaCl)

Figure 2.2 Ionic reaction. **a.** During the formation of sodium chloride, an electron is transferred from the sodium atom to the chlorine atom. At the completion of the reaction, each atom has eight electrons in the outer shell, but each also carries a charge as shown. **b.** In a sodium chloride crystal, bonding between ions creates a three-dimensional lattice in which each Na$^+$ ion is surrounded by six Cl$^-$ ions, and each Cl$^-$ is surrounded by six Na$^+$.

b. 1 mm

Na$^+$
Cl$^-$

the outer shell, reacts with a single chlorine atom. Why? Because once the reaction is finished and sodium loses one electron to chlorine, its outer shell will have eight electrons. Similarly, a chlorine atom, which has seven electrons already, needs only to acquire one more electron to have a stable outer shell.

Ions are particles that carry either a positive (+) or negative (−) charge. When the reaction between sodium and chlorine is finished, the sodium ion carries a positive charge because it now has 11 positively charged protons and only 10 negatively charged electrons (11^+, 10^-; net charge: 1^+). Meanwhile, the negatively charged chloride ion has 18 protons and 19 electrons (18^+, 19^-; net charge: 1^-). The attraction between oppositely charged sodium ions and chloride ions forms an ionic bond. The resulting compound, sodium chloride, is table salt, which we use to enliven the taste of foods. Salts characteristically form an **ionic lattice** that **dissociates** (separates into ions) after dissolving in water (Fig. 2.2b).

In contrast to sodium, why would calcium, with two electrons in the outer shell, react with two chlorine atoms? Because whereas calcium needs to lose two electrons, each chlorine, with seven electrons already, requires only one more electron to have a stable outer shell. The resulting salt ($CaCl_2$) is called calcium chloride.

The balance of various ions in the body is important to our health. Too much sodium in the blood can contribute to **hypertension** (high blood pressure); not enough calcium leads to **rickets** (a bowing of the legs) in children; too much or too little potassium results in **arrhythmia** (heartbeat irregularities). Bicarbonate, hydrogen, and hydroxide ions are all involved in maintaining the acid-base balance of the body (see p. 25).

Covalent Bonds

As a result of other reactions, atoms share electrons in covalent bonds instead of losing or gaining them. The overlapping outermost shells in Figure 2.3 indicate that the atoms are sharing electrons. Just as two hands participate in a handshake, each atom contributes one electron to the pair that is shared. These electrons spend part of their time in the outer shell of each atom; therefore, they are counted as belonging to both bonded atoms.

Covalent bonds can be represented in a number of ways. In contrast to the diagrams in Figure 2.3, structural formulas use straight lines to show the covalent bonds between the atoms. Each line represents a pair of shared electrons. Molecular formulas indicate only the number of each type of atom making up a molecule. A comparison follows:

Structural formula: Cl—Cl
Molecular formula: Cl_2

Double and Triple Bonds Besides a single bond, in which atoms share only a pair of electrons, a double or a triple bond can form. In a double bond, atoms share two pairs of electrons, and in a triple bond, atoms share three pairs of electrons between them. For example, in Figure 2.3, each nitrogen atom

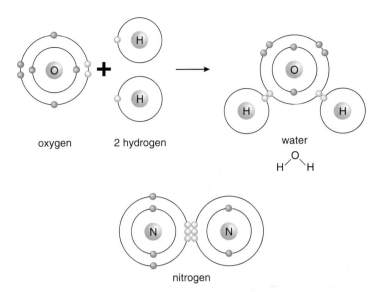

Figure 2.3 Covalent reactions. After a covalent reaction, each atom will have filled its outer shell by sharing electrons. To determine this, it is necessary to count the shared electrons as belonging to both bonded atoms. Oxygen and nitrogen are most stable with eight electrons in the outer shell. Hydrogen is most stable with two electrons in the outer shell.

(N) requires three electrons to achieve a total of eight electrons in the outer shell. Notice that six electrons are placed in the outer overlapping shells in the diagram and that three straight lines are in the structural formula for nitrogen gas (N_2).

What would be the structural and molecular formulas for carbon dioxide? Carbon, with four electrons in the outer shell, requires four more electrons to complete its outer shell. Each oxygen, with six electrons in the outer shell, needs only two electrons to complete its outer shell. Therefore, carbon shares two pairs of electrons with each oxygen atom, and the formulas are as follows:

Structural formula: O=C=O
Molecular formula: CO_2

Polar Covalent Bonds

In atoms that form a covalent molecule, custody of the shared electrons is not always equal. One of the elements often keeps the electrons for a longer period of time. Such elements are called **electronegative** (think "electron-grabbers"). When electronegative elements such as oxygen are teamed with smaller, weaker elements like hydrogen, the resulting compound (water, H_2O) forms a **polar covalent bond.** Because the electrons spend most of their time with the electronegative element (oxygen), this element acquires a partial negative charge. The weaker hydrogen atoms gain a partial positive charge because they surrender their electrons for a longer period of time. Because polar covalent bonds help determine the characteristics of water and other molecules, they are very important in living organisms.

1. One particular isotope of the element potassium has the following symbol: ^{39}K. The atomic number (number of protons) for potassium is 19. How many neutrons are contained in this isotope?

2. Fluorine gas contains two atoms of fluorine that share two electrons. Draw the structural and molecular formulas for fluorine gas.

3. The world's worst nuclear accident, at Chernobyl, Russia in 1986, released radioactive strontium into the air. Strontium is chemically similar to calcium. What body tissue might be damaged by radioactive strontium?

2.2 Water, Acids, and Bases

Water is the most abundant molecule in living organisms, usually making up about 60–70% of the total body weight. Even so, water is an **inorganic molecule** because it does not contain carbon atoms. Carbon atoms are common to **organic molecules.**

As mentioned previously, in water the electrons spend more time circling the larger oxygen (O) atom than the smaller hydrogen (H) atoms. This imparts a slight negative charge (symbolized as δ^-) to the oxygen and a slight positive charge (symbolized as δ^+) to the hydrogen atoms. Therefore, water is a **polar molecule** with negative and positive ends:

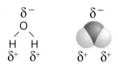

The diagram on the left shows the structural formula of water, and the one on the right is called a space-filling model.

Hydrogen Bonds

A **hydrogen bond** occurs whenever a covalently bonded hydrogen has a slight positive charge, and is attracted to a slight negatively charged atom nearby. A hydrogen bond is represented by a dotted line because it is relatively weak and can be broken rather easily. In Figure 2.4, you can see that each hydrogen atom, being slightly positive, bonds to the slightly negative oxygen atom of another water molecule nearby.

Properties of Water

Polarity and hydrogen bonding cause water to have many properties beneficial to life, including the three to be mentioned here.

1. Water is a **solvent** for polar (charged) molecules and thereby facilitates chemical reactions both outside and within our bodies.

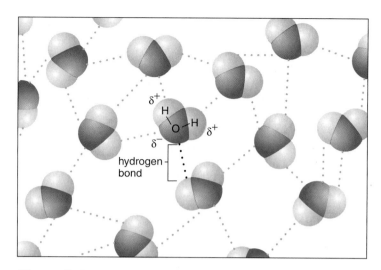

Figure 2.4 Hydrogen bonding between water molecules. The polarity of the water molecules causes hydrogen bonds (dotted lines) to form between the molecules.

When ions and molecules disperse in water, they move about and collide, allowing reactions to occur. Therefore, water is a solvent that facilitates chemical reactions. For example, when a salt such as sodium chloride (NaCl) is put into water, the negative ends of the water molecules are attracted to the sodium ions, and the positive ends of the water molecules are attracted to the chloride ions. This causes the sodium ions and the chloride ions to dissolve and to separate in water:

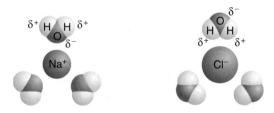

The salt NaCl dissolves in water.

Ions and molecules that are water-soluble are said to be **hydrophilic** ("water loving"). Nonionized and nonpolar molecules that do not interact with water are said to be **hydrophobic** ("water fearing").

2. Water molecules are **cohesive. Water-based solutions** fill vessels, such as blood vessels.

Water molecules cling together because of hydrogen bonding, and yet water flows freely. This property allows dissolved and suspended molecules to be evenly distributed throughout a system. Therefore, water is an excellent transport medium. Within our bodies, the blood that fills our arteries and veins is 92% water. Blood transports oxygen and nutrients to the cells and removes wastes such as carbon dioxide from the cells.

3. Water has a high **specific heat capacity** and a high **heat of vaporization.**

Specific heat capacity is the amount of heat energy needed to change an object's temperature by exactly 1°C. Water has the highest specific heat of almost any substance in nature. Thus, water can absorb a tremendous amount of heat energy without changing its temperature. This is due to the hydrogen bonds that link water molecules together. Maintaining a fairly constant body temperature is possible, in part, because the body is roughly 60% water. High water content helps to insulate the body from extreme heat and cold in the environment. Hydrogen bonding also accounts for water's high heat of vaporization—the amount of energy needed to turn water into steam. As water evaporates during sweating, the body can be cooled rapidly and effectively.

Acids and Bases

When water molecules dissociate (break up), they release an equal number of hydrogen ions (H^+) and hydroxide ions (OH^-):

$$H-O-H \rightleftharpoons H^+ + OH^-$$

water · · · · hydrogen ion · · · · hydroxide ion

Only a few water molecules at a time dissociate, and the actual number of H^+ and OH^- is very small (1×10^{-7} moles/liter).

Acids are substances that dissociate in water, releasing hydrogen ions (H^+). For example, an important inorganic acid is hydrochloric acid (HCl), which dissociates in this manner:

$$HCl \longrightarrow H^+ + Cl^-$$

Dissociation is almost complete; therefore, HCl is called a **strong acid.** If hydrochloric acid is added to a beaker of water, the number of hydrogen ions (H^+) increases greatly. Lemon juice, vinegar, tomatoes, and coffee are all acidic solutions.

Bases are substances that either take up hydrogen ions (H^+) or release hydroxide ions (OH^-). For example, an important inorganic base is sodium hydroxide (NaOH), which dissociates in this manner:

$$NaOH \longrightarrow Na^+ + OH^-$$

Dissociation is almost complete; therefore, sodium hydroxide is called a **strong base.** If sodium hydroxide is added to a beaker of water, the number of hydroxide ions increases. Milk of magnesia and ammonia are common basic solutions.

pH Scale

In chemistry, pH is a chemist's "shorthand"; it is defined as the negative of the base 10 logarithm of free hydrogen ion concentration. Suppose a chemist stirs up a solution containing 0.00001 moles per liter of free hydrogen ions. The solution's concentration can be written (using scientific notation) as 1×10^{-5} moles per liter. The base 10 logarithm is the exponent, -5. Using the pH system, concentration is expressed as negative times (-5), or pH 5. (Recall that a negative number multiplied by a negative number equals a positive number.)

The **pH scale,** which ranges from 0 to 14, is used to indicate the acidity and basicity (alkalinity) of a solution. pH 7, which is the pH of water, is neutral pH because water releases an equal number of hydrogen ions (H^+) and hydroxide ions (OH^-). Notice in Figure 2.5 that any pH above 7 is a base, with more hydroxide ions than hydrogen ions. Any pH below 7 is an acid, with more hydrogen ions than hydroxide ions. As we move toward a higher pH, each unit has 10 times the basicity of the previous unit, and as we move toward a lower pH, each unit has 10 times the acidity of the previous unit. This means that even a small change in pH represents a large change in the proportional number of hydrogen and hydroxide ions in the body.

In chemistry, concentration is measured in units of **moles per liter** of solution. Just as one dozen equals 12 of something, one **mole** in chemistry is a specific, very large number: 6.023×10^{23} atoms, ions, or molecules. This number, termed Avogadro's number, is the number of carbon atoms in exactly 12 grams of ^{12}C.

The pH of body fluids needs to be maintained within a narrow range, or else health suffers. The pH of our blood when we are healthy is always about 7.4—that is, just slightly basic (alkaline). If the pH value drops below 7.35, the person is said to have **acidosis**; if it rises above 7.45, the condition is called **alkalosis**. The pH stability is normally possible because the body has built-in mechanisms to

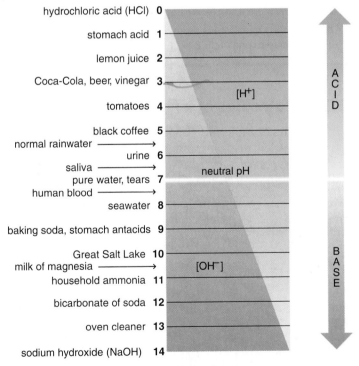

Figure 2.5 The pH scale. The proportionate amount of hydrogen ions to hydroxide ions is indicated by the diagonal line. Any solution with a pH above 7 is basic, while any solution with a pH below 7 is acidic.

prevent pH changes. The respiratory and urinary systems co-operate to maintain normal pH (see Chapters 14 and 16). Additionally, **buffers** in blood and body fluids help to keep pH within its normal range. Buffers are chemicals, or pairs of chemicals, that take up excess hydrogen ions (H^+) or hydroxide ions (OH^-). For example, the combination of carbonic acid (H_2CO_3) and the bicarbonate ion (HCO_3^-) helps keep the pH of the blood relatively constant because carbonic acid can dissociate to release hydrogen ions, while the bicarbonate ion can take them up!

Electrolytes

As we have seen, salts, acids, and bases are molecules that dissociate; that is, they ionize in water. For example, when a salt such as sodium chloride is put in water, the Na^+ ion separates from the Cl^- ion.

Substances that release ions when put into water are called **electrolytes** because the ions can conduct an electrical current. The electrolyte balance in the blood and body tissues is important for good health because it affects the functioning of vital organs such as the heart and the brain.

> *Content* **CHECK-UP!**
>
> **4.** A soda has a pH of 4. How many free hydrogen ions are in the soda in moles per liter?
>
> **5.** Depending on what you've eaten recently (as well as other factors), the pH of your urine can vary from 5 to 8.
>
> a. Which is more acidic—urine at pH 5 or pH 8?
>
> b. How many more free hydrogen ions are in a solution of pH 5, compared to one of pH 8?

2.3 Molecules of Life

Four categories of molecules called carbohydrates, lipids, proteins, and nucleic acids are unique to cells. They are called *macromolecules*, or *polymers*, because each is composed of many smaller subunits, or *monomers*.

Category	Polymer	Monomer(s) (subunits)
Carbohydrates	Polysaccharide	Monosaccharide
Lipids	Fat	Glycerol and fatty acids
Proteins	Polypeptide	Amino acid
Nucleic acid	DNA, RNA	Nucleotide

During synthesis of macromolecules, the cell uses a **dehydration reaction,** so called because an —OH (hydroxyl group) and an —H (hydrogen atom)—the equivalent of a water molecule—are removed as the molecule forms (Fig. 2.6a). The result is reminiscent of a train whose length is determined by how many boxcars are hitched together. To break up macromolecules, the cell reverses dehydration, using a

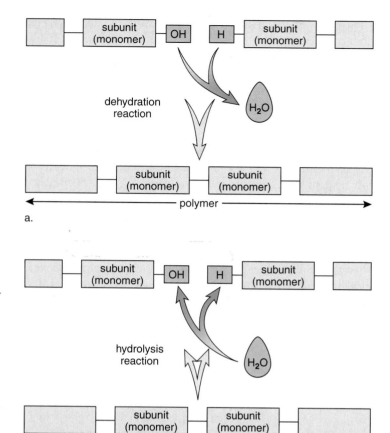

a.

b.

Figure 2.6 Synthesis and decomposition of macromolecules. **a.** In cells, synthesis often occurs when subunits bond following a dehydration reaction (removal of H_2O). **b.** Decomposition occurs when the subunits in a macromolecule separate after a hydrolysis reaction (addition of H_2O).

hydrolysis reaction. In hydrolysis ("splitting with water"), the components of water are added (Fig. 2.6b).

2.4 Carbohydrates

Carbohydrates, like all organic molecules, always contain carbon (C) and hydrogen (H) atoms. Carbohydrate molecules are characterized by the presence of the atomic grouping H—C—OH, in which the ratio of hydrogen atoms (H) to oxygen atoms (O) is approximately 2:1. Because this ratio is the same as the ratio in water, the name "hydrates of carbon" seems appropriate. **Carbohydrates** first and foremost function for quick, short-term cellular energy in all organisms, including humans. The role of carbohydrates and other nutrients in maintaining good health is discussed in Medical Focus, p. 28.

Simple Carbohydrates

If the number of carbon atoms in a carbohydrate is low (between three and seven), it is called a simple sugar, or **monosaccharide.** The designation **pentose** means a 5-carbon sugar, and the designation **hexose** means a 6-carbon sugar. **Glucose,** the hexose our bodies use as an immediate source of energy, can be written in any one of these ways:

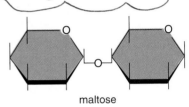

$C_6H_{12}O_6$

Other common hexoses are fructose, found in fruits, and galactose, a constituent of milk. A **disaccharide** (*di*, two; *saccharide*, sugar) is made by joining only two monosaccharides together by a dehydration reaction (see Fig. 2.6a). **Maltose** is a disaccharide that contains two glucose molecules:

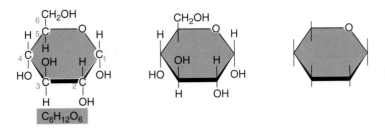

maltose

When glucose and fructose join, the disaccharide **sucrose** forms. Sucrose—ordinary table sugar—is derived from sugar cane or sugar beets. **Lactose,** or milk sugar, is formed from glucose and galactose. (Individuals who are *lactose intolerant* cannot digest this sugar. If they consume dairy products, they experience intestinal cramping and diarrhea).

Complex Carbohydrates (Polysaccharides)

Macromolecules such as starch, glycogen, and cellulose are **polysaccharides** that contain many glucose units. Although polysaccharides can contain other sugars, we will study the ones that use glucose.

Starch and Glycogen

Starch and **glycogen** are ready storage forms of glucose in plants and animals, respectively. Some of the macromolecules in starch are long chains of up to 4,000 glucose units. Starch has fewer side branches, or chains of glucose that branch off from the main chain, than does glycogen, as shown in Figures 2.7 and 2.8. Flour, usually acquired by grinding wheat and used for baking, is high in starch, and so are potatoes.

After we eat starchy foods such as potatoes, bread, and pasta, glucose enters the bloodstream and the liver stores excess glucose as glycogen. In between meals, the liver releases glucose so that the blood glucose concentration is always about 0.1%. Maintaining a fairly constant blood

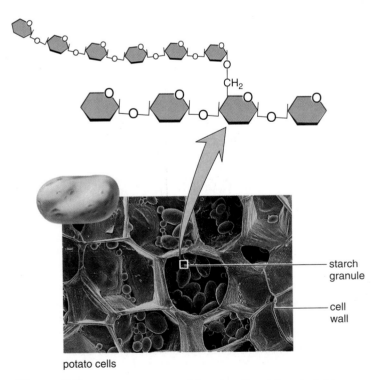

starch granule

cell wall

potato cells

Figure 2.7 Starch structure and function. Starch has straight chains of glucose molecules. Some chains are also branched, as indicated. The electron micrograph shows starch granules in potato cells. Starch is the storage form of glucose in plants.

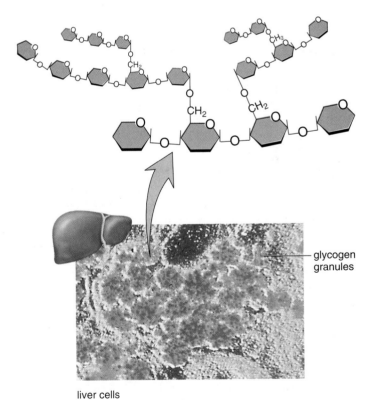

glycogen granules

liver cells

Figure 2.8 Glycogen structure and function. Glycogen is more branched than starch. The electron micrograph shows glycogen granules in liver cells. Glycogen is the storage form of glucose in humans.

Nutrition Labels

No, it's not quite as interesting as your morning newspaper, but tucked into the corner of your *Aqua Puffs* (and all other packaged food) is a piece of important information. Reading nutrition labels, like the one depicted in Figure 2A, should become habit for anyone interested in health and wellness. These labels contain important information about the chemistry of foods we normally consume, providing the nutritional breakdown for a single serving of the food (in the Aqua Puffs example, 1¼ cup, or 57 grams). Labels also tell us what share of daily nutrients and Calories are represented in that serving, based on a daily diet of 2,000 Calories. Calories are a measurement of food energy. One serving of the cereal provides 220 Calories (20 from fat).

Carbohydrates

When you study nutrition labels, first check the carbohydrate content. Carbohydrates (simple sugars and complex carbohydrates, or polysaccharides) are the most readily available source of body energy and the optimal energy source for brain and nerve tissue. Breads, cereals, vegetables, and fruits contain complex carbohydrates, as well as protein, minerals, and vitamins. Complex carbohydrates also contain fiber, and our diets should contain 25 to 30 grams of fiber daily. *Soluble fiber* combines with cholesterol in food, preventing cholesterol from being absorbed from the digestive tract into the body. *Insoluble fiber* has a laxative effect. Breads and cereals made from whole grains (whole wheat, for example) and fruits and vegetables have the highest fiber content. Current dietary guidelines recommend that at least half of one's daily intake of breads, cereals, pasta, etc., should consist of whole-grain varieties.

Fats

Dietary fat is needed for energy, for absorption of specific vitamins, and for manufacturing cell lipids and steroids. The body

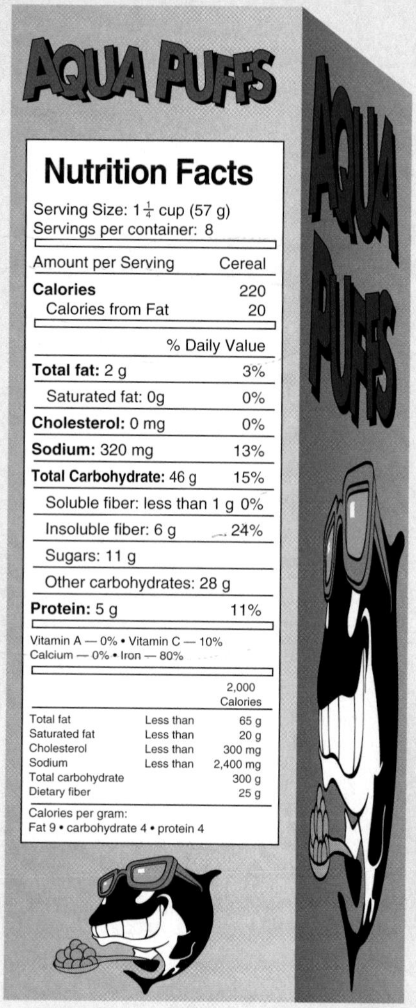

Nutrition Facts	
Serving Size: 1¼ cup (57 g)	
Servings per container: 8	
Amount per Serving	Cereal
Calories	220
Calories from Fat	20
	% Daily Value
Total fat: 2 g	3%
Saturated fat: 0g	0%
Cholesterol: 0 mg	0%
Sodium: 320 mg	13%
Total Carbohydrate: 46 g	15%
Soluble fiber: less than 1 g	0%
Insoluble fiber: 6 g	24%
Sugars: 11 g	
Other carbohydrates: 28 g	
Protein: 5 g	11%
Vitamin A — 0% • Vitamin C — 10%	
Calcium — 0% • Iron — 80%	

		2,000 Calories
Total fat	Less than	65 g
Saturated fat	Less than	20 g
Cholesterol	Less than	300 mg
Sodium	Less than	2,400 mg
Total carbohydrate		300 g
Dietary fiber		25 g

Calories per gram:
Fat 9 • carbohydrate 4 • protein 4

Figure 2A Nutrition label on the side panel of a cereal box.

glucose concentration is important because the brain and nerve tissues work best when they use glucose as their energy source.

Cellulose

The polysaccharide **cellulose** is found in plant cell walls. In cellulose, the glucose units are joined by a slightly different type of linkage from that in starch or glycogen. Although this might seem to be a technicality, actually it is important because humans are unable to digest foods containing this type of linkage. Cellulose largely passes through our digestive tract as fiber, or roughage. It is believed that fiber in the diet is

necessary to good health, and some researchers have suggested it may even help prevent colon cancer.

2.5 Lipids

Lipids contain more energy per gram than other biological molecules, and some function as long-term energy storage molecules in organisms. Other lipids form cell membranes that enclose individual cells, as well as membranes that surround the organelles found inside cells. **Steroids** are a large class of lipids that includes, among other molecules, the sex hormones.

Lipids are diverse in structure and function, but they have a common characteristic: They do not dissolve in water. Their

stores any excess nutrient energy as fat. An overconsumption of daily Calories can lead to obesity and adverse effects on health. High levels of saturated fat have been implicated in cancer of the colon, pancreas, ovary, prostate, and breast. Cholesterol and saturated fat contribute to atherosclerosis, the narrowing of arteries caused by formation of fat-filled plaques. Atherosclerosis is the major cause of hypertension (high blood pressure), heart attack, and stroke in the Western world. A 2,000 Calorie per day diet should contain no more than 65 grams (585 Calories) of fat. Knowing how specific foods contribute to the maximum recommended daily amount of fat, saturated fat, and cholesterol is important for long-term health.

Proteins

The recommended daily adult allowance for protein is approximately 5 to 6 ounces per day—a serving about the same size as a deck of playing cards. This small portion is usually adequate because protein is needed only for tissue growth and repair. Proteins from animal sources, although a staple of the Western diet, may be high in saturated fat. Therefore, it is considered good health sense to rely on lower-fat protein from plant origins (e.g., whole-grain cereals, dark breads, rice, and legumes such as beans).

Minerals and Vitamins

The amount of dietary sodium (as in table salt) in foods can be a concern because excessive sodium intake seems to further elevate blood pressure in people with hypertension. Sodium intake should be no more than 2,400 milligrams (mg) per day for people with hypertension.

Vitamins are organic molecules required in small amounts in the diet for good health. Each vitamin has a recommended daily intake, and nutrition labels tell the percentage of the recommended amount provided by one food serving. The antioxidant vitamins C, E, and beta-carotene (which is converted to vitamin A in the body) seem to have a special importance to cells. These forms of vitamins can neutralize free radicals, which are unstable molecules generated during cellular metabolism. Free radicals damage cell components and are linked to cancer and atherosclerosis.

So how nutritious is that bowl of Aqua Puffs? A single serving provides 15% of recommended daily carbohydrate—but closer study will reveal that almost one-fourth of the total carbohydrate is simple sugar. Sugar added to food tastes good, but adds calories with few nutrients. A serving does contribute 6 grams of beneficial insoluble fiber, as well as 5 grams of protein. Total fat content is acceptably small, with no saturated fat. And while the cereal provides no vitamin A or calcium, one serving does contain 10% of the suggested daily intake of vitamin C and 80% of the recommended daily intake of iron. Add a cup of low-fat milk, and this cereal becomes acceptable for an occasional quick breakfast.

A better choice for a high-fiber complex carbohydrate at breakfast might be fresh fruit (perhaps with a dairy protein source, like low-fat yogurt). In addition to carbohydrate energy and fiber, fruits and vegetables contribute antioxidant vitamins to the diet. Current recommendations call for 2 cups of fresh fruit and 2½ cups of vegetables daily. While dietary vitamin supplements may provide a potential safeguard against cancer and cardiovascular disease, taking supplements instead of improving your intake of fruits and vegetables is not a good approach. Fruits and vegetables provide hundreds of beneficial compounds that cannot be obtained from a vitamin pill.

Current nutritional guidelines from the United States Department of Agriculture are discussed in Chapter 15.

low solubility in water is due to an absence of polar groups. They contain little oxygen and consist mostly of carbon and hydrogen atoms.

Fats and Oils

The most familiar lipids are those found in fats and oils. **Fats,** which are usually of animal origin (e.g., lard and butter), are solid at room temperature. **Oils,** which are usually of plant origin (e.g., corn oil and soybean oil), are liquid at room temperature. Fat has several functions in the body: It is used for long-term energy storage, it insulates against heat loss, and it forms a protective cushion around major organs.

Fats and oils form when one **glycerol** molecule reacts with three fatty acid molecules (Fig. 2.9). A fat is sometimes called a **triglyceride** because of its three-part structure, or a **neutral fat** because the molecule is nonpolar and carries no charge.

Emulsification

Emulsifiers can cause fats to mix with water. They contain molecules with a nonpolar end and a polar end. The molecules position themselves about an oil droplet so that their nonpolar ends project. Now the droplet disperses in water, which means that **emulsification** has occurred.

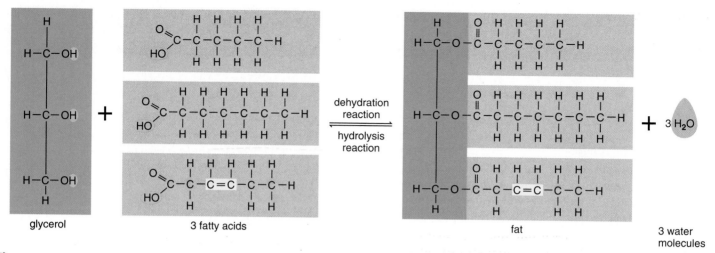

Figure 2.9 Synthesis and degradation of a fat molecule. Fatty acids can be saturated (no double bonds between carbon atoms) or unsaturated (have double bonds, colored yellow, between carbon atoms). When a fat molecule forms, three fatty acids combine with glycerol, and three water molecules are produced.

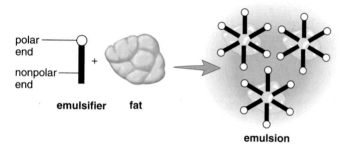

Emulsification takes place when dirty clothes are washed with soaps or detergents. Also, prior to the digestion of fatty foods, fats are emulsified by bile. The gallbladder stores bile for emulsifying fats prior to the digestive process.

Saturated and Unsaturated Fatty Acids

A **fatty acid** is a carbon–hydrogen chain that ends with the acidic group —COOH (Fig. 2.9). Most of the fatty acids in cells contain 16 or 18 carbon atoms per molecule, although smaller ones with fewer carbons are also known.

Fatty acids are either saturated or unsaturated. **Saturated fatty acids** have only single covalent bonds because the carbon chain is saturated, so to speak, with all the hydrogens it can hold. Saturated fatty acids account for the solid nature at room temperature of fats such as lard and butter. **Unsaturated fatty acids** have double bonds between carbon atoms wherever fewer than two hydrogens are bonded to a carbon atom. Unsaturated fatty acids account for the liquid nature of vegetable oils at room temperature. Hydrogenation of vegetable oils can convert them to margarine and products such as Crisco.

Phospholipids

Phospholipids, as their name implies, contain a phosphate group (Fig. 2.10). Essentially, they are constructed like fats, except that in place of the third fatty acid, there is a phosphate

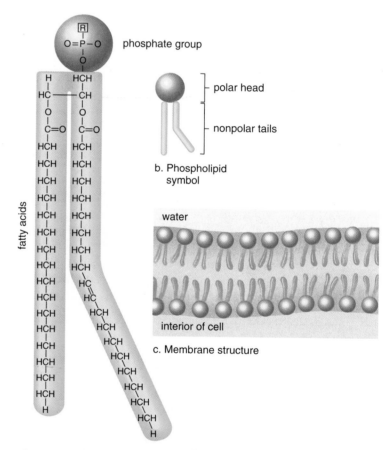

a. Phospholipid structure

Figure 2.10 Phospholipid structure and function.
a. Phospholipids are structured like fats, but one fatty acid is replaced by a polar phosphate group. **b.** Therefore, the head is polar while the tails are nonpolar. **c.** This causes the molecule to arrange itself as shown when exposed to water.

group or a grouping that contains both phosphate and nitrogen. Phospholipid molecules are not electrically neutral, as are fats, because the phosphate and nitrogen-containing groups are ionized. They form the so-called hydrophilic head of the molecule, while the rest of the molecule becomes the hydrophobic tails. Phospholipids are the backbone of cellular membranes; they spontaneously form a bilayer in which the hydrophilic heads face outward toward watery solutions and the tails form the hydrophobic interior.

Steroids

Steroids are lipids that have an entirely different structure from those of fats. Steroid molecules have a backbone of four fused carbon rings. Each one differs primarily by the side-chain molecules, called **functional groups,** attached to the rings.

Cholesterol is a component of an animal cell's outer membrane and is the precursor of several other steroids, such as the sex hormones estrogen and testosterone. The male sex hormone, testosterone, is formed primarily in the testes, and the female sex hormone, estrogen, is formed primarily in the ovaries. Testosterone and estrogen differ only by the functional groups attached to the same carbon backbone, yet they have a profound effect on the body and on our sexuality (Fig. 2.11a,b). Testosterone is a steroid that causes males to have greater muscle strength than females. Taking synthetic testosterone for this purpose, however, is dangerous to your health, as will be discussed in Chapter 10.

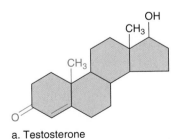

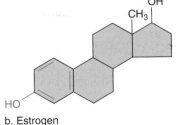

a. Testosterone   b. Estrogen

Figure 2.11 Steroids. All steroids have four rings, but they differ by attached groups. The effects of **(a)** testosterone and **(b)** estrogen on the body largely depend on the difference in the attached groups shown in red.

We know that a diet high in saturated fats and cholesterol can cause fatty material to accumulate inside the lining of blood vessels, thereby reducing blood flow. As discussed in the Medical Focus on page 28, nutrition labels are now required to list the calories from fat per serving and the percent daily value from saturated fat and cholesterol.

2.6 Proteins

Proteins perform a myriad of functions, including the following:

- Collagen (found in connective tissue and skin) and keratin (found in skin, hair, and nails) are fibrous structural proteins.
- Many hormones, which are messengers that influence cellular metabolism, are proteins.
- The proteins actin and myosin account for the movement of cells and the ability of our muscles to contract.
- Some proteins transport molecules in the blood; for example, hemoglobin is a complex protein in our blood that transports oxygen.
- Antibodies in blood and other body fluids are proteins that combine with pathogens or their toxins.
- Enzymes are globular (round) proteins that speed chemical reactions.

Structure of Proteins

Proteins are macromolecules composed of amino acid subunits. An **amino acid** has a central carbon atom bonded to a hydrogen atom and three groups. The name of the molecule is appropriate because one of these groups is an amino group and another is an acidic group. The third group is called an *R* group because it is the *remainder* of the molecule (Fig. 2.12a).

Amino acids differ from one another by the R group, which varies in structure from a single hydrogen atom to a complicated ring. When two amino acids join, a **dipeptide** results. **Polypeptides** contain three or more amino acids (Fig. 2.12b) Proteins are large polypeptides. Twenty different amino acids compose all human polypeptides. The sequence of amino acids in a polypeptide is called its **primary structure** (Fig. 2.12b).

The bond between amino acids is termed a **peptide bond.** The atoms of a peptide bond share electrons unevenly, making hydrogen bonding possible in a polypeptide. Due to hydrogen bonding, the polypeptide often twists to form a coil or folds into a fan shape. Coiling or folding creates the **secondary structure** of the protein (Fig. 2.12c) Finally, the coil bends and twists into a particular shape because of bonding between R groups. Hydrogen, ionic, and covalent bonding all occur in polypeptides, creating the **tertiary structure** (Fig. 2.12d) Hydrophobic portions of a polypeptide are found inside the molecule, while the hydrophilic portions are outside where they can contact water.

Some proteins have just one polypeptide, while others have more than one polypeptide, each with its own primary, secondary, and tertiary structures. If a protein has more than

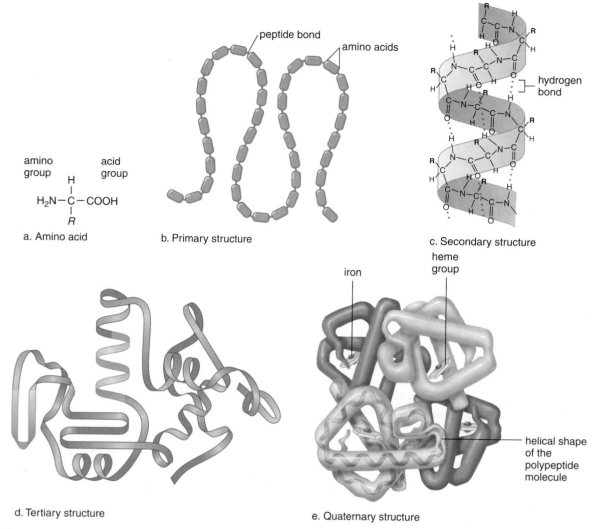

amino group   acid group

$$H_2N-\overset{\overset{\textstyle H}{|}}{\underset{\underset{\textstyle R}{|}}{C}}-COOH$$

a. Amino acid

peptide bond

amino acids

b. Primary structure

hydrogen bond

c. Secondary structure

iron

heme group

d. Tertiary structure

helical shape of the polypeptide molecule

e. Quaternary structure

Figure 2.12 Levels of polypeptide structure. **a.** Amino acids are the subunits of polypeptides. Note that an amino acid contains nitrogen. **b.** Polypeptides differ by the sequence of their amino acids, which are joined by peptide bonds. **c.** A polypeptide often twists to become a coil due to hydrogen bonding between members of the peptide bonds. **d.** The tertiary structure of polypeptide structure is due to various types of bonding between the *R* groups of the amino acids. **e.** The grouping of two or more small proteins creates the quaternary structure of a large protein molecule.

one polypeptide, arrangement of individual polypeptides gives a protein its **quaternary structure.**

The final three-dimensional shape of a protein is very important to its function. When proteins are exposed to extremes in heat and pH, they undergo an irreversible change in shape called **denaturation.** For example, addition of acid to milk (as when milk goes "sour") causes curdling. Heating causes egg white protein, called *albumin,* to coagulate (form clumps). Denaturation occurs because the normal bonding between the R groups has been disturbed. Once a protein loses its normal shape, it is no longer able to function normally. Researchers hypothesize that an alteration in protein organization may be the cause of Alzheimer disease (see the Medical Focus in Chapter 8) and variant Creutzfeldt-Jacob disease (see the Medical Focus, on page 33).

Enzymatic Reactions

Metabolism is the sum of all of the chemical reactions that occur in a cell. Most cellular reactions cannot happen unless an **enzyme** is present. Enzymes are protein *catalysts*—molecules that enable a particular metabolic reaction to occur at the body's normal temperature. Atoms and molecules frequently do not react unless they are first activated in some way. *Energy of activation* is the energy needed to start a reaction. In the lab, heat is often used to supply the energy needed for a reaction. For example, sugar can be broken down to form carbon dioxide and water, but doing so in a laboratory requires tremendous heat. In the body, enzymes lower the energy of activation by forming a complex with particular molecules. Just as a mutual friend can cause particular people to interact at a crowded party, enzymes in cells bring together atoms or molecules and cause

Prions: Malicious Proteins?

Infectious diseases are known to be caused by minute organisms that successfully dodge the body's defense mechanisms. These invaders include parasites such as liver flukes and parasitic worms; smaller, one-celled parasites like disease-causing strains of *Amoeba*, and the smallest unicellular organisms, the bacteria. Viruses are disease-causing particles that are even smaller than bacteria. All of these disease agents share a common trait: All contain the nucleic acids DNA or RNA, or both. Nucleic acids are large molecules that comprise the genetic material of parasites, bacteria, and viruses. Until recently, scientists believed that infectious organisms had to contain genetic material.

However, disease outbreaks in England, Canada, and the United States have raised concerns that even smaller agents, consisting only of a protein molecule, are capable of causing disease. These proteinaceous infectious particles, called **prions** (pronounced pree-ahns), were probably responsible for an outbreak of variant Creutzfeld-Jakob (croits-feld yay-kob) disease, or "mad-cow" disease. Twelve British people contracted the illness between 1994 and 1996 after eating meat from infected cattle, and all died. Upon autopsy, scientists noted that holes pockmarked the brains of all human and animal victims, giving the tissue a spongy, Swiss-cheese appearance. Prions are now generally accepted to be the cause for a set of diseases called transmissible

spongiform encephalopathies (TSEs), diseases that cause progressive death of brain tissue. *Scrapie* (a disease of sheep and goats) and *kuru* in humans are also TSE diseases.

How can something as small as a protein molecule cause disease? Scientists theorize that prions cause normal brain cell proteins to change into prion proteins. Accumulating prions kill brain cells and spread throughout brain tissue. Symptoms indicate brain destruction: "Mad" cows wobble, stagger, refuse food, and ultimately starve. Sheep with scrapie will frantically rub wool off their bodies. Infected humans lose memory, muscle control, and finally stop breathing. At this time, there is no treatment—all TSEs are fatal.

Preventing the spread of prions requires improved methods of raising beef cattle. Researchers believe that prions were spread from sheep to cattle, and later among infected cattle, when the animals ate feed that contained waste tissue from slaughterhouses. The waste tissue, which consists of brains and spinal cords, had been ground up and mixed with the animal feed. (It is interesting to note that kuru in humans was spread by cannibalism among the Fore' tribe of New Guinea, who honored dead relatives by eating their brains!) The addition of waste tissue to animal feed has been outlawed in the United States, Canada, and Europe, and infected animals have been destroyed. However, many puzzling questions about prions remain, and much additional research is needed.

reactions. Our bodies use enzymes to convert sugar into CO_2 and water, and the reaction can occur at body temperature.

Enzyme-Substrate Complex

In any reaction, the molecules that interact are called *reactants*, while the substances that form as a result of the reaction are the *products*. The reactants in an enzymatic reaction are its **substrate(s)**. Enzymes are often named for their substrate(s); for example, maltase is the enzyme that digests maltose. Enzymes have a specific region, called an *active site*, where the reaction occurs. An enzyme's specificity is caused by the shape of the active site, where the enzyme and its substrate(s) fit together, much like pieces of a jigsaw puzzle (Fig. 2.13). After a reaction is complete and the products are released, the enzyme is ready to catalyze its reaction again:

$$E + S \rightarrow ES \rightarrow E + P$$

(where E = enzyme, S = substrate, ES = enzyme-substrate complex, and P = product).

Many enzymes require **cofactors.** Cofactors assist an enzyme and may even accept or contribute atoms to the reaction. Some cofactors are inorganic metals, such as copper, zinc, and iron. Other cofactors are organic, nonprotein molecules called coenzymes. Vitamins are often components of coenzymes.

Types of Reactions

Certain types of chemical reactions are common to metabolism.

Synthesis Reactions During **synthesis reactions,** two or more reactants combine to form a larger and more complex product (Fig. 2.13*a*). The dehydration synthesis reaction we have already studied (i.e., the joining of subunits to form a macromolecule) is an example of a synthesis reaction (see Fig. 2.8). When glucose molecules join in the liver, forming glycogen, dehydration synthesis has occurred. Notice that synthesis reactions always involve bond formation and require energy.

Degradation Decomposition Reactions During **degradation reactions** (also called decomposition reactions), a larger and more complex molecule breaks down into smaller, simpler products (Fig. 2.13*b*). The hydrolysis reactions that break down macromolecules into their subunits are examples of degradation reactions. When protein is digested to amino acids in the stomach, a degradation reaction has occurred.

Replacement Reactions Replacement reactions involve both degradation decomposition and synthesis. For example, when ADP joins with inorganic phosphate, ℗, and ATP forms, the last hydrogen in ADP is replaced by a ℗. The ℗ loses a hydroxyl group. The hydrogen and hydroxyl group join to become water.

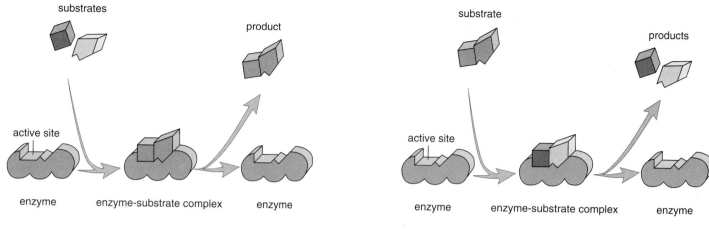

a. Synthesis

b. Degradation/Decomposition

Figure 2.13 Enzymatic action. An enzyme has an active site where the substrates come together and react. The products are released, and the enzyme is free to act again. **a.** In synthesis, the substrates join to produce a larger product. **b.** In degradation/decomposition, the substrate breaks down to smaller products.

Content CHECK-UP!

6. The sequence of amino acids found in a protein is that protein's _____ structure.
   a. primary
   c. tertiary
   b. secondary
   d. quaternary

7. Which of the following *cannot* be digested by humans?
   a. cellulose
   c. glycogen
   b. starch
   d. maltose

8. What type of reaction is the digestion of glucose to form carbon dioxide and water?
   a. synthesis
   c. replacement
   b. degradation/decomposition

2.7 Nucleic Acids

Nucleic acids are huge macromolecules composed of nucleotides. Every **nucleotide** is a molecular complex of three types of subunit molecules—a phosphate (phosphoric acid), a pentose sugar, and a nitrogen-containing base:

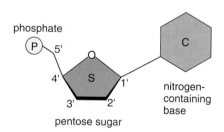

phosphate
P 5'
O
4' S 1'
3' 2'
nitrogen-containing base
pentose sugar

Nucleic acids contain hereditary information that determines which proteins a cell will have. Two classes of nucleic acids are in cells: **DNA (deoxyribonucleic acid)** and **RNA (ribonucleic acid)**. DNA makes up the hereditary units called **genes**. Genes pass on from generation to generation the instructions for replicating DNA, making RNA, and joining amino acids to form the proteins of a cell. RNA is an intermediary in the process of protein synthesis, conveying information from DNA regarding the amino acid sequence in proteins.

The nucleotides in DNA contain the 5-carbon sugar deoxyribose; the nucleotides in RNA contain the sugar ribose. This difference accounts for their respective names. As indicated in Figure 2.14, there are four different types of bases in DNA: A = adenine, T = thymine, G = guanine, and C = cytosine. The base can have two rings (like adenine or guanine) or one ring (like thymine or cytosine). In RNA, the base uracil replaces the base thymine.

The bases in DNA and RNA are nitrogen-containing bases—that is, a nitrogen atom is a part of the ring. Like other bases, the presence of the nitrogen-containing base in DNA and RNA raises the pH of a solution.

The nucleotides in DNA and RNA form a linear molecule called a *strand*. A strand has a backbone made up of phosphate-sugar-phosphate-sugar, with the bases projecting to one side of the backbone. Because the nucleotides occur in a definite order, so do the bases. Any particular DNA or RNA has a definite sequence of bases, although the sequence can vary between molecules. RNA is usually single-stranded, while DNA is usually double-stranded, with the two strands twisted about each other in the form of a double helix (like a spiral staircase). The molecular differences between DNA and RNA are listed in Table 2.1.

In DNA, the two strands are held together by hydrogen bonds between the bases (see Fig. 2.14). When unwound, DNA resembles a stepladder. The sides of the ladder are made entirely of phosphate and sugar molecules, and the rungs of the ladder are made only of complementary paired bases. Thymine (T) always pairs with adenine (A), and guanine (G)

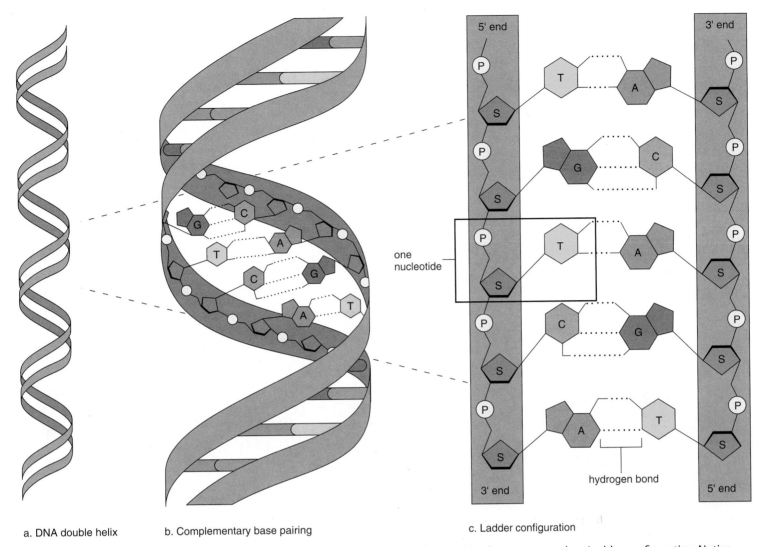

5' end   3' end

one
nucleotide

hydrogen bond

3' end   5' end

a. DNA double helix   b. Complementary base pairing   c. Ladder configuration

Figure 2.14 Overview of DNA structure. **a.** Double helix. **b.** Complementary base pairing between strands. **c.** Ladder configuration. Notice that the uprights are composed of phosphate and sugar molecules and that the rungs are complementary paired bases.

always pairs with cytosine (C) (see Fig. 2.14). This is called **complementary base pairing.**

Complementary bases pair because they have shapes that fit together. We shall see that complementary base pairing allows DNA to replicate in a way that ensures the sequence of bases will remain the same. When RNA is produced, comple-mentary base pairing occurs between DNA and RNA, but uracil takes the place of thymine. Then, the sequence of the bases in RNA determines the sequence of amino acids in a protein because every three bases code for a particular amino acid (see Chapter 3, pp. 53–54). The code is nearly universal and is the same in other organisms as it is in humans.

ATP (Adenosine Triphosphate)

Individual nucleotides can have metabolic functions in cells. Some nucleotides are important in energy transfer. When adenosine (adenine plus ribose) is modified by the addition of three phosphate groups, it becomes **ATP (adenosine triphosphate),** the primary energy carrier in cells.

Cells require a constant supply of ATP. To obtain it, they break down glucose and convert the energy that is released into ATP molecules. The amount of energy in ATP is just right for more chemical reactions in cells. As an analogy, the energy in glucose is like a $100 bill, and the energy in ATP is like a $20 bill.

TABLE 2.1	DNA Structure Compared to RNA Structure	
	DNA	**RNA**
Sugar	Deoxyribose	Ribose
Bases	Adenine, guanine, thymine, cytosine	Adenine, guanine, uracil, cytosine
Strands	Double-stranded	Single-stranded
Helix	Yes	No

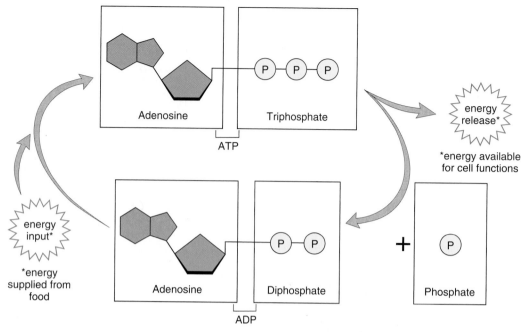

Figure 2.15 Breakdown and Formation of ATP.

Just as you might go to the bank to change a $100 bill (glucose) into $20 bills (ATP molecules), in order to spend money, cells "spend" ATP when cellular reactions require energy. Therefore, ATP is called the energy currency of cells. Cells use ATP when macromolecules such as carbohydrates and proteins are synthesized. In muscle cells, ATP is used for muscle contraction, and in nerve cells, it is used for the conduction of nerve impulses.

ATP is a high-energy molecule because the last two phosphate bonds are unstable and easily broken. In cells, the terminal phosphate bond is usually hydrolyzed, releasing energy for cell reactions. The terminal bond is sometimes called a high-energy bond, symbolized by a wavy line. But this terminology is misleading—the breakdown of ATP *releases* energy because the products of hydrolysis are *more* stable than ATP. Breakdown of ATP leaves one molecule of ADP (adenosine diphosphate) and a molecule of inorganic phosphate, Ⓟ (Fig. 2.15).

After ATP breaks down and the energy is used for a cellular purpose, ATP is rebuilt by the addition of Ⓟ to ADP again (Fig. 2.15). There is enough energy in one glucose molecule to build 36 ATP molecules in this way. Homeostasis is only possible because cells continually produce and use ATP molecules. The use of ATP as the energy currency of cells also occurs in other organisms, ranging from bacteria to humans.

SELECTED NEW TERMS

Basic Key Terms

acid (as′id), p. 25

amino acid (uh-me′no as′id), p. 31

ATP (adenosine triphosphate) (uh-den′o-sēn tri-fos′fāt), p. 35

base (bās), p. 25

buffer (buf′er), p. 26

carbohydrate (kar″bo-hi′drāt), p. 26

cellulose (sĕl′yə-lōs′), p. 28

cofactors (kō′făk′tərz), p. 33

cohesive (kō-hē′-sĭv), p. 24

complementary base pairing (kŏm′plə-mĕn′tə-rē bās pâr-ing), p. 35

compound (kŏm-pound′), p. 22

covalent bond (ko-va′lent bond), p. 22

degradation decomposition reactions (dĕg′rə-dā′shən dē-kŏm′pə-zish′ən rē-ăk′shənz), p. 33

dehydration reaction (dē′hī-drā′shən rē-ăk′shənz), p. 26

denaturation (dē-nā′chər-ā′shən), p. 32

disaccharide (di-sak′uh-rīd), p. 27

DNA (deoxyribonucleic acid) (de-oks′e-ri″-bo-nu-kla″ikas′id), p. 34

electrolyte (e-lek′tro-līt), p. 26

electronegative (ĭ-lĕk′trō-nĕg′ə-tĭv), p. 23

electrons (ĭ-lĕk′tron′z), p. 21

elements (ĕl′ə-məntz), p. 21

emulsification (ĭ-mŭl′sə-fə-cā′-shən), p. 29

enzyme (en′zīm), p. 32

fats (fătz), p. 29

fatty acid (fat′e as′id), p. 30

functional groups (fŭngk′shə-nəl groopz), p. 31

gene (jēn), p. 34

glycerol (glis′er-ol), p. 29

glycogen (gli′ko-jen), p. 27

heat of vaporization (hēt ŭv vā′pə-rī-zā′shən), p. 24

hexose (hĕk′sōs′), p. 27

hydrogen bond (hi′dro-jen bond), p. 24

hydrolysis reaction (hi-drol′ĭ-sis re-ak′shun), p. 26

hydrophilic (hī′drə-fīl′ĭk), p. 24

hydrophobic (hī′drə-fō′bĭk), p. 24

inorganic molecule (in-or-gan′ik mol′e-kyūl), p. 24

ion (i′on), p. 22

SUMMARY

2.1 Basic Chemistry

A. All matter is composed of elements, each made up of just one type of atom. An atom has an atomic symbol, atomic number (number of protons and, therefore, electrons when neutral), and atomic weight (number of protons and neutrons). The isotopes of some atoms are radioactive and have biological and medical applications.

B. Atoms react with one another to form molecules. Following an ionic reaction, charged ions are attracted to one another. Following a covalent reaction, atoms share electrons. Polar covalent bonding results from unequal sharing of electrons.

2.2 Water, Acids, and Bases

A. In water, the electrons are shared unequally, and the result is a polar molecule. Hydrogen bonding can occur between polar molecules.

B. Water is a polar molecule and acts as a solvent; it dissolves various chemical substances and facilitates chemical reactions. Because of hydrogen bonding, water molecules are cohesive, and also,

water heats up and cools down slowly. This helps keep body temperature within normal limits.

C. Substances such as salts, acids, and bases that dissociate in water are called electrolytes. The electrolyte balance in the blood and body tissues is important for good health.

D. Acids have a pH less than 7, and bases have a pH greater than 7. The actions of the lungs and kidneys, and the presence of buffers, help to keep the pH of body fluids around pH 7.

2.3 Molecules of Life

A. Carbohydrates, lipids, proteins, and nucleic acids are the molecules of life.

B. A monosaccharide, such as glucose, is a subunit (monomer) for larger carbohydrates. Glycerol and fatty acids are monomers for fat. Amino acids are monomers for proteins, and nucleotides are monomers for nucleic acids.

2.4 Carbohydrates

Glucose is an immediate source of energy in cells. Glycogen stores energy in the body, starch is a dietary source of energy, and cellulose is fiber in the diet.

2.5 Lipids

Lipids include neutral fat (a long-term, energy-storage molecule that forms from glycerol and three fatty acids) and the related phospholipids, which have a charged group. Fatty acids can be saturated or unsaturated. Steroids have an entirely different structure from that of fats.

2.6 Proteins

A. Proteins, which are composed of one or more long polypeptides, have both structural and physiological functions. Polypeptides have several levels of structure. Their three-dimensional shape is necessary to their function.

B. Enzymes are proteins necessary to metabolism. The reaction occurs at the active site of an enzyme.

2.7 Nucleic Acids

A. Both DNA and RNA are polymers of nucleotides; only DNA is double-stranded. DNA makes up the genes, and along with RNA, specifies protein synthesis.

B. ATP is the energy "currency" of cells because its breakdown supplies energy for many cellular processes.

STUDY QUESTIONS

1. Name the subatomic particles of an atom; describe their charge, atomic mass unit, and location in the carbon atom. (p. 21)

2. What is an isotope? A radioactive isotope? Discuss the clinical uses of radioactive isotopes. (p. 21)

3. Give an example of an ionic reaction, and explain it. (p. 22)

4. Give an example of a covalent reaction, and explain it. (p. 23)

5. Relate three characteristics of water to its polarity and hydrogen bonding between water molecules (p. 24)
6. What is an acid? A base? (p. 25)
7. On the pH scale, which numbers indicate a basic solution? An acidic solution? Why? (p. 25)
8. What are buffers, and how do they function? (p. 26)
9. Name the four categories of macromolecules (polymers) in cells; give an example for each category, and name the subunits (monomers) of each. (p. 26)

10. Tell how macromolecules are built up and broken down. (p. 26)
11. Name some monosaccharides, disaccharides, and polysaccharides, and give the functions for each. (pp. 26–27)
12. What is a lipid? A saturated fatty acid? An unsaturated fatty acid? What is the function of fats? (p. 28)
13. Relate the structure of a phospholipid to that of a neutral fat. What is the function of a phospholipid? (p. 30)
14. Name two steroids that function as sex hormones in humans. (p. 31)

15. What are some functions of proteins? Why do proteins stop functioning if exposed to the wrong pH or high temperature? (p. 31)
16. Discuss the levels of protein structure. (p. 31)
17. How do enzymes function? Name three types of metabolic reactions. (p. 32)
18. Discuss the structure and function of the nucleic acids DNA and RNA. (pp. 34–35)

LEARNING OBJECTIVE QUESTIONS

Fill in the blanks.

1. _____ are the smallest units of matter nondivisible by chemical means.
2. Isotopes differ by the number of _____ in the nucleus.
3. The two primary types of reactions and bonds are _____ and _____.
4. A type of weak bond, called a _____ bond, exists between water molecules.

5. Acidic solutions contain more _____ ions than basic solutions, but they have a _____ pH.
6. Glycogen is a polymer of _____, molecules that serve to give the body immediate _____.
7. A fat hydrolyzes to give one _____ molecule and three _____ molecules.

8. A polypeptide has levels of structure. The first level is the sequence of _____; the second level is very often a _____; the third level is its final _____.
9. _____ speed chemical reactions in cells.
10. Genes are composed of _____, a nucleic acid made up of _____ joined together.

MEDICAL TERMINOLOGY EXERCISE

Consult Appendix B for help in pronouncing and analyzing the meaning of the terms that follow.

1. anisotonic (an-i″so-ton′ik)
2. dehydration (de″hi-dra′shun)
3. hypokalemia (hi″po-kă-le′me-uh)
4. hypovolemia (hi″po-vo-le′me-uh)

5. nonelectrolyte (non″ē-lek′tro-līt)
6. lipometabolism (lip″o-mĕ-tab′o-lizm)
7. hyperlipoproteinemia (hi″per-lip″o-pro″te-in-e′me-uh)
8. hyperglycemia (hi″per-gli-se′me-uh)
9. hypoxemia (hi″pok-se′me-uh)
10. hydrostatic pressure (hi″dro-stat′-ik-presh′ur)

11. galactosemia (guh-lak-to-se′me-uh)
12. hypercalcemia (hi″per-kal-se′me-uh)
13. hyponatremia (hi″po-nuh-tre′me-uh)
14. gluconeogenesis (glu″ko-ne-o-jen′uh-sis)
15. edema (uh-de′muh)

WEBSITE LINK

Visit this text's website at http://www.mhhe.com/maderap6 for additional quizzes, interactive learning exercises, and other study tools.

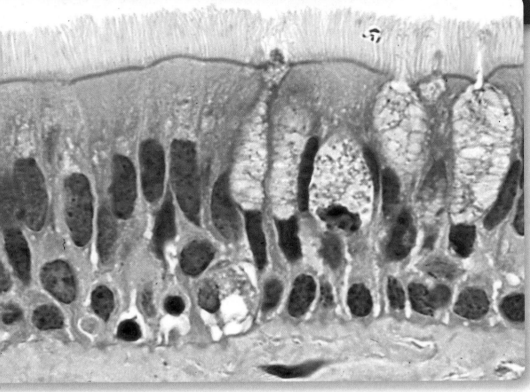

CHAPTER

3

Cell Structure and Function

Cellular organelles are exquisitely tailored to each cell in the body. Note the fine, hairlike cilia on these cells lining the trachea, or windpipe. These cilia trap inhaled pollutants and prevent them from entering the lungs.

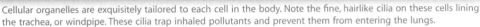

Chapter Outline & Learning Objectives

After you have studied this chapter, you should be able to:

3.1 Cellular Organization (p. 40)
- Name the three main parts of a human cell.
- Describe the structure and function of the plasma membrane.
- Describe the structure and function of the nucleus.
- Describe the structures and roles of the endoplasmic reticulum and the Golgi apparatus in the cytoplasm.
- Describe the structures of lysosomes and the role of these organelles in the breakdown of molecules.
- Describe the structure of mitochondria and their role in producing ATP.
- Describe the structures of centrioles, cilia, and flagella and their roles in cellular movement.

- Describe the structures and function of the cytoskeleton.

3.2 Crossing the Plasma Membrane (p. 48)
- Describe how substances move across the plasma membrane, and distinguish between passive and active transport.

3.3 The Cell Cycle (p. 52)
- Describe the phases of the cell cycle.
- As a part of interphase, describe the process of DNA replication.

- As a part of interphase, also describe how cells carry out protein synthesis.
- Describe the phases of mitosis, and explain the function of mitosis.

Medical Focus
Dehydration and Water Intoxication (p. 51)

Focus on Forensics
DNA Fingerprinting (p. 57)

3.1 Cellular Organization

Every human cell has a plasma membrane, a nucleus, and cytoplasm. (Some exceptions to this rule exist. A mature erythrocyte, or red blood cell, eliminates its nucleus once development is complete. Thus, erythrocytes are *anucleate*. Cells of skeletal muscle, liver, and other tissues may have up to 50 nuclei and are *multinucleate*.) The **plasma membrane**, which surrounds the cell and keeps it intact, regulates what enters and exits a cell. The plasma membrane is a phospholipid bilayer that is said to be *semipermeable* because it allows certain molecules but not others to enter the cell. Proteins present in the plasma membrane play important roles in allowing substances to enter the cell.

The **nucleus** is a large, centrally located structure that can often be seen with a light microscope. The nucleus contains the chromosomes and is the control center of the cell. It controls the metabolic functioning and structural characteristics of the cell. The **nucleolus** is a region inside the nucleus.

The **cytoplasm** is the portion of the cell between the nucleus and the plasma membrane. Cytoplasm is a gelatinous, semifluid medium that contains water and various types of molecules suspended or dissolved in the medium. The presence of proteins accounts for the semifluid nature of cytoplasm.

The cytoplasm contains various **organelles** (Table 3.1 and Fig. 3.1). Organelles are small, usually membranous structures that are best seen with an electron microscope.[1] Each type of organelle has a specific function. For example, one type of organelle transports substances, and another type produces ATP for the cell. *Organelles* compartmentalize the cell, keeping the various cellular activities separated from one another. Just as the rooms in your house have particular pieces of furniture that serve a particular purpose, organelles have a structure that suits their function.

Cells also have a **cytoskeleton,** a network of interconnected filaments and microtubules in the cytoplasm. The name cytoskeleton is convenient in that it allows us to compare the cytoskeleton to our bones and muscles. Bones and muscles give us structure and produce movement. Similarly, the elements of the cytoskeleton maintain cell shape and allow the cell and its contents to move. Some cells move by using cilia and flagella, which are made up of microtubules.

The Plasma Membrane

Our cells are surrounded by an outer plasma membrane. The plasma membrane separates the inside of the cell, termed the

---

[1]Electron microscopes are high-powered instruments that are used to generate detailed photographs of cellular contents. The photographs are called electron micrographs. Scanning electron micrographs have depth while transmission electron micrographs are flat (see Fig. 3.3). Light microscopes are used to generate photomicrographs that are often simply called micrographs.

TABLE 3.1 Structures in Human Cells

Name	Composition	Function
MEMBRANOUS STRUCTURES		
Plasma membrane	Phospholipid bilayer with embedded proteins	Cell border; selective passage of molecules into and out of cell; location of cell markers, cell receptors
Nucleus	Nuclear membrane (envelope) surrounding nucleoplasm, chromatin, and nucleolus	Storage of genetic information; control center of cell; cell replication
Nucleolus	Concentrated area of chromatin, RNA, and proteins	Ribosomal formation
Ribosome	Two subunits composed of protein and RNA	Protein synthesis
Endoplasmic reticulum	Complex system of tubules, vesicles, and sacs	Synthesis and/or modification of proteins and other substances; transport by vesicle formation
Rough endoplasmic reticulum	Endoplasmic reticulum studded with ribosomes	Protein synthesis for export
Smooth endoplasmic reticulum	Endoplasmic reticulum without ribosomes	Varies: lipid and/or steroid synthesis; calcium storage
Golgi apparatus	Stacked, concentrically folded membranes	Processing, packaging, and distribution of molecules
Vacuole	Small membranous sac	Isolates substances inside cell
Vesicle	Small membranous sac	Storage and transport of substances in/out of cell
Lysosome	Vesicle containing digesting enzymes	Intracellular digestion; self-destruction of the cell
Peroxisome	Vesicle containing oxidative enzymes	Detoxifies drugs, alcohol, etc.; breaks down fatty acids
Mitochondrion	Inner membrane within outer membrane	Cellular respiration
Cytoskeleton	Microtubules, actin filaments	Shape of cell and movement of its parts
Cilia and flagella	9 + 2 pattern of microtubules	Movement by cell; movement of substances inside a tube
Centriole	9 + 0 pattern of microtubules	Formation of basal bodies for cilia and flagella; formation of spindle in cell division

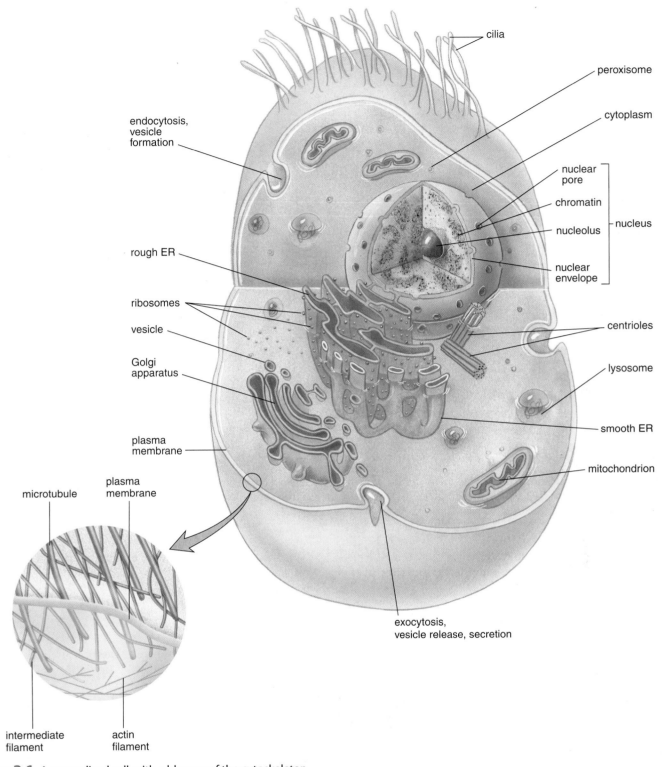

cilia

peroxisome

cytoplasm

endocytosis,
vesicle
formation

nuclear
pore

chromatin

nucleolus ⎫ nucleus

nuclear
envelope

rough ER

ribosomes

vesicle

Golgi
apparatus

centrioles

lysosome

smooth ER

mitochondrion

plasma
membrane

microtubule

plasma
membrane

exocytosis,
vesicle release, secretion

intermediate
filament

actin
filament

Figure 3.1 A generalized cell, with a blowup of the cytoskeleton.

cytoplasm, from the outside. Plasma membrane integrity is necessary to the life of the cell.

The plasma membrane is a phospholipid bilayer with attached (also called *peripheral*) or embedded (also called *integral*) proteins. The phospholipid molecule has a polar head and nonpolar tails (Fig. 3.2*a*). Because the polar heads are charged, they are *hydrophilic* (water-loving) and face outward, where they are likely to encounter a watery environment. The nonpolar tails are *hydrophobic* (water-fearing) and face inward, where there is no water. When

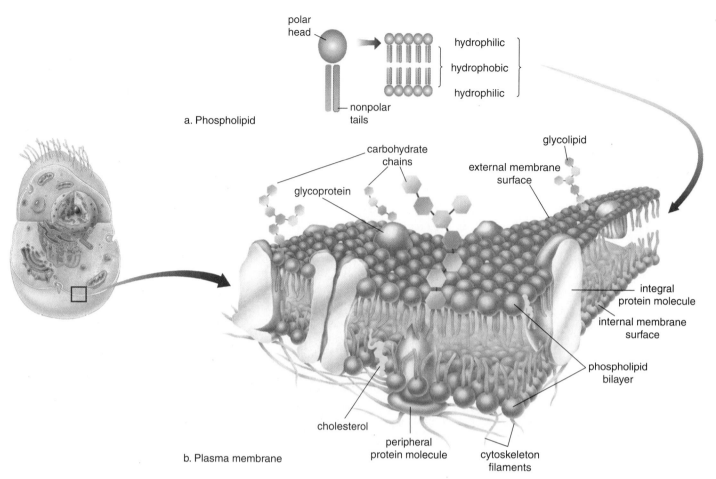

a. Phospholipid

polar head

nonpolar tails

hydrophilic
hydrophobic
hydrophilic

carbohydrate chains

glycoprotein

glycolipid

external membrane surface

integral protein molecule

internal membrane surface

phospholipid bilayer

cholesterol

peripheral protein molecule

cytoskeleton filaments

b. Plasma membrane

Figure 3.2 Fluid-mosaic model of the plasma membrane. **a.** In the phospholipid bilayer, the polar (hydrophilic) heads project outward and the nonpolar (hydrophobic) tails project inward. **b.** Proteins are embedded in the membrane. Glycoproteins have attached carbohydrate chains, as do glycolipids.

phospholipids are placed in water, they naturally form a spherical bilayer because of the chemical properties of the heads and the tails.

At body temperature, the phospholipid bilayer is a liquid; it has the consistency of olive oil, and the proteins are able to change their positions by moving laterally. The *fluid-mosaic model,* a working description of membrane structure, suggests that the protein molecules have a changing pattern (form a mosaic) within the fluid phospholipid bilayer (Fig. 3.2b). Our plasma membranes also contain a substantial number of *cholesterol* molecules. These molecules stabilize the phospholipid bilayer and prevent a drastic decrease in fluidity at low temperatures.

Short chains of sugars are attached to the outer surfaces of some protein and lipid molecules (called *glycoproteins* and *glycolipids,* respectively). These carbohydrate chains, specific to each cell, mark the cell as belonging to a particular individual. Such cell markers account for such characteristics as blood type or why a patient's system sometimes rejects an organ transplant. Some glycoproteins have a special configuration that allows them to act as a receptor for a chemical messenger such as a hormone. Some integral plasma membrane proteins

form channels through which certain substances can enter cells, while others are carriers involved in the passage of molecules through the membrane.

The Nucleus

The nucleus is a prominent structure in human cells. The nucleus is of primary importance because it stores the genetic information that determines the characteristics of the body's cells and their metabolic functioning. The unique chemical composition of each person's DNA forms the basis for DNA fingerprinting (see Focus on Forensics, p. 57). Every cell contains a copy of genetic information, but each cell type has certain genes turned on and others turned off. Activated DNA, with messenger RNA (mRNA) acting as an intermediary, controls protein synthesis (see page 53). The proteins of a cell determine its structure and the functions it can perform.

When you look at the nucleus, even in an electron micrograph, you cannot see DNA molecules, but you can see chromatin (Fig. 3.3). Chemical analysis shows that **chromatin** contains DNA and much protein, as well as some RNA. Chromatin undergoes coiling into rodlike structures called

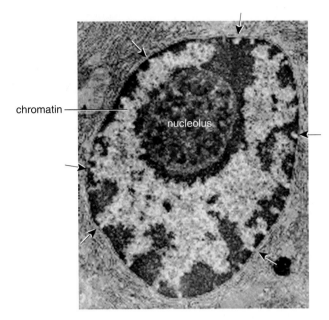

chromatin

nucleolus

Figure 3.3 The nucleus. The nuclear envelope with pores (arrows) surrounds the chromatin. Chromatin has a special region called the nucleolus where rRNA is produced and ribosomal subunits are assembled.

chromosomes just before the cell divides. Chromatin is immersed in a semifluid medium called nucleoplasm.

Most likely, too, when you look at an electron micrograph of a nucleus (Fig. 3.3), you will see one or more regions that look darker than the rest of the chromatin. These are nucleoli (sing., nucleolus) where another type of RNA, called ribosomal RNA (rRNA), is produced and where rRNA joins with proteins to form the subunits of ribosomes. (Ribosomes are small bodies in the cytoplasm that contain rRNA and proteins.)

The nucleus is separated from the cytoplasm by a double membrane known as the **nuclear envelope,** which is continuous with the endoplasmic reticulum (Fig. 3.4) discussed on this page. The nuclear envelope has **nuclear pores** of sufficient size to permit the passage of proteins into the nucleus and ribosomal subunits out of the nucleus. Additionally, the double membrane of the nuclear envelope surrounds and contains cellular DNA, protecting the vital genetic information contained within its molecules.

Ribosomes

Ribosomes are composed of two subunits, one large and one small. Each subunit has its own mix of proteins and rRNA. Protein synthesis occurs at the ribosomes.

Ribosomes are found free within the cytoplasm either singly or in groups called **polyribosomes** (called polysomes for short). Ribosomes are often attached to the endoplasmic reticulum, a membranous system of saccules and channels discussed next (Fig. 3.4).

Proteins synthesized by cytoplasmic ribosomes are used inside the cell for various purposes. Those produced by

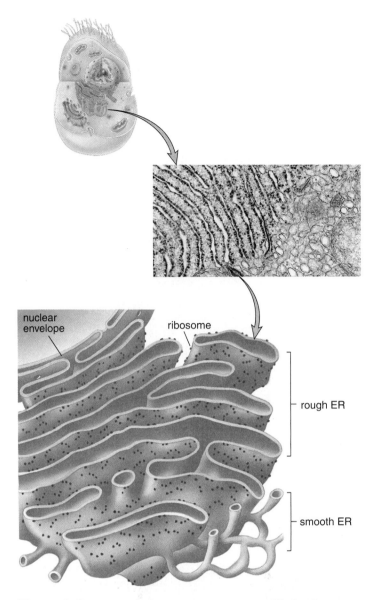

nuclear envelope

ribosome

rough ER

smooth ER

Figure 3.4 Rough endoplasmic reticulum is studded with ribosomes where protein synthesis occurs. Smooth endoplasmic reticulum, which has no attached ribosomes, produces lipids and often has other functions as well in particular cells.

ribosomes attached to endoplasmic reticulum may eventually be secreted from the cell.

Endomembrane System

The **endomembrane system** consists of the nuclear envelope, the endoplasmic reticulum, the Golgi apparatus, lysosomes, and **vesicles** (tiny membranous sacs) (Fig. 3.5). These components of the cell work together to produce and secrete a product.

The Endoplasmic Reticulum

The **endoplasmic reticulum (ER),** a complicated system of membranous channels and saccules (flattened vesicles), is

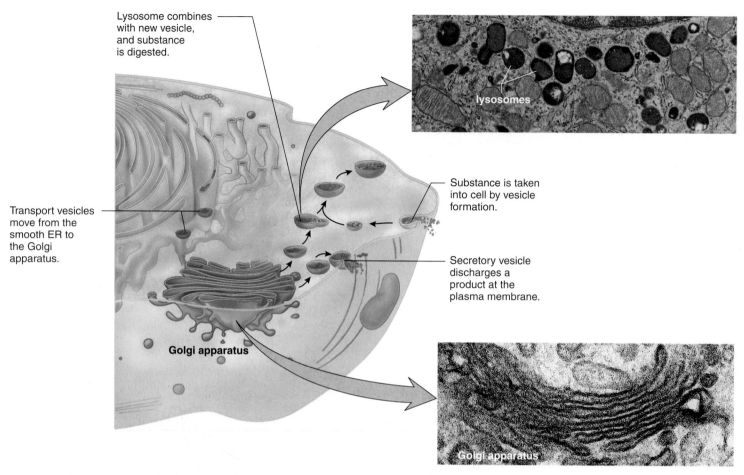

Figure 3.5 The endomembrane system. Transport vesicles from the ER bring proteins and lipids to the Golgi apparatus where they are modified and repackaged into secretory vesicles. Secretion occurs when vesicles fuse with the plasma membrane. Lysosomes made at the Golgi apparatus digest macromolecules after fusing with incoming vesicles.

physically continuous with the outer membrane of the nuclear envelope. **Rough ER** is studded with ribosomes on the side of the membrane that faces the cytoplasm. Here proteins are synthesized and enter the ER interior where processing and modification begin. Some of these proteins are incorporated into membrane, and some are for export. Smooth ER, which is continuous with rough ER, does not have attached ribosomes. **Smooth ER** synthesizes the phospholipids that occur in membranes and has various other functions, depending on the particular cell. In the testes, it produces testosterone, and in the liver it helps detoxify drugs.

Regardless of any specialized function, ER also forms vesicles in which large molecules are transported to other parts of the cell. Often these vesicles are on their way to the plasma membrane or the Golgi apparatus.

The Golgi Apparatus

The **Golgi apparatus** is named for Camillo Golgi, who discovered its presence in cells in 1898. The Golgi apparatus consists of a stack of three to twenty slightly curved saccules whose appearance can be compared to a stack of pancakes

(Fig. 3.5). In animal cells, one side of the stack (the inner face) is directed toward the ER, and the other side of the stack (the outer face) is directed toward the plasma membrane. Vesicles can frequently be seen at the edges of the saccules.

The Golgi apparatus receives protein and/or lipid-filled vesicles that bud from the ER. Some biologists believe that these fuse to form a saccule at the inner face and that this saccule remains a part of the Golgi apparatus until the molecules are repackaged in new vesicles at the outer face. Others believe that the vesicles from the ER proceed directly to the outer face of the Golgi apparatus, where processing and packaging occur within its saccules. The Golgi apparatus contains enzymes that modify proteins and lipids. For example, it can add a chain of sugars to proteins and lipids, thereby making them glycoproteins and glycolipids, which are molecules found in the plasma membrane.

The vesicles that leave the Golgi apparatus move to other parts of the cell. Some vesicles proceed to the plasma membrane where they discharge their contents. Because this is secretion, note that the Golgi apparatus is involved in processing, packaging, and secretion. Other vesicles that leave the Golgi apparatus are lysosomes.

Lysosomes

Lysosomes, membranous sacs produced by the Golgi apparatus, contain hydrolytic digestive enzymes. Sometimes macromolecules are brought into a cell by vesicle formation at the plasma membrane (Fig. 3.5). When a lysosome fuses with such a vesicle, its contents are digested by lysosomal enzymes into simpler subunits that then enter the cytoplasm. Even parts of a cell are digested by its own lysosomes (called autodigestion). Normal cell rejuvenation most likely takes place in this manner, but autodigestion is also important during development. For example, when a tadpole becomes a frog, lysosomes digest away the cells of the tail. The fingers of a human embryo are at first webbed, but they are freed from one another as a result of lysosomal action.

Occasionally, a child is born with **Tay-Sachs disease,** a metabolic disorder involving a missing or inactive lysosomal enzyme in nerve cells. In these cases, the lysosomes fill to capacity with macromolecules that cannot be broken down. The nerve cells become so full of these lysosomes that the child dies. Someday soon, it may be possible to provide the missing enzyme for these children.

Peroxisomes and Vacuoles

Although **peroxisomes** are not part of the endomembranous system, they are similar to lysosomes in structure. Like lysosomes, they are vesicles that contain enzymes. Peroxisome enzymes function to detoxify drugs, alcohol, and other potential toxins. The liver and kidneys contain large numbers of peroxisomes because these organs help to cleanse the blood. Peroxisomes also break down fatty acids from fats so that fats can be metabolized by the cell.

Vacuoles are occasionally found within human cells, where they isolate substances captured inside the cell. For example, vacuoles may contain parasites that are awaiting digestion by lysosomes.

Mitochondria

Although the size and shape of mitochondria (sing., **mitochondrion**) can vary, all are bounded by a double membrane. The inner membrane is folded to form little shelves called *cristae*, which project into the *matrix*, an inner space filled with a gel-like fluid (Fig. 3.6).

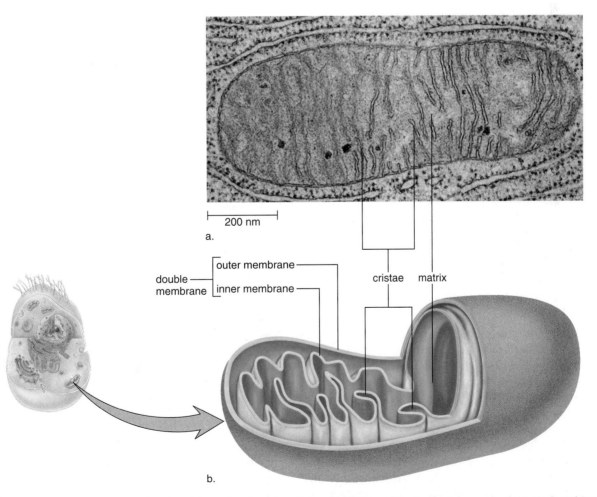

200 nm

a.

double membrane

outer membrane

inner membrane

cristae   matrix

b.

Figure 3.6 Mitochondrion structure. Mitochondria are involved in cellular respiration. **a.** Electron micrograph of a mitochondrion. **b.** Generalized drawing in which the outer membrane and portions of the inner membrane have been cut away to reveal the cristae.

Mitochondria are the site of ATP (adenosine triphosphate) production involving complex metabolic pathways. As you know, ATP molecules are the common carrier of energy in cells. A shorthand way to indicate the chemical transformation that involves mitochondria is as follows:

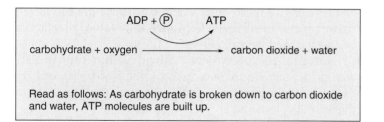

Read as follows: As carbohydrate is broken down to carbon dioxide and water, ATP molecules are built up.

Mitochondria are often called the powerhouses of the cell: Just as a powerhouse burns fuel to produce electricity, the mitochondria convert the chemical energy of carbohydrate molecules into the chemical energy of ATP molecules. In the process, mitochondria use up oxygen and give off carbon dioxide and water. The oxygen you breathe in enters cells and then mitochondria; the carbon dioxide you breathe out is released by mitochondria. Because oxygen is used up and carbon dioxide is released, we say that mitochondria carry on **cellular respiration.**

Fragments of digested carbohydrate, protein, and lipid enter the mitochondrial matrix from the cytoplasm. The matrix contains enzymes for metabolizing these fragments to carbon dioxide and water. Energy released from metabolism is used for ATP production, which occurs at the cristae. The protein complexes that aid in the conversion of energy are located in an assembly-line fashion on these membranous shelves.

Every cell uses a certain amount of ATP energy to synthesize molecules, but many cells use ATP to carry out their specialized functions. For example, muscle cells use ATP for muscle contraction, which produces movement, and nerve cells use it for the conduction of nerve impulses, which make us aware of our environment.

The Cytoskeleton

Several types of filamentous protein structures form a cytoskeleton that helps maintain the cell's shape and either anchors the organelles or assists their movement as appropriate. The cytoskeleton includes microtubules, intermediate filaments, and actin filaments (see Fig. 3.1).

Microtubules are hollow cylinders whose wall is made up of 13 longitudinal rows of the globular protein tubulin. Remarkably, microtubules can assemble and disassemble. Microtubule assembly is regulated by the centrosome, which lies near the nucleus. The centrosome is the region of the cell that contains the centrioles. Microtubules radiate from the centrosome, helping to maintain the shape of the cell and acting as tracks along which organelles move. It is well known that

during cell division, microtubules form spindle fibers, which assist the movement of chromosomes.

Intermediate filaments differ in structure and function. Because they are tough and resist stress, intermediate filaments often form cell-to-cell junctions. Intermediate filaments join skin cells in the outermost skin layer, the epidermis. **Actin filaments** are long, extremely thin fibers that usually occur in bundles or other groupings. Actin filaments have been isolated from various types of cells, especially those in which movement occurs. Microvilli, which project from certain cells and can shorten and extend, contain actin filaments. Actin filaments, like microtubules, can assemble and disassemble.

Centrioles

Centrioles are short cylinders with a 9 + 0 pattern of microtubules, meaning that there are nine outer microtubule triplets and no center microtubules (see Fig. 3.8b). Each cell has a pair of centrioles in the centrosome, a region near the nucleus. The members of each pair of centrioles are at right angles to one another. Before a cell divides, the centrioles duplicate, and the members of the new pair are also at right angles to one another. During cell division, the pairs of centrioles separate so that each daughter cell gets one centrosome.

Centrioles are involved in the formation of the spindle apparatus, which functions during cell division. Their role in forming the spindle is described later in the Mitosis/Meiosis section of this chapter on pages 55–56. A single centriole forms the anchor point, or *basal body*, for each individual cilium or flagellum. Basal bodies direct the formation of cilia and flagella as well.

Cilia and Flagella

Cilia and flagella (sing., **cilium, flagellum**) are projections of cells that can move either in an undulating fashion, like a whip, or stiffly, like an oar. Cilia are shorter than flagella (Fig. 3.7). Cells that have these organelles are capable of self-movement or moving material along the surface of the cell. For example, sperm cells, carrying genetic material to the egg, move by means of flagella. The cells that line our respiratory tract are ciliated. These cilia sweep debris trapped within mucus back up the throat, and this action helps keep the lungs clean. Within a woman's uterine tubes, ciliated cells move the ovum toward the uterus (womb), where a fertilized ovum grows and develops.

Recall that each cilium and flagellum is anchored by its basal body, which lies in the cytoplasm of the cell. Basal bodies, like centrioles, have a 9 + 0 pattern of microtubule triplets (Fig. 3.8a). They are believed to organize the structure of cilia and flagella even though cilia and flagella have a 9 + 2 pattern of microtubules. In cilia and flagella, nine microtubule doublets surround two central microtubules. This arrangement is believed to be necessary to their ability to move (Fig. 3.8b).

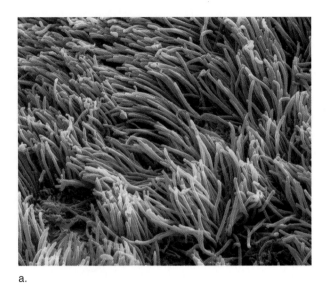
a.

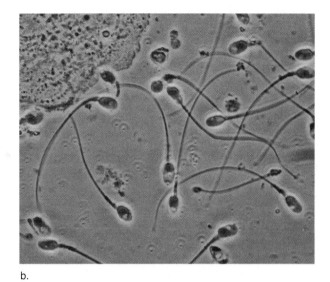
b.

Figure 3.7 Cilia and flagella. **a.** Cilia are common on the surfaces of certain tissues, such as the one that forms the inner lining of the respiratory tract. **b.** Flagella form the tails of human sperm cells.

The basal body of a flagellum has a ring of nine microtubule triplets with no central microtubules.

basal body

triplets

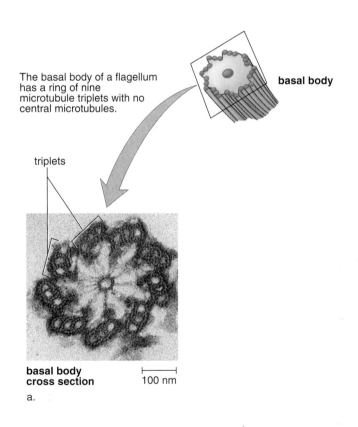

basal body cross section

100 nm

a.

The shaft of the flagellum has a ring of nine microtubule doublets anchored to a central pair of microtubules.

plasma membrane

outer microtubule doublet

side arms

central microtubules

radial spoke

flagellum cross section

25 nm

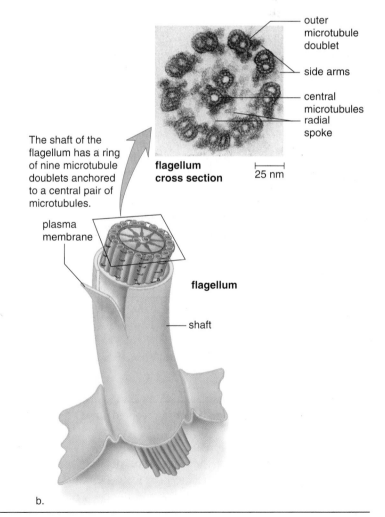

flagellum

shaft

b.

Figure 3.8 Structure of basal bodies and centrioles. **a.** Basal bodies have a 9 + 0 pattern of microtubule triplets. **b.** Centrioles have a 9 + 2 pattern of microtubules.

3.2 Crossing the Plasma Membrane

The plasma membrane keeps a cell intact. It allows only certain molecules and ions to enter and exit the cytoplasm freely; therefore, the plasma membrane is said to be **selectively permeable.** Both passive and active methods are used to cross the plasma membrane (see Table 3.2).

Simple Diffusion

Simple diffusion is the random movement of simple atoms or molecules from area of higher concentration to an area of lower concentration until they are equally distributed. To illustrate diffusion, imagine putting a tablet of dye into water.

The water eventually takes on the color of the dye as the dye molecules diffuse (Fig. 3.9).

The chemical and physical properties of the plasma membrane allow only a few types of molecules to enter and exit a cell by simple diffusion. Lipid-soluble molecules such as alcohols can diffuse through the membrane because lipids are the membrane's main structural components. Gases can also diffuse through the lipid bilayer; this is the mechanism by which oxygen enters cells and carbon dioxide exits cells. For example, consider the movement of oxygen from the lungs to the bloodstream. When you inhale, oxygen fills the tiny air sacs, or *alveoli*, within your lungs. Neighboring lung capillaries contain red blood cells with a very low oxygen concentration. Oxygen diffuses from the area of highest concentration to the area of lowest concentration: first through alveolar cells, then lung capillary cells, and finally into the red blood cells.

When atoms or molecules diffuse from areas of higher to lower concentration across plasma membranes, no cellular energy is involved. Instead, kinetic or thermal energy of matter is the energy source for diffusion.

Osmosis

Osmosis is the diffusion of water across a plasma membrane. Osmosis occurs whenever an unequal concentration of water exists on either side of a selectively permeable membrane. (Recall that a selectively permeable membrane allows water to pass freely, but not most dissolved substances.) In a solution, water is *more* concentrated when it contains fewer dissolved substances, or **solutes,** (and thus is closest to pure water). Water is *less* concentrated as solute concentration increases. **Osmotic pressure** is the force exerted on a selectively permeable membrane because water has moved from the area of higher water concentration to the area of lower water concentration (higher concentration of solute).

TABLE 3.2 Crossing the Plasma Membrane

Name	Direction	Requirement	Examples
PASSIVE METHODS			
Simple diffusion	High to low concentration	Concentration gradient	Lipid-soluble molecules, gases
Osmosis	High to low concentration	Semipermeable membrane, water concentration gradient	Absorption of water from digestive tract to bloodstream
Facilitated diffusion	High to low concentration	Carrier molecule plus concentration gradient	Sugars and amino acids
ACTIVE METHODS			
Active transport	Low to high concentration	Carrier molecule plus cell energy	Ions, sugars, amino acids
Endocytosis			
Phagocytosis	Into the cell ("cell eating")	Vesicle formation	Bacterial cells, viruses, cell debris
Pinocytosis	Into the cell ("cell drinking")	Vesicle formation	Breast milk absorption in infants
Exocytosis	Out of the cell	Vesicle fuses with plasma membrane	Hormones, messenger chemicals, other macromolecules

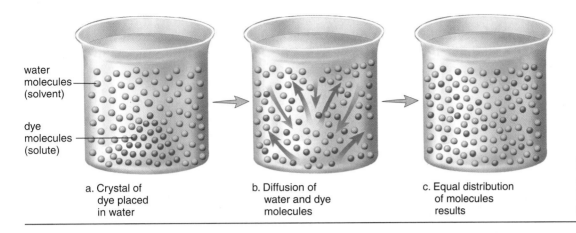

water
molecules
(solvent)

dye
molecules
(solute)

a. Crystal of
dye placed
in water

b. Diffusion of
water and dye
molecules

c. Equal distribution
of molecules
results

Figure 3.9 Diffusion. As dye molecules move from areas of higher to lower concentration, the water takes on the color of the dye.

Tonicity is the degree to which a solution's concentration of solute-versus-water causes water to move into or out of cells. Normally, body fluids are **isotonic** to cells (Fig. 3.10*a*)—that is, there is an equal concentration of solutes (dissolved substances) and solvent (water) on both sides of the plasma membrane, and cells maintain their usual size and shape. Medically administered intravenous solutions usually have this tonicity. Body fluids which are *not* isotonic to body cells are the result of dehydration or water intoxication (see Medical Focus, p. 51).

Solutions (solute plus solvent) that cause cells to swell or even to burst due to an intake of water are said to be **hypotonic** solutions. If red blood cells are placed in a hypotonic solution, which has a higher concentration of water (lower concentration of solute) than do the cells, water enters the cells and they swell to bursting (Fig. 3.10*b*). The term **lysis** refers to disrupted cells: **hemolysis,** then, is disrupted red blood cells.

Solutions that cause cells to shrink or to shrivel due to a loss of water are said to be **hypertonic** solutions. If red blood cells are placed in a hypertonic solution, which has a lower concentration of water (higher concentration of solute) than do the cells, water leaves the cells and they shrink (Fig. 3.10*c*). The term **crenation** refers to red blood cells in this condition.

Filtration

Because capillary walls are only one cell thick, small molecules (e.g., water or small solutes) tend to passively diffuse across these walls, from areas of higher concentration to those of lower

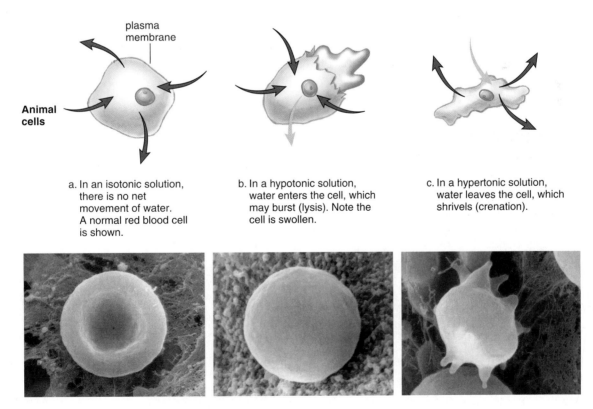

plasma
membrane

Animal cells

a. In an isotonic solution, there is no net movement of water. A normal red blood cell is shown.

b. In a hypotonic solution, water enters the cell, which may burst (lysis). Note the cell is swollen.

c. In a hypertonic solution, water leaves the cell, which shrivels (crenation).

Figure 3.10 Tonicity. The arrows indicate the movement of water.

concentration. However, blood pressure aids matters by pushing water and dissolved solutes out of the capillary, through tiny pores between capillary cells. This process, called **filtration,** is the movement of liquid from high pressure to low pressure.

You can observe filtration in a drip coffeemaker. Water moves from the area of high pressure (the water reservoir) to the area of low pressure (the coffee pot). Large substances (the coffee grounds) remain behind in the coffee filter, but small molecules (caffeine, flavor) and water pass through.

Transport by Carriers

Most solutes do not simply diffuse across a plasma membrane; rather, they are transported by means of protein carriers within the membrane. During **facilitated diffusion** (facilitated transport), a molecule (e.g., an amino acid or glucose) is transported across the plasma membrane from the side of higher concentration to the side of lower concentration. The cell does not need to expend energy for this type of transport because the molecules are moving down their concentration gradient.

During **active transport,** a molecule is moving contrary to the normal direction—that is, from lower to higher concentration (Fig. 3.11). For example, iodine collects in the cells of the thyroid gland; sugar is completely absorbed from the gut by cells that line the digestive tract; and sodium (Na^+) is sometimes almost completely withdrawn from urine by cells lining kidney tubules. Active transport requires a protein carrier and the use of cellular energy obtained from the breakdown of ATP. When ATP is broken down, energy is released, and in this case the energy is used by a carrier to carry out active transport. Therefore, it is not surprising that cells involved in active transport have a large number of mitochondria near the plasma membrane at which active transport is occurring.

Proteins involved in active transport often are called pumps because just as a water pump uses energy to move water against the force of gravity, proteins use energy to move substances against their concentration gradients. One type of pump that is active in all cells but is especially associated with nerve and muscle cells moves sodium ions (Na^+) to the outside of the cell and potassium ions (K^+) to the inside of the cell.

The passage of salt (NaCl) across a plasma membrane is of primary importance in cells. First, sodium ions are pumped across a membrane; then, chloride ions simply diffuse through channels that allow their passage. Chloride ion channels malfunction in persons with **cystic fibrosis,** and this leads to the symptoms of this inherited (genetic) disorder.

Endocytosis and Exocytosis

During **endocytosis,** a portion of the plasma membrane forms an inner pocket to envelop a substance, and then the membrane pinches off to form an intracellular vesicle (see Fig. 3.1, *top*). Two forms of endocytosis exist: **phagocytosis,** or "cell eating," is a mechanism whereby the cell can ingest solid particles. White blood cells consume bacterial cells by phagocytosis. Once inside the cell, the bacterial cell can be destroyed. **Pinocytosis,** or "cell drinking," allows the cell to consume solutions. An infant's intestinal lining ingests breast milk by pinocytosis, allowing the mother's protective antibodies to enter the baby's bloodstream.

During exocytosis, a vesicle fuses with the plasma membrane as secretion occurs (see Fig. 3.1, *bottom*). This is the way insulin leaves insulin-secreting cells, for instance. Table 3.2 summarizes the various ways molecules cross the plasma membrane.

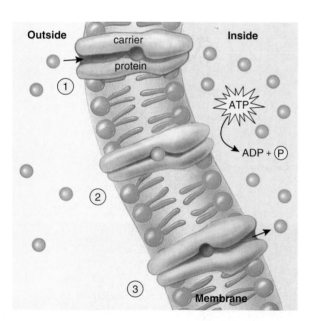

Figure 3.11 Active transport through a plasma membrane. Active transport allows a molecule to cross the membrane from lower concentration to higher concentration. ① Molecule enters carrier. ② Breakdown of ATP induces a change in shape that ③ drives the molecule across the membrane.

Content **CHECK-UP!**

4. Which process requires cellular ATP energy?
   a. osmosis
   b. facilitated diffusion (facilitated transport)
   c. active transport
   d. simple diffusion

5. A researcher studying the white blood cells of a patient infected with tuberculosis (TB) bacteria notices the bacteria are in *vesicles* in the cytoplasm. How did the bacteria come to be inside the cell?
   a. pinocytosis   c. exocytosis
   b. phagocytosis

6. The cell organelle that is needed to destroy the TB bacterium discussed in question 5 is a:
   a. ribosome.   c. centrosome.
   b. lysosome.

Dehydration and Water Intoxication

Dehydration is due to a loss of water. The solute concentration in extracellular fluid increases—that is, tissue fluid becomes hypertonic to cells, and water leaves the cells. Common causes of dehydration are excessive sweating, perhaps during exercise, without any replacement of the water lost. Dehydration can also be a side effect of any illness that causes prolonged vomiting or diarrhea. The signs of moderate dehydration are a dry mouth, sunken eyes, and skin that will not bounce back after light pinching. If dehydration becomes severe, the pulse and breathing rate are rapid, the hands and feet are cold, and the lips are blue. Although dehydration leads to weight loss, it is never a good idea to dehydrate on purpose for this reason.

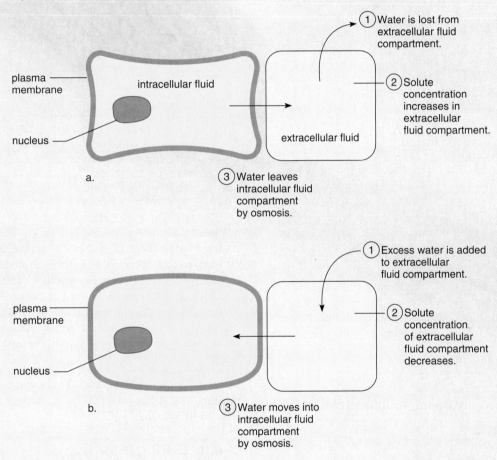

Figure 3A Dehydration versus water intoxication. **a.** If extracellular fluid loses too much water, cells lose water by osmosis and become dehydrated. **b.** If extracellular fluid gains too much water, cells gain water by osmosis and water intoxication occurs.

Water intoxication is due to excessive consumption of pure water. The tissue fluid becomes hypotonic to the cells, and water enters the cells. Water intoxication can lead to pulmonary edema (excess tissue fluid in the lungs) and swelling in the brain. In extreme cases, it is fatal. Water intoxication is not nearly as common in adults as is dehydration. It can result from a mental disorder termed *psychogenic polydipsia*. Another cause can be the intake of too much pure water during vigorous exercise: for example, a marathon race. Marathoners who collapse and have nausea and vomiting after a race may be suffering from water intoxication. The cure, an intravenous solution containing high amounts of sodium, is the opposite of that for dehydration. Therefore, it is important that physicians be able to diagnose water intoxication in athletes who have had an opportunity to drink fluids over a period of a few hours. To prevent both dehydration and water intoxication, athletes should replace lost fluids continuously. Pure water is a good choice if the exercise period is short. Low-sodium solutions, such as sports drinks, are a good choice for longer-duration events like marathons.

3.3 The Cell Cycle

The **cell cycle** is an orderly set of stages that take place between the time a cell divides and the time the resulting daughter cells also divide. The cell cycle is controlled by internal and external signals. A signal is a molecule that stimulates or inhibits a metabolic event. For example, *growth factors* are external signals received at the plasma membrane that cause a resting cell to undergo the cell cycle. When blood platelets release a growth factor, skin fibroblasts in the vicinity finish the cell cycle, thereby repairing an injury. Other signals ensure that the stages follow one another in the normal sequence and that each stage is properly completed before the next stage begins.

The cell cycle has a number of checkpoints, places where the cell cycle stops if all is not well. Any cell that did not successfully complete mitosis and is abnormal undergoes apoptosis at the *restriction checkpoint*. **Apoptosis** is often defined as *programmed cell death* because the cell progresses through a series of events that bring about its destruction. The cell rounds up and loses contact with its neighbors. The nucleus fragments and the plasma membrane develops blisters. Finally, the cell fragments and its bits and pieces are engulfed by white blood cells and/or neighboring cells. The enzymes that bring about apoptosis are ordinarily held in check by inhibitors, but are unleashed by either internal or external signals.

Following a certain number of cell cycle revolutions, cells are apt to become specialized and no longer go through the cell cycle. Muscle cells and nerve cells typify specialized cells that rarely, if ever, go through the cell cycle. At the other extreme, some cells in the body, called stem cells, are always immature and go through the cell cycle repeatedly. There is a great deal of interest in stem cells today because it may be possible to control their future development into particular tissues and organs.

Cell Cycle Stages

The cell cycle has two major portions: **interphase** and the **mitotic stage** (Fig. 3.12).

Interphase

The cell in Figure 3.1 is in interphase because it is not dividing. During interphase, the cell carries on its regular activities, and it also gets ready to divide if it is going to complete the cell cycle. For these cells, interphase has three stages, called G_1 phase, S phase, and G_2 phase.

G_1 Phase Early microscopists named the phase before DNA replication G_1, and they named the phase after DNA replication G_2. G stood for "gap." Now that we know how metabolically active the cell is, it is better to think of G as standing for "growth." Protein synthesis is very much a part of these growth phases.

During G_1, a cell doubles its organelles (such as mitochondria and ribosomes) and accumulates materials that will be used for DNA synthesis.

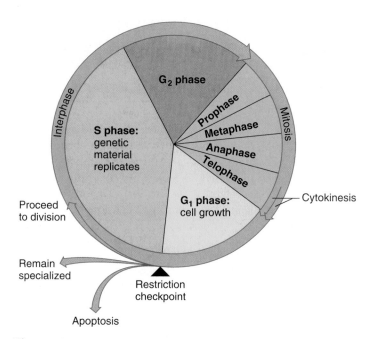

Figure 3.12 The cell cycle consists of interphase, during which cellular components duplicate, and a mitotic stage, during which the cell divides. Interphase consists of two so-called "growth" phases (G_1 and G_2) and a DNA synthesis (S) phase. The mitotic stage consists of the phases noted plus cytokinesis.

S Phase Following G_1, the cell enters the S (for "synthesis") phase. During the S phase, DNA replication occurs. At the beginning of the S phase, each chromosome is composed of one DNA double helix, which is equal to a chromatid. At the end of this phase, each chromosome has two identical DNA double helix molecules, and therefore is composed of two sister chromatids. Another way of expressing these events is to say that DNA replication has resulted in duplicated chromosomes.

G_2 Phase During this phase, the cell synthesizes proteins that will assist cell division, such as the protein found in microtubules. The role of microtubules in cell division is described later in this section.

Events During Interphase

Two significant events during interphase are replication of DNA and protein synthesis.

Replication of DNA

During **replication,** an exact copy of a DNA helix is produced. (DNA and RNA structure are described on pp. 34–35.) The double-stranded structure of DNA aids replication because each strand serves as a template for the formation of a complementary strand. A **template** is most often a mold used to produce a shape opposite to itself. In this case, each old (parental) strand is a template for each new (daughter) strand.

Figures 3.13 and 3.14 show how replication is carried out. Figure 3.14 uses the ladder configuration of DNA for easy viewing.

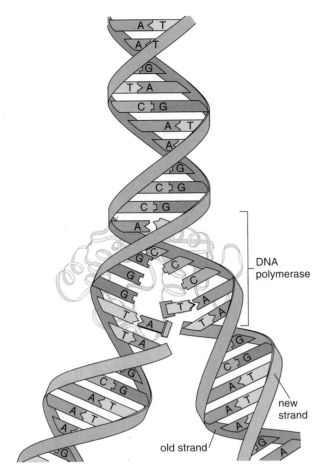
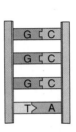

DNA
polymerase

new
strand

old strand

Figure 3.13 Overview of DNA replication. During replication, an old strand serves as a template for a new strand. The new double helix is composed of an old (parental) strand and a new (daughter) strand.

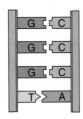
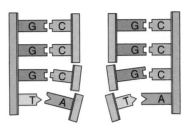
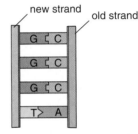

Parental DNA molecule contains so-called old strands hydrogen-bonded by complementary base pairing.

Region of replication. Parental DNA is unwound and unzipped. New nucleotides are pairing with those in old strands.

new strand   old strand

Replication is complete. Each double helix is composed of an old (parental) strand and a new (daughter) strand.

Figure 3.14 Ladder configuration and DNA replication. Use of the ladder configuration better illustrates how complementary nucleotides available in the cell pair with those of each old strand before they are joined together to form a daughter strand.

1. Before replication begins, the two strands that make up parental DNA are hydrogen-bonded to one another.
2. During replication, the old (parental) DNA strands unwind and "unzip" (i.e., the weak hydrogen bonds between the two strands break).
3. New complementary nucleotides, which are always present in the nucleus, pair with the nucleotides in the old strands. A pairs with T and C pairs with G. The enzyme DNA polymerase joins the new nucleotides forming new (daughter) complementary strands.
4. When replication is complete, two double helix molecules are identical.

Each strand of a double helix is equal to a **chromatid**, which means that at the completion of replication each chromosome is composed of two sister chromatids. They are called **sister chromatids** because they are identical. The chromosome is called a **duplicated chromosome** (see Fig. 3.16).

Protein Synthesis

DNA not only serves as a template for its own replication, but is also a template for RNA formation and for the construction of protein by the cell. Protein synthesis requires two steps: transcription and translation. During **transcription**, a messenger RNA (mRNA) molecule is produced, using DNA as the template. During **translation**, this mRNA specifies the order of amino acids in a particular polypeptide (Fig. 3.15). A gene (i.e., DNA) contains coded information for the sequence of amino acids in a particular polypeptide. The code is a **triplet code:** Every three bases in DNA (and therefore in mRNA) stand for a particular amino acid.

Transcription and Translation

During transcription, DNA unwinds and unzips, as though it is preparing for replication. Complementary RNA nucleotides from an RNA nucleotide pool in the nucleus pair with the DNA nucleotides of one strand. The RNA nucleotides are joined by an enzyme called RNA polymerase, and an RNA molecule results (Fig. 3.15, *step 1*). There are three forms of RNA: messenger, transfer, and ribosomal; however, only messenger RNA determines amino acid sequence. When mRNA forms, it has a sequence of bases complementary to those of DNA. A sequence of three bases that is complementary to the DNA triplet code is a **codon** (Fig. 3.15, *step 3*).

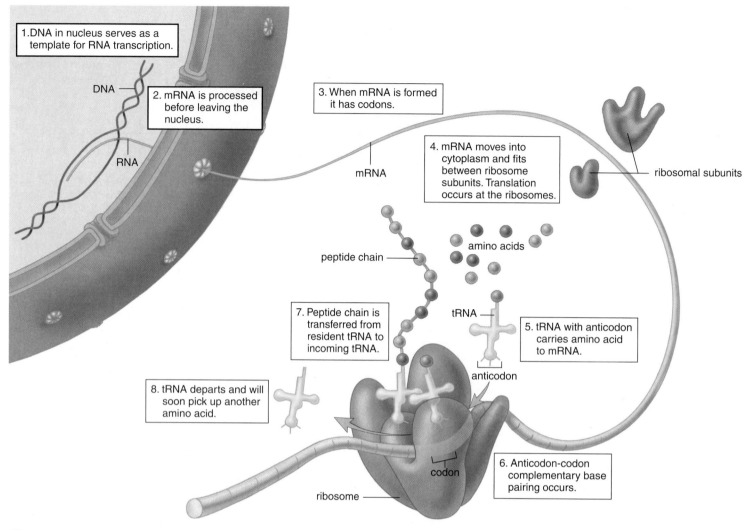

1. DNA in nucleus serves as a template for RNA transcription.

DNA

RNA

2. mRNA is processed before leaving the nucleus.

3. When mRNA is formed it has codons.

mRNA

4. mRNA moves into cytoplasm and fits between ribosome subunits. Translation occurs at the ribosomes.

ribosomal subunits

peptide chain

amino acids

tRNA

5. tRNA with anticodon carries amino acid to mRNA.

7. Peptide chain is transferred from resident tRNA to incoming tRNA.

anticodon

8. tRNA departs and will soon pick up another amino acid.

codon

6. Anticodon-codon complementary base pairing occurs.

ribosome

Figure 3.15 Protein synthesis. The two steps required for protein synthesis are transcription, which occurs in the nucleus and translation, which occurs in the cytoplasm at the ribosomes.

Translation requires several enzymes and all three types of RNA. Messenger RNA, containing the polypeptide's code, is sandwiched between the two ribosome subunits (Fig. 3.15, *step 4*). **Transfer RNA (tRNA)** molecules deliver amino acids to the ribosomes, which are composed of **ribosomal RNA (rRNA)** and protein (Fig. 3.15, *step 5*). There is at least one tRNA molecule for each of the 20 amino acids found in proteins. The amino acid binds to one end of the molecule, and the entire complex is designated as tRNA-amino acid.

Recall that the primary sequence of a protein is determined by the ordering of its amino acids (see Chapter 2, pages 31–32). Transfer RNA must deliver the correct amino acid so that the primary sequence is also correct. At the other end of each tRNA molecule is a specific **anticodon,** a group of three bases that is complementary to an mRNA codon. A tRNA molecule comes to the ribosome, where its anticodon pairs with an mRNA codon (Fig. 3.15, *step 6*). For example, if the codon is ACC, the anticodon is UGG and the amino acid is threonine. (The codes for each of the 20 amino acids are known.) Notice that the order of the codons of the mRNA determines the order that tRNA-amino acids come to a ribosome, and therefore the final sequence of amino acids in a polypeptide.

Mitotic Stage

Following interphase, the cell enters the M (for mitotic) stage. This cell division stage includes **mitosis** (division of the nucleus) and **cytokinesis** (division of the cytoplasm). The **parental cell** is the cell that divides, and the **daughter cells** are the cells that result. During mitosis, chromosomes are distributed to two separate nuclei. When cytokinesis is complete, two daughter cells are present.

As mitosis begins (Fig. 3.16, *upper left*), the centrioles have doubled in preparation for mitosis. Each chromosome is duplicated—it is composed of two chromatids held together at a central region, called the centromere. Mitosis is divided into four phases: prophase, metaphase, anaphase, and telophase (Fig. 3.16).

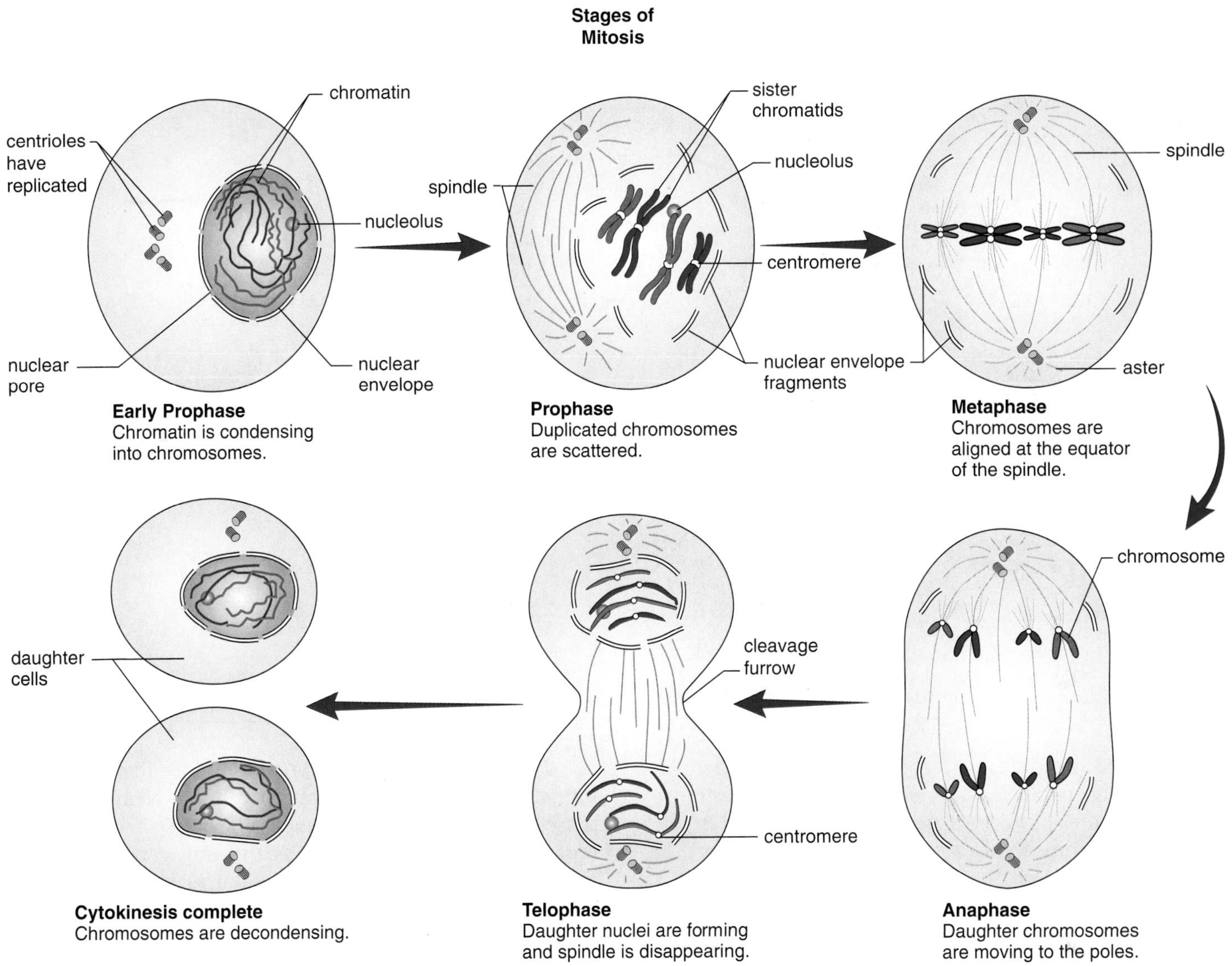

Stages of Mitosis

Early Prophase
Chromatin is condensing into chromosomes.
centrioles have replicated — chromatin — nucleolus — nuclear pore — nuclear envelope

Prophase
Duplicated chromosomes are scattered.
spindle — sister chromatids — nucleolus — centromere — nuclear envelope fragments

Metaphase
Chromosomes are aligned at the equator of the spindle.
spindle — aster

Anaphase
Daughter chromosomes are moving to the poles.
chromosome

Telophase
Daughter nuclei are forming and spindle is disappearing.
cleavage furrow — centromere

Cytokinesis complete
Chromosomes are decondensing.
daughter cells

Figure 3.16 The mitotic stage of the cell cycle. Humans have 46 chromosomes; four are shown here. The blue chromosomes were originally inherited from the father, and the red were originally inherited from the mother.

Prophase

Several events occur during **prophase** that visibly indicate the cell is about to divide. The two pairs of centrioles outside the nucleus begin moving away from each other toward opposite ends of the nucleus. Spindle fibers appear between the separating centriole pairs, the nuclear envelope begins to fragment, and the nucleolus begins to disappear.

The chromosomes are now fully visible. Although humans have 46 chromosomes, only four are shown in Figure 3.16 for ease in following the phases of mitosis. Spindle fibers attach to the centromeres as the chromosomes continue to shorten and thicken. During prophase, chromosomes are randomly placed in the nucleus.

Structure of the Spindle At the end of prophase, a cell has a fully formed spindle. A **spindle** has poles, asters, and fibers. The **asters** are arrays of short microtubules that radiate from the **poles** and the **fibers** are bundles of microtubules that stretch between the poles. (A spindle resembles a lopsided bicycle wheel; the asters are the "spokes.") Centrioles are located in centrosomes, at opposite poles of the cell. Centrosomes are believed to organize the spindle.

Metaphase

During **metaphase,** the nuclear envelope is fragmented, and the spindle occupies the region formerly occupied by the nucleus. The paired chromosomes are now at the equator (center) of the spindle. Metaphase is characterized by a fully formed spindle, and the chromosomes, each with two sister chromatids, are aligned at the equator (Fig. 3.17).

Anaphase

At the start of **anaphase,** the sister chromatids separate. *Once separated, the chromatids are called chromosomes.* Separation of the sister chromatids ensures that each cell receives a copy of

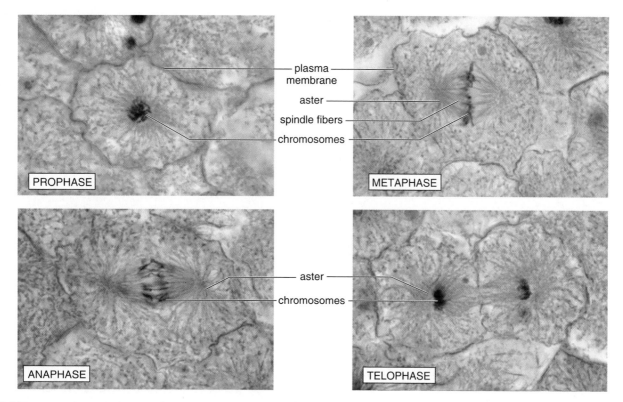

Figure 3.17 Micrographs of mitosis occurring in a whitefish embryo.

each type of chromosome and thereby has a full complement of genes. During anaphase, the daughter chromosomes move to the poles of the spindle. Anaphase is characterized by the movement of chromosomes toward each pole and thus, to opposite sides of the cell.

Function of the Spindle The spindle brings about chromosome movement. Two types of spindle fibers are involved in the movement of chromosomes during anaphase. One type extends from the poles to the equator of the spindle; there, they overlap. As mitosis proceeds, these fibers increase in length, and this helps push the chromosomes apart. The chromosomes themselves are attached to other spindle fibers that simply extend from their centromeres to the poles. These fibers get shorter and shorter as the chromosomes move toward the poles. Therefore, they pull the chromosomes apart.

Spindle fibers, as stated earlier, are composed of microtubules. Microtubules can assemble and disassemble by the addition or subtraction of tubulin (protein) subunits. This is what enables spindle fibers to lengthen and shorten, and it ultimately causes the movement of the chromosomes.

Telophase and Cytokinesis

Telophase begins when the chromosomes arrive at the poles. During telophase, the chromosomes become indistinct chromatin again. The spindle disappears as nucleoli appear, and nuclear envelope components reassemble in each cell. Telophase is characterized by the presence of two daughter nuclei.

Cytokinesis is division of the cytoplasm and organelles. In human cells, a slight indentation called a **cleavage furrow** passes around the circumference of the cell. Actin filaments form a contractile ring, and as the ring gets smaller and smaller, the cleavage furrow pinches the cell in half. As a result, each cell becomes enclosed by its own plasma membrane.

Importance of Mitosis

Because of mitosis, each cell in our body is genetically identical, meaning that it has the same number and kinds of chromosomes. Mitosis is important to the growth and repair of multicellular organisms. When a baby develops in the mother's womb, mitosis occurs as a component of growth. As a wound heals, mitosis occurs, and the damage is repaired.

Meiosis: Reduction-Division

Mitosis is the process for growing new body cells, whereas meiosis is the process for producing a person's *gametes*, or sex cells—sperm cells in males and ova in females—are produced by meiosis. In meiosis, the stages of mitosis—prophase, metaphase, anaphase, and telophase—are repeated twice. When meiosis is complete, the sperm or ova that result have half the normal chromosome number, or 23 (instead of 46 chromosomes in a body cell). When the ovum is joined by a sperm at conception, 23 ovum chromosomes join with 23 sperm chromosomes to form the new individual. (See Chapter 17 for a complete discussion of meiosis.)

FOCUS on FORENSICS

DNA Fingerprinting

It's almost a cliché of modern life: Look no further than the evening news if you want to view evidence of human cruelty. Tragic scenarios are played out daily in emergency rooms and morgues across North America, as increasing numbers of innocent people become victims of violent crimes—beatings, shootings, stabbings, sexual assault. However, recent advancements in biotechnology make the likelihood of catching and convicting criminals greater than ever before. DNA identification technology, also called DNA fingerprinting, identifies the DNA samples recovered from a crime scene. Like actual fingerprints, the composition of DNA is unique to each individual. Only identical twins share the same DNA.

The process of identifying human DNA begins with sample collection at the crime scene. Blood, hair, bone, tissue, nail clippings, saliva, even vomit—all contain human cells. An individual's DNA is contained within the nucleus and the mitochondrion of the cell, although investigators use nuclear DNA whenever possible because of its larger sample size. Exceedingly small samples of only a nanogram (one-billionth of a gram, approximately 160 human cells) are sufficient, making it possible to identify human DNA from saliva on a cigarette butt or a single human hair.

DNA is first extracted from the sample; if necessary, it can be reproduced from a very small sample to create adequate amounts for testing. The DNA is digested into fragments of differing sizes using restriction enzymes, which act as a molecular "scissors." Each fragment is polar—possessing either a positive or negative charge. A technique called *gel electrophoresis* separates the fragments by size and electrical charge, spreading the fragments throughout a gel medium. Identifying the exact nucleotide sequence for each fragment is never completed—to do so would be extremely expensive and time consuming. Instead, the best DNA analysis technique identifies STRs, or "short tandem repeats." STRs are segments of repeating nucleotides (for example, A-A-A-A or T-T-T-T). They can be found in all human DNA, but their exact location in the molecule is unique to each individual. Each STR is tagged with radioactive molecules. Finally, the radioactively labeled gel is exposed to X-ray film. The result is a pattern similar to a grocery store "bar code," with alternating light and dark bands.

If a suspect's DNA matches the crime scene DNA, the probability of finding another person on earth with the same DNA pattern is 1 in 300 billion (a number roughly equal to the number of human beings found on earth!) Thus, DNA fingerprinting is recognized as a very powerful tool in the courtroom for convicting a guilty individual. However, the same technique can also be used to clear the innocent. If there is no match between the crime scene DNA and that of the suspect, the suspect can be declared innocent of the crime.

Crime labs at the state, national, and international levels now maintain computerized databases of DNA fingerprints. These databases allow rapid electronic comparison of DNA samples from crime scenes to the DNA fingerprints of known and suspected felons. Laws currently exist in all 50 states that require felons convicted of murder, assault, sexual offenses, etc., to submit DNA samples on demand to the Combined DNA Index System (CODIS), which is maintained by the FBI.

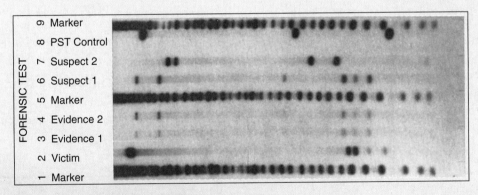

Figure 3B Actual DNA fingerprint, from a rape trial. Lanes 1, 2, 5, 8, and 9 are controls. Lane 2 is the victim's DNA, taken from a blood sample. Lanes 3 and 4 were created using semen samples taken from the victim's body after the rape was committed. Study lanes 6 and 7 carefully—who committed the crime—suspect 1 or suspect 2?

Content CHECK-UP!

7. How many chromosomes are found in a human cell after the S phase of the cell cycle is complete?

  a. 46

  c. 92 (46 × 2)

  b. 23

8. The nuclear envelope fragments and the nucleolus begins to disappear during:

  a. anaphase.

  c. telophase.

  b. prophase.

  d. metaphase.

9. Formation of a cleavage furrow signals the beginning of which stage?

  a. anaphase

  c. telophase

  b. metaphase

  d. cytokinesis

SELECTED NEW TERMS

Basic Key Terms

actin filaments (ăk′tĭn fĭl′ə-məntz), p. 46

active transport (ak′tiv trans′port), p. 50

anaphase (ăn′ə-fāz), p. 55

anticodon (ăn′tī′-kō′dŏn′), p. 54

apoptosis (ap″o-to′-sis), p. 52

asters (ăs′tərz), p. 55

cell cycle (sel sī′-kl), p. 52

cellular respiration (sĕl′yə-lər rĕs′pə-rā′shən), p. 46

centriole (sen′tre-ōl), p. 46

chromatid (krō′mə-tĭd), p. 53

chromatin (kro′muh-tin), p. 42

chromosome (kro′mo-sōm), p. 43

cilium (sĭl′ē-əm), p. 46

cleavage furrow (klēv′ij fur′o), p. 56

codon (kō′dŏn′), p. 53

crenation (krē-nā′-shun), p. 49

cytokinesis (si′to-kī-ne′sis), p. 54

cytoplasm (si′to-plazm), p. 40

cytoskeleton (si′to-skel″ĕ-tun), p. 40

dehydration (dē′hī-drā′shən), p. 51

endocytosis (ĕn′dō-sī′tō-sĭs), p. 50

endomembrane system (en″do-mem′brān sis′tem), p. 43

endoplasmic reticulum (en-do-plaz′mic rĕ-tik′yū-lum), p. 43

facilitated diffusion (fuh-sil′-ĭ-tāt′id di-fyū-zhun), p. 50

filtration (fil-tra′shun), p. 50

flagellum (flə-jĕl′əm), p. 46

Golgi apparatus (gol′je ap″uh-rā′tus), p. 44

intermediate filaments (ĭn′tər-mē′dē-ĭt fĭl′ə-məntz), p. 46

interphase (ĭn′tər-fāz′), p. 52

lysis (lī′sĭs), p. 49

lysosome (li′so-sōm), p. 45

metaphase (mĕt′ə-fāz), p. 55

microtubule (mi″kro-tu′byūl), p. 46

mitochondrion (mi″to-kon′dre-on), p. 45

mitosis (mi-to′sis), p. 54

nuclear envelope (nu′kle-er en′vě-lōp), p. 43

nuclear pore (nu′kle-er por), p. 43

nucleolus (nu-kle′o-lus), p. 40

nucleus (nu′kle-us), p. 40

organelle (or′guh-nel), p. 40

osmosis (oz-mo′sis), p. 48

osmotic pressure (ŏz-mŏt′ĭk prĕsh′ər), p. 48

peroxisome (per-ok′-si-sōm), p. 45

phagocytosis (făg′ə-sĭ-tō′sĭs), p. 50

pinocytosis (pĭn′ə-sī-tō′sĭs), p. 50

plasma membrane (plaz′muh mem′bran), p. 40

poles (pōlz), p. 55

prophase (prō′fāz′), p. 55

replication (rĕp′lĭ-kā′shən), p. 52

ribosomal RNA (ri′bo-sōm′al RNA), p. 54

ribosome (ri′bo-sōm), p. 43

selectively permeable (se-lĕk′tiv-le per′me-uh-bl), p. 48

simple diffusion (sĭm′păl dĭ-fyo͞o′zhən), p. 48

solute (sol′ūt), p. 48

spindle (spin′dl), p. 55

telophase (tĕl′ə-fāz), p. 56

template (tĕm′plĭt), p. 52

tonicity (tō-nĭs′ĭ-tē), p. 49

transcription (trans-krip′shun), p. 53

transfer RNA (trans′fer RNA), p. 54

translation (trans-la′shun), p. 53

triplet code (trip′let cōd), p. 53

vesicle (ves′ĭ-kl), p. 43

water intoxication (wôltər ĭn-tŏk′sĭ-kāt′shən), p. 51

Clinical Key Terms

cystic fibrosis (sis′tik fĭ-brō′-sis), p. 50

Tay-Sachs (tā saks), p. 45

SUMMARY

Cells differ in shape and function, but even so, a generalized cell can be described.

3.1 Cellular Organization
All human cells, despite varied shapes and sizes, have a plasma membrane. Most have a single nucleus, although some are multinucleate. Mature red blood cells are anucleate. Cell cytoplasm contains organelles and a cytoskeleton.

A. The plasma membrane, composed of phospholipid and protein molecules, regulates the entrance and exit of other molecules into and out of the cell.

B. The nucleus contains chromatin, which condenses into chromosomes just prior to cell division. Genes, composed of DNA, are on the chromosomes, and they code for the production of proteins in the cytoplasm. The nucleolus is involved in ribosome formation.

C. Ribosomes are small organelles where protein synthesis occurs. Ribosomes occur in the cytoplasm, both singly and in groups. Numerous ribosomes are attached to the endoplasmic reticulum.

D. The endomembrane system consists of the endoplasmic reticulum (ER), the Golgi apparatus, and the lysosomes and various transport vesicles.

E. The ER is involved in protein synthesis (rough ER) and various other processes such as lipid synthesis (smooth ER). Molecules produced or modified in the ER are eventually enclosed in vesicles that take them to the Golgi apparatus.

F. The Golgi apparatus processes and packages molecules, distributes them within the cell, and transports them out of the cell. It is also involved in secretion.

G. Lysosomes are produced by the Golgi apparatus, and their hydrolytic enzymes digest macromolecules from various sources. Peroxisomes digest toxins and fatty acids.

H. Mitochondria are involved in cellular respiration, a metabolic pathway that provides ATP molecules to cells.

I. Notable among the contents of the cytoskeleton are microtubules, intermediate filaments, and actin filaments. The cytoskeleton maintains the shape of the cell and also directs the movement of cell parts.

J. Centrioles lie near the nucleus and are involved in the production of the spindle during cell division and in the formation of cilia and flagella.

3.2 **Crossing the Plasma Membrane**
When substances enter and exit cells by simple diffusion, osmosis, or filtration, no carrier is required. Facilitated diffusion (facilitated transport) and active transport do require a carrier.

A. Some substances can simply diffuse across a plasma membrane. The diffusion of water is called osmosis. In an isotonic solution, cells neither gain nor lose water. In a hypotonic solution, cells swell. In a hypertonic solution, cells shrink.

B. During filtration, movement of water and small molecules out of a blood vessel is aided by blood pressure.

C. During facilitated diffusion (facilitated transport), a carrier is required, but energy is not because the substance is moving from higher to lower concentration. Active transport, which requires a carrier and ATP energy, moves substances from lower to higher concentration.

D. Endocytosis involves the uptake of substances by a cell through vesicle formation. Phagocytosis allows solids to be taken up, whereas pinocytosis is ingestion of a liquid. Exocytosis involves the release of substances from a cell as vesicles within the cell cytoplasm fuse with the plasma membrane.

3.3 **The Cell Cycle**
The cell cycle consists of interphase (G_1 phase, S phase, G_2 phase) and the mitotic stage, which includes mitosis and cytokinesis.

A. During interphase, DNA replication and protein synthesis take place. DNA serves as a template for its own replication: The DNA parental molecule unwinds and unzips, and new (daughter) strands form by complementary base pairing. Protein synthesis consists of transcription and translation. During transcription, DNA serves as a template for the formation of RNA. During translation, mRNA, rRNA, and tRNA are involved in polypeptide synthesis.

B. Mitosis consists of a number of phases, during which each newly formed cell receives a copy of each kind of chromosome. Later, the cytoplasm divides by furrowing. Mitosis occurs during tissue growth and repair.

STUDY QUESTIONS

1. What are the three main parts to any human cell? (p. 40)
2. Describe the fluid-mosaic model of membrane structure. (p. 42)
3. Describe the nucleus and its contents, and include the terms *DNA* and *RNA* in your description. (p. 42)
4. Describe the structure and function of ribosomes. (p. 43)
5. What is the endomembrane system? What organelles belong to this system? (p. 43)
6. Describe the structure and function of endoplasmic reticulum (ER). Include the terms smooth ER, rough ER, and ribosomes in your description. (p. 43)
7. Describe the structure and function of the Golgi apparatus. Mention vesicles and lysosomes in your description. (p. 44)
8. Describe the structure and function of mitochondria. Mention the energy molecule ATP in your description. (p. 45)
9. What is the cytoskeleton? What role does the cytoskeleton play in cells? (p. 46)
10. Describe the structure and function of centrioles. Mention the mitotic spindle in your description. (p. 46)
11. Contrast passive transport (diffusion, osmosis, filtration) with active transport of molecules across the plasma membrane. (pp. 48–49)
12. Define osmosis, and discuss the effects of placing red blood cells in isotonic, hypotonic, and hypertonic solutions. (p. 48–49)
13. What is the cell cycle? What stages occur during interphase? What happens during the mitotic stage? (p. 52)
14. Describe the structure of DNA. How does this structure contribute to the process of DNA replication? (p. 52–53)
15. Briefly describe the events of protein synthesis. (p. 53)
16. List the phases of mitosis. What happens during each phase? (pp. 54–55)
17. Discuss the importance of mitosis in humans. (p. 56)

LEARNING OBJECTIVE QUESTIONS

I. Match the organelles in the key to the functions listed in questions 1–5.

Key:

    a. mitochondria
    b. nucleus
    c. Golgi apparatus
    d. rough ER
    e. centrioles

1. packaging and secretion
2. cell division
3. powerhouse of the cell
4. protein synthesis
5. control center for the cell

II. Fill in the blanks.

6. The fluid-mosaic model of membrane structure says that _____ molecules drift about within a double layer of _____ molecules.
7. Rough ER has _____, but smooth ER does not.

8. Basal bodies that organize the microtubules within cilia and flagella are believed to be derived from _____.
9. Water will enter a cell when it is placed in a _____ solution.
10. Active transport requires a protein _____ and _____ for energy.
11. Vesicle formation occurs when a cell takes in material by _____.
12. At the conclusion of mitosis, each newly formed cell in humans contains _____ chromosomes.
13. The _____, which is the substance outside the nucleus of a cell, contains bodies called _____, each with a specific structure and function.

III. Match the organelles in the key to the functions listed in questions 14–17.

Key:

    a. DNA
    b. mRNA
    c. tRNA
    d. rRNA

14. Joins with proteins to form subunits of a ribosome.
15. Contains codons that determine the sequence of amino acids in a polypeptide.
16. Contains a code and serves as a template for the production of RNA.
17. Brings amino acids to the ribosomes during the process of transcription.

MEDICAL TERMINOLOGY EXERCISE

Consult Appendix B for help in pronouncing and analyzing the meaning of the terms that follow.

1. hemolysis (he″mol′ĭ-sis)
2. cytology (si-tol′o-je)
3. cytometer (si-tom′ĕ-ter)

4. nucleoplasm (nu′kle-o-plazm)
5. pancytopenia (pan″si-to-pe′ne-uh)
6. cytogenic (si-to-jen′ik)
7. erythrocyte (ĕ-rith′ro-sit)
8. atrophy (at′ro-fe)
9. hypertrophy (hi-per′tro-fe)

10. oncotic pressure, colloid osmotic pressure (ong-kot′ik presh′er)(kol′oyd oz-mah′-tik presh′er)
11. hyperplasia (hi-per-pla′zhe-uh)

WEBSITE LINK

Visit this text's website at http://www.mhhe.com/maderap6 for additional quizzes, interactive learning exercises, and other study tools.

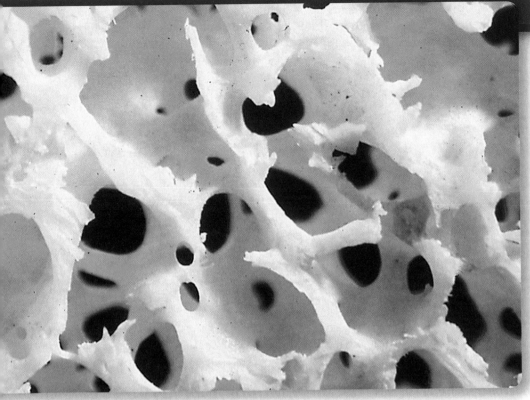

CHAPTER

4

Body Tissues and Membranes

Spongy bone consists of bars and plates separated by irregular spaces, but it is still quite strong.

Chapter Outline & Learning Objectives

After you have studied this chapter, you should be able to:

4.1 Epithelial Tissue (p. 62)

- Describe the general characteristics and functions of epithelial tissue.
- Name the major types of epithelial tissue, and relate each one to a particular organ.

4.2 Connective Tissue (p. 65)

- Describe the general characteristics and functions of connective tissue.
- Name the major types of connective tissue, and relate each one to a particular organ.

4.3 Muscular Tissue (p. 69)

- Describe the general characteristics and functions of muscular tissue.

- Name the major types of muscular tissue, and relate each one to a particular organ.

4.4 Nervous Tissue (p. 72)

- Describe the general characteristics and functions of nervous tissue.

4.5 Extracellular Junctions, Glands, and Membranes (p. 73)

- Describe the structure and function of three types of extracellular junctions.
- Describe the difference between an exocrine and an endocrine gland with examples.

- Describe the way the body's membranes are organized.
- Name and describe the major types of membranes in the body.

Medical Focus

Cancer—The Traitor Inside (p. 74)

What's New

Blocking Blood Supply to Treat Cancer (p. 76)

A tissue is composed of specialized cells of similar structure that perform a common function in the body. There are four major types of tissues: (1) Epithelial tissue, also called epithelium, covers body surfaces and organs, and lines body cavities; (2) connective tissue binds and supports body parts; (3) muscular tissue contracts; and (4) nervous tissue responds to stimuli and transmits impulses from one body part to another. Body organs are typically composed of all four tissue types. For example, the heart is covered by *epithelium*; its valves are supported by *connective tissue*; cardiac *muscle* pumps blood, and *nerves* regulate how rapidly the heart beats.

4.1   Epithelial Tissue

Epithelial tissues share a set of common characteristics. In all epithelial tissues, the cells are tightly packed, with little space between them. Externally, this tissue protects the body from drying out, injury, and bacterial invasion. On internal surfaces, epithelial tissue protects, but it also may have an additional function. For example, in the respiratory tract, epithelial tissue sweeps up impurities by means of cilia. Along the digestive tract, it secretes mucus, which protects the lining from digestive enzymes. In kidney tubules, its absorptive function is enhanced by the presence of fine, cellular extensions called microvilli.

Epithelial cells readily divide to produce new cells that replace lost or damaged ones. Skin cells as well as those that line the stomach and intestines are continually being replaced. To support its very high rate of reproduction, epithelial tissue must get its nutrients from underlying connective tissues. Epithelial tissues are *avascular*—they lack blood vessels. Thus, they can be shed safely without the risk of bleeding.

Because epithelial tissue covers surfaces and lines cavities, it always has a *free surface*. The other surface is attached to underlying tissue by a layer of carbohydrates and proteins called the *basement membrane*.

Epithelial tissues are classified according to the shape of the cells and the number of cell layers (Table 4.1). **Simple** epithelial tissue is composed of a single layer, and **stratified** epithelial tissue is composed of two or more layers. **Squamous** epithelium has flattened cells; **cuboidal** epithelium has cube-shaped cells; and **columnar** epithelium has elongated cells.

Squamous Epithelium

Simple squamous epithelium is composed of a single layer of flattened cells, and therefore its protective function is not as significant as that of other epithelial tissues (Fig. 4.1). It is found in areas where simple diffusion occurs. (Recall that simple diffusion is movement of a substance from areas of high to low concentration.) For example, simple squamous epithelium forms the tiny air sacs (called alveoli) of the lungs, where oxygen and carbon dioxide are exchanged. Capillaries (the smallest blood vessels) are tubes of simple squamous epithelium. Nutrients and wastes are exchanged between capillaries and neighboring cells by diffusion.

Stratified squamous epithelium has many cell layers and does play a protective role (Fig. 4.2). While the deeper cells may be cuboidal or columnar, the outer layer is composed of squamous-shaped cells (Note that stratified epithelial tissues are named according to the cells on their outer layer.). The outer portion of skin is stratified squamous epithelium. New cells produced in a basal layer become reinforced by keratin, a protein that waterproofs and provides strength, as they move toward the skin's surface. Aside from skin, stratified squamous epithelium is found lining the various orifices of the body.

Cuboidal Epithelium

Simple cuboidal epithelium (Fig. 4.3) consists of a single layer of cube-shaped cells attached to a basement membrane. This type of epithelium is frequently found in glands, such as salivary glands, the thyroid gland, and the pancreas, where its function is secretion. Simple cuboidal epithelium also covers the ovaries and lines most of the kidney tubules, the portion of the kidney where urine is formed. In one part of the kidney

TABLE 4.1 Epithelial Tissues		
Type of Tissue	**Description**	**Location**
Simple squamous	One layer of flattened cells	Blood capillaries; air sacs (alveoli) of lungs
Stratified squamous	Many layers; cells flattened at free surface	Skin and body orifices
Simple cuboidal	One layer of cube-shaped cells	Secreting glands, ovaries, linings of kidney tubules
Stratified cuboidal	Two or more layers of cube-shaped cells	Linings of salivary gland and mammary gland ducts
Simple columnar	One layer of elongated cells	Lining of digestive organs; lining of uterine tubes
Stratified columnar	Two or more layers of elongated cells	Pharynx (back of throat); male urethra (tube that carries urine out of body)
Pseudostratified columnar	One layer of elongated, tapered cells; appear stratified	Air passages of the respiratory system
Transitional	Many layers; when tissue stretches, layers become fewer	Urinary bladder (stores urine); ureters and urethra (tubes carrying urine)

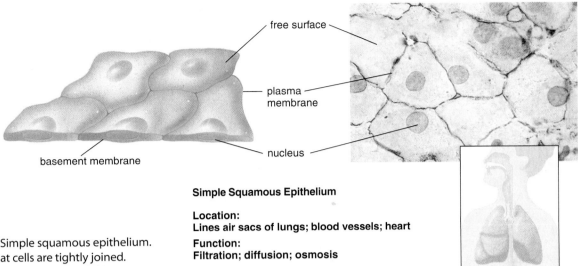

Simple Squamous Epithelium

Location:
Lines air sacs of lungs; blood vessels; heart

Function:
Filtration; diffusion; osmosis

Figure 4.1 Simple squamous epithelium. The thin and flat cells are tightly joined. The nuclei tend to be broad and thin.

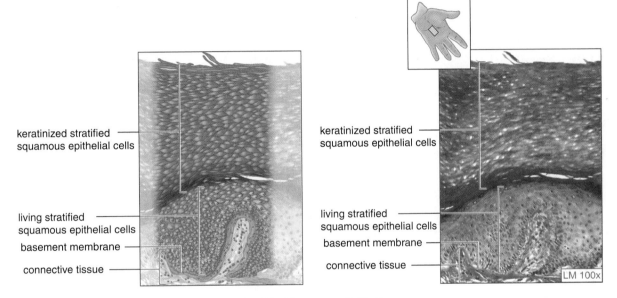

Stratified Squamous Epithelium

Location:
Lining of esophogus, mouth, vagina; epidermis
Function:
Protection from abrasion & desication in keratinized form

Figure 4.2 Stratified squamous epithelial tissue. This type of tissue is found in the skin.

tubule, it absorbs substances from the tubule, and in another part it secretes substances into the tubule. Tubular absorption and secretion are both forms of active transport. Thus, the cuboidal epithelial cells contain many mitochondria, which supply the ATP needed for active transport. Additionally, where the cells function in reabsorption, microvilli (tiny, fanlike folds in the plasma membrane) increase the surface area of the cells.

Stratified cuboidal epithelium is mostly found lining the larger ducts of certain glands, such as the mammary glands and the salivary glands. Often this tissue has only two layers.

Columnar Epithelium

Simple columnar epithelium (Fig. 4.4) has cells that are longer than they are wide. They are modified to perform particular functions. Some of these cells are goblet cells that secrete mucus onto the free surface of the epithelium.

This tissue is well known for lining digestive organs, including the small intestine, where microvilli expand the surface area and aid in absorbing the products of digestion. Simple columnar epithelium also lines the uterine tubes. Here, many cilia project from the cells and propel the egg toward the uterus, or womb.

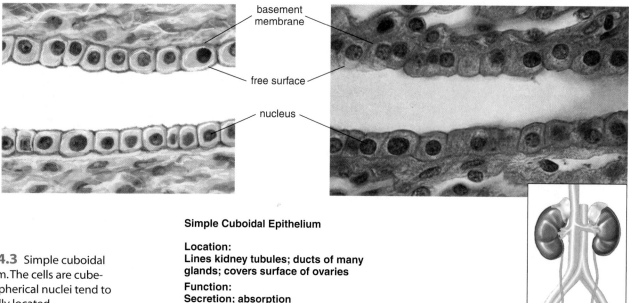

basement membrane

free surface

nucleus

Simple Cuboidal Epithelium

Location:
Lines kidney tubules; ducts of many glands; covers surface of ovaries

Function:
Secretion; absorption

Figure 4.3 Simple cuboidal epithelium. The cells are cube-shaped. Spherical nuclei tend to be centrally located.

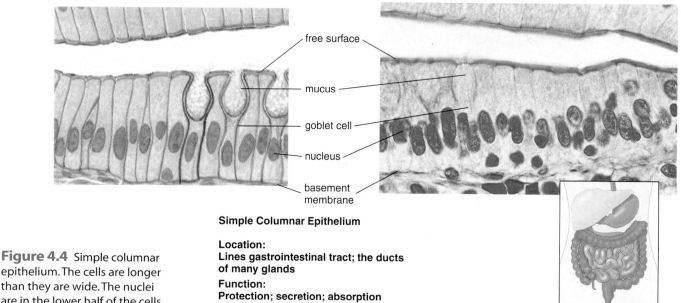

free surface

mucus

goblet cell

nucleus

basement membrane

Simple Columnar Epithelium

Location:
Lines gastrointestinal tract; the ducts of many glands

Function:
Protection; secretion; absorption

Figure 4.4 Simple columnar epithelium. The cells are longer than they are wide. The nuclei are in the lower half of the cells.

Stratified columnar epithelium is not very common but does exist in parts of the pharynx (back of the throat) and the male urethra.

Pseudostratified Columnar Epithelium

Pseudostratified columnar epithelium is so named because it appears to be layered (*pseudo*, false; *stratified*, layers). However, true layers do not exist because each cell touches the basement membrane. Each cell is tapered and narrow at one end; the opposite end contains the nucleus. The irregular placement of the nuclei creates the appearance of several layers where only one exists.

Pseudostratified ciliated columnar epithelium (Fig. 4.5) lines parts of the reproductive tract as well as the air passages of the respiratory system, including the nasal cavities and the trachea (windpipe) and its branches. Mucous-secreting goblet cells are scattered among the ciliated epithelial cells. A surface covering of mucus traps foreign particles, and upward ciliary motion carries the mucus to the back of the throat, where it may be either swallowed or expectorated.

Transitional Epithelium

The term **transitional epithelium** implies changeability, and this tissue changes in response to tension. It forms the lining

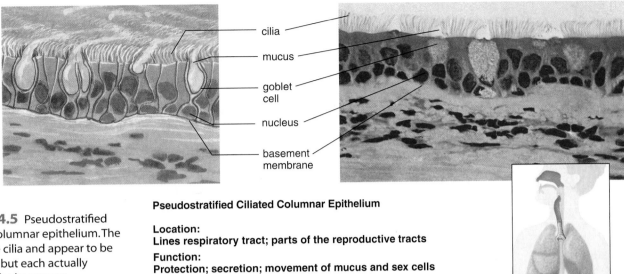

Figure 4.5 Pseudostratified ciliated columnar epithelium. The cells have cilia and appear to be stratified, but each actually touches the basement membrane.

Pseudostratified Ciliated Columnar Epithelium

Location:
Lines respiratory tract; parts of the reproductive tracts

Function:
Protection; secretion; movement of mucus and sex cells

of the urinary bladder, the ureters (tubes that carry urine from the kidneys to the bladder), and part of the urethra (the single tube that carries urine to the outside). All are organs that may need to stretch. When the walls of the bladder are relaxed, the transitional epithelium consists of several layers of cuboidal cells. When the bladder is distended with urine, the epithelium stretches, and the outer cells take on a squamous appearance. It's interesting to observe that the cells in transitional epithelium of the bladder are physically able to slide in relation to one another while at the same time forming a barrier that prevents any part of urine from diffusing into the internal environment.

Content **CHECK-UP!**

1. Which of the following is not a characteristic of epithelial tissues?
   a. possess a blood supply
   b. rapid rate of mitosis
   c. very tightly packed
   d. attached to underlying basement membrane
2. A single layer of tall, skinny epithelial cells would be best described as:
   a. stratified columnar epithelium.
   b. simple cuboidal epithelium.
   c. simple squamous epithelium.
   d. simple columnar epithelium.
3. A student studying slides of epithelial tissues under a microscope observes the presence of microvilli. She guesses that the function of the tissue is probably:
   a. secreting mucus.
   b. moving substances through a tube.
   c. absorbing something inside a tube.

4.2 Connective Tissue

Connective tissue binds structures together, provides support and protection, fills spaces, produces blood cells, and stores fat. The body uses this stored fat for energy, insulation, and organ protection. As a rule, connective tissue cells are widely separated by a nonliving, extracellular **matrix** composed of an *organic ground substance* that contains *fibers* and varies in consistency from solid to semifluid to fluid. Whereas the functional and physical properties of epithelial tissues are derived from its cells, connective tissue properties are largely derived from the characteristics of the matrix (Table 4.2).

The fibers within the matrix are of three types. **Collagen fibers** contain the fibrous protein *collagen*, a substance that gives the fibers flexibility and tremendous strength. **Elastic fibers** contain the protein *elastin*, which is not as strong as collagen but is more elastic. **Reticular fibers** are very thin, highly branched, collagenous fibers that form delicate supporting networks.

Fibrous Connective Tissue

Fibrous connective tissue includes loose connective tissue and dense connective tissue. The body's membranes are composed of an epithelium and fibrous connective tissue (see page 75).

Loose (areolar) connective tissue commonly lies between other tissues or between organs, binding them together. The cells of this tissue are mainly **fibroblasts**—large, star-shaped cells that produce extracellular fibers (Fig. 4.6). The cells are located some distance from one another because they are separated by a matrix with a jellylike ground substance that contains many collagen and elastin fibers. The collagen fibers occur in bundles and are strong and flexible. The elastin fibers form a highly elastic network that returns to its original length after stretching. *Adipose*

TABLE 4.2 Classification of Connective Tissue

Type	Structure of Matrix	Example of Location
FIBROUS CONNECTIVE TISSUE		
Loose (areolar) connective tissue	Collagen, elastin, and reticular fibers	Between tissues and organs
Adipose tissue	Fibroblasts enlarge and store fat; very little matrix	Beneath skin; around organs
Dense connective tissue		
Regular	Bundles of parallel collagen fibers	Tendons; ligaments; aponeuroses
Irregular	Bundles of nonparallel collagen fibers	Dermis of skin; joint capsules
Reticular connective tissue	Reticular fibers	Lymphatic organs and liver
CARTILAGE		
Hyaline cartilage	Fine collagen fibers	Ends of long bones; rib cartilages; nose
Elastic cartilage	Many elastin fibers	External ear
Fibrocartilage	Strong collagen fibers	Between vertebrae of spine
BONE		
Compact	Collagen plus calcium salts; arranged in *osteons*	Skeleton
Spongy	Collagen plus calcium salts; arranged in *trabeculae*	Ends of long bones
BLOOD	Plasma plus red and white blood cells, platelets	Inside blood vessels

tissue (Fig. 4.7) is a type of loose connective tissue in which the fibroblasts enlarge and store fat, and there is limited extracellular matrix.

Dense connective tissue (Fig. 4.8) has a matrix produced by fibroblasts that contains thick bundles of collagen fibers. In *dense regular connective tissue,* the bundles are parallel as in **tendons** (which connect muscles to bones), **ligaments** (which connect bones to other bones at joints), and aponeuroses, (which join muscle to muscle). In *dense irregular connective tissue,* the bundles run in different directions. This type of tissue is found in the inner portion of the skin (called the *dermis*) and in joint capsules (Fig. 4.9).

The fibroblasts of *reticular connective tissue* are called reticular cells, and the matrix contains only reticular fibers. This tissue, also called **lymphatic tissue,** is found in lymph nodes, the spleen, thymus, and red bone marrow. These organs are a part of the immune system because they store and/or produce white blood cells, particularly lymphocytes. All types of blood cells are produced in red bone marrow.

Cartilage

In **cartilage,** the cells (*chondrocytes*), which lie in small chambers called **lacunae,** are separated by a matrix that is solid yet flexible. Unfortunately, because this tissue lacks

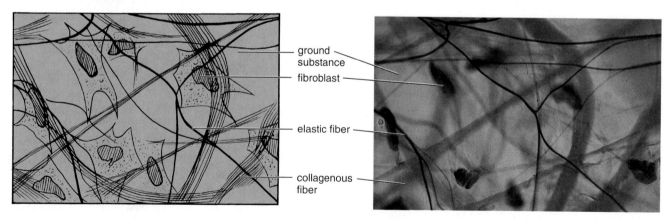

ground substance

fibroblast

elastic fiber

collagenous fiber

Loose (Areolar) Connective Tissue

Location:
Between muscles; beneath the skin;
beneath most epithelial layers
Function:
Binds organs together

Figure 4.6 Loose (areolar) connective tissue. This tissue has a loose network of fibers.

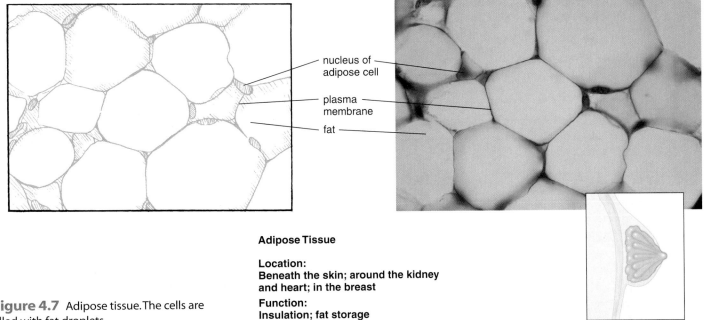

nucleus of adipose cell

plasma membrane

fat

Adipose Tissue

Location:
Beneath the skin; around the kidney
and heart; in the breast

Function:
Insulation; fat storage

Figure 4.7 Adipose tissue. The cells are filled with fat droplets.

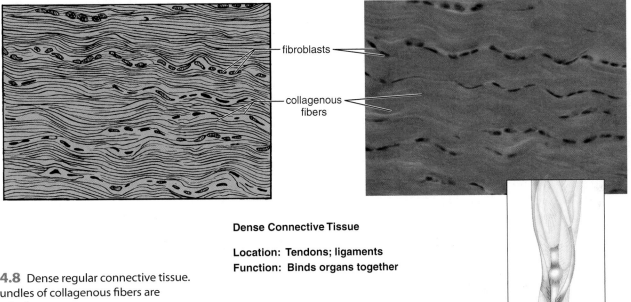

fibroblasts

collagenous fibers

Dense Connective Tissue

Location: Tendons; ligaments
Function: Binds organs together

Figure 4.8 Dense regular connective tissue. Parallel bundles of collagenous fibers are closely packed.

a direct blood supply, it heals very slowly. The three types of cartilage are classified according to the type of fiber in the matrix.

Hyaline cartilage (Fig. 4.10) is the most common type of cartilage. The matrix, which contains only very fine collagen fibers, has a glassy, white, opaque appearance. This type of cartilage is found in the nose, at the ends of the long bones and ribs, and in the supporting rings of the trachea. The fetal skeleton is also made of this type of cartilage, although the cartilage is later replaced by bone.

Elastic cartilage has a matrix containing many elastic fibers, in addition to collagen fibers. For this reason, elastic cartilage is more flexible than hyaline cartilage. Elastic cartilage is found, for example, in the framework of the outer ear.

Fibrocartilage has a matrix containing strong collagen fibers. This type of cartilage absorbs shock and reduces friction between joints. Fibrocartilage is found in structures that withstand tension and pressure, such as the disks between the vertebrae in the backbone and the pads in the knee joint.

Bone

Bone is the most rigid of the connective tissues. It has an extremely hard matrix of mineral salts, notably calcium salts,

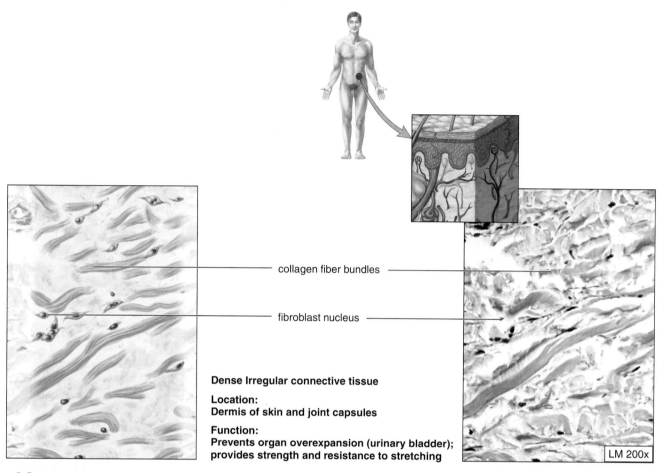

collagen fiber bundles

fibroblast nucleus

Dense Irregular connective tissue

Location:
Dermis of skin and joint capsules

Function:
Prevents organ overexpansion (urinary bladder);
provides strength and resistance to stretching

LM 200x

Figure 4.9 Dense irregular connective tissue. Bundles of collagen fibers run in different directions enabling the tissue to resist or withstand multi-directional forces.

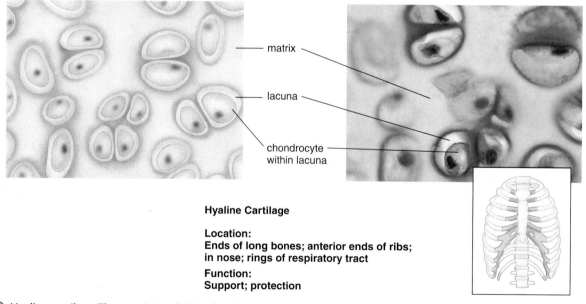

matrix

lacuna

chondrocyte
within lacuna

Hyaline Cartilage

Location:
Ends of long bones; anterior ends of ribs;
in nose; rings of respiratory tract

Function:
Support; protection

Figure 4.10 Hyaline cartilage. The matrix is solid but flexible.

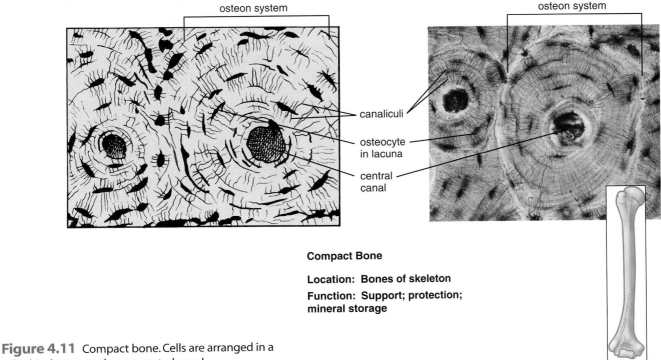

osteon system

osteon system

canaliculi

osteocyte
in lacuna

central
canal

Compact Bone

Location: Bones of skeleton

**Function: Support; protection;
mineral storage**

Figure 4.11 Compact bone. Cells are arranged in a
cylindrical manner about a central canal.

deposited around collagen fibers. The minerals give bone rigidity, and the collagen fibers provide elasticity and strength, much as steel rods do in reinforced concrete.

The outer portion of a long bone contains compact bone. **Compact bone** consists of many cylindrical-shaped units called *osteons*, or *Haversian systems* (Fig. 4.11). In an osteon, matrix is deposited in thin layers called *lamellae* that form a concentric pattern around tiny tubes called *central canals*. The canals contain nerve fibers and blood vessels. The blood vessels bring nutrients to bone cells (called *osteocytes*) that are located in small hollows called *lacunae* between the lamellae. The nutrients can reach all of the cells because minute canals (*canaliculi*) containing thin extensions of the osteocytes connect the osteocytes with one another and with the central canals.

The ends of a long bone contain spongy bone, which has an entirely different structure. **Spongy bone** contains numerous bony bars and plates called **trabeculae** separated by irregular spaces. Although lighter than compact bone, spongy bone is still designed for strength. Like braces used for support in buildings, the solid plates of spongy bone follow lines of stress. Blood cells are formed within red marrow found in spongy bone at the ends of certain long bones.

Blood

Blood (Fig. 4.12) is a connective tissue composed of **formed elements** suspended in a liquid matrix called **plasma**. There are three types of formed elements: **red blood cells (erythrocytes)**, which carry oxygen, **white blood cells (leukocytes)**,

which aid in fighting infection; and **platelets (thrombocytes)**, which are important to the initiation of blood clotting. Platelets are not complete cells; rather, they are fragments of giant cells called **megakaryocytes**, which are found in the bone marrow.

In red bone marrow, stem cells continually divide to produce new cells that mature into the different types of blood cells. The rate of cell division is high because blood cells have a relatively short life span and must be replaced constantly.

Blood is unlike other types of connective tissue in that the extracellular matrix (*plasma*) is not made by the cells of the tissue. Plasma is a mixture of different types of molecules that enter blood at various organs.

4.3 Muscular Tissue

Muscular (contractile) tissue is composed of cells called muscle fibers (Table 4.3). Muscle fibers contain actin and myosin, which are protein filaments whose interaction accounts for movement. The three types of vertebrate muscles are skeletal, smooth, and cardiac.

Skeletal Muscle

Skeletal muscle, also called *voluntary muscle* (Fig. 4.13), is attached by tendons to the bones of the skeleton. When skeletal muscle contracts, the muscle shortens, and body parts such as arms and legs move. Contraction of skeletal muscle is generally

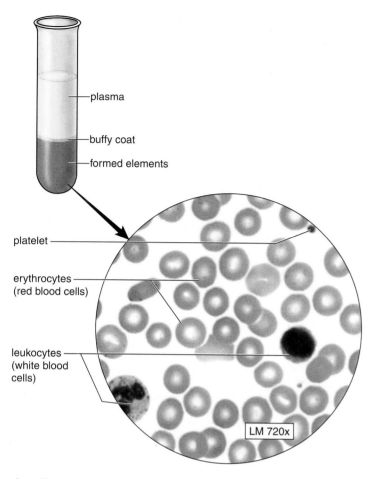

plasma

buffy coat

formed elements

platelet

erythrocytes
(red blood cells)

leukocytes
(white blood
cells)

LM 720x

Location:
In the blood vessels

Function:
Supplies cells with nutrients and oxygen
and takes away their wastes; fights infection

Figure 4.12 Blood. When a blood sample is centrifuged, the formed elements settle out below the plasma. White blood cells form a layer termed the "buffy coat" on top of red blood cells. Plasma is the liquid portion of the blood. Red blood cells, white blood cells, and platelets are called the formed elements.

TABLE 4.3	Classification of Muscular Tissue		
Type	**Fiber Appearance**	**Location**	**Control**
Skeletal	Striated	Attached to skeleton	Voluntary
Smooth	Spindle-shaped	Wall of hollow organs (e.g., intestine, urinary bladder, uterus, and blood vessels)	Involuntary
Cardiac	Striated and branched	Heart	Involuntary

under one's conscious control. Skeletal muscle fibers are cylindrical and quite long—sometimes they run the length of the muscle. They arise during development when several cells fuse, resulting in one fiber with multiple nuclei. The nuclei are located at the periphery of the cell, just inside the plasma membrane. The fibers have alternating light and dark bands that give them a *striated* (striped) appearance. These bands are due to the placement of actin filaments and myosin filaments in the fiber.

Smooth Muscle

Smooth (visceral) muscle is so named because the arrangement of actin and myosin does not give the appearance of cross-striations. The *spindle-shaped cells* form layers in which the thick middle portion of one cell is opposite the thin ends of adjacent cells. (A *spindle* is a long, pointed oval structure.) Consequently, the nuclei form an irregular pattern in the tissue (Fig. 4.14).

Smooth muscle is not under conscious control and therefore is said to be *involuntary*. Smooth muscle is found in the walls of hollow viscera, such as the intestines, stomach, uterus, urinary bladder, and blood vessels. Smooth muscle contracts more slowly than skeletal muscle but can remain

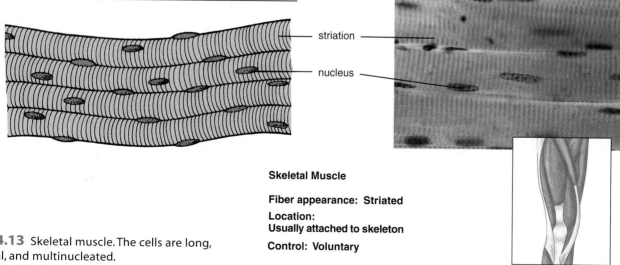

striation

nucleus

Skeletal Muscle

Fiber appearance: Striated

Location:
Usually attached to skeleton

Control: Voluntary

Figure 4.13 Skeletal muscle. The cells are long, cylindrical, and multinucleated.

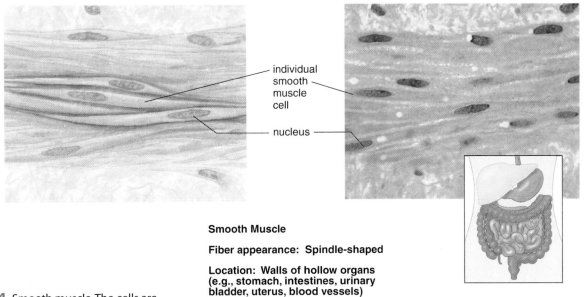

Smooth Muscle

Fiber appearance: Spindle-shaped

Location: Walls of hollow organs (e.g., stomach, intestines, urinary bladder, uterus, blood vessels)

Control: Involuntary

individual smooth muscle cell

nucleus

Figure 4.14 Smooth muscle. The cells are spindle-shaped.

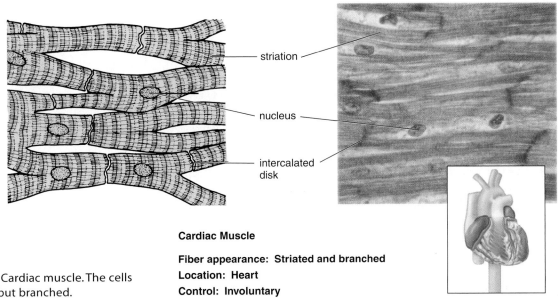

striation

nucleus

intercalated disk

Cardiac Muscle

Fiber appearance: Striated and branched

Location: Heart

Control: Involuntary

Figure 4.15 Cardiac muscle. The cells are cylindrical but branched.

contracted for a longer time. Contractility is inherent in this type of muscle, and it contracts rhythmically on its own. Even so, its contraction can be modified by the nervous system. Smooth muscle of the small intestine contracts in waves, thereby moving food along its lumen (central cavity). When the smooth muscle of blood vessels contracts, blood vessels constrict, helping to regulate blood flow.

Cardiac Muscle

Cardiac muscle (Fig. 4.15) is found only in the walls of the *heart*. Its contraction pumps blood and accounts for the heartbeat. Cardiac muscle combines features of both smooth muscle and skeletal muscle. Like skeletal muscle, it has striations, but the contraction of the heart is *involuntary* (although the use of relaxation therapy does enable some to consciously slow the heart). Also, like skeletal muscle, its contractions are strong, but like smooth muscle, the contraction of the heart is inherent and rhythmical. Cardiac muscle contraction can be modified by the nervous system.

Even though cardiac muscle fibers are *striated*, the cells differ from skeletal muscle fibers in that they have a single, centrally placed nucleus. The cells are branched and seemingly fused one with the other, and the heart appears to be

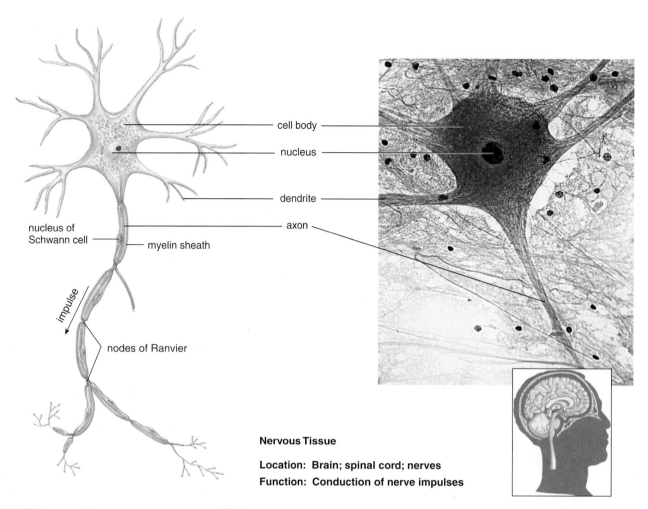

cell body

nucleus

dendrite

axon

nucleus of
Schwann cell

myelin sheath

impulse

nodes of Ranvier

Nervous Tissue

Location: Brain; spinal cord; nerves

Function: Conduction of nerve impulses

Figure 4.16 Nervous tissue. Neurons are surrounded by neuroglia, such as Schwann cells, which envelope axons. Only neurons conduct nerve impulses.

composed of one large, interconnecting mass of muscle cells. Actually, cardiac muscle cells are separate and individual, but they are bound end-to-end at **intercalated disks,** areas where folded plasma membranes between two cells contain adhesion junctions and gap junctions (see page 73). These permit extremely rapid spread of contractile stimuli so that the fibers contract almost simultaneously.

4.4   Nervous Tissue

Nervous tissue, found in the brain and spinal cord, contains specialized cells called neurons that conduct nerve impulses. A **neuron** (Fig. 4.16) has three parts: (1) A *dendrite* receives signals that may result in a nerve impulse; (2) the *cell body* contains the nucleus and most of the cytoplasm of the neuron; and (3) the *axon* conducts nerve impulses.

Long axons are called *fibers.* In the brain and spinal cord, fibers form *tracts.* Outside the brain and spinal cord, fibers are bound together by connective tissue to form **nerves.** Nerves conduct impulses from sense organs to the spinal cord and brain, where the phenomenon called *sensation* occurs. They also conduct nerve impulses away from the spinal cord and brain to muscles, glands, and organs.

In addition to neurons, nervous tissue contains neuroglia.

Neuroglia

Neuroglia are cells that outnumber neurons nine to one and take up more than half the volume of the brain. The primary function of neuroglia is to support and nourish neurons. For example, types of neuroglia found in the brain are microglia, astrocytes, oligodendrocytes, and ependymal cells. *Microglia,* in addition to supporting neurons, engulf bacterial and cellular debris. *Astrocytes* provide nutrients to neurons and produce a hormone known as glia-derived growth factor, which someday might be used as a cure for Parkinson disease and other diseases caused by neuron degeneration. *Oligodendrocytes* form myelin, a protective layer of fatty insulation. *Ependymal cells* form a plasma-like solution called cerebrospinal fluid (CSF), which supports and nourishes the brain and spinal cord.

Schwann cells are the type of neuroglia that enclose all long nerve fibers located outside the brain or spinal cord. Each Schwann cell covers only a small section of a nerve fiber. The gaps between Schwann cells are called **nodes of Ranvier.** Collectively, the Schwann cells provide nerve fibers with a myelin sheath separated by the nodes. The **myelin sheath** speeds conduction because the nerve impulse jumps from node to node. Because the myelin sheath is white, all myelinated nerve fibers appear white.

Content **CHECK-UP!**

4. Bundles of nonparallel collagen fibers can be found in:
   a. dense regular connective tissue.
   b. dense irregular connective tissue.
   c. reticular connective tissue.
   d. adipose tissue.

5. Gap junctions and intercalated disks allow this type of muscle to contract as a unit:
   a. skeletal muscle   c. cardiac muscle
   b. smooth muscle

6. Complete this statement: In a nerve cell, a(n) _____ receives signals while a(n) _____ conducts impulses.
   a. axon, dendrite   c. dendrite, cell body
   b. dendrite, axon

4.5 Extracellular Junctions, Glands, and Membranes

Extracellular Junctions

The cells of a tissue can function in a coordinated manner when the plasma membranes of adjoining cells interact. The junctions that occur between cells help cells function as a tissue.

A **tight junction** forms an impermeable barrier because adjacent plasma membrane proteins actually join, producing a zipperlike fastening (Fig. 4.17*a*). In the stomach, digestive secretions are contained, and in the kidneys, the urine stays within kidney tubules because epithelial cells are joined by tight junctions. A **gap junction** forms when two adjacent plasma membrane channels join (Fig. 4.17*b*). This lends strength, but it also allows ions, sugars, and small molecules to pass between the two cells. Gap junctions in heart and smooth muscle ensure synchronized contraction. In an **adhesion junction** (desmosome), the adjacent plasma membranes do not touch but are held together by extracellular filaments firmly attached to cytoplasmic plaques, composed of dense protein material (Fig. 4.17*c*). Desmosomes that join heart muscles cells prevent the cells from tearing apart during contraction. Similarly, desmosomes in the *cervix*, the opening to the uterus (womb), prevent the cervix from ripping when a woman gives birth.

Glands

A **gland** consists of one or more cells that produce and secrete a product. Most glands are composed primarily of epithelium in which the cells secrete their product by exocytosis. During secretion, the contents of a vesicle are released when the vesicle fuses with the plasma membrane.

The mucous-secreting goblet cells within the columnar epithelium lining the digestive tract are single cells (see Fig. 4.4). Glands with ducts that secrete their product onto the outer surface (e.g., sweat glands and mammary glands) or into a cavity (e.g., pancreas) are called *exocrine glands.* Ducts can be simple or compound, as illustrated in Figure 4.18.

Glands that no longer have a duct are appropriately known as the ductless glands, or endocrine glands. *Endocrine glands* (e.g., pituitary gland and thyroid) secrete their products internally so they are transported by the bloodstream. Endocrine glands produce hormones that help promote homeostasis. Each type of hormone influences the metabolism of a particular target organ or cells.

Glands are composed of epithelial tissue, but they are supported by connective tissue, as are other epithelial tissues.

Membranes

Membranes line the internal spaces of organs and tubes that open to the outside, and they also line the body cavities discussed on page 6.

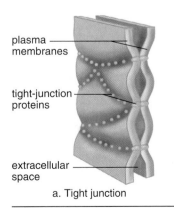

a. Tight junction

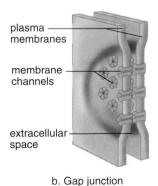

b. Gap junction

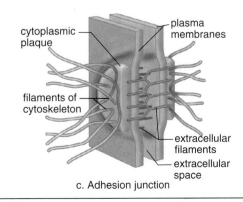

c. Adhesion junction

Figure 4.17 Extracellular junctions. Tissues are held together by **(a)** tight junctions that are impermeable; **(b)** gap junctions that allow materials to pass from cell to cell; and **(c)** adhesion junctions that allow tissues to stretch.

Cancer—The Traitor Inside

The life of almost every person has been touched, either directly or indirectly, by the specter of cancer. **Cancer** is not one disease, but perhaps several hundred diseases, all sharing a common characteristic: rapid, uncontrolled, and disorganized growth of tissue cells. Thus, any cell in any of the body's tissues can be the starting point for cancer. Cancers are classified according to the type of tissue from which they arise. **Carcinomas,** the most common type, are cancers of epithelial tissues (skin and linings); **sarcomas** are cancers that arise in connective tissue (muscle, bone, and cartilage); **leukemias** are cancers of the blood; and **lymphomas** are cancers of reticular connective tissue. The chance of cancer occurring in a particular tissue is related to the rate of cell division. Because epithelial cells reproduce at a high rate, carcinomas account for 90% of all human cancers.

In the body, a cancer cell divides to form a **malignant neoplasm** ("new tissue"), or a **malignant tumor,** that invades and destroys neighboring tissue. Cancer cells can also detach and spread to other sites by invading the blood vessels or the lymphatic vessels. Through this process, called **metastasis,** cancer tumors colonize healthy tissue elsewhere in the body. By contrast, noncancerous, or **benign tumors** are encapsulated and stay in one place. To support their growth, cancer cells release a growth factor that causes neighboring blood vessels to branch into the cancerous tissue. This phenomenon is called **vascularization,** and some modes of cancer treatment are aimed at preventing vascularization (see What's New on page 76).

Cancer development seems to occur by a two-step process involving (1) initiation and (2) promotion. Initiation of cancer is caused by a change, or a **mutation,** in the DNA (genes) of a cell, which results in runaway cell growth. Agents that are known to cause DNA mutations are called **carcinogens.** Known carcinogens include viruses, excessive radiation, and certain chemicals. For example, cigarette smoke contains chemical carcinogens that may initiate cancers of the lung, throat, mouth, and urinary bladder. A cancer **promoter** is any influence that causes a mutated cell to start growing in an uncontrolled manner. A promoter might cause a second mutation or provide the environment for cells to form a tumor. Some evidence suggests that a diet rich in saturated fats and cholesterol promotes colon cancer. Considerable time may elapse between initiation and promotion, and this is one reason why cancer is seen more often in older people.

Cancer can be detected by physical examination, assisted by various means of viewing the internal organs. Mammograms can detect early breast cancer using low-level X ray, and thyroid cancer is diagnosed using radioactive iodine (see the Medical Focus on page 16). Specific blood tests exist for tumors that secrete a particular chemical in the blood. For example, the level of prostate-specific antigen (PSA) appears to increase in the blood according to the size of a prostate tumor. Tissue samples can also detect early malignancy. During a **Pap smear** (named for George Papanicolaou, the Greek doctor who first described the test), epithelial tissue lining the cervix at the opening of the uterus is obtained using a cotton swab. A **biopsy** is the removal of sample tissue using a plunger-like device. A **pathologist** is skilled at recognizing the abnormal characteristics that allow for cancer **diagnosis.** If cancer is found and treated before metastasis occurs, chances for a complete cure are greatly increased.

Mucous Membranes

Mucous membranes line the interior walls of the organs and tubes that open to the outside of the body, such as those of the digestive, respiratory, urinary, and reproductive systems. These membranes consist of an epithelium overlying a layer of loose connective tissue. The epithelium contains goblet cells that secrete mucus.

The mucus secreted by mucous membranes ordinarily protects interior walls from invasion by bacteria and viruses; for example, more mucus is secreted when a person has a

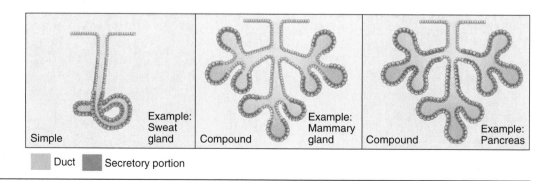

Figure 4.18 Multicellular exocrine glands. Exocrine glands have ducts that can be simple or compound. Compound glands vary according to the placement of secretory portions.

Simple — Example: Sweat gland

Compound — Example: Mammary gland

Compound — Example: Pancreas

☐ Duct ■ Secretory portion

Tumors can often be removed surgically, but there is always the danger that they have metastasized. For this reason, surgery is often preceded or followed by radiation therapy and/or chemotherapy. The goal of both is to destroy rapidly dividing cancer cells. Radiation destroys the more rapidly dividing cancer cells, but may also damage the more slowly dividing normal cells. The use of radioactive protons is preferred over X ray because proton beams can be aimed directly at the tumor, like an automatic rifle hitting the bull's-eye of a target.

Chemotherapy is the use of drugs to kill actively growing cancer cells. Sometimes, cancer cells become resistant to chemotherapy (even when several drugs are used in combination). The plasma membrane in resistant cells contains a carrier that pumps toxic chemicals out of the cell. Researchers are testing drugs known to poison the pump in an effort to restore sensitivity to chemotherapy. Unfortunately, both chemotherapy and radiation therapy kill normal cells in addition to the cancer. The patient will suffer the negative side effects of therapy: nausea, vomiting, hair loss, weight loss, anemia, etc. Thus, the use of chemotherapy and radiation must be balanced carefully: It must be strong enough to kill cancer, but not so strong as to cause the person's death.

Immunotherapy and gene therapy are new, experimental ways of treating cancer. Immunotherapy is the use of an immune system component to treat a disease. For example, cancer patients are sometimes given *cytotoxins*—chemicals released by lymphocytes, a type of white blood cell. Gene therapy is the substitution of "good genes" for defective or missing genes to treat a disease. The hope is that, one day, cancer can be cured by replacing a normal gene for the mutated gene in cancerous cells.

Individuals should be aware of the seven danger signals for cancer (Table 4A) and inform their doctor when any one of these are observed. Further, the evidence is clear that the risk of certain types of cancer can be reduced by adopting certain behaviors. For example, avoiding excessive sunlight reduces the risk of skin cancer, and abstaining from cigarettes, cigars, and chewing tobacco reduces the risk of oral, throat, and lung cancers, as well as other types of cancer. Exercise and a healthy diet are also believed to be important. Recommendations include:

1. Lowering total fat intake
2. Eating more high-fiber foods
3. Increasing consumption of foods rich in vitamins A and C
4. Reducing consumption of salt-cured and smoked foods
5. Including vegetables of the cabbage family in the diet
6. Consuming only moderate amounts of alcohol

TABLE 4.a Danger Signals for Cancer

C hange in bowel or bladder habits
A sore that does not heal
U nusual bleeding or discharge
T hickening or lump in breast or elsewhere
I ndigestion or difficulty in swallowing
O bvious change in wart or mole
N agging cough or hoarseness

cold, resulting in a "runny nose." In addition, mucus usually protects the walls of the stomach and small intestine from digestive juices, but this protection breaks down when a person develops an *ulcer*.

Serous Membranes

As also discussed on page 6, **serous membranes** line cavities, including the thoracic and abdominopelvic cavities, and cover internal organs such as the intestines. The term **parietal** refers to the wall of the body cavity, while the term visceral pertains to the internal organs. Therefore, parietal membranes line the interior of the thoracic and abdominopelvic cavities, and visceral membranes cover the organs.

Serous membranes consist of a layer of simple squamous epithelium overlying a layer of loose connective tissue. They secrete serous fluid that keeps the membranes lubricated. Serous membranes support the internal organs and tend to compartmentalize the large thoracic and abdominopelvic cavities. This helps hinder the spread of any infection.

In the thorax, the **pleurae** are serous membranes that form a double layer around the lungs. The parietal pleura lines the inside of the thoracic wall, while the visceral pleura adheres to the surface of the lungs. Similarly a double-layered serous membrane is a part of the **pericardium**, a covering for the heart (see Fig. 12.2).

The **peritoneum** is the serous membrane within the abdomen. The parietal peritoneum lines the abdominopelvic wall, and the visceral peritoneum covers the organs. In between the organs, the visceral peritoneum comes together to form a double-layered **mesentery** that supports these organs (see Fig. 15.5).

Synovial Membranes

Synovial membranes line freely movable joint cavities and are composed of connective tissues. They secrete synovial fluid into the joint cavity; this fluid lubricates the ends of the

What's New

Blocking Blood Supply to Treat Cancer

All tissue cells share some basic requirements for survival: a constant supply of oxygen and nutrients, and a mechanism to eliminate wastes. Epithelial tissues meet these needs by simple diffusion from underlying tissues, but the rest of the body's tissues must have a continuous blood supply. The basic needs of cancer cells are further increased by their furious pace of growth and metabolism. Cancers guarantee their own continuous blood supply by stimulating nearby blood vessels to sprout new capillaries, in a process called *angiogenesis* (*angio,* blood vessel; *genesis,* to create).

Angiogenesis occurs in normal tissues as well as cancerous growths. Blood vessels proliferate and grow during embryonic and fetal development, as a child grows, whenever a wound heals, and as a woman's uterus goes through the early stage of her menstrual cycle. Blood vessel growth requires that cells produce specific chemicals called growth factors. Cancer cells have been shown to produce large quantities of a specific growth factor, termed vascular endothelial growth factor, or VEGF. This chemical will induce the blood vessels that neighbor a tumor to branch toward the tumor, creating a vascular network. The completed network not only supplies the tumor's metabolic needs—it also serves as a "super-highway" that allows tumor cells to metastasize to other parts of the body.

New medications, currently in clinical trials, have successfully blocked the action of VEGF and other angiogenesis growth factors. These medications are antibodies, small immune system proteins, that combine with VEGF. Thus, VEGF is prevented from signaling the growth of blood vessels to supply a tumor—essentially starving the tumor. Researchers studying angiogenesis inhibitors were also pleasantly surprised to find that using the drugs boosted the effectiveness of traditional cancer-fighting methods like chemotherapy and radiation.

Blocking a tumor's blood supply by using immune system proteins is one aspect of *immunotherapy.* Research in immunotherapy is promising because this type of treatment allows cancer cells to be specifically targeted, without causing collateral damage to healthy tissue.

---

bones so that they can move freely (see Fig. 6.21). In *rheumatoid arthritis,* the synovial membrane becomes inflamed and grows thicker. Fibrous tissue then invades the joint and may eventually become bony so that the bones of the joint are no longer capable of moving.

Meninges

The **meninges** are membranes found within the posterior cavity (see Fig. 1.5). They are composed only of connective tissue and serve as a protective covering for the brain and spinal cord. *Meningitis* is a life-threatening infection of the meninges (see the Medical Focus on page 9).

Cutaneous Membrane

The **cutaneous membrane,** or skin, forms the outer covering of the body. It consists of an outer portion of keratinized stratified squamous epithelium attached to a thick underlying layer of dense irregular connective tissue. The skin is discussed in detail in Chapter 5.

Content **CHECK-UP!**

7. Match each type of extracellular junction to its function:

a. desmosome _____   1. allows ions and small molecules to pass between cells

b. tight junction _____   2. prevents adjacent cells from tearing apart

c. gap junction _____   3. forms an impermeable barrier

8. Match each type of membrane to its function:

a. synovial membrane _____   1. forms outer covering of the body

b. cutaneous membrane _____   2. lines walls of organs that open to the outside of the body

c. mucous membrane _____   3. lines the interior of a joint capsule

SELECTED NEW TERMS

Basic Key Terms

blood (blŭd), p. 69
bone (bōn), p. 67
cartilage (kar′tĭ-lij), p. 66

collagen fibers (kŏl′ə-jən fī′bərz), p. 65
columnar (kə-lūm′nər), p. 62
connective tissue (kŏ-nek′tiv tish′u), p. 65
cuboidal (kyōō-boid′l), p. 62

cutaneous membrane (kyū-ta′ne-us mem′brān), p. 76
elastic cartilage (ĭ-lăs′tĭk kär′tl-ĭj), p. 67
elastic fibers (ĭ-lăs′tĭk fī′bərz), p. 65

epithelial tissue (epi″ĭ-the′le-al tish′u), p. 62

fibroblasts (fī′brə-blăst′), p. 65

fibrocartilage (fī′brō-kär′tl-ĭj), p. 67

gap junction (găp jŭngk′shən), p. 73

hyaline cartilage (hī′ə-lĭn kär′tl-ĭj), p. 67

lacuna (luh-ku′na), p. 66

ligaments (lĭg′a-məntz), p. 66

lymphatic tissue (lĭm-făt′ĭk tĭsh′oͦo), p. 66

matrix (ma′triks), p. 65

meninges (mĕ-nin′jēz), p. 76

mesentery (mes′en-tĕr″e), p. 75

mucous membrane (myŭ′kus mem′brān), p. 74

muscular tissue (mus′kyū-ler tish′u), p. 69

myelin sheath (mi′ĕ-lin shēth), p. 73

nervous tissue (ner′vus tish′u), p. 72

neuroglia (nu-rog′le-uh), p. 72

neuron (nu′ron), p. 72

pericardium (pĕr′ĭ-kär′dē-əm), p. 75

promoter (prə-mō′tər), p. 74

pseudostratified (su″do-strat′ĭ-fīd), p. 64

reticular fibers (rĭ-tĭk′yə-lər fī′bərz), p. 65

serous membrane (sĕr′us mem′brān), p. 75

simple (sĭm′pəl), p. 62

squamous (skwā′məs), p. 62

stratified (strat′ĭ-fīd), p. 62

synovial membrane (sĭ-no′ve-al mem′brān), p. 75

tendons (tĕn′dənz), p. 66

tight junction (tīt jŭnāk′shən), p. 73

transitional epithelium (trăn-zĭsh′ən-əl ĕp′ə-thē′lē-əm), p. 64

Clinical Key Terms

benign (bē-nīn′) tumor, p. 74

biopsy (bi′op-se), p. 74

cancer (kăn′sər), p. 74

carcinogen (kar-sin′ō-jen), p. 74

carcinoma (kar-sĭ-no′muh), p. 74

diagnosis (di-ahg-no′sis), p. 74

leukemia (lu-ke′me-uh), p. 74

lymphoma (lim-fo′muh), p. 74

malignant (mă-lig′-nant) tumor, p. 74

metastasis (mĕ-tas′tă-sis), p. 74

mutation (myōo-tā′shən), p. 74

neoplasm (nē′ə-plăz′əm), p. 74

Pap smear (pap smēr), p. 74

pathologist (puh-thol′uh-jist), p. 74

sarcoma (sar-ko′muh), p. 74

vascularization (văs′kyə-lər-ĭ-zā′shən), p. 74

SUMMARY

4.1 Epithelial Tissue
A. Body tissues are categorized into four types: epithelial, connective, muscular, and nervous.
B. Epithelial tissue. This tissue is classified according to cell shape and number of layers. The cell shape may be squamous, cuboidal, or columnar. Simple tissues have one layer of cells, and stratified tissues have several layers.

4.2 Connective Tissue
A. In connective tissue, cells are separated by a matrix (organic ground substance plus fibers).
B. Fibrous connective tissue can be loose connective tissue, in which fibroblasts are separated by a jellylike ground substance, or dense connective tissue, which contains bundles of collagenous fibers. Adipose tissue is a type of loose connective tissue in which the fibroblasts enlarge and store fat.

C. Cartilage and bone are support tissues. Cartilage is more flexible than bone because the matrix is rich in protein, rather than the mineral salts found in bone.
D. Blood is a connective tissue in which the matrix is plasma.

4.3 Muscular Tissue
Muscular tissue contains actin and myosin protein filaments. These form a striated pattern in skeletal and cardiac muscle, but not in smooth muscle. Cardiac and smooth muscle are under involuntary control. Skeletal muscle is under voluntary control.

4.4 Nervous Tissue
Nervous tissue contains conducting cells called neurons. Neurons have processes called axons and dendrites. In the brain and spinal cord, axons are organized into tracts. Outside the brain and spinal cord, axons (fibers) are found in nerves.

4.5 Extracellular Junctions, Glands, and Membranes
A. In a tissue, cells can be joined by tight junctions, gap junctions, or adhesion junctions (desmosomes).
B. Glands are composed of epithelial tissue that produces and secretes a product, usually by exocytosis. Glands can be unicellular or multicellular. Multicellular exocrine glands have ducts and secrete onto surfaces; endocrine glands are ductless and secrete into the bloodstream.
C. Mucous membranes line the interior of organs and tubes that open to the outside. Serous membranes line the thoracic and abdominopelvic cavities, and cover the organs within these cavities. Synovial membranes line certain joint cavities. Meninges are membranes that cover the brain and spinal cord. The skin forms a cutaneous membrane.

STUDY QUESTIONS

1. What is a tissue? (p. 62)
2. Name the four major types of tissues. (p. 62)
3. What are the functions of epithelial tissue? Name the different kinds of epithelial tissue, and give a location for each. (pp. 62–64)

4. What are the functions of connective tissue? Name the different kinds of connective tissue, and give a location for each type. (pp. 65–69)
5. Contrast the structure of cartilage with that of bone, using the words

lacunae and *central canal* in your description. (p. 66)
6. Describe the composition of blood, and give a function for each type of blood cell. (p. 69)
7. What are the functions of muscular tissue? Name the different kinds of

muscular tissue, and give a location for each. (pp. 69–71)
8. What types of cells does nervous tissue contain? Which organs in the body are made up of nervous tissue? (p. 72)
9. Name three types of junctions, and state the function of each with examples. (p. 72)
10. Describe the structure of a gland. What is the difference between an
exocrine gland and an endocrine gland? (p. 73)
11. Name the different types of body membranes, and associate each type with a particular location in the body. (pp. 73–76)

LEARNING OBJECTIVE QUESTIONS

I. Fill in the blanks.

1. Most organs contain several different types of _____.
2. Pseudostratified ciliated columnar epithelium contains cells that appear to be _____, have projections called _____, and are _____ in shape.
3. Connective tissue cells are widely separated by a _____ that usually contains _____.
4. Both cartilage and blood are classified as _____ tissue.
5. A mucous membrane contains

_____ tissue overlying _____ tissue.

II. Match the organs in the key to the epithelial tissues listed in questions 6–9.

Key:
   a. kidney tubules
   b. small intestine
   c. air sacs of lungs
   d. trachea (windpipe)

6. simple squamous
7. simple cuboidal
8. simple columnar

9. pseudostratified ciliated columnar

III. Match the muscle tissues in the key to the descriptions listed in questions 10–12.

Key:
   a. skeletal muscle
   b. smooth muscle
   c. cardiac muscle

10. striated and branched, involuntary
11. striated and voluntary
12. visceral and involuntary

MEDICAL TERMINOLOGY EXERCISE

Consult Appendix B for help in pronouncing and analyzing the meaning of the terms that follow.

1. epithelioma (ep″ĭ the″le-o′muh)
2. fibrodysplasia (fi″bro-dis-pla′se-uh)
3. meningoencephalopathy (mĕ-ning″go-en-sef′ul-lop′uh-the)
4. pericardiocentesis (per″i-kar″de-o-sen-te′sis)

5. peritonitis (per″ĭ-to-ni′tis)
6. intrapleural (in″tra-plūr′al)
7. neurofibromatosis (nu″ro-fi″bro″muh-to′sis)
8. submucosa (sub″myū-ko′suh)
9. polyarthritis (pol″e-ar-thri′tis)
10. cardiomyopathy (kar′de-o-mi-ah′puh-the)

11. encephalitis (en-sef′-uh-li-tis)
12. glioma (gle-o′-muh)
13. pleurisy (plūr′ĭ-se)
14. chondroblast (kon′-dro-blast)
15. osteology (os′te-ol′-o-je)

WEBSITE LINK

Visit this text's website at http://www.mhhe.com/maderap6 for additional quizzes, interactive learning exercises, and other study tools.

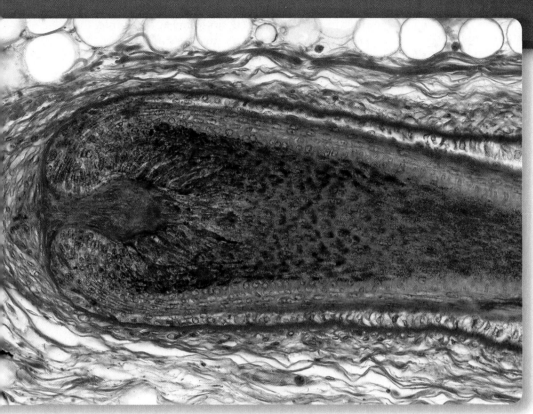

PART II

CHAPTER

5

The Integumentary System

We cut it, shave it, pluck and dye it—hair is an important accessory structure of the skin. This photomicrograph shows the base of a hair follicle, including the region of cell division.

Chapter Outline & Learning Objectives

After you have studied this chapter, you should be able to:

5.1 Structure of the Skin (p. 80)
- Describe the regions of the skin and the hypodermis.
- Name two main epidermal layers, and describe their structure and function.

5.2 Accessory Structures of the Skin (p. 82)
- Describe the structure and growth of hair and nails.
- Name three glands of the skin, and describe their structure and function.

5.3 Disorders of the Skin (p. 84)
- Name the three types of skin cancer, and state their risk factor.

- Name and describe four types of burns with regard to depth.
- Describe how the "rule of nines" may be used to estimate the extent of a burn.
- Describe the steps by which a skin wound heals.

5.4 Effects of Aging (p. 87)
- Describe the anatomical and physiological changes that occur in the integumentary system as we age.

5.5 Homeostasis (p. 88)
- List and discuss four functions of the skin that contribute to homeostasis.

Medical Focus
Decubitus Ulcers (p. 82)

What's New
Pills for Sun Protection? (p. 88)

Human Systems Work Together
Integumentary System (p. 90)

Medical Focus
Body Art: Buyer Beware! (p. 91)

5.1 Structure of the Skin

The skin, sometimes called the **cutaneous membrane** or the **integument,** covers the entire surface of the human body. In an adult, the skin has a surface area of about 1.8 square meters (20.83 square feet). Thus, it is the largest organ in the human body. It is important to note that the skin is an organ, comprised of all four tissue types: epithelial, connective, muscle, and nervous tissue. Because the skin has several accessory organs, it is also technically an organ system sometimes referred to as the **integumentary system.**

The skin (Fig. 5.1) has two regions: the epidermis and the dermis. The **hypodermis,** a subcutaneous tissue, is found between the skin and any underlying structures, such as muscle. Usually, the hypodermis is only loosely attached to underly-ing muscle tissue, but where no muscles are present, the hypodermis attaches directly to bone. For example, there are *flexion creases* where the skin attaches directly to the joints of the fingers.

Epidermis

The **epidermis** is the outer and thinner region of the skin. It is made up of stratified squamous epithelium divided into five separate layers, or *strata* (sing., *stratum*). From deepest to most superficial, the layers are *stratum basale, stratum spinosum, stratum granulosum, stratum lucidum,* and *stratum corneum.* Only the stratum basale, stratum lucidum, and stratum corneum will be discussed here. Like all epithelial tissues, epidermis lacks blood vessels and has tightly packed cells.

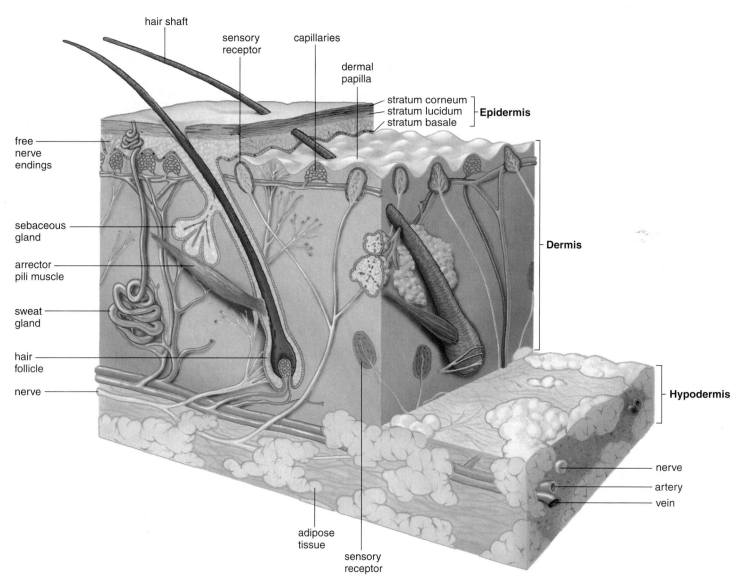

Figure 5.1 Skin anatomy. Skin is composed of two regions: the epidermis and the dermis. The hypodermis, or subcutaneous layer, is located beneath the skin.

Stratum Basale

The basal cells of the **stratum basale** lie just superior to the dermis and are constantly dividing and producing new cells that are pushed to the surface of the epidermis in two to four weeks. As the cells move away from the dermis, they get progressively farther away from the blood vessels in the dermis. Because these cells are not being supplied with nutrients and oxygen (since epidermis lacks blood vessels), they eventually die and are sloughed off.

Langerhans cells are macrophages found deep in the epidermis. Macrophages are a type of white blood cell (leukocyte). These cells phagocytize microbes and then travel to lymphatic organs, where they stimulate the immune system to react.

Melanocytes are another type of specialized cell located in the deeper epidermis. Melanocytes produce **melanin,** the pigment primarily responsible for skin color. Because the number of melanocytes is about the same in all individuals, variation in skin color is due to the amount of melanin produced and its distribution. When skin is exposed to the sun, melanocytes produce more melanin to protect the skin from the damaging effects of the ultraviolet (UV) radiation in sunlight. The melanin is passed to other epidermal cells, and the result is tanning, or in some people, the formation of patches of melanin called freckles. A hereditary trait characterized by the lack of ability to produce melanin is known as **albinism.** Individuals with this disorder lack pigment not only in the skin, but also in the hair and eyes. Another pigment, called carotene, is present in epidermal cells and in the dermis and gives the skin of certain Asians its yellowish hue. The pinkish color of fair-skinned people is due to the pigment hemoglobin in the red blood cells in the capillaries of the dermis.

Sensory nerves also supply the stratum basale. **Free nerve endings** supply pain and temperature sensations to the brain (Fig. 5.1). **Tactile cells** (also called **Merkel cells**) signal the brain that an object has touched the skin.

Stratum Lucidum

As you examine your skin, you can likely identify areas where constant abrasion has created calluses. These are areas where the epidermis has formed **stratum lucidum,** just deep to the stratum corneum. This additional layer is found only in thick skin: the palms of the hands, soles of the feet, elbows, etc. The stratum lucidum provides protection from constant friction.

Stratum Corneum

As cells are pushed toward the surface of the skin, they become flat and hard, forming the tough, uppermost layer of the epidermis, the **stratum corneum.** Hardening is caused by keratinization, the cellular production of a fibrous, waterproof protein called **keratin.** Over much of the body, keratinization

is minimal. However, in areas containing an underlying stratum lucidum, a particularly thick layer of dead, keratinized cells affords extra protection.

The waterproof nature of keratin protects the body from water loss and water gain. The stratum corneum allows us to live in a desert or a tropical rain forest without damaging our inner cells. The stratum corneum also serves as a mechanical barrier against microbe invasion. This protective function of skin is assisted by the secretions of *sebaceous glands* (discussed in section 5.2).

Dermis

The **dermis,** a deeper and thicker region than the epidermis, is composed of dense irregular connective tissue. The upper layer of the dermis has fingerlike projections called *dermal papillae.* Dermal papillae project into and anchor the epidermis. In the overlying epidermis, dermal papillae cause ridges, resulting in spiral and concentric patterns commonly known as "fingerprints." The function of the epidermal ridges is to increase friction and thus provide a better gripping surface. Because they are unique to each person, fingerprints and footprints can be used for identification purposes.

The dermis contains collagen and elastic fibers. The *collagen fibers* are flexible but offer great resistance to overstretching; they prevent the skin from being torn. The *elastic fibers* stretch to allow movement of underlying muscles and joints, but they maintain normal skin tension. The dermis also contains blood vessels that supply oxygen and nutrients to its cells, and those of the epidermis as well. Blood rushes into these vessels when a person blushes; *pallor* (pale skin) develops when blood flow to dermal vessels is reduced. If blood is not adequately supplied with oxygen (perhaps because of lung disease), the person turns *cyanotic,* or "blue."

Extended periods of diminished blood flow to the dermis can cause the formation of decubitus ulcers, or bedsores (see Medical Focus, p. 82).

There are also numerous sensory nerve fibers in the dermis that take nerve impulses to and from the accessory structures of the skin, which are discussed in section 5.2.

Hypodermis

Hypodermis, or **subcutaneous tissue,** lies below the dermis. From the names for this layer, we get the terms **subcutaneous injection,** performed with a **hypodermic needle.**

The hypodermis is composed of loose connective tissue, including adipose (fat) tissue. **Fat** is an energy storage form that can be called upon when necessary to supply the body with molecules for cellular respiration. Adipose tissue also helps insulate the body. A well-developed hypodermis gives the body a rounded appearance and provides protective padding against external assaults. Excessive development of adipose tissue in the hypodermis layer results in obesity.

Decubitus Ulcers

Decubitus ulcers, or *bedsores*, are a critical problem that should concern *anyone* who is a care provider for patients requiring long-term or extended care. Bedsores develop when there is constant, unrelieved pressure on a single area of the skin. Blood supply to the dermis is blocked by the continuous pressure. Because the epidermis relies on diffusion of oxygen from underlying dermis, epidermal cells will begin to die. Areas of the body where bone is close to the skin are common areas for development of bedsores. These include the sacral and coxal areas (see Fig. 1.3), as well as the ankle, heel, shoulder, and elbow.

Populations at risk for bedsores include elderly patients, quadriplegics and paraplegics, and brain-injured individuals. Additional risk factors include anemia (decreased red blood cell count), urinary/fecal incontinence, and malnutrition. Signs of injury to the skin begin with a sunburn-like redness; as injury worsens, blisters develop. Progressive injury will cause skin loss, first involving the epidermis and subsequently spreading to the dermis and hypodermis (Fig. 5A). Full-thickness bedsores may penetrate to underlying muscle tissue or even to the bone itself. Patients will have a fever and increased white blood cell count. If the patient is able to feel pain, the sore is very painful; however, many quadriplegics and paraplegics have no pain sensation. Infection of the bedsore can slow the healing process, and in severe cases will be fatal.

Once formed, decubitus ulcers are very difficult to cure. Keeping pressure off the affected area is critical. Moist sterile dressings are applied to the wound to protect growing skin. Therapy using high-pressure oxygen is sometimes helpful. Severe cases may require a skin graft to replace lost tissue.

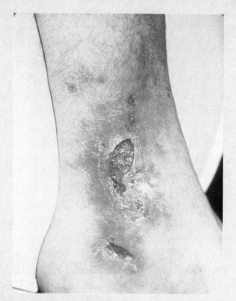

Figure 5A A decubitus ulcer (bedsore). The most frequent sites for bedsores are in the skin overlying a bony projection, such as on the hip, ankle, heel, shoulder, or elbow.

The best treatment for bedsores is prevention. Bedridden patients must be turned and repositioned hourly. Wet, soiled clothing must be changed immediately to prevent further irritation to the skin. Bony areas at risk for bedsore development should be well padded. Massaging the skin gently to stimulate blood flow is helpful. Proper nutrition is important to provide the nutrients needed by rapidly growing skin cells.

Content **CHECK-UP!**

1. Blood vessels can be found in:
   a. all three layers of the skin.
   b. the dermis only.
   c. the hypodermis and the dermis.

2. Which of the following are white blood cells found in the epidermis that phagocytize microbes and stimulate the immune system?
   a. Merkel cells   c. melanocytes
   b. Langerhans cells

3. The deepest layer of the epidermis is the:
   a. stratum corneum.   c. stratum basale.
   b. stratum lucidum.

5.2 Accessory Structures of the Skin

Hair, nails, and glands are structures of epidermal origin, even though some parts of hair and glands are largely in the dermis.

Hair and Nails

Hair is found on all body parts except the palms, soles, lips, nipples, and portions of the external reproductive organs. Most of this hair is fine and downy, but the hair on the head includes stronger types as well. After puberty, when sex hormones are made in quantity, there is noticeable hair in the axillary and pelvic regions of both sexes. In the male, a beard develops, and other parts of the body may also become quite hairy. When women produce more male sex hormone than usual, they can develop **hirsutism,** a condition characterized

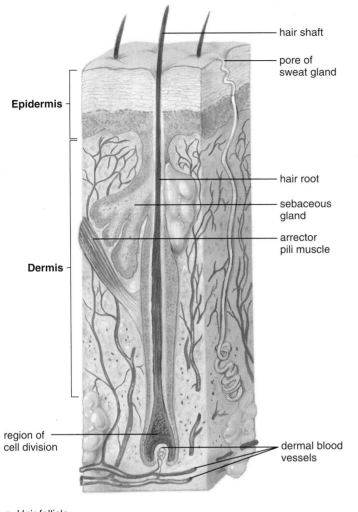

a. Hair follicle

Labels for figure a:
- hair shaft
- pore of sweat gland
- Epidermis
- hair root
- sebaceous gland
- arrector pili muscle
- Dermis
- region of cell division
- dermal blood vessels

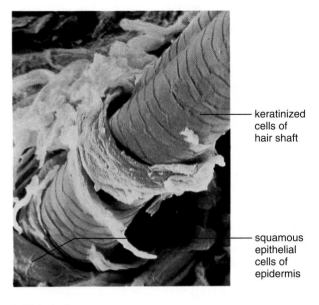

b. Hair shaft

Labels for figure b:
- keratinized cells of hair shaft
- squamous epithelial cells of epidermis

Figure 5.2 Hair follicle and hair shaft. **a.** A hair grows from the base of a hair follicle where epidermal cells produce new cells as older cells move outward and become keratinized. **b.** A hair shaft penetrating the outer squamous epithelial cells of the epidermis.

by excessive body and facial hair. Hormonal injections and procedures to kill hair roots are possible treatments.

Hairs project from complex structures called **hair follicles.** These hair follicles are formed from epidermal cells but are located in the dermis of the skin (Fig. 5.2). Certain hair follicle cells continually divide, producing new cells that form a hair. At first, the cells are nourished by dermal blood vessels, but as the hair grows up and out of the follicle, they are pushed farther away from this source of nutrients, become keratinized, and die. The portion of a hair within the follicle is called the *root,* and the portion that extends beyond the skin is called the *shaft.*

The life span of any particular hair is usually three to four months for an eyelash and three to four years for a scalp hair; then it is shed and regrows. In males, baldness occurs when the hair on the head fails to regrow. **Alopecia,** meaning hair loss, can have many causes. Male pattern baldness, or *androgenic alopecia,* is an inherited condition. *Alopecia areata* is characterized by the sudden onset of patchy hair loss. It is most common among children and young adults, and can affect either sex.

Each hair has one or more oil, or sebaceous, glands, whose ducts empty into the follicle. A smooth muscle, the **arrector pili,** attaches to the follicle in such a way that con-

traction of the muscle causes the hair to stand on end. If a person has had a scare or is cold, "goose bumps" develop due to contraction of these muscles.

Nails grow from special epithelial cells at the base of the nail in the region called the *nail root* (Fig. 5.3). These cells become keratinized as they grow out over the nail bed. The visible portion of the nail is called the *nail body.* The cuticle is a fold of skin that hides the nail root. Ordinarily, nails grow only about 1 millimeter per week.

The pink color of nails is due to the vascularized dermal tissue beneath the nail. The whitish color of the half-moon-shaped base, or **lunula,** results from the thicker layer of rapidly reproducing cells in this area.

Glands

The *glands* in the skin are groups of cells specialized to produce and secrete a substance into ducts.

Sweat Glands

Sweat glands, or sudoriferous glands, are present in all regions of the skin. There can be as many as 90 glands per

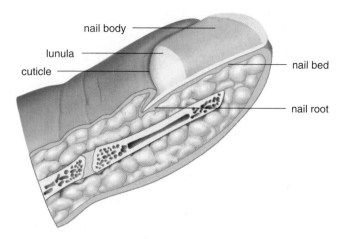

Figure 5.3 Sagittal section of a nail. Cells produced by the nail root become keratinized, forming the nail body.

square centimeter on the leg, 400 glands per square centimeter on the palms and soles, and an even greater number on the fingertips. A sweat gland is tubular. The tubule is coiled, particularly at its origin within the dermis. These glands become active when a person is under stress.

Two types of sweat glands are shown in Figure 5.4. Both types secrete their products by exocytosis (see Chapter 3). *Apocrine glands* open into hair follicles in the anal region,

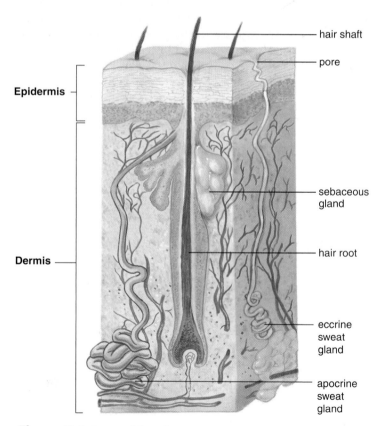

Figure 5.4 Types of skin glands. Apocrine glands and eccrine glands are types of sweat glands.

groin, and armpits. These glands begin to secrete at puberty, and a component of their secretion may act as a sex attractant. *Eccrine glands* open onto the surface of the skin. They become active when a person is hot, helping to lower body temperature as sweat evaporates. The sweat (perspiration) produced by these glands is mostly water, but it also contains salts and some urea, a waste substance. Therefore, sweat is a form of excretion. The opening into the ear, called the external ear canal, contains modified sweat glands, called ceruminous glands, which produce cerumen, or earwax.

Sebaceous Glands

Most **sebaceous glands** are associated with a hair follicle. These glands secrete an oily substance called **sebum** that flows into the follicle and then out onto the skin surface. This secretion lubricates the hair and skin, and helps waterproof them. Sebum also weakens or kills bacteria on the skin surface.

Particularly on the face and back, the sebaceous glands may fail to discharge sebum, and the secretions collect, forming whiteheads or blackheads. If pus-inducing bacteria are also present, a boil or pimple may result.

Acne vulgaris, the most common form of acne, is an inflammation of the sebaceous glands that most often occurs during adolescence. Hormonal changes during puberty cause the sebaceous glands to become more active at this time.

Mammary Glands

Mammary glands are modified apocrine sweat glands that produce milk only after childbirth. The anatomy and physiology of mammary glands are discussed in Chapter 17.

Content **CHECK-UP!**

4. Which glands begin secretion at puberty and contain a sex attractant?

   a. eccrine glands   c. ceruminous glands

   b. apocrine glands   d. sebaceous glands

5. Which part of the hair contains actively growing cells?

   a. base of hair follicle   c. hair shaft

   b. hair root

6. Which of the following pathologies causes women to develop excessive body and facial hair?

   a. alopecia   c. acne vulgaris

   b. hirsutism

5.3 Disorders of the Skin

The skin is subject to many disorders, some of which are more annoying than life-threatening. For example, **athlete's foot** is caused by a fungal infection that usually involves the skin of the toes and soles. **Impetigo** is a highly contagious disease occurring most often in young children. It is caused by a bacterial infection that results in pustules that crust over. **Psoriasis** is a chronic condition, possibly hereditary, in which the skin develops pink or reddish patches covered by silvery scales due to overactive cell division. **Eczema**, an inflammation of the skin, is caused by sensitivity to various chemicals (e.g., soaps or detergents), to certain fabrics, or even to heat or dryness. **Dandruff** is a skin disorder not caused by a dry scalp, as is commonly thought, but by an accelerated rate of keratinization in certain areas of the scalp, producing flaking and itching. **Urticaria**, or hives, is an allergic reaction characterized by the appearance of reddish, elevated patches and often by itching.

Individuals who choose to pierce or tattoo their skin must take care to avoid skin damage or life threatening infection (see Medical Focus, p. 91).

Skin Cancer

Skin cancer is categorized as either melanoma or non-melanoma. Like all cancers, it begins with mutation of the skin cell DNA.

Nonmelanoma cancers, which include basal cell carcinoma and squamous cell carcinoma, are much less likely to metastasize than melanoma cancer. **Basal cell carcinoma** (Fig. 5.5a), the most common type of skin cancer, begins when ultraviolet (UV) radiation causes epidermal basal cells to form a tumor, while at the same time suppressing the immune system's ability to detect the tumor. The signs of a tumor are varied. They include an open sore that will not heal; a recurring reddish patch; a smooth, circular growth with a raised edge; a shiny bump; or a pale mark. About 95% of patients are easily cured by surgical removal of the tumor, but recurrence is common.

Squamous cell carcinoma (Fig. 5.5b) begins in the epidermis proper. While five times less common than basal cell carcinoma, it is more likely to spread to nearby organs, and death occurs in about 1% of cases. It, too, is triggered by excessive UV exposure. The signs of squamous cell carcinoma are the same as those for basal cell carcinoma, except that it may also show itself as a wart or scaly growth that bleeds and scabs.

TABLE 5.1	**Warning Signs for Melanoma: The ABCDE Rule**

WHEN EXAMINING A MOLE ON THE BODY, NOTE THE FOLLOWING SIGNS OF MELANOMA:

A **Asymmetrical**—instead of being perfectly round, the mole has an oval or irregular shape.

B **Borders** of the mole are irregular and have notches or indentation in them.

C **Color** is uneven and several colors may be present: black, brown, tan, red or blue.

D **Diameter** of the mole is greater than 6 mm. (larger than a pencil eraser).

E **Elevation:** the mole is elevated above the surface of the skin.

Melanoma (Fig. 5.5c), the type that is more likely to be malignant starts in the melanocytes and has the appearance of an unusual **mole** (Table 5.1). Unlike a normal mole, which is dark, circular, and confined, a melanoma mole looks like a spilled ink spot, and a single melanoma mole may display a variety of shades. A melanoma mole can also itch, hurt, or feel numb. The skin around the mole turns gray, white, or red. Melanoma is most common in fair-skinned persons, particularly if they have suffered occasional severe burns as children. Melanoma risk increases with the number of moles a person has. Most moles appear before the age of 14, and their appearance is linked to sun exposure. Melanoma rates have risen since the turn of the century, but the incidence has doubled in the last decade. Each year, an estimated 62,000 new cases of melanoma are diagnosed, and melanoma is responsible for approximately 8,000 fatalities annually. Protecting the skin from the harmful effects of ultraviolet radiation, as detailed in the What's New reading on page 88, is the best strategy to prevent all forms of skin cancer.

Kaposi's sarcoma is a form of skin cancer that is most commonly seen in patients with AIDS, and in others whose immune system defenses are weakened or non-functional. The tumors of Kaposi's sarcoma appear as red, blue or black

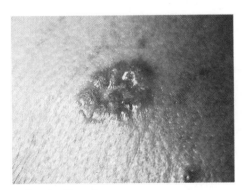
a. Basal cell carcinoma

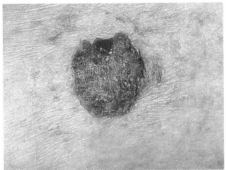
b. Squamous cell carcinoma

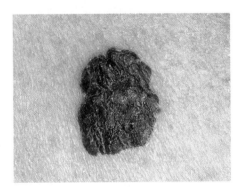
c. Melanoma

Figure 5.5 Skin cancer. In each of the three types shown, the skin clearly has an abnormal appearance.

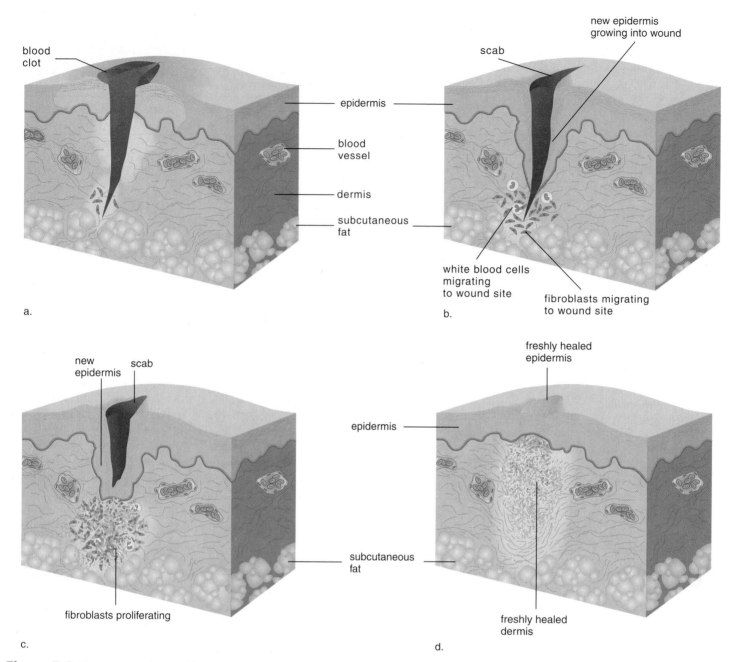

Figure 5.6 The process of wound healing. **a.** Tissue injury causes inflammation, with redness, swelling, heat, and pain. Deep wounds rupture blood vessels, and blood fills the wound. **b.** After a blood clot forms, a protective scab develops. Fibroblasts and white blood cells migrate to the wound site. **c.** New epidermis forms, and fibroblasts promote tissue regeneration. **d.** Freshly healed skin.

spots on the skin. The tumors respond to treatment with a combined set of drugs commonly referred to as the "AIDS cocktail."

Raised growths on the skin, such as moles and **warts,** usually are not cancerous. Moles are due to an overgrowth of melanocytes, and warts are due to a viral infection.

Wound Healing

Injury to the skin will cause an inflammatory response, characterized by redness, swelling, heat, and pain (inflammation

is discussed fully in Chapter 13). A wound that punctures a blood vessel will fill with blood. Chemicals released by damaged tissue cells will cause the blood to clot. The clot prevents pathogens and toxins from spreading to other tissues (Fig. 5.6a). The part of the clot exposed to air will dry and harden, gradually becoming a scab. White blood cells and fibroblasts move into the area. White blood cells help fight infection and fibroblasts are able to pull the margins of the wound together (Fig. 5.6b). Fibroblasts promote tissue regeneration: The basal layer of the epidermis begins to produce new cells at a faster than usual rate. The proliferating fibroblasts bring about scar formation; the scar may or may not be visible from the surface (Fig. 5.6c). A scar is a tissue composed of many collagen fibers arranged to provide maximum strength. A scar does not contain the accessory organs of the skin and is usually devoid of feeling. In any case, epidermis and dermis have now healed (Fig. 5.6d).

Burns

The epidermal injury known as a burn is usually caused by heat but can also be caused by radioactive, chemical, or electrical agents. Two factors affect burn severity: the depth of the burn and the extent of the burned area.

A useful technique for estimating the extent of a burn, called the "rule of nines," is often employed (Fig. 5.7). In this method, the total body surface is divided into regions as follows: the head and neck, 9% of the total body surface; each upper limb, 9%; each lower limb, 18%; the front and back portions of the trunk, 18% each; and the perineum, which includes the anal and urogenital regions, 1%. Physicians use the Lund-Browder chart to estimate the extent of burns in children. This system adjusts for the fact that a child's head is proportionally larger than an adult's.

One way to classify burns is according to the depth of the burned area. In *first-degree burns,* only the epidermis is affected. The person experiences redness and pain, but no blisters or swelling. A classic example of a first-degree burn is a moderate sunburn. The pain subsides within 48–72 hours, and the injury heals without further complications or scarring. The damaged skin peels off in about a week.

A *second-degree burn* extends through the entire epidermis and part of the dermis. The person experiences not only redness and pain, but also blistering in the region of the damaged tissue. The deeper the burn, the more prevalent the blisters, which enlarge during the hours after the injury. Unless they become infected, most second-degree burns heal without complications and with little scarring in 10–14 days. If the burn extends deep into the dermis, it heals more slowly over a period of 30–105 days. The healing epidermis is extremely fragile, and scarring is common. First- and second-degree burns are sometimes referred to as partial-thickness burns.

Third-degree burns, or full-thickness burns, destroy the entire thickness of the skin. The surface of the wound is leathery and may be brown, tan, black, white, or red. The patient feels no pain because the pain receptors have been destroyed, as have blood vessels, sweat glands, sebaceous glands, and hair follicles.

Fourth-degree burns involve tissues down to the bone. Obviously, the chances of a person surviving fourth-degree burns are not good unless a very limited area of the body is affected.

A burn is a critical injury if any of the following conditions are met: (1) Second-degree burns cover 25% or more of the patient's body; (2) third-degree burns cover 10% or more of the patient's body; (3) any portion of the body has a fourth-degree burn; or (4) third-degree burns occur on the face, hands, or feet. Facial burns may be accompanied by damage to the lungs from smoke inhalation. Burns to the hands or feet may result in loss of joint movement because of scar tissue formation.

The major concerns with severe burns are fluid loss, heat loss, and bacterial infection. Fluid loss is counteracted by intravenous administration of a balanced salt solution. Heat loss is minimized by placing the burn patient in a warm environment. Bacterial infection is treated by isolation and the application of an antibacterial dressing.

As soon as possible, the damaged tissue is removed, and skin grafting is begun. The skin needed for grafting is usually taken from other parts of the patient's body. This is called *autografting,* as opposed to heterografting, in which the graft is received from another person. Autografting is preferred because rejection rates are very low. However, if the burned area is quite extensive, it may be difficult to acquire enough skin for autografting. In that case, skin can be grown in the laboratory from only a few cells taken from the patient.

5.4 Effects of Aging

As aging occurs, the epidermis maintains its thickness, but the rate of cell mitosis decreases. The dermis becomes thinner, the dermal papillae flatten, and the epidermis is held less tightly to the dermis so that the skin is looser. Adipose tissue in the hypodermis of the face and hands also decreases, which means that older people are more likely to feel cold.

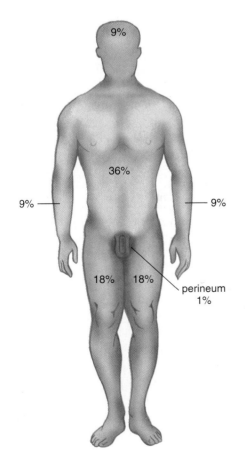

Figure 5.7 The "rule of nines" for estimating the extent of burns.

Now you may be able to protect yourself from the harmful effects of ultraviolet light—just by taking a pill. New research has shown that a plant extract, taken orally, provides protection from the harmful effects of sun exposure. Previous studies had shown the plant, the fern *Polypodium leucotomos*, to have antioxidant and antitumor characteristics.

In a small study, scientists at the Harvard School of Medicine fed volunteers an extract of *P. leucotomos*. Subjects were then exposed to ultraviolet radiation, and their skin was studied carefully for signs of sun damage. Participants in the study showed less reddening of the skin and less damage from UV light after taking the extract.

If further studies come to the same conclusion, sun block pills could give the advantage of protecting the entire body, not just the parts where a lotion is applied.

If an individual has experienced severe sunburns as a child, the chance of having skin cancer as an adult is greater. The sun gives off two types of UV rays: UV-A rays and UV-B rays. UV-A rays penetrate the skin deeply, affect connective tissue, and cause the skin to sag and wrinkle, and UV-A rays may help cause skin cancer. At any rate, UV-A rays are believed to increase the effects of UV-B rays, which are the primary cancer-causing rays. UV-B rays are more prevalent at midday.

Until medications that protect us from UV light are widely available, it is important that we shelter ourselves from the sun. Regardless of where you live, or your racial or ethnic group, take the following steps to protect yourself:

- Use a broad-spectrum sunscreen that protects you from both UV-A and UV-B radiation, and has a sun protection factor (SPF) of at least 15. (This means that if you usually burn, for example, after a 20-minute exposure, it will take 15 times longer, or 5 hours, before you will burn.) Children should use a higher SPF such as 30 or 45 (a sun block).

- Wear protective clothing. Choose fabrics with a tight weave, and wear a wide-brimmed hat. A baseball cap does not protect the rims of the ears, which often burn and then get infected. Wherever the ozone layer is thinner than usual, even more protection is required. In Australia, because of a thin ozone layer due especially to the Earth's rotation, school children are allowed outside for recess only if they wear a wide-brimmed hat and long sleeves.

- Stay out of the sun altogether between the hours of 10 A.M. and 3 P.M. Some authorities believe this action will reduce annual exposure to the sun's rays by as much as 60%.

- Wear sunglasses that have been treated to absorb both UV-A and UV-B radiation. Otherwise, darkened sunglasses can expose the eyes to more damage than usual because the pupils dilate in the shade. For this reason, do not let children wear "fun" sunglasses outside in the sun. Purchase children's sunglasses only if there is a tag indicating UV-ray protection.

- Avoid tanning machines unless prescribed by a physician for Seasonal Affective Disorder (SAD). Although most tanning devices use only high levels of UV-A radiation, the deep layers of the skin become more vulnerable to UV-B radiation upon later exposure to the sun.

The fibers within the dermis change with age. The collagenous fibers become coarser, thicker, and farther apart; therefore, there is less collagen than before. Elastic fibers in the upper layer of the dermis are lost, and those in the lower dermis become thicker, less elastic, and disorganized. The skin wrinkles because (1) the epidermis is loose, (2) the fibers are fewer and those remaining are disorganized, and (3) the hypodermis has less padding.

With aging, homeostatic adjustment to heat is limited due to less vasculature (fewer blood vessels) and fewer sweat glands. The number of hair follicles decreases, causing the hair on the scalp and extremities to thin. Because of a reduced number of sebaceous glands, the skin tends to crack.

As a person ages, the number of melanocytes decreases. This causes the hair to turn gray and the skin to become paler. In contrast, some of the remaining pigment cells are larger, and pigmented blotches appear on the skin.

Many of the changes that occur in the skin as a person ages appear to be due to sun damage. Ultraviolet radiation causes rough skin, mottled pigmentation, fine lines and wrinkles, deep furrows, numerous benign skin growths, and the various types of skin cancer discussed in section 5.3.

5.5 Homeostasis

The illustration on page 90, called Human Systems Work Together, tells how the functions of the skin assist the other systems of the body and how the other systems help the skin carry out these functions.

Functions of the Skin

Skin has a protective function. First and foremost, the skin forms a protective covering over the entire body, safeguarding

MEDICAL FOCUS

Body Art: Buyer Beware!

There's no doubt about it: Body art has gone mainstream. Piercing and tattooing the body are practices that have existed for thousands of years in many different human cultures. In the past decade, both practices have become commonplace in Western society as well. It is no longer surprising to see actors, sports figures, and other celebrities with visible tattoos, pierced navels, multiple earrings, or other forms of wearable art. As the trend continues, more and more individuals are tempted to get their own tattoo or piercing. Tattooing is the process of injecting inks into the dermal layer of the skin. The most common sites for piercing remain the earlobes and ear cartilage, but lips, eyebrows, navels, tongues, and genitals are also common sites.

Individuals must consider the decision to tattoo or pierce very carefully, and health-care providers must be prepared to give accurate information about both practices. With the increase in both tattooing and piercing, reports of complications have increased as well.

Tattooing and piercing can result in infections because both practices can potentially introduce bacteria into the skin, blood vessels, or lymphatic vessels. Complications from infection can range from minor skin irritations to life-threatening blood poisoning. Several cases on record have involved infection with *resistant* bacteria, which are extremely dangerous because they are not killed by common antibiotics. Fatalities have been reported from infections caused by piercing.

Further, the reaction of a person's skin to tattoo ink is unpredictable. Tattoo inks are not regulated by any form of federal agency, and many color additives have not been approved for contact with the skin. Some are inks used in paper printing or automobile paint. It is hardly surprising that many tattooed individuals develop allergies to ingredients in tattoo dyes. Tattooed skin may develop unattractive granulomas: hard connective tissue nodules under the skin. Keloids—scars that grow beyond normal boundaries—may also develop as the body reacts to chemicals in the ink.

Metals (especially nickel) used in body jewelry are another common cause for allergy. Once an allergic reaction develops, it typically persists for life. In such cases, the jewelry must be permanently removed if at all possible. Additionally, the jewelry may damage nerves or blood vessels as piercing occurs. Jewelry in and around the mouth may damage gums, break teeth or dentures, and interfere with chewing or swallowing. Mouth jewelry that becomes loose may be aspirated (sucked into the airways) where it can obstruct an airway and interfere with breathing.

Anyone considering body art must proceed with caution. If local, state, or provincial law regulates tattoo and piercing artists, use an approved artist or shop. Never try a "do-it-yourself" approach. Check the cleanliness of the shop and make sure that all equipment is steam-sterilized using a device called an *autoclave*. Insist that the artist or piercer use sterile gloves at all times. Make sure that jewelry is appropriate: not too heavy, and made of inert metals such as gold, titanium, or surgical stainless steel, which are less likely to cause allergic reactions. Keep the affected area scrupulously clean after piercing or tattooing and touch jewelry as little as possible. Allow adequate time for healing. Pierced navels, for example, may require up to a year to heal completely. Finally, if complications develop (excessive pain, redness, swelling, fever, etc.), seek medical attention promptly.

Most important, anyone considering body art must remember that the decision may very likely be *permanent*. Piercings may never reclose completely and they may leave a large scar. Although tattoos can be removed by laser therapy, removal is expensive and may leave large scars—and it often cannot completely remove the tattoo.

SELECTED NEW TERMS

Basic Key Terms

arrector pili (ah-rek′tor pil′i), p. 83

dermis (der′mis), p. 81

epidermis (ep″ĭ-der′mis), p. 80

free nerve endings (frē nûrv ĕn′dĭng), p. 81

hair follicle (hār fol′ĭ-kl), p. 83

hypodermis (hi″po-der′mis), p. 80

integument (in-teg′yū-ment), p. 80

integumentary system (in-teg″yū-men′tar-e sis′tem), p. 80

keratin (kĕr′uh-tin), p. 81

Langerhans cells (lăhng′hăhnz sĕlz), p. 81

lunula (lu′nu-luh), p. 83

melanin (mel′uh-nin), p. 81

melanocyte (mel′uh-no-sīt), p. 81

Merkel cells (mer′kelz sĕlz), p. 81

sebaceous gland (sĕ-ba′shus gland), p. 84

sebum (se′bum), p. 84

stratum basale (strā′təm bā′səlē′), p. 81

stratum corneum (strā′təm kor′nē-um), p. 81

stratum lucidum (strā′təm lōō′sĭd-um), p. 81

subcutaneous tissue (sŭb′kyōō-tā′nē-əs tĭsh′ōō), p. 81

sweat gland (swet gland), p. 83

tactile cells (tăk′təl sĕlz), p. 81

Clinical Key Terms

acne vulgaris (ak′ne vul-ga′-ris), p. 84

albinism (al′bī-nizm), p. 81

alopecia (al-o-pe′she-uh), p. 83

athlete's foot (ath′lĕts fŭt), p. 84

basal cell carcinoma (bās′al sel kar-sĭ-no′muh), p. 85

dandruff (dan′druf), p. 85

decubitus ulcer (de-kyū'bĭ-tus ul'ser), p. 82
eczema (ek'zĕ-muh), p. 85
hirsutism (her'suh-tizm), p. 82
hyperthermia (hi"per-ther'me-uh), p. 89
hypodermic needle (hi-po-der'mik ne'dl), p. 81

hypothermia (hi"po-ther'me-uh), p. 89
impetigo (im"pĕ-ti'go), p. 84
melanoma (mel-uh-no'muh), p. 85
mole (mōl), p. 85
psoriasis (so-ri'uh-sis), p. 85

squamous cell carcinoma (skwa'mus sel kar-sĭ-no'muh), p. 85
subcutaneous injection (sub"kyū-ta'ne-us in-jek'shun), p. 81
urticaria (ur"tĭ-kār'e-uh), p. 85

SUMMARY

5.1 Structure of the Skin
The skin has two regions: the epidermis and the dermis. The hypodermis lies below the skin.
A. The epidermis, the outer region of the skin, is made up of stratified squamous epithelium. New cells continually produced in the stratum basale of the epidermis are pushed outward and become the keratinized cells of the stratum corneum. Stratum lucidum is found in thick skin of palms and soles.
B. The dermis, which is composed of dense irregular connective tissue, lies beneath the epidermis. It contains collagenous and elastic fibers, blood vessels, and nerve fibers.
C. The hypodermis is made up of loose connective tissue and adipose tissue, which insulates the body from heat and cold.

5.2 Accessory Structures of the Skin
Accessory structures of the skin include hair, nails, and glands.
A. Both hair and nails are produced by the division of epidermal cells and consist of keratinized cells.
B. Sweat glands are numerous and present in all regions of the skin. Sweating helps lower the body temperature.
C. Sebaceous glands are associated with a hair follicle and secrete sebum, which lubricates the hair and skin.

D. Mammary glands located in the breasts produce milk after childbirth.

5.3 Disorders of the Skin
A. Skin cancer. Skin cancer, which is associated with ultraviolet radiation, occurs in three forms. Basal cell carcinoma and squamous cell carcinoma can usually be removed surgically. Melanoma is the most dangerous form of skin cancer.
B. Wound healing. The skin has regenerative powers and can grow back on its own if a wound is not too extensive.
C. Burns. The severity of a burn depends on its depth and extent. First-degree burns affect only the epidermis. Second-degree burns affect the entire epidermis and a portion of the dermis. Third-degree burns affect the entire epidermis and dermis. The "rule of nines" provides a means of estimating the extent of a burn injury.

5.4 Effects of Aging
Skin wrinkles with age because the epidermis is held less tightly, fibers in the dermis are fewer, and the hypodermis has less padding. The skin has fewer blood vessels, sweat glands, and hair follicles. Although pigment cells are fewer and the hair turns gray, pigmented blotches appear on the skin. Exposure to the sun results in many of the skin changes we associate with aging.

5.5 Homeostasis
A. Skin protects the body from physical trauma and bacterial invasion.
B. Skin helps regulate water loss and gain, which helps the urinary system. Also, sweat glands excrete some urea.
C. The skin produces a precursor molecule that is converted to vitamin D following exposure to UV radiation. A hormone derived from vitamin D helps regulate calcium and phosphorus metabolism involved in bone development.
D. The skin contains sensory receptors for touch, pressure, pain, hot, and cold, which help people to be aware of their surroundings. These receptors send information to the nervous system.
E. The skin helps regulate body temperature. When the body is too hot, surface blood vessels dilate, and the sweat glands are active. When the body is cold, surface blood vessels constrict, and the sweat glands are inactive.
F. Hyperthermia and hypothermia are two conditions that can result when the body's temperature regulatory mechanism is overcome. With hyperthermia, the body temperature rises above normal, and with hypothermia, the body temperature falls below normal.

STUDY QUESTIONS

1. In general, describe the two regions of the skin. (pp. 80–81)
2. Describe the process by which epidermal tissue continually renews itself. (pp. 80–81)
3. What function does the dermis have in relation to the epidermis? (p. 81)
4. What primary role does adipose tissue play in the hypodermis? (p. 81)
5. Describe in general the structure of a hair follicle and a nail. How do hair follicles and nails grow? (pp. 82–83)
6. Describe the structure and function of sweat glands and sebaceous glands. (pp. 83–84)
7. Name the three types of skin cancer, and cite the most frequent cause of skin cancer. (p. 85)

8. Describe how a wound heals and how a scar forms. (pp. 86–87)
9. Explain how to determine the severity of a burn. Describe the proper treatment for burns. (p. 87)
10. Explain three changes that happen to the skin with age. (pp. 87–88)
11. Name five functions of the skin, and tell what system of the body is assisted by these functions and how they contribute to homeostasis. (pp. 88–89)

LEARNING OBJECTIVE QUESTIONS

I. Match the terms in the key to the items listed in questions 1–5.

Key:
   a. epidermis
   b. dermis
   c. hypodermis

1. blood vessels and nerve fibers
2. fat cells
3. basal cells
4. location of sweat glands
5. many collagenous and elastic fibers

II. Fill in the blanks.

6. Sebaceous glands are associated with _____ in the dermis, and they secrete an oily substance called _____.
7. Sweat glands are involved in body _____ regulation.
8. Skin protects against _____ trauma, _____ invasion, and _____ gain or loss.
9. Skin cells produce vitamin _____, which is needed for strong bones.
10. The severity of a burn is determined by _____ and _____.
11. The type of skin cancer with the highest death rate is _____, while the most common form is _____.

MEDICAL TERMINOLOGY EXERCISE

Consult Appendix B for help in pronouncing and analyzing the meaning of the terms that follow.

1. epidermomycosis (ep″ĭ-der″mo-mi-ko′sis)
2. melanogenesis (mel″uh-no-jen′ĕ-sis)
3. acrodermatosis (ak″ro-der″muh-to′sis)
4. pilonidal cyst (pi″lo-ni′dal sist)
5. mammoplasty (mam′o-plas″te)
6. antipyretic (an″ti-pi-ret′ik)
7. dermatome (der′muh-tōm)
8. hypodermic (hy″po-der′mik)
9. onychocryptosis (on″ĭ-ko-krip-to′sis)
10. hyperhydrosis (hi″per-hi-dro′sis)
11. scleroderma (sklēr-o-der′muh)
12. piloerection (pi′lo-e-rek′shun)
13. cellulitis (sel′yū-li′tis)
14. dermatitis (der-muh-ti′tis)
15. rhytidoplasty (rit′ĭ-do-plas-te)
16. trichopathy (tri-kop′uh-the)

WEBSITE LINK

Visit this text's website at \underline{\text{http://www.mhhe.com/maderap6}} for additional quizzes, interactive learning exercises, and other study tools.

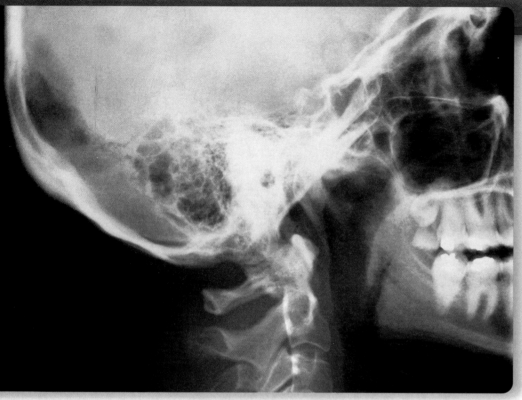

CHAPTER 6

The Skeletal System

An Xray showing a lateral view of the skull. When you have finished this chapter, see how many structures you recognize!

Chapter Outline & Learning Objectives

After you have studied this chapter, you should be able to:

6.1 Skeleton: Overview

The skeletal system consists of the bones (206 in adults) and joints, along with the cartilage and ligaments that occur at the joints.

Functions of the Skeleton

The skeleton has the following functions:

The skeleton supports the body. The bones of the lower limbs support the entire body when we are standing, and the pelvic girdle supports the abdominal cavity.

The skeleton protects soft body parts. The bones of the skull protect the brain; the rib cage protects the heart and lungs.

The skeleton produces blood cells. All bones in the fetus have red bone marrow that produces blood cells. In the adult, only certain bones produce blood cells.

The skeleton stores minerals and fat. All bones have a matrix that contains calcium phosphate, a source of calcium ions and phosphate ions in the blood. Fat is stored in yellow bone marrow.

The skeleton, along with the muscles, permits flexible body movement. While articulations (joints) occur between all the bones, we associate body movement in particular with the bones of the limbs.

Anatomy of a Long Bone

Bones are classified according to their shape. Long bones are longer than they are wide. Short bones are cube shaped—that is, their lengths and widths are about equal. Flat bones, such as those of the skull, are platelike with broad surfaces. Irregular bones have varied shapes that permit connections with other bones. Round bones are circular in shape (Fig. 6.1).

A long bone, such as the one in Figure 6.2*a* can be used to illustrate certain principles of bone anatomy. The bone is enclosed in a tough, fibrous, connective tissue covering called the **periosteum**, which is continuous with the ligaments and tendons that anchor bones. The periosteum contains blood vessels that enter the bone and supply its cells. At both ends of a long bone is an expanded portion called an **epiphysis** (pl.: **epiphyses**); the portion between the epiphyses is called the **diaphysis.**

As shown in the section of an adult bone in Figure 6.2, the diaphysis, or shaft, of a long bone, is not solid but has a **medullary cavity** containing yellow marrow. Yellow marrow contains large amounts of fat. The medullary cavity is bounded at the sides by compact bone. The epiphyses contain spongy bone. Beyond the spongy bone is a thin shell of compact bone and, finally, a layer of hyaline cartilage called the **articular cartilage.** Articular cartilage is so named because it occurs where bones articulate (join). **Articulation** is the joining together of bones at a joint. The medullary cavity and the spaces of spongy bone are lined with *endosteum*, a thin, fibrous membrane.

In infants, **red bone marrow,** a specialized tissue that produces blood cells, is found in the medullary cavities of

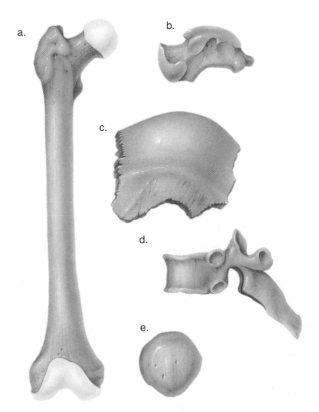

Figure 6.1 Classification of bones. **a.** Long bones are longer than they are wide. **b.** Short bones are cube shaped; their lengths and widths are about equal. **c.** Flat bones are platelike and have broad surfaces. **d.** Irregular bones have varied shapes with many places for connections with other bones. **e.** Round bones are circular.

most bones. In adults, red blood cell formation, called **hematopoiesis,** occurs in the spongy bone of the skull, ribs, sternum (breastbone), and vertebrae, and in the ends of the long bones.

Compact Bone

Compact bone, or dense bone, contains many cylinder-shaped units called osteons. The osteocytes (bone cells) are in tiny chambers called *lacunae* that occur between concentric layers of matrix called *lamellae*. The matrix contains collagenous protein fibers and mineral deposits, primarily of calcium and phosphorus salts.

In each osteon, the lamellae and lacunae surround a single central canal. Blood vessels and nerves from the periosteum enter the central canal. The osteocytes have extensions that extend into passageways called *canaliculi,* and thereby the osteocytes are connected to each other and to the central canal.

Spongy Bone

Spongy bone, or cancellous bone, contains numerous bony bars and plates, called *trabeculae*. Although lighter than compact bone, spongy bone is still designed for strength. Like braces used for support in buildings, the trabeculae of spongy bone follow lines of stress.

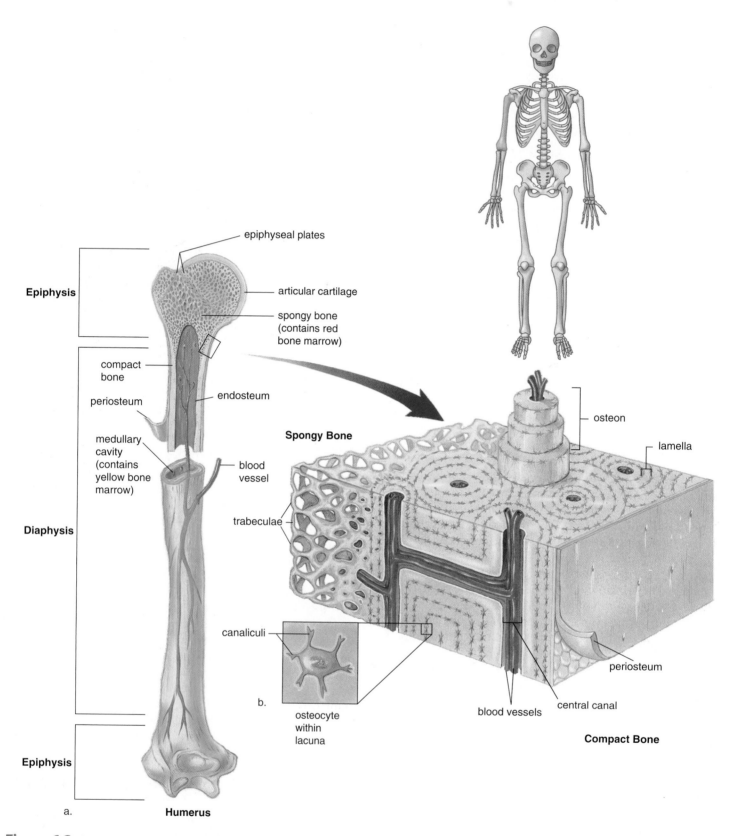

Figure 6.2 Anatomy of a long bone. **a.** A long bone is encased by the periosteum except at the epiphyses, which are covered by articular cartilage. Spongy bone of the epiphyses contains red bone marrow. The diaphysis contains yellow bone marrow and is bordered by compact bone. **b.** The detailed anatomy of spongy bone and compact bone is shown in the enlargement, along with a blowup of an osteocyte in a lacuna.

Bone Growth and Repair

Bones are composed of living tissues, as exemplified by their ability to grow and undergo repair. Several different types of cells are involved in bone growth and repair:

Osteoprogenitor cells are unspecialized cells present in the inner portion of the periosteum, in the endosteum, and in the central canal of compact bone.

Osteoblasts are bone-forming cells derived from osteoprogenitor cells. They are responsible for secreting the matrix characteristic of bone.

Osteocytes are mature bone cells derived from osteoblasts. Once the osteoblasts are surrounded by matrix, they become the osteocytes in bone.

Osteoclasts are thought to be derived from monocytes, a type of white blood cell present in red bone marrow. Osteoclasts perform bone resorption; that is, they break down bone and assist in depositing calcium and phosphate in the blood. The work of osteoclasts is important to the growth and repair of bone.

Bone Development and Growth

The term **ossification** refers to the formation of bone. The bones of the skeleton form during embryonic development in two distinctive ways—intramembranous ossification and endochondral ossification.

In **intramembranous ossification,** bone develops between sheets of fibrous connective tissue. Cells derived from connective tissue become osteoblasts that form a matrix resembling the trabeculae of spongy bone. Other osteoblasts associated with a periosteum lay down compact bone over the surface of the spongy bone. The osteoblasts become osteocytes when they are surrounded by a mineralized matrix. The bones of the skull develop in this manner.

Most of the bones of the human skeleton form by **endochondral ossification.** Hyaline cartilage models, which appear during fetal development, are replaced by bone as development continues. During endochondral ossification of a long bone, the cartilage begins to break down in the center of the diaphysis, which is now covered by a periosteum (Fig. 6.3). Osteoblasts invade the region and begin to lay down spongy bone in what is called a *primary ossification center.* Other osteoblasts lay down compact bone beneath the periosteum. As the compact bone thickens, the spongy bone of the diaphysis is broken down by osteoclasts, and the cavity created becomes the medullary cavity.

After birth, the epiphyses of a long bone continue to grow, but soon secondary ossification centers appear in these regions. Here spongy bone forms and does not break down. A band of cartilage called an **epiphyseal plate** remains between the primary ossification center and each secondary center. The limbs keep increasing in length and width as long as epiphyseal plates are still present. The rate of growth is controlled by hormones, such as growth hormones and the sex hormones. Eventually, the epiphyseal plates become ossified, and the bone stops growing in length.

Long bone growth ends when the epiphyseal plates are ossified, but it is possible for bones to increase their diameter by **appositional growth.** In this process, osteoprogenitor cells in the inner periosteum convert to osteoblast cells, which in turn

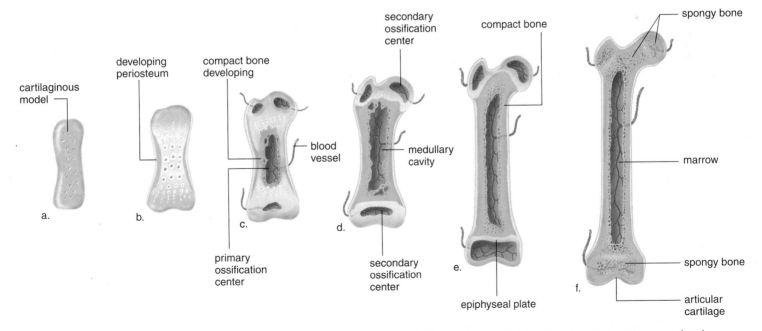

Figure 6.3 Endochondral ossification of a long bone. **a.** A cartilaginous model forms during fetal development. **b.** A periosteum develops. **c.** A primary ossification center contains spongy bone surrounded by compact bone. **d.** The medullary cavity forms in the diaphysis, and secondary ossification centers develop in the epiphyses. **e.** After birth, growth is still possible as long as cartilage remains at the epiphyseal plates. **f.** When the bone is fully formed, the remnants of the epiphyseal plates become a thin line.

add more matrix to the outer surface of the bone. Osteoblasts in the matrix are then converted to osteocytes. Appositional growth causes the bone to become thicker and stronger.

Remodeling of Bones

In the adult, bone is continually being broken down and built up again. Osteoclasts derived from monocytes in red bone marrow break down bone, remove worn cells, and assist in depositing calcium in the blood. After a period of about three weeks, the osteoclasts disappear, and the bone is repaired by the work of osteoblasts. As they form new bone, osteoblasts take calcium from the blood. Eventually some of these cells get caught in the matrix they secrete and are converted to osteocytes, the cells found within the lacunae of osteons.

Strange as it may seem, adults apparently require at least as much calcium in the diet (about 1,000 to 1,500 mg daily) as do actively growing children. Calcium promotes the work of osteoblasts in adults and children. Although adults no longer experience growth in the long bones, high levels of calcium are necessary to prevent **osteoporosis**. In osteoporosis, bones are thin, weak, and fracture easily (see the Medical Focus on page 99).

Growth of bone is a complex process involving over 20 different known hormones and other messenger chemicals. Three of the most important hormones that regulate bone growth are parathyroid hormone, calcitonin, and growth hormone. Their effects on bone are discussed in Chapter 10, The Endocrine System.

Bone Repair

Repair of a bone is required after it breaks, or **fractures.** Bone repair occurs in a series of four steps (Fig. 6.4):

1. *Hematoma.* Within six to eight hours after a fracture, blood escapes from ruptured blood vessels and forms a hematoma (mass of clotted blood) in the space between the broken bones (Fig. 6.4*a*).
2. *Fibrocartilaginous callus.* Tissue repair begins, and fibrocartilage fills the space between the ends of the broken bone for about three weeks (Fig. 6.4*b*).
3. *Bony callus.* Osteoblasts produce trabeculae of spongy bone and convert the fibrocartilaginous callus to a bony callus that joins the broken bones together and lasts about three to four months (Fig. 6.4*c*).
4. *Remodeling.* Osteoblasts build new compact bone at the periphery, and osteoclasts reabsorb the spongy bone, creating a new medullary cavity (Fig. 6.4*d*).

In some ways, bone repair parallels the development of a bone except that the first step, hematoma, indicates that injury has occurred, and then fibrocartilage instead of hyaline cartilage precedes the production of compact bone.

The naming of fractures describes what kind of break occurred. A fracture is *complete* if the bone is broken clear through and *incomplete* if the bone is not separated into two parts. A fracture is *simple* if it does not pierce the skin and *compound* if it does pierce the skin. *Impacted* means that the broken ends are wedged into each other, and a *spiral fracture* occurs when the break is ragged due to twisting of a bone. Repair of a fracture is called **reduction. Closed reduction** involves realigning the bone fragments into their normal position without surgery. **Open reduction** requires surgical repair of the bone using plates, screws, or pins.

Surface Features of Bones

As we study the various bones of the skeleton, refer to Table 6.1, which lists and explains the surface features of bones.

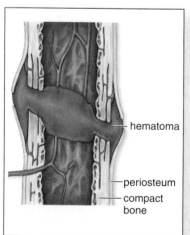
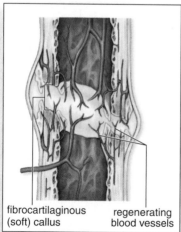
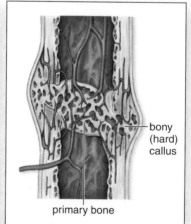
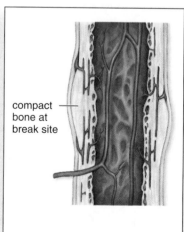

Figure 6.4 Repair of a broken bone. **a.** A hematoma forms between the broken bones. **b.** Fibrocartilage fills the space between bones for about 3 weeks. **c.** Bony callus is formed by osteoblast cells. **d.** Osteoclasts reabsorb the callus and create a new medullary cavity.

Osteoporosis

Osteoporosis is a condition in which the bones are weakened due to a decrease in the bone mass that makes up the skeleton. Throughout life, bones are continuously remodeled. While a child is growing, the rate of bone formation is greater than the rate of bone breakdown. The skeletal mass continues to increase until ages 20 to 30. After that, the rates of formation and breakdown of bone mass are equal until ages 40 to 50. Then, reabsorption begins to exceed formation, and the total bone mass slowly decreases.

Over time, men are apt to lose 25% and women 35% of their bone mass. But we have to consider that men tend to have denser bones than women anyway, and their testosterone (male sex hormone) level generally does not begin to decline significantly until after age 65. In contrast, the estrogen (female sex hormone) level in women begins to decline at about age 45. Because sex hormones play an important role in maintaining bone strength, this difference means that women are more likely than men to suffer fractures, involving especially the hip, vertebrae, long bones, and pelvis. Although osteoporosis may at times be the result of various disease processes, it is essentially a disease of aging.

Everyone can take measures to avoid having osteoporosis when they get older. Adequate dietary calcium throughout life is an important protection against osteoporosis. The U.S. National Institutes of Health recommend a calcium intake of 1,200–1,500 mg per day during puberty. Males and females require 1,000 mg per day until age 65 and 1,500 mg per day after age 65, because the intestinal tract has fewer vitamin D receptors in the elderly.

A small daily amount of vitamin D is also necessary to absorb calcium from the digestive tract. Exposure to sunlight is required to allow skin to synthesize vitamin D. If you reside on or north of a "line" drawn from Boston to Milwaukee, to Minneapolis, to Boise, chances are, you're not getting enough vitamin D during the winter months. Therefore, you should avail yourself of the vitamin D in fortified foods such as low-fat milk and cereal.

Postmenopausal women should have an evaluation of their bone density. Presently, bone density is measured by a method called dual energy X-ray absorptiometry (DEXA). This test measures bone density based on the absorption of photons generated by an X-ray tube. Soon, a blood and urine test may be able to detect the biochemical markers of bone loss, making it possible for physicians to screen all older women and at-risk men for osteoporosis.

If the bones are thin, it is worthwhile to take measures to gain bone density because even a slight increase can significantly reduce fracture risk. Regular, moderate, weight-bearing exercise such as walking or jogging is a good way to maintain bone strength (Fig. 6A). A combination of exercise and drug treatment, as recommended by a physician, may yield the best results.

Figure 6A Preventing osteoporosis. **a.** Exercise can help prevent osteoporosis, but when playing golf, you should carry your own clubs and walk instead of using a golf cart. **b.** Normal bone growth compared to bone from a person with osteoporosis.

Medications for treatment of osteoporosis can slow or reverse the patient's bone loss. Medications that inhibit the action of bone-resorbing osteoclast cells are called bisphosphonates (Fosamax, Actonel). Calcitonin and parathyroid hormone are the body's two naturally occurring hormones for calcium homeostasis. Both are available as medications for people with osteoporosis. Calcitonin is administered as a nasal spray or an injection. Like the bisphosphonates, it inhibits osteoclasts and slows the rate of bone thinning. Parathyroid hormone is given by injection to patients at high risk of bone fracture. It stimulates osteoblast cells to build new bone. The breast cancer drugs tamoxifen and raloxifene also stimulate the growth of new bone tissue.

The sex hormones are also used occasionally for treatment of osteoporosis. Estrogen therapy is used in women after menopause to slow the rate of bone loss. Testosterone can be given to men; it, too, will slow the rate of bone loss. However, therapy with sex hormones must be carefully monitored because sex hormones may trigger the growth of certain reproductive tissue cancers.

TABLE 6.1 Surface Features of Bones

ARTICULATIONS AND PROJECTIONS

Term	Definition	Example
ARTICULATING SURFACES		
Condyle (kon'dīl)	A large, rounded, articulating knob	Mandibular condyle of the mandible (Fig 6.7b)
Head	A prominent, rounded, articulating proximal end of a bone	Head of the femur (Fig. 6.18)
Process	Any bony prominence or knob; often forms a joint with a fossa	Olecranon process of the ulna (Fig. 6.15)
PROJECTIONS FOR MUSCLE ATTACHMENT		
Crest	A narrow, ridgelike projection	Iliac crest of the coxal bone (Fig. 6.17)
Spine	A sharp, slender process	Spine of the scapula (Fig. 6.13)
Trochanter (tro-kan'ter)	A massive process found only on the femur	Greater trochanter and lesser trochanter of the femur (Fig. 6.18)
Tubercle (tu'ber-kl)	A small, rounded process	Greater tubercle of the humerus (Fig. 6.14)
Tuberosity (tu"bĕ-ros'ĭ-te)	A large, roughened process	Radial tuberosity of the radius (Fig. 6.15)
DEPRESSIONS AND OPENINGS		
Foramen (fo-ra'men)	A rounded opening through a bone	Foramen magnum of the occipital bone (Fig. 6.8a)
Fossa (fos'uh)	A flattened or shallow surface	Mandibular fossa of the temporal bone (Fig. 6.8a)
Meatus (me-a'tus)	A tubelike passageway through a bone	External auditory meatus of the temporal bone (Fig. 6.7b)
Sinus (si'nus)	A cavity or hollow space in a bone	Frontal sinus of the frontal bone (Fig. 6.6)

Source: Data from Kent M. Van De Graaff and Stuart Ira Fox, *Concepts of Human Anatomy and Physiology,* 5th ed., 1999, p. 187.

Content CHECK-UP!

1. The term for the expanded portions at the ends of a long bone is:
   a. diaphysis.
   c. periosteum.
   b. epiphysis.
   d. articular cartilage.
2. Which type of bone cell breaks down bone and deposits calcium into the blood?
   a. osteoblast
   c. osteoprogenitor
   b. osteocyte
   d. osteoclast
3. The term for a rounded opening through a bone is:
   a. foramen.
   c. trochanter.
   b. tuberosity.
   d. condyle.

6.2 Axial Skeleton

The skeleton is divided into the axial skeleton and the appendicular skeleton. The tissues of the axial and appendicular skeletons are bone (both compact and spongy), cartilage (hyaline, fibrocartilage, and elastic cartilage), and dense connective tissue, a type of fibrous connective tissue. (The various types of connective tissues were extensively discussed in Chapter 4.)

In Figure 6.5, the bones of the axial skeleton are colored orange, and the bones of the appendicular skeleton are colored yellow for easy distinction. Notice that the **axial skeleton** lies in the midline of the body and contains the bones of the skull, the hyoid bone, the vertebral column, and the thoracic cage. Six tiny middle ear bones (three in each ear) are also in the axial skeleton; we will study them in Chapter 9 in connection with the ear.

Skull

The skull is formed by the cranium and the facial bones. These bones contain **sinuses** (Figure 6.6), air spaces lined by mucous membranes, that reduce the weight of the skull and give the voice a resonant sound. The paranasal sinuses empty into the nose and are named for their locations. They include the maxillary, frontal, sphenoidal, and ethmoidal sinuses. **Sinusitis** is infection or inflammation of the paranasal sinuses. The two mastoid sinuses drain into the middle ear. **Mastoiditis,** a condition that can lead to deafness, is an inflammation of these sinuses.

Bones of the Cranium

The cranium protects the brain and is composed of eight bones. These bones are separated from each other by

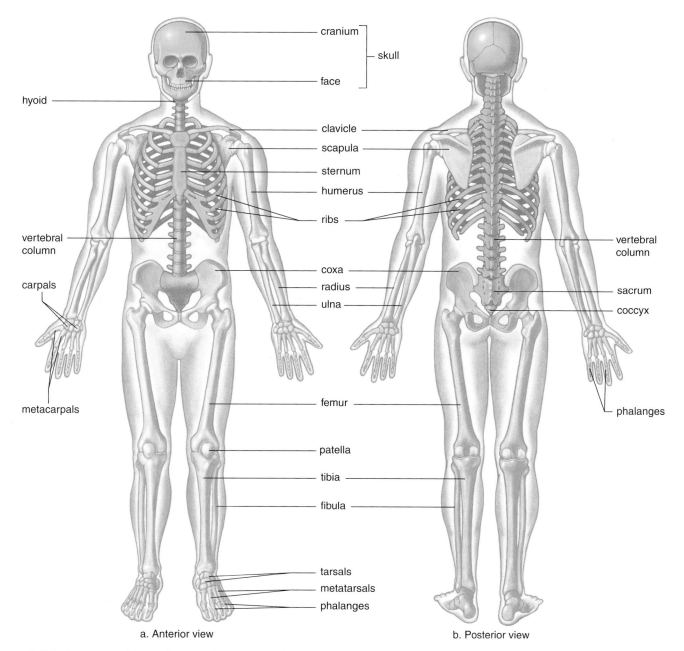

Figure 6.5 Major bones of the skeleton. **a.** Anterior view. **b.** Posterior view. The bones of the axial skeleton are shown in orange, and those of the appendicular skeleton are shown in yellow.

immovable joints called **sutures.** Newborns have membranous regions called **fontanels,** where the bones of the cranial vault have not yet fused together. The fontanels permit the bones of the skull to shift during birth as the head passes through the birth canal. The largest fontanel is the anterior fontanel (often called the "soft spot"), which is located where the two parietal bones meet the two unfused parts of the frontal bone. The anterior fontanel usually closes by the age of two years. Besides the frontal bone, the cranium is composed of two parietal bones, one occipital bone, two temporal bones, one sphenoid bone, and one ethmoid bone (Figs. 6.7 and 6.8).

Frontal Bone One frontal bone forms the forehead, a portion of the nose, and the superior portions of the orbits (bony sockets of the eyes).

Parietal Bones Two parietal bones are just posterior to the frontal bone. They form the roof of the cranium and also help form its sides.

Occipital Bones One occipital bone forms the most posterior part of the skull and the base of the cranium. The spinal cord joins the brain by passing through a large opening in the occipital bone called the foramen magnum. The **occipital condyles** (Fig. 6.8a) are rounded processes on either side of

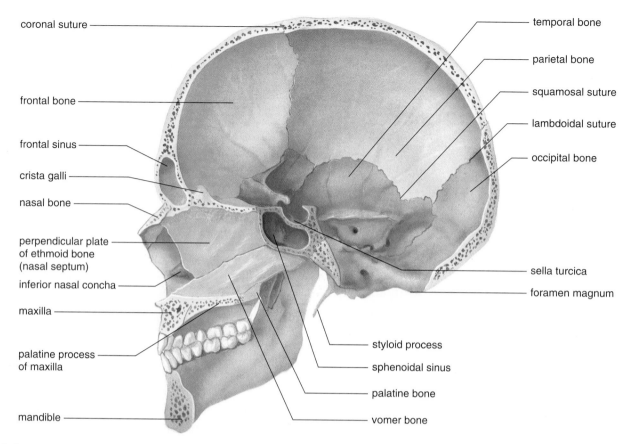

coronal suture

frontal bone

frontal sinus

crista galli

nasal bone

perpendicular plate
of ethmoid bone
(nasal septum)

inferior nasal concha

maxilla

palatine process
of maxilla

mandible

temporal bone

parietal bone

squamosal suture

lambdoidal suture

occipital bone

sella turcica

foramen magnum

styloid process

sphenoidal sinus

palatine bone

vomer bone

Figure 6.6 Sagittal section of the skull.

the foramen magnum that articulate with the first vertebra of the spinal column.

Temporal Bones Two temporal bones are just inferior to the parietal bones on the sides of the cranium. They also help form the base of the cranium (Figs. 6.6 and 6.7*a*). Each temporal bone has the following:

external auditory meatus, a canal that leads to the middle ear;
mandibular fossa, which articulates with the mandible;
mastoid process, which provides a place of attachment for certain neck muscles;
styloid process, which provides a place of attachment for muscles associated with the tongue and larynx;
zygomatic process, which projects anteriorly and helps form the cheekbone.

Sphenoid Bone The sphenoid bone helps form the sides and floor of the cranium and the rear wall of the orbits. The sphenoid bone has the shape of a bat and this shape means that it articulates with and holds together the other cranial bones (Fig. 6.8). Within the cranial cavity, the sphenoid bone has a saddle-shaped midportion called the **sella turcica** (Fig. 6.8*b*), which houses the pituitary gland in a depression.

Ethmoid Bone The ethmoid bone is anterior to the sphenoid bone and helps form the floor of the cranium. It

contributes to the medial sides of the orbits and forms the roof and sides of the nasal cavity (Figs. 6.7 and 6.8*b*). Important features of the ethmoid bone include:

crista galli (cock's comb), a triangular process that serves as an attachment for membranes that enclose the brain;
cribriform plate with tiny holes that serve as passageways for nerve fibers from the olfactory receptors;
perpendicular plate (Fig. 6.6), which projects downward to form the nasal septum;
superior and **middle nasal conchae,** which project toward the perpendicular plate. These projections support mucous membranes that line the nasal cavity.

Bones of the Face

Maxillae The two maxillae form the upper jaw. Aside from contributing to the floors of the orbits and to the sides of the floor of the nasal cavity, each maxilla has the following processes:

alveolar process (Fig. 6.7*a*). The alveolar processes contain the tooth sockets for teeth: incisors, canines, premolars, and molars.
palatine process (Fig. 6.8*a*). The left and right palatine processes form the anterior portion of the **hard palate** (roof of the mouth).

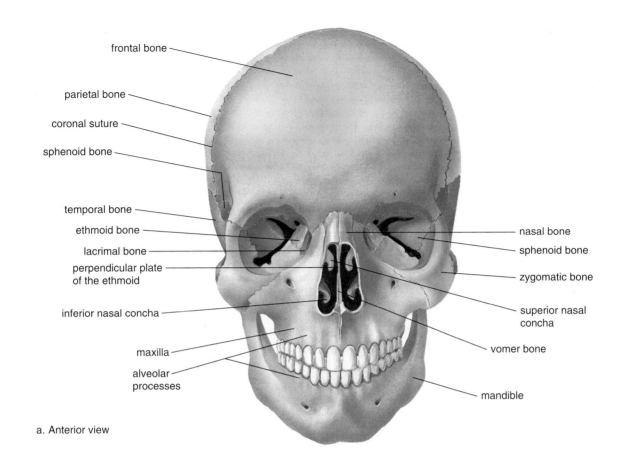

frontal bone

parietal bone

coronal suture

sphenoid bone

temporal bone

ethmoid bone

lacrimal bone

perpendicular plate
of the ethmoid

inferior nasal concha

maxilla

alveolar
processes

nasal bone

sphenoid bone

zygomatic bone

superior nasal
concha

vomer bone

mandible

a. Anterior view

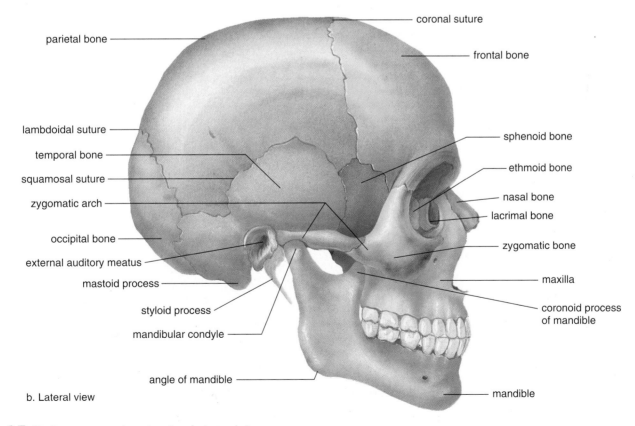

coronal suture

parietal bone

frontal bone

lambdoidal suture

temporal bone

squamosal suture

zygomatic arch

occipital bone

external auditory meatus

mastoid process

styloid process

mandibular condyle

sphenoid bone

ethmoid bone

nasal bone

lacrimal bone

zygomatic bone

maxilla

coronoid process
of mandible

angle of mandible

mandible

b. Lateral view

Figure 6.7 Skull anatomy. **a.** Anterior view. **b.** Lateral view.

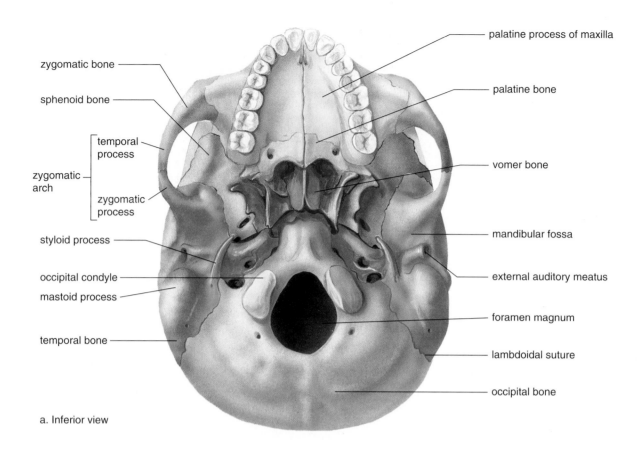

zygomatic bone

sphenoid bone

temporal process

zygomatic arch

zygomatic process

styloid process

occipital condyle

mastoid process

temporal bone

palatine process of maxilla

palatine bone

vomer bone

mandibular fossa

external auditory meatus

foramen magnum

lambdoidal suture

occipital bone

a. Inferior view

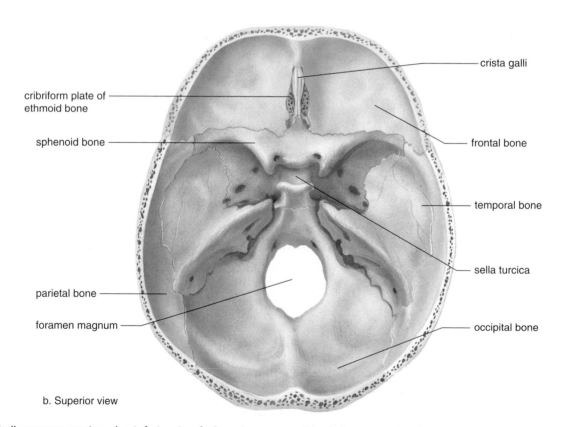

cribriform plate of ethmoid bone

sphenoid bone

parietal bone

foramen magnum

crista galli

frontal bone

temporal bone

sella turcica

occipital bone

b. Superior view

Figure 6.8 Skull anatomy continued. **a.** Inferior view. **b.** Superior cross-sectional view.

Palatine Bones The two palatine bones contribute to the floor and lateral wall of the nasal cavity (Fig. 6.6). The horizontal plates of the palatine bones form the posterior portion of the hard palate (Fig. 6.8a).

Notice that the hard palate consists of (1) portions of the maxillae (i.e., the palatine processes) and (2) horizontal plates of the palatine bones. A cleft palate results when either (1) or (2) have failed to fuse.

Zygomatic Bones The two zygomatic bones form the sides of the orbits (Fig. 6.8a). They also contribute to the "cheekbones." Each zygomatic bone has a **temporal process.** A **zygomatic arch,** the most prominent feature of a cheekbone consists of a temporal process connected to a zygomatic process (a portion of the temporal bone).

Lacrimal Bones The two small, thin lacrimal bones are located on the medial walls of the orbits (Fig. 6.7). A small opening between the orbit and the nasal cavity serves as a pathway for a duct that carries tears from the eyes to the nose.

Nasal Bones The two nasal bones are small, rectangular bones that form the bridge of the nose (Fig. 6.6). The ventral portion of the nose is cartilage, which explains why the nose is not seen on a skull.

Vomer Bone The vomer bone joins with the perpendicular plate of the ethmoid bone to form the nasal septum (Figs. 6.6 and 6.7a).

Inferior Nasal Conchae The two inferior nasal conchae are thin, curved bones that form a part of the inferior lateral wall of the nasal cavity (Fig. 6.7a). Like the superior and middle nasal conchae, they project into the nasal cavity and support the mucous membranes that line the nasal cavity.

Mandible The mandible, or lower jaw, is the only movable portion of the skull. The horseshoe-shaped front and horizontal sides of the mandible, referred to as the *body,* form the chin. The body has an **alveolar process** (Fig. 6.7a), which contains tooth sockets for 16 teeth. Superior to the left and right angle of the mandible are upright projections called *rami.* Each ramus has the following:

mandibular condyle (Fig. 6.7b), which articulates with a temporal bone;
coronoid process (Fig. 6.7b), which serves as a place of attachment for the muscles used for chewing.

Hyoid Bone

The U-shaped hyoid bone (Fig. 6.5) is located superior to the larynx (voice box) in the neck. It is the only bone in the body that does not articulate with another bone. Instead, it is suspended from the styloid processes of the temporal bones by the stylohyoid muscles and ligaments. It anchors the tongue and serves as the site for the attachment of several muscles associated with swallowing.

Vertebral Column (Spine)

The **vertebral column** extends from the skull to the pelvis. It consists of a series of separate bones, the **vertebrae,** separated by pads of fibrocartilage called the **intervertebral disks** (Fig. 6.9). The vertebral column is located in the middorsal region and forms the vertical axis. The skull rests on the superior end of the vertebral column, which also supports the rib cage and serves as a point of attachment for the pelvic

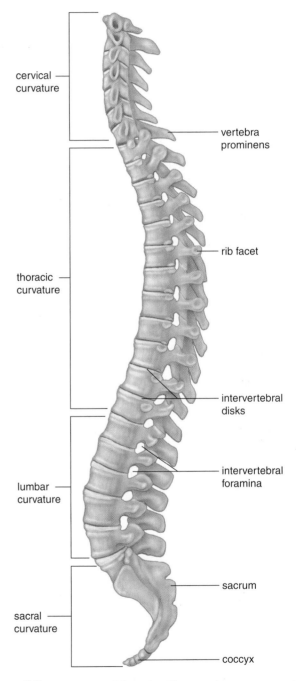

Figure 6.9 Curvatures of the spine. The vertebrae are named for their location in the body. Note the presence of the coccyx, also called the tailbone.

girdle. The vertebral column also protects the spinal cord, which passes through a vertebral canal formed by the vertebrae. The vertebrae are named according to their location: seven *cervical* (neck) *vertebrae,* twelve *thoracic* (chest) *vertebrae,* five *lumbar* (lower back) *vertebrae,* five *sacral vertebrae* fused to form the sacrum, and three to five *coccygeal vertebrae* fused into one coccyx.

When viewed from the side, the vertebral column has four normal curvatures, named for their location (Fig. 6.9). The cervical and lumbar curvatures are convex anteriorly, and the thoracic and sacral curvatures are concave anteriorly. In the fetus, the vertebral column has but one curve, and it is concave anteriorly. The cervical curve develops three to four months after birth, when the child begins to hold the head up. The lumbar curvature develops when a child begins to stand and walk, around one year of age. The curvatures of the vertebral column provide more support than a straight column would, and they also provide the balance needed to walk upright.

The curvatures of the vertebral column are subject to abnormalities (Fig. 6.10). An abnormally exaggerated lumbar curvature is called **lordosis,** or "swayback." People who are balancing a heavy midsection, such as pregnant women or men with "potbellies," may have swayback. An increased roundness of the thoracic curvature is **kyphosis,** or "hunchback." This abnormality sometimes develops in older people as the center sections of thoracic vertebrae become compressed. An abnormal lateral (side-to-side) curvature is called **scoliosis.** Occurring most often in the thoracic region, scoliosis is usually first seen during late childhood.

Intervertebral Disks

The fibrocartilaginous intervertebral disks located between the vertebrae act as a cushion. The disks are filled with gelatinous material, which prevents the vertebrae from grinding against one another and absorbs shock caused by such movements as running, jumping, and even walking. The disks also allow motion between the vertebrae so that a person can bend forward, backward, and from side to side. Unfortunately, these disks become weakened with age, and can slip or even rupture (called a **herniated disk**). A damaged disk pressing against the spinal cord or the spinal nerves causes pain. Such a disk may need to be removed surgically. If a disk is removed, the vertebrae are fused together, limiting the body's flexibility.

Vertebrae

Figure 6.11*a* shows that a typical vertebra has an anteriorly placed *body* and a posteriorly placed vertebral *arch.* The vertebral arch forms the wall of a *vertebral foramen* (pl., *foramina*). The foramina become a canal through which the spinal cord passes.

The structure of a single vertebra can be likened to a house (the vertebral arch) sitting on a hill (the body of the vertebra). *Pedicles* are the upright walls of the house and the *laminae* form a slanting roof. The single *spinous process,* or *spine,* is like a flagpole, and the *transverse processes* are gutters projecting sideways at the corners where pedicles join laminae. Each of these bony projections on an actual vertebra serves as a site for muscle attachment. Additionally, articular processes (superior and inferior) function to join vertebrae.

The vertebrae have regional differences. For example, as the vertebral column descends, the bodies get bigger and are better able to carry more weight. In the cervical region, the spines are short and tend to have a split, or bifurcation. The thoracic spines are long and slender and project downward. The lumbar spines are massive and square and project posteriorly. The transverse processes of thoracic vertebrae have articular facets for connecting to ribs.

Atlas and Axis The first two cervical vertebrae are not typical (Fig. 6.11*b*). The **atlas** supports and balances the head. It has two depressions that articulate with the occipital condyles, allowing movement of the head up and down (as though nodding "yes"). The **axis** has an *odontoid process* (also called the dens) that projects into the ring of the atlas. When the head

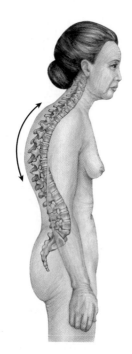 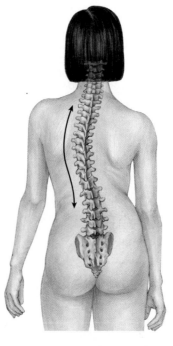

Lordosis ("swayback")   Kyphosis ("hunchback")   Scoliosis

Figure 6.10 Abnormal curvatures of the vertebral column. **a.** Lordosis, or "swayback." **b.** Kyphosis, or "hunchback." **c.** Scoliosis.

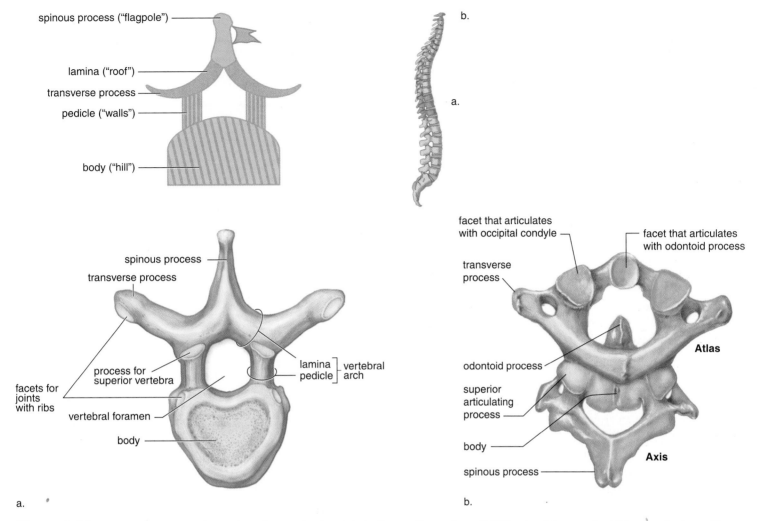

spinous process ("flagpole")

lamina ("roof")

transverse process

pedicle ("walls")

body ("hill")

b.

a.

spinous process

transverse process

process for superior vertebra

facets for joints with ribs

vertebral foramen

body

lamina
pedicle | vertebral arch

a.

facet that articulates with occipital condyle

transverse process

facet that articulates with odontoid process

odontoid process

superior articulating process

body

spinous process

Atlas

Axis

b.

Figure 6.11 Vertebrae. **a.** A typical vertebra. The vertebral arch is similar to a "house" on a "hill" (body of the vertebra). The vertebral canal is formed by adjacent vertebral foramina and contains the spinal cord. **b.** Atlas and axis, showing their interaction. The interaction/articulating facets between $C_1 + C_2$ are shaded blue. The odontoid process of the axis is the pivot around which the atlas turns, as when we shake our head and say "no." The superior articulating processes on the axis forms a joint with the inferior articular processes on the atlas.

moves from side to side, the atlas pivots around the odontoid process (as though shaking the head "no").

Sacrum and Coccyx The five sacral vertebrae are fused to form the **sacrum**. The sacrum articulates with the pelvic girdle and forms the posterior wall of the pelvic cavity (see Fig. 6.17). The **coccyx**, or tailbone, is the last part of the vertebral column. It is formed from a fusion of three to five vertebrae.

The Rib Cage

The **rib cage** (Fig. 6.12), sometimes called the thoracic cage, is composed of the thoracic vertebrae, ribs and associated cartilages, and sternum.

The rib cage demonstrates how the skeleton is protective but also flexible. The rib cage protects the heart and lungs; yet it swings outward and upward upon inspiration and then downward and inward upon expiration. The rib cage also provides support for the bones of the pectoral girdle (see page 109).

The Ribs

There are 12 pairs of ribs. All 12 pairs connect directly to the thoracic vertebrae in the back. After connecting with thoracic vertebrae, each rib first curves outward and then forward and downward. The first pair of ribs attaches to the body of the first thoracic vertebra, or T1. It also attaches to the transverse process of T1, at the facet for the joint with the rib. The next eight pairs of ribs (ribs 2 through 9) have three points of attachment to the vertebrae: to the body of the same numbered vertebra; to the body of the vertebra immediately superior; and to the transverse process of the same numbered vertebra (see Fig. 6.11a). For example, rib 2 attaches to the bodies of thoracic vertebrae 1 and 2, and to the transverse process of vertebra 2. Rib pairs 10 through 12 attach only to their respective vertebrae.

The upper seven pairs of ribs connect directly to the sternum by means of costal cartilages. These are called the "true ribs," or the *vertebrosternal* ribs. The next five pairs of ribs are called the "false ribs" because they attach indirectly to the sternum or are not attached at all. Ribs 8, 9, and 10 are called

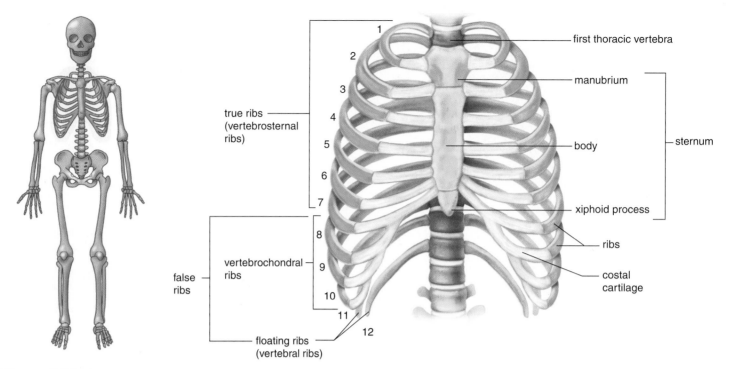

Figure 6.12 The rib cage. This structure includes the thoracic vertebrae, the ribs, and the sternum. The three bones that make up the sternum are the manubrium, body, and xiphoid process. The ribs numbered 1–7 are true ribs; those numbered 8–10 are false ribs; ribs 11 and 12 are "floating" ribs.

vertebrochondral ribs; each attaches its costal cartilage to the cartilage of the rib superior to it. All three ribs attach indirectly to the sternum using the costal cartilage of rib 7. Ribs 11 and 12 are *vertebral*, or "floating," ribs. These are short ribs with no attachment to the sternum (Fig. 6.12).

The Sternum

The **sternum,** or breastbone, is a flat bone that has the shape of a blade. The sternum, along with the ribs, helps protect the heart and lungs. During surgery the sternum may be split to allow access to the organs of the thoracic cavity.

The sternum is composed of three bones that fuse during fetal development (Fig. 6.12). These bones are the manubrium, the body, and the xiphoid process. The *manubrium* is the superior portion of the sternum. The *body* is the middle and largest part of the sternum, and the *xiphoid process* is the inferior and smallest portion of the sternum. The manubrium joins with the body of the sternum at an angle. This joint is an important anatomical landmark because it occurs at the level of the second rib, and therefore allows the ribs to be counted. Counting the ribs is sometimes done to determine where the apex of the heart is located—usually between the fifth and sixth ribs.

The manubrium articulates with the costal cartilages of the first and second ribs; the body articulates with costal cartilages of the second through seventh ribs; and the xiphoid process doesn't articulate with any ribs.

The xiphoid process is the third part of the sternum. Composed of hyaline cartilage in the child, it becomes ossified in the adult. The variably shaped xiphoid process serves as an attachment site for the diaphragm, which separates the thoracic cavity from the abdominal cavity.

6.3 Appendicular Skeleton

The **appendicular skeleton** contains the bones of the pectoral girdle, upper limbs, pelvic girdle, and lower limbs.

Pectoral Girdle

The **pectoral girdle** (shoulder girdle) contains four bones: two clavicles and two scapulae (Fig. 6.13). It supports the arms and serves as a place of attachment for muscles that move the arms. The bones of this girdle are not held tightly

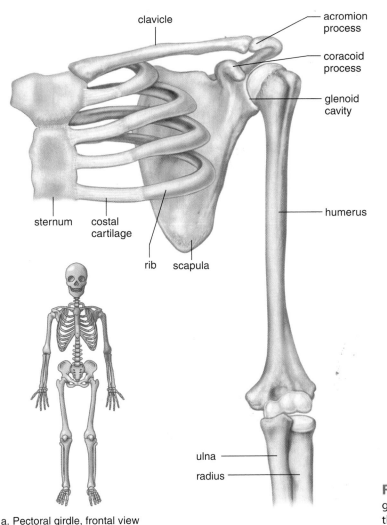

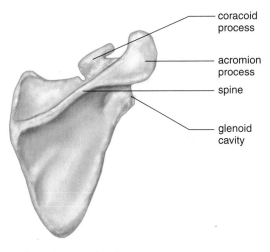

b. Scapula, posterior view

a. Pectoral girdle, frontal view

Figure 6.13 The pectoral girdle. **a.** Frontal view of the pectoral girdle (left side) with the upper limb attached. **b.** Posterior view of the right scapula.

together; rather, they are weakly attached and held in place by ligaments and muscles. This arrangement allows great flexibility but means that the arm is prone to dislocation at the shoulder joint.

Clavicles

The **clavicles** (collarbones) are slender and S-shaped (Fig. 6.13). Each clavicle articulates medially with the manubrium of the sternum. This is the only place where the pectoral girdle is attached to the axial skeleton.

Each clavicle also articulates with a scapula. The clavicle serves as a brace for the scapula and helps stabilize the shoulder. It is structurally weak, however, and if undue force is applied to the shoulder, the clavicle will fracture.

Scapulae

The **scapulae** (sing., scapula), also called the shoulder blades, are broad bones that somewhat resemble triangles (Fig. 6.13b). One reason for the pectoral girdle's flexibility is that the scapulae are not joined to each other (see Fig. 6.5).

Each scapula has a spine. Note the following features as well:

acromion process, which articulates with a clavicle and provides a place of attachment for arm and chest muscles;
coracoid process, which serves as a place of attachment for arm and chest muscles;
glenoid cavity, which articulates with the head of the arm bone (humerus). The arm joint's flexibility is also a result of the glenoid cavity being smaller than the head of the humerus.

Upper Limb

The upper limb includes the bones of the arm (humerus), the forearm (radius and ulna), and the hand (carpals, metacarpals, and phalanges).[1]

---

[1]The term *upper extremity* is used to include a clavicle and scapula (of the pectoral girdle), an arm, forearm, wrist, and hand.

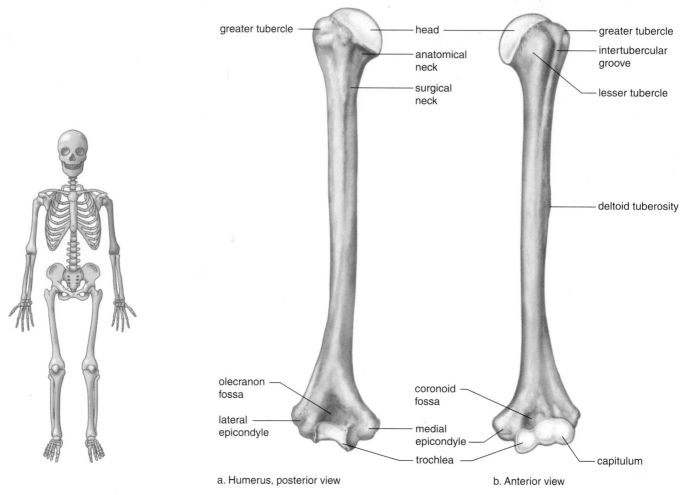

greater tubercle — head — greater tubercle

anatomical neck — intertubercular groove

surgical neck — lesser tubercle

deltoid tuberosity

olecranon fossa — coronoid fossa

lateral epicondyle — medial epicondyle — trochlea — capitulum

a. Humerus, posterior view   b. Anterior view

Figure 6.14 Left humerus. **a.** Posterior surface view. **b.** Anterior surface view.

Humerus

The **humerus** (Fig. 6.14) is the bone of the arm. It is a long bone with the following features at the proximal end:

head, which articulates with the glenoid cavity of the scapula;

greater and **lesser tubercles,** which provide attachments for muscles that move the arm and shoulder;

intertubercular groove, which holds the tendon from the biceps brachii, a muscle of the arm;

deltoid tuberosity, which provides an attachment for the deltoid, a muscle that covers the shoulder joint.

The humerus has the following features at the distal end:

capitulum, a lateral condyle that articulates with the head of the radius;

trochlea, a spool-shaped condyle that articulates with the ulna;

coronoid fossa, a depression for a process of the ulna when the elbow is flexed;

olecranon fossa, a depression for a process of the ulna when the elbow is extended.

Radius

The **radius** and **ulna** (see Figs. 6.13a and 6.15) are the bones of the forearm. The radius is on the lateral side of the forearm (the thumb side). When you turn your hand from the "palms up" position to the "palms down" position, the radius crosses over the ulna, so the two bones are criss-crossed. Proximally, the radius has the following features:

head, which articulates with the capitulum of the humerus and fits into the radial notch of the ulna;

radial tuberosity, which serves as a place of attachment for a tendon from the biceps brachii;

Distally, the radius has the following features:

ulnar notch, which articulates with the head of the ulna;

styloid process, which serves as a place of attachment for ligaments that run to the wrist.

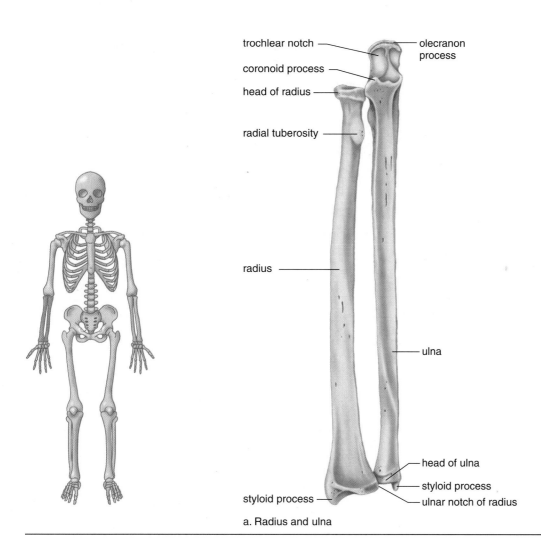

trochlear notch — olecranon process

coronoid process —

head of radius —

radial tuberosity —

radius —

ulna

head of ulna

styloid process —

styloid process — ulnar notch of radius

a. Radius and ulna

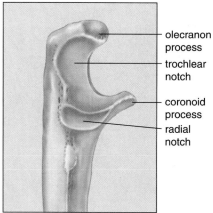

— olecranon process
— trochlear notch
— coronoid process
— radial notch

b. Ulna, proximal, anterior view

Figure 6.15 Right radius and ulna. **a.** The head of the radius articulates with the radial notch of the ulna. The head of the ulna articulates with the ulnar notch of the radius. **b.** Anterior view of the proximal end of the ulna.

Ulna

The ulna is the longer bone of the forearm. Proximally, the ulna has the following features:

coronoid process, which articulates with the coronoid fossa of the humerus when the elbow is flexed;
olecranon process, the point of the elbow, articulates with the olecranon fossa of the humerus when the elbow is extended;
trochlear notch, which articulates with the trochlea of the humerus at the elbow joint;
radial notch, which articulates with head of the radius.

Distally, the ulna has the following features:

head, which articulates with the ulnar notch of the radius;
styloid process, which serves as a place of attachment for ligaments that run to the wrist.

Hand

Each hand (Fig. 6.16) has a wrist, a palm, and five fingers, or digits.

The wrist, or carpus, contains eight small carpal bones, tightly bound by ligaments in two rows of four each. Where we wear a "wrist watch" is the distal forearm—the true wrist is the proximal part of what we generally call the hand. Only two of the carpals (the scaphoid and lunate) articulate with the radius. Anteriorly, the concave region of the wrist is covered by a ligament called the flexor retinaculum, forming the so-called "carpal tunnel." Inflammation of the tendons running through this area is usually caused by abuse or overuse of the wrist area. The inflamed tendons compress a nerve, and the resulting pain and numbness is *carpal tunnel syndrome.*

Five metacarpal bones, numbered 1 to 5 from the thumb side of the hand toward the little finger, fan out to form the palm. When the fist is clenched, the heads of the metacarpals, which articulate with the phalanges, become obvious. The first metacarpal is more anterior than the others, and this allows the thumb to touch each of the other fingers.

The fingers, including the thumb, contain bones called the *phalanges.* The thumb has only two phalanges (proximal and distal), but the other fingers have three each (proximal, middle, and distal).

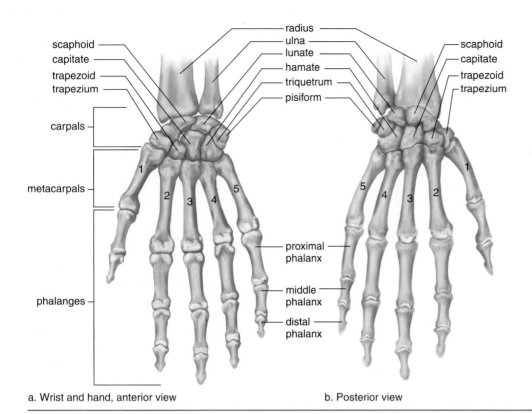

scaphoid
capitate
trapezoid
trapezium

carpals

metacarpals

phalanges

radius
ulna
lunate
hamate
triquetrum
pisiform

scaphoid
capitate
trapezoid
trapezium

1
2 3 4 5

5 4 3 2

1

proximal phalanx

middle phalanx

distal phalanx

a. Wrist and hand, anterior view

b. Posterior view

Figure 6.16 Right wrist and hand.
a. Anterior view. **b.** Posterior view.

Pelvic Girdle

The **pelvic girdle** contains two coxal bones (hipbones), as well as the sacrum and coccyx (Fig. 6.17a,b; see Fig. 6.9). The strong bones of the pelvic girdle are firmly attached to one another and bear the weight of the body. The pelvis also serves as the place of attachment for the lower limbs and protects the urinary bladder, the internal reproductive organs, and a portion of the large intestine.

Coxal Bones

Each **coxal bone** has the following three parts:

1. **ilium** (Fig. 6.17). The ilium, the largest part of a coxal bone, flares outward to give the hip prominence. The margin of the ilium is called the **iliac crest** (when you put your hands on your hips, you are resting them on the iliac crests). The end points of the iliac crest are called the **anterior superior iliac spine** and the **posterior superior iliac spine** (Fig. 6.17c). The **greater sciatic notch** is the site where blood vessels and the large sciatic nerve pass posteriorly into the leg. Each ilium connects posteriorly with the sacrum at a **sacroiliac joint.**
2. **ischium** (Fig. 6.17c). The ischium is the most inferior part of a coxal bone. Its posterior region, the **ischial tuberosity,** allows a person to sit. Near the junction of the ilium and ischium is the **ischial spine,** which projects into the pelvic cavity. The distance between the ischial spines is the size of the pelvic cavity.

3. **pubis** (Fig. 6.17c). The pubis is the anterior part of a coxal bone. The two pubic bones are joined by a fibrocartilage disk at the **pubic symphysis.** Posterior to where the pubis and the ischium join together is a large opening, the **obturator foramen,** through which blood vessels and nerves pass anteriorly into the leg.

Where the three parts of each coxal bone meet is a depression called the **acetabulum,** which receives the rounded head of the femur.

False and True Pelvises

The false pelvis is the portion of the trunk bounded laterally by the flared parts of the ilium. This space is much larger than that of the true pelvis. The true pelvis, which is inferior to the false pelvis, is the portion of the trunk bounded by the sacrum, lower ilium, ischium, and pubic bones. The true pelvis is said to have an upper inlet (also called the pelvic brim), and a lower outlet. The dimensions of these outlets are important for females because the outlets must be large enough to allow a baby to pass through during the birth process.

Gender Differences

Female and male pelvises (Fig. 6.17) usually differ in several ways, including the following:

1. Female iliac bones are more flared than those of the male; therefore, the female has broader hips.
2. The female pelvis is wider between the ischial spines and the ischial tuberosities.

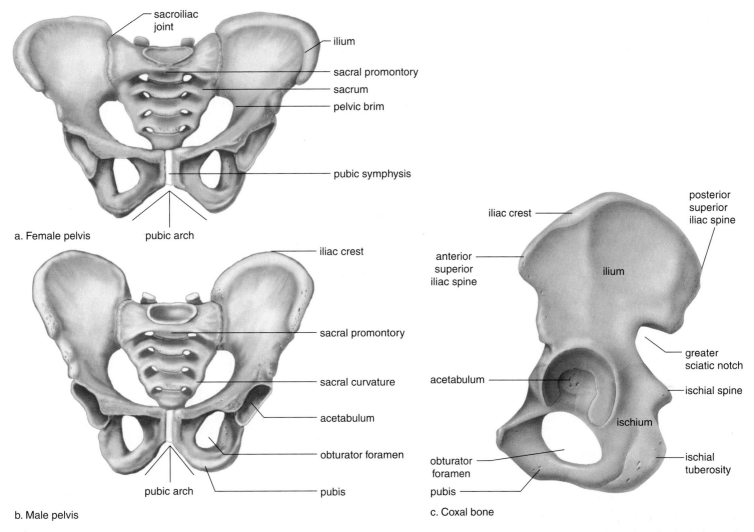

sacroiliac joint

ilium

sacral promontory

sacrum

pelvic brim

pubic symphysis

a. Female pelvis   pubic arch

iliac crest

sacral promontory

sacral curvature

acetabulum

obturator foramen

pubic arch   pubis

b. Male pelvis

iliac crest

posterior superior iliac spine

anterior superior iliac spine

ilium

acetabulum

greater sciatic notch

ischial spine

ischium

obturator foramen

pubis

ischial tuberosity

c. Coxal bone

Figure 6.17 The female pelvis is usually wider in all diameters and roomier than that of the male. **a.** Female pelvis. **b.** Male pelvis. **c.** Left coxal bone, lateral view.

3. The female inlet and outlet of the true pelvis are wider.
4. The female pelvic cavity is more shallow, while the male pelvic cavity is more funnel shaped.
5. Female bones are lighter and thinner.
6. The female pubic arch (angle at the pubic symphysis) is wider. In females, the pubic arch resembles an inverted letter U; in males, the pubic arch resembles an inverted V.

In addition to these differences in pelvic structure, male pelvic bones are larger and heavier, the articular ends are thicker, and the points of muscle attachment may be larger.

Lower Limb

The lower limb includes the bones of the thigh (femur), the kneecap (patella), the leg (tibia and fibula), and the foot (tarsals, metatarsals, and phalanges).[2]

---

[2]The term *lower extremity* is used to include a coxal bone (of the pelvic girdle), the thigh, kneecap, leg, ankle, and foot.

Femur

The **femur** (Fig. 6.18), or thighbone, is the longest and strongest bone in the body. Proximally, note these structures on the femur:

head, which fits into the acetabulum of the coxal bone;
greater and **lesser trochanters,** which provide a place of attachment for the muscles of the thighs and buttocks;
linea aspera, a crest that serves as a place of attachment for several muscles.

Distally, the femur has the following features:

medial and **lateral epicondyles,** which serve as sites of attachment for muscles and ligaments;
lateral and **medial condyles,** which articulate with the tibia;
patellar surface, which is located between the condyles on the anterior surface, articulates with the **patella,** a small triangular bone that protects the knee joint.

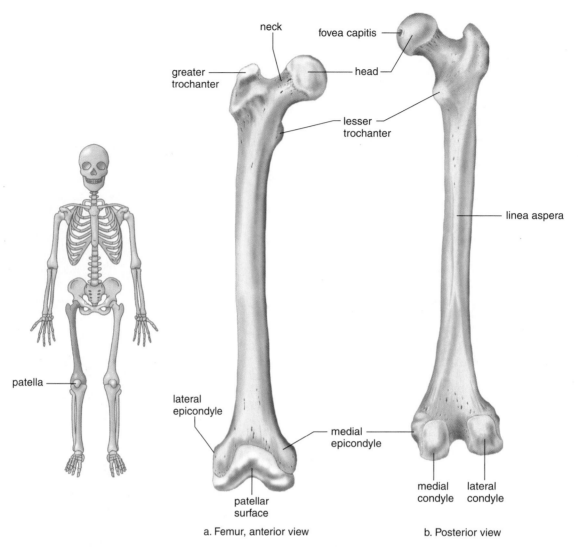

Figure 6.18 Right femur. **a.** Anterior view. **b.** Posterior view.

Tibia

The **tibia** and **fibula** (Fig. 6.19) are the bones of the leg. The tibia, or shinbone, is medial to the fibula. It is thicker than the fibula and bears the weight from the femur, with which it articulates. Observe these structures on the tibia:

medial and **lateral condyles,** which articulate with the femur;

tibial tuberosity, where the patellar (kneecap) ligaments attach;

anterior crest, commonly called the shin;

medial malleolus, the bulge of the inner ankle, which articulates with the talus in the foot.

Fibula

The fibula is lateral to the tibia and is more slender. It has a head that articulates with the tibia just below the lateral condyle. Distally, the **lateral malleolus** articulates with the talus and forms the outer bulge of the ankle. Its role is to stabilize the ankle.

Foot

Each foot (Fig. 6.20) has an ankle, an instep, and five toes (also called digits).

The ankle has seven **tarsal bones;** together, they are called the tarsus. Only one of the seven bones, the **talus,** can move freely where it joins the tibia and fibula. The largest of the ankle bones is the **calcaneus,** or heel bone. Along with the talus, it supports the weight of the body.

The instep has five elongated **metatarsal bones.** The distal ends of the metatarsals form the ball of the foot. Along with the tarsals, these bones form the arches of the foot (longitudinal and transverse), which give spring to a person's step. If the ligaments and tendons holding these bones together weaken, fallen arches, or "flat feet," can result.

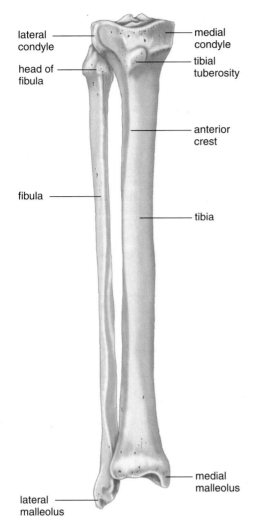

Leg bones, anterior view

Figure 6.19 Bones of the right leg, viewed anteriorly.

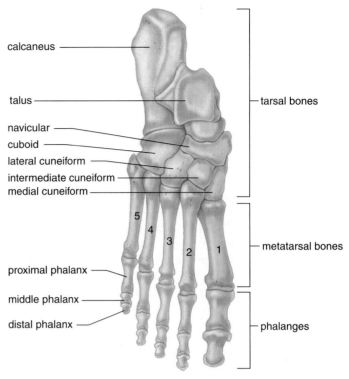

Right foot, superior view

Figure 6.20 The right foot, viewed superiorly.

The toes contain the **phalanges**. The big toe has only two phalanges, but the other toes have three each.

Content **CHECK-UP!**

7. The glenoid process is a part of which bone?

a. coxal bone   c. humerus

b. scapula   d. femur

8. The acetabulum is a part of which bone?

a. coxal bone   c. humerus

b. scapula   d. femur

9. Which bone forms the heel?

a. talus   c. calcaneus

b. fibula   d. tibia

6.4   Joints (Articulations)

Bones articulate at the joints. There are two systems for classifying joints: (1) according to the amount of movement they allow; and (2) according to their structure (the convention followed here). A joint called a **synarthrosis** is immovable; an **amphiarthrosis** allows slight movement; and a **diarthrosis** is freely movable.

Fibrous joints occur where fibrous connective tissue joins bone to bone. These joints are typically immovable (thus, synarthrosis joints) although exceptions exist.

Cartilaginous joints occur where fibrocartilage or hyaline cartilage joins bones. These are generally slightly movable (amphiarthrosis joints), with a few exceptions.

Synovial joints are formed when bone ends do not contact each other, but are enclosed in a capsule. These are usually freely movable (diarthrosis joints). Again, there are a few exceptions.

Fibrous Joints

Some bones, such as those that make up the adult cranium, are sutured together by a thin layer of fibrous connective tissue and are immovable. Review Figures 6.7 and 6.8, and note the following immovable *sutures*:

coronal suture, between the parietal bones and the frontal bone;

lambdoidal suture, between the parietal bones and the occipital bone;
squamosal suture, between each parietal bone and each temporal bone;
sagittal suture, between the parietal bones (not shown).

The joints formed by each tooth in its tooth socket are also fibrous joints.

Cartilaginous Joints

Where bones are joined by hyaline cartilage or fibrocartilage, the joint that forms is usually slightly movable. The ribs are joined to the sternum by costal cartilages, which are hyaline cartilage (see Fig. 6.12). The *bodies* of adjacent vertebrae are separated by fibrocartilage intervertebral disks, which increase vertebral flexibility. The **pubic symphysis,** the joint between the two pubic bones (see Fig. 6.17), consists largely of fibrocartilage. Due to hormonal changes, this joint becomes more flexible during late pregnancy, allowing the pelvis to expand during childbirth.

Synovial Joints

Synovial joints are generally freely movable because, unlike the joints discussed so far, the two bones are separated by a *joint cavity* (Figs. 6.21 and 6.22). The joint cavity is lined by a **synovial membrane,** which produces **synovial fluid,** a lubricant for the joint. The absence of tissue between the articulating bones allows them to be freely movable but means that the joint has to be stabilized in some way.

The joint is stabilized by the joint capsule, a sleevelike extension of the periosteum of each articulating bone. **Ligaments,** which are composed of dense regular connective tissue, bind the

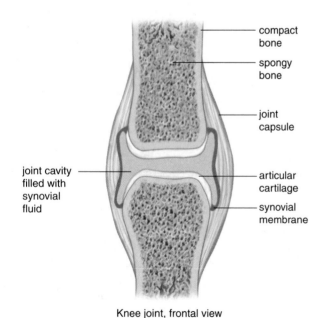

Knee joint, frontal view

Figure 6.21 Generalized anatomy of a synovial joint.

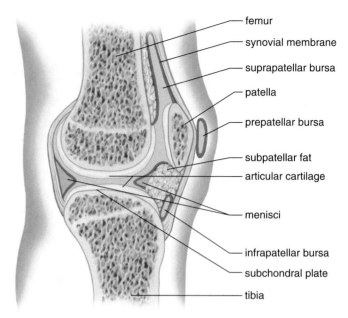

Knee joint, lateral view

Figure 6.22 The knee joint. Notice the menisci and bursae associated with the knee joint.

two bones to one another and add even more stability. Tendons, which are cords of dense regular connective tissue that connect muscle to bone, also help stabilize a synovial joint.

The articulating surfaces of the bones are protected in several ways. The bones are covered by a layer of articular (hyaline) cartilage. In addition, the joint, such as the knee, contains **menisci** (sing., *meniscus*), crescent-shaped pieces of cartilage, and fluid-filled sacs called **bursae,** which ease friction between all parts of the joint. Inflammation of the bursae is called **bursitis.** Tennis elbow is a form of bursitis. Articular cartilage can be worn away by years of constant use, such as when performing a sport. The What's New on p. 118 details a method for repairing this cartilage.

Types of Synovial Joints

Different types of freely movable joints are listed here and depicted in Figure 6.23.

Saddle joint. Each bone is saddle-shaped and fits into the complementary regions of the other. A variety of movements are possible. *Example*: the joint between the carpal and metacarpal bones of the thumb.
Ball-and-socket joint. The ball-shaped head of one bone fits into the cup-shaped socket of another. Movement in all planes, as well as rotation, are possible. *Examples*: the shoulder and hip joints.
Pivot joint. A small, cylindrical projection of one bone pivots within the ring formed of bone and ligament of another bone. Only rotation is possible. *Examples*: the joint between the proximal ends of the radius

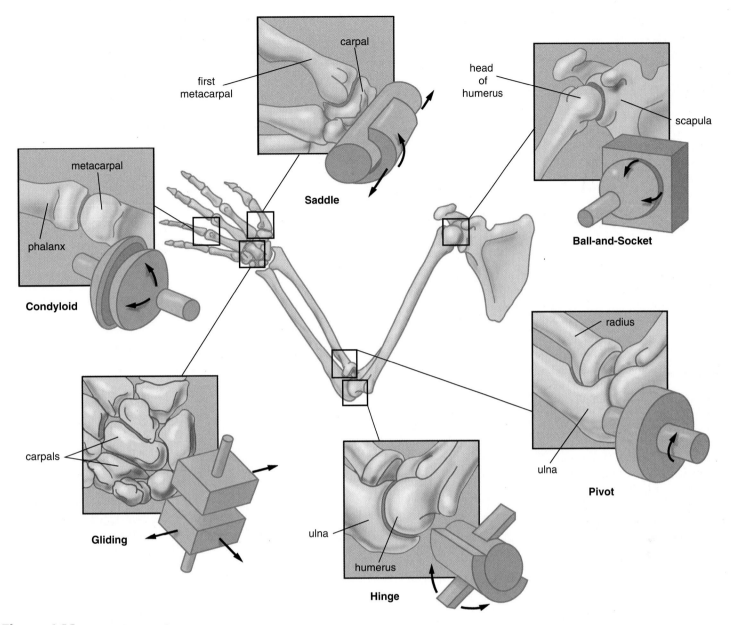

Figure 6.23 Types of synovial joints.

and ulna, and the joint between the atlas and axis (see Fig. 6.11).

Hinge joint. The convex surface of one bone articulates with the concave surface of another. Up-and-down motion in one plane is possible. *Examples*: the elbow and knee joints.

Gliding joint. Flat or slightly curved surfaces of bones articulate. Sliding or twisting in various planes is possible. *Examples*: the joints between the bones of the wrist and between the bones of the ankle.

Condyloid joint. The oval-shaped condyle of one bone fits into the elliptical cavity of another. Movement in different planes is possible, but rotation is not. *Examples*: the joints between the metacarpals and phalanges.

Movements Permitted by Synovial Joints

Skeletal muscles are attached to bones by tendons that cross joints. When a muscle contracts, one bone moves in relation to another bone. The more common types of movements are described here.

Angular Movements (Fig. 6.24*a*):

Flexion decreases the joint angle. Flexion of the elbow moves the forearm toward the arm; flexion of the knee moves the leg toward the thigh. *Dorsiflexion* is flexion of the foot upward, as when you stand on your heels; *plantar flexion* is flexion of the foot downward, as when you stand on your toes.

Extension increases the joint angle. Extension of the flexed elbow straightens the upper limb. *Hyperextension* occurs

What's New

Coaxing the Chondrocytes for Knee Repair

To the young, otherwise healthy, 30-something athlete on the physician's exam table, the diagnosis must seem completely unfair. Perhaps he's a former football player, or she's a trained dancer. Whatever the sport or activity, the patient is slender and fit, but knee pain and swelling are this athlete's constant companions. Examination of the knee shows the result of decades of use and abuse while performing a sport: The hyaline cartilage, also called articular cartilage, of the knee joint has degenerated. Hyaline cartilage (see page 95) is the "Teflon coating" for the bones of freely movable joints such as the knee. Hyaline cartilage allows easy, frictionless movement between the bones of the joint. Once repeated use has worn it away, hyaline cartilage does not grow back naturally. Exposed bone ends can grind against one another, resulting in pain, swelling, and restricted movements that can cripple the athlete. In severe cases, total knee replacement with a prosthetic joint is the athlete's only option (Fig. 6B).

Now the technique of tissue culture (growing cells outside of the patient's body in a special medium) can help young athletes with cartilage injuries regenerate their own hyaline cartilage. In an autologous chondrocyte implantation (ACI) surgery, a piece of healthy hyaline cartilage from the patient's knee is first removed surgically. This piece of cartilage, about the size of a pencil eraser, is typically taken from an undamaged area at the top edge of the knee. The chondrocytes, living cells of hyaline cartilage, are grown outside the body in tissue culture medium. Millions of the patient's own cells can be grown to create a "patch" of living cartilage. Growing these cells takes two to three weeks. Once the chondrocytes have grown, a pocket is created over the damaged area using the patient's own periosteum, the connective tissue that surrounds the bone (see page 95). The periosteum pocket will hold the hyaline cartilage cells in place. The cells are injected into the pocket and left to grow.

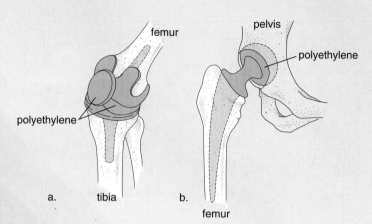

Figure 6B Artificial joints in which polyethylene replaces articular cartilage. **a.** Knee. **b.** Hip.

As with all injuries to the knee, once the cartilage cells are firmly established, the patient still faces a lengthy rehabilitation. The patient must use crutches or a cane for three to four months to protect the joint. Physical therapy will stimulate cartilage growth without overstressing the area being repaired. In six months, the athlete can return to light-impact training and jogging. Full workouts can be resumed in about one year after surgery. However, most patients regain full mobility and a pain-free life after ACI surgery and do not have to undergo total knee replacement.

ACI surgery can't be used for the elderly or for overweight patients with osteoarthritis. Muscle or bone defects in the knee joint must be corrected before the surgery can be attempted. As with all surgeries, there is a risk for postoperative complications, such as bleeding or infection. However, ACI may offer young athletes the chance to restore essential hyaline cartilage and regain a healthy, functional knee joint.

---

when a portion of the body part is extended beyond 180°. It is possible to hyperextend the head and the trunk of the body, and also the shoulder and wrist (arm and hand).

Adduction is the movement of a body part toward the midline. For example, adduction of the arms or legs moves them back to the sides, toward the body.

Abduction is the movement of a body part laterally, away from the midline. Abduction of the arms or legs moves them laterally, away from the body.

Circular Movements (Fig. 6.24*b*):

Circumduction is the movement of a body part in a wide circle, as when a person makes arm circles. Careful observation of the motion reveals that, because the

proximal end of the arm is stationary, the shape outlined by the arm is actually a cone.

Rotation is the movement of a body part around its own axis, as when the head is turned to answer "no" or when the arm is twisted toward the trunk (medial rotation) and away from the trunk (lateral rotation).

Supination is the rotation of the forearm so that the palm is upward; **pronation** is the opposite—the movement of the forearm so that the palm is downward.

Special movements (Fig. 6.24*c*):

Inversion and **eversion** apply only to the feet. Inversion is turning the foot so that the sole faces inward, and eversion is turning the foot so that the sole faces outward.

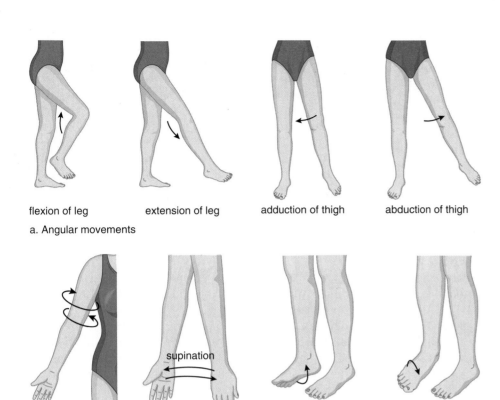

flexion of leg   extension of leg   adduction of thigh   abduction of thigh

a. Angular movements

rotation of arm   supination / pronation of hand   inversion   eversion

b. Circular movements   c. Special movements

Figure 6.24 Joint movements. **a.** Angular movements increase or decrease the angle between the bones of a joint. **b.** Circular movements describe a circle or part of a circle. **c.** Special movements are unique to certain joints.

Elevation and **depression** refer to the lifting up and down, respectively, of a body part, as when you shrug your shoulders or move your jaw up and down.

Content **CHECK-UP!**

10. A joint that is slightly movable is called:

    a. a synarthrosis.   c. a diarthrosis.

    b. an amphiarthrosis.

11. The joint found at the elbow is what type of synovial joint?

    a. hinge   c. saddle

    b. ball-and-socket   d. condyloid

12. The joint found between the atlas and the axis is what type of synovial joint?

    a. hinge   c. ball-and-socket

    b. pivot   d. condyloid

6.5   Effects of Aging

Both cartilage and bone tend to deteriorate as a person ages. The chemical nature of cartilage changes, and the bluish color typical of young cartilage changes to an opaque, yellowish color. The chondrocytes die, and reabsorption occurs as the cartilage undergoes calcification, becoming hard and brittle. Calcification interferes with the ready diffusion of nutrients and waste products through the matrix. The articular cartilage may no longer function properly, and symptoms of arthritis can appear. There are three common types of arthritis: (1) **Osteoarthritis** is accompanied by deterioration of the articular cartilage. (2) In **rheumatoid arthritis**, the synovial membrane becomes inflamed and grows thicker cartilage. This is due to an autoimmune reaction, in which the body's immune system mistakenly attacks the synovial membrane. Although rheumatoid arthritis is more common with age, it can affect children and young adults as well. (3) **Gout,** or gouty arthritis, is caused by an excessive buildup of uric acid (a metabolic waste) in the blood. Rather than being excreted in the urine, the acid is deposited as crystals in the joints, where it causes inflammation and pain.

Osteoporosis, discussed in the Medical Focus on page 99, is present when weak and thin bones cause aches and pains. Such bones tend to fracture easily.

6.6   Homeostasis

The illustration in Human Systems Work Together on page 122 tells how the skeletal system assists other systems and how other systems assist the skeletal system. Let's review again the functions of the skeletal system, but this time as they relate to the other systems of the body.

Functions of the Skeletal System

The bones protect the internal organs. The rib cage protects the heart and lungs; the skull protects the brain; and the vertebrae protect the spinal cord. The endocrine organs, such as the pituitary gland, pineal gland, thymus, and thyroid gland, are also protected by bone. The nervous system and the endocrine system work together to control the other organs and, ultimately, homeostasis.

The bones assist all phases of respiration (Fig. 6.25). The rib cage assists the breathing process, enabling oxygen to enter the blood, where it is transported by red blood cells to the tissues. Red bone marrow produces the blood cells, including the red blood cells that transport oxygen. Without a supply of oxygen, the cells of the body could not efficiently produce ATP. ATP is needed for muscle contraction and for nerve conduction as well as for the many synthesis reactions that occur in cells.

The bones store and release calcium. The storage of calcium in the bones is under hormonal control. A dynamic equilibrium is maintained between the concentrations of calcium in the bones and in the blood. Calcium ions play a major role in muscle contraction and nerve conduction. Calcium ions also help regulate cellular metabolism. Protein hormones, which cannot enter cells, are called the first messenger, and a second messenger such as calcium ions jump-starts cellular metabolism, directing it to proceed in a particular way.

The bones assist the lymphatic system and immunity. Red bone marrow produces not only the red blood cells but also the white blood cells. The white cells, which congregate in the lymphatic organs, are involved in defending the body against pathogens and cancerous cells. Without the ability to withstand foreign invasion, the body may quickly succumb to disease and die.

The bones assist digestion. The jaws contain sockets for the teeth, which chew food, and a place of attachment for the muscles that move the jaws. Chewing breaks food into pieces small enough to be swallowed and chemically digested. Without digestion, nutrients would not enter the body to serve as building blocks for repair and a source of energy for the production of ATP.

The skeleton is necessary to locomotion. Locomotion is efficient in human beings because they have a jointed skeleton for the attachment of muscles that move the bones. Our jointed skeleton allows us to seek out and move to a more suitable external environment in order to maintain the internal environment within reasonable limits.

Functions of Other Systems

How do the other systems of the body help the skeletal system carry out its functions?

The integumentary system and the muscles help the skeletal system protect internal organs. For example, anteriorly, the abdominal organs are only protected by muscle and skin.

The digestive system absorbs the calcium from food so that it enters the body. The plasma portion of blood transports calcium from the digestive system to the bones and any other organs that need it. The endocrine system regulates the storage of calcium in the bones.

The thyroid gland, a lymphatic organ, is instrumental in growth of bone and other tissues, and in the maturity of certain white blood cells produced by the red bone marrow. The cardiovascular system transports the red blood cells as they deliver oxygen to the tissues and as they return to the lungs where they pick up oxygen.

Movement of the bones would be impossible without contraction of the muscles. In these and other ways, the systems of the body help the skeletal system carry out its functions.

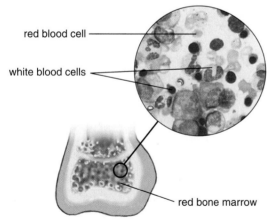

a. Production of blood cells

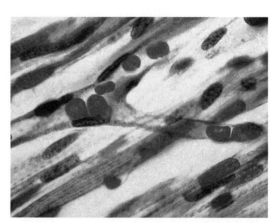

b. Red blood cells in capillaries

Figure 6.25 The skeletal system and cardiovascular system work together. **a.** Red bone marrow produces red and white blood cells. **b.** As the red blood cells pass through the capillaries, they deliver oxygen to the body's cells. Some white blood cells exit blood vessels and enter the tissues at capillaries where they phagocytize pathogens. Others stay in the blood (and lymph) where they produce antibodies against invaders.

Skeletal Remains

"JOHN/JANE DOE, SKELETAL REMAINS, AGE UNKNOWN" is the initial identification given by law enforcement officials to the bones of an unidentified human being. The bones may have been found in the woods by a hiker or a hunter, or in a field after a farmer harvests his crops. Bones may be uncovered when a building is demolished, or if natural events such as floods or earthquakes disrupt the soil. Regardless of how human bones are found, questions must be answered. Who was this person? Was this a male or a female, and how old? What was the person's ethnicity? How did the person die, and how long ago? Was this person murdered, or did death come from natural causes?

It is the job of a *forensic anthropologist* to collect, analyze, and ultimately identify the remains. Forensic anthropologists typically have extensive training in the structure of the human skeleton and are able to examine the features of the recovered bones. These scientists rely on a national forensic analysis data bank that contains measurements and observations from thousands of skeletons. In addition, forensic anthropologists are routinely called upon to testify in criminal cases as to a victim's time and cause of death.

Clues about the identity and history of a deceased person can be found throughout the skeleton. Age is approximated by *dentition*—the structure of the teeth in the upper jaw (maxilla) and lower jaw (mandible). For example, infants aged 0–4 months have no teeth present; children aged approximately 6 through 10 have missing deciduous, or "baby teeth"; young adults acquire their last molars, or "wisdom teeth," around age 20. The age of older adults can be approximated by the number and location of missing or broken teeth.

In addition, ossification of bones—that is, replacement of a baby/child's incomplete cartilage skeleton with bone—continues in an orderly fashion until about the age of 20. Studying areas of bone ossification also gives clues to the age of the deceased at the time of death. In older adults, signs of joint breakdown provide additional information about age. Hyaline cartilage becomes worn, yellowed, and brittle with age, and the hyaline cartilages covering bone ends wear down over time. The amount of yellowed, brittle, or missing cartilage helps scientists to estimate the person's age.

If skeletal remains include the individual's pelvic bones, these provide the best method for determining an adult's gender (see pages 112–113). The long bones, particularly the humerus and femur, give information about gender as well. Long bones are thicker and denser in males, and points of muscle attachment are bigger and more prominent. The skull of a male has a

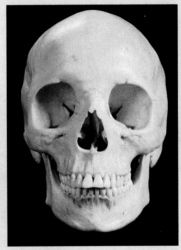
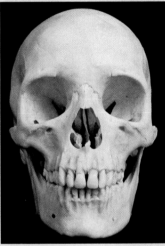

a. Female adult skull   b. Male adult skull

Figure 6C Gender differences of the skull. **a.** Note that the female skull is smaller, more delicate, and has a pointed chin. **b.** The male skull is large, bulky, and has a squared-off chin.

square chin and more prominent ridges above the eye sockets or orbits.

Determining ethnic origin of skeletal remains can be difficult because so many people have a mixed racial heritage. Forensic anatomists rely on observed racial characteristics of the skull. In general, individuals of African or African American descent have a greater distance between the eyes, eye sockets that are roughly rectangular, and a jaw that is large and prominent. Skulls of Native Americans typically have round eye sockets, prominent cheek (zygomatic) bones, and a rounded palate. Caucasian skulls usually have a U-shaped palate and a visible suture line between the frontal bones. Additionally, the external ear canals in Caucasians are long and straight, so that the auditory ossicles (tiny bones built inside the temporal bone and used for hearing) are visible.

Once the identity of the individual has been determined, the skeletal remains can be returned to the victim's family for proper burial. Although this can be a sorrowful event, return of physical remains provides closure and solace to many families. For this reason, special teams of forensic anthropologists employed by the U.S. military are currently researching the identities of bones from soldiers who fought in World War II, as well as the Korean, Vietnam, and Gulf wars. Ancestral remains from Native Americans are protected by the Native American Grave Protection and Repatriation Act, and must be returned to the leadership of the tribe.

Human Systems Work Together

Integumentary System

Bones provide support for skin.

Skin protects bones; helps provide vitamin D for Ca^{2+} absorption.

Muscular System

Bones provide attachment sites for muscles; store Ca^{2+} for muscle function.

Muscular contraction causes bones to move joints; muscles help protect bones.

Nervous System

Bones protect sense organs, brain, and spinal cord; store Ca^{2+} for nerve function.

Receptors send sensory input from bones to joints.

Endocrine System

Bones provide protection for glands; store Ca^{2+} used as second messenger.

Growth hormone regulates bone development; parathyroid hormone and calcitonin regulate Ca^{2+} content.

Cardiovascular System

Rib cage protects heart; red bone marrow produces blood cells; bones store Ca^{2+} for blood clotting.

Blood vessels deliver nutrients and oxygen to bones, carry away wastes.

How the Skeletal System works with other body systems

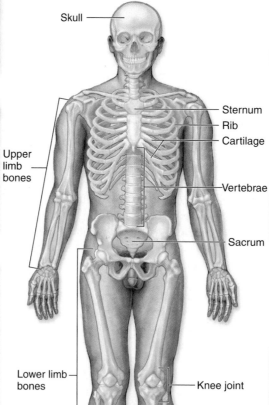

Skull

Upper limb bones

Sternum

Rib

Cartilage

Vertebrae

Sacrum

Lower limb bones

Knee joint

Lymphatic System/Immunity

Red bone marrow produces white blood cells involved in immunity.

Lymphatic vessels pick up excess tissue fluid; immune system protects against infections.

Respiratory System

Rib cage protects lungs and assists breathing; bones provide attachment sites for muscles involved in breathing.

Gas exchange in lungs provides oxygen and rids body of carbon dioxide.

Digestive System

Jaws contain teeth that chew food; hyoid bone assists swallowing.

Digestive tract provides Ca^{2+} and other nutrients for bone growth and repair.

Urinary System

Bones provide support and protection.

Kidneys provide active vitamin D for Ca^{2+} absorption and help maintain blood level of Ca^{2+} needed for bone growth and repair.

Reproductive System

Bones provide support and protection of reproductive organs.

Sex hormones influence bone growth and density in males and females.

SELECTED NEW TERMS

Basic Key Terms

abduction (ab-duk'shun), p. 118

adduction (uh-duk'shun), p. 118

amphiarthrosis (am-fee-ar-thrō-sis), p. 115

appendicular skeleton (ap"en-dik'yū-ler skel'ĕ-ton), p. 108

appositional growth (ăp'ə-zĭsh'ən-əl grŏth), p. 97

articular cartilage (ar-tik'yū-ler kar'tĭ-lij), p. 95

articulation (ar-tik"yū-la'shun), p. 95

atlas (ăt-ləs), p. 106

axial skeleton (ak'se-al skel'ĕ-ton), p. 100

axis (āk'sĭs), p. 106

ball-and-socket joint (bôl and sŏk'ĭt joint), p. 116

bursa (bur'suh), p. 116

calcaneus (kăl-kā'nē-əas), p. 114

cartilaginious joints (kär'tl-ăj'ə-nas jointz), p. 115

circumduction (ser"kum-duk'shun), p. 118

clavicles (klăv'ĭ-kəls), p. 109

coccyx (kŏk'sĭks), p. 107

compact bone (kom'pakt bōn), p. 95

condyloid joint (kŏn'dĭl-oid joint), p. 117

coxal bone (kŏk-săl bōn), p. 112

depression (dĭ-prĕsh'ən), p. 119

diaphysis (di-af'ĭ-sis), p. 95

diarthrosis (dī-ar-thrō'-sis), p. 115

elevation (ĕl'ə-vā'shən), p. 119

endochondral ossification (ĕn'dō-kŏn'drəl ŏs'ə-fĭ-kā'shən), p. 97

epiphyseal plate (ep"ĭ-fĭz'e-al plāt), p. 97

epiphysis (ĕ-pif'ĭ-sis), p. 95

eversion (e-ver'zhun), p. 118

extension (ek-sten'shun), p. 117

femur (fē'mər), p. 113

fibrous joints (fī'brəs joint), p. 115

fibula (fĭb'yə-lə), p. 114

flexion (flek'shun), p. 117

fontanel (fon"tuh-nel'), p. 101

gliding joint (glīd'-ing joint), p. 117

hematopoiesis (hem"ah-to-poi-e'sis), p. 95

hinge joint (hĭnj joint), p. 117

humerus (hyōō'mər-ās), p. 110

ilium (ĭl'ē-əm), p. 112

intervertebral disk (in"ter-ver'tĕ-bral disk), p. 105

intramembranous ossification (ĭn'trə-mĕm'brān-əs ŏs'ə-fĭ-kā'shən), p. 97

inversion (in-ver'zhun), p. 118

ischium (ĭs'kē-əm), p. 112

ligament (lig'uh-ment), p. 116

medullary cavity (med'u-lār"e kav'ĭ-te), p. 95

meniscus (mĕ-nis'kus), p. 116

metatarsal bones (mĕt'ə-tär'səl bōn), p. 114

ossification (os'-ĭ-fĭ-ka'shun), p. 97

osteoblast (os'te-o-blast"), p. 97

osteoclast (os'te-o-klast"), p. 97

osteocyte (os'te-o-sīt), p. 97

osteoprogenitor cells (ŏs'tē-ə-prō-jĕn'ĭ-tər sĕlz), p. 97

patella (pə-tĕl'ə), p. 113

pectoral girdle (pek'tor-al ger'dl), p. 108

pelvic girdle (pel'vik ger'dl), p. 112

periosteum (per"e-os'te-um), p. 95

phalanges (fə-lăn'jēz), p. 115

pivot joint (pĭv'ət joint), p. 116

pronation (pro-na'shun), p. 118

pubic symphysis (pyōō'bĭk sĭm'fĭ-sĭs), p. 116

pubis (pyōō'bĭs), p. 112

radius (rā'dē-əs), p. 110

red bone marrow (red bōn mār'o), p. 95

rib cage (rĭb kāj), p. 107

rotation (ro-ta'shun), p. 118

sacrum (sā'krəm), p. 107

saddle joint (săd'l joint), p. 116

scapulae (skăp'yə-lə), p. 109

sinus (si'nus), p. 100

sinusitis (sī'nə-sī'tĭs), p. 100

spongy bone (spunj'e bōn), p. 95

sternum (stûr'nəm), p. 108

supination (su"pĭ-na'shun), p. 118

suture (su'cher), p. 101

synarthrosis (sĭn-ar-thrō'-sis), p. 115

synovial fluid (si-no've-al flu'id), p. 116

synovial joint (si-no've-al joint), p. 115

talus (tā'ləs), p. 114

tarsal bones (tär'səl bōns), p. 114

tibia (tĭb'ē-ə), p. 114

ulna (ŭl'nə), p. 110

vertebrae (vûr'tə-brē), p. 105

vertebral column (ver'tĕ-bral kah'lum), p. 105

Clinical Key Terms

bursitis (ber-si'tis), p. 116

fracture (frak'cher), p. 98

gout (gowt), p. 119

herniated disk (her'ne-a-ted disk), p. 106

kyphosis (ki-fo'sis), p. 106

lordosis (lor-do'sis), p. 106

mastoiditis (mas"toi-di'tis), p. 100

osteoarthritis (os"te-o-ar-thri'tis), p. 119

osteoporosis (os"te-o-po-ro'sis), p. 119

reduction (rĭ-duk'shən), p. 98

rheumatoid arthritis (ru'muh-toid ar-thri'tis), p. 119

scoliosis (sko"le-o'sis), p. 106

SUMMARY

6.1 Skeleton: Overview

A. The skeleton supports and protects the body; produces red blood cells; serves as a storehouse for inorganic calcium and phosphate ions and fat; and permits flexible movement.

B. A long bone has a shaft (diaphysis) and two ends (epiphyses), which are covered by articular cartilage. The diaphysis contains a medullary cavity with yellow marrow and is bounded by compact bone. The epiphyses contain spongy bone with red bone marrow that produces red blood cells.

C. Bone is a living tissue. It develops, grows, remodels, and repairs itself. In all these processes, osteoclasts break down bone, and osteoblasts build bone.

D. Fractures are of various types, but repair requires four steps: (1) hematoma, (2) fibrocartilaginous callus, (3) bony callus, and (4) remodeling.

6.2 Axial Skeleton

The axial skeleton lies in the midline of the body and consists of the skull, the hyoid bone, the vertebral column, and the thoracic cage.

A. The skull is formed by the cranium and the facial bones. The cranium includes the frontal bone, two parietal bones, one occipital bone, two temporal bones, one sphenoid bone, and one ethmoid bone. The facial bones include two maxillae, two palatine bones, two zygomatic bones, two lacrimal bones, two nasal bones, the vomer bone, two inferior nasal conchae, and the mandible.

B. The U-shaped hyoid bone is located in the neck. It anchors the tongue and does not articulate with any other bone.

C. The typical vertebra has a body, a vertebral arch surrounding the vertebral foramen, and a spinous process. The first two vertebrae are the atlas and axis. The vertebral column has four curvatures and contains the cervical, thoracic, lumbar, sacral, and coccygeal vertebrae. Cervical, thoracic, and lumbar vertebrae are separated by intervertebral disks.

D. The rib cage contains the thoracic vertebrae, ribs and associated cartilages, and the sternum.

6.3 **Appendicular Skeleton**
The appendicular skeleton consists of the bones of the pectoral girdle, upper limbs, pelvic girdle, and lower limbs.

A. The pectoral (shoulder) girdle contains two clavicles and two scapulae.

B. The upper limb contains the humerus, the radius, the ulna, and the bones of the hand (the carpals, metacarpals, and phalanges).

C. The pelvic girdle contains two coxal bones, as well as the sacrum and coccyx. The female pelvis is generally wider and more shallow than the male pelvis.

D. The lower limb contains the femur, the patella, the tibia, the fibula, and the bones of the foot (the tarsals, metatarsals, and phalanges).

6.4 **Joints (Articulations)**

A. Joints are regions of articulation between bones. They are classified according to their structure and/or degree of movement. Some joints are immovable, some are slightly movable, and some are freely movable (synovial). The different kinds of synovial joints are ball-and-socket, hinge, condyloid, pivot, gliding, and saddle.

B. Movements at joints are broadly classified as angular (flexion, extension, adduction, abduction); circular (circumduction, rotation, supination, and pronation); and special (inversion, eversion, elevation, and depression).

6.5 **Effects of Aging**
Two fairly common effects of aging on the skeletal system are arthritis and osteoporosis.

6.6 **Homeostasis**

A. The bones protect the internal organs: The rib cage protects the heart and lungs; the skull protects the brain; and the vertebrae protect the spinal cord.

B. The bones assist all phases of respiration. The rib cage assists the breathing process, and red bone marrow produces the red blood cells that transport oxygen.

C. The bones store and release calcium. Calcium ions play a major role in muscle contraction and nerve conduction. Calcium ions also help regulate cellular metabolism.

D. The bones assist the lymphatic system and immunity. Red bone marrow produces not only the red blood cells but also the white blood cells.

E. The bones assist digestion. The jaws contain sockets for the teeth, which chew food, and a place of attachment for the muscles that move the jaws.

F. The skeleton is necessary for locomotion. Locomotion is efficient in human beings because they have a jointed skeleton for the attachment of muscles that move the bones.

STUDY QUESTIONS

1. What are five functions of the skeleton? (p. 95)
2. What are five major categories of bones based on their shapes? (p. 95)
3. What are the parts of a long bone? What are some differences between compact bone and spongy bone? (p. 95)
4. How does bone grow in children, and how is it remodeled in all age groups? (pp. 97–98)
5. What are the various types of fractures? What four steps are required for fracture repair? (p. 98)
6. List the bones of the axial and appendicular skeletons. (Fig. 6.5, p. 101)
7. What are the bones of the cranium and the face? What are the special features of the temporal bones, sphenoid bone, and ethmoid bone? (pp. 100–105)
8. What are the parts of the vertebral column, and what are its curvatures? Distinguish between the atlas, axis, sacrum, and coccyx. (pp. 105–107)
9. What are the bones of the rib cage, and what are several of its functions? (p. 107)
10. What are the bones of the pectoral girdle? Give examples to demonstrate the flexibility of the pectoral girdle. What are the special features of a scapula? (p. 108–109)
11. What are the bones of the upper limb? What are the special features of these bones? (pp. 109–111)
12. What are the bones of the pelvic girdle, and what are their functions? (p. 112)
13. What are the false and true pelvises, and what are several differences between the male and female pelvises? (pp. 112–113)
14. What are the bones of the lower limb? Describe the special features of these bones. (pp. 113–115)
15. How are joints classified? Give examples of each type of joint. (p. 115)
16. How can joint movements permitted by synovial joints be categorized? Give an example of each category. (pp. 117–119)
17. How does aging affect the skeletal system? (p. 119)
18. What functions of the skeletal system are particularly helpful in maintaining homeostasis? (p. 120)

LEARNING OBJECTIVE QUESTIONS

I. Match the items in the key to the bones listed in questions 1–6.

Key:
- a. forehead
- b. chin
- c. cheekbone
- d. elbow
- e. shoulder blade
- f. hip
- g. ankle

1. temporal and zygomatic bones
2. tibia and fibula
3. frontal bone
4. ulna
5. coxal bone
6. scapula

II. Match the items in the key to the bones listed in questions 7–13.

Key:
- a. external auditory meatus
- b. cribriform plate
- c. xiphoid process
- d. glenoid cavity
- e. olecranon process
- f. acetabulum
- g. greater and lesser trochanters

7. scapula
8. sternum
9. femur
10. temporal bone
11. coxal bone
12. ethmoid bone
13. ulna

III. Fill in the blanks.

14. Long bones are _____ than they are wide.
15. The epiphysis of a long bone contains _____ bone, where red blood cells are produced.
16. The _____ are the air-filled spaces in the cranium.
17. The sacrum is a part of the _____, and the sternum is a part of the _____.
18. The pectoral girdle is specialized for _____, while the pelvic girdle is specialized for _____.
19. The term *phalanges* is used for the bones of both the _____ and the _____.
20. The knee is a freely movable (synovial) joint of the _____ type.

MEDICAL TERMINOLOGY EXERCISE

Consult Appendix B for help in pronouncing and analyzing the meaning of the terms that follow.

1. chondromalacia (kon″dro-muh-la′she-uh)
2. osteomyelitis (os″te-o-mi″e-li′tis)
3. craniosynostosis (kra″ne-o-sin″os-to′sis)
4. myelography (mi″ĕ-log′ruh-fe)
5. acrocyanosis (ak″ro-si″uh-no′sis)
6. syndactylism (sin-dak′tĭ-lizm)
7. orthopedist (or″tho-pe′dist)
8. prognathism (prog′nah-thizm)
9. micropodia (mi″kro-po′de-uh)
10. arthroscopic (ar″thro-skop′ik)
11. bursectomy (ber-sek′to-me)
12. synovitis (sin-o-vi′tis)
13. acephaly (a-sef′uh-le)
14. sphenoidostomy (sfe-noy-dos′to-me)
15. acetabuloplasty (as-ĕ-tab′yū-lo-plas-te)

WEBSITE LINK

Visit this text's website at http://www.mhhe.com/maderap6 for additional quizzes, interactive learning exercises, and other study tools.

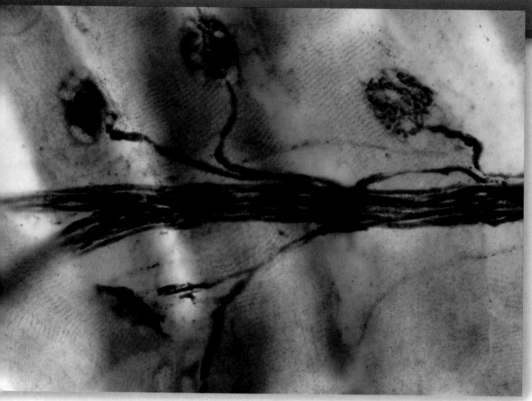

Photomicrograph of skeletal muscle motor neurons at their junction, or synapse, with skeletal muscle fibers. A muscle fiber receives the stimulus to contract at this neuromuscular junction.

Chapter Outline & Learning Objectives

After you have studied this chapter, you should be able to:

7.1 Functions and Types of Muscles

All muscles, regardless of the particular type, can contract—that is, shorten. When muscles contract, some part of the body or the entire body moves. Humans have three types of muscles: smooth, cardiac, and skeletal (Fig. 7.1). The contractile cells of these tissues are elongated and therefore are called **muscle fibers.**

Smooth Muscle

Smooth muscle is located in the walls of hollow internal organs and blood vessels. Its involuntary contraction moves materials through organs, and regulates blood flow in blood vessels. Smooth muscle fibers are narrow, tapered rod-shaped cells, and each cell is uninucleate (has a single nucleus). The cells are usually arranged in parallel lines, forming sheets. Smooth muscle does not have the striations (bands of light and dark) seen in cardiac and skeletal muscle. Although smooth muscle is slower to contract than skeletal muscle, it can sustain prolonged contractions and does not fatigue easily.

Cardiac Muscle

Cardiac muscle forms the heart wall. Its fibers are uninucleated, striated, tubular, and branched, which allows the fibers to interlock at *intercalated disks.* Intercalated disks contain gap junctions, which permit contractions to spread quickly throughout the heart. Cardiac fibers relax completely between contractions, which prevents fatigue. Contraction of cardiac muscle fibers is rhythmical and occurs without requiring outside nervous stimulation. Thus, cardiac muscle contraction is involuntary. However, keep in mind the nerves that supply the heart do affect heart rate and strength of contraction.

Skeletal Muscle

Skeletal muscle fibers are tubular, multinucleated, and striated. They make up the skeletal muscles attached to the skeleton. Skeletal muscle fibers can run the length of a muscle and therefore can be quite long. Skeletal muscle is voluntary because its contraction is always stimulated and controlled by the nervous system. In this chapter, we will explore why skeletal muscle (and cardiac muscle) is striated.

Connective Tissue Coverings

Muscles are organs, and as such they contain other types of tissues, such as nervous tissue, blood vessels, and connective tissue. Connective tissue is essential to the organization of the fibers within a muscle (Fig. 7.2). First, each fiber is surrounded by a thin layer of areolar (loose) connective tissue called the *endomysium.* Blood capillaries and nerve fibers reach each muscle fiber by way of the endomysium. Second, the muscle fibers are grouped into bundles called *fascicles.* The fascicles have a sheath of connective tissue called the

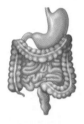
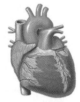

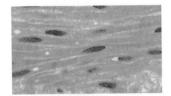

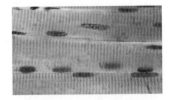

Smooth muscle
- has narrow, tapered rod-shaped cells.
- has nonstriated, uninucleated fibers.
- occurs in walls of internal organs and blood vessels.
- is involuntary.

Cardiac muscle
- has striated, tubular, branched, uninucleated fibers.
- occurs in walls of heart.
- is involuntary.

Skeletal muscle
- has striated, tubular, multinucleated fibers.
- is usually attached to skeleton
- is voluntary.

Figure 7.1 Types of muscles. The three types of muscles in the body have the appearance and characteristics shown here.

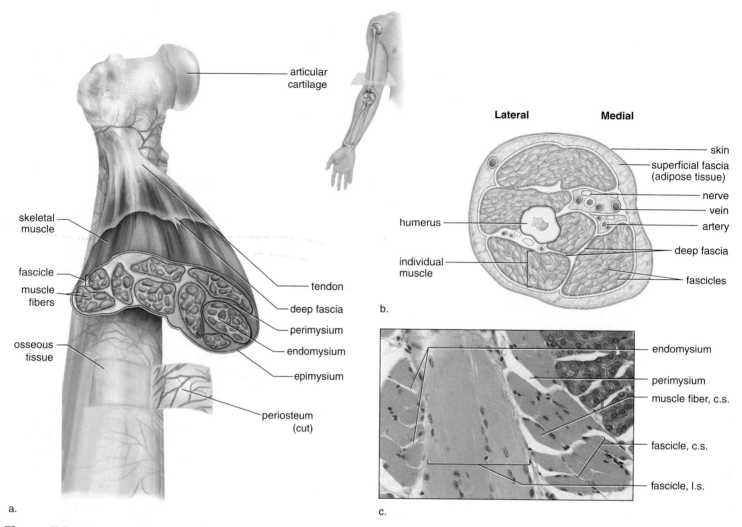

Figure 7.2 Connective tissue of a skeletal muscle. **a.** Trace the connective tissue of a muscle from the endomysium to the perimysium to the epimysium, which becomes a part of the deep fascia and from which the tendon extends to attach a muscle to the periosteum of a bone. **b.** Cross section of the arm showing the arrangement of the muscles, which are separated from the skin by fascia. The superficial fascia contains adipose tissue. **c.** Photomicrograph of muscle fascicles from the tongue where the fascicles run in different directions. (c.s. = cross section; l.s. = longitudinal section.)

perimysium. Finally, the muscle itself is covered by a connective tissue layer called the *epimysium.* The epimysium becomes a part of the *fascia,* a layer of fibrous tissue that separates muscles from each other (deep fascia) and from the skin (superficial fascia). Collagen fibers of the epimysium continue as a strong, fibrous **tendon** that attaches the muscle to a bone. The epimysium merges with the periosteum of the bone.

Functions of Skeletal Muscles

This chapter concerns the skeletal muscles, and therefore it is fitting to consider their functions independent of the other types of muscles:

Skeletal muscles support the body. Skeletal muscle contraction opposes the force of gravity and allows us to remain upright. Some skeletal muscles are serving this purpose even when you think you are relaxed.

Skeletal muscles make bones and other body parts move. Muscle contraction accounts not only for the movement of limbs but also for eye movements, facial expressions, and breathing.

Skeletal muscles help maintain a constant body temperature. Skeletal muscle contraction causes ATP to break down, releasing heat that is distributed about the body.

Skeletal muscle contraction assists movement in cardiovascular and lymphatic vessels. The pressure of skeletal muscle contraction keeps blood moving in cardiovascular veins and lymph moving in lymphatic vessels.

Skeletal muscles help protect bones and internal organs and stabilize joints. Muscles pad the bones, and the muscular wall in the abdominal region protects the internal organs. Muscle tendons help hold bones together at the joints.

1. Which of the following is not a characteristic of skeletal muscle?
   a. uninucleate cells
   c. under voluntary control
   b. can be very long
   d. striated
2. Which of the following is not a characteristic of smooth muscle?
   a. uninucleate cells
   c. no striations
   b. cells arranged in sheets
   d. fatigues easily
3. Select the connective tissue covering of a muscle that is most superficial.
   a. perimysium
   c. epimysium
   b. endomysium

7.2 Microscopic Anatomy and Contraction of Skeletal Muscle

We have already examined the structure of skeletal muscle as seen with the light microscope. As you know, skeletal muscle tissue has alternating light and dark bands, giving it a striated appearance. The electron microscope shows that these bands are due to the arrangement of protein filaments, called *myofilaments,* in a muscle fiber.

Muscle Fiber

A muscle fiber contains the usual cellular components, but special names have been assigned to some of these components (note that the terms used to describe muscle start with the prefixes *myo-* and *sarco-;* see Table 7.1 and Fig. 7.3).

TABLE 7.1 Microscopic Anatomy of a Muscle

Name	Function
Sarcolemma	Plasma membrane of a muscle fiber that forms T tubules
Sarcoplasm	Cytoplasm of a muscle fiber that contains organelles, including myofibrils
Glycogen	A polysaccharide that stores energy for muscle contraction
Myoglobin	A red pigment that stores oxygen for muscle contraction
T tubule	Extension of the sarcolemma that extends into the muscle fiber and conveys nerve signals that cause Ca^{2+} to be released from the sarcoplasmic reticulum into the sarcoplasm.
Sarcoplasmic reticulum	The smooth ER of a muscle fiber that stores Ca^{2+}
Myofibril	A bundle of myofilaments that contracts
Myofilament	Thick and thin filaments whose structure and functions account for muscle striations and contractions

The plasma membrane is called the *sarcolemma;* the cytoplasm is the *sarcoplasm;* and the endoplasmic reticulum is the *sarcoplasmic reticulum.* A muscle fiber also has some unique anatomical characteristics. One feature is its T (for transverse) system; the sarcolemma forms **T (transverse) tubules** that penetrate, or dip down, into the cell so that they come into contact—but do not fuse—with expanded portions of the sarcoplasmic reticulum. The expanded portions of the sarcoplasmic reticulum are calcium storage sites. Calcium ions (Ca^{2+}), as we shall see, are essential for muscle contraction.

The sarcoplasmic reticulum encases hundreds and sometimes even thousands of **myofibrils,** which are bundles of myofilaments. Each myofibril is about 1 micrometer in diameter. Myofibrils are the contractile portions of the muscle fibers. Any other organelles, such as mitochondria, are located in the sarcoplasm between the myofibrils. The sarcoplasm also contains glycogen, which provides stored energy for muscle contraction, and the red pigment **myoglobin,** which binds oxygen until it is needed for muscle contraction.

Myofibrils and Sarcomeres

Myofibrils are cylindrical in shape and run the length of the muscle fiber. Each myofibril is composed of numerous **sarcomeres** (Fig. 7.3). As an analogy, imagine each myofibril as a single freight train. Each "boxcar" of the "train" is a sarcomere; thus, the myofibril "train" is composed of linked sarcomeres. Each sarcomere extends between two dark, vertical lines called **Z lines.** The horizontal stripes, or *striations,* of skeletal muscle fibers are formed by the placement of myofilaments within the sarcomeres. A sarcomere contains two types of protein myofilaments. The **thick filaments** are made up of a single protein called **myosin.** Thin filaments are made up of three proteins: a globular protein called **actin,** plus **tropomyosin** and **troponin.** In each sarcomere, the **I band** is light-colored because it contains only thin filaments attached to a Z line. Note that I bands overlap adjacent sarcomeres. The dark regions of the **A band,** found in the center of the sarcomere, contain overlapping thick and thin filaments. In the center of the A band is the lighter **H zone,** which contains only myosin filaments (Fig. 7.3*b* and *c*).

Myofilaments

The thick and thin filaments differ in the following ways:

Thick Filaments A thick filament is composed of several hundred molecules of the protein myosin. Each myosin molecule is composed of two protein strands, each shaped like a golf club. The straight portions of each strand coil around each other. Each myosin molecule ends in a double globular head, or *cross-bridge.* Cross-bridges are slanted away from the middle of a sarcomere, toward the thin filaments surrounding them. One portion of the cross-bridge will bind to actin in thin filaments during muscle contraction.

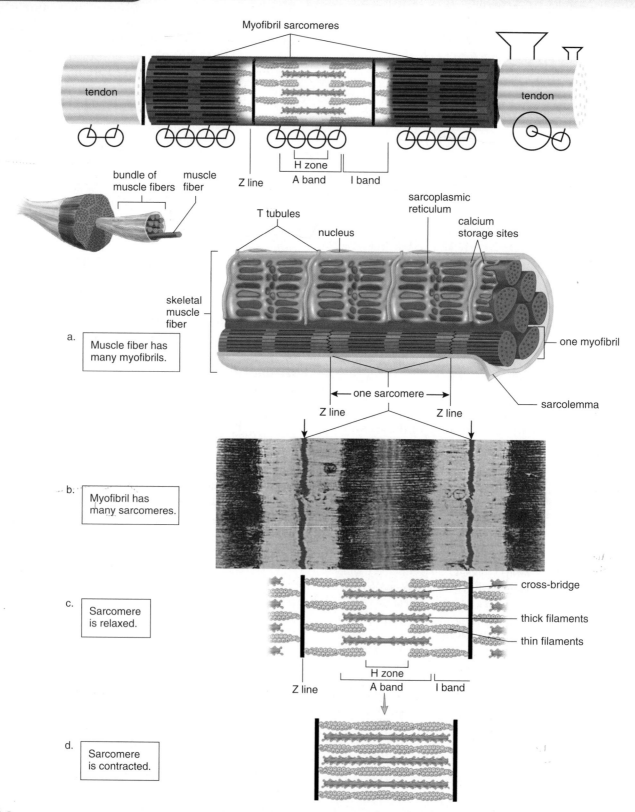

Myofibril sarcomeres

tendon
tendon

H zone
A band   I band
Z line

bundle of   muscle
muscle fibers   fiber

T tubules
nucleus
sarcoplasmic reticulum
calcium storage sites

skeletal muscle fiber

a. Muscle fiber has many myofibrils.

one myofibril

sarcolemma

one sarcomere
Z line   Z line

b. Myofibril has many sarcomeres.

c. Sarcomere is relaxed.

cross-bridge
thick filaments
thin filaments

H zone
Z line   A band   I band

d. Sarcomere is contracted.

Figure 7.3 Anatomy of a muscle fiber. **a.** A muscle fiber contains many myofibrils with the components shown. **b.** A myofibril has many sarcomeres that contain myosin and actin filaments whose arrangement gives rise to the striations so characteristic of skeletal muscle. **c, d.** Muscle contraction occurs when sarcomeres contract and thin filaments slide past thick filaments.

It's one of the deadliest toxins known to mankind: the group of protein chemicals collectively referred to as **botulism toxin.** Each poison is produced by different strains of the bacterial organism *Clostridium botulinum*. If swallowed or inhaled, the toxins cause progressive muscle paralysis. Symptoms begin with blurred vision and difficulty swallowing, then gradual paralysis moves down the body. Death is due to respiratory muscle failure.

As a biological weapon, botulism toxin could be catastrophic: one U.S. bioweapons expert has calculated that releasing the toxin in a powdered, aerosol form could kill 10% of the people who inhale it. Yet, this same toxin, in the form of a preparation called *Botox®*; is injected daily in doctor's offices, hospitals, and clinics around the country, into the muscles of thousands of people seeking a more youthful appearance.

Botox is a diluted form of type A botulism toxin. The toxin works by affecting the motor nerves that supply skeletal muscle. It enters the axon terminals (the ends of motor nerves, at their synapse with a muscle; see Fig. 7.4). After entering the nerve terminals, botox blocks the release of acetylcholine, the neuro-transmitter that signals muscle to contract. By blocking neurotransmitter release, Botox temporarily paralyzes the muscles of facial expression. This muscle paralysis reduces the appearance of wrinkles in the face and neck.

Botox is not used only for cosmetic purposes, however. Patients with brain and spinal cord injuries may suffer from painful and debilitating muscle *contractures* caused by their injury. (Muscle contractures cause muscles to remain in a fixed, contracted position; the patient cannot relax the muscle.) Contractures can be relaxed by injecting the affected muscle with Botox. For example, patients whose cervical spinal cord has been injured may suffer from *torticollis*, or "wryneck," caused by spasm of the neck muscles. Botox has been used to successfully relieve this condition, as well as other conditions caused by muscle spasm.

When used appropriately by trained medical personnel, this deadly toxin is surprisingly safe and effective. Side effects include minor bruising or inflammation at the site of injection. Some patients develop allergies and must discontinue Botox treatment. In other patients, the treatment must be stopped because it loses its effectiveness.

Thin Filaments A thin filament consists primarily of two strands of the globular protein actin, twisted around each other like intertwined bead necklaces. Double strands of tropomyosin coil over each actin strand. Troponin occurs at intervals on the tropomyosin strand (see Fig. 7.5a).

Sliding Filaments When muscles are activated by motor nerves, impulses travel down a T tubule, and calcium is released from the sarcoplasmic reticulum. Now the muscle fiber contracts as the sarcomeres within the myofibrils shorten. When a sarcomere shortens, the actin (thin) filaments slide past the myosin (thick) filaments and approach one another. This causes the I band to shorten and the H zone to almost or completely disappear (Fig. 7.3c and d). The movement of actin filaments in relation to myosin filaments is called the **sliding filament theory** of muscle contraction. During the sliding process, each sarcomere shortens even though the thick and thin filaments themselves remain the same length. Although the actin filaments slide past the myosin filaments, it is the myosin filaments that do the work. Myosin filaments break down ATP, which supplies the energy for muscle contraction. Myosin molecules have cross-bridges that pull the actin filaments toward the center of the sarcomere. Returning to the train analogy, when each sarcomere ("boxcar") shortens during a muscle contraction, the myofibril ("train") becomes shorter as well.

Skeletal Muscle Contraction

Muscle fibers are innervated—that is, they are stimulated to contract by motor neurons whose axons are found in nerves. Recall from Chapter 4 that an axon is the single, long extension of a nerve cell or neuron. The axon of one motor neuron has several branches and can stimulate from a few to several muscle fibers of a particular muscle. Each branch of the axon ends in an axon terminal that lies in close proximity to the sarcolemma of a muscle fiber. A small gap, called a *synaptic cleft*, separates the axon terminal from the sarcolemma. This entire region is called a **neuromuscular junction** (Fig. 7.4).

Axon terminals contain synaptic vesicles that are filled with the neurotransmitter acetylcholine (ACh). When nerve signals traveling down a motor neuron arrive at an axon terminal, the synaptic vesicles release neurotransmitter into the synaptic cleft, which quickly diffuses across the cleft and binds to receptors on the sarcolemma. Now the sarcolemma generates signals that spread over the sarcolemma and down T tubules to the sarcoplasmic reticulum, triggering calcium release. The release of calcium from the sarcoplasmic reticulum causes the filaments within the sarcomeres to slide past one another. Sarcomere contraction results in myofibril contraction, which in turn results in muscle fiber, and finally muscle, contraction.

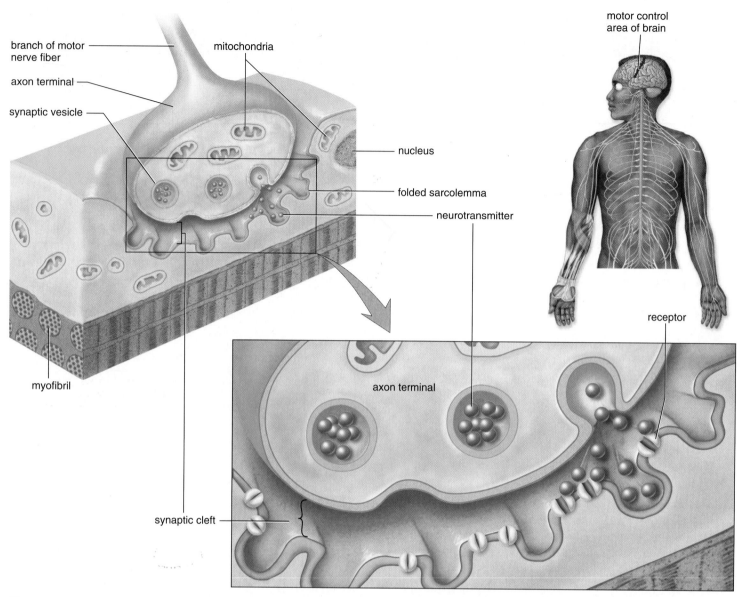

Figure 7.4 Neuromuscular junction. The branch of a nerve axon ends in an axon terminal that meets but does not touch a muscle fiber. A synaptic cleft separates the axon terminal from the sarcolemma of the muscle fiber. Nerve impulses traveling down an axon cause synaptic vesicles to discharge acetylcholine, which diffuses across the synaptic cleft. When the neurotransmitter is received by the sarcolemma of a muscle fiber, impulses begin and lead to muscle fiber contractions. The motor control area of the brain signals muscles via motor nerves.

The Role of Actin and Myosin

Figure 7.5a shows the placement of the three proteins that comprise a thin filament. Thin filaments are composed of a double row of twisted actin molecules. Threads of tropomyosin wind around each actin filament, and troponin occurs at intervals along the tropomyosin threads. A myosin binding site occurs on each actin molecule; when muscle is relaxed, these binding sites are covered by tropomyosin. Calcium ions (Ca^{2+}) that have been released from the sarcoplasmic reticulum combine with troponin. After binding occurs, the tropomyosin threads shift their position, and myosin binding sites are exposed (Fig. 7.5b).

Each of the cross-bridges of a myosin thick filament (recall that these are the paired globular heads of the filament) has two binding sites. One site binds to ATP and then functions as an ATPase enzyme, splitting ATP into ADP and $\circledP$ (see Fig. 7.5c, step 1). This reaction activates the second binding site so that it will bind to actin. The ADP and $\circledP$ remain on the myosin heads until the heads attach to actin (Fig. 7.5c, step 2). Now, ADP and $\circledP$ are released, and this causes the cross-bridges to bend sharply (Fig. 7.5c, step 3). This is the **power stroke** that pulls the thin filaments toward the middle of the sarcomere. When another ATP molecule binds to a myosin head, the head detaches from actin (Fig. 7.5c, step 4). The

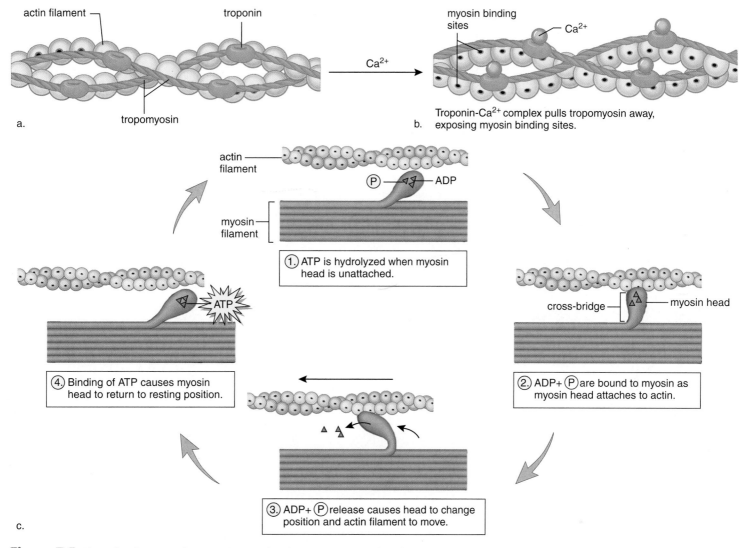

a.

actin filament — troponin

tropomyosin

b.

myosin binding sites — Ca²⁺

Troponin-Ca²⁺ complex pulls tropomyosin away, exposing myosin binding sites.

c.

actin filament

(P) ADP

myosin filament

① ATP is hydrolyzed when myosin head is unattached.

cross-bridge — myosin head

② ADP+ (P) are bound to myosin as myosin head attaches to actin.

③ ADP+ (P) release causes head to change position and actin filament to move.

ATP

④ Binding of ATP causes myosin head to return to resting position.

Figure 7.5 The role of actin and myosin in muscle contraction. **a.** In relaxed muscle, tropomyosin covers myosin binding sites on actin. **b.** Upon release, calcium binds to troponin, exposing myosin binding sites. **c.** After breaking down ATP, myosin cross-bridges bind to actin filaments and, later, a power stroke causes the actin filament to move.

cycle begins again; the thin filaments move nearer the center of the sarcomere each time the cycle is repeated. Note that ATP has two roles in this process: first to energize the myosin cross-bridge, and then to break the link between myosin and actin.

Contraction continues until nerve impulses cease and calcium ions are returned to their storage sites. The membranes of the sarcoplasmic reticulum contain active transport proteins that pump calcium ions back into the sarcoplasmic reticulum. Of course, this active transport process also requires ATP energy.

Contraction of Smooth Muscle

You will recall that smooth muscle cells are uninucleate, cylindrical cells with pointed ends, and that smooth muscle control is involuntary. Like skeletal muscle, smooth muscle contains thick and thin filaments. However, in smooth muscle these filaments are not arranged into myofibrils that create visible striations. Instead, thin filaments in smooth muscle are anchored directly to the sarcolemma or to protein molecules called *dense bodies*. Dense bodies are scattered through the sarcoplasm. When smooth muscle contracts, the elongated cells become shorter and wider, much like an oval balloon being inflated. Smooth muscle contraction occurs very slowly, but can last for long periods of time without fatigue.

Energy for Muscle Contraction

ATP produced previous to strenuous exercise lasts a few seconds, and then muscles acquire new ATP in three different ways: creatine phosphate breakdown, cellular respiration, and fermentation (Fig. 7.6). Creatine phosphate breakdown and

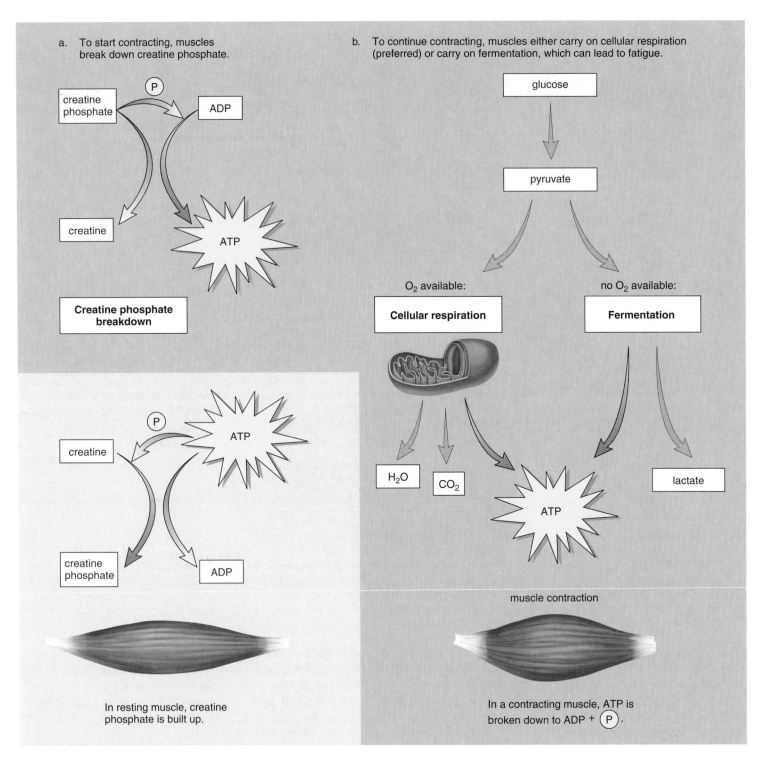

a. To start contracting, muscles break down creatine phosphate.

Creatine phosphate breakdown

In resting muscle, creatine phosphate is built up.

b. To continue contracting, muscles either carry on cellular respiration (preferred) or carry on fermentation, which can lead to fatigue.

glucose

pyruvate

O_2 available:

Cellular respiration

no O_2 available:

Fermentation

H_2O CO_2

lactate

ATP

muscle contraction

In a contracting muscle, ATP is broken down to ADP + Ⓟ.

Figure 7.6 Energy sources for muscle contraction. **a.** Creatine phosphate can transfer a phosphate group to ADP to form ATP. **b.** Cellular respiration and fermentation also form ATP in muscle tissue.

fermentation are anaerobic, meaning that they do not require oxygen.

Creatine Phosphate Breakdown

Creatine phosphate is a high-energy compound built up when a muscle is resting. Creatine phosphate cannot participate directly in muscle contraction. Instead, it can regenerate ATP by transferring its phosphate to ADP, using the following reaction:

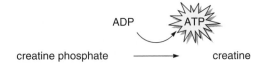

This reaction occurs in the midst of sliding filaments, and therefore is the speediest way to make ATP available to muscles. Creatine phosphate provides enough energy for only about eight seconds of intense activity, and then it is spent. Creatine phosphate is rebuilt when a muscle is resting by transferring a phosphate group from ATP to creatine (see Fig. 7.6a).

Cellular Respiration

Cellular respiration completed in mitochondria usually provides most of a muscle's ATP. Glycogen and fat are stored in muscle cells. Therefore, a muscle cell can use glucose from glycogen and fatty acids from fat as fuel to produce ATP if oxygen is available:

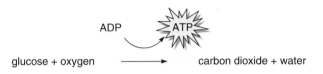

Myoglobin, an oxygen carrier similar to hemoglobin in red blood cells, is synthesized in muscle cells. Its presence accounts for the reddish-brown color of skeletal muscle fibers. Myoglobin has a higher affinity for oxygen than does hemoglobin. Therefore, myoglobin can pull oxygen out of blood and make it available to muscle mitochondria that are carrying on cellular respiration. Then, too, the ability of myoglobin to temporarily store oxygen reduces a muscle's immediate need for oxygen when cellular respiration begins. The end products of cellular respiration (carbon dioxide and water) are usually no problem. Carbon dioxide leaves the body at the lungs, and water simply enters the extracellular space. The by-product, heat, keeps the entire body warm.

Fermentation

Fermentation, like creatine phosphate breakdown, supplies ATP without consuming oxygen. Thus, it is an *anaerobic* process. Fermentation produces the ATP necessary for short bursts of exercise—for example, a 50-yard dash or a run around the bases in a baseball game. During fermentation, glucose is broken down to lactate (lactic acid):

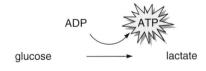

The accumulation of lactate in a muscle fiber makes the cytoplasm more acidic, and eventually enzymes cease to function well. If fermentation continues longer than two or three minutes, cramping and fatigue set in. Cramping seems to be due to lack of ATP. Recall that ATP is needed to pump calcium ions back into the sarcoplasmic reticulum and to break the linkages between the actin and myosin filaments so that muscle fibers can relax.

Oxygen Debt

When a muscle uses fermentation to supply its energy needs, it incurs an **oxygen debt.** Oxygen debt is obvious when a person continues to breathe heavily after exercising. The ability to run up an oxygen debt is one of muscle tissue's greatest assets. Brain tissue cannot last nearly as long without oxygen as muscles can.

Repaying an oxygen debt requires replenishing creatine phosphate supplies and disposing of lactic acid. Lactic acid can be changed back to a compound called pyruvic acid and metabolized completely in mitochondria, or it can be sent to the liver to reconstruct glycogen. A marathon runner who has just crossed the finish line is not exhausted due to oxygen deficit. Instead, the runner has used up all the muscles', and probably the liver's, glycogen supply. It takes about two days to replace glycogen stores on a high-carbohydrate diet.

People who train rely more heavily on cellular respiration than do people who do not train. In people who train, the number of muscle mitochondria increases, and so fermentation is not needed to produce ATP. Their mitochondria can start consuming oxygen as soon as the ADP concentration starts rising during muscle contraction. Because mitochondria can break down fatty acid, instead of glucose, blood glucose is spared for the activity of the brain. (The brain, unlike other organs, ordinarily utilizes only glucose to produce ATP.) Because less lactate is produced in people who train, the pH of the blood remains steady, and there is less of an oxygen debt.

Content **CHECK-UP!**

4. The extension of the muscle cell membrane that extends into the muscle and conveys nerve signals is called the:
   a. sarcoplasmic reticulum.   c. transverse (T) tubules.
   b. sarcolemma.   d. sarcoplasm.

5. Thin filaments contain all of the following proteins except:
   a. myosin.   c. tropomyosin.
   b. actin.   d. troponin.

6. Which of the following is the process in a muscle cell that requires oxygen to produce ATP energy?
   a. fermentation   c. cellular respiration
   b. creatine phosphate breakdown

7.3 Muscle Responses

Muscles can be studied in the laboratory in an effort to understand how they respond when in the body.

In the Laboratory

When a muscle fiber is isolated, placed on a microscope slide, and provided with ATP plus the various electrolytes it requires, it contracts completely along its entire length. This

For everything there is a season, and a time for every matter under heaven: a time to be born, and a time to die . . .
—Ecclesiastes 3:1,2 KJV

When a person dies, the physiologic events that accompany death occur in an orderly progression. Respiration ceases, the heart ultimately stops beating, and tissue cells begin to die. The first tissues to die are those with the highest oxygen requirement. Brain and nervous tissues have an extremely high requirement for oxygen. Deprived of oxygen, these cells typically die after only six minutes, as the ATP energy reserve within the cell is used up. However, tissues that can produce ATP by fermentation (which does not require oxygen) can "live" for an hour or more before ATP is completely depleted. Muscle is capable of generating ATP by both fermentation and creatine phosphate breakdown; thus, muscle cells can survive for a time after clinical death occurs. Muscle death is signaled by a process termed *rigor mortis*—the "stiffness of death." Rigor mortis develops and reverses itself within a known time period of hours to days after death occurs. Forensic pathologists use this information to study the condition of a body to approximate the time of death.

Rigor mortis occurs as dying muscle cells deplete the last of their ATP energy reserve. Relaxing the muscle becomes impossible because ATP energy is needed to break the bond between an actin binding site and the myosin cross-bridge. In addition, ATP energy is required for muscle relaxation to return calcium ions from the sarcoplasm back to the sarcoplasmic reticulum. Thus, without ATP energy, the muscle remains fixed in the same state of contraction that preceded that person's death. If, for example, a murder victim dies while sitting at a desk, the body in rigor mortis will be frozen in the sitting position.

Forensic pathologists use body temperature and the presence or absence of rigor mortis to estimate time of death. For example, the body of someone who has been dead for 3 hours or less will still be warm (close to normal body temperature, 98.6°F or 37°C) and rigor mortis will be absent. After approximately 3 hours, the body will be significantly cooler than normal, and rigor mortis will begin to develop. The corpse of an individual who has been dead at least 8 hours will be in full rigor mortis, and the temperature of the body will be the same as the surroundings. Rigor mortis resolves approximately 24–36 hours after death. Muscles lose their stiffness because lysosomes inside the cell eventually rupture, releasing enzymes that break the myosin-actin bonds. Forensic pathologists know that a person has been dead for more than 24 hours if the body temperature is the same as the environment and there is no longer a trace of rigor mortis.

observation has resulted in the **all-or-none law:** A muscle fiber contracts completely or not at all. In contrast, a whole muscle (made up of many muscle fibers) shows degrees of contraction. To study whole muscle contraction in the laboratory, an isolated muscle is stimulated electrically, and the mechanical force of contraction is recorded as a visual pattern called a *myogram*. When the strength of the stimulus is above a threshold level, the muscle contracts and then relaxes. This action—a single contraction that lasts only a fraction of a second—is called a **muscle twitch.** Figure 7.7 is a myogram of a muscle twitch, which is customarily divided into three stages: the latent period, or the period of time between stimulation and initiation of contraction; the contraction period, when the muscle shortens; and the relaxation period, when the muscle returns to its former length. It's interesting to use our knowledge of muscle fiber contraction to understand these events. From our study thus far, we know that a muscle fiber in an intact muscle contracts when calcium leaves storage sacs in the sarcoplasmic reticulum and relaxes when calcium returns to storage sacs.

But unlike the contraction of a muscle fiber, a muscle has degrees of contraction, and a twitch can vary in height (strength) depending on the degree of stimulation. Why should that be? Obviously, a stronger stimulation causes more individual fibers to contract than before.

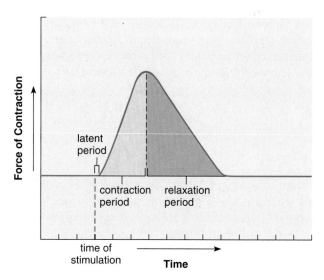

Figure 7.7 A myogram showing a single muscle twitch.

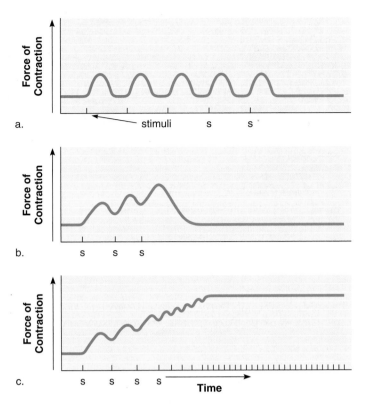

Figure 7.8 Myograms showing **a.** a series of twitches, **b.** summation, and **c.** a tetanic contraction. Note that an increased frequency of stimulations has resulted in these different responses.

If a whole muscle is given a rapid series of stimuli, it can respond to the next stimulus without relaxing completely. **Summation** is increased muscle contraction until maximal sustained contraction, called a **tetanic contraction,** is achieved (Fig. 7.8). The myogram no longer shows individual twitches; rather, the twitches are fused and blended completely into a straight line. Tetanus continues until the muscle *fatigues.*

Fatigue in muscles is apparent when the muscle relaxes even though stimulation continues. In the body, fatigue is a gradual weakening that occurs after repetitive use. There are several reasons why muscles become fatigued. First, ATP is depleted during constant use of a muscle; the muscle essentially "runs out of energy." Repetitive use causes production of lactic acid by fermentation, which lowers the pH of the sarcoplasm and inhibits muscle function. Additionally, the motor nerves that supply muscle can run out of their neurotransmitter, acetylcholine. Finally, the brain itself, through mechanisms not well understood, may signal a person to stop exercising even if the muscles are not truly fatigued. People who train can exercise for longer periods without experiencing fatigue.

In the Body

In the body, muscles are innervated to contract by nerves. As mentioned, each axon within a nerve stimulates a number of muscle fibers. A nerve fiber together with all of the muscle fibers it innervates is called a **motor unit.** A motor unit obeys the all-or-none law. Why? Because all the muscle fibers in a motor unit are stimulated at once, and they all either contract or do not contract. A variable of interest is the number of muscle fibers within a motor unit. For example, in the ocular muscles that move the eyes, the innervation ratio is one motor axon per 23 muscle fibers, while in the gastrocnemius muscle of the lower leg, the ratio is about one motor axon per 1,000 muscle fibers. No doubt, moving the eyes requires finer control than moving the legs.

Tetanic contractions ordinarily occur in the body because, as the intensity of nervous stimulation increases, more and more motor units are activated. This phenomenon, known as **recruitment,** results in stronger and stronger muscle contractions. But while some muscle fibers are contracting, others are relaxing. Because of this, intact muscles rarely fatigue completely. Even when muscles appear to be at rest, they exhibit **tone,** in which some of their fibers are always contracting. Muscle tone is particularly important in maintaining posture. If all the fibers within the muscles of the neck, trunk, and legs were to suddenly relax, the body would collapse.

Athletics and Muscle Contraction

Athletes who excel in a particular sport, and much of the general public as well, are interested in staying fit by exercising. The Medical Focus on page 149 gives suggestions for exercise programs according to age.

Exercise and Size of Muscles Muscles that are not used or that are used for only very weak contractions decrease in size, or atrophy. **Atrophy** can occur when a limb is placed in a cast or when the nerve serving a muscle is damaged. If nerve stimulation is not restored, muscle fibers are gradually replaced by fat and fibrous tissue. Unfortunately, atrophy can cause muscle fibers to shorten progressively, leaving body parts contracted in contorted positions.

Forceful muscular activity over a prolonged period causes muscle to increase in size as the number of myofibrils within the muscle fibers increases. Increase in muscle size, called **hypertrophy,** occurs only if the muscle contracts to at least 75% of its maximum tension.

Some athletes take anabolic steroids, either testosterone or related chemicals, to promote muscle growth. This practice has many undesirable side effects as discussed in the Medical Focus on page 220.

Slow-Twitch and Fast-Twitch Muscle Fibers We have seen that all muscle fibers metabolize both aerobically (using oxygen during cellular respiration) and anaerobically (without oxygen, using fermentation or creatine phosphate breakdown). Some muscle fibers, however, utilize one method more than the other to provide myofibrils with ATP. Slow-twitch fibers tend to be aerobic, and fast-twitch fibers tend to

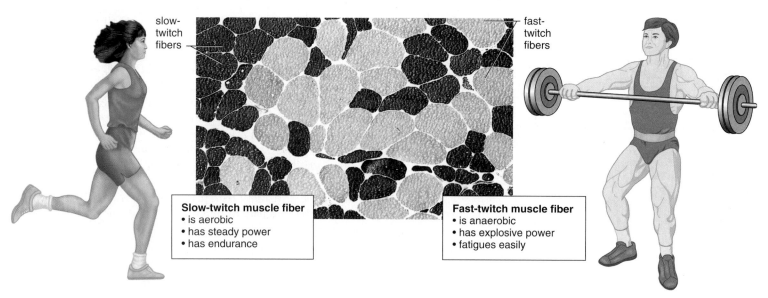

Figure 7.9 Slow- and fast-twitch fibers. If your muscles contain many slow-twitch fibers (dark color), you would probably do better at a sport like cross-country running. But if your muscles contain many fast-twitch fibers (light color), you would probably do better at a sport like weight lifting.

be anaerobic (Fig. 7.9). Slow-twitch fibers are also referred to as type I fibers and fast-twitch fibers are called type II fibers.

Slow-twitch fibers have a steadier tug and more endurance, despite having motor units with a smaller number of fibers. These muscle fibers are most helpful in sports such as long-distance running, biking, jogging, and swimming. Because they produce most of their energy aerobically, they tire only when their fuel supply is gone. Slow-twitch fibers have many mitochondria and are dark in color because they contain myoglobin, the respiratory pigment found in muscles. They are also surrounded by dense capillary beds and draw more blood and oxygen than fast-twitch fibers. Slow-twitch fibers have a low maximum tension, which develops slowly, but these muscle fibers are highly resistant to fatigue. Because slow-twitch fibers have a substantial reserve of glycogen and fat, their abundant mitochondria can maintain a steady, prolonged production of ATP when oxygen is available.

Fast-twitch fibers tend to be anaerobic and seem to be designed for strength because their motor units contain many fibers. They provide explosions of energy and are most helpful in sports activities such as sprinting, weight lifting, swinging a golf club, or throwing a shot. Fast-twitch fibers are light in color because they have fewer mitochondria, little or no myoglobin, and fewer blood vessels than slow-twitch fibers do. Fast-twitch fibers can develop maximum tension more rapidly than slow-twitch fibers can, and their maximum tension is greater. However, their dependence on anaerobic energy leaves them vulnerable to an accumulation of lactic acid that causes them to fatigue quickly.

Content **CHECK-UP!**

7. A motor unit is:
   a. a group of myofibrils.
   b. a group of skeletal muscles that perform a specific function.
   c. a group of motor nerves in the brain.
   d. a motor nerve and all of the muscle fibers it innervates.

8. Which of the following is not a characteristic of slow-twitch muscle fibers?
   a. produce most of their energy aerobically
   b. contain many mitochondria
   c. fatigue easily
   d. have an extensive blood supply

9. An increase in muscle size, caused by increasing the number of myofibrils in individual muscle cells, is called:
   a. recruitment.   c. atrophy.
   b. hypertrophy.   d. tetanic contraction.

7.4 Skeletal Muscles of the Body

The human body has some 600 skeletal muscles, but this text will discuss only some of the most significant of these. First, let us consider certain basic principles of muscle contraction.

Basic Principles

When a muscle contracts at a joint, one bone remains fairly stationary, and the other one moves. The **origin** of a muscle is

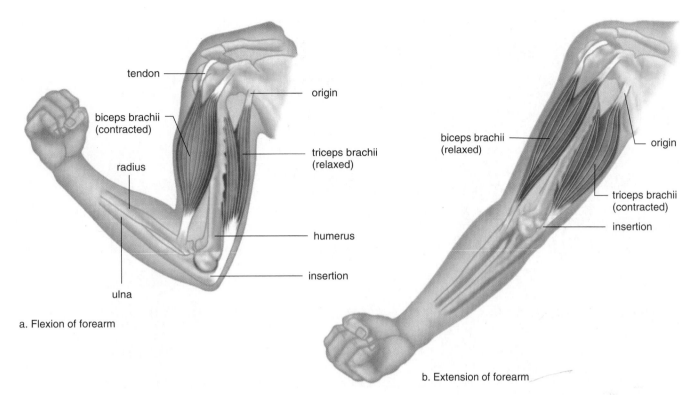

tendon

origin

biceps brachii (contracted)

triceps brachii (relaxed)

radius

humerus

insertion

ulna

a. Flexion of forearm

biceps brachii (relaxed)

origin

triceps brachii (contracted)

insertion

b. Extension of forearm

Figure 7.10 The origin of a muscle is on a bone that remains stationary, and the insertion of a muscle is on a bone that moves when a muscle contracts. Two of the muscles shown here are antagonistic. **a.** When the biceps brachii contracts, the lower arm flexes. **b.** When the triceps brachii contracts, the lower arm extends.

on the stationary bone, and the **insertion** of a muscle is on the bone that moves.

Frequently, a body part is moved by a group of muscles working together. Even so, one muscle does most of the work, and this muscle is called the **prime mover.** For example, in flexing the elbow, the prime mover is the biceps brachii (Fig. 7.10) The assisting muscles are called the **synergists.** The brachialis (see Fig. 7.16) is a synergist that helps the biceps brachii flex the elbow. A prime mover can have several synergists.

When muscles contract, they shorten. Therefore, muscles can only pull; they cannot push. However, muscles have **antagonists,** and antagonistic pairs work opposite one another to bring about movement in opposite directions. For example, the biceps brachii and the triceps brachii are antagonists; one flexes the forearm, and the other extends the forearm (Fig. 7.10). Later on in our discussion, we will encounter other antagonistic pairs.

Naming Muscles

When learning the names of muscles, considering what the name means will help you remember it. The names of the various skeletal muscles are often combinations of the following terms used to characterize muscles:

1. **Size.** For example, the *gluteus maximus* is the largest muscle that makes up the buttocks. The *gluteus minimus* is the smallest of the gluteal muscles. Other terms used

to indicate size are vastus (huge), longus (long), and brevis (short).

2. **Shape.** For example, the *deltoid* is shaped like a delta, or triangle, while the *trapezius* is shaped like a trapezoid. Other terms used to indicate shape are latissimus (wide) and teres (round).

3. **Direction of fibers.** For example, the *rectus abdominis* is a longitudinal muscle of the abdomen (*rectus* means straight). The *orbicularis oculi* is a circular muscle around the eye. Other terms used to indicate direction are transverse (across) and oblique (diagonal).

4. **Location.** For example, the *frontalis* overlies the frontal bone. The *external obliques* are located outside the internal obliques. Other terms used to indicate location are pectoralis (chest), gluteus (buttock), brachii (arm), and sub (beneath). You should also review these directional terms: anterior, posterior, lateral, medial, proximal, distal, superficial, and deep.

5. **Attachment.** For example, the *sternocleidomastoid* is attached to the sternum, clavicle, and mastoid process. The *brachioradialis* is attached to the brachium (arm) and the radius.

6. **Number of attachments.** For example, the biceps brachii has two attachments, or origins (and is located on the arm). The quadriceps femoris has four origins (and is located on the anterior femur).

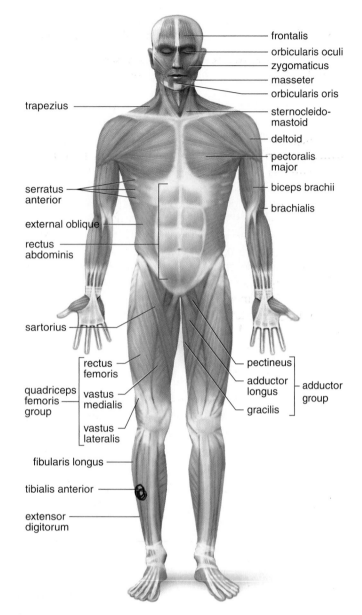

Figure 7.11 Anterior view of the body's superficial skeletal muscles.

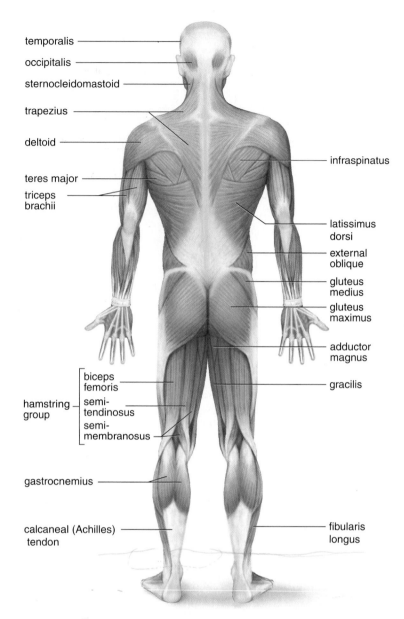

Figure 7.12 Posterior view of the body's superficial skeletal muscles.

7. **Action.** For example, the *extensor digitorum* extends the fingers or digits. The *adductor magnus* is a large muscle that adducts the thigh. Other terms used to indicate action are flexor (to flex), masseter (to chew), and levator (to lift).

With every muscle you learn, try to understand its name.

Skeletal Muscle Groups

In our discussion, the muscles of the body (Figs. 7.11 and 7.12) will be grouped according to their location and their action. To better understand muscle groups, recall from Chapter 6 that the term *arm* refers to the humeral area, while the radius/ulna area is referred to as the *forearm*. Likewise, the femur is the *thigh* area, while the tibia/fibular area is referred to as the *leg*.

After you understand the meaning of a muscle's name, try to correlate its name with the muscle's location and the action it performs. Knowing the origin and insertion will also help you remember what the muscle does. Why? Because the insertion is on the bone that moves. You should review the various body movements listed and illustrated in Chapter 6 (see page 119). Only then will you be able to understand the actions of the muscles listed in Tables 7.2–7.5. Scientific terminology is necessary because it allows all persons to know the exact action being described for that muscle.

Muscles of the Head

The muscles of the head and neck are the first group of muscles we will study. The muscles of the head and neck are illustrated

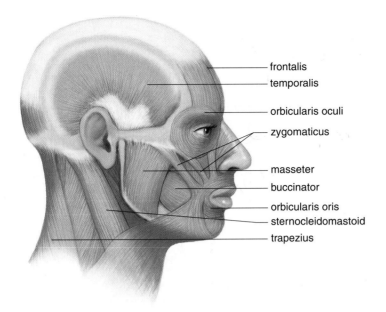

frontalis
temporalis
orbicularis oculi
zygomaticus
masseter
buccinator
orbicularis oris
sternocleidomastoid
trapezius

Figure 7.13 Muscles of the head and neck. Some of these muscles account for our facial expressions and the ability to chew our food; others move the head.

in Figure 7.13 and listed in Table 7.2. The muscles of the head are responsible for facial expression and mastication (chewing). One muscle of the head and several muscles of the neck allow us to swallow. The muscles of the neck also move the head.

Muscles of Facial Expression

The muscles of facial expression are located on the scalp and face. These muscles are unusual in that they insert into and move the skin. Therefore, we expect them to move the skin and

not a bone. The use of these muscles communicates to others whether we are surprised, angry, fearful, happy, and so forth.

Frontalis lies over the frontal bone; it raises the eyebrows and wrinkles the brow. Frequent use results in furrowing of the forehead.

Orbicularis oculi is a ringlike band of muscle that encircles (forms an orbit about) the eye. It causes the eye to close or blink, and is responsible for "crow's feet" at the eye corners.

Orbicularis oris encircles the mouth and is used to pucker the lips, as in forming a kiss. Frequent use results in lines about the mouth.

Buccinator muscles are located in the cheek areas. When a buccinator contracts, the cheek is compressed, as when a person whistles or blows out air. Therefore, this muscle is called the "trumpeter's muscle." Important to everyday life, the buccinator helps hold food in contact with the teeth during chewing. Babies use this muscle for suckling. It is also used in swallowing, as discussed next.

Zygomaticus extends from each zygomatic arch (cheekbone) to the corners of the mouth. It raises the corners of the mouth when a person smiles.

Muscles of Mastication

We use the muscles of mastication when we chew food or bite something. Although there are four pairs of muscles for chewing, only two pairs are superficial and shown in Figure 7.13. As you might expect, both of these muscles insert on the mandible.

Each **masseter** has its origin on the zygomatic arch and its insertion on the mandible. The masseter is a muscle of mastication (chewing) because it is a prime mover for elevating the mandible.

TABLE 7.2 Muscles of the Head and Neck		
Name	**Function**	**Origin/Insertion**
MUSCLES OF FACIAL EXPRESSION		
Frontalis (frun-ta′lis)	Raises eyebrows	Cranial fascia/skin and muscles around eye
Orbicularis oculi (or-bik′yū-lā-ris ok′yū-li)	Closes eye	Maxillary and frontal bones/skin around eye
Orbicularis oris (or-bik′yū-lā-ris o′ris)	Closes and protrudes lips	Muscles near the mouth/skin around mouth
Buccinator (buk′si-na″tor)	Compresses cheeks inward	Outer surfaces of maxilla and mandible/orbicularis oris
Zygomaticus (zi″go-mat′ik-us)	Raises corner of mouth	Zygomatic bone/skin and muscle around mouth
MUSCLES OF MASTICATION		
Masseter (mas-se′ter)	Closes jaw	Zygomatic arch/mandible
Temporalis (tem-po-ra′lis)	Closes jaw	Temporal bone/mandibular coronoid process
MUSCLES THAT MOVE THE HEAD		
Sternocleidomastoid (ster″no-kli″do-mas′toid)	Flexes head and rotates head	Sternum and clavicle/mastoid process of temporal bone
Trapezius (truh-pe′ze-us)	Extends head and adducts scapula	Occipital bone C_7 vertebra, all thoracic vertebrae/spine of scapula and clavicle

Each **temporalis** is a fan-shaped muscle that overlies the temporal bone. It is also a prime mover for elevating the mandible. The masseter and temporalis are synergists.

Muscles of the Neck

Deep muscles of the neck (not illustrated) are responsible for swallowing. Superficial muscles of the neck move the head (see Table 7.2 and Figure 7.13).

Swallowing

Swallowing is an important activity that begins after we chew our food. First, the tongue (a muscle) and the buccinators squeeze the food back along the roof of the mouth toward the pharynx. An important bone that functions in swallowing is the hyoid (see page 105). The hyoid is the only bone in the body that does not articulate with another bone.

Muscles that lie superior to the hyoid, called the suprahyoid muscles, and muscles that lie inferior to the hyoid, called the infrahyoid muscles, move the hyoid. Because these muscles lie deep in the neck, they are not illustrated in Figure 7.13. The suprahyoid muscles pull the hyoid forward and upward toward the mandible. Because the hyoid is attached to the larynx, this pulls the larynx upward and forward. The epiglottis now lies over the glottis and closes the respiratory passages. Small palatini muscles (not illustrated) pull the soft palate backward, closing off the nasal passages. Pharyngeal constrictor muscles (not illustrated) push the bolus of food into the pharynx, which widens when the suprahyoid muscles move the hyoid. The hyoid bone and larynx are returned to their original positions by the infrahyoid muscles. Notice that the suprahyoid and infrahyoid muscles are antagonists.

Muscles That Move the Head

Two muscles in the neck are of particular interest: The sternocleidomastoid and the trapezius are listed in Table 7.2 and illustrated in Figure 7.13. Recall that *flexion* is a movement that closes the angle at a joint and *extension* is a movement that increases the angle at a joint. Recall that *abduction* is a movement away from the midline of the body, while *adduc-*tion is a movement toward the midline. Also, *rotation* is the movement of a part around its own axis.

Sternocleidomastoid muscles ascend obliquely from their origin on the sternum and clavicle to their insertion on the mastoid process of the temporal bone. Which part of the body do you expect them to move? When both sternocleidomastoid muscles contract, flexion of the head occurs. When only one contracts, the head turns to the opposite side. If you turn your head to the right, you can see how the left sternocleidomastoid shortens, pulling the head to the right.

Each **trapezius** muscle is triangular, but together, they take on a diamond or trapezoid shape. The origin of a trapezius is at the base of the skull. Its insertion is on a clavicle and scapula. You would expect the trapezius muscles to move the scapulae, and they do. They adduct the scapulae when the shoulders are shrugged or pulled back. The trapezius muscles also help extend the head, however. The prime movers for head extension are actually deep to the trapezius and not illustrated in Figure 7.13.

Muscles of the Trunk

The muscles of the trunk are listed in Table 7.3 and illustrated in Figure 7.14. The muscles of the thoracic wall are primarily involved in breathing. The muscles of the abdominal wall protect and support the organs within the abdominal cavity.

Muscles of the Thoracic Wall

External intercostal muscles occur between the ribs; they originate on a superior rib and insert on an inferior rib. These muscles elevate the rib cage during the inspiration phase of breathing.

The **diaphragm** is a dome-shaped muscle that, as you know, separates the thoracic cavity from the abdominal cavity (see Fig. 1.5). Contraction of the diaphragm also assists inspiration.

Internal intercostal muscles originate on an inferior rib and insert on a superior rib. These muscles depress the rib cage and contract only during a forced expiration. Normal expiration does not require muscular action.

TABLE 7.3 Muscles of the Trunk		
Name	**Function**	**Origin/Insertion**
External intercostals	Elevate rib cage for inspiration	Superior rib/inferior rib
Internal intercostals	Depress rib cage for forced expiration	Inferior rib/superior rib
External oblique	Tenses abdominal wall; lateral rotation of trunk	Lower eight ribs/iliac crest
Internal oblique	Tenses abdominal wall; lateral rotation of trunk	Iliac crest/lower three ribs
Transversus abdominis	Tenses abdominal wall	Lower six ribs/pubis
Rectus abdominis	Flexes and rotates the vertebral column	Pubis, pubic symphysis/xiphoid process of sternum, fifth to seventh costal cartilages

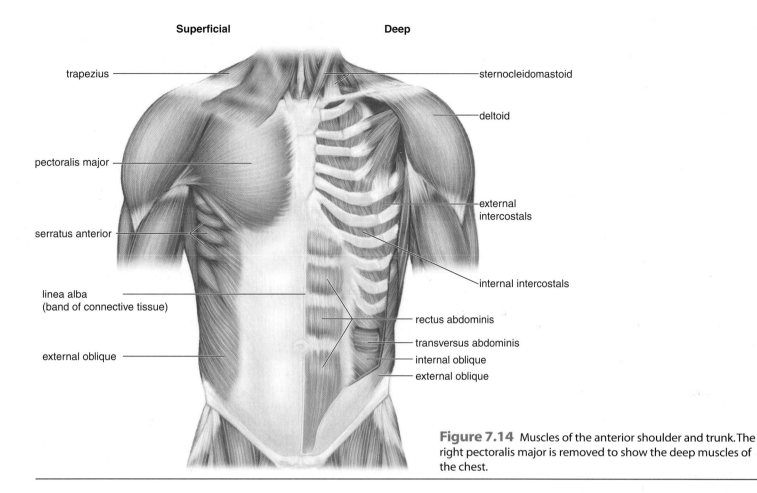

Superficial   Deep

trapezius ——————————   —————— sternocleidomastoid

                                                  —————— deltoid

pectoralis major ——————

                                                  —— external
                                                     intercostals

serratus anterior ——————

                                                  —— internal intercostals

linea alba
(band of connective tissue) ——————   —— rectus abdominis

                                                  —— transversus abdominis
external oblique ——————   —— internal oblique
                                                  —— external oblique

Figure 7.14 Muscles of the anterior shoulder and trunk. The right pectoralis major is removed to show the deep muscles of the chest.

Muscles of the Abdominal Wall

The abdominal wall has no bony reinforcement (Fig. 7.14). The wall is strengthened by four pairs of muscles that run at angles to one another. The external and internal obliques and the transversus abdominis occur laterally, but the fasciae of these muscle pairs meet at the midline of the body, forming a tendinous area called the linea alba. The rectus abdominis is a superficial medial pair of muscles.

All of the muscle pairs of the abdominal wall compress the abdominal cavity and support and protect the organs within the abdominal cavity.

External and **internal obliques** occur on a slant and are at right angles to one another. They are located between the lower ribs and the pelvic girdle. The internal obliques are deep to the external obliques. These muscles also aid trunk rotation and lateral flexion.

Transversus abdominis, deep to the obliques, extends horizontally across the abdomen. The obliques and the transversus abdominis are synergistic muscles.

Rectus abdominis has a straplike appearance but takes its name from the fact that it runs straight (*rectus* means straight) up from the pubic bones to the ribs and sternum. These muscles also help flex and rotate the lumbar portion of the vertebral column.

Muscles of the Shoulder

Muscles of the shoulder are shown in Figures 7.14 and 7.15. They are also listed in Table 7.4 on page 144. The muscles of the shoulder attach the scapula to the thorax and move the scapula; they also attach the humerus to the scapula and move the arm.

Muscles That Move the Scapula

Of the muscles that move the scapula, we have already discussed the trapezius (see page 142).

Serratus anterior is located below the axilla (armpit) on the lateral chest. It runs between the upper ribs and the scapula. It depresses the scapula and pulls it forward, as when we push something. Because this muscle causes a fast-forward jab of the arm, it is often called the boxer's muscle. It also helps to elevate the arm above the horizontal level.

Muscles That Move the Arm

Deltoid is a large, fleshy, triangular muscle (*deltoid* in Greek means triangular) that covers the shoulder and causes a bulge in the arm where it meets the shoulder. It runs

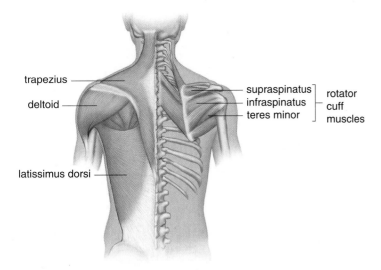

trapezius

deltoid

latissimus dorsi

supraspinatus
infraspinatus
teres minor

rotator
cuff
muscles

Figure 7.15 Muscles of the posterior shoulder. The right trapezius is removed to show deep muscles that move the scapula and the rotator cuff muscles.

from both the clavicle and the scapula of the pectoral girdle to the humerus. This muscle abducts the arm to the horizontal position.

Pectoralis major (Fig. 7.14) is a large anterior muscle of the upper chest. It originates from a clavicle, but also from the sternum and ribs. It inserts on the humerus. The pectoralis major flexes the arm (raises it anteriorly) and adducts the arm, pulling it toward the chest.

Latissimus dorsi (Fig. 7.15) is a large, wide, triangular muscle of the back. This muscle originates from the lower spine and sweeps upward to insert on the humerus. The latissimus dorsi extends and adducts the arm (brings it down from a raised position). This muscle is very important for swimming, rowing, and climbing a rope.

Rotator cuff (Figs. 7.15 and 7.16). This group of muscles is so named because their tendons help form a cuff over the proximal humerus. There are four rotator cuff muscles. Three are located on the posterior scapula: supraspinatus, infraspinatus, and teres minor. The last rotator cuff muscle is the subscapularis muscle (not shown) located on the anterior surface of the scapula. These muscles lie deep to those already mentioned, and they are synergists to them.

Muscles of the Arm

The muscles of the arm move the forearm. They are illustrated in Figure 7.16 and listed in Table 7.4.

Biceps brachii is a muscle of the proximal anterior arm (Fig. 7.16a) that is familiar because it bulges when the forearm is flexed. It also supinates the hand when a doorknob is turned or the cap of a jar is unscrewed. The name of the muscle refers to its two heads that attach to the scapula, where it originates. The biceps brachii inserts on the radius.

Brachialis originates on the humerus and inserts on the ulna. It is a muscle of the distal anterior humerus and

TABLE 7.4 Muscles of the Shoulder and Upper Limb		
Name	**Function**	**Origin/Insertion**
MUSCLES THAT MOVE THE SCAPULA AND ARM		
Serratus anterior	Depresses scapula and pulls it forward; elevates arm above horizontal	Upper nine ribs/vertebral border of scapula
Deltoid	Abducts arm to horizontal	Acromion process, spine of scapula, and clavicle/deltoid tuberosity of humerus
Pectoralis major	Flexes and adducts arm	Clavicle, sternum, second to sixth costal cartilages/intertubular groove of humerus
Latissimus dorsi	Extends or adducts arm	Iliac crest/intertubular groove of humerus
Rotator cuff	Angular and rotational movements of arm	Scapula/humerus
MUSCLES THAT MOVE THE FOREARM		
Biceps brachii	Flexes forearm, and supinates forearm	Scapula/radial tuberosity
Triceps brachii	Extends forearm	Scapula, proximal humerus/olecranon process of ulna
Brachialis	Flexes forearm	Anterior humerus/coronoid process of ulna
MUSCLES THAT MOVE THE HAND AND FINGERS		
Flexor carpi and extensor carpi	Move wrist and hand	Humerus/carpals and metacarpals
Flexor digitorum and extensor digitorum	Move fingers	Humerus, radius, ulna/phalanges

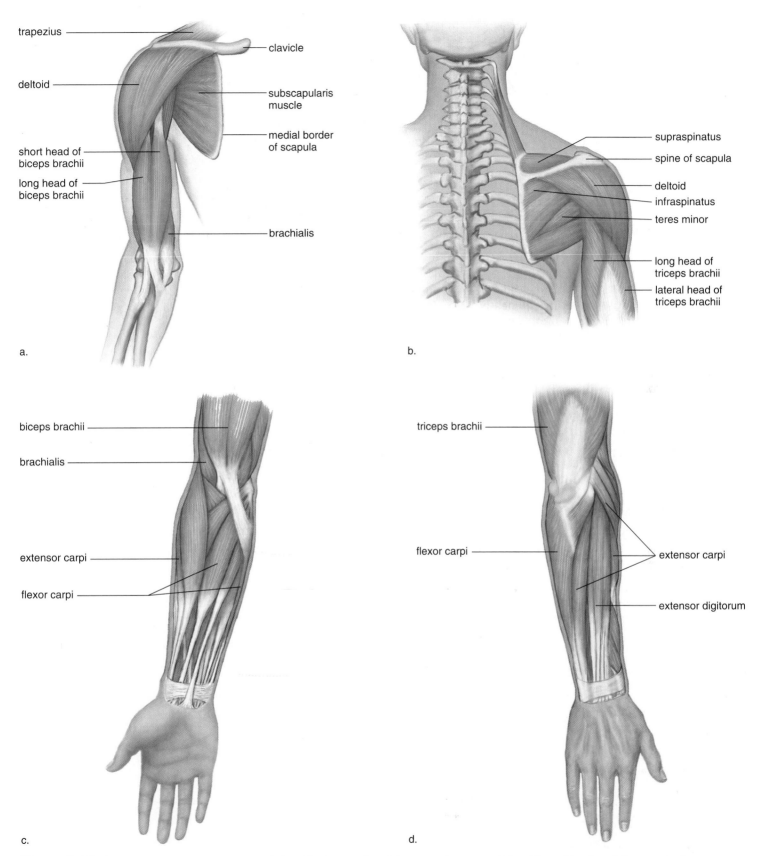

trapezius

clavicle

deltoid

subscapularis muscle

short head of biceps brachii

medial border of scapula

long head of biceps brachii

brachialis

a.

supraspinatus

spine of scapula

deltoid

infraspinatus

teres minor

long head of triceps brachii

lateral head of triceps brachii

b.

biceps brachii

brachialis

extensor carpi

flexor carpi

c.

triceps brachii

flexor carpi

extensor carpi

extensor digitorum

d.

Figure 7.16 **a.** Muscles of the anterior arm and shoulder. **b.** Muscles of the posterior arm and shoulder. **c.** Muscles of the anterior forearm. **d.** Muscles of the posterior forearm.

lies deep to the biceps brachii. It is synergistic to the biceps brachii in flexing the forearm.

- **Triceps brachii** is the only muscle of the posterior arm (Fig. 7.16*b*). It has three heads that attach to the scapula and humerus, and it inserts on the ulna. The triceps extends the forearm. The triceps is also used in tennis to do a backhand volley.

Muscles of the Forearm

The muscles of the forearm move the hand and fingers. They are illustrated in Figure 7.16*c,d* and listed in Table 7.4. Note that in anatomical position, extensors of the wrists and fingers are on the posterior and lateral forearm and flexors are on the anterior and lateral forearm.

- **Flexor carpi** and **extensor carpi** muscles originate on the bones of the forearm and insert on the bones of the hand. The flexor carpi flex the wrists and hands, and the extensor carpi extend the wrists and hands.
- **Flexor digitorum** and **extensor digitorum** muscles also originate on the bones of the forearm and insert on the bones of the hand. The flexor digitorum (not shown) flexes the wrist and fingers, and the extensor digitorum extends the wrist and fingers (i.e., the digits).

Muscles of the Hip and Lower Limb

The muscles of the hip and lower limb are listed in Table 7.5 and shown in Figures 7.17 to 7.20. These muscles, particularly those of the hips and thigh, tend to be large and heavy because they are used to move the entire weight of the body and to resist the force of gravity. Therefore, they are important for movement and balance.

Muscles That Move the Thigh

The muscles that move the thigh have at least one origin on the pelvic girdle and insert on the femur. Notice that the iliopsoas is an anterior muscle that moves the thigh, while the gluteal muscles ("gluts") are posterior muscles that move the thigh. The adductor muscles are medial muscles (Fig. 7.17 and Fig. 7.18). Before studying the action of these muscles, review the movement of the hip joint when the thigh flexes, extends, abducts, and adducts (see Chapter 6, page 117).

- **Iliopsoas** (includes psoas major and iliacus) originates at the ilium and the bodies of the lumbar vertebrae, and inserts on the femur anteriorly (Fig. 7.17). This muscle is the prime mover for flexing the thigh and also the trunk, as when we bow. As the major flexor of the thigh, the iliopsoas is important to the process of walking. It also helps prevent the trunk from falling backward when a person is standing erect.

The gluteal muscles form the buttocks. We will consider only the gluteus maximus and the gluteus medius, both of which are illustrated in Figure 7.18.

- **Gluteus maximus** is the largest muscle in the body and covers a large part of the buttock (*gluteus* means buttocks in Greek). It originates at the ilium and sacrum, and inserts on the femur. The gluteus maximus is a prime mover of thigh extension, as when a person is walking, climbing stairs, or jumping from a crouched position. Notice that the iliopsoas and the gluteus maximus are antagonistic muscles.
- **Gluteus medius** lies partly behind the gluteus maximus (Fig. 7.18). It runs between the ilium and the femur, and

TABLE 7.5 Muscles of the Hip and Lower Limb

Name	Function	Origin/Insertion
MUSCLES THAT MOVE THE THIGH		
Iliopsoas (il′e-o-so′us)	Flexes thigh	Lumbar vertebrae, ilium/lesser trochanter of femur
Gluteus maximus	Extends thigh	Posterior ilium, sacrum/proximal femur
Gluteus medius	Abducts thigh	Ilium/greater trochanter of femur
Adductor group	Adducts thigh	Pubis, ischium/femur and tibia
MUSCLES THAT MOVE THE LEG		
Quadriceps femoris group	Extends leg	Ilium, femur/patellar tendon that continues as a ligament to tibial tuberosity
Sartorius	Flexes, abducts, and rotates leg laterally	Ilium/medial tibia
Hamstring group	Flexes and rotates leg medially, and extends thigh	Ischial tuberosity/lateral and medial tibia
MUSCLES THAT MOVE THE ANKLE AND FOOT		
Gastrocnemius (gas″trok-ne′me-us)	Plantar flexion and eversion of foot	Condyles of femur/calcaneus by way of Achilles tendon
Tibialis anterior (tib″e-a′lis an-te′re-or)	Dorsiflexion and inversion of foot	Condyles of tibia/tarsal and metatarsal bones
Fibularis group	Plantar flexion and eversion of foot	Fibula/tarsal and metatarsal bones
Flexor and extensor digitorum longus	Moves toes	Tibia, fibula/phalanges

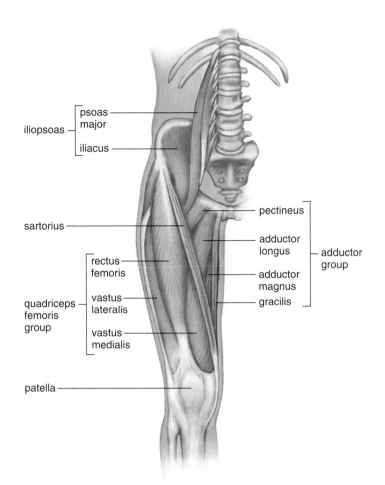

Figure 7.17 Muscles of the anterior right hip and thigh.

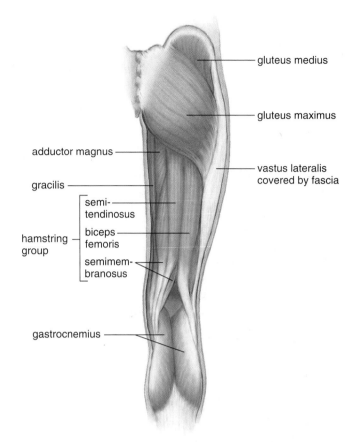

Figure 7.18 Muscles of the posterior right hip and thigh.

functions to abduct the thigh. The gluteus maximus assists the gluteus medius in this function. Therefore, they are synergistic muscles.

Adductor group muscles (pectineus, adductor longus, adductor magnus, gracilis) are located on the medial thigh (Fig. 7.17). All of these muscles originate from the pubis and ischium, and insert on the femur; the deep adductor magnus is shown in Figure 7.17. Adductor muscles adduct the thigh—that is, they lower the thigh sideways from a horizontal position. Because they squeeze the thighs together, these are the muscles that keep a rider on a horse. Notice that the gluts and the adductor group are antagonistic muscles.

Muscles That Move the Leg

The muscles that move the leg originate from the pelvic girdle or femur and insert on the tibia. They are listed in Table 7.5 and illustrated in Figures 7.17 and 7.18. Before studying these muscles, review the movement of the knee when the leg extends and when it flexes (see Chapter 6, page 117).

Quadriceps femoris group (rectus femoris, vastus lateralis, vastus medialis, vastus intermedius), also known as the "quads," is found on the anterior and medial thigh. The rectus femoris, which originates from the ilium, is external to the vastus intermedius, and therefore the vastus intermedius is not shown in Figure 7.17. These muscles are the primary extensors of the leg, as when you kick a ball by straightening your knee.

Sartorius is a long, straplike muscle that has its origin on the iliac spine and then goes across the anterior thigh to insert on the medial side of the knee (Fig. 7.17). Because this muscle crosses both the hip and knee joint, it acts on the thigh in addition to the leg. The insertion of the sartorius is such that it flexes both the leg and the thigh. It also rotates the thigh laterally, enabling us to sit cross-legged, as tailors were accustomed to do in another era. Therefore, it is sometimes called the "tailor's muscle," and in fact, *sartor* means tailor in Latin.

Hamstring group (biceps femoris, semimembranosus, semitendinosus) is located on the posterior thigh (Fig. 7.18). Notice that these muscles also cross the hip and knee joint because they have origins on the ischium and insert on the tibia. They flex and rotate the leg medially, but they also extend the thigh. Their strong tendons can be felt behind the knee. These same tendons are present in hogs and were used by butchers as strings to hang up hams for smoking—hence, the name. Notice that the quadriceps femoris group and the hamstring group are antagonistic muscles in that the quads extend the leg and the hamstrings flex the leg.

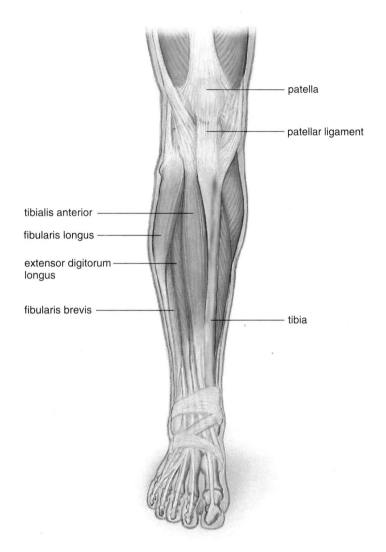

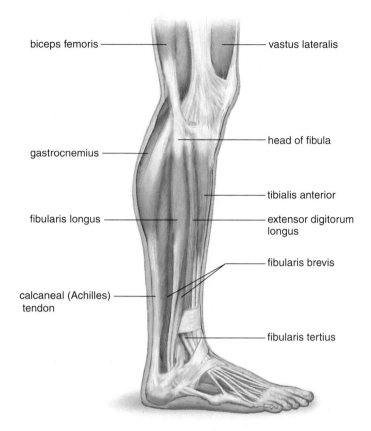

Figure 7.20 Muscles of the lateral right leg.

Figure 7.19 Muscles of the anterior right leg.

Muscles That Move the Ankle and Foot

Muscles that move the ankle and foot are shown in Figures 7.19 and 7.20.

Gastrocnemius is a muscle of the posterior leg, where it forms a large part of the calf. It arises from the femur; distally, the muscle joins the strong calcaneal tendon, which attaches to the calcaneus bone (heel). The gastrocnemius is a powerful plantar flexor of the foot that aids in pushing the body forward during walking or running. It is sometimes called the "toe dancer's muscle" because it allows a person to stand on tiptoe.

Tibialis anterior is a long, spindle-shaped muscle of the anterior leg. It arises from the surface of the tibia and attaches to the bones of the ankle and foot. Contraction of this muscle causes dorsiflexion and inversion of the foot.

Fibularis muscles (fibularis longus, fibularis brevis) are found on the lateral side of the leg, connecting the fibula to the metatarsal bones of the foot. These muscles evert the foot and also help bring about plantar flexion.

Flexor (not shown) and **extensor digitorum longus** muscles are found on the lateral and posterior portion of the leg. They arise mostly from the tibia and insert on the toes. They flex and extend the toes, respectively, and assist in other movements of the feet.

Content **CHECK-UP!**

10. Which muscle is named for its shape?
    a. latissimus dorsi   c. gluteus maximus
    b. triceps brachii   d. frontalis
11. Which muscle of facial expression is also used for swallowing?
    a. frontalis   c. zygomaticus
    b. masseter   d. buccinator
12. Which arm muscle extends the forearm?
    a. biceps brachii   c. triceps brachii
    b. brachialis   d. biceps femoris

7.5 Effects of Aging

Muscle mass and strength tend to decrease as people age. How much of this is due to lack of exercise and a poor diet has yet to be determined. Deteriorated muscle elements are replaced

Benefits of Exercise

Exercise programs improve muscular strength, muscular endurance, and flexibility. Muscular strength is the force a muscle group (or muscle) can exert against a resistance in one maximal effort. Muscular endurance is judged by the ability of a muscle to contract repeatedly or to sustain a contraction for an extended period. Flexibility is tested by observing the range of motion about a joint.

As muscular strength improves, the overall size of the muscle, as well as the number of muscle fibers and myofibrils in the muscle, increases. The total amount of protein, the number of capillaries, and the amounts of connective tissue, including tissue found in tendons and ligaments, also increase. Physical training with weights can improve muscular strength and endurance in all adults, regardless of their age. Over time, increased muscle strength promotes strong bones.

A surprising finding, however, is that health benefits also accompany less strenuous programs, such as those described in Table 7A. A study of 12,000 men by Dr. Arthur Leon at the University of Minnesota showed that even moderate exercise lowered the risk of a heart attack by one-third. People with arthritis reported much less pain, swelling, fatigue, and depression after only four months of attending a twice-weekly, low-impact aerobics class. Increasing daily activity by walking to the corner store instead of driving and by taking the stairs instead of the elevator can improve a person's health.

The benefits of exercise are most apparent with regard to cardiovascular health. Brisk walking for 2.5–4 hours a week can raise the blood levels of high-density lipoprotein (HDL), a chemical that promotes healthy blood vessels (see Chapter 12). Exercise also helps prevent osteoporosis, a condition in which the bones are weak and tend to break. The stronger the bones are when a person is young, the less chance of osteoporosis as a person ages. Exercise promotes the activity of osteoblasts (as opposed to osteocytes) in young people, as well as older people. An increased activity level can also keep off unwanted pounds, which is a worthwhile goal because added body weight contributes to numerous conditions, such as type II diabetes (see page 217). Increased muscle activity is also helpful by causing glucose to be transported into muscle cells and making the body less dependent on the presence of insulin.

Cancer prevention and early detection involve eating properly, not smoking, avoiding cancer-causing chemicals and radiation, undergoing appropriate medical screening tests, and knowing the early warning signs of cancer. However, evidence indicates that because it helps prevent obesity, exercise also helps prevent certain kinds of cancer. Studies show that people who exercise are less likely to develop colon, breast, cervical, uterine, and ovarian cancer.

TABLE 7.A A Checklist for Staying Fit			
Children, 7–12	**Teenagers, 13–18**	**Adults, 19–55**	**Seniors, 56 and Up**
Vigorous activity 1–2 hours daily	Vigorous activity 1 hour 3–5 days a week; otherwise, $\frac{1}{2}$ hour daily moderate activity	Vigorous activity 1 hour 3 days a week; otherwise, $\frac{1}{2}$ hour daily moderate activity	Moderate exercise 1 hour daily 3 days a week; otherwise, $\frac{1}{2}$ hour daily moderate activity
Free play	Build muscle with calisthenics	Exercise to prevent lower back pain: aerobics, stretching, yoga	Take a daily walk
Build motor skills through team sports, dance, swimming	Do aerobic exercise to control buildup of fat cells	Take active vacations: hike, bicycle, cross-country ski	Do daily stretching exercises
Encourage more exercise outside of physical education classes	Pursue tennis, swimming, horseback riding—sports that can be enjoyed for a lifetime	Find exercise partners: join a running club, bicycle club, outing group	Learn a new sport or activity: golf, fishing, ballroom dancing
Initiate family outings: bowling, boating, camping, hiking	Continue team sports, dancing, hiking, swimming		Try low-impact aerobics. Before undertaking new exercises, consult your doctor

initially by connective tissue and, eventually, by fat. With age, degenerative changes take place in the mitochondria, and endurance decreases. Also, changes in the nervous and cardiovascular systems adversely affect the structure and function of muscles.

Muscle mass and strength can improve remarkably if elderly people undergo a training program. Exercise at any age appears to stimulate muscle buildup. As discussed in the Medical Focus above, exercise has many other benefits as well. For example, exercise improves the

MEDICAL FOCUS

Muscular Disorders and Neuromuscular Disease

During the course of a lifetime, nearly everyone will suffer from some type of muscular disorder. Muscular disorders cover a wide spectrum in terms of severity. Minor muscle irritation, inflammation, or injury may resolve without any medical care. However, many diseases affecting the neuromuscular system are extremely serious and eventually prove to be fatal.

Spasms are sudden, involuntary muscular contractions, most often accompanied by pain. Spasms can occur in both smooth and skeletal muscles. A spasm of the intestinal tract is a "bellyache"; most such spasms are not serious. Multiple spasms of skeletal muscles are called a **convulsion. Cramps** are strong, painful spasms, especially of the leg and foot, usually due to strenuous athletic activity. Cramps typically occur after a strenuous workout, and may even occur when sleeping. Facial **tics,** such as periodic eye blinking or grimacing, are spasms that can be controlled voluntarily but only with great effort.

Muscles, joints, and their connective tissues are often subject to overuse injuries: strains, sprains, and tendinitis. A **strain** is caused by stretching or tearing of a muscle. A **sprain** is the twisting of a joint, leading to swelling and injury not only of muscles but also of ligaments, tendons, blood vessels and nerves. The ankle and knee are two areas often subject to sprains. **Tendinitis** is inflammation of a tendon due to repeated athletic activity. The tendons most commonly affected are those associated with the shoulder, elbow, hip, and knee.

Overuse injuries are often minor and can be treated with pain medication and rest. However, one should seek medical attention if the area is extremely painful, hot or swollen, or if accompanied by a fever.

Neuromuscular Diseases

Neuromuscular disease may result from pathologic changes to the muscle itself. It may also result from excessive motor nerve stimulation, or from damage or destruction of the motor neurons that supply the muscle.

Tetanus develops in persons who have not been properly immunized against the toxin of the tetanus bacterium. Tetanus toxin shuts down brain areas which normally inhibit unnecessary muscle contractions. As a result, excessive brain stimulation causes muscles to lock in a tetanic contraction (from which the disease

gets its name.) A rigidly locked jaw is one of the first signs of bacterial infection and toxin production. Though antibiotics will kill the bacteria, once the toxin is circulating in the bloodstream, it cannot be removed or neutralized. Because muscles can't relax, the patient cannot breathe or swallow, and death may occur due to respiratory failure. Immunization and periodic booster shots will prevent the toxin's effects (see Immunization: The Great Protector, p. 290.)

Fibromyalgia is a chronic condition whose symptoms include achy pain, tenderness, and stiffness of muscles. Its precise cause is not known, though 80–90% of sufferers are women. Substance P, a neurotransmitter (messenger chemical) of pain pathways in the brain, has been found in the bloodstream of affected individuals. Exercise seems to decrease blood levels of substance P. Therapeutic massage, over-the-counter pain medication, and muscle relaxants are also recommended.

Muscular dystrophy is a broad term applied to a group of disorders that causes progressive degeneration and weakening of muscles. As muscle fibers die, fat and connective tissue take their place. **Duchenne muscular dystrophy,** the most common type, is inherited through a flawed gene carried by the mother. It is now known that the lack of a protein called *dystrophin* causes the condition. When dystrophin is absent, calcium leaks into the cell and activates an enzyme that dissolves muscle fibers. Treatment includes muscle injections with immature muscle cells that do produce dystrophin.

Myasthenia gravis is an autoimmune disease characterized by weakness that especially affects the muscle of the eyelids, face, neck, and extremities. Muscle contraction is impaired because the immune system mistakenly produces antibodies that destroy acetylcholine receptors on the sarcolemma. (Recall that acetylcholine is the neurotransmitter released by motor neurons.) In many cases, the first signs of the disease are drooping eyelids and double vision. Treatment includes drugs that inhibit the enzyme that digests acetylcholine, thus allowing it to accumulate.

Amyotrophic Lateral Sclerosis (ALS) is often called Lou Gehrig's disease, after its most famous victim, the 1930s-era baseball player. ALS sufferers experience the gradual death of their motor neurons, thus losing the ability to walk, talk, chew, swallow, etc. Intellect and sensation are not affected, however. Drugs can slow the disease's progression, but ALS is always fatal.

---

cardiovascular system and reduces the risk of diabetes and glycation. During glycation, excess glucose molecules stick to body proteins so that the proteins no longer have their normal structure and cannot function properly. Exercise burns glucose and, in this way, helps prevent muscle deterioration.

7.6 Homeostasis

The illustration in Human Systems Work Together on page 151 tells how the muscular system works with other systems of the body to maintain homeostasis.

Cardiac muscle contraction accounts for the heartbeat, which creates blood pressure, the force that propels blood in

Human Systems Work Together

MUSCULAR SYSTEM

Integumentary System

Muscle contraction provides heat to warm skin. Muscle moves skin of face.

Skin protects muscles; rids the body of heat produced by muscle contraction.

Skeletal System

Muscle contraction causes bones to move joints; muscles help protect bones.

Bones provide attachment sites for muscles; store Ca^{2+} for muscle function.

Nervous System

Muscle contraction moves eyes, permits speech, creates facial expressions.

Brain controls nerves that innervate muscles; receptors send sensory input from muscles to brain.

Endocrine System

Muscles help protect glands.

Androgens promote growth of skeletal muscle; epinephrine stimulates heart and constricts blood vessels.

Cardiovascular System

Muscle contraction keeps blood moving in heart and blood vessels.

Blood vessels deliver nutrients and oxygen to muscles, carry away wastes.

How the Muscular System works with other body systems

Lymphatic System/Immunity

Skeletal muscle contraction moves lymph; physical exercise enhances immunity.

Lymphatic vessels pick up excess tissue fluid; immune system protects against infections.

Respiratory System

Muscle contraction assists breathing; physical exercise increases respiratory capacity.

Lungs provide oxygen for, and rid the body of, carbon dioxide from contracting muscles.

Digestive System

Smooth muscle contraction accounts for peristalsis; skeletal muscles support and help protect abdominal organs.

Digestive tract provides glucose for muscle activity; liver metabolizes lactic acid following anaerobic muscle activity.

Urinary System

Smooth muscle contraction assists voiding of urine; skeletal muscles support and help protect urinary organs.

Kidneys maintain blood levels of Na^+, K^+, and Ca^{2+}, which are needed for muscle innervation, and eliminate creatinine,

Reproductive System

Muscle contraction occurs during orgasm and moves gametes; abdominal and uterine muscle contraction occurs during childbirth.

Androgens promote growth of skeletal muscle.

the arteries and arterioles. The walls of the arteries and arterioles contain smooth muscle. Constriction of arteriole walls is regulated to help maintain blood pressure. Arterioles branch into the capillaries where exchange takes place that creates and cleanses tissue fluid. Blood and tissue fluid are the internal environment of the body, and without cardiac and smooth muscle contraction, blood would never reach the capillaries for exchange to take place. Blood is returned to the heart in cardiovascular veins, and excess tissue fluid is returned to the cardiovascular system within lymphatic vessels. Skeletal muscle contraction presses on the cardiovascular veins and lymphatic vessels, and this creates the pressure that moves fluids in both types of vessels. Without the return of blood to the heart, circulation would stop, and without the return of lymph to the blood vessels, normal blood pressure could not be maintained.

The contraction of sphincters composed of smooth muscle fibers temporarily prevents the flow of blood into a capillary. This is an important homeostatic mechanism because in times of emergency it is more important, for example, for blood to be directed to the skeletal muscles than to the tissues of the digestive tract. Smooth muscle contraction also accounts for peristalsis, the process that moves food along the digestive tract. Without this action, food would never reach all the organs of the digestive tract where digestion releases nutrients that enter the bloodstream. Smooth muscle contraction assists the voiding of urine, which is necessary for ridding the body of metabolic wastes and for regulating the blood volume, salt concentration, and pH of internal fluids.

Skeletal muscles protect internal organs, and their strength protects joints by stabilizing their movements. Skeletal muscle contraction raises and lowers the rib cage and diaphragm during the active phases of breathing. As we breathe, oxygen enters the blood and is delivered to the tissues, including the muscles, where ATP is produced in mitochondria with heat as a by-product. The heat produced by skeletal muscle contraction allows the body temperature to remain within the normal range for human beings.

Finally, skeletal muscle contraction moves bones and allows us to perform those daily activities necessary to our health and benefit. Although it may seem as if movement of our limbs does not affect homeostasis, it does so by allowing us to relocate our bodies to keep the external environment within favorable limits for our existence.

SELECTED NEW TERMS

Basic Key Terms

a band (ā bănd), p. 129
actin (ak′tin), p. 129
all-or-none law (ôl-ôr-nŭn lô), p. 136
antagonist (an-tag′o-nist), p. 139
cardiac muscle (kar′de-ak mus′el), p. 127
creatine phosphate (kre′uh-tin fos′fāt), p. 134
H zone (H zōn), p. 129
I band (ī bănd), p. 129
insertion (in-ser′shun), p. 139
motor unit (mo′tor yū′nit), p. 137
muscle fiber (mus′el fi′ber), p. 127
muscle twitch (mus′el twich), p. 136
myofibril (mi″o-fi′bril), p. 129
myoglobin (mi″o-glo′bin), p. 129
myosin (mi′o-sin), p. 129
neuromuscular junction (nu″ro-mus′kyū-ler junk′shun), p. 131
origin (or′ĭ-jin), p. 138

oxygen debt (ok′sĭ-jen debt), p. 135
power stroke (pou′ər strōk), p. 132
prime mover (prĭm mu′ver), p. 139
recruitment (re-krūt′ment), p. 137
sarcomere (sar′ko-mĕr), p. 129
skeletal muscle (skel′ĕ-tal mus′el), p. 127
sliding filament theory (sli′ding fil′uh-ment the′o-re), p. 131
smooth muscle (smūth mus′el), p. 127
summation (sə-mā′shən), p. 137
synergist (sin′er-jist), p. 139
tendon (ten′don), p. 128
tetanic contraction (tē-tăn′ĭk kən-trăk′shən), p. 137
thick filaments (thĭk fĭl′ə-mənts), p. 129
thin filaments (thĭn fĭl′ə-mənts), p. 129
tone (tōn), p. 137
T (transverse) tubules (tranz-vers′ tu′byūl), p. 129
tropomyosin (trō′pə-mĭ′ə-sĭn), p. 129
troponin (trō′pə-nĭn), p. 129

Clinical Key Terms

amyotrophic lateral sclerosis (ALS) (ă′mī-ə-trō′fĭk lăt′ər-əl sklə-rō′sĭs), p. 150
atrophy (at′ro-fe), p. 137
convulsion (kən-vŭl′shən), p. 150
cramps (krămpz), p. 150
fatigue (fə-tēg′), p. 137
fibromyalgia (fī′brō-mī-ăl′je-ə), p. 150
hypertrophy (hi-per′tro-fe), p. 137
muscular dystrophy (mus′kyū-ler dis′trĕ-fe), p. 150
myasthenia gravis (mi″as-the′ne-uh grah′vis), p. 150
spasm (spazm), p. 150
sprain (sprān), p. 150
strain (strān), p. 150
tendinitis (ten″(dĕ-ni′tis), p. 150
tetanus (tet′uh-nus), p. 150
tics (tĭks), p. 150
z lines (z līnz), p. 129

SUMMARY

7.1 **Functions and Types of Muscles**
 A. Muscular tissue is either smooth, cardiac, or skeletal. Skeletal muscles have tubular,
multinucleated, and striated fibers that contract voluntarily.
 B. Skeletal muscles support the body, make bones move, help
maintain a constant body temperature, assist movement in cardiovascular and lymphatic vessels, and help protect

internal organs and stabilize joints.

7.2 **Microscopic Anatomy and Contraction of Skeletal Muscle**
   A. The sarcolemma, which extends into a muscle fiber, forms T tubules; the sarcoplasmic reticulum has calcium storage sites. The placement of actin and myosin in the contractile myofibrils accounts for the striations of skeletal muscle fibers.
   B. Skeletal muscle innervation occurs at neuromuscular junctions. Impulses travel down the tubules of the T system and cause the release of calcium from calcium storage sites. The presence of calcium and ATP in muscle cells prompts actin myofilaments to slide past myosin myofilaments, shortening the length of the sarcomere.
   C. ATP, required for muscle contraction, can be generated by way of creatine phosphate breakdown and fermentation. Lactic acid from fermentation represents an oxygen deficit, because oxygen is required to metabolize this product. Cellular respiration, an aerobic process, is the best source of ATP.

7.3 **Muscle Responses**
   A. In the laboratory, muscle fibers obey the all-or-none law, but whole muscles in the body do not. The occurrence of a muscle twitch, summation, or tetanic contraction depends on the frequency with which a muscle is stimulated.
   B. In the body, muscle fibers belong to motor units that obey the all-or-none law. The strength of muscle contraction depends on the recruitment of motor units. A muscle has tone because some fibers are always contracting.

7.4 **Skeletal Muscles of the Body**
   A. When muscles cooperate to achieve movement, some act as prime movers, others as synergists, and still others as antagonists.
   B. The skeletal muscles of the body are divided into those that move: the head and neck (see Table 7.2); the trunk (see Table 7.3); the shoulder and arm (see Table 7.4); the forearm (see Table 7.4); the hand and fingers (see Table 7.4); the thigh (see Table 7.5); the leg (see Table 7.5); and the ankle and foot (see Table 7.5).

7.5 **Effects of Aging**
   As we age, muscles become weaker, but exercise can help retain vigor.

7.6 **Homeostasis**
   Smooth muscle contraction helps move the blood; cardiac muscle contraction pumps the blood. Skeletal muscle contraction produces heat and is needed for breathing.

STUDY QUESTIONS

1. Name and describe the three types of muscles, and give a general location for each type. (p. 127)
2. List and discuss five functions of muscles. (p. 128)
3. Describe the anatomy of a muscle, from the whole muscle to the myofilaments within a sarcomere. Name the layers of fascia that cover a skeletal muscle and divide the muscle interior. (pp. 129–131)
4. List the sequential events that occur after nerve impulse reaches a muscle. (pp. 131–133)
5. How is ATP supplied to muscles? What is oxygen debt? (pp. 132–135)
6. What is the all-or-none law? What is the difference between a single muscle twitch, summation, and a tetanic contraction? (pp. 136–137)
7. What is muscle tone? How does muscle contraction affect muscle size? (p. 137)
8. Describe how muscles are attached to bones. Define the terms prime mover, synergist, and antagonist. (p. 139)
9. How do muscles get their names? Give an example for each characteristic used in naming muscles. (pp. 139–140)
10. Which of the muscles of the head are used for facial expression? Which are used for chewing? (p. 141)
11. Which muscles of the neck flex and extend the head? (p. 142)
12. What are the muscles of the thoracic wall? What are the muscles of the abdominal wall? (pp. 142–143)
13. Which of the muscles of the shoulder and upper limb move the arm and forearm, and what are their actions? Name the muscles that move the hand and fingers. (pp. 144–146)
14. Which of the muscles of the hip move the thigh, and what are their actions? Which of the muscles of the thigh move the leg, and what are their actions? Which of the muscles of the leg move the feet? (pp. 146–148)

LEARNING OBJECTIVE QUESTIONS

I. Fill in the blanks.

1. _____ muscle is uninucleated, nonstriated, and located in the walls of internal organs.
2. The fascia called _____ separates muscle fibers from one another within a fascicle.
3. When a muscle fiber contracts, an _____ myofilament slides past a myosin myofilament.
4. The energy molecule _____ is needed for muscle fiber contraction.
5. Whole muscles have _____, a condition in which some fibers are always contracted.
6. When muscles contract, the _____ does most of the work, but the _____ help.
7. The _____ is a muscle in the arm that has two origins.
8. The _____ acts as the origin of the latissimus dorsi, and the

_____ acts as the insertion during most activities.

II. **For questions 9–12, name the muscle indicated by the combination of origin and insertion shown.**

	Origin	Insertion
9.	temporal bone	mandibular coronoid process
10.	scapula, clavicle	humerus
11.	scapula, proximal humerus	olecranon process of ulna
12.	posterior ilium, sacrum	proximal femur

III. **Match the muscles in the key to the actions listed in questions 13–18.**

Key:

a. orbicularis oculi
b. zygomaticus
c. deltoid
d. serratus anterior
e. rectus abdominis
f. iliopsoas
g. gluteus maximus
h. gastrocnemius

13. Allows a person to stand on tiptoe
14. Tenses abdominal wall
15. Abducts arm
16. Flexes thigh
17. Raises corner of mouth
18. Closes eyes

MEDICAL TERMINOLOGY EXERCISE

Consult Appendix B for help in pronouncing and analyzing the meaning of the terms that follow.

1. hyperkinesis (hi″per-ki-ne′sis)
2. dystrophy (dis′tro-fe)
3. electromyogram (e-lek″tro-mi′-o-gram)
4. menisectomy (men″i-sek′to-me)
5. tenorrhaphy (te-nor′uh-fe)
6. myatrophy (mi-at′ro-fe)
7. leiomyoma (li″o-mi-o′muh)
8. kinesiotherapy (ki-ne″se-o-thĕr′uh-pe)
9. myocardiopathy (mi″o-kar″de-op′ uh-the)
10. myasthenia (mi″as-the′ne-uh)

WEBSITE LINK

Visit this text's website at http://www.inhhe.com/maderap6 for additional quizzes, interactive learning exercises, and other study tools.

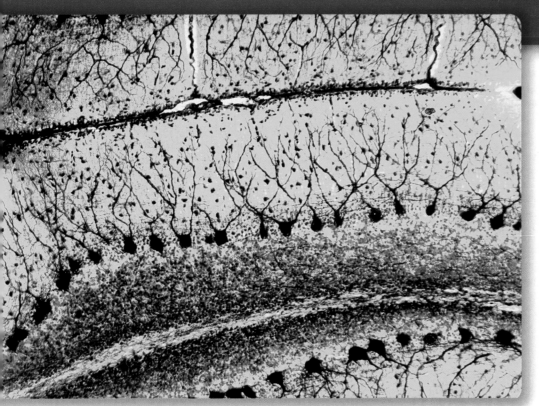

PART III

C H A P T E R

8

The Nervous System

Cells of the cerebral cortex, the superficial layer of brain tissue, function to control sensation, voluntary movements, and conscious thought.

Chapter Outline & Learning Objectives
After you have studied this chapter, you should be able to:

8.1 Nervous System

The nervous system has three specific functions:

1. *Sensory input.* Sensory receptors in skin and organs respond to external and internal stimuli by generating nerve signals that travel to the brain and spinal cord. For example, temperature sensors in the skin may signal to the brain that the air surrounding the body is cold.
2. *Integration.* The brain and spinal cord interpret the data received from sensory receptors all over the body, and signal the appropriate nerve responses. To continue the example of body temperature, sensory information from temperature receptors is sent to the *hypothalamus,* the brain center that controls body temperature.
3. *Motor output.* The nerve impulses from the brain and spinal cord go to the effectors, which are muscles, glands, and organs. Muscle contractions, gland secretions, and changes in organ function are responses to stimuli received by sensory receptors. To adjust body temperature, the hypothalamus triggers shivering—skeletal muscles contract rhythmically, producing heat that warms the body.

It is important to stress that the nervous system maintains **homeostasis** by receiving sensory information, integrating that information, and making an appropriate response.

Divisions of the Nervous System

The nervous system has two major divisions: the central nervous system and the peripheral nervous system (Fig. 8.1). The **central nervous system (CNS)** includes the brain and spinal cord, which have a *central* location—they lie in the midline of the body. The **peripheral nervous system (PNS),** which is further divided into the *afferent* (sensory) and *efferent* (motor) divisions, includes all the cranial and spinal nerves. Nerves have a *peripheral* location in the body, meaning that they project out from the central nervous system. The division between the central nervous system and the peripheral nervous system is arbitrary; the two systems work together, as we shall see.

Nervous Tissue

Although exceedingly complex, nervous tissue is made up of just two principal types of cells: (1) **neurons,** also called nerve cells, which transmit nerve impulses; and (2) **neuroglia,** which support and nourish neurons (see Chapter 4, page 72).

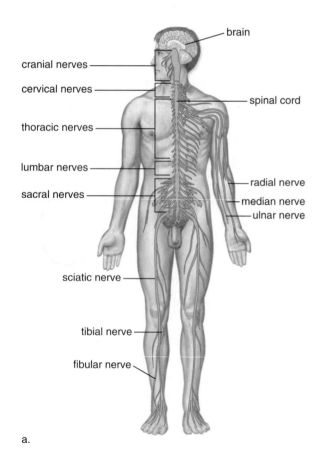

a.

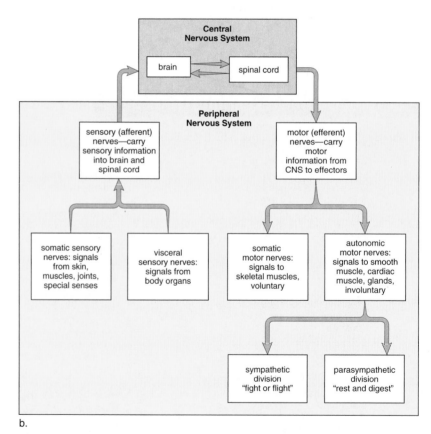

b.

Figure 8.1 Organization of the nervous system in humans. **a.** This pictorial representation shows the central nervous system (CNS; composed of brain and spinal cord) and some of the nerves of the peripheral nervous system (PNS). **b.** The CNS and PNS communicate with each other. Somatic sensory and visceral sensory nerves carry information to the brain and spinal cord. Somatic motor nerves signal skeletal muscles and autonomic motor nerves signal smooth muscle, cardiac muscle, and glands.

Neuron Structure

Neurons vary in appearance, but all of them have just three parts: a cell body, dendrite(s), and an axon. As shown in Figure 8.2, the **cell body** of each type of neuron contains the nucleus, as well as other organelles. **Dendrites** are shorter, branched extensions that receive signals from sensory receptors or other neurons. At the dendrites, signals can result in nerve impulses that are then conducted away from the cell body by an axon. The **axon** is the portion of a neuron that conducts nerve signals.

Axons may be grouped together in bundles. A bundle of parallel axons in the peripheral nervous system is called a **nerve,** whereas a similar axon bundle in the central nervous system is a **tract.** Axons found in nerves or in tracts may be covered by myelin, a lipid coating that insulates the nerve.

The myelin covering of axons in the PNS is formed by neuroglial cells called *Schwann cells* or *neurolemmocytes.* (Recall from Chapter 4 that neuroglial cells are support cells for the brain and nervous system.) *Oligodendrocytes,* another type of neuroglial cell, perform a similar function in the CNS. Gaps in the myelin sheath are called *nodes of Ranvier* (*neurofibril nodes*). These gaps greatly increase the speed of conduction for a nerve signal, as will be discussed shortly.

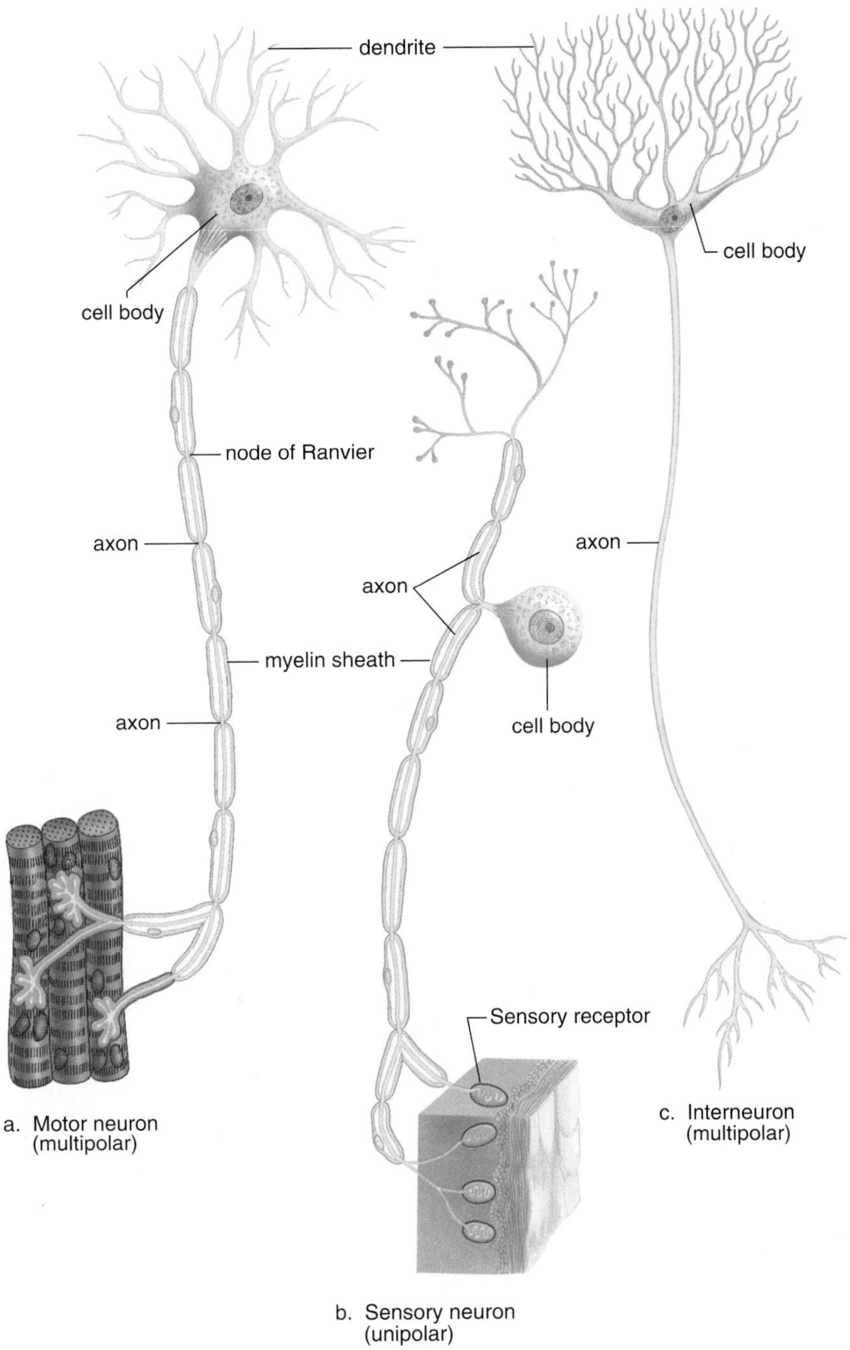

Types of Neurons

Neurons can be classified according to their function and structure. **Motor neurons** take nerve impulses from the CNS to muscles, organs, or glands. Motor neurons are said to be multipolar because they have many dendrites and a single axon (Fig. 8.2*a*). Motor neurons cause muscle fibers to contract, organs to modify their function, or glands to secrete, and therefore they are said to *innervate* these structures.

Sensory neurons take nerve impulses from sensory receptors to the CNS. The sensory receptor, which is the distal end of the long axon of a sensory neuron, may be as simple as a naked nerve ending (a pain receptor), or it may be a part of a highly complex organ, such as the eye or ear. Almost all sensory neurons have a structure that is termed unipolar (Fig. 8.2*b*). In unipolar neurons, the extension from the cell body divides into a branch that comes to the periphery and another that goes to the CNS. Because both branches are long and transmit nerve impulses, it is now generally accepted to refer to them collectively as an axon.

Interneurons, also known as association neurons, occur entirely within the CNS. Interneurons, which are typically multipolar (Fig. 8.2*c*), convey nerve impulses between various parts of the CNS. Some lie between sensory neurons and motor neurons, and some take messages from one side of the spinal cord to the other or from the brain to the cord, and vice versa. They also form complex pathways in the brain where processes accounting for thinking, memory, and language occur.

Nerve Signal Conduction

The function of neurons is to conduct nerve signals and, thus, to transmit information. When neurons are resting, they are not conducting nerve signals,

Figure 8.2 Neuron anatomy. **a.** Motor neuron. Note the branched dendrites and the single, long axon, which branches only near its tip. **b.** Sensory neuron with dendritelike structures projecting from the peripheral end of the axon. **c.** Interneuron (from the cortex of the cerebellum) with highly branched dendrites. Note that the myelin sheath is absent.

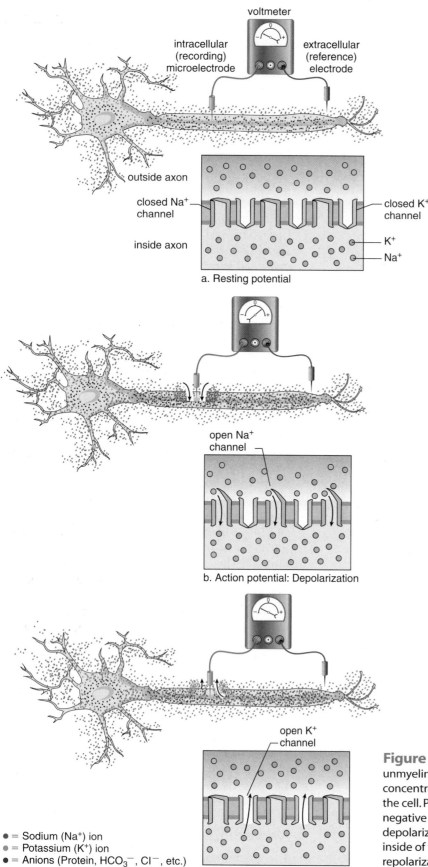

but they are constantly prepared to do so. A nerve signal is also called an *action potential*; it is conducted by the axon of the neuron.

Resting Potential

Anyone who has ever used a battery has employed an energy source that is manufactured by separating positively charged ions across a membrane from negative ions. The battery's *potential energy* can be used to perform work—lighting a flashlight, for example. When a neuron is resting, it too possesses potential energy, much like a fully charged battery. This energy, called the **resting potential**, exists because the cell membrane is *polarized*—positively charged outside the cell and negatively charged inside. The outside of the cell is positive because positively charged sodium ions (Na^+) cluster around the cell. In addition, positively charged potassium ions (K^+) are constantly leaking out of the cell and contribute to the positive charge. The inside of the cell is negative because of the presence of large, negatively charged proteins and other molecules that are stuck inside the cell because of their size.

Like a battery, the neuron's resting potential energy can be measured in volts. A D-size flashlight battery has 1.5 volts; a nerve cell typically has 0.07 volts, or 70 millivolts, of stored energy (Fig. 8.3*a*). By convention, the voltage measurement is assigned a negative value. This is because scientists compare the inside of the cell, (where negatively charged proteins and other large molecules cluster) to the outside of the cell (where positively charged sodium ions are gathered). Just like rechargeable batteries, neurons must maintain their resting potential to be able to work effectively. To do so, neurons continually transport sodium ions out of the cell and return potassium to the cytoplasm. A protein carrier in the membrane, called the sodium-potassium pump, pumps sodium (Na^+) out of the neuron and potassium (K^+) into the neuron. This action effectively "recharges" the cell, so, like a fresh battery, it can perform work.

Figure 8.3 Resting potential and action potential in an unmyelinated axon. **a.** Resting potential. Sodium ions are concentrated outside the cell; potassium and large anions are inside the cell. Potential is approximately 70 mV; the inside of the cell is negative compared to the outside. **b.** Action potential: depolarization. Na^+ gates open and sodium ions flow into cell; the inside of the cell becomes positive. **c.** Action potential: repolarization. K^+ gates open and potassium ions flow out; the inside of the cell becomes negative again.

Action Potential

The resting potential energy of the neuron can be used to perform the work of the neuron: conduction of nerve signals. The process of conduction, called an **action potential**, occurs in the axons of neurons. An action potential begins with a **stimulus,** which activates the neuron. (For example, poking the skin with a sharp pin would be a stimulus for pain neurons in the skin.) The stimulus causes protein channels for sodium located in the cell membrane to open, and sodium ions rush into the cell. Adding positively charged sodium ions causes the inside of the axon to become positive, compared to the outside (Fig. 8.3b). This change is called **depolarization.**

Immediately after depolarization, the channels for sodium close and a separate set of protein channels for potassium open. Potassium rapidly leaves the cell; as positively charged potassium ions exit the cell, the inside of the cell becomes negative again (due to the presence of large negatively charged ions stuck inside the cell). This change in polarity is called **repolarization.**

Conduction of Action Potentials

If an axon is unmyelinated, an action potential at one locale stimulates an adjacent part of the axon membrane to produce an action potential. Conduction along the entire axon in this fashion can be rather slow (approximately 1 meter/second in thin axons) because each section of the axon must be stimulated. In myelinated fibers, an action potential at one node of Ranvier causes an action potential at the next node, jumping over the entire myelin-coated portion of the axon (Fig. 8.4). This type of conduction, called **saltatory conduction,** is much faster: In thick, myelinated fibers, the rate is more than 100 m/sec.

The conduction of an action potential is an all-or-none event—either an axon conducts its action potential or it does not. The intensity of a message is determined by how many action potentials are generated within a given time span. A neuron can conduct a volley of action potentials quickly because only a small number of ions are exchanged with each action potential. Once the action potential is complete, the ions are rapidly restored to their proper place through the action of the sodium-potassium pump. The cell is once more at resting potential, and ready to discharge another action potential if necessary.

As soon as the action potential has passed by each successive portion of an axon, that portion undergoes a short **refractory period** during which it is unable to conduct an action potential. This ensures the one-way direction of an impulse from the cell body down the length of the axon to the axon terminal.

It is interesting to observe that all functions of the nervous system, from our deepest emotions to our highest reasoning abilities, are dependent on the conduction of action potentials.

Transmission Across a Synapse

Every axon branches into many fine endings, each tipped by a small swelling called an **axon terminal.** Each swelling lies very close to either the dendrite or the cell body of another neuron. This region of close proximity is called a **synapse** (Fig. 8.5). At a synapse, the membrane of the first neuron is called the *pre*synaptic membrane, and the membrane of the next neuron is called the *post*synaptic membrane. The small gap between is the **synaptic cleft.**

Transmission across a synapse is carried out by molecules called **neurotransmitters,** which are stored in synaptic vesicles in the axon terminals. When nerve impulses traveling along an axon reach an axon terminal, channels for calcium ions (Ca^{2+}) open, and calcium enters the terminal. This sudden rise in Ca^{2+} stimulates synaptic vesicles to merge with the presynaptic membrane, and neurotransmitter molecules are released into the synaptic cleft. They diffuse across the cleft to the postsynaptic membrane, where they bind with specific receptor proteins.

Depending on the type of neurotransmitter and the type of receptor, the response of the postsynaptic neuron can be toward excitation or toward inhibition. After excitatory neurotransmitters combine with a receptor, a sodium ion channel opens, and Na^+ enters the neuron (Fig. 8.5). Other neurotransmitters have an inhibitory effect as described in the next section.

Graded Potentials and Synaptic Integration

A single neuron can have many dendrites plus the cell body, and both can synapse with many other neurons. Typically, a neuron is on the receiving end of many excitatory and inhibitory synapses (Fig. 8.5). Each of the small signals from a synapse is called a **graded potential.** An excitatory neurotransmitter opens the neuron's sodium channels, producing a graded potential that drives the polarity of a neuron closer to an action potential. An inhibitory neurotransmitter produces a graded potential that makes it harder for a neuron to have

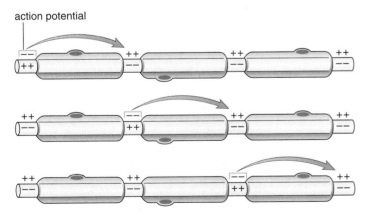

action potential

Figure 8.4 Conduction of an action potential in a myelinated axon. The action potential jumps from one node of Ranvier (neurofibril node) to the next along the axon. This makes the speed of a nerve impulse much faster than in unmyelinated axons. Almost all axons are myelinated in humans.

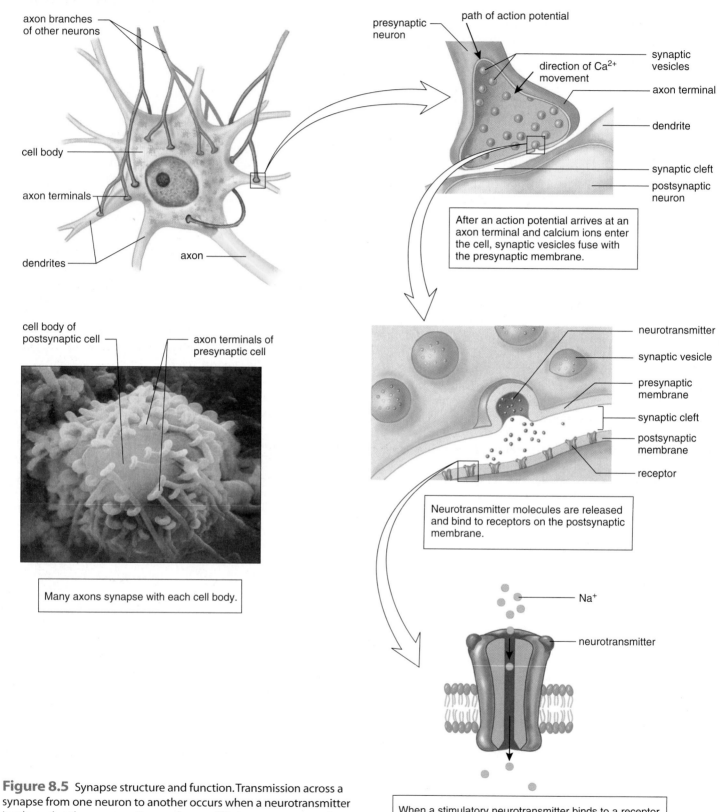

axon branches of other neurons

cell body

axon terminals

dendrites

axon

presynaptic neuron

path of action potential

synaptic vesicles

direction of Ca²⁺ movement

axon terminal

dendrite

synaptic cleft

postsynaptic neuron

After an action potential arrives at an axon terminal and calcium ions enter the cell, synaptic vesicles fuse with the presynaptic membrane.

cell body of postsynaptic cell

axon terminals of presynaptic cell

Many axons synapse with each cell body.

neurotransmitter

synaptic vesicle

presynaptic membrane

synaptic cleft

postsynaptic membrane

receptor

Neurotransmitter molecules are released and bind to receptors on the postsynaptic membrane.

Na⁺

neurotransmitter

When a stimulatory neurotransmitter binds to a receptor, Na⁺ diffuses into the postsynaptic neuron.

Figure 8.5 Synapse structure and function. Transmission across a synapse from one neuron to another occurs when a neurotransmitter is released at the presynaptic membrane, diffuses across a synaptic cleft, and binds to a receptor in the postsynaptic membrane.

Alzheimer Disease

Alzheimer disease (AD) is a disorder characterized by a gradual loss of reason that begins with memory lapses and ends with the inability to perform any type of daily activity. Personality changes signal the onset of AD. A normal 50- to 60-year-old might forget the name of a friend not seen for years. People with AD, however, forget the name of a neighbor who visits daily. People afflicted with AD become confused and tend to repeat the same question. Signs of mental disturbance eventually appear, and patients gradually become bedridden and die of a complication, such as pneumonia.

Researchers have discovered that in some families whose members have a 50% chance of AD, a genetic defect exists on chromosome 21. This is of extreme interest because Down syndrome results from the inheritance of three copies of chromosome 21, and people with Down syndrome tend to develop AD.

AD is characterized by the presence of abnormally structured neurons and a reduced amount of ACh. The AD neuron has two features: (1) Bundles of fibrous protein, called neurofibrillary tangles, surround the nucleus in the cells, and (2) Protein-rich accumulations, called amyloid plaques, envelop the axon branches. These abnormal neurons are especially seen in the portions of the brain involved in reason and memory.

Treatment for Alzheimer Disease

Treatment for AD involves using one of two categories of drugs. Cholinesterase inhibitors (Aricept, Cognex, Exelon, Reminyl) work at neuron synapses in the brain, slowing the activity of the enzyme that breaks down acetylcholine (ACh). Allowing ACh to accumulate in synapses keeps memory pathways in the brain functional for a longer period of time. A second drug, memantine (Namenda), blocks *excitotoxicity*—the tendency of diseased neurons to self-destruct. This recently approved medication is used only in moderately to severely affected patients. The drug allows neurons involved in memory pathways to survive longer in affected patients. However, it is important to note that neither type of medication cures AD—both merely slow the progress of disease symptoms, allowing the patient to function independently for a longer period of time. Additional research is currently underway to test the effectiveness of anti-cholesterol *statin* drugs, as well as anti-inflammatory medications, in slowing the progress of the disease.

Much of current research on AD focuses on the prevention and cure of the disease. Scientists believe that curing AD will require an early diagnosis because it is thought that the disease may begin in the brain 15–20 years before symptoms ever develop. Currently, diagnosis cannot be made with absolute certainty until the brain is examined at autopsy. A new test on the cerebrospinal fluid may allow early detection of amyloid proteins and a much earlier diagnosis of the disease. Researchers are also testing vaccines for AD that would target the patient's immune system to destroy amyloid.

Early findings have shown that risk factors for cardiovascular disease—heart attacks and stroke—also contribute to an increased incidence of AD. Risk factors for cardiovascular disease include elevated blood cholesterol and blood pressure, smoking, obesity, sedentary lifestyle, and diabetes mellitus. Thus, evidence suggests that a lifestyle tailored for good cardiovascular health may also prevent AD.

---

an action potential, typically by opening potassium channels and allowing potassium ions to leave the cell.

Neurons integrate these incoming signals. Integration is the summing up of excitatory and inhibitory signals. If a neuron receives many excitatory signals (either from different synapses, or at a rapid rate from one synapse), chances are the axon will transmit an action potential. On the other hand, if a neuron receives both inhibitory and excitatory signals, the summing up of these signals may prohibit the axon from firing.

Sensory receptors (for example, light receptors in the eye) have special graded potentials called **receptor potentials.** These will be discussed in Chapter 9.

Neurotransmitter Molecules

At least 50 different neurotransmitters have been identified, but two very well-known ones are **acetylcholine (ACh)** and **norepinephrine (NE).**

Once a neurotransmitter has been released into a synaptic cleft and has initiated a response, it is removed from the cleft. In some synapses, the postsynaptic membrane contains enzymes that rapidly inactivate the neurotransmitter. For example, the enzyme **acetylcholinesterase (AChE)** breaks down acetylcholine. In other synapses, the presynaptic membrane rapidly reabsorbs the neurotransmitters, possibly for repackaging in synaptic vesicles or for molecular breakdown. The short existence of neurotransmitters at a synapse prevents continuous stimulation (or inhibition) of postsynaptic membranes.

The Medical Focus above discusses Alzheimer disease, which may be due in part to a lack of ACh in the brain. It is also of interest to note that many drugs are available that enhance or block the release of a neurotransmitter, mimic the action of a neurotransmitter or block the receptor, or interfere with the removal of a neurotransmitter from a synaptic cleft.

8.2 Central Nervous System

The CNS, consisting of the brain and spinal cord, is composed of gray matter and white matter. **Gray matter** is gray because it contains cell bodies and short, nonmyelinated fibers. **White matter** is white because it contains myelinated axons that run together in bundles called tracts. The myelin covering on these axons gives them a shiny, white appearance.

Meninges and Cerebrospinal Fluid

Both the spinal cord and the brain are wrapped in protective membranes known as **meninges** (sing., meninx). The outer meninx, the **dura mater,** is tough, white, fibrous connective tissue that lies next to the skull and vertebrae. The dura mater is constructed of two separate membrane layers that are fused. However, in several areas the layers separate to form the **dural venous sinuses.** The dural venous sinuses collect venous blood and excess cerebrospinal fluid, returning both to the cardiovascular system. Deep to the dura mater is the **arachnoid mater** ("spider-like"), so-called because it consists of spider-web-like connective tissue. Thin strands of the arachnoid mater attach it to the **pia mater,** the deepest meninx. The pia mater is very thin and closely follows the contours of the brain and spinal cord (Fig. 8.6). Between the arachnoid mater and the pia mater is the subarachnoid space, which is filled with **cerebrospinal fluid.** This clear tissue fluid, similar to blood plasma, forms a protective cushion around and within the CNS.

Cerebrospinal fluid (CSF) is produced by **ependymal cells,** a type of neuroglial cell. These cells are found in hollow, interconnecting cavities of the brain called **ventricles.** CSF fills brain ventricles, as well as the hollow **central canal** of the spinal cord. Normally, any excess CSF is drained into the dural venous sinuses to return to the cardiovascular system. However, blockages can occur. In an infant, the brain can enlarge due to cerebrospinal fluid accumulation, resulting in a condition called **hydrocephalus** ("water on the brain").

The Spinal Cord

The **spinal cord** is a cylinder of nervous tissue that begins at the base of the brain and extends through a large opening in the skull called the **foramen magnum** (see Chapter 6, Fig. 6.8).

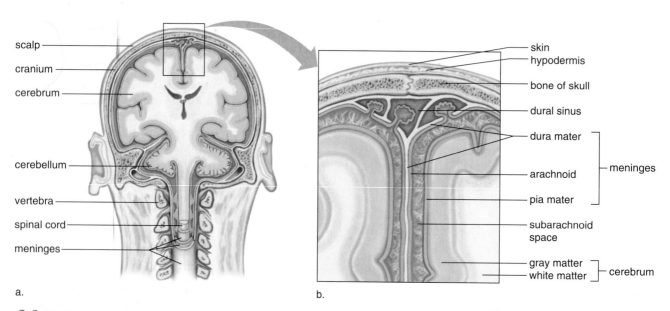

a.

b.

Figure 8.6 Meninges. **a.** Meninges are protective membranes that enclose the brain and spinal cord. **b.** The meninges include three layers: the dura mater, the arachnoid, and the pia mater.

The spinal cord is protected by the vertebral column, which is composed of individual vertebrae. The cord passes through the vertebral canal formed by openings in the vertebrae. It ends at the first lumbar vertebra (see Fig. 6.8).

Structure of the Spinal Cord

Figure 8.7*a* shows how an individual vertebra protects the spinal cord. The spinal nerves extend from the cord between the vertebrae. Intervertebral disks, which are composed of tough fibrocartilage and filled with gelatinous material, separate each vertebra. If a disk is torn open, this **herniated disk** may compress spinal nerves. Pain and loss of function result.

A cross section of the spinal cord shows a central canal, gray matter, and white matter (Fig. 8.7*b,c*). The central canal contains cerebrospinal fluid, as do the meninges that protect the spinal cord. The gray matter is centrally located and shaped like the letter H. Portions of sensory neurons and motor neurons are found there, as are interneurons that communicate with these two types of neurons. The posterior (dorsal) root of a spinal nerve contains sensory fibers entering the gray matter, and the anterior (ventral) root of a spinal nerve contains motor fibers exiting the gray matter. The posterior and anterior roots join, forming a spinal nerve that leaves the vertebral canal. Spinal nerves are a part of the PNS.

The white matter of the spinal cord contains *ascending tracts*, which take sensory information to the spinal cord and brain, and *descending tracts*, which take motor information from the brain. Ascending tracts are generally located in the posterior white matter; descending tracts are found in the anterior white matter. Because the tracts typically cross just after they enter and exit the brain, the left brain controls the right side of the body and the right brain controls the left side of the body.

Functions of the Spinal Cord

The spinal cord provides a means of communication between the brain and the peripheral nerves that leave the cord.

When someone touches your hand, sensory receptors generate action potentials that travel by way of sensory nerve axons to the spinal cord. One of several ascending tracts next carries the information to the sensory area of the brain. When you voluntarily move your limbs, action potentials originating in the motor control area of the brain pass down one of several descending tracts to the spinal cord and out to your muscles by way of motor nerve axons. The "What's New" on page 165 discusses promising new therapies for patients whose spinal cord is injured.

We will see that the spinal cord is also the center for thousands of reflex arcs (see Fig. 8.13): A stimulus causes sensory receptors to generate action potentials that travel in sensory neurons to the spinal cord. Interneurons integrate the incoming data and relay signals to motor neurons. A response to the stimulus occurs when motor axons cause skeletal muscles to

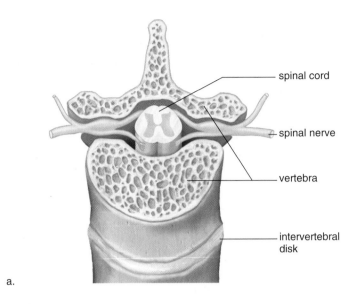

a.

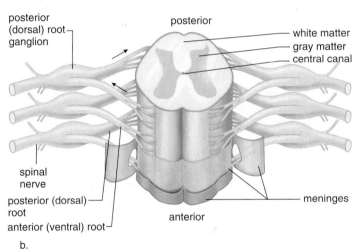

b.

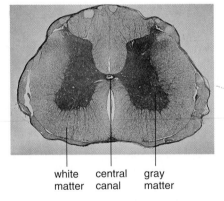

c.

Figure 8.7 Spinal cord. **a.** The spinal cord passes through the vertebral canal formed by the vertebrae. Spinal nerves branch off the spinal cord and project through openings between the vertebrae. **b.** The spinal cord has a central canal filled with cerebrospinal fluid, gray matter in an H-shaped configuration, and white matter elsewhere. The white matter contains tracts that take nerve impulses to and from the brain. **c.** Photomicrograph of a cross section of the spinal cord.

contract. Each interneuron in the spinal cord has synapses with many other neurons, and therefore they send signals to several other interneurons (as well as to the brain) in addition to motor neurons.

The Brain

We will discuss the parts of the brain with reference to the cerebrum, the diencephalon, the cerebellum, and the brain stem. The brain's four ventricles are called, in turn, the two lateral ventricles, the third ventricle, and the fourth ventricle. It will be helpful for you to associate the cerebrum with the two lateral ventricles, the diencephalon with the third ventricle, and the brain stem and the cerebellum with the fourth ventricle (Fig. 8.8a).

The electrical activity of the brain can be recorded in the form of an **electroencephalogram (EEG)**. Electrodes are taped to different parts of the scalp, and an instrument records the so-called brain waves. The EEG is a diagnostic tool; for example, an irregular pattern can signify epilepsy or a brain tumor. Absence of electrical activity on an EEG signifies brain death.

The Cerebrum

The **cerebrum** is the largest portion of the brain in humans. The cerebrum is the last center to receive sensory input and carry out integration before commanding voluntary motor responses. It communicates with and coordinates the activities of the other parts of the brain. The cerebrum carries out the higher thought processes required for learning and memory and for language and speech.

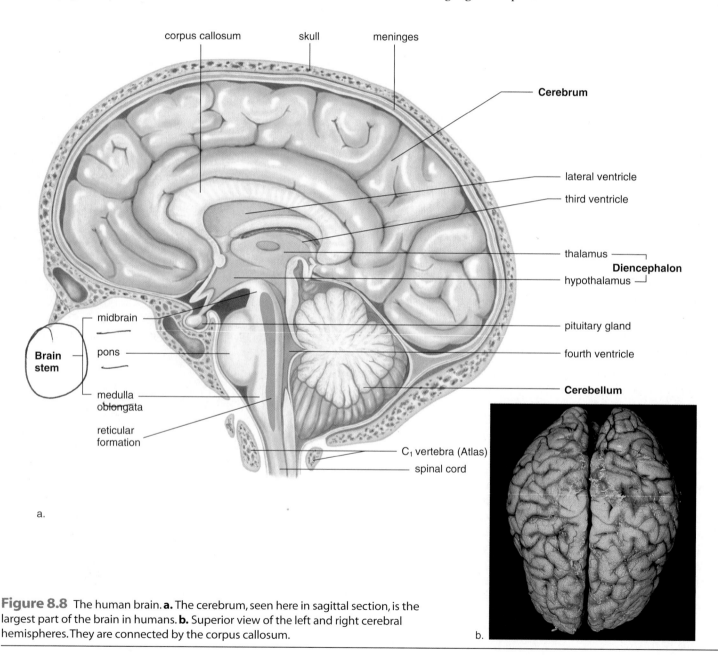

Figure 8.8 The human brain. **a.** The cerebrum, seen here in sagittal section, is the largest part of the brain in humans. **b.** Superior view of the left and right cerebral hemispheres. They are connected by the corpus callosum.

Spinal Cord Injuries—Cause for Hope *In Memoriam: Christopher Reeve, 1952–2004*

It happened in a split second: The tall, athletic actor, who achieved worldwide recognition in his role as "Superman," was thrown from his horse in 1995. In the accident, Christopher Reeve shattered both atlas and axis, the first two cervical vertebrae, and became a ventilator-dependent quadriplegic. Although he spent the remainder of his life in a wheelchair, Reeve redefined the recovery process for spinal cord injury with tremendous courage and determination.

Spinal cord injuries can result from accidents or other trauma. It is rare for the cord to be completely cut across (transection). Commonly, the damage is caused by compression or bruising to one side. Injured cells in the center of the cord will die, leaving a rim of functional cells. The location and extent of the damage produce a variety of effects, depending on the partial or complete stoppage of impulses passing up and down the spinal cord. If the spinal cord is completely transected, no sensations or somatic motor impulses traveling in the cord are able to pass the point where the cord is cut. If the injury is between the first thoracic vertebra (T1) and the second lumbar vertebra (L2), paralysis of the lower body and legs occurs. This condition is known as **paraplegia.** If the injury is between the cervical vertebrae and the first thoracic vertebra (T1), the entire body and all four limbs are usually affected. This condition is called **quadriplegia.** If the injury is a unilateral hemisection (half-cut), motor loss will occur on the same side as the injury because motor neuron crossover occurs in the medulla oblongata. At the same time, loss of sensation will vary and the pattern and type of such loss can be analyzed to locate the lesion. Other serious consequences of spinal cord injury include exaggerated reflexes, spasms in skeletal muscles, loss of breathing ability and sexual function, loss of bowel and bladder control, and pain. *Autonomic dysreflexia* is a life-threatening complication that causes elevated blood pressure, heart rate, and temperature.

After his injury, Reeve continued to pursue recovery, despite expert medical opinions that his case was hopeless. He pioneered *functional electrical stimulation* (FES), the technique of using computer-directed muscle activation to move the limbs in exercise. FES therapy has since been shown to improve muscle tone and bone density in spinal cord-injured patients, and several studies have shown that such exercise seems to stimulate spinal cord recovery. In Reeve's case, it worked: His was the **first** documented case of significant improvement in a quadriplegic occurring more than 2 years after injury. Reeve regained movement of his fingers and toes, as well as sensation over most of his body. He was also able to finally breathe independently for several hours at a time. Ironically, Reeve died in 2004 of complications resulting from an infected *decubitus ulcer* (bedsore; see page 82), a common affliction of quadriplegics.

The Christopher Reeve Foundation, and others dedicated to spinal cord research, have made significant advances. Emergency care of the spinal cord-injured patient has improved, and new drugs allow for improved survival of neural tissue in the injured cord. Prosthetics for the spinal cord-injured include stair-climbing and voice-activated wheelchairs, electronic devices that allow independent breathing, and devices to restore control of bladder and bowel. Studies using grafts of neural tissue, neural stem cells, and nerve growth factors—hormone-like chemicals that stimulate growth—have shown promise in regenerating spinal cord tissue. The goal of the Christopher Reeve Paralysis Foundation (www.christopherreeve.org) is to one day produce a cure for spinal cord paralysis.

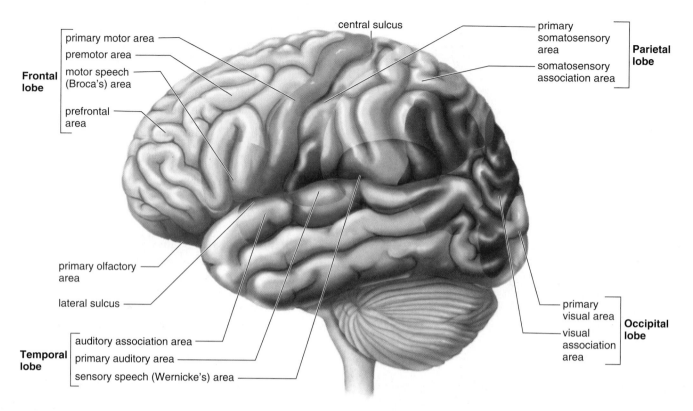

primary motor area

premotor area

motor speech (Broca's) area

prefrontal area

Frontal lobe

central sulcus

primary somatosensory area

somatosensory association area

Parietal lobe

primary olfactory area

lateral sulcus

primary visual area

visual association area

Occipital lobe

auditory association area

primary auditory area

sensory speech (Wernicke's) area

Temporal lobe

Figure 8.9 The lobes of a cerebral hemisphere. Each cerebral hemisphere is divided into four lobes: frontal, parietal, temporal, and occipital. These lobes contain centers for reasoning and movement (frontal lobe), somatic sensing including taste (parietal lobe), hearing (temporal lobe), and vision (occipital lobe). Broca's area is only in the left lobe.

The Cerebral Hemispheres The cerebrum has two halves called the left and right **cerebral hemispheres** (Fig. 8.8b). A deep groove, the longitudinal *fissure*, divides the left and right cerebral hemispheres. Still, the two cerebral hemispheres are connected by a bridge of white matter called the **corpus callosum.**

Ridges called *gyri* are separated by shallow grooves called *sulci* (sing., sulcus). Specific sulci divide each hemisphere into lobes (Fig. 8.9). Note that each lobe of the brain is located underneath the skull bone that shares its name. The *frontal lobe* lies under the frontal bone, anterior to the *parietal lobe* and bone. The *occipital lobe* is deep to the occipital bone, in the posterior area of the cranial vault. The *temporal lobe* is the lateral portion of the cerebral hemisphere. A fifth, very small lobe called the *insula* lies directly deep to the lateral sulcus (not shown on Fig. 8.9).

The **cerebral cortex** is a thin but highly convoluted outer layer of gray matter that covers the cerebral hemispheres. The cerebral cortex contains over one billion cell bodies and is the region of the brain that accounts for sensation, voluntary movement, and all the thought processes we associate with consciousness.

Motor and Sensory Areas of the Cortex The **primary motor area** is in the frontal lobe just anterior to the central sulcus (Fig. 8.9). Voluntary commands to skeletal muscles begin in the primary motor area, and each part of the body is controlled by a certain section (see Fig. 8.10a).

The **primary somatosensory area** is just posterior to the central sulcus in the parietal lobe. Sensory information from the skin and skeletal muscles arrives here, where each part of the body is sequentially represented (see Fig. 8.10b). As you study Figure 8.10, notice that the areas of the body with the greatest voluntary control—the face and hands—have the largest area of motor cortex in the brain dedicated to them. Similarly, the face and hands are among the most sensitive areas of the body, and the large area of the sensory cortex receiving information from them corresponds to that fact.

A primary taste area, located within adjacent areas of the parietal lobe and insula, accounts for taste sensations. A primary visual area in the occipital lobe receives information from our eyes, and a primary auditory area in the temporal lobe receives information from our ears.

Association Areas **Association areas** are places where integration occurs and where memories are stored. Anterior to the primary motor area is a premotor area. The premotor area organizes motor functions for skilled motor activities, and then the primary motor area sends signals to the cerebellum and the basal nuclei, which integrate them. A momentary lack of oxygen during birth can damage the motor areas of the cerebral cortex so that **cerebral palsy,** a

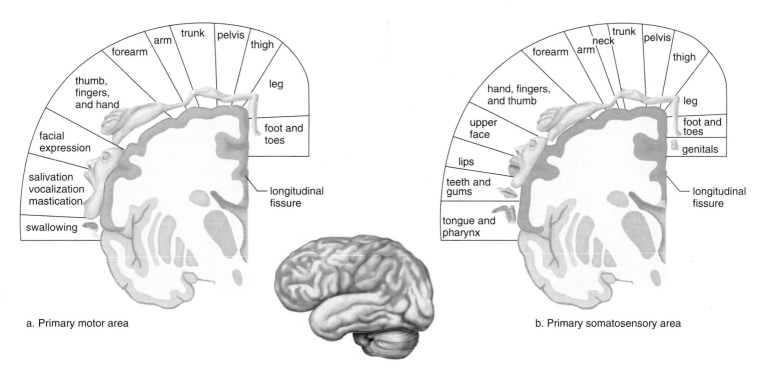

a. Primary motor area

b. Primary somatosensory area

Figure 8.10 Portions of the body controlled by the primary motor area and the primary somatosensory area of the cerebrum. Notice that the size of the body part in the diagram reflects the amount of cerebral cortex devoted to that body part.

condition characterized by a spastic weakness of the arms and legs, develops.

The somatosensory association area, located just posterior to the primary somatosensory area, processes and analyzes sensory information from the skin and muscles. The visual association area associates new visual information with memories of previously received visual information. It might "decide", for example, whether we have seen this face, tool, or whatever before. The auditory association area performs the same functions with regard to sounds.

Processing Centers There are a few areas of the cortex that receive information from the other association areas and perform higher-level analytical functions. The **prefrontal area**, a processing area in the frontal lobe, receives information from the other association area and uses this information to reason and plan our actions. Integration in this area accounts for our most cherished human abilities to think critically and to formulate appropriate behaviors.

The unique ability of humans to speak is partially dependent upon the **motor speech area**, also called **Broca's area**, a processing area usually located in the left frontal lobe. Signals originating here pass to the premotor area before reaching the primary motor area. Damage to this area can interfere with a person's ability to control the muscles of the face and neck that allow speech.

Wernicke's area, also called the **general interpretive area**, receives information from all of the other sensory asso-

ciation areas. Damage to this area hinders a person's ability to interpret written and spoken messages, even though the individual understands the words. Wernicke's area and Broca's area cooperate to allow human communication.

The Medical Focus (p. 168) describes additional ways in which the brain uses separate areas to carry out specific functions.

Central White Matter Much of the rest of the cerebrum beneath the cerebral cortex is composed of white matter. Tracts within the cerebrum take information between the different sensory, motor, and association areas pictured in Figure 8.9. The corpus callosum, previously mentioned, contains tracts that join the two cerebral hemispheres. Descending tracts from the primary motor area communicate with various parts of the brain, and ascending tracts from lower brain centers send sensory information up to the primary somatosensory area (Fig. 8.10).

Basal Nuclei While the bulk of the cerebrum is composed of tracts, there are masses of gray matter located deep within the white matter. These so-called **basal nuclei** (formerly termed basal ganglia) integrate motor commands, ensuring that proper muscle groups are activated or inhibited. Huntington disease and Parkinson disease, which are both characterized by uncontrollable movements, are believed to be due to an imbalance of neurotransmitters in the basal nuclei.

Left and Right Brain

Current research indicates that the right side of the cerebral hemisphere handles emotion and holistic thoughts ("the big picture"), and is more intuitive than the left side. The left side appears to handle language, math, and music, and is said to be the "rational" side of the brain. Brain imaging techniques illustrate more activity in the right hemisphere for artists and navigators. The motor cortex, cerebellum, and basal ganglia are more organized in dancers and other athletes, while individuals who work with people, such as psychologists, use their limbic system more efficiently.

From ages 7–10 years to adulthood, males are observed to excel at visual-spatial skills, whereas females during the same time period are more generalists. In general, males use the left hemisphere (including Broca's area) more while females use both hemispheres equally. This explains why males tend to have more speaking difficulties after a stroke affects the brain's left side than females in the same situation. Females have an analogous region to Broca's area in their right side that can take over speech functions.

Limbic System The **limbic system** (blue in the following figure) is a collection of structures that lies just inferior to the cerebral cortex and contains neural pathways that connect portions of the cerebral cortex and the temporal lobes with the thalamus and the hypothalamus:

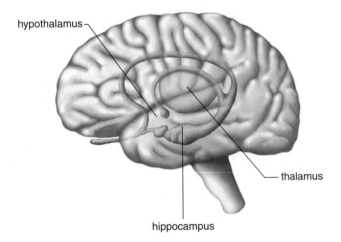

Stimulation of different areas of the limbic system causes the subject to experience rage, pain, pleasure, or sorrow. By causing pleasant or unpleasant feelings about experiences, the limbic system apparently guides the individual into behavior that is likely to increase the chance of survival.

The limbic system is also involved in learning and memory. In particular, the most inferior structure of the limbic system, the **hippocampus,** is vital in processing of short-term memory to become long-term memory. Learning requires memory and memory is stored in the sensory regions of the cerebrum, but just what permits memory development is not definitely known. The involvement of the limbic system in memory explains why emotionally charged events result in our most vivid memories. The fact that the limbic system communicates with the sensory areas for touch, smell, vision, and so forth accounts for the ability of any particular sensory stimulus to awaken a complex memory.

The Diencephalon

The hypothalamus and the thalamus are both in the **diencephalon,** a region that encircles the third ventricle (see Fig. 8.8a). The **hypothalamus** forms the floor of the third ventricle. The hypothalamus is an integrating center that helps maintain homeostasis by regulating hunger, sleep, thirst, body temperature, and water balance. The hypothalamus produces the hormones secreted by the posterior pituitary gland and secretes hormones that control the anterior pituitary. Therefore, it is a link between the nervous and endocrine systems.

The **thalamus** consists of two masses of gray matter located in the sides and roof of the third ventricle. The thalamus is on the receiving end for all sensory input except smell; it functions as a sensory "relay center." Visual, auditory, and somatosensory information arrive at the thalamus via the cranial nerves and tracts from the spinal cord. The thalamus integrates this information and sends it on to the appropriate portions of the cerebrum. The thalamus is involved in arousal of the cerebrum, and it also participates in higher mental functions such as memory and emotions.

The pineal gland, which secretes the hormone melatonin and regulates our body's daily rhythms, is located in the diencephalon.

The Cerebellum

The **cerebellum** is separated from the brain stem by the fourth ventricle (see Fig. 8.8a). The cerebellum has two hemispheres, which are joined by a narrow median portion. Each portion is primarily composed of white matter, which in longitudinal section has a treelike pattern. Overlying the white matter is a thin layer of gray matter that forms a series of complex folds.

The cerebellum receives sensory input from the eyes, ears, joints, and muscles about the present position of body parts. It also receives motor output from the cerebral cortex about where these parts should be located. After integrating this information, the cerebellum sends motor impulses by way of the brain stem to the skeletal muscles. In this way, the cerebellum maintains posture and balance. It also ensures that all of the muscles work together to produce smooth, coordinated voluntary movements. In addition, the cerebellum assists the learning of new motor skills, such as playing the piano or hitting a baseball.

The Brain Stem

The **brain stem** contains the midbrain, the pons, the medulla oblongata, and the reticular formation (see Fig. 8.8*a*). The **midbrain** acts as a relay station for tracts passing between the cerebrum and the spinal cord or cerebellum. It also has reflex centers for visual, auditory, and tactile responses. The word *pons* means "bridge" in Latin, and true to its name, the **pons** contains bundles of axons traveling between the cerebellum and the rest of the CNS. In addition, the pons functions with the medulla oblongata to regulate breathing rate and has reflex centers concerned with head movements in response to visual and auditory stimuli.

The **medulla oblongata** contains a number of reflex centers for regulating heartbeat, breathing, and vasoconstriction. It also contains the reflex centers for vomiting, coughing, sneezing, hiccuping, and swallowing. The medulla oblongata lies just superior to the spinal cord, and it contains tracts that ascend or descend between the spinal cord and higher brain centers.

The **reticular formation** assists the cerebellum in maintaining muscle tone; it also assists the pons and medulla in regulating respiration, heart rate, and blood pressure. The sensory component of the reticular formation processes sensory stimuli—sounds, sights, touch—and uses these signals to keep us mentally alert. Additionally, this portion of the brain helps to rouse a sleeping person.

Content CHECK-UP!

4. The deepest of the three meninges is the:
   a. dura mater.
   c. pia mater.
   b. arachnoid mater.

5. Which of the following statements is not correct?
   a. The motor cortex is in the parietal lobe of the brain.
   b. The primary somatosensory area is in the parietal lobe.
   c. The primary visual area is in the occipital lobe.
   d. The primary auditory area is in the temporal lobe.

6. Which structure integrates motor commands to make sure that proper muscle groups are activated or inhibited?
   a. Wernicke's area
   c. limbic system
   b. corpus callosum
   d. basal nuclei

8.3 Peripheral Nervous System

The peripheral nervous system (PNS) lies outside the central nervous system and is composed of nerves and ganglia. Nerves are bundles of axons, both myelinated and unmyelinated, that travel together. Ganglia (sing., **ganglion**) are swellings associated with nerves that contain collections of cell bodies. As with muscles, connective tissue separates axons at various levels of organization:

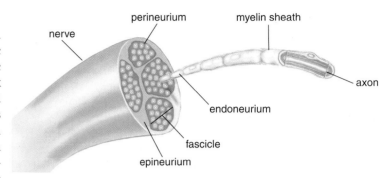

The PNS is subdivided into the **afferent,** or **sensory, system** and the **efferent,** or **motor, system** (see Fig. 8.1). The **somatic sensory system** serves the skin, skeletal muscles, joints, and tendons. The special senses (vision, hearing, taste, smell) are also part of the somatic sensory system. The **visceral sensory system** supplies the internal organs. Nerves from both sensory systems take information from peripheral sensory receptors to the CNS.

The **somatic motor system** carries commands away from the CNS to the skeletal muscles. The **autonomic motor system,** with a few exceptions, regulates the activity of cardiac and smooth muscles and glands.

Types of Nerves

The **cranial nerves** are attached to the brain, and the **spinal nerves** are attached to the spinal cord.

Cranial Nerves

Humans have 12 pairs of cranial nerves (Table 8.1). By convention, the pairs of cranial nerves are referred to by Roman numerals (Fig. 8.11*a*). Some cranial nerves are sensory nerves— that is, they contain only sensory fibers; some are *motor nerves,* containing only motor fibers; and others are *mixed nerves,* so called because they contain both sensory and motor fibers. Cranial nerves are largely concerned with the head, neck, and facial regions of the body. However, the *vagus nerve* (cranial nerve X), has sensory and motor branches to the face and most of the internal organs. This nerve, whose name aptly derives from a Latin word that means "traveler," contains both somatic and visceral sensory nerves. Its efferent nerves belong to the somatic and the autonomic motor systems.

Spinal Nerves

Humans have 31 pairs of spinal nerves; one of each pair is on either side of the spinal cord (Fig. 8.11*b*). The spinal nerves are grouped as shown in Table 8.2 because they are at either the cervical, thoracic, lumbar, or sacral regions of the vertebral column. The spinal nerves are designated according to their

TABLE 8.1 Cranial Nerves

Nerve	Type		Brain Pathway	Transmits Nerve Impulses to (Motor) or from (Sensory)
Olfactory (I)	Sensory		I: Mucous membrane of nose to olfactory bulbs	Olfactory receptors for sense of smell
Optic (II)	Sensory		II: Retina → optic nerve → thalamus → occipital lobe	Retina for sense of sight
Oculomotor (III)	Motor		III: Midbrain → eye and eyelid	Eye muscles (including eyelids and lens); pupil (parasympathetic division)
Trochlear (IV)	Motor		IV: Midbrain → eye	Eye muscles
Trigeminal (V)	Mixed	Sensory	V: Sensory: Teeth, eye, skin, Tongue → pons	Teeth, eyes, skin, and tongue
		Motor	Motor: Pons → jaw muscles	Jaw muscles (chewing)
Abducens (VI)	Motor		VI: Pons → eye	Eye muscles
Facial (VII)	Mixed	Sensory	VII: Sensory: Tongue → pons	Taste buds of anterior tongue
		Motor	Motor: Pons → facial muscles, Salivary glands, tear glands	Facial muscles (facial expression) and glands (tear and salivary)
Vestibulocochlear (VIII)	Sensory		VIII: Inner ear → pons and medulla	Inner ear for sense of balance and hearing
Glossopharyngeal (IX)	Mixed	Sensory	IX: Sensory: Tongue, throat → pons	Pharynx
		Motor	Motor: Pons → Salivary gland, Throat muscles	Pharyngeal muscles (swallowing), salivary glands
Vagus (X)	Sensory Motor		X: Sensory: Eardrum, ear canal, throat, heart, lungs, abdominal organs → medulla	Internal organs, external ear canal, eardrum, back of throat
			Motor: Medulla → throat and larynx, heart, lungs, abdominal organs	Internal organs (parasympathetic division), throat muscles (somatic motor division)
Spinal accessory (XI)	Motor		XI: Medulla → muscles of throat, neck, shoulder	Neck and back muscles
Hypoglossal (XII)	Motor		XII: Medulla → tongue muscles	Tongue muscles

location in relation to the vertebrae because each passes through an intervertebral foramen as it leaves the spinal cord. This organizational principle is illustrated in Figure 8.12.

Spinal nerves are called mixed nerves because they contain both sensory fibers that conduct impulses to the spinal cord from sensory receptors and motor fibers that conduct impulses away from the cord to effectors. The sensory fibers enter the cord via the posterior root, and the motor fibers exit by way of the anterior root. The cell body of a sensory neuron is in a **posterior (dorsal)-root ganglion.** Each spinal nerve serves the particular region of the body in which it is located.

Somatic Motor Nervous System and Reflexes

Most actions in the somatic motor nervous system are voluntary. These actions, such as when we decide to move a limb, always originate in the motor cortex. Recall that the motor

cortex is in the posterior part of the frontal lobe. Other actions in the somatic motor nervous system are due to **reflexes**— automatic involuntary responses to changes occurring inside or outside the body. A reflex occurs quickly; we don't even have to think about it. Reflexes are protective mechanisms that are essential to homeostasis. They keep the internal organs functioning within normal bounds and protect the body from external harm.

Some reflexes, called *cranial reflexes,* involve the brain, as when we automatically blink our eyes when an object nears the eye suddenly. Figure 8.13 illustrates the path of a reflex within the somatic motor nervous system that involves only the spinal cord (called a *spinal reflex*). If your skin is poked with a sharp pin, a pain sensory receptor in the skin generates action potentials that move along a sensory fiber through the posterior root ganglia toward the spinal cord. Sensory neurons enter the cord posteriorly and pass signals

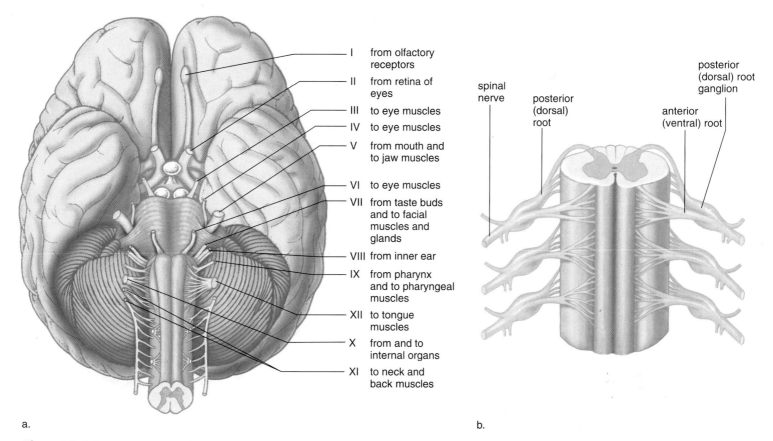

I — from olfactory receptors
II — from retina of eyes
III — to eye muscles
IV — to eye muscles
V — from mouth and to jaw muscles
VI — to eye muscles
VII — from taste buds and to facial muscles and glands
VIII — from inner ear
IX — from pharynx and to pharyngeal muscles
XII — to tongue muscles
X — from and to internal organs
XI — to neck and back muscles

spinal nerve   posterior (dorsal) root   posterior (dorsal) root ganglion   anterior (ventral) root

a.

b.

Figure 8.11 Cranial and spinal nerves. **a.** Ventral surface of the brain showing the attachment of the 12 pairs of cranial nerves. **b.** Cross section of the spinal cord, showing three pairs of spinal nerves. Each spinal nerve has a posterior root and an anterior root that join shortly beyond the cord.

TABLE 8.2 Spinal Nerves		
Name	**Spinal Nerves Involved***	**Function**
Musculocutaneous nerves	C_5–T_1	Supply muscles of the arms on the anterior sides, and skin of the forearms
Radial nerves	C_5–T_1	Supply muscles of the arms on the posterior sides, and skin of the forearms and hands
Median nerves	C_5–T_1	Supply muscles of the forearms, and muscles and skin of the hands
Ulnar nerves	C_5–T_1	Supply muscles of the forearms and hands, and skin of the hands
Phrenic nerves	C_3–C_5	Supply the diaphragm
Intercostal nerves	T_2–T_{12}	Supply intercostal muscles, abdominal muscles, and skin of the trunk
Femoral nerves	L_2–L_4	Supply muscles and skin of the anterior thighs and legs
Sciatic nerves	L_4–S_3	Supply muscles and skin of the posterior thighs, legs, and feet
*C = cervical; T = thoracic; L = lumbar		

on to many interneurons. Some of these interneurons synapse with motor neurons whose short dendrites and cell bodies are in the spinal cord. Nerve action potentials travel along a motor fiber to an effector, which brings about a response to the stimulus. In this case, the effector is a skeletal muscle, which contracts so that you withdraw your hand from the pain stimulus.

Various other reactions are also possible—you will most likely look at the pin, wince, and cry out in pain. This whole series of responses occurs because certain interneurons carry nerve impulses to the brain via tracts in the spinal cord and brain. The brain makes you aware of the stimulus and directs your other reactions to it. You don't feel pain until the brain receives the information and interprets it.

plantar flexion due to contraction of the gastrocnemius and soleus muscles.

Some reflexes are important for avoiding injury, but the knee-jerk and ankle-jerk reflexes are important for normal physiological functions. For example, the knee-jerk reflex helps a person stand erect. If the knee begins to bend slightly when a person stands still, the quadriceps femoris is stretched, and the leg straightens.

Autonomic Motor Nervous System and Visceral Reflexes

The autonomic motor nervous system (ANS) is composed of the sympathetic and parasympathetic divisions (Fig. 8.14). These two divisions have several features in common: (1) They function automatically and usually in an involuntary manner; (2) they innervate all internal organs; and (3) they utilize two motor neurons and one ganglion to transmit an action potential. (By contrast, a somatic motor neuron travels directly to its effector, without synapsing at a ganglion.) The first neuron has a cell body within the CNS and a preganglionic axon fiber. The second neuron has a cell body within the ganglion and a postganglionic axon fiber.

Visceral reflex actions, such as those that regulate blood pressure and breathing rate, are especially important to maintenance of homeostasis. These reflexes begin when the sensory neurons in contact with internal organs send messages via spinal nerves to the CNS. They are completed when motor neurons within the autonomic system stimulate smooth muscle, cardiac muscle, or a gland. These structures are also effectors.

Sympathetic Division: "Fight or Flight"

Most preganglionic fibers of the **sympathetic division** arise from the middle, or thoracic-lumbar, portion of the spinal cord and almost immediately terminate in ganglia that lie near the cord. Therefore, in this division, the preganglionic fiber is short, but the postganglionic fiber that makes contact with an organ is long:

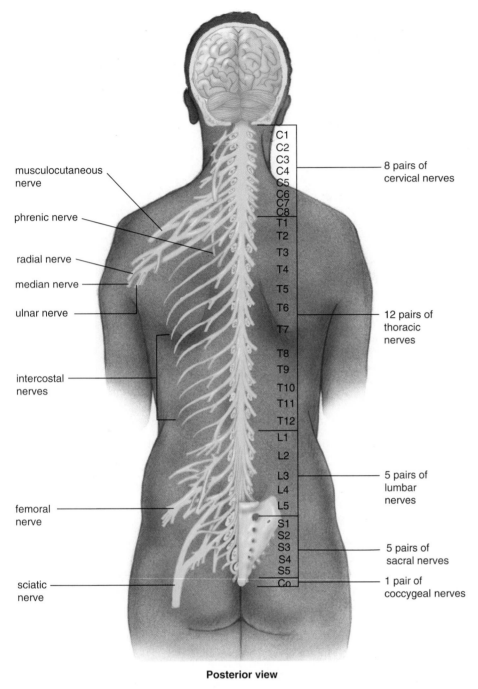

musculocutaneous nerve

phrenic nerve

radial nerve

median nerve

ulnar nerve

intercostal nerves

femoral nerve

sciatic nerve

C1
C2
C3
C4
C5
C6
C7
C8 — 8 pairs of cervical nerves
T1
T2
T3
T4
T5
T6
T7 — 12 pairs of thoracic nerves
T8
T9
T10
T11
T12
L1
L2
L3
L4 — 5 pairs of lumbar nerves
L5
S1
S2
S3
S4 — 5 pairs of sacral nerves
S5
Co — 1 pair of coccygeal nerves

Posterior view

Figure 8.12 Spinal nerves. The number and kinds of spinal nerves are given on the right. The location of major peripheral nerves is given on the left. Table 8.2 lists the functions of these nerves.

Reflexes can also be used to determine if the nervous system is reacting properly. Two of these types of reflexes are:

knee-jerk reflex (patellar reflex), initiated by striking the patellar ligament just below the patella. The response is contraction of the quadriceps femoris muscles, which causes the lower leg to extend;

ankle-jerk reflex, initiated by tapping the Achilles tendon just above its insertion on the calcaneus. The response is

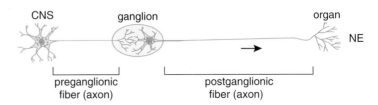

CNS
ganglion
organ
NE
preganglionic fiber (axon)
postganglionic fiber (axon)

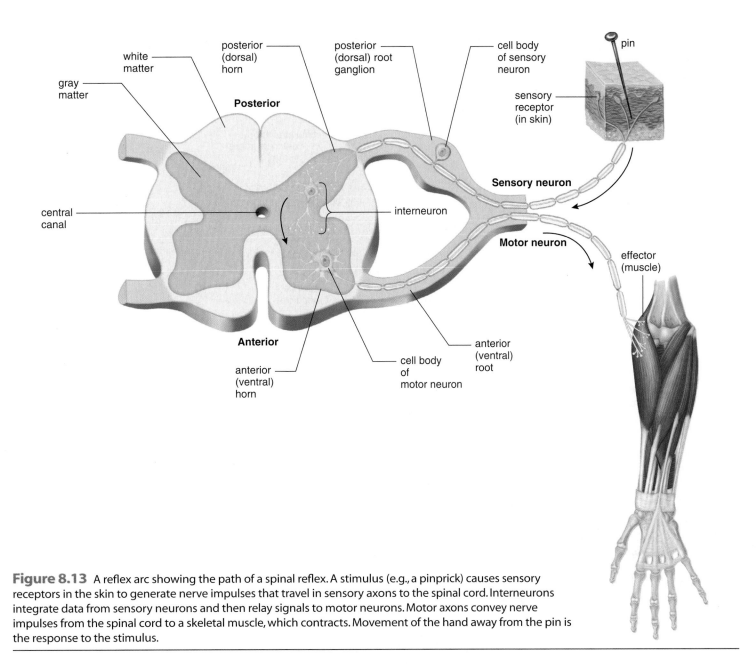

Figure 8.13 A reflex arc showing the path of a spinal reflex. A stimulus (e.g., a pinprick) causes sensory receptors in the skin to generate nerve impulses that travel in sensory axons to the spinal cord. Interneurons integrate data from sensory neurons and then relay signals to motor neurons. Motor axons convey nerve impulses from the spinal cord to a skeletal muscle, which contracts. Movement of the hand away from the pin is the response to the stimulus.

The sympathetic division is especially important during emergency situations when a person might be required to fight or take flight. It accelerates the heartbeat and dilates the bronchi—active muscles, after all, require a ready supply of glucose and oxygen. On the other hand, the sympathetic division inhibits the digestive tract—digestion is not an immediate necessity if you are under attack. The neurotransmitter released by the postganglionic axon is primarily norepinephrine (NE). The structure of NE is like that of epinephrine (adrenaline), an adrenal medulla hormone that usually increases heart rate and contractility.

Parasympathetic Division: "Rest and Digest"

The **parasympathetic division** includes several cranial nerves (e.g., the vagus nerve) as well as fibers that arise from the sacral (bottom) portion of the spinal cord. Therefore, this division is often referred to as the craniosacral portion of the autonomic system. In the parasympathetic division, the preganglionic fiber (axon) is long, and the postganglionic fiber (axon) is short because the ganglia lie near or within the organ:

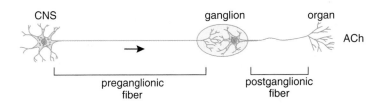

The parasympathetic division, sometimes called the "housekeeper" division, promotes all of the internal responses

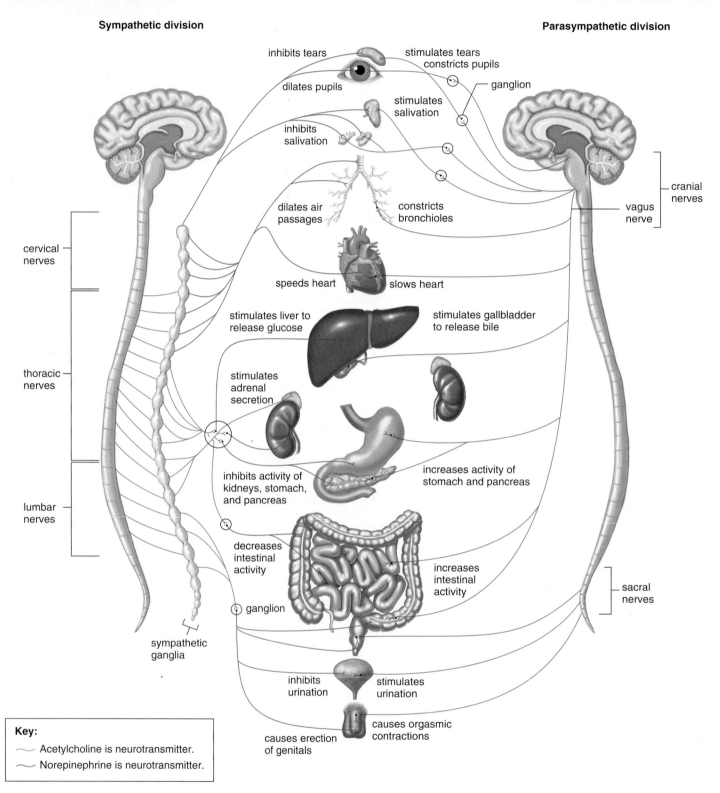

Sympathetic division

Parasympathetic division

inhibits tears

stimulates tears
constricts pupils

dilates pupils

ganglion

stimulates
salivation

inhibits
salivation

cranial
nerves

dilates air
passages

constricts
bronchioles

vagus
nerve

cervical
nerves

speeds heart

slows heart

stimulates liver to
release glucose

stimulates gallbladder
to release bile

thoracic
nerves

stimulates
adrenal
secretion

increases activity of
stomach and pancreas

inhibits activity of
kidneys, stomach,
and pancreas

lumbar
nerves

decreases
intestinal
activity

increases
intestinal
activity

ganglion

sacral
nerves

sympathetic
ganglia

inhibits
urination

stimulates
urination

causes orgasmic
contractions

causes erection
of genitals

Key:

Acetylcholine is neurotransmitter.

Norepinephrine is neurotransmitter.

Figure 8.14 Autonomic system structure and function. Sympathetic preganglionic fibers (*left*) arise from the cervical, thoracic, and lumbar portions of the spinal cord; parasympathetic preganglionic fibers (*right*) arise from the cranial and sacral portions of the spinal cord. Each system innervates the same organs but has contrary effects.

TABLE 8.3 Autonomic Motor Pathways

	Sympathetic	Parasympathetic
Type of control	Involuntary	Involuntary
Number of neurons per message	Two (preganglionic *shorter* than postganglionic)	Two (preganglionic *longer* than postganglionic)
Location of motor fiber	Thoracolumbar spinal nerves	Cranial (e.g., vagus) and sacral spinal nerves
Neurotransmitter	Norepinephrine	Acetylcholine
Effectors	Smooth and cardiac muscle, glands	Smooth and cardiac muscle, glands

we associate with a relaxed state. Parasympathetic function can be referred to as "rest and digest." For example, parasympathetic nerves cause the pupil of the eye to contract, promote digestion of food, slow heart rate, and decrease the strength of cardiac contraction. The neurotransmitter utilized by the parasympathetic division is acetylcholine (ACh).

Table 8.3 contrasts the two divisions of the autonomic system.

Content **CHECK-UP!**

7. Which of the following cranial nerves does not control the eye muscles?

   a. oculomotor   c. trochlear

   b. abducens   d. optic

8. Which spinal nerve supplies the diaphragm?

   a. radial   c. median

   b. phrenic   d. intercostal

9. Put the following parts of a reflex in the correct order:
   1. effector   2. sensory receptor   3. motor nerve
   4. sensory nerve   5. interneuron

   a. 2—5—4—3—1   c. 4—2—5—3—1

   b. 2—4—5—3—1   d. 5—2—4—1—3

8.4 Effects of Aging

After age 60, the brain begins to lose thousands of neurons a day. When these cells die, they are not replaced. By age 80, the brain weighs about 10% less than when the person was a young adult. The cerebral cortex shrinks more than other areas of the brain, losing as much as 45% of its cells. Therefore, such mental activities as learning, memory, and reasoning decline.

Neurotransmitter production also decreases, resulting in slower synaptic transmission. As a person ages, thought processing and translating a thought into action take longer. This partly explains why younger athletes tend to outshine older athletes in sports. However, it is important to note that although structural changes occur, mental impairment is *not* an automatic consequence of getting older. Maintaining the health of the cardiovascular system is fundamental to retaining mental function—after all, the heart and blood vessels supply oxygen and nutrients to brain cells. In addition, older individuals can stay mentally alert by challenging the brain: taking courses, reading, solving puzzles, etc. Avoiding depression is important because it is a contributor to mental impairment; thus, the elderly should try to maintain a social network and interact with others in meaningful ways. Exercise is also an important ingredient in good health—including mental health—at every age.

Neurological disorders, especially Alzheimer disease which is discussed in the Medical Focus on page 161, are more apt to occur in the elderly. The Medical Focus reading on page 177 describes a new procedure for the treatment of Parkinson disease.

8.5 Homeostasis

The nervous system detects, interprets, and responds to changes in internal and external conditions to keep the internal environment relatively constant. Together with the endocrine system, it coordinates and regulates the functioning of the other systems in the body to maintain homeostasis.

The everyday regulation of internal organs that maintains the composition of blood and tissue fluid usually takes place below the level of consciousness. Subconscious control is dependent on reflex actions that involve the hypothalamus and the medulla oblongata. The hypothalamus and the medulla oblongata act through the autonomic motor nervous system to control such important parameters as the heart rate, the constriction of the blood vessels, and the breathing rate.

The illustration in Human Systems Work Together on page 176 tells how the nervous system works with other systems in the body to maintain homeostasis. The hypothalamus works closely with the endocrine system and even produces the hormone ADH, which causes the kidneys to reabsorb water. Other hormones also influence the work of the kidneys in maintaining blood volume and pressure.

Because the nervous system stimulates skeletal muscles to contract, it controls the major movements of the body. When we are in a "fight-or-flight" mode, the nervous system stimulates the adrenal glands and voluntarily controls the skeletal muscles to keep us from danger. On a daily basis, you might think that voluntary movements don't play a role in homeostasis, but actually we usually take all necessary actions to stay in as moderate an environment as possible. Otherwise, we are testing the ability of the nervous system to maintain homeostasis despite extreme conditions.

Integumentary System

Brain controls nerves that regulate size of cutaneous blood vessels; activates sweat glands and arrector pili muscles.

Skin protects nerves, helps regulate body temperature; skin receptors send sensory input to brain.

Skeletal System

Receptors send sensory input from bones and joints to brain.

Bones protect sense organs, brain, and spinal cord; store Ca^{2+} for nerve function.

Muscular System

Brain controls nerves that innervate muscles; receptors send sensory input from muscles to brain.

Muscle contraction moves eyes, permits speech, and creates facial expressions.

Endocrine System

Hypothalamus is part of endocrine system; nerves innervate certain glands of secretion.

Sex hormones affect development of brain.

Cardiovascular System

Brain controls nerves that regulate the heart and dilation of blood vessels.

Blood vessels deliver nutrients and oxygen to neurons; carry away wastes.

How the Nervous System works with other body systems.

Brain

Spinal cord

Lymphatic System/Immunity

Microglial cells engulf and destroy pathogens.

Lymphatic vessels pick up excess tissue fluid; immune system protects against infections of nerves.

Respiratory System

Respiratory centers in brain regulate breathing rate.

Lungs provide oxygen for neurons and rid the body of carbon dioxide produced by neurons.

Digestive System

Brain controls nerves that innervate smooth muscle and permit digestive tract movements.

Digestive tract provides nutrients for growth, maintenance, and repair of neurons and neuroglial cells.

Urinary System

Brain controls nerves that innervate muscles that permit urination.

Kidneys maintain blood levels of Na^+, K^+, and Ca^{2+}, which are needed for nerve conduction.

Reproductive System

Brain controls onset of puberty; nerves are involved in erection of penis and clitoris, contraction of ducts that carry gametes, and contraction of uterus.

Sex hormones masculinize or feminize the brain, exert feedback control over the hypothalamus, and influence sexual behavior.

Pacemakers for Parkinson Disease

"My body is completely out of control. That's the hardest thing about this disease. Sometimes I can't move at all, or I move so slowly that it takes forever just to cross the room. Next thing you know, I'm jerking around like a puppet."

Your patient has just described the classic symptoms of **Parkinson disease,** a progressive central nervous system disorder. The Parkinson patient is usually a person age 60 or older. However, the disease is seen increasingly in younger people as well, making headlines when 38-year-old actor Michael J. Fox announced publicly in 1999 that he suffered from Parkinson disease. If the facial muscles are involved, the person's face may not be able to show emotion, resulting in a fixed, masklike appearance. Routine tasks such as dressing and bathing become very difficult. The sufferer has an increased risk of falling and injuring himself because balance and coordination are also affected. The disease takes its toll on the patient psychologically; most suffer depression as their activities and independence become more and more limited.

Parkinson disease is caused by destruction of specific areas of the brain called the basal nuclei (see page 167). Researchers have determined that these basal nuclei nerve cells produce the neurotransmitter dopamine. The lack of this neurotransmitter seems to cause the signs and symptoms of the disorder. Treatment for the disease has, until recently, focused on ways to replace dopamine in the brain. Drug treatment produces temporary dopamine replacement and relieves the symptoms completely for a few weeks to months. However, as the disease progresses, patients need increasingly stronger medications in higher dosages to relieve the symptoms. These stronger medications produce undesirable side effects, such as dizziness, sleepiness, and memory loss.

Implants of dopamine-producing cells have also been placed into the brain. These implants have had low to moderate success rates in relieving symptoms. Because the cells are often obtained from human embryos, scientists have also raised ethical concerns about the source of the implanted cells.

A novel approach to therapy involves the use of deep-brain stimulation. Similar to a cardiac pacemaker, this "pacemaker for the brain" consists of a set of electrodes implanted into precise centers in the brain. The electrodes are connected to a wire extension, threaded under the skin from the head to the upper chest. The extension is connected to an electrical neurostimulator implanted into the chest near the clavicle, or collarbone. The stimulator delivers continuous electrical signals into the patient's brain. The electrical impulses block the signals that cause Parkinsonian movement. Once implanted, the stimulator can be adjusted from outside the patient's body. Using radio waves, the stimulator can be set to achieve maximum control and symptom relief. Additional surgery is only necessary to replace the stimulator after its three-year life span. The "brain pacemaker" can achieve up to 85% improvement in symptoms and may allow patients to resume normal activities.

SELECTED NEW TERMS

Basic Key Terms

acetylcholine (as″ē-til-ko′lēn), p. 161

acetylcholinesterase (as″ē-til-ko″lin-es′ter-ās), p. 161

action potential (ak′-shun po-ten′shul), p. 159

arachnoid mater (uh-rak′noyd mem′brān), p. 162

association area (uh-so″se-a′shun a′re-uh), p. 166

autonomic motor system (aw″to-nom′ik mō′tər sis′tem), p. 169

axon (ak′son), p. 157

basal nuclei (bas′al nu′kle-i), p. 167

brain stem (brān′ stĕm′), p. 169

Broca's area (brō′kəz âr′ē-ə), p. 167

cell body (sel bod′e), p. 157

central canal (sĕn′trəl kə-năl′), p. 162

central nervous system (sen′tral ner′vus sis′tem), p. 156

cerebellum (sĕr″ĕ-bel′um), p. 168

cerebral cortex (sĕr′ĕ-bral kor′teks), p. 166

cerebral hemisphere (sĕr″ĕ-bral hem′ĭ-sfēr), p. 166

cerebrospinal fluid (sĕr′ĕ-bro-spi′nal flu′id), p. 162

cerebrum (sĕr′ĕ-brum), p. 164

corpus callosum (kôr′pəs kə-lō′səm), p. 166

cranial nerve (kra′ne-al nerve), p. 169

dendrite (den′drīt), p. 157

depolarization (dē-pō′lə-rīz-ā-shən), p. 159

diencephalon (di″en-sef′uh-lon), p. 168

dura mater (dŏŏr′ ə mă′tər), p. 162

ependymal cells (ĕ-pen′dĭ-mal selz), p. 162

ganglion (gang′gle-on), p. 169

graded potential (grād-ed pə-tĕn′shəl), p. 159

gray matter (grā mat′er), p. 162

herniated disk (hûr′nē-āt-ed dĭsk), p. 163

hippocampus (hĭp′ə-kăm′pəs), p. 168

hypothalamus (hi″po-thal′uh-mus), p. 168

interneuron (in″ter-nu′ron), p. 157

limbic system (lim′bik sis′tem), p. 167

medulla oblongata (mĭ-dŭl′ə ŏb′lông-gä′tə), p. 169

midbrain (mid′brān), p. 169

nerve (nerv), p. 157

neurotransmitter (nu″ro-trans′mit-er), p. 159

norepinephrine (nor″ep-ĭ-nef′rin), p. 161

parasympathetic division (pār″uh-sim″ puh-thet′ik dĭ-vizh′un), p. 173

peripheral nervous system (pĕ-rif'er-al ner'vus sis'tem), p. 156

pia mater (pī'ə mă'tər), p. 162

pons (ponz), p. 169

posterior-root ganglion (pos-tēr'e-or-rut gang'gle-on), p. 170

prefrontal area (prē-frŭn'tl âr'ē-ə), p. 167

primary motor area (pri'ma-re mo'tor a're-uh), p. 166

primary somatosensory area (pri'ma-re so"măto-sen'so-re a're-uh), p. 166

receptor potential (rĭ-sĕp'tər pə-tĕn'shəl), p. 161

reflex (re'fleks), p. 170

refractory period (rĭ-frăk'tə-rē pîr'ē-əd), p. 159

repolarization (rē-pō'lər-ĭ-zā'shən), p. 159

resting potential (rĕst-ing pə-tĕn'shəl), p. 158

reticular formation (rĭ-tĭk'yə-lər fôr-mā'-shən), p. 169

saltatory conduction (săl'tə-tôr'ē kən-dŭk'shən), p. 159

somatic sensory system (so-mat'ik sĕn'sə-rē sis'tem), p. 169

spinal cord (spi'nal kord), p. 162

spinal nerve (spi'nal nerv), p. 169

stimulus (stĭm'yə-ləs), p. 159

sympathetic division (sim"puh-thet'ik dĭ-vizh'un), p. 172

synapse (sin'aps), p. 159

synaptic cleft (sĭ-nap'tik kleft), p. 159

thalamus (thal'uh-mus), p. 168

tract (trakt), p. 157

ventricle (ven'trĭ-kl), p. 162

Wernicke's area (ver'nĭ-kez âr'ē-ə), p. 167

white matter (whĭt mat'er), p. 162

Clinical Key Terms

Alzheimer disease (altz'hi-mer dĭ-zēz'), p. 161

ankle-jerk reflex (an'kl-jerk re'fleks), p. 172

cerebral palsy (sĕr'ē-bral pal'ze), p. 166

electroencephalogram (e-lek"tro-en-sef'uh-lo-gram), p. 164

hydrocephalus (hi"dro-sĕ'fuh-lus), p. 162

knee-jerk reflex (ne'jerk re'fleks), p. 172

paraplegia (par-uh-ple'je-uh), p. 165

Parkinson disease (par'kin-sun dĭ-zēz'), p. 177

quadriplegia (kwah-druh-ple'je-uh), p. 165

SUMMARY

8.1 Nervous System

A. The nervous system permits sensory input, performs integration, and stimulates motor output.

B. The nervous system is divided into the central nervous system (brain and spinal cord) and the peripheral nervous system (afferent and efferent nervous systems). The CNS lies in the midline of the body, and the PNS is located peripherally to the CNS.

C. Nervous tissue contains neurons and neuroglia. Each type of neuron (motor, sensory, and interneuron) has three parts (dendrites, cell body, and axon). Neuroglia support, protect, and nourish the neurons.

D. All axons transmit the same type of nerve impulse: a change in polarity (called an action potential) that moves along the membrane of a nerve fiber. Saltatory conduction in myelinated axons is a faster type of conduction.

E. Transmission of a nerve impulse across a synapse is dependent on the release of a neurotransmitter into a synaptic cleft.

8.2 Central Nervous System

A. The CNS, consisting of the spinal cord and brain, is protected by the meninges and the cerebrospinal fluid.

B. The spinal cord, located in the vertebral column, is composed of white matter and gray matter. White matter contains bundles of nerve fibers, called tracts, which conduct nerve impulses to and from the higher centers of the brain. Gray matter is mainly made up of short fibers and cell bodies. The spinal cord is a center for reflex action and allows communication between the brain and the peripheral nerves leaving the spinal cord.

C. The brain has four ventricles. The lateral ventricles are found in the left and right cerebral hemispheres. The third ventricle is found in the diencephalon. The fourth ventricle is found in the brain stem.

D. The cerebrum is divided into the left and right hemispheres. The cerebral cortex, a thin layer of gray matter, has four lobes in each hemisphere. The frontal lobe initiates motor output. The parietal lobe is the final receptor for sensory input from the skin and muscles. The other lobes receive specific sensory input.

Various association areas integrate sensory data. Processing centers integrate data from other association areas: The prefrontal area carries out higher mental processes; Broca's area and Wernicke's area are concerned with speech.

E. The limbic system includes portions of the cerebrum, the thalamus, and the hypothalamus. It is involved in learning and memory and in causing the emotions that guide behavior.

F. The hypothalamus helps control the functioning of most internal organs and controls the secretions of the pituitary gland. The thalamus receives sensory impulses from all parts of the body and channels them to the cerebrum.

G. The cerebellum controls balance and complex muscular movements.

H. The brain stem contains the medulla oblongata, pons, midbrain, and reticular formation. The medulla oblongata contains vital centers for regulating heartbeat, breathing, and blood pressure. The pons assists the medulla oblongata in regulating the breathing rate. The midbrain contains tracts that conduct impulses to and from the higher parts of the brain. The reticular formation maintains alertness and assists other brainstem centers.

8.3 **Peripheral Nervous System**
  A. A nerve contains bundles of long fibers covered by fibrous connective tissue layers.
  B. Cranial nerves take impulses to and/or from the brain. Spinal nerves take impulses to and from the spinal cord.
  C. Reflexes (automatic reactions to internal and external stimuli) depend on the reflex arc. Some reflexes are important for avoiding injury, and others are necessary for normal physiological functions.
  D. The autonomic nervous system controls the functioning of internal organs.

1. The divisions of the autonomic nervous system: (1) function automatically and usually subconsciously in an involuntary manner; (2) innervate all internal organs; and (3) utilize two motor neurons and one ganglion for each impulse.
2. The sympathetic division brings about the responses associated with the "fight-or-flight" response.
3. The parasympathetic division "rest and digest" brings about the responses associated with normally restful activities.

8.4 **Effects of Aging**
  A. The brain loses nerve cells, and this affects learning, memory, and reasoning. Mental impairment is not an automatic consequence of aging.
  B. Alzheimer disease is more often seen among the elderly.

8.5 **Homeostasis**
  A. The nervous system, along with the endocrine system, regulates and coordinates the other systems to maintain homeostasis.
  B. Skeletal muscle contraction also plays a role because movement helps us take precautions or stay in a moderate environment.

STUDY QUESTIONS

1. What are the functions of the nervous system? (p. 156)
2. What are the two main divisions of the nervous system? How are these divisions subdivided? (p. 156)
3. What is the general structure of a neuron, and what are the functions of three different types of neurons? (p. 157)
4. What constitutes a nerve impulse (action potential)? Describe the resting potential. Why do myelinated fibers have a faster speed of conduction? (pp. 158–159)
5. How is the nerve impulse transmitted across a synapse? Name two well-known neurotransmitters. (p. 159)

6. Name the meninges, and describe their locations. Where do you find cerebrospinal fluid? (p. 162)
7. Describe the structure and function of the spinal cord. (p. 163)
8. What is the difference between the cerebrum and the cerebral cortex? Name the lobes of the cerebral cortex, and state their function. Describe the primary motor area and the primary somatosensory area. (pp. 164, 166)
9. What is the limbic system, and what is its function? (pp. 167–168)
10. Name the other parts of the brain, and give a location and function for each part. (pp. 168–169)

11. Describe the structure of a nerve. In general, discuss the location and function of the cranial nerves and the spinal nerves. (p. 169)
12. Describe a spinal reflex, including the role played by a sensory nerve fiber, interneurons, and a motor fiber. (pp. 170–171)
13. Contrast the actions of the sympathetic and the parasympathetic divisions of the autonomic system. (pp. 172–173)
14. What role does the nervous system play in homeostasis? (p. 175)

LEARNING OBJECTIVE QUESTIONS

Fill in the blanks.

1. A(n) _____ carries nerve impulses away from the cell body.
2. During the depolarization portion of an action potential, _____ ions are moving to the _____ of the nerve fiber.
3. The space between the axon ending of one neuron and the dendrite of another is called the _____.
4. ACh is broken down by the enzyme _____ after it has initiated an

action potential on a neighboring neuron.
5. Motor nerves stimulate _____.
6. In a reflex arc, only the _____ is completely within the CNS.
7. The _____ is the part of the brain responsible for coordinating body movements.
8. The _____ is the part of the brain responsible for consciousness.

9. The brain and spinal cord are covered by protective layers called _____.
10. The vagus nerve is a _____ nerve that controls _____.
11. Whereas the central nervous system is composed of the _____ and _____, the peripheral nervous system is composed of the _____.

12. The limbic system records emotions and also is involved in _____ and _____ .

13. While the _____ division of the autonomic nervous system brings about organ responses that are part of the "fight-or-flight" response, the _____ division brings about responses associated with normal restful conditions.

14. The electrical activity of the brain can be recorded in the form of a(n) _____ .

15. Label the following diagram.

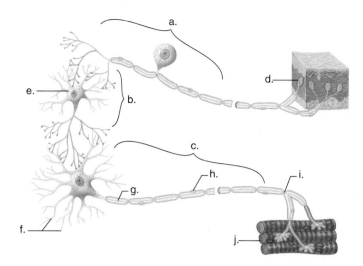

MEDICAL TERMINOLOGY EXERCISE

Consult Appendix B for help in pronouncing and analyzing the meaning of the terms that follow.

1. neuropathogenesis (nu″ro-path″o-jen″ĕ-sis)
2. anesthesia (an″es-the′ze-uh)
3. encephalomyeloneuropathy (en-sef″uh-lo-mi″ĕ-lo-nu-rop′uh-the)
4. hemiplegia (hem″ĭ-ple′je-uh)
5. glioblastoma (gli″o-blas-to′muh)
6. subdural hemorrhage (sub-du′ral hem′or-ij)
7. cephalometer (sef″uh-lom′ĕ-ter)
8. meningoencephalocele (me-ning″go-en-sef″ uh-lo-sēl)
9. neurorrhaphy (nu-rōr′uh-fe)
10. ataxiaphasia (uh-tak″se-uh-fa′ze-uh)
11. cerebrovascular accident (sĕr′-e-bro′-vas-kyū-ler ak′suh-dent)
12. duraplasty (du′ruh-plas-te)
13. brachycephalic (brak′e-sef-al′ik)
14. arachnoiditis (uh-rak″noy-di′tis)

WEBSITE LINK

Visit this text's website at http://www.mhhe.com/maderap6 **for additional quizzes, interactive learning exercises, and other study tools.**

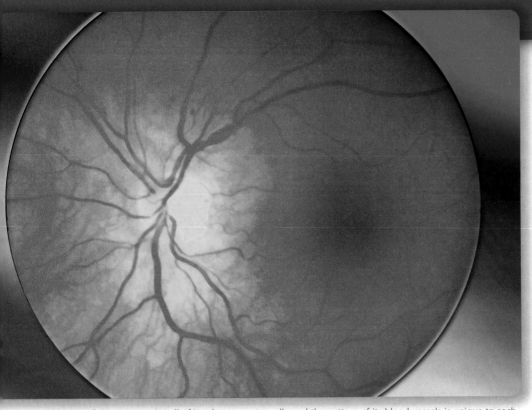

CHAPTER 9

The Sensory System

The retina of the eye contains all of its photoreceptor cells, and the pattern of its blood vessels is unique to each individual—just like a fingerprint!

Chapter Outline & Learning Objectives
After you have studied this chapter, you should be able to:

9.1 General Senses (p. 182)
- Categorize sensory receptors according to five types of stimuli.
- Discuss the function of proprioceptors.
- Relate specific sensory receptors in the skin to particular senses of the skin.
- Discuss the phenomenon of referred pain.

9.2 Senses of Taste and Smell (p. 184)
- Name the chemoreceptors, and state their location, anatomy, and mechanism of action.

9.3 Sense of Vision (p. 186)
- Describe the anatomy and function of the accessory organs of the eye.
- Describe the anatomy of the eye, and give a function of each part.
- Describe the sensory receptors for sight, their mechanism of action, and the mechanism for stereoscopic vision.
- Describe some common disorders of sight.

9.4 Sense of Hearing (p. 196)
- Describe the anatomy of the ear, and give a function of each part.
- Describe the sensory receptors for hearing and their mechanism of action.

9.5 Sense of Equilibrium (p. 197)
- Describe the sensory receptors for equilibrium and their mechanism of action.

9.6 Effects of Aging (p. 200)
- Describe the anatomical and physiological changes that occur in the sensory system as we age.

Medical Focus
Corrective Lenses (p. 191)

What's New
A Bionic Cure for Macular Degeneration (pp. 192–193)

Focus on Forensics
Retinal Hemorrhage in Shaken Baby Syndrome (p. 194)

Medical Focus
Hearing Damage and Deafness (p. 200)

When a sensory receptor is stimulated, it generates nerve signals that travel to your brain. Interpretation of these signals is the function of the brain, which has a special region for receiving information from each of the sense organs. Signals arriving at a particular sensory area of the brain can be interpreted in only one way; for example, those arriving at the olfactory area result in smell sensation, and those arriving at the visual area result in sight sensation.

Specialized sensory receptors throughout the body start signal transmission to the brain by using a system called a **receptor potential.** Receptor potentials begin with a stimulus (for example, a light stimulus for receptors in the eye). Unlike an action potential (which is an all-or-nothing event), receptor potentials can be weak or strong. Like the signals that occur when nerves synapse, receptor potentials can add together. Although receptors do not generate action potentials, they are parts of neurons or they synapse with other neurons that do create action potentials. In this way, sensory information is transmitted to the brain. The brain integrates data from various sensory receptors in order to perceive whatever caused the stimulation of olfactory and visual receptors—for example, a flower.

Sensory receptors may be categorized into five types based on their stimuli:

Mechanoreceptors, such as pressure receptors in the skin and proprioceptors (specialized stretch receptors) in muscle, are stimulated by changes in pressure or body movement. Other mechanoreceptors in the inner ear also detect body movements; these will be discussed in Section 9.5.

Thermoreceptors in the skin and in the internal organs are stimulated by changes in the external or internal temperature.

Pain receptors, such as those in the skin, are stimulated by damage or oxygen deprivation to the tissues.

Chemoreceptors are stimulated by changes in the chemical concentration of substances. Taste buds in the tongue and olfactory (smell) receptors in the nose are chemoreceptors. Other chemoreceptors can sense the concentration of oxygen, carbon dioxide, and hydrogen ions in the blood.

Photoreceptors, which are located only in the eye, are stimulated by light energy.

9.1 General Senses

Sensory receptors in the muscles, joints, tendons, some internal organs, and skin send action potentials to the spinal cord. From there, the action potentials travel up the spinal cord in tracts to the **thalamus,** the sensory relay center of the brain. The information from these receptors is then relayed to the somatosensory areas of the cerebral cortex, located in the parietal lobe of the brain. These general sensory receptors can be categorized into three types: proprioceptors, cutaneous receptors, and pain receptors.

Proprioceptors

Proprioceptors are mechanoreceptors involved in reflex actions that maintain muscle tone and thereby the body's

equilibrium and posture. They help us know the position of our limbs in space by detecting the degree of muscle relaxation, the stretch of tendons, and the movement of ligaments. Muscle spindles act to increase the degree of muscle contraction, and Golgi tendon organs act to decrease it. The result is a muscle that has the proper length and tension, or **muscle tone.**

Figure 9.1 illustrates the activity of a muscle spindle. In a muscle spindle, sensory nerve endings are wrapped around thin muscle cells within a connective tissue sheath. When the muscle relaxes and undue stretching of the muscle spindle occurs, nerve action potentials are generated. The rapidity of the action potentials generated by the muscle spindle is proportional to the stretching of a muscle. A reflex action then occurs, which results in contraction of muscle fibers adjoining the muscle spindle. The knee-jerk reflex, which involves muscle spindles, offers an opportunity for physicians to test a reflex action. By contrast, a Golgi tendon organ, which is built into the tough connective tissue of a tendon, is activated by excessive muscle contraction. When Golgi tendon organs generate action potentials, inhibitory signals are sent to the muscle. The muscle is prevented from contracting too forcefully, which could possibly injure both muscle and tendon.

The information sent by muscle spindles and Golgi tendon organs to the CNS is used to maintain the body's equilibrium and posture despite the force of gravity always acting upon the skeleton and muscles.

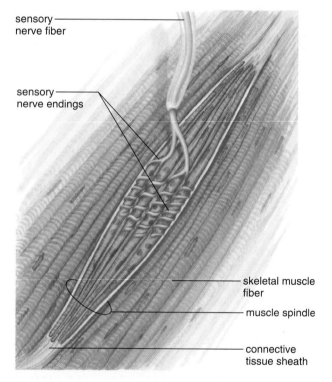

Figure 9.1 Muscle spindle. When a muscle is stretched, a muscle spindle sends sensory nerve impulses to the spinal cord. Motor nerve impulses from the spinal cord result in muscle fiber contraction so that muscle tone is maintained.

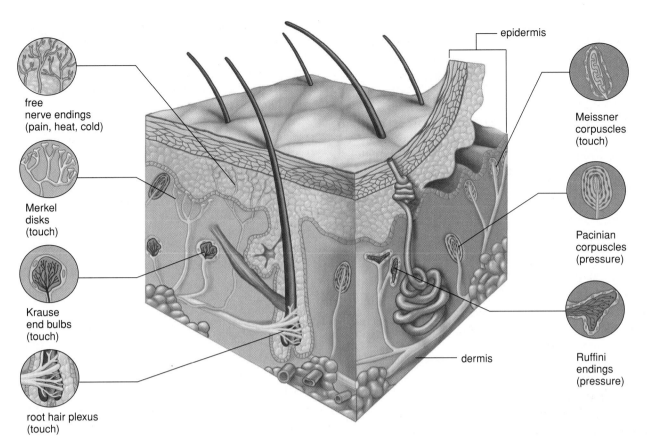

Figure 9.2 Sensory receptors in human skin. The classical view is that each sensory receptor has the main function shown here. However, investigators report that matters are not so clear-cut. For example, microscopic examination of the skin of the ear shows only free nerve endings (pain receptors), and yet the skin of the ear is sensitive to all sensations. Therefore, it appears that the receptors of the skin are somewhat, but not completely, specialized.

Cutaneous Receptors

The skin is composed of two layers: the epidermis and the dermis. In Figure 9.2, the artist has dramatically indicated these two layers by separating the epidermis from the dermis in one location. The epidermis is stratified squamous epithelium in which cells become keratinized as they rise to the surface. Once at the surface, cells are sloughed off. The dermis is a thick connective tissue layer. The deepest layer of epidermis, and the entire dermis layer, contain **cutaneous receptors,** which make the skin sensitive to touch, pressure, pain, and temperature (warmth and cold). The dermis is a mosaic of these tiny sensory receptors, as you can determine by slowly passing a metal probe over your skin. At certain points, you will feel touch or pressure, and at others, you will feel heat or cold (depending on the probe's temperature).

Three types of cutaneous receptors are sensitive to fine touch. *Meissner corpuscles* are concentrated in the fingertips, the palms, the lips, the tongue, the nipples, the penis, and the clitoris. *Merkel disks* are found in the deepest epidermal layer, where the epidermis meets the dermis. A free nerve ending called a *root hair plexus* winds around the base of a hair follicle and fires if the hair is touched.

The three different types of cutaneous receptors that are sensitive to pressure are Pacinian corpuscles, Ruffini endings, and Krause end bulbs. *Pacinian corpuscles* are onion-shaped sensory receptors that lie deep inside the dermis. *Ruffini endings* and *Krause end bulbs* are encapsulated by sheaths of connective tissue and contain lacy networks of nerve fibers.

Temperature receptors are simply free nerve endings in the superficial dermis and epidermis. Some free nerve endings are responsive to cold; others are responsive to warmth. Cold receptors are far more numerous than warmth receptors, but the two types have no known structural differences.

Pain Receptors

The skin and many (but not all) internal organs have pain receptors, also called **nociceptors.** *Somatic nociceptors* from the skin and skeletal muscles respond to mechanical, thermal, electrical, or chemical damage to these tissues. Skinning a knee, burning a finger, or straining a muscle all stimulate somatic nociceptors. *Visceral nociceptors* react in response to excessive stretching of the internal organs, oxygen deprivation when blood supply is reduced, or chemicals released by damaged tissues. Visceral nociceptors create the pain sensation when the stomach is too full, as well as the crushing pain of a heart attack when blood supply to the heart is reduced.

Stimulation of internal pain receptors is sometimes felt as pain from the skin as well as the internal organs; this is called **referred pain.** Referred pain occurs because the somatic pain nociceptors travel in the same spinal cord pathway as the internal visceral nociceptors. Signals from both sets of neurons converge on the same nervous pathway, and the brain cannot distinguish between the two. For example, pain from the heart that occurs during a heart attack is often accompanied by referred pain in the left shoulder and arm, especially in males.

Content CHECK-UP!

1. Sensory receptors for sensing pain are called:
   a. proprioceptors.
   c. photoreceptors.
   b. nociceptors.
   d. thermoreceptors.

2. Which of the following are not pressure receptors?
   a. Meissner corpuscles
   c. Krause end bulbs
   b. Ruffini endings
   d. Pacinian corpuscles

3. Following are pairs of sensory receptors and stimuli to which they respond. Choose the incorrect pair.
   a. Golgi tendon organs—excessive muscle contraction
   b. visceral nociceptors—oxygen deprivation to an organ
   c. Merkel disks—pressure
   d. thermoreceptors—temperature changes

9.2 Senses of Taste and Smell

Taste and smell are called chemical senses because their receptors are sensitive to molecules in the food we eat and the air we breathe. The body also has other chemoreceptors that govern respiratory rate.

Chemoreceptors in the carotid arteries and in the aorta are primarily sensitive to the hydrogen ion concentration of the blood. These receptors communicate with the respiratory center in the medulla oblongata. Similar chemoreceptors located directly in the medulla respond to increases in both carbon dioxide content and hydrogen ion concentration of the blood. When the hydrogen ion or carbon dioxide concentration of blood increases, both sets of chemoreceptors signal an increase in breathing rate. Exhaling carbon dioxide lowers hydrogen ion concentration, and thus, raises the pH of the blood.

Sense of Taste

The sensory receptors for the sense of taste are located in **taste buds.** Taste buds are embedded in epithelium primarily on the tongue (Fig. 9.3). Many lie along the walls of the papillae, the small elevations on the tongue that are visible to the naked eye. Isolated taste buds are also present on the hard palate, the pharynx, and the epiglottis. We have at least five primary types of taste sensations: sweet, sour, salty, bitter, and umami (pronounced "yoo-mommy"). Umami sensation is named after the Japanese

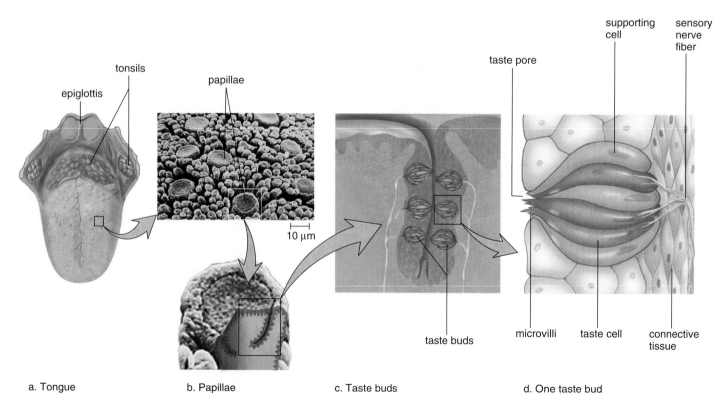

a. Tongue   b. Papillae   c. Taste buds   d. One taste bud

Figure 9.3 Taste buds. **a.** Papillae on the tongue contain taste buds that are sensitive to sweet, sour, salty, and bitter tastes. **b.** Enlargement of papillae. **c.** Taste buds occur along the walls of the papillae. **d.** Taste cells end in microvilli that bear receptor proteins for certain molecules. When molecules bind to the receptor proteins, nerve impulses are generated that go to the brain where the sensation of taste occurs.

word for "delicious"; it can best be described as the pleasant, savory taste of well-seasoned meat. The taste buds for each taste sensation are scattered throughout the tongue (Fig. 9.3a).

How the Brain Receives Taste Information

Taste buds open at a taste pore. They have supporting cells and a number of elongated taste cells that end in microvilli. The microvilli of taste cells project through the taste pore. These microvilli have receptor proteins for molecules that cause the brain to distinguish between sweet, sour, salty, umami, and bitter tastes. When these molecules bind to receptor proteins, nerve impulses are generated in associated sensory nerve fibers. These nerve impulses go to the brain, including the cortical areas, which interpret them as tastes. Sensory receiving and memory areas for taste are located in overlapping areas of the parietal lobe and the insula, the small inner lobe of the brain (see Section 8.2, p. 166).

Since we can respond to a range of sweet, sour, salty, umami, and bitter tastes, the brain appears to survey the overall pattern of incoming sensory impulses and to take a "weighted average" of their taste messages as the perceived taste. Again, we can note that even though our senses are dependent on sensory receptors, the brain integrates the incoming information and gives us our sense perceptions.

Sense of Smell

Our sense of smell is dependent on **olfactory cells** located within olfactory epithelium high in the roof of the nasal cavity (Fig. 9.4). Olfactory cells are modified neurons. Each cell ends in a tuft of about five olfactory cilia, which bear receptor proteins for odor molecules. The brain distinguishes odors after odor molecules bind to the receptor proteins.

How the Brain Receives Odor Information

Each olfactory cell has only one type out of 1,000 different types of receptor proteins. Nerve fibers from like olfactory cells lead to the same neuron in the olfactory bulb, an extension of the brain. An odor contains many odor molecules, which activate a characteristic combination of receptor proteins. For example, a rose might stimulate olfactory cells, designated by purple and green in Figure 9.4, while a hyacinth might stimulate a different combination. An odor's signature in the olfactory bulb is determined by which neurons are stimulated. When the neurons

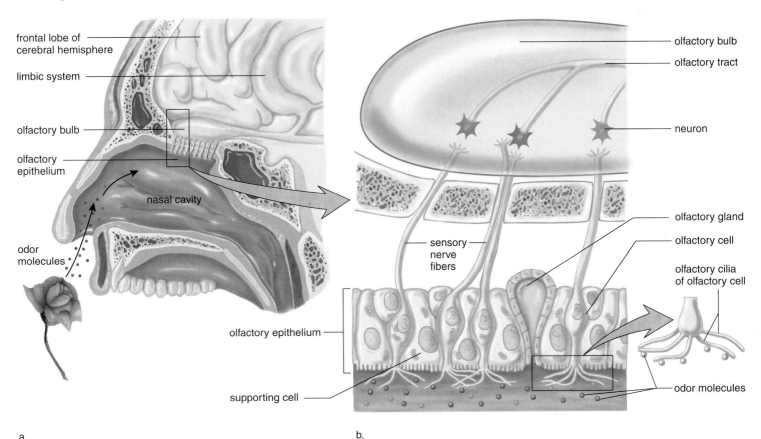

a.

b.

Figure 9.4 Olfactory cell location and anatomy. **a.** The olfactory epithelium in humans is located in the nasal cavity. **b.** Olfactory cells end in cilia that bear receptor proteins for specific odor molecules. The cilia of each olfactory cell can bind to only one type of odor molecule (signified here by color). For example, if a rose causes olfactory cells sensitive to "purple" and "green" odor molecules to be stimulated, then neurons designated by purple and green in the olfactory bulb are activated. The primary olfactory area of the cerebral cortex interprets the pattern of stimulation as the scent of a rose.

communicate this information via the olfactory tract to the olfactory areas of the cerebral cortex, we know we have smelled a rose or a hyacinth. The olfactory cortex is located in the temporal lobe. Some areas of the olfactory cortex receive smell sensations; other areas contain olfactory memories.

Have you ever noticed that a certain aroma vividly brings to mind a certain person or place, and can recreate emotions you feel about that person or place? The smell of a certain perfume may depress you by reminding you of a failed relationship, or the smell of boxwood might create happier emotions because it reminds you of your grandfather's farm. The olfactory bulbs have direct connections with the limbic system and its centers for emotion and memory. One investigator showed that when subjects smelled an orange while viewing a painting, they later more vividly recalled memories of the painting.

Sense of Taste and Sense of Smell

Actually, the sense of taste and the sense of smell work together to create a combined effect when interpreted by the cerebral cortex. For example, when you have a cold, you think food has lost its taste, but most likely you have lost the ability to sense its smell. This method works in reverse also. When you smell something, some of the molecules move from the nose down into the mouth region and stimulate the taste buds there. Therefore, part of what we refer to as smell may in fact be taste.

> ## Content CHECK-UP!
>
> 4. Select one incorrect statement about the sensation of taste:
>   a. Taste buds are located in the back of the throat (pharynx).
>   b. Taste buds respond to five primary taste sensations.
>   c. Taste buds are a type of chemoreceptor.
>   d. Taste cells in taste buds end in cilia.
> 5. Select one incorrect statement about the sense of smell:
>   a. Taste and smell work separately.
>   b. Olfactory epithelium is located in the upper part of the nasal cavity.
>   c. Olfactory cells are modified neurons.
>   d. Olfactory cells end in olfactory cilia.
> 6. Complete this statement correctly: The gustatory (taste) control center is located in the _____ lobe, and the olfactory (smell) area is located in the _____ lobe of the brain.
>   a. insula, temporal   c. frontal, temporal
>   b. temporal, parietal   d. occipital, frontal

9.3   Sense of Vision

The photoreceptors for sight are in the eyes. The eyes are located in orbits formed by seven of the skull's bones (frontal, lacrimal, ethmoid, zygomatic, maxilla, sphenoid, and palatine). The bony ridge superior to the orbits, called the *supraorbital ridge,* protects the eye from blows, and serves as a location for the eyebrows. The eye has certain accessory organs.

Accessory Organs of the Eye

Accessory organs of the eye include: (1) the eyebrows, eyelids, and eyelashes; (2) the lacrimal apparatus, which produces tears; and (3) the extrinsic muscles that move the eye.

Eyebrows, Eyelids, and Eyelashes

Eyebrows have short, thick hairs positioned transversely above the eye along the supraorbital ridge (Fig. 9.5a). Eyebrows shade the eyes from the sun and prevent perspiration or debris from falling into the eye.

Eyelids are a continuation of the skin. The eyelashes of the eye can trap debris and keep it from entering the eyes. Sebaceous glands associated with each eyelash produce an oily secretion that lubricates the eye. Inflammation of one of the glands is called a **sty.**

Blinking of eyelids keeps the eye lubricated and free of debris. The eyelids are operated by the orbicularis oculi muscle, which closes the lid, and by the levator palpebrae superioris muscle, which raises the lid. A person with **myasthenia gravis** has weakness in these muscles due to destruction of receptors for the neurotransmitter acetylcholine, and their eyelids often have to be taped open.

The inner surface of an eyelid is lined by a transparent mucous membrane, called the **conjunctiva.** The conjunctiva folds back to cover the anterior of the eye, except for the cornea which is covered by a delicate epithelium.

Lacrimal Apparatus

A **lacrimal apparatus** consists of the lacrimal gland and the lacrimal sac with its ducts (Fig. 9.5b). The lacrimal gland, which lies in the orbit above the eye, produces tears that flow over the eye when the eyelids are blinked. The tears, collected by two small ducts, pass into the lacrimal sac before draining into the nose by way of the nasolacrimal duct.

Extrinsic Muscles

Within an orbit, the eye is anchored in place by the **extrinsic muscles,** whose contractions move the eyes. Each of these muscles originates from the bony orbit and inserts by tendons to the outer layer of the eyeball. There are three pairs of antagonistic extrinsic muscles (Fig. 9.6):

First pair:

Superior rectus	Rolls eye upward
Inferior rectus	Rolls eye downward

Second pair:

Lateral rectus	Turns eye outward, away from midline
Medial rectus	Turns eye inward, toward midline

Third pair:

Superior oblique	Rotates eye counterclockwise
Inferior oblique	Rotates eye clockwise

Although stimulation of each muscle causes a precise movement of the eyeball, most movements of the eyeball involve the combined contraction of two or more muscles. For

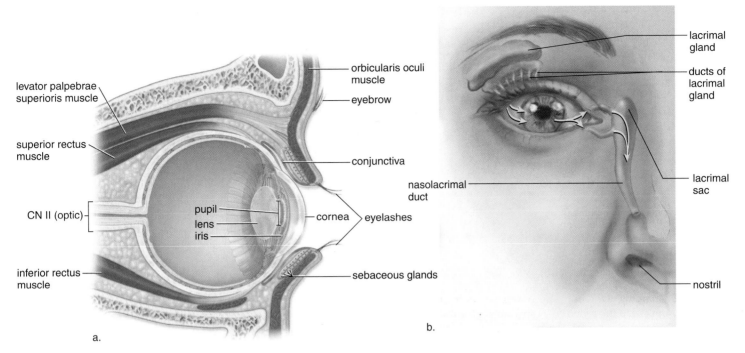

Figure 9.5 Accessory structures of the orbit. **a.** Sagittal section of the eye and orbit. **b.** The lacrimal apparatus.

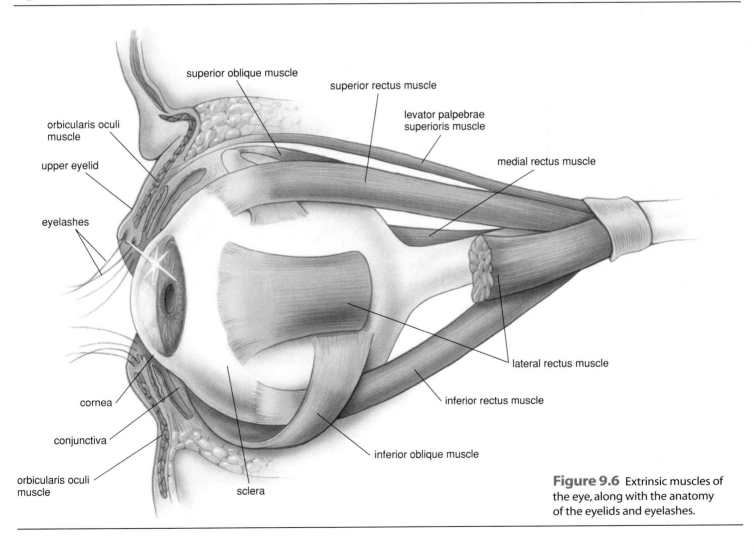

Figure 9.6 Extrinsic muscles of the eye, along with the anatomy of the eyelids and eyelashes.

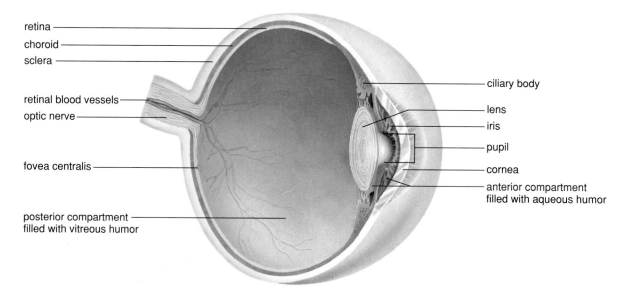

retina
choroid
sclera
retinal blood vessels
optic nerve
fovea centralis
posterior compartment
filled with vitreous humor

ciliary body
lens
iris
pupil
cornea
anterior compartment
filled with aqueous humor

Figure 9.7 Anatomy of the human eye. Notice that the sclera, the outer layer of the eye, becomes the cornea and that the choroid, the middle layer, is continuous with the ciliary body and the iris. The retina, the inner layer, contains the photoreceptors for vision; the fovea centralis is the region where vision is most acute. The retina is supplied with a network of tiny blood vessels.

example, if your left eye is directed upward toward your nose, which muscles are required? The answer is the superior and medial rectus muscles.

Three cranial nerves—the oculomotor, abducens, and trochlear—control these muscles. The oculomotor nerve innervates the superior, inferior, and medial rectus muscles, as well as the inferior oblique muscles; the abducens nerve innervates the lateral rectus muscle; and the trochlear nerve innervates the superior oblique muscle. The motor units of these muscles are the smallest in the body. A single motor axon serves only about 10 muscle fibers, allowing eyeball movements to be very precise.

Anatomy and Physiology of the Eye

The eyeball, which is an elongated sphere about 2.5 cm in diameter, has three layers or coats: the sclera, the choroid, and the retina (Fig. 9.7). Only the retina contains photoreceptors for light energy. Table 9.1 gives the functions of the parts of the eye.

The outer layer, the **sclera,** is white and fibrous except for the **cornea,** which is made of transparent collagen fibers. The cornea is the window of the eye. The middle, thin, darkly pigmented layer, the **choroid,** is vascular and absorbs stray light rays that photoreceptors have not absorbed. Toward the front, the choroid becomes the donut-shaped **iris.** The iris regulates the size of the **pupil,** a hole in the center of the iris through which light enters the eyeball. The color of the iris (and therefore the color of your eyes) correlates with its pigmentation. Heavily pigmented eyes are brown, while lightly pigmented eyes are green or blue. Behind the iris, the choroid thickens and forms the circular ciliary body. The **ciliary body** contains the **ciliary muscle,** which controls the shape of the lens for near and far vision. Changing the shape of the lens is a process called accommodation.

TABLE 9.1 Functions of the Parts of the Eye

Part	Function
Sclera	Tough outermost connective tissue layer; protects and supports eyeball
Cornea	Refracts (bends) light rays
Choroid	Blood vessel layer; absorbs stray light
Ciliary body	Holds the lens in place
Ciliary muscle	Accommodation: changes the shape of the lens for near or far vision
Iris	Regulates entrance of light into retina
Pupil	Opening in iris; admits light into retina
Retina	Contains sensory receptors for light
Rods	Receptors for black and white, dim light vision; peripheral vision
Cones	Receptors for color vision; bright-light vision
Fovea centralis	Largest concentration of cone cells; makes acute vision possible
Optic nerve	Transmits visual signals to brain
Lens	Refracts (bends) and focuses light rays
Suspensory ligaments	Support lens; attach lens to ciliary body
Aqueous humor	Transmits light rays; supports anterior chamber
Vitreous humor	Transmits light rays; supports posterior chamber

The **lens,** attached to the ciliary body by the suspensory ligaments, divides the eye into two compartments; the one in front of the lens is the anterior compartment, and the one behind the lens is the posterior compartment. The anterior

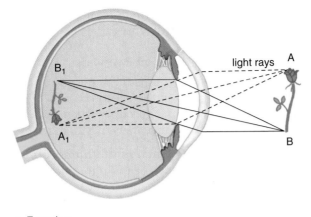

a. Focusing

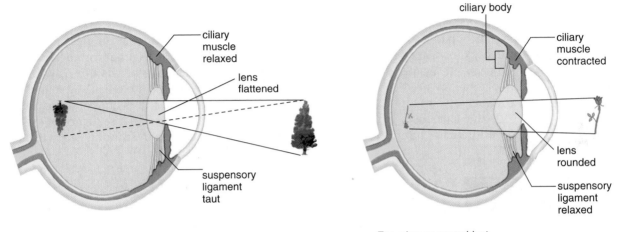

b. Focusing on distant object

c. Focusing on near object

Figure 9.8 Focusing. **a.** Light rays from each point on an object are bent by the cornea and the lens in such a way that an inverted and reversed image of the object forms on the retina. **b.** When focusing on a distant object, the lens is flat because the ciliary muscle is relaxed and the suspensory ligament is taut. **c.** When focusing on a near object, the lens accommodates; it becomes rounded because the ciliary muscle contracts, causing the suspensory ligament to relax.

compartment is filled with a clear, watery fluid called the **aqueous humor.** A small amount of aqueous humor is continually produced each day. Normally, it leaves the anterior compartment by way of tiny ducts. When a person has **glaucoma,** these drainage ducts are blocked, and aqueous humor builds up. If glaucoma is not treated, the resulting pressure compresses the arteries that serve the nerve fibers of the retina where photoreceptors are located. The nerve fibers begin to die due to lack of nutrients, and the person becomes partially blind. Eventually, total blindness can result.

The third layer of the eye, the **retina,** is located in the posterior compartment, which is filled with a clear, gelatinous material called the **vitreous humor.** The retina contains photoreceptors called rod cells and cone cells. The rods are very sensitive to light, but they do not see color; therefore, at night or in a darkened room, we see only shades of gray. Rods also give us peripheral vision, as when you sense movement to your side. The cones, which require bright light, are sensitive to different wavelengths of light, and therefore we have the ability to distinguish colors. The retina has a very special

region called the **fovea centralis** where cone cells are densely packed. Light is normally focused on the fovea when we look directly at an object. This is helpful because vision is most acute in the fovea centralis. Sensory fibers from the retina form the **optic nerve,** which takes nerve impulses to the brain.

The delicate retina is easily dislodged from the choroid layer underneath. A sudden blow to the eye (for example, being struck by the ball during a baseball game) may detach the retina. Because detachment interrupts blood supply, retinal cells will die. Permanent blindness will result unless prompt surgical reattachment can correct the condition. Tragically, retinal detachment, hemorrhage, blindness, and severe brain damage occurs in very young children who are victims of shaken baby syndrome (see Focus on Forensics, p. 194.)

Function of the Lens

The lens, assisted by the cornea and the humors, focuses images on the retina (Fig. 9.8*a*). Focusing starts with the cornea and continues as the rays pass through the lens and the humor. The image produced is much smaller than the object because light rays are bent (refracted) when they are brought into focus. Notice that the image on the retina is inverted (upside down) and reversed from left to right.

Accommodation Imagine taking a walk in the country: as you stroll along, you focus first on a distant tree, then on a nearby flower. To maintain focus as you view distant and then near objects, the lens of the eye must change its shape—a process called **visual accommodation**. The shape of the lens is controlled by the ciliary muscle within the ciliary body. When we view a distant object, the ciliary muscle is relaxed, causing the suspensory ligaments attached to the ciliary body to be taut; therefore, the lens remains relatively flat (Fig. 9.8*b*). When we view a near object, the ciliary muscle contracts, releasing the tension on the suspensory ligaments, and the lens rounds up due to its natural elasticity (Fig. 9.8*c*). Because close work requires continuous contraction of the ciliary muscle, it very often causes muscle fatigue, known as eyestrain. As discussed in the Medical Focus on page 191, if a person's eyeball is too long or too short, he or she may need corrective lenses to focus the image on the retina.

Usually after the age of 40, the lens loses some of its elasticity and is unable to accommodate. Bifocal lenses may then be necessary for those who already have corrective lenses.

Vision Pathway

The pathway for vision begins once light has been focused on the photoreceptors in the retina. Some integration occurs in the retina where nerve impulses begin before the optic nerve transmits them to the brain.

Function of Photoreceptors Figure 9.9 illustrates the structure of the photoreceptors called **rod cells** and **cone cells.** Both rods and cones have an outer segment joined to an inner segment by a stalk. Pigment molecules are embedded in the membrane of the many disks present in the outer segment. Synaptic vesicles are located at the synaptic endings of the inner segment.

The visual pigment in rods is a deep purple pigment called rhodopsin. **Rhodopsin** is a complex molecule made up of the protein opsin and a light-absorbing molecule called **retinal,** which is a derivative of vitamin A. When a rod absorbs light, rhodopsin splits into opsin and retinal, leading to a cascade of reactions and the closure of ion channels in the rod cell's plasma membrane. In a sequence of events unique to photoreceptors, the light stimulus *stops* the release of

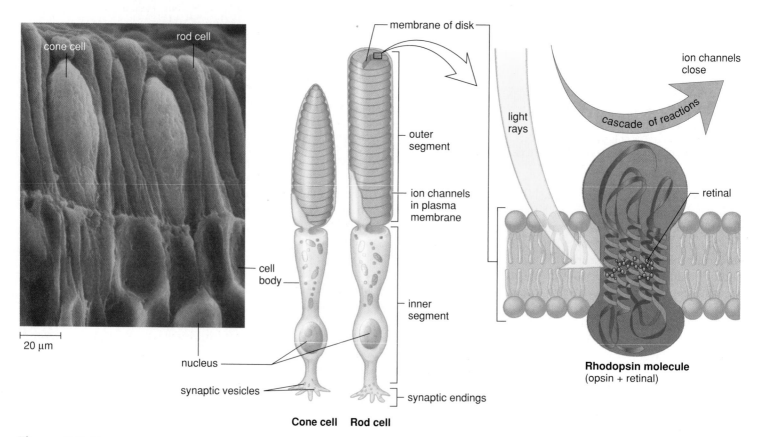

Figure 9.9 Photoreceptors in the eye. The outer segment of rods and cones contains stacks of membranous disks, which contain visual pigments. In rods, the membrane of each disk contains rhodopsin, a complex molecule containing the protein opsin and the pigment retinal. When retinal absorbs light energy, it splits, releasing opsin, which sets in motion a cascade of reactions that cause ion channels in the plasma membrane to close. Thereafter, nerve impulses go to the brain.

Corrective Lenses

Normal vision is commonly designated as 20/20. This ratio is determined using a **Snellen chart,** which uses letters of different sizes to test visual acuity (alternative charts for children or illiterate individuals use symbols). The numerator of the ratio is always the distance from the chart: 20 feet. The denominator is the distance at which a normal individual can read the letter: 20 feet. The larger this number, the poorer the subject's vision. Thus, a person with 20/200 vision must stand 20 feet away to read a letter that the normal person can read at 200 feet. (In countries that use metric measures, normal vision is 6/6.) Younger people may actually have vision that is better than 20/20; 20/15 vision is not uncommon in teenagers and young adults. Persons who can see close objects but cannot see the letters from this distance have **myopia**—that is, nearsightedness. Nearsighted people can see close objects better than they can see objects at a distance. These individuals have an elongated eyeball, and when they attempt to look at a distant object, the image is brought to focus in front of the retina (Fig. 9A*a*). They can see close objects because they can adjust the lens to allow the image to focus on the retina, but to see distant objects, these people must wear concave lenses, which diverge the light rays so that the image can be focused on the retina.

Rather than wear glasses or contact lenses, many nearsighted people are now choosing to undergo laser surgery. First, specialists determine how much the cornea needs to be flattened to achieve visual acuity. Controlled by a computer, the laser then removes this amount of the cornea. Most patients achieve at least 20/40 vision, but a few complain of glare and varying visual acuity.

Persons who can easily see the optometrist's chart but cannot see close objects well have **hyperopia**—that is, farsightedness. These individuals can see distant objects better than they can see close objects. They have a shortened eyeball, and when they try to see close objects, the image is focused behind the retina (Fig. 9A*b*). When the object is distant, the lens can compensate for the short eyeball, but when the object is

close, these persons must wear a convex lens to increase the bending of light rays so that the image can be focused on the retina.

In some individuals, the cornea assumes an oval shape along one axis (imagine a football-shaped cornea, instead of the normal round "tennis ball" shape). In rare cases, the lens itself may take on this oblong shape. The light rays cannot be evenly focused on the retina; thus, the image is blurred. This condition, called **astigmatism,** can be corrected by an unevenly ground lens to compensate for the uneven cornea (Fig. 9A*c*).

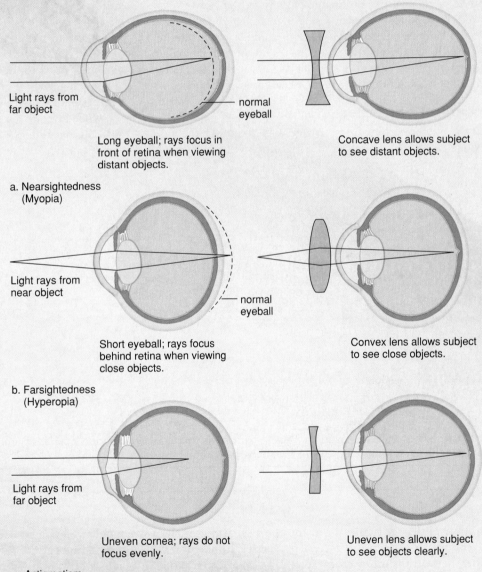

Light rays from far object

normal eyeball

Long eyeball; rays focus in front of retina when viewing distant objects.

Concave lens allows subject to see distant objects.

a. Nearsightedness (Myopia)

Light rays from near object

normal eyeball

Short eyeball; rays focus behind retina when viewing close objects.

Convex lens allows subject to see close objects.

b. Farsightedness (Hyperopia)

Light rays from far object

Uneven cornea; rays do not focus evenly.

Uneven lens allows subject to see objects clearly.

c. Astigmatism

Figure 9A Common abnormalities of the eye, with possible corrective lenses. **a.** A concave lens in nearsighted persons focuses light rays on the retina. **b.** A convex lens in farsighted persons focuses light rays on the retina. **c.** An uneven lens in persons with astigmatism focuses light rays on the retina.

A Bionic Cure for Macular Degeneration

As described in Figure 9.10, the retina is a three-layered tissue. The ganglion cells are the outermost layer, and light passing through the eye strikes these retinal cells first. The axons of ganglion cells form the optic nerve. Ganglion cells connect to the middle layer of bipolar cells. Bipolar cells then connect to rod and cone cells. The rod and cone cells are the actual photoreceptor cells, forming the deepest layer of the retina. When light enters the eye, it must penetrate the three layers—ganglion cells, bipolar cells, and finally the rods and cones. Recall that rods and cones contain the photochemicals that can respond to light. Rods respond to movement and changes in light intensity, and cones can respond to color. Once the rods or cones have responded, the nerve signal is sent backward through the retinal layers: from rod or cone, to bipolar cell, to ganglion cell, to the optic nerve, and from there to the visual cortex of the brain.

Macular Degeneration

If the photoreceptors—rods or cones—are destroyed, the individual will be blind, even if the rest of the visual pathway is undamaged. The most common cause of blindness in the Western world is age-related **macular degeneration**, which results in destruction of the macula lutea, a yellowish area in the central region of the retina. The macula lutea contains a concentration of cones, especially in the fovea centralis. Individuals with this condition have a distorted visual field: Blurriness or a blind spot is present, straight lines may look wavy, objects may appear larger or smaller than they are, and colors may look faded (Fig. 9B).

There are two main forms of age-related macular degeneration. "Wet" macular degeneration means that abnormal growth of new blood vessels is evident in the region of the macula. The blood vessels leak serum and blood, and the retina becomes distorted, leading to severe scarring that completely destroys the macula. "Dry" macular degeneration is not accompanied by the growth of blood vessels, and visual loss is less dramatic.

Heredity plays a role in the development of age-related macular degeneration: 15% of people with a family history of the condition develop the disease after age 60. Also, light-eyed

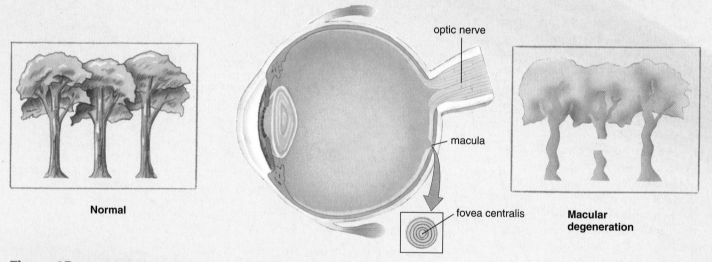

Normal

optic nerve

macula

fovea centralis

Macular degeneration

Figure 9B Macular degeneration. When a person with macular degeneration looks at a clump of trees, the trees may appear larger or smaller than they really are, the trunks may look wavy, details may be absent, and the colors may be dim.

neurotransmitter molecules from the rod's synaptic vesicles. (By contrast, other sensory receptors react to stimuli by releasing neurotransmitter from synaptic vesicles.) Neurons that synapse with the photoreceptors react to the *absence* of neurotransmitter by creating action potentials. Thereafter, nerve impulses go to the visual area of the cerebral cortex. Rods are very sensitive to light and therefore are suited to night vision. (Because carrots are rich in vitamin A, it is true that eating carrots can improve your night vision.) Rod cells are plentiful throughout the entire retina; therefore, they also provide us with peripheral vision and perception of motion.

The cones, on the other hand, are located primarily in the fovea and are activated by bright light. They allow us to detect the fine detail and the color of an object. Therefore, the people tend to be afflicted more frequently than dark-eyed people. Smoking, hypertension, and excessive sun exposure are possible contributing factors.

A yearly eye examination assists in the early detection of many eye diseases, including macular degeneration, cataracts, and glaucoma. When an ophthalmologist presents an Amsler grid (a crosshatched pattern of straight lines) to someone with macular degeneration, the grid looks blurred, distorted, or discolored. Signs of the "wet" form can be detected by an examination of the retina and confirmed by a fluorescein angiogram. In this test, a number of pictures are taken of the macula lutea after an orange dye has been injected into a vein in the patient's arm.

Currently, the treatment for the "dry" form of macular degeneration is the use of vitamin and mineral supplements, which may help stem the disease. For example, research indicates that consumption of zinc may prevent further loss of vision. On the other hand, when the "wet" form of the disease is diagnosed early, laser treatment can sometimes stop the growth of blood vessels.

Although people with age-related macular degeneration are classified as blind, they still have normal peripheral vision (outside the macula), which they can learn to use effectively. Because the periphery of the retina contains a high concentration of rods, vision there is less acute, and colors are not detected. But high-powered eyeglasses, magnifying devices, closed-circuit television, and special lamps can help patients see details more clearly.

Accumulating evidence suggests that both macular degeneration and cataracts, which tend to occur in the elderly, are caused by long-term exposure to the ultraviolet rays of the sun. Therefore, everyone—especially people who live in sunny climates or work outdoors—should wear sunglasses that absorb ultraviolet light. Large lenses worn close to the eyes offer further protection. The Sunglass Association of America has devised a helpful system for categorizing sunglasses.

Beginnings for Bionic Eyes

It has long been a dream of biomedical engineers to be able to craft a device to restore sight to the blind. Actual pioneering studies of retinal implant prosthetic devices began in the early 1990s, and continue to the present. The devices show some promise of being able to restore limited vision to individuals with retinal destruction. Although the cone cells are useless, the ganglion cells in the retinas of these patients can still send nerve signals. Two types of "bionic eyes" are currently being studied: a *subretinal implant* and an *epiretinal implant*. Both devices are designed to directly stimulate the ganglion cells of the retina. The subretinal system is surgically placed below the retina. It is a simple and very tiny, solar-powered silicon chip. Electricity from the solar chip produces nerve signals in ganglion cells. The epiretinal implant, which sits on top of the ganglion cells of the retina, consists of several parts. A miniature digital camera and computer are mounted in special glasses worn by the user. The glasses can transmit information to a silicon microchip placed on top of the ganglion cells. A battery pack worn at the belt transmits power to the implanted microchip.

Currently, clinical research has shown that subretinal implanted silicon chips do indeed stimulate the ganglion cells. Blind human volunteers have reported return of some vision after receiving these implants. In the most remarkable case, a totally blind patient was able to see his wife's face for the first time in decades. Epiretinal implants have also triggered visual sensations in blind human volunteers. More important, these tiny silicon chips seem to be stable after surgery. They do not cause infection, irritation, or breakdown of the retinal tissue.

Neither the subretinal implant nor the epiretinal implant is currently approved by the Food and Drug Administration for widespread use in patients. Both require further study and experimentation to ensure that they are totally safe and effective for use in humans. It is also important to note that these implants can't restore perfect vision at present. However, as the technology allowing miniaturization of electronics continues to improve, the blind may soon be able to obtain a device that restores some useful vision. Future research may result in even better vision.

condition called macular degeneration, which affects the fovea, is particularly devastating. The What's New reading above describes the condition and tells of a promising treatment that may soon be available.

Color vision depends on three different kinds of cones, which contain pigments called the B (blue), G (green), and R (red) pigments. Each pigment is made up of retinal and opsin, but there is a slight difference in the opsin structure of each, which accounts for their individual absorption patterns. Various combinations of cones are believed to be stimulated by in-between shades of color. Inherited absence of the color pigments in the cones of the eye causes **color blindness.** Red-green color blindness, which occurs primarily in men, is the most common form of color blindness.

Function of the Retina The retina has three layers of neurons (Fig. 9.10). The deepest layer (closest to the choroid) contains the rod cells and cone cells; the middle layer contains bipolar cells; and the innermost layer contains ganglion cells, whose sensory fibers become the *optic nerve.* Only the rod cells and the cone cells are sensitive to light, and therefore light must penetrate to the back of the retina before they are stimulated.

The rod cells and the cone cells synapse with the bipolar cells, which in turn synapse with ganglion cells that initiate nerve impulses. Notice in Figure 9.10 that there are many more rod cells and cone cells than ganglion cells. In fact, the retina has as many as 150 million rod cells and 6 million cone cells but only one million ganglion cells. The sensitivity of cones versus rods is mirrored by how directly they connect to ganglion cells. As many as 150 rods may activate the same ganglion cell. No wonder stimulation of rods results in vision that is blurred and indistinct. In contrast, some cone cells in the fovea centralis activate only one ganglion cell. This explains why cones, especially in the fovea, provide us with a sharper, more detailed image of an object.

As signals pass to bipolar cells and ganglion cells, integration occurs. Each ganglion cell receives signals from rod cells covering about one square millimeter of retina (about the size of a thumbtack hole). This region is the ganglion cell's receptive field. Some time ago, scientists discovered that a ganglion cell is stimulated only by nerve impulses received from the center of its receptive field; otherwise, it is inhibited. If all the rod cells in the receptive field receive light, the ganglion cell responds in a neutral way—that is, it reacts only weakly or perhaps not at all. This supports the hypothesis that considerable processing occurs in the retina before nerve impulses are sent to the brain. Additional integration occurs in the visual areas of the cerebral cortex.

Blind Spot Figure 9.10 provides an opportunity to point out that there are no rods and cones where the optic nerve exits the retina. Therefore, no vision is possible in this area. You can prove this to yourself by putting a dot to the right of center on a piece of paper. Use your right hand to move the paper slowly toward your right eye while you look straight ahead. The dot will disappear at one point—when its image falls on the retina where receptors are absent. This is your **blind spot.**

From the Retina to the Visual Cortex As stated, sensory fibers from the ganglion cells in the retina assemble to form the optic nerves. The **optic nerves** carry nerve impulses from the eyes to the optic chiasma. The **optic chiasma** has an X-shape formed by a crossing over of some of the optic nerve fibers. At the chiasma, fibers from the right half of each retina converge and continue on together in the *right optic tract,* and fibers from the left half of each retina converge and continue on together in the *left optic tract.*

The optic tracts sweep around the hypothalamus, and most fibers synapse with neurons in nuclei (masses of neuron cell bodies) in the thalamus. Axons from the thalamic nuclei form *optic radiations* that take nerve impulses to the primary visual areas of the occipital lobes (Fig. 9.11). The occipital lobes are a part of the cerebral cortex (see Fig. 8.9).

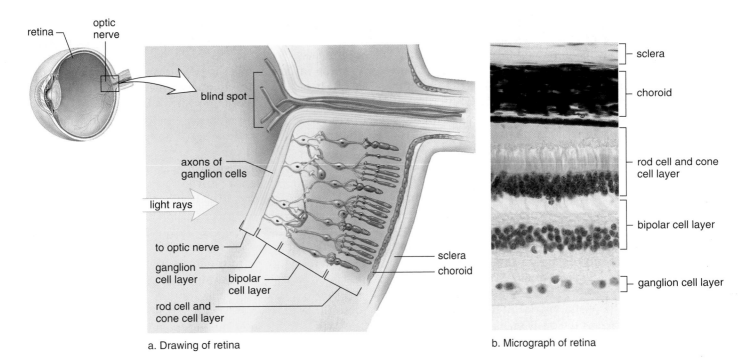

a. Drawing of retina

b. Micrograph of retina

Figure 9.10 Structure and function of the retina. **a.** The retina is the inner layer of the eyeball. Rod cells and cone cells, located at the back of the retina nearest the choroid, synapse with bipolar cells, which synapse with ganglion cells. Integration of signals occurs at these synapses; therefore, much processing occurs in bipolar and ganglion cells. Further, notice that many rod cells share one bipolar cell, but cone cells do not. Certain cone cells synapse with only one bipolar cell. Cone cells, in general, distinguish more detail than do rod cells. **b.** This micrograph shows that the sclera and choroid are relatively thin compared to the retina, which is composed of several layers of cells.

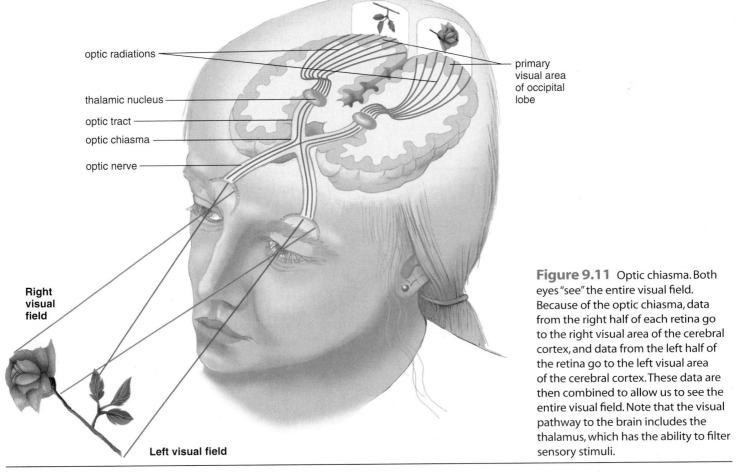

Figure 9.11 Optic chiasma. Both eyes "see" the entire visual field. Because of the optic chiasma, data from the right half of each retina go to the right visual area of the cerebral cortex, and data from the left half of the retina go to the left visual area of the cerebral cortex. These data are then combined to allow us to see the entire visual field. Note that the visual pathway to the brain includes the thalamus, which has the ability to filter sensory stimuli.

The *visual cortex* consists of the primary visual area and the visual association areas of the occipital lobes. Notice that the image arriving at the thalamus, and therefore the primary visual areas, has been split because the left optic tract carries information about the right portion of the visual field and the right optic tract carries information about the left portion of the visual field. Therefore, the right and left visual cortex must communicate with each other for us to see the entire visual field. Also, because the image is inverted and reversed (see Figs. 9.8 and 9.11) it must be righted for us to correctly perceive the visual field.

The most surprising finding has been that each primary visual area of the cerebral cortex acts like a post office, parceling out information regarding color, form, motion, and possibly other attributes to different portions of the adjoining visual association areas. In other words, the visual field has been taken apart even though we see a unified field. The visual association areas are believed to rebuild the field and give us an understanding of it. Visual association areas also store visual memories of objects previously seen.

Content **CHECK-UP!**

7. Which of the following structures is not part of the choroid layer of the eye?

   a. iris   c. ciliary body

   b. ciliary muscle   d. cornea

8. The posterior compartment of the eye is filled with:

   a. aqueous humor.   c. blood.

   b. tears.   d. vitreous humor.

9. Which statement is false?

   a. Rod cells contain the pigment rhodopsin.

   b. Light stimulus to rod cells causes release of neurotransmitter from the rods.

   c. Rod cells provide peripheral vision.

   d. Rod cells are found throughout the retina.

9.4 Sense of Hearing

The ear has two sensory functions: hearing and equilibrium (balance). The sensory receptors for both of these are located in the inner ear, and each consists of **hair cells** with *stereocilia* (long microvilli) that are sensitive to mechanical stimulation. The hair cells are mechanoreceptors, which respond to pressure or body movement.

Anatomy of the Ear

Figure 9.12 shows that the ear has three divisions: outer, middle, and inner. The **outer ear** consists of the **pinna** (external flap) and the external **auditory canal.** The opening of the auditory canal is lined with fine hairs and sweat glands. Modified sweat glands are located in the upper wall of the canal;

they secrete cerumen, or earwax, a substance that helps guard the ear against the entrance of foreign materials, such as air pollutants.

The **middle ear** is carved into the temporal bone of the skull. It begins at the **tympanic membrane** (eardrum) and ends at a bony wall containing two small openings covered by membranes. These openings are called the **oval window** and the **round window.** Three small bones connect the tympanic membrane to the oval window. Collectively called the **ossicles,** individually they are the **malleus** (hammer), the **incus** (anvil), and the **stapes** (stirrup) because their shapes resemble these objects. The malleus adheres to the tympanic membrane, and the stapes touches the oval window. An **auditory tube** (Eustachian tube), which extends from each middle ear to the nasopharynx (area at the back of the throat, which joins the nasal cavity), permits equalization of air pressure. Chewing gum, yawning, and swallowing in elevators and airplanes help move air through the auditory tubes upon ascent and descent. As this occurs, we often hear the ears "pop."

Whereas the outer ear and the middle ear contain air, the inner ear is filled with fluid. Anatomically speaking, the **inner ear** has three areas: The **semicircular canals** and the **vestibule** are both concerned with equilibrium; the **cochlea** is concerned with hearing. The cochlea resembles the shell of a snail because it spirals.

Sound Pathway

Sound waves pass through the auditory canal and middle ear to the cochlea in the inner ear, which transforms them into nerve impulses conducted in the auditory nerve to the brain.

Through the Auditory Canal and Middle Ear The process of hearing begins when sound waves enter the auditory canal. Just as ripples travel across the surface of a pond, sound waves travel by the successive vibrations of air molecules. Sound waves do not carry much energy, but when a large number of waves strike the tympanic membrane, it moves back and forth (vibrates) ever so slightly. The malleus then takes the pressure from the inner surface of the tympanic membrane, vibrating the incus and then the stapes, in such a way that the pressure is multiplied about 20 times as it moves. The stapes strikes the membrane of the oval window, causing it to vibrate, and in this way, the pressure is passed to the fluid within the cochlea of the inner ear.

From the Cochlea to the Auditory Cortex If the cochlea is unwound and examined in cross section (Fig. 9.13), you can see that it has three canals: the vestibular canal, the **cochlear canal**, and the tympanic canal. The sense organ for hearing, called the **spiral organ** (organ of Corti), is located in the cochlear canal. The spiral organ consists of little hair cells and a gelatinous material called the tectorial membrane. The hair cells sit on the basilar membrane and their stereocilia are embedded in the tectorial membrane.

When the stapes strikes the membrane of the oval window, pressure waves move from the vestibular canal to the tympanic canal across the basilar membrane. The basilar

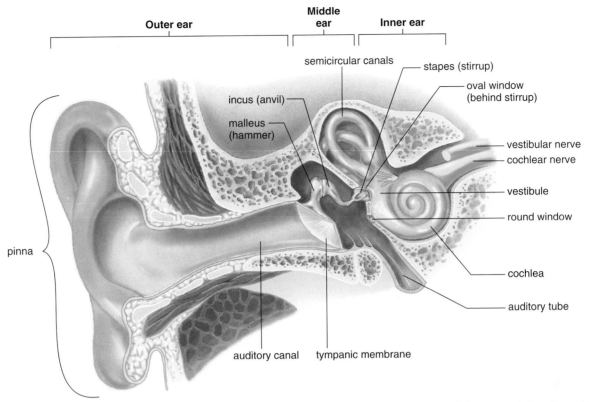

Figure 9.12 Anatomy of the human ear. In the middle ear, the malleus (hammer), the incus (anvil), and the stapes (stirrup) amplify sound waves. In the inner ear, the mechanoreceptors for equilibrium are in the semicircular canals and the vestibule, and the mechanoreceptors for hearing are in the cochlea.

membrane moves up and down, and the stereocilia of the hair cells embedded in the tectorial membrane bend. Bending the stereocilia causes action potentials, which begin in the cochlear nerve and travel to the brain stem. When they reach the auditory cortex of the cerebral cortex in the temporal lobe, they are interpreted as a sound.

Each part of the spiral organ is sensitive to different wave frequencies, or *pitch*. Near the tip, the spiral organ responds to low pitches, such as a tuba, and near the base, it responds to higher pitches, such as a bell or a whistle. The nerve fibers from each region along the length of the spiral organ lead to slightly different areas in the brain. The pitch sensation we experience depends upon which region of the basilar membrane vibrates and which area of the brain is stimulated.

Volume is a function of the amplitude (height) of sound waves. Loud noises cause the fluid within the vestibular canal to exert more pressure and the basilar membrane to vibrate farther and faster. The resulting increased stimulation is interpreted by the brain as volume.

9.5 Sense of Equilibrium

Mechanoreceptors in the semicircular canals detect rotational and/or angular movement of the head (**rotational equilibrium**), while mechanoreceptors in the vestibule detect move-

ment of the head in the vertical or horizontal planes (**gravitational equilibrium**) (Fig. 9.14).

Through their communication with the brain, these mechanoreceptors help us achieve equilibrium, but other structures in the body are also involved. The cerebellum must integrate information from mechanoreceptors in the inner ear, proprioceptors in muscles and joints, and photoreceptors in the eye to maintain balance and equilibrium.

Rotational Equilibrium Pathway

Rotational equilibrium involves the three semicircular canals, which are arranged so that there is one in each dimension of space. The base of each of the three canals, called the **ampulla,** is slightly enlarged. Little hair cells, whose stereocilia are embedded within a gelatinous material called a *cupula*, are found within the ampullae. Because of the way the semicircular canals are arranged, each ampulla responds to head rotation in a different plane of space. As fluid within a semicircular canal flows over and displaces a cupula, the stereocilia of the hair cells bend, and the pattern of impulses carried by the vestibular nerve to the brain changes. The cerebellum and other brain centers use information from the hair cells within the ampullae of the semicircular canals to maintain balance. The sensory input from semicircular canals tells the brain that you are rotating; motor output to various skeletal muscles can

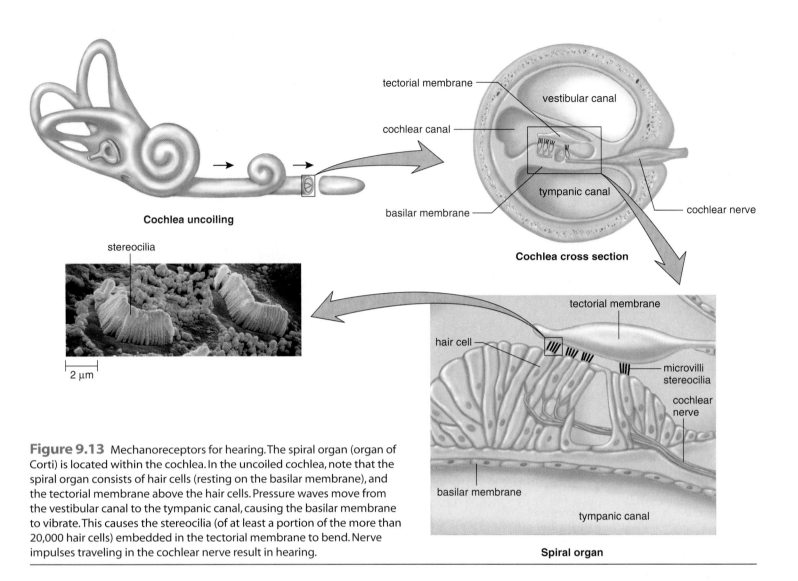

Cochlea uncoiling

tectorial membrane
vestibular canal
cochlear canal
tympanic canal
basilar membrane
cochlear nerve

Cochlea cross section

stereocilia

2 µm

tectorial membrane
hair cell
microvilli
stereocilia
cochlear
nerve
basilar membrane
tympanic canal

Spiral organ

Figure 9.13 Mechanoreceptors for hearing. The spiral organ (organ of Corti) is located within the cochlea. In the uncoiled cochlea, note that the spiral organ consists of hair cells (resting on the basilar membrane), and the tectorial membrane above the hair cells. Pressure waves move from the vestibular canal to the tympanic canal, causing the basilar membrane to vibrate. This causes the stereocilia (of at least a portion of the more than 20,000 hair cells) embedded in the tectorial membrane to bend. Nerve impulses traveling in the cochlear nerve result in hearing.

right your present position in space as need be (so that you don't fall).

Sometimes data regarding rotational equilibrium bring about unfortunate results. For example, continuous movement of fluid in the semicircular canals causes one form of motion sickness. **Vertigo** is dizziness and a sensation of rotation. It is possible to simulate a feeling of vertigo by spinning rapidly and stopping suddenly. Inflammation of cranial nerve VIII, the vestibulocochlear nerve from the ear, can cause vertigo as well. Motion sickness is also possible if the sensory input from the inner ear is different from visual sensation. For example, a person standing *inside* a large ship on rough seas will sense movement with the inner ear (as the ship tosses back and forth on the waves), but the visual input will register no movement (because the ship's walls don't move relative to the person's body). The person is *seasick*, with severe nausea and vomiting. People who are seasick sometimes find relief if they stand outdoors and focus on the horizon, so that both visual and inner ear inputs to the brain signal movement.

Gravitational Equilibrium Pathway

Gravitational equilibrium depends on the **utricle** and **saccule,** two membranous sacs located in the vestibule. Both of these sacs contain little hair cells, whose stereocilia are embedded within a gelatinous material called an otolithic membrane. Calcium carbonate ($CaCO_3$) granules, or **otoliths,** rest on this membrane. The utricle is especially sensitive to horizontal (back-forth) movements and the bending of the head, while the saccule responds best to vertical (up-down) movements.

When the body is still, the otoliths in the utricle and the saccule rest on the otolithic membrane above the hair cells. When the head bends or the body moves in the horizontal and vertical planes, the otoliths are displaced and the otolithic membrane sags, bending the stereocilia of the hair cells beneath. If the stereocilia move toward the largest stereocilium, called the *kinocilium,* nerve impulses increase in the vestibular nerve. If the stereocilia move away from the kinocilium, nerve impulses decrease in the vestibular nerve

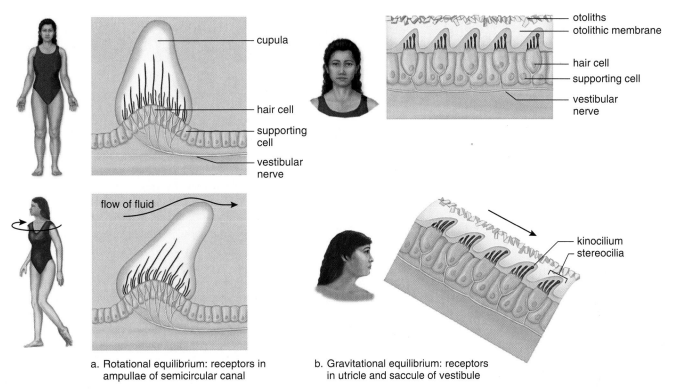

a. Rotational equilibrium: receptors in ampullae of semicircular canal

b. Gravitational equilibrium: receptors in utricle and saccule of vestibule

Figure 9.14 Mechanoreceptors for equilibrium. **a.** Rotational equilibrium. The ampullae of the semicircular canals contain hair cells with stereocilia embedded in a cupula. When the head rotates, the cupula is displaced, bending the stereocilia. Thereafter, nerve impulses travel in the vestibular nerve to the brain. **b.** Gravitational equilibrium. The utricle and the saccule contain hair cells with stereocilia embedded in an otolithic membrane. When the head bends, otoliths are displaced, causing the membrane to sag and the stereocilia to bend. If the stereocilia bend toward the kinocilium, the longest of the stereocilia, nerve impulses increase in the vestibular nerve. If the stereocilia bend away from the kinocilium, nerve impulses decrease in the vestibular nerve. The difference tells the brain in which direction the head moved.

TABLE 9.2 Functions of the Parts of the Ear			
Part	**Medium**	**Function**	**Mechanoreceptor**
OUTER EAR	**AIR**		
Pinna		Collects sound waves	—
Auditory canal		Filters air; directs sound waves to tympanic membrane	—
MIDDLE EAR	**AIR**		
Tympanic membrane and ossicles		Amplify sound waves	—
Auditory tube (Eustachian tube)		Equalizes air pressure	—
INNER EAR	**FLUID**		
Cochlea (contains spiral organ)		Hearing	Stereocilia embedded in tectorial membrane
Semicircular canals		Rotational equilibrium	Stereocilia embedded in cupula
Vestibule (contains utricle and saccule)		Gravitational equilibrium	Stereocilia embedded in otolithic membrane

(see Fig. 9.14*b*). If you are upside down, nerve impulses in the vestibular nerve cease. These data tell the brain the direction of the movement of the head at the moment. The cerebellum and other brain centers use this sensory information to main-tain gravitational equilibrium and balance. Appropriate motor output from the brain to various skeletal muscles can right our present position in space.

Table 9.2 summarizes the functions of the parts of the ear.

Hearing Damage and Deafness

There are two major types of deafness: conduction deafness and sensorineural, or nerve, deafness. **Conduction deafness** occurs when a mechanical blockage keeps sound waves from reaching the oval window, the membrane at the beginning of the inner ear. Conductive deafness can result from the presence of foreign objects in the external ear canal, impacted ear wax, or cancerous tumors in the external ear canal or middle ear. Conduction deafness can also be due to repeated infections or **otosclerosis.** With otosclerosis, the normal bone of the middle ear is replaced by vascular, spongy bone.

Nerve deafness most often occurs when cilia on the receptors within the cochlea have worn away. Because this may happen with normal aging, older people are more likely to have trouble hearing. However, studies also suggest that age-associated hearing loss can be prevented if ears are protected from loud noises, starting even during infancy. Hospitals are now aware of the problem and are taking steps to ensure that neonatal intensive care units and nurseries are as quiet as possible. Neural deafness may also result from congenital defects, particularly when a pregnant woman contracts German measles (rubella) during the first trimester of pregnancy. For this reason, every female should be immunized against rubella before her childbearing years.

In today's society, exposure to the types of noises listed in Table 9A is common. Everyone should consider three aspects of noise to prevent hearing loss: (1) how loud is the noise, (2) how long is the noise heard, and (3) how close is the noise to the ear. Loudness is measured in decibels, and any level above 80 decibels could damage the hair cells of the organ of Corti. Exposure to intense sounds of short duration, such as a burst of gunfire, can result in an immediate hearing loss. Since the butt of a rifle offers some protection, hunters may have a significant hearing reduction in the ear opposite the shoulder they use for support while firing their gun. Because even listening to city traffic for extended periods can damage hearing, frequent attendance at rock concerts and constant listening to loud music from a stereo are obviously dangerous. Noisy indoor or outdoor equipment, such as a rug-cleaning machine or a chain saw, is also troublesome. Even motorcycles and recreational vehicles, such as snowmobiles and motocross bikes, can contribute to a gradual hearing loss.

The first hint of a problem could be temporary hearing loss, a "full" feeling in the ears, muffled hearing, or **tinnitus** (ringing in the ears). If you have any of these symptoms, modify your listening habits immediately to prevent further damage. If exposure to noise is unavoidable, use specially designed noise-reduction earmuffs or purchase earplugs made from a compressible, sponge-like material at a drugstore or sporting goods store. These earplugs are not the same as those worn for swimming, and they should not be used interchangeably.

Finally, people need to be aware that some medicines are **ototoxic** (damaging to any of the elements of hearing or balance).

Content **CHECK-UP!**

10. Choose the pathway for a sound wave to travel to the inner ear:

    a. auditory canal; oval window; malleus, stapes, incus; tympanic membrane

    b. auditory canal; tympanic membrane; stapes, incus, malleus; auditory tube

    c. auditory canal; tympanic membrane; malleus, incus, stapes; oval window

    d. auditory canal; tympanic membrane; incus, stapes, malleus; oval window

11. Which structure allows air pressure to equalize in the middle ear?

    a. auditory canal   c. cochlear canal

    b. auditory tube   d. semicircular canal

12. Which of the following receptors are found in the semicircular canals, utricle, and saccule?

    a. mechanoreceptors   c. chemoreceptors

    b. pressure receptors   d. nociceptors

9.6   Effects of Aging

As we age, assistance is likely required to improve our sight and hearing. The lens of the eye does not accommodate as well, a condition called presbyopia. Therefore, eyeglasses, contact lenses, or corrective surgery will most likely be needed to improve vision. Also, three serious visual disorders are seen more frequently in older persons: (1) Possibly due to exposure to the sun, the lens is subject to **cataracts.** The lens becomes opaque and therefore incapable of transmitting rays of light. Today, the lens is usually surgically replaced with an artificial lens or the lens is removed entirely and replaced by special glasses. In the future, it may be possible to restore the original configuration of the proteins making up the lens. (2) Age-related macular degeneration (see the What's New reading on pages 192–193) is the most frequent cause of blindness in older people. (3) Glaucoma is more likely to develop because the anterior compartment of the eye (see Fig. 9.7) undergoes a reduction in size.

The need for a hearing aid also increases with age. Atrophy of the organ of Corti can lead to **presbycusis** (age-related hearing decline). First, people tend to lose the ability to detect

Anticancer drugs—most notably, cisplatin—and certain antibiotics (for example, streptomycin, kanamycin, and gentamicin) make the ears especially susceptible to a hearing loss. People taking such medications should protect their ears from any excessive noises.

Cochlear implants that directly stimulate the auditory nerve are available for persons with nerve deafness. However, they are costly, and people wearing these electronic devices report that the speech they hear is like that of a robot.

TABLE 9A Sound Intensity and Hearing Damage

Type of Noise	Sound Level (decibels)	Effect
Rock concert, shotgun, jet engine	Over 125	Beyond threshold of pain; potential for hearing loss is high.
Nightclub, boom box, thunderclap	Over 120	Hearing loss is likely.
Chain saw, pneumatic drill, jackhammer, symphony orchestra, snowmobile, garbage truck, cement mixer	100–200	Regular exposure of longer than 1 minute risks permanent hearing loss.
Farm tractor, newspaper press, subway, motorcycle	90–100	Fifteen minutes of unprotected exposure is potentially harmful.
Lawnmower, food blender	85–90	Continuous daily exposure for more than 8 hours can cause hearing damage.
Diesel truck, average city traffic noise	80–85	Annoying; constant exposure may cause hearing damage.

Source: National Institute on Deafness and Other Communication Disorders, National Institutes of Health, January 1990.

high-frequency tones, and later the lower tones are affected. Eventually, they can hear speech but cannot understand the words being said.

Otosclerosis, an overgrowth of bone that causes the stapes to adhere to the oval window, is the most frequent cause of conduction deafness in adults (see the Medical Focus on page 200). The condition actually begins during youth but may not become evident until later in life. Dizziness and the inability to maintain balance may also occur in older people due to changes in the inner ear.

SELECTED NEW TERMS

Basic Key Terms

ampulla (am-pul′uh), p. 197

aqueous humor (a′kwe-us hyū′mer), p. 189

auditory canal (aw′dĭ-to″re kuh-nal′), p. 196

auditory tube (aw′dĭ-to″re tūb), p. 196

blind spot (blīnd spot), p. 194

chemoreceptor (ké′mō-rĭ-sĕp′tar), p. 182

choroid (ko′royd), p. 188

ciliary body (sĭl′ē-ĕr′ē bŏd′ē), p. 188

ciliary muscle (sil′e-ĕr″e mus′l), p. 188

cochlea (kōk′le-uh), p. 196

cochlear canal (kōk′le-er kuh-nal′), p. 196

cone cell (kōn sel), p. 190

conjunctiva (kŏn′jŭngk-tī′və), p. 186

cornea (kor′ne-uh), p. 188

fovea centralis (fō′vē-ə sĕn-trā′lĭs), p. 189

incus (ing′kus), p. 196

iris (i′ris), p. 188

lacrimal apparatus (lak′rĭ-mul ap″uh-rắ′tus), p. 186

lens (lenz), p. 188

malleus (mal′e-us), p. 196

mechanoreceptor (mĕk′ə-nō-rĭ-sĕp′tər), p. 182

nociceptor (no″sē-sep′tor), p. 183

olfactory cell (ol-fak′to-re sel), p. 185

optic chiasma (ŏp′tĭk kī-ăz′mə), p. 194

optic nerve (op′tik nerv), p. 189

ossicle (os′ĭ-kl), p. 196

otolith (ō′tō-lith), p. 198

oval window (ō′vəl wĭn′dō), p. 196

pain receptors (pān rĭ-sĕp′tər), p. 183

photoreceptor (fō′tō-rĭ-sĕp′tər), p. 182

pinna (pin′uh), p. 196

proprioceptor (pro″pre-o-sep′tor), p. 182

pupil (pyū′pl), p. 188

receptor potential (rĭ-sĕp′tər pə-tĕn′shəl), p. 182

retina (rĕt′n-ə), p. 189
retinal (rĕt′n-ôl′), p. 190
rhodospin (rō-dŏp′sĭn), p. 190
rod cell (rod sel), p. 190
round window (round wĭn′dō), p. 196
saccule (sak′yūl), p. 198
sclera (sklēr′uh), p. 188
semicircular canal
    (sem″e-ser′kyū-ler kuh-nal′), p. 196
spiral organ (spi′rul or′gun), p. 196
stapes (sta′pēz), p. 196
taste bud (tāst bud), p. 184
thermoreceptor (thûr′mō-rĭ-sĕp′tər), p. 182
tympanic membrane
    (tim-pan′ik mem′brān), p. 196

utricle (u′trĭ-kl), p. 198
vestibule (vĕs′tə-byōol), p. 196
visual accommodation
    (vizh′ū-ul uh-kom″o-da′shun), p. 190
vitreous humor (vit′re-us hyū′mor), p. 189

Clinical Key Terms

astigmatism (ə-stĭg′mə-tĭz′əm), p. 191
cataract (kat′uh-rakt), p. 200
cochlear implant (kōk′le-er im′plant), p. 201
color blindness (kŭl′ər-blīnd nəss), p. 193
conduction deafness
    (kon-duk′shun def′nes), p. 200
glaucoma (glaw-ko′muh), p. 189

hyperopia (hi″per-o′pe-uh), p. 191
macular degeneration
    (mă′kyū-ler de″jen-er-a′shun), p. 192
myopia (mi-o′pe-uh), p. 191
nerve deafness (nerv def′nes), p. 200
otosclerosis (ō″tō-sklĕ-ro′sis), p. 200
ototoxic (ō″tō-tok′sik), p. 200
presbycusis (prez″be-ku′sis), p. 200
referred pain (rĭ-fūr′əd pān), p. 184
sty (sti), p. 186
tinnitus (tĭn′ĭ-təs), p. 200
vertigo (vûr′tĭ-gō), p. 198

SUMMARY

9.1 General Senses
Each type of sensory receptor detects a particular kind of stimulus. When stimulation occurs, sensory receptors initiate graded potentials. Action potentials are generated and transmitted to the spinal cord and/or brain. Sensation occurs when nerve impulses reach the cerebral cortex. Perception is an interpretation of the meaning of sensations.

9.2 Senses of Taste and Smell
A. Taste and smell are due to chemoreceptors that are stimulated by molecules in the environment. The taste buds contain taste cells that communicate with sensory fibers, while the chemoreceptors for smell are olfactory neurons.
B. After molecules bind to plasma membrane receptor proteins on the microvilli of taste cells and the cilia of olfactory cells, nerve action potentials eventually reach the cerebral cortex, which determines the taste and odor according to the pattern of stimulation.

9.3 Sense of Vision
A. Vision is dependent on the eye, the optic nerves, and the visual areas of the cerebral cortex. The eye has three layers. The outer layer, the sclera, can be seen as the white of the eye; it also becomes the transparent bulge in the front of the eye called the cornea. The middle pigmented layer, called the choroid, absorbs stray light rays. The rod cells (sensory receptors for dim light)

and the cone cells (sensory receptors for bright light and color) are located in the retina, the inner layer of the eyeball. The cornea, the humors, and especially the lens bring the light rays to focus on the retina. To see a close object, accommodation occurs as the lens rounds up.
B. When light strikes rhodopsin within the membranous disks of rod cells, rhodopsin splits into opsin and retinal. A cascade of reactions leads to the closing of ion channels in a rod cell's plasma membrane. Neurotransmitter molecules are no longer released. Nerve cells, which synapse with photoreceptors, transmit action potentials, which are carried in the optic nerve to the brain.
C. Integration occurs in the retina, which is composed of three layers of cells: the rod and cone layer, the bipolar cell layer, and the ganglion cell layer. Integration also occurs in the brain. The visual field is taken apart by the optic chiasma and by the primary visual area in the cerebral cortex, which parcels out signals for color, form, and motion to the visual association area. Then the cortex rebuilds the field.

9.4 Sense of Hearing
A. Hearing is dependent on the ear, the cochlear nerve, and the auditory areas of the cerebral cortex.
B. The ear is divided into three parts: outer, middle, and inner.

The outer ear consists of the pinna and the auditory canal, which direct sound waves to the middle ear. The middle ear begins with the tympanic membrane and contains the ossicles (malleus, incus, and stapes). The malleus is attached to the tympanic membrane, and the stapes is attached to the oval window, which is covered by a membrane. The inner ear contains the cochlea and the semicircular canals, plus the utricle and the saccule.
C. Hearing begins when the outer ear receives and the middle ear amplifies the sound waves that then strike the oval window membrane. Its vibrations set up pressure waves across the cochlear canal, which contains the spiral organ, consisting of hair cells whose stereocilia are embedded within the tectorial membrane. When the basilar membrane vibrates, the stereocilia of the hair cells bend. Nerve impulses begin in the cochlear nerve and are carried to the brain.

9.5 Sense of Equilibrium
The ear also contains mechano-receptors for our sense of equilibrium. Rotational equilibrium is dependent on the stimulation of hair cells within the ampullae of the semicircular canals. Gravitational equilibrium relies on the stimulation of hair cells within the utricle and the saccule.

9.6 **Effects of Aging**

As we age, assistance is likely needed to improve our failing senses of sight and hearing. Three more serious visual disorders—cataracts, age-related macular degeneration, and glaucoma—may occur, making medical intervention necessary.

STUDY QUESTIONS

1. Name the three different types of general sensory receptors. (p. 182)
2. Name the receptors for the senses of taste and smell. Which brain areas receive their sensory information? (pp. 184–185)
3. Describe the relationship between taste and smell. (p. 186)
4. Explain the accommodation process. Describe the shape of the lens when a person is viewing a faraway object; then when he or she is viewing a near object. (p. 190)
5. Describe sight in dim light. What chemical reaction is responsible for vision in dim light? (pp. 190–194)
6. Compare and contrast rods and cones. Explain what role each plays in vision; what chemical each contains; where each is located on the retina; and how each reacts to light. (pp. 190–191)
7. Place the structures of the visual pathway (optic radiations, optic nerve, optic chiasma, optic tracts, thalamus, occipital lobes) in the correct order as an action potential signal travels from the eye to the brain. How does the retina integrate visual information? How does the brain process visual information? (p. 189)
8. Place the structures of the outer, middle, and inner ears in the correct order as a sound wave travels through the ear. How does the spiral organ (organ of Corti) in the cochlea translate a sound wave into an action potential? Which cranial nerve carries the action potential into the brain? (p. 196)
9. Compare and contrast the utricle and saccule and the semicircular canals. How are they alike? How are they different? (p. 198)
10. Define conduction deafness and sensorineural deafness. Why do young people frequently suffer hearing loss? (p. 200)

LEARNING OBJECTIVE QUESTIONS

Fill in the blanks.

1. The sensory organs for position and movement are called _____.
2. Taste buds and olfactory cells are termed _____ because they are sensitive to chemicals in the air and food.
3. The sensory receptors for sight, the _____ and _____, are located in the _____, the inner layer of the eye.
4. The cones give us _____ vision and work best in _____ light.
5. The lens _____ for viewing close objects.
6. People who are nearsighted cannot see objects that are _____. A _____ lens will restore this ability.
7. The ossicles are the _____, _____, and _____.
8. The semicircular canals are involved in the sense of _____.
9. The spiral organ is located in the _____ canal of the _____.
10. Vision, hearing, taste, and smell do not occur unless nerve impulses reach the proper portion of the _____.

MEDICAL TERMINOLOGY EXERCISE

Consult Appendix B for help in pronouncing and analyzing the meaning of the terms that follow.

1. ophthalmologist (of″thal-mol′o-jist)
2. presbyopia (pres″be-o′pe-uh)
3. blepharoptosis (blef″uh-ro-to′sis)
4. keratoplasty (ker′uh-to-plas″te)
5. optometrist (op-tom′ĕ-trist)
6. lacrimator (lak′rĭ-ma″tor)
7. otitis media (o-ti′tis me′de-uh)
8. tympanocentesis (tim″puh-no-sen-te′sis)
9. microtia (mi″kro′she-uh)
10. myringotome (mi-ring′go-tōm)
11. iridomalacia (ir′ĭ-do-muh′la′she-uh)
12. hypogeusia (hi-po′go′sē-uh)

WEBSITE LINK

Visit this textbook's website at http://www.mhhe.com/maderap6 **for additional quizzes, interactive learning exercises, and other study tools.**

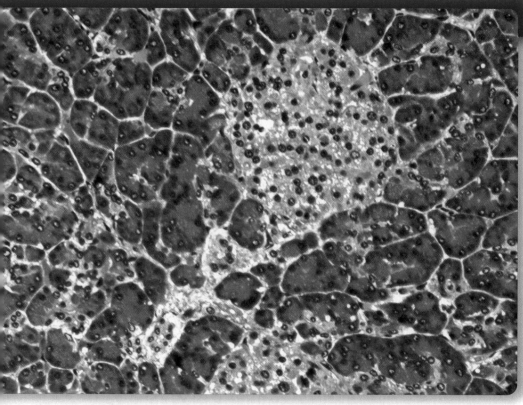

Pancreatic islets (light pink areas) are shown in this photomicrograph of the pancreas.

CHAPTER

10

The Endocrine System

Chapter Outline & Learning Objectives

After you have studied this chapter, you should be able to:

10.1 Endocrine Glands

The endocrine system consists of glands and tissues that secrete hormones. This chapter will give many examples of the close association between the endocrine and nervous systems. Like the nervous system, the endocrine system is intimately involved in homeostasis.

Hormones are chemical signals that affect the behavior of other glands or tissues. Hormones influence the metabolism of cells, the growth and development of body parts, and homeostasis. **Endocrine glands** are ductless; they secrete their hormones directly into tissue fluid. From there, the hormones diffuse into the bloodstream for distribution throughout the body. Endocrine glands can be contrasted with exocrine glands, which have ducts and secrete their products into these ducts. For example, the salivary glands send saliva into the mouth by way of the salivary ducts.

Figure 10.1 depicts the locations of the major endocrine glands in the body; Table 10.1 lists the hormones they release. Each type of hormone has a unique composition. Even so, hormones can be categorized as either peptides (which include proteins, glycoproteins, and modified amino acids) or steroids. All steroid hormones have the same four-carbon ring complex, but each has different side chains.

How Hormones Function

Along with fundamental differences in structure, peptide and steroid hormones also function differently. Most peptide hormones bind to a receptor protein in the plasma membrane and activate a "second messenger" system (Fig. 10.2). The "second messenger" causes the cellular changes for which the hormone is credited. As an analogy, suppose you are the person in charge of a crew assigned to redecorate a room. As such, you stand outside the room and direct the workers inside the room. The workers clean, paint, apply wallpaper, etc. Like the "boss" in this analogy, the peptide hormone stays outside the cell and directs activities within. The peptide hormone, or "first messenger," activates a "second messenger"—the crew workers inside the cell. Common second messengers found in many body cells include **cyclic AMP** (made from ATP, and abbreviated cAMP) and calcium. The second messenger sets in motion an enzyme cascade, so-called because each enzyme in turn activates several others, and so on. These intracellular enzymes cause the changes in the cell that are associated with the hormone. Because of the second messenger system, the binding of a single peptide hormone can result in as much as a thousandfold response. The cellular response can be a change in cellular behavior or the formation of an end product that leaves the cell. For example, by activating a second messenger, insulin causes the facilitated diffusion of glucose into body cells, while thyroid-stimulating hormone causes thyroxine release from the thyroid gland.

Steroid hormones are lipids, and therefore they diffuse across the plasma membrane and other cellular membranes (Fig. 10.3). Only after they are inside the cell do steroid hormones, such as estrogen and progesterone, bind to receptor proteins. The hormone-receptor complex then binds to DNA, activating particular genes. Activation leads to production of a cellular enzyme in varying quantities. Again, it is intracellular enzymes that cause the cellular changes for which the hormone receives credit. For example, estrogen

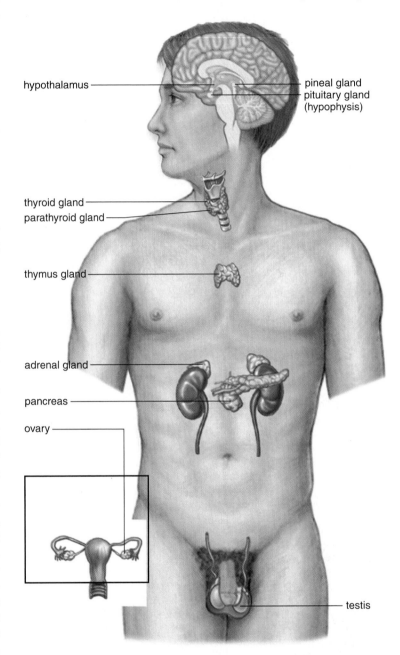

hypothalamus

pineal gland
pituitary gland
(hypophysis)

thyroid gland
parathyroid gland

thymus gland

adrenal gland

pancreas

ovary

testis

Figure 10.1 The endocrine system. Anatomical location of major endocrine glands in the body. The hypothalamus and pituitary gland are in the brain, the thyroid and parathyroids are in the neck, and the adrenal glands and pancreas are in the pelvic cavity. The gonads include the ovaries in females, located in the pelvic cavity, and the testes in males, located outside this cavity in the scrotum. Also shown are the pineal gland, located in the brain, and the thymus gland, which lies within the thoracic cavity.

TABLE 10.1 Principal Endocrine Glands and Hormones

Endocrine Gland	Hormone Released	Chemical Class	Target Tissues/Organs	Chief Function(s) of Hormone
Hypothalamus	Hypothalamic-releasing and inhibiting hormones	Peptide	Anterior pituitary	Regulate anterior pituitary hormones
Produced by hypothalamus, released from posterior pituitary	Antidiuretic (ADH)	Peptide	Kidneys	Stimulates water reabsorption by kidneys
	Oxytocin	Peptide	Uterus, mammary glands	Stimulates uterine muscle contraction, release of milk by mammary glands
Anterior pituitary	Thyroid-stimulating (TSH)	Glycoprotein	Thyroid	Stimulates thyroid
	Adrenocorticotropic (ACTH)	Peptide	Adrenal cortex	Stimulates adrenal cortex
	Gonadotropic Follicle-stimulating Hormone (FSH) Luteinizing Hormone (LH)	Glycoprotein	Gonads	Egg and sperm production; sex hormone production
	Prolactin (PRL)	Protein	Mammary glands	Milk production
	Growth (GH)	Protein	Soft tissues, bones	Cell division, protein synthesis, and bone growth
	Melanocyte-stimulating (MSH)	Peptide	Melanocytes in skin	Unknown function in humans; regulates skin color in lower vertebrates
Thyroid	Thyroxine (T_4) and triiodothyronine (T_3)	Iodinated amino acid	All tissues	Increases metabolic rate; regulates growth and development
	Calcitonin	Peptide	Bones, kidneys, intestine	Lowers blood calcium level
Parathyroids	Parathyroid (PTH)	Peptide	Bones, kidneys, intestine	Raises blood calcium level
Adrenal gland				
Adrenal cortex	Glucocorticoids (cortisol)	Steroid	All tissues	Raise blood glucose level; stimulate breakdown of protein
	Mineralocorticoids (aldosterone)	Steroid	Kidneys	Reabsorb sodium and excrete potassium
	Sex hormones	Steroid	Gonads, skin, muscles, bones	Stimulate reproductive organs and bring about sex characteristics
Adrenal medulla	Epinephrine and norepinephrine	Modified amino acid	Cardiac and other muscles	Released in emergency situations; raise blood glucose level
Pancreas	Insulin	Protein	Liver, muscles, adipose tissue	Lowers blood glucose level; promotes formation of glycogen
	Glucagon	Protein	Liver, muscles, adipose tissue	Raises blood glucose level
Gonads				
Testes	Androgens (testosterone)	Steroid	Gonads, skin, muscles, bones	Stimulate male sex characteristics
Ovaries	Estrogens and progesterone	Steroid	Gonads, skin, muscles, bones	Stimulate female sex characteristics
Thymus	Thymosins	Peptide	T lymphocytes	Stimulate production and maturation of T lymphocytes
Pineal gland	Melatonin	Modified amino acid	Brain	Controls circadian and circannual rhythms; possibly involved in maturation of sexual organs

directs cellular enzymes that cause the growth of axillary and pubic hair in an adolescent female.

When protein hormones such as insulin are used for medical purposes, they must be administered by injection. If these hormones were taken orally, they would be acted on by digestive enzymes. Once digested, insulin cannot carry out its functions. Steroid hormones, such as those in birth control pills, can be taken orally because they can pass through the plasma membrane without prior digestion.

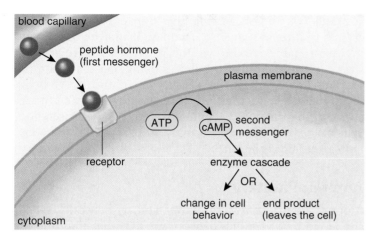

Figure 10.2 The binding of a peptide hormone leads to cAMP and then to activation of an enzyme cascade.

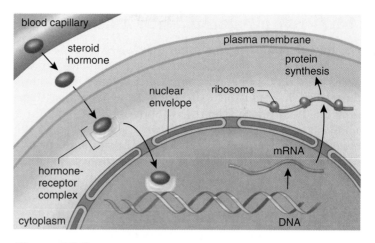

Figure 10.3 A steroid hormone results in a hormone-receptor complex that activates DNA and protein synthesis.

Hormones and Homeostasis

The release of hormones is usually controlled by one or more of three different mechanisms: 1) Negative feedback controls hormone release; 2) hormone release is controlled by the actions of other hormones; and/or 3) hormone release is controlled by the nervous system.

For the majority of hormones, release is controlled by a negative feedback system. The result is that the activity of the hormone is maintained within normal limits and homeostasis is ensured. Recall that in a negative feedback system, a stimulus causes a body response. The body response, in turn, corrects the initial stimulus. For example, when the blood glucose level rises following a meal (stimulus), the pancreas secretes insulin (response). Insulin causes the liver to store glucose and the cells to take up glucose. Thus, blood glucose is lowered back to normal (stimulus corrected). Once the initial stimulus is corrected, the pancreas stops producing insulin. In much the same way, a low level of calcium ions in the blood stimulates secretion of parathyroid hormone (PTH) from the parathyroid gland. When blood calcium rises to a normal level, secretion of PTH stops. These examples illustrate regulation by negative feedback.

Hormone release may also be controlled by specific stimulating or inhibiting hormones. Thyroid-stimulating hormone (TSH) from the pituitary gland (also called **thyrotropin**) does exactly what its name implies: It stimulates the thyroid gland to produce thyroid hormone. By contrast, the release of insulin is inhibited by the production of glucagon by the pancreas. Insulin lowers the blood glucose level, whereas glucagon raises it. In subsequent sections of this chapter, we will point out other instances in which pairs of hormones work opposite to one another, and thereby bring about the regulation of a substance in the blood.

The nervous system is an important controller of the endocrine system. Upon receiving sensory information from the body, the brain can make appropriate adjustments to hormone secretion, to ensure homeostasis. For example, while you eat a meal, sensory information is relayed to the brain; in turn, the brain signals parasympathetic motor neurons to release the hormone insulin from the pancreas. (Recall that the parasympathetic neurons control "rest and digest" functions.) Insulin will allow body cells to take up glucose and amino acids from digested food.

It's important to stress that many hormones are influenced by more than one control mechanism. In the previous examples, you can see that insulin release is influenced by all three controllers: negative feedback, hormonal, and neural control. In addition, note that in a few instances *positive* feedback controls the release of a hormone: for example, release of oxytocin during labor and delivery (discussed on page 208.)

Content CHECK-UP!

1. Antidiuretic hormone (ADH), a peptide hormone, works by:
   a. binding to a receptor outside the cell and activating a second messenger.
   b. diffusing into the cell, binding to a receptor inside the cell, and activating a second messenger.
   c. diffusing into the cell, binding to a receptor inside the cell, and activating genes in DNA.

2. Testosterone, a steroid hormone, works by:
   a. binding to a receptor outside the cell and activating a second messenger.
   b. diffusing into the cell, binding to a receptor inside the cell, and activating a second messenger.
   c. diffusing into the cell, binding to a receptor inside the cell, and activating genes in DNA.

3. Antidiuretic hormone stimulates the kidneys to reabsorb water and return it to the blood plasma. ADH release is controlled by a negative feedback system. Which action causes ADH to be released?
   a. drinking a big bottle of water
   b. finishing a marathon race and becoming dehydrated

10.2 Hypothalamus and Pituitary Gland

The **hypothalamus** regulates the internal environment. For example, through the autonomic system, it helps control heartbeat, body temperature, and water balance (by creating thirst). The hypothalamus also controls the glandular secretions of the **pituitary gland (hypophysis).** The pituitary, a small gland about 1 cm in diameter, is connected to the hypothalamus by a stalklike structure. The pituitary has two portions: the **posterior pituitary (neurohypophysis)** and the **anterior pituitary (adrenohypophysis).**

Posterior Pituitary

Neurons in the hypothalamus called neurosecretory cells produce the hormones antidiuretic hormone (ADH) and oxytocin (Fig. 10.4, *left*). These hormones pass through axons into the posterior pituitary (neurohypophysis) where they are stored in axon endings. Thus, the hypothalamic hormones antidiuretic hormone and oxytocin are produced in the hypothalamus, but are released into the bloodstream from the posterior pituitary.

Antidiuretic Hormone and Oxytocin

Certain neurons in the hypothalamus are sensitive to the water-salt balance of the blood. When these cells determine that the blood is too concentrated, **antidiuretic hormone (ADH,** also called *vasopressin*) is released from the posterior pituitary. In your body, blood becomes concentrated if you have just finished exercising heavily and body water has been lost as sweat. Upon reaching the kidneys, ADH causes more water to be reabsorbed into kidney capillaries. As the blood becomes dilute once again, ADH is no longer released. This is an example of control by negative feedback because the effect of the hormone (to dilute blood) acts to shut down the release of the hormone. An additional effect of ADH is to raise blood pressure, by vasoconstriction of blood vessels throughout the body (hence, the hormone's additional name of vasopressin). This mechanism also illustrates negative feedback: Blood pressure falls because body water is lost as sweat (stimulus); vasopressin is released (response); blood vessels constrict and blood pressure rises to normal (stimulus corrected). Negative feedback maintains stable conditions and homeostasis.

Inability to produce ADH causes **diabetes insipidus** (watery urine), in which a person produces copious amounts of urine with a resultant loss of ions from the blood. The condition can be corrected by the administration of ADH.

Oxytocin, the other hormone made in the hypothalamus, causes uterine contraction during childbirth and milk letdown when a baby is nursing. The more the uterus contracts during labor, the more nerve impulses reach the hypothalamus, causing oxytocin to be released. Similarly, the more a baby suckles, the more oxytocin is released. In both instances, the release of oxytocin from the posterior pituitary is controlled by **positive feedback**—that is, the stimulus continues to bring about an effect that ever increases in intensity. Positive feedback is not the best way to maintain stable conditions and homeostasis. However, it works during childbirth and nursing because external mechanisms interrupt the process. In childbirth, the delivery of the baby and afterbirth (the placenta and membranes surrounding the baby) eventually stops oxytocin secretion. When a baby with a full tummy stops nursing, that, too, halts oxytocin secretion.

Anterior Pituitary

A portal system, consisting of two capillary networks or beds connected by a vein, lies between the hypothalamus and the anterior pituitary (Fig. 10.4, *right*). The hypothalamus controls the anterior pituitary by producing **hypothalamic-releasing hormones** and **hypothalamic-inhibiting hormones.** For example, there is a thyrotropin-releasing hormone (TRH) and a prolactin-inhibiting hormone (PIH). TRH stimulates the anterior pituitary to secrete thyroid-stimulating hormone, and PIH inhibits the pituitary from secreting prolactin.

Hormones That Affect Other Glands

Three of the hormones produced by the anterior pituitary have an effect on other glands: **Thyroid-stimulating hormone (TSH)** stimulates the thyroid to produce the thyroid hormones; **adrenocorticotropic hormone (ACTH)** stimulates the adrenal cortex to produce its hormones; and **gonadotropic hormones** (follicle-stimulating hormone—FSH and luteinizing hormone—LH) stimulate the gonads—the testes in males and the ovaries in females—to produce gametes and sex hormones. The hypothalamus, the anterior pituitary, and other glands controlled by the anterior pituitary are all involved in self-regulating negative feedback mechanisms that maintain stable conditions. In each instance, the blood level of the last hormone in the sequence exerts negative feedback control over the secretion of the first two hormones:

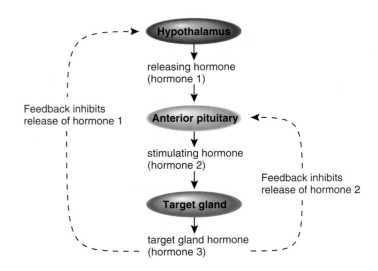

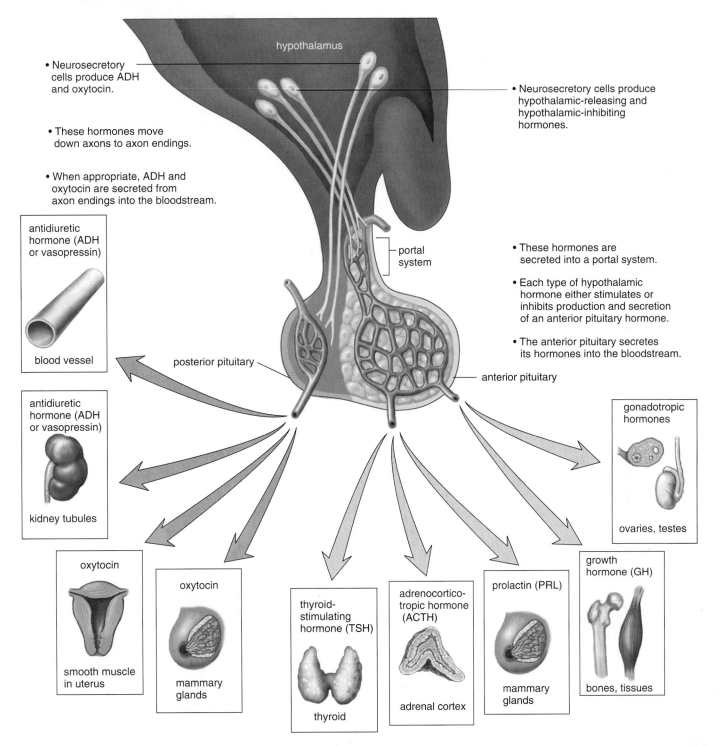

- Neurosecretory cells produce ADH and oxytocin.

- These hormones move down axons to axon endings.

- When appropriate, ADH and oxytocin are secreted from axon endings into the bloodstream.

hypothalamus

- Neurosecretory cells produce hypothalamic-releasing and hypothalamic-inhibiting hormones.

portal system

- These hormones are secreted into a portal system.

- Each type of hypothalamic hormone either stimulates or inhibits production and secretion of an anterior pituitary hormone.

- The anterior pituitary secretes its hormones into the bloodstream.

antidiuretic hormone (ADH or vasopressin)

blood vessel

posterior pituitary

anterior pituitary

antidiuretic hormone (ADH or vasopressin)

kidney tubules

oxytocin

smooth muscle in uterus

oxytocin

mammary glands

thyroid-stimulating hormone (TSH)

thyroid

adrenocortico-tropic hormone (ACTH)

adrenal cortex

prolactin (PRL)

mammary glands

gonadotropic hormones

ovaries, testes

growth hormone (GH)

bones, tissues

Figure 10.4 The hypothalamus and the pituitary. *Left:* The hypothalamus produces two hormones, ADH and oxytocin, which are stored and secreted by the posterior pituitary. *Right:* The hypothalamus controls the secretions of the anterior pituitary, and the anterior pituitary controls the secretions of the thyroid, adrenal cortex, and gonads, which are also endocrine glands. It also secretes growth hormone and prolactin.

Effects of Other Hormones

Other hormones produced by the anterior pituitary do not affect other endocrine glands. **Prolactin (PRL)** is produced beginning at about the fifth month of pregnancy, and is produced in quantity after childbirth. It causes the mammary glands in the breasts to develop and produce milk. It also plays a role in carbohydrate and fat metabolism.

Growth hormone (GH), or somatotropic hormone, stimulates protein synthesis within cartilage, bone, and muscle. It stimulates the rate at which amino acids enter cells and protein synthesis occurs. It also promotes fat metabolism as opposed to glucose metabolism.

Effects of Growth Hormone

The amount of GH produced by the anterior pituitary affects the height of the individual. The quantity of GH produced is greatest during childhood and adolescence, when most body growth is occurring (Fig. 10.5). If too little GH is produced during childhood, the individual has **pituitary dwarfism,** characterized by perfect proportions but small stature. If too much GH is secreted, a person can become a giant (Fig. 10.5). Giants usually have poor health, primarily because GH has a secondary effect on the blood sugar level, promoting an illness called diabetes mellitus (see page 216).

On occasion, GH is overproduced in the adult and a condition called **acromegaly** results. Because long bone growth is no longer possible in adults, only the feet, hands, and face (particularly the chin, nose, and eyebrow ridges) can respond, and these portions of the body become overly large (Fig. 10.6).

Figure 10.5 Effect of growth hormone. The amount of growth hormone produced by the anterior pituitary during childhood affects the height of an individual. Too much growth hormone can lead to giantism, while an insufficient amount results in limited stature and even pituitary dwarfism.

Age 9

Age 16

Age 33

Age 52

Figure 10.6 Acromegaly. Acromegaly is caused by overproduction of GH in the adult. It is characterized by enlargement of the bones in the face, the fingers, and the toes as a person ages.

Content CHECK-UP!

4. Which of the following anterior pituitary hormones does not cause the release of another hormone?

    a. thyroid-stimulating hormone   c. adrenocorticotropic hormone

    b. prolactin   d. gonadotropic hormones

5. Oxytocin release from the hypothalamus during labor and delivery is a mechanism that works by:

    a. positive feedback.   b. negative feedback.

6. Which of the following is an effect of growth hormone?

    a. It promotes fat metabolism.

    b. It stimulates protein synthesis in bone and cartilage.

    c. It causes a person to grow taller.

    d. all of the above

10.3 Thyroid and Parathyroid Glands

The **thyroid gland** is a large gland located in the neck, where it is attached to the trachea just below the larynx (see Fig. 10.1). The parathyroid glands are embedded in the posterior surface of the thyroid gland.

Thyroid Gland

The thyroid gland is composed of a large number of follicles, each a small spherical structure made of thyroid cells filled with triiodothyronine (T_3), which contains three iodine atoms, and **thyroxine** (T_4), which contains four iodine atoms. These are the two forms of thyroid hormone; T_3 is thought to have the greatest effect on the body.

Effects of Thyroid Hormones

To produce triiodothyronine and thyroxine, the thyroid gland actively acquires iodine. The concentration of iodine in the thyroid gland can increase to as much as 25 times that of the blood. If iodine is lacking in the diet, the thyroid gland is unable to produce the thyroid hormones. In response to constant stimulation by the anterior pituitary, the thyroid enlarges, resulting in a **simple,** or **endemic, goiter** (Fig. 10.7). Some years ago, it was discovered that the use of iodized salt allows the thyroid to produce the thyroid hormones, and therefore helps prevent simple goiter.

Thyroid hormones increase the metabolic rate. They do not have a single target organ; instead, they stimulate all cells of the body to metabolize at a faster rate. More glucose is broken down, and more energy is utilized.

If the thyroid fails to develop properly, a condition called **congenital hypothyroidism** results (Fig. 10.8). Individuals with this condition are short and stocky and have had extreme hypothyroidism (undersecretion of thyroid hormone) since infancy or childhood. Thyroid hormone therapy can initiate growth, but unless treatment is begun within the first two months of life, mental retardation results. The occurrence of hypothyroidism in adults produces the condition known as **myxedema,** which is characterized by lethargy, weight gain, loss of hair, slower pulse rate, lowered body temperature, and

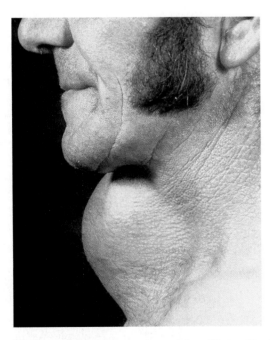

Figure 10.7 Simple goiter. An enlarged thyroid gland is often caused by a lack of iodine in the diet. Without iodine, the thyroid is unable to produce its hormones, and continued anterior pituitary stimulation causes the gland to enlarge.

Figure 10.8 Congenital hypothyroidism. Individuals who have hypothyroidism since infancy or childhood do not grow and develop as others do. Unless medical treatment is begun, the body is short and stocky; mental retardation is also likely.

thickness and puffiness of the skin. The administration of adequate doses of thyroid hormones restores normal function and appearance.

In the case of hyperthyroidism (oversecretion of thyroid hormone), as seen in **Graves' disease,** the thyroid gland is overactive, and a goiter forms. This type of goiter is called **exophthalmic goiter.** The eyes protrude because of edema in eye socket tissues and swelling of the muscles that move the eyes. The patient usually becomes hyperactive, nervous, and irritable, and suffers from insomnia. Removal or destruction of a portion of the thyroid by means of radioactive iodine is sometimes effective in curing the condition. Hyperthyroidism can also be caused by a thyroid tumor, which is usually detected as a lump during physical examination. Again, the treatment is surgery in combination with administration of radioactive iodine. The prognosis for most patients is excellent.

Calcitonin

Calcium (Ca^{2+}) plays a significant role in both nervous conduction and muscle contraction. It is also necessary for coagulation (clotting) of blood. The blood calcium level is regulated in part by **calcitonin,** a hormone secreted by the thyroid gland when the blood calcium level rises (Fig. 10.9). The primary effect of calcitonin is to bring about the deposit of calcium in the bones. It does this by temporarily reducing the activity and number of osteoclasts. Recall from Chapter 6 that these cells break down bone. When the blood calcium lowers to normal, the release of calcitonin by the thyroid is inhibited, but a low calcium level stimulates the release of parathyroid hormone (PTH) by the parathyroid glands. Calcitonin is an important hormone in children, whose skeleton is undergoing rapid growth. It is of minor importance in adults. Parathyroid hormone is the major controller of calcium homeostasis. However, calcitonin can be used therapeutically in adults to reduce the effects of osteoporosis (see the Medical Focus on page 99).

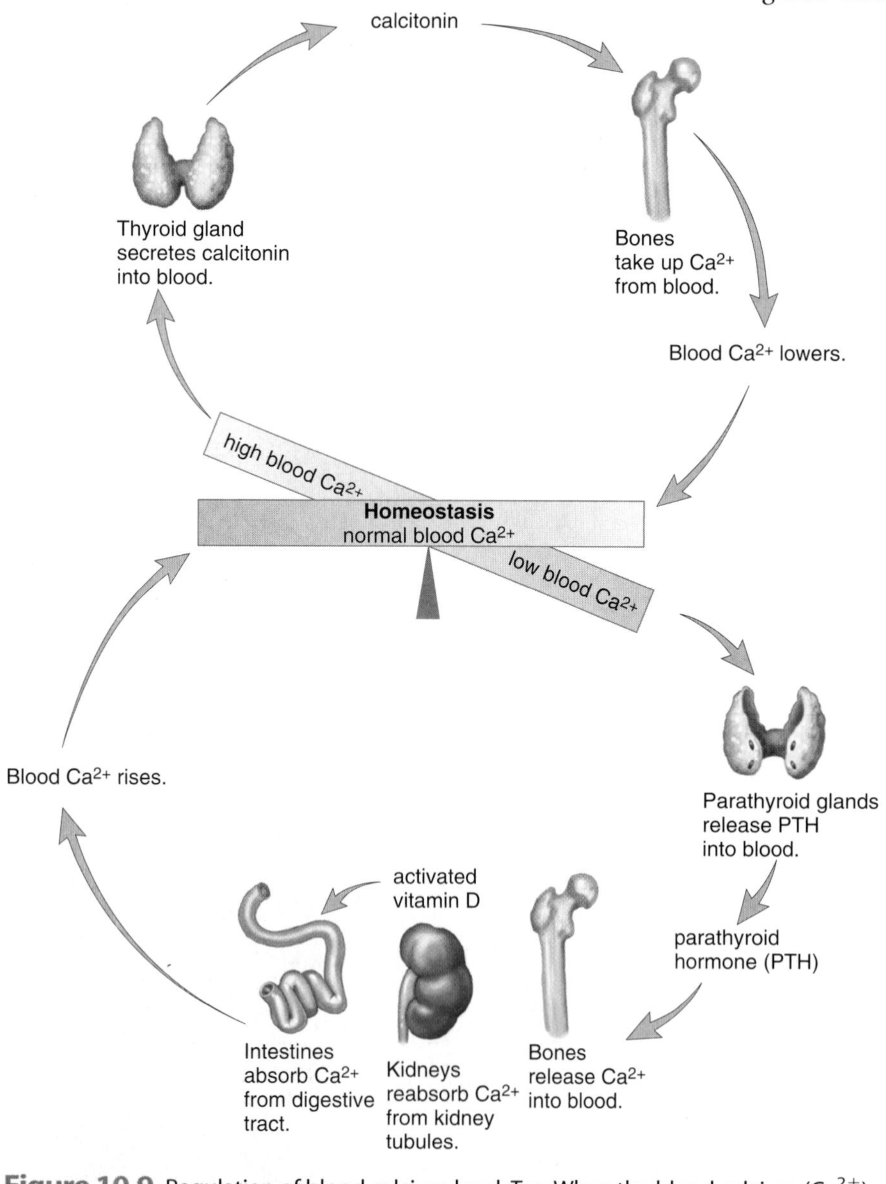

Parathyroid Glands

Parathyroid hormone (PTH), the hormone produced by the **parathyroid glands,** causes the blood phosphate (HPO_4^{2-}) level to decrease and the ionic blood calcium (Ca^{2+}) level to increase. The antagonistic actions of calcitonin, from the thyroid gland, and parathyroid hormone, from the parathyroid glands, maintain the blood calcium level within normal limits.

Note in Figure 10.9 that after a low blood calcium level stimulates the release of PTH, it promotes release of calcium from the bones. (It does this by promoting the activity of osteoclasts.) PTH promotes the reabsorption of calcium by the kidneys, where it also activates vitamin D. Vitamin D, in turn, stimulates the absorption of calcium from the intestine. These effects bring the blood calcium level back to the normal range so that the parathyroid glands no longer secrete PTH.

Many years ago, the four parathyroid glands were sometimes mistakenly removed during thyroid surgery because of their size and location in the thyroid. When insufficient parathyroid hormone production leads to a dramatic drop in the blood calcium level, hypocalcemic tetany results.

Figure 10.9 Regulation of blood calcium level. *Top:* When the blood calcium (Ca^{2+}) level is high, the thyroid gland secretes calcitonin. Calcitonin promotes the uptake of Ca^{2+} by the bones, and therefore the blood Ca^{2+} level returns to normal. *Bottom:* When the blood Ca^{2+} level is low, the parathyroid glands release parathyroid hormone (PTH). PTH causes the bones to release Ca^{2+}. It also causes the kidneys to reabsorb Ca^{2+} and activate vitamin D; thereafter, the intestines absorb Ca^{2+}. Therefore, the blood Ca^{2+} level returns to normal.

Labels in figure: calcitonin · Thyroid gland secretes calcitonin into blood. · Bones take up Ca^{2+} from blood. · Blood Ca^{2+} lowers. · high blood Ca^{2+} · **Homeostasis** normal blood Ca^{2+} · low blood Ca^{2+} · Blood Ca^{2+} rises. · Parathyroid glands release PTH into blood. · parathyroid hormone (PTH) · activated vitamin D · Intestines absorb Ca^{2+} from digestive tract. · Kidneys reabsorb Ca^{2+} from kidney tubules. · Bones release Ca^{2+} into blood.

In **tetany,** the body shakes from continuous muscle contraction. This effect is brought about by increased excitability of the nerves, which initiate nerve impulses spontaneously and without rest. In severe cases, hypocalcemic tetany is fatal.

10.4 Adrenal Glands

The **adrenal glands** sit atop the kidneys (see Fig. 10.1). Each adrenal gland consists of an inner portion called the **adrenal medulla** and an outer portion called the **adrenal cortex.** These portions, like the anterior pituitary and the posterior pituitary, have no physiological connection with one another. The adrenal medulla is under nervous control, and the adrenal cortex is under the control of ACTH (also called **corticotropin**), an anterior pituitary hormone. Stress of all types, including emotional and physical trauma, prompts the hypothalamus to stimulate the adrenal glands (Fig. 10.10).

Adrenal Medulla

The hypothalamus initiates nerve impulses that travel by way of the brain stem, spinal cord, and sympathetic nerve fibers to the adrenal medulla, which then secretes its hormones.

   Epinephrine (adrenaline) and **norepinephrine** (noradrenaline) produced by the adrenal medulla rapidly bring about all the body changes that occur when an individual reacts to an emergency situation. The release of epinephrine

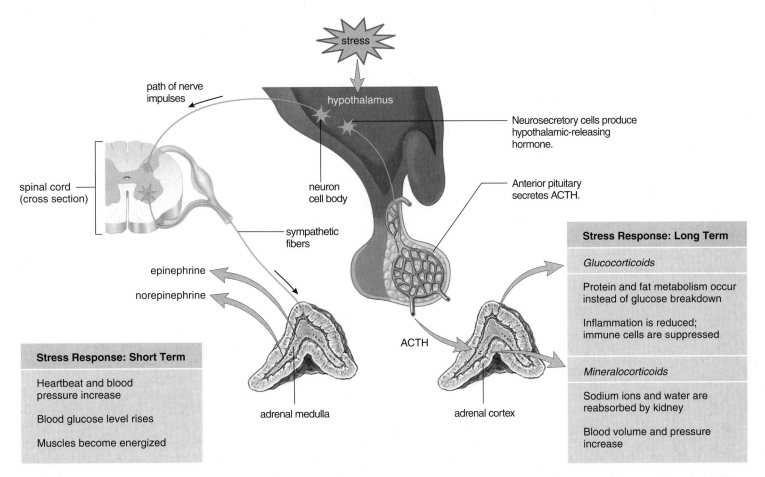

Figure 10.10 Adrenal glands. Both the adrenal medulla and the adrenal cortex are under the control of the hypothalamus when they help us respond to stress. *Left:* The adrenal medulla provides a rapid, but short-term stress response. *Right:* The adrenal cortex provides a slower, but long-term stress response.

and norepinephrine achieves the same results as sympathetic stimulation—the "fight-or-flight" responses: increased heart rate, rapid respiration, dilation of the pupils, etc. Thus, these hormones assist sympathetic nerves in providing a short-term response to stress.

Adrenal Cortex

In contrast, the hormones produced by the adrenal cortex provide a long-term response to stress (Fig. 10.10). The two major types of hormones produced by the adrenal cortex are the mineralocorticoids and the glucocorticoids. The **mineralocorticoids** regulate salt and water balance, leading to increases in blood volume and blood pressure. The **glucocorticoids** regulate carbohydrate, protein, and fat metabolism, leading to an increase in blood glucose level. Cortisone, the medication often administered for inflammation of joints, is a glucocorticoid.

The adrenal cortex also secretes small amounts of both male and female sex hormones—regardless of one's gender. Male and female sex hormones promote skeletal growth in adolescents. The male hormones from the adrenal gland stimulate the growth of axillary and pubic hair at puberty. In addition, male hormones help to sustain the sex drive, or *libido*, in both men and women.

Glucocorticoids

Cortisol is a biologically significant glucocorticoid produced by the adrenal cortex. Cortisol raises the blood glucose level in at least two ways: (1) It promotes the breakdown of muscle proteins to amino acids, which are taken up by the liver from the bloodstream. The liver then converts these excess amino acids to glucose, which enters the blood. (2) Cortisol promotes the metabolism of fatty acids rather than carbohydrates, and this spares glucose for the brain.

Cortisol also counteracts the inflammatory response that leads to the pain and swelling of joints in arthritis and bursitis. The administration of cortisol aids these conditions because it reduces inflammation. Very high levels of glucocorticoids in the blood can suppress the body's defense system, including the inflammatory response that occurs at infection sites. Cortisone (the glucocorticoid used as a medication) and other glucocorticoids can relieve swelling and pain from inflammation, but by suppressing pain and immunity, they can also make a person highly susceptible to injury and infection. (See the Medical Focus on page 224).

Mineralocorticoids

Aldosterone is the most important of the mineralocorticoids. Aldosterone primarily targets the kidney where it promotes renal absorption of sodium (Na^+) and water, and renal excretion of potassium (K^+).

As one might expect, secretion of mineralocorticoids from the adrenal cortex is influenced by ACTH (adrenocorticotropic hormone or corticotropin) from the pituitary gland. However, the pituitary hormone is not the primary controller for aldosterone secretion. When the blood sodium level and therefore the blood pressure are low, the kidneys secrete **renin** (Fig. 10.11). Renin is an enzyme that converts the plasma protein angiotensinogen to angiotensin I. Angiotensin I is fully activated to angiotensin II by a converting enzyme found in lung capillaries. Angiotensin II stimulates the adrenal cortex to release aldosterone. The effect of this system, called the renin-angiotensin-aldosterone system, is to raise blood pressure in two ways: Angiotensin II constricts arterioles, and aldosterone causes the kidneys to reabsorb sodium. When the blood sodium level rises, water is reabsorbed, in part because the hypothalamus secretes ADH (see page 208). Reabsorption means that water enters kidney capillaries and thus returns to the blood. Then blood pressure increases to normal.

There is an antagonistic hormone to aldosterone, as you might suspect. When the atria of the heart are stretched due to increased blood volume, cardiac cells release a hormone called **atrial natriuretic hormone (ANH,** or **atriopeptide),** which inhibits the secretion of aldosterone from the adrenal cortex. The effect of ANH is the excretion of sodium—that is, *natriuresis*. When sodium is excreted, so is water, and therefore blood pressure lowers to normal.

Malfunction of the Adrenal Cortex

Malfunction of the adrenal cortex can lead to a **syndrome,** a set of symptoms that occur together. The syndromes commonly associated with the adrenal cortex are Addison disease and Cushing's syndrome.

Addison Disease and Cushing's Syndrome

When the level of adrenal cortex hormones is low due to hyposecretion, a person develops **Addison disease.** The presence of excessive but ineffective ACTH causes a bronzing of the skin because ACTH can lead to a buildup of melanin (Fig. 10.12). Without cortisol, glucose cannot be replenished when a stressful situation arises. Even a mild infection can lead to death. The lack of aldosterone results in a loss of sodium and water, the development of low blood pressure, and possibly severe dehydration. Left untreated, Addison disease can be fatal.

When the level of adrenal cortex hormones is high due to hypersecretion, a person develops **Cushing syndrome** (Fig. 10.13). The excess cortisol results in a tendency toward diabetes mellitus as muscle protein is metabolized and subcutaneous fat is deposited in the midsection. The trunk is obese, while the arms and legs remain a normal size. An excess of aldosterone and reabsorption of sodium and water by the kidneys leads to a basic blood pH and hypertension. The face is moon-shaped due to edema. Masculinization may occur in women because of excess adrenal male sex hormones.

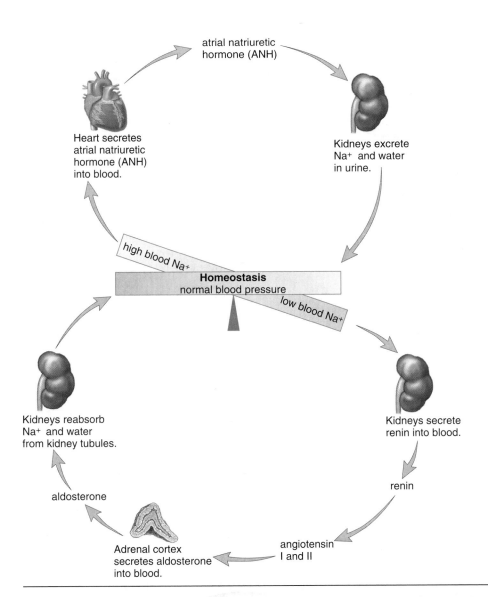

atrial natriuretic
hormone (ANH)

Heart secretes
atrial natriuretic
hormone (ANH)
into blood.

Kidneys excrete
Na⁺ and water
in urine.

high blood Na⁺

Homeostasis
normal blood pressure

low blood Na⁺

Kidneys reabsorb
Na⁺ and water
from kidney tubules.

Kidneys secrete
renin into blood.

aldosterone

renin

Adrenal cortex
secretes aldosterone
into blood.

angiotensin
I and II

Figure 10.11 Regulation of blood pressure and volume. *Bottom:* When the blood sodium (Na^+) level is low, low blood pressure causes the kidneys to secrete renin. Renin leads to the secretion of aldosterone from the adrenal cortex. Aldosterone causes the kidneys to reabsorb Na^+, and water follows, so that blood volume and pressure return to normal. *Top:* When the blood Na^+ is high, a high blood volume causes the heart to secrete atrial natriuretic hormone (ANH). ANH causes the kidneys to excrete Na^+, and water follows. The blood volume and pressure return to normal.

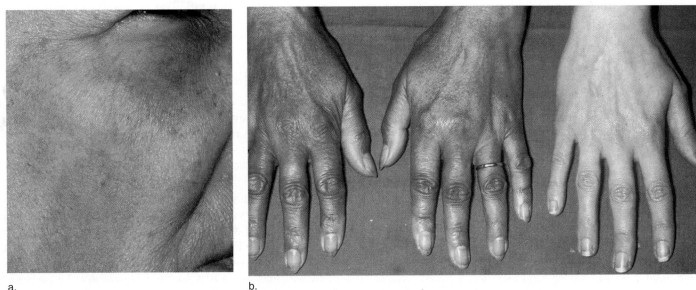

a.

b.

Figure 10.12 Addison disease. Addison disease is characterized by a peculiar bronzing of the skin, particularly noticeable in these light-skinned individuals. Note the color of **(a)** the face and **(b)** the hands compared to the hand of an individual without the disease.

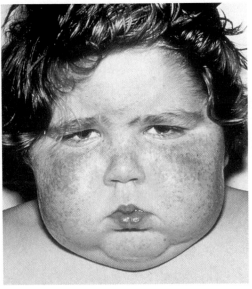

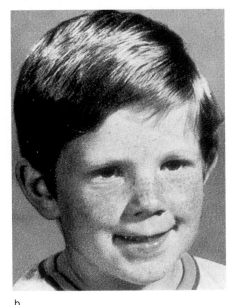

a.   b.

Figure 10.13 Cushing syndrome. Cushing syndrome results from hypersecretion of adrenal cortex hormones. **a.** Patient first diagnosed with Cushing syndrome. **b.** Four months later, after therapy.

Content CHECK-UP!

10. From the following list of hormones of the adrenal cortex and their corresponding effects, choose the pair that is incorrect:

    a. cortisol → decreases blood glucose

    b. male hormones → stimulate sex drive

    c. aldosterone → increases sodium concentration in the blood

    d. female hormones → promote long bone growth in adolescents

11. Which hormone opposes the effect of aldosterone in the body?

    a. renin

    b. angiotensin I

    c. atrial natriuretic hormone

    d. cortisol

12. Aldosterone returns blood pressure to normal by causing the kidneys to reabsorb water and sodium. Because it works by a negative feedback mechanism, which of the following actions could cause aldosterone to be released?

    a. giving a unit of blood

    b. drinking a big bottle of a sports drink

    c. running a marathon and becoming dehydrated

    d. both a and c are correct

10.5 Pancreas

The **pancreas** is a long organ that lies transversely in the abdomen between the kidneys and near the duodenum of the small intestine. It is composed of two types of tissue.

Exocrine tissue produces and secretes digestive juices that go by way of ducts to the small intestine. Endocrine tissue, called the **pancreatic islets** (islets of Langerhans), produces and secretes the hormones **insulin** and **glucagon** directly into the blood (Fig. 10.14).

The two antagonistic hormones insulin and glucagon, both produced by the pancreas, help maintain the normal level of glucose in the blood. Insulin is secreted when the blood glucose level is high, which usually occurs just after eating. Insulin stimulates the uptake of glucose by most body cells. Insulin is not necessary for the transport of glucose into brain or red blood cells, but muscle cells and adipose tissue cells require insulin for glucose transport. In liver and muscle cells, insulin stimulates enzymes that promote the storage of glucose as glycogen. In muscle cells, the glucose supplies energy for muscle contraction, and in fat cells, glucose enters the metabolic pool and thereby supplies glycerol for the formation of fat. In these ways, insulin lowers the blood glucose level. As discussed in the What's New reading on page 218, individuals who do not produce insulin have a condition called diabetes mellitus type I.

Glucagon is secreted from the pancreas, usually between meals, when the blood glucose level is low. The major target tissues of glucagon are the liver and adipose tissue. Glucagon stimulates the liver to break down glycogen to glucose and to use fat and protein in preference to glucose as energy sources. Adipose tissue cells break down fat to glycerol and fatty acids. The liver takes these up and uses them as substrates for glucose formation. In these ways, glucagon raises the blood glucose level.

Diabetes Mellitus

Diabetes mellitus is a fairly common hormonal disease in which insulin-sensitive body cells are unable to take up and/or metabolize glucose. Therefore, the blood glucose level is elevated—a condition called **hyperglycemia**. Because body cells cannot access glucose, starvation occurs at the cell level. The person becomes extremely hungry—a condition called **polyphagia**. As the blood glucose level rises, glucose will be lost in the urine (**glycosuria**). Glucose in the urine causes excessive water loss through urination (**polyuria**). The loss of water in this way causes the diabetic to be extremely thirsty (**polydipsia**). Glucose is not being metabolized, so the body turns to the breakdown of protein and fat for energy. Fat metabolism leads to the buildup of ketones in the blood, and excretion of ketones in the urine (**ketonuria**). Because ketones are acidic, their buildup in the

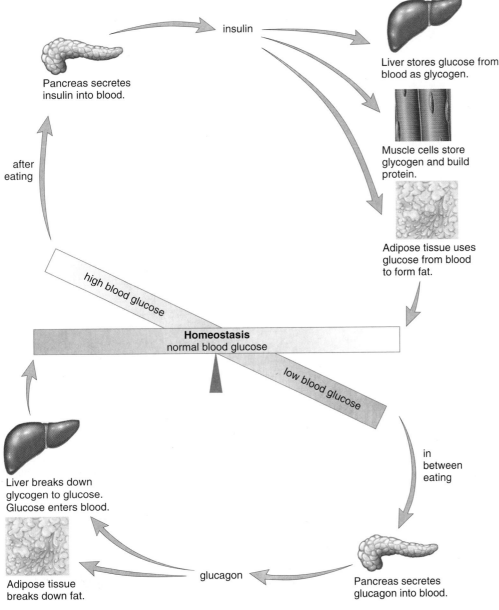

Figure 10.14 Regulation of blood glucose level. *Top:* When the blood glucose level is high, the pancreas secretes insulin. Insulin promotes the storage of glucose as glycogen in the liver and muscles and the use of glucose to form fat in adipose tissue. Therefore, insulin lowers the blood glucose level. *Bottom:* When the blood glucose level is low, the pancreas secretes glucagon. Glucagon acts opposite to insulin; therefore, glucagon raises the blood glucose level to normal.

Labels within figure:

insulin

Pancreas secretes insulin into blood.

Liver stores glucose from blood as glycogen.

Muscle cells store glycogen and build protein.

Adipose tissue uses glucose from blood to form fat.

after eating

high blood glucose

Homeostasis
normal blood glucose

low blood glucose

in between eating

Liver breaks down glycogen to glucose. Glucose enters blood.

Adipose tissue breaks down fat.

glucagon

Pancreas secretes glucagon into blood.

---

blood causes **acidosis** (acid blood), which can lead to coma and death.

We now know that diabetes mellitus exists in two forms. In type I, more often called **insulin-dependent diabetes mellitus (IDDM),** the pancreas does not produce insulin. This condition is believed to be brought on, at least in part, by exposure to an environmental agent. This agent—very likely a virus—causes immune cells to destroy the pancreatic islets. As a result, the individual must have daily insulin injections. Daily injections control the diabetic symptoms but can still cause inconveniences. Either an overdose of insulin or missing a meal can bring on the symptoms of **hypoglycemia** (low blood sugar). These symptoms include perspiration, pale skin, shallow breathing, and anxiety. The cure is quite simple: Immediate ingestion of a sugar cube or fruit juice can quickly counteract hypoglycemia.

Of the 16 million people who now have diabetes in the United States, most have type II, more often called **noninsulin-dependent diabetes (NIDDM).** This type of diabetes mellitus usually occurs in people of any age who tend to be obese. Researchers theorize that perhaps adipose tissue produces a substance that interferes with the transport of glucose into cells. The amount of insulin in the blood of these patients is normal or elevated, but the insulin receptors on the cells do not respond to it. It is possible to prevent, or at least control, type II diabetes by adhering to a low-fat, low-sugar diet, maintaining a healthy weight, and exercising regularly. If these attempts fail, oral drugs are available to stimulate the pancreas to secrete more insulin. Other oral medications enhance the metabolism of glucose in the liver and muscle cells. It is projected that as many as 7 million Americans may have type II diabetes without being aware of it. Yet, the effects of untreated type II diabetes are as serious as those of type I diabetes. In addition, without stringent control the NIDDM diabetic will ultimately require insulin injections, thus becoming insulin-dependent.

Long-term complications of both types of diabetes are blindness, kidney disease, and circulatory disorders, including atherosclerosis, heart disease, stroke, and reduced circulation. The latter can lead to gangrene in the arms and legs. Pregnancy carries an increased risk of diabetic coma, and the child of a diabetic is somewhat more likely to be stillborn or to die shortly after birth. However, these complications of diabetes are not expected to appear if the mother's blood glucose level is carefully regulated and kept within normal limits during the pregnancy.

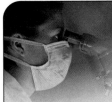
"I can remember getting sick with the flu just before I was diagnosed. I was eleven, and I was sick enough to miss two or three days of school. Then I just never got my strength back. I ate and drank constantly because I was thirsty and hungry all the time. I was always in the bathroom. It was so embarrassing. I started wetting the bed—can you imagine, at age 11? I fell asleep in school, and the teacher could barely get me to wake up. That's when my doctor diagnosed my diabetes for the first time."

The patient, age 25, is a typical type I, juvenile-onset or insulin-dependent diabetic. Her symptoms are classic for insulin-dependent diabetes mellitus (IDDM) (see page 217).

In insulin-dependent diabetes, the insulin-producing islet cells of the pancreas have been destroyed. Researchers think this is due to a malfunction of the immune system that causes the body's own immune cells to target the pancreas. Thus, insulin-dependent diabetes is considered an autoimmune disease. As the name suggests, insulin must be taken by injection. The diabetic patient's life then revolves around two to three daily insulin injections or monitoring by an insulin pump device that injects insulin automatically. Four or more daily blood tests are used to check blood glucose levels, and the patient must also monitor diet, activity level, exercise, and stress.

"My insulin pump saves me from those three-a-day shots, but boy, do I hate finger sticks to test my blood," the patient says with a wistful smile. "I know how carefully I have to manage this disease. Diabetics lose their sight, or go into kidney failure, or wind up having an early heart attack or stroke. I wish I could be placed on a transplant list for a pancreas, but everybody wants a pancreas. There aren't enough human donors to go around."

Pancreatic transplantation has been available to IDDM sufferers since 1966, but it suffers from the same limitations of all transplant technology. Transplanting an entire organ is major surgery, and there is always a shortage of available donors. Strong drugs must be taken for the rest of the patient's life in order to suppress the immune system. These antirejection drugs can have toxic effects on normal body cells. Moreover, with a weakened immune system, the patient has an increased risk of developing life-threatening infections or cancer.

The technique of *pancreatic islet cell transplantation* seems to hold promise for solving the problems of the traditional pancreas transplant. The islet cells are first isolated from a donor pancreas. The cells are then directly injected through the hepatic portal vein into the liver, where they form colonies and begin to produce insulin. This technique is much simpler than whole-pancreas transplantation and does not involve major surgery. Following a procedure developed at the University of Alberta in Edmonton, centers in the United States, Canada, and Europe have been successful in transplanting islets, allowing recipients to be freed from their daily insulin injections. Key to the success of the technique is a new combination of antirejection drugs that are less toxic to both the transplanted islet cells and the body's own cells.

It is estimated that 700,000 islets will be needed to produce enough insulin for an adult. Several donor pancreases are needed to harvest sufficient islet cells for a single transplant. Embryonic stem cell research has not yet yielded a reliable source for islets, and debate persists about using human embryo cells. If an animal cell source could be used, unlimited islet cells would be available. Heart valves from pigs have been used for decades, and insulin for injection into humans was first isolated from pigs. Tissue engineers are now experimenting with islet cells from pigs. These islet cells have been isolated and surrounded by a semipermeable plastic membrane, a process called microencapsulation. These capsules are so small that they can be placed into the abdomen, where they will float freely and produce insulin as needed (Fig. 10A). The membrane of the capsule contains pores large enough to allow oxygen and nutrients to flow in and wastes and insulin to flow out by diffusion. But the membrane prevents immune cells from coming into contact with the enclosed pancreatic cells. Unless immune cells actually come in contact with transplanted cells, they cannot destroy them. Therefore, the patient does not need to take harsh antirejection drugs, and the immune system can function normally to suppress infection and cancer. Researchers are optimistic that prepared microencapsulated islet cells could soon be available for clinical trials.

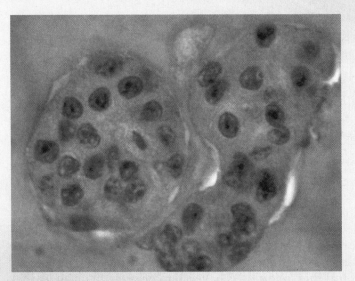

Figure 10A Encapsulated insulin-producing pancreatic islet cells from pigs can be transplanted into patients without the need for immune system-suppressing drugs.

13. Insulin-sensitive cells in the human body include:

    a. muscle cells.

    b. adipose tissue cells.

    c. brain and nerve cells.

    d. a and b.

    e. all of the above.

14. Which of the following is not an effect of glucagon?

    a. causes the liver to break down stored glycogen

    b. causes adipose tissue to break down fat

    c. lowers blood glucose level

    d. causes the liver to use protein and fat as energy sources instead of glucose

15. Glucagon release is controlled by a negative feedback system. Which action causes glucagon to be released?

    a. skipping breakfast and going to morning classes with an empty stomach

    b. eating a big holiday meal

    c. running a marathon race for several hours without pausing for food

    d. a and c

    e. b and c

10.6 Other Endocrine Glands

The body has a number of other endocrine glands, including the **gonads** (testes in males and the ovaries in females). Other lesser-known glands, such as the thymus gland and the pineal gland, also produce hormones. Some tissues within organs produce hormones and/or growth factors. Individual body cells produce local messenger chemicals termed prostaglandins.

Testes and Ovaries

The **testes** are located in the scrotum, and the **ovaries** are located in the pelvic cavity. The testes produce **androgens** (e.g., **testosterone**), which are the male sex hormones, and the ovaries produce **estrogens** and **progesterone**, the female sex hormones. The hypothalamus and the pituitary gland control the hormonal secretions of these organs in the manner previously described on page 208.

Androgens

Puberty is the time of life when sexual maturation occurs. Greatly increased testosterone secretion during puberty stimulates the growth of the penis and the testes. Testosterone also brings about and maintains the male secondary sex characteristics that develop during puberty, including the growth of a beard, axillary (underarm) hair, and pubic hair. It prompts the larynx and the vocal cords to enlarge, causing the voice to change. It is partially responsible for the muscular strength of males. This is why some athletes take supplemental amounts of **anabolic steroids,** which are either testosterone or related chemicals. The contraindications of taking anabolic steroids

are discussed in the Medical Focus on page 220. Testosterone also stimulates oil and sweat glands in the skin; therefore, it is largely responsible for acne and body odor. Another side effect of testosterone is baldness. Genes for baldness are probably inherited by both sexes, but baldness is seen more often in males because of the presence of testosterone.

Estrogen and Progesterone

The female sex hormones, estrogens and progesterone, have many effects on the body. In particular, estrogens secreted during puberty stimulate the growth of the uterus and the vagina. Estrogen is necessary for ovum maturation and is largely responsible for the secondary sex characteristics in females, including female body hair and fat distribution. In general, females have a more rounded appearance than males because of a greater accumulation of fat beneath the skin. Also, the pelvic girdle is wider in females than in males, resulting in a larger pelvic cavity. Both estrogen and progesterone are required for breast development and for regulation of the uterine cycle, which includes monthly menstruation (discharge of blood and mucosal tissues from the uterus).

Thymus Gland

The lobular **thymus gland,** which lies just beneath the sternum (see Fig. 10.1), reaches its largest size and is most active during childhood. **Lymphocytes** are white blood cells that originate in the bone marrow and are responsible for specific defenses against a particular invader. When lymphocytes pass complete development in the thymus, they are transformed into **thymus-derived lymphocytes,** or **T-lymphocytes.** The lobules of the thymus are lined by epithelial cells that secrete hormones called **thymosins.** These hormones aid in the differentiation of lymphocytes packed inside the lobules. Although the hormones secreted by the thymus ordinarily work only in the thymus, researchers hope that these hormones could be injected into AIDS or cancer patients where they would enhance T-lymphocyte function.

Pineal Gland

The **pineal gland,** which is located in the brain (see Fig. 10.1), produces the hormone **melatonin,** primarily at night. Melatonin is involved in our daily sleep-wake cycle; normally we grow sleepy at night when melatonin levels increase and awaken once daylight returns and melatonin levels are low (Fig. 10.15). Daily 24-hour cycles such as this are called **circadian rhythms.** Circadian rhythms are controlled by an internal timing mechanism called a biological clock.

Based on animal research, it appears that melatonin also regulates sexual development. It has also been noted that children whose pineal gland has been destroyed due to a brain tumor experience early puberty.

Hormones from Other Tissues

We have already mentioned that the heart produces atrial natriuretic hormone (see page 214). The kidney also influences

Side Effects of Anabolic Steroids

The story rocked major league baseball fans across the country: Baseball's sluggers—home run hitters who had set new records for the sport—were implicated in a scandal involving abuse of anabolic steroids. In testimony before the U.S. House of Representatives in March 2005, Mark McGwire of the St. Louis Cardinals denied using steroids during his career in baseball and stood by his hitting records. However, other players—most notably Jose Canseco of the Oakland Athletics and Jason Giambi of the New York Yankees—admitted using performance-enhancing drugs.

During the hearings, Congressional officials denounced the Major League Baseball's punishment: a ten-game, "slap on the wrist" suspension for players who test positive for steroids. Many legislators questioned whether league records created by those testing positive should be allowed to stand. Of tremendous concern to lawmakers, educators, and parents is the increased use of steroids by teens wishing to bulk up quickly, perhaps seeking to be just like the sports heroes they admire.

Anabolic steroids are synthetic forms of the male sex hormone testosterone. Taking doses 10 to 100 times the amount prescribed by doctors for various illnesses promotes larger muscles when the person also exercises. Trainers may have been the first to acquire anabolic steroids for weight lifters, bodybuilders, and other athletes, such as professional baseball players. However, being a steroid user can have serious detrimental effects. Men often experience decreased sperm counts and decreased sexual desire due to atrophy of the testicles. Some develop an enlarged prostate gland or grow breasts. On the other hand, women can develop male sexual characteristics. They grow hair on their chests and faces, and lose hair from their heads; many experience abnormal enlargement of the clitoris. Some cease ovulating or menstruating, sometimes permanently.

Some researchers predict that two or three months of high-dosage use of anabolic steroids as a teen can cause death by age 30 or 40. Steroids have even been linked to heart disease in both sexes and implicated in the deaths of young athletes from liver cancer and one type of kidney tumor. Steroids can cause the body to retain fluid, which results in increased blood pressure. Users then try to get rid of "steroid bloat" by taking large doses of diuretics. A young California weight lifter had a fatal heart attack after using steroids, and the postmortem showed a lack of **electrolytes,** salts that help regulate the heart. Finally, steroid abuse has psychological effects, including depression, hostility, aggression, and eating disorders. Unfortunately, these drugs make a person feel invincible. One abuser even had his friend videotape him as he drove his car at 40 miles an hour into a tree!

The many harmful effects of anabolic steroids are given in Figure 10B. The Federal Food and Drug Administration now bans most steroids, and steroid use has also been banned by the National Collegiate Athletic Association (NCAA), the National Football League (NFL), and the International Olympic Committee (IOC).

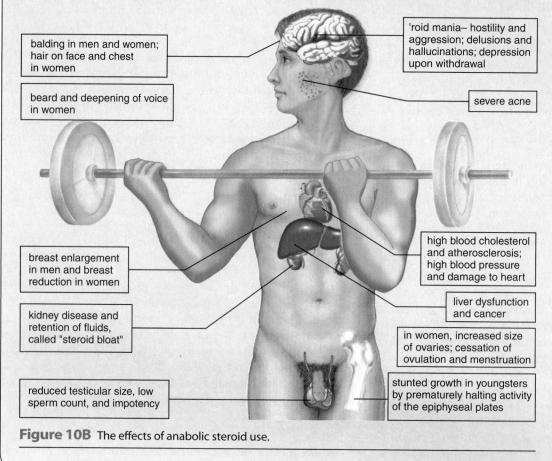

balding in men and women; hair on face and chest in women

beard and deepening of voice in women

'roid mania– hostility and aggression; delusions and hallucinations; depression upon withdrawal

severe acne

breast enlargement in men and breast reduction in women

kidney disease and retention of fluids, called "steroid bloat"

reduced testicular size, low sperm count, and impotency

high blood cholesterol and atherosclerosis; high blood pressure and damage to heart

liver dysfunction and cancer

in women, increased size of ovaries; cessation of ovulation and menstruation

stunted growth in youngsters by prematurely halting activity of the epiphyseal plates

Figure 10B The effects of anabolic steroid use.

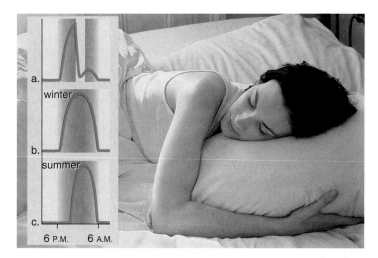

Figure 10.15 Melatonin production. Melatonin production is greatest at night when we are sleeping. Light suppresses melatonin production **(a)** so its duration is longer in the winter **(b)** than in the summer **(c)**.

cardiovascular system function by producing the hormone erythropoietin (EPO), which stimulates red blood cell production by the bone marrow. And you will see in Chapter 15 that the stomach and small intestine produce peptide hormones that regulate digestive secretions.

Leptin

Leptin is a protein hormone produced by adipose tissue. Leptin acts on the hypothalamus, where it signals satiety—that the individual has had enough to eat. Strange to say, the blood of obese individuals may be rich in leptin. It is possible that the leptin they produce is ineffective because of a genetic mutation, or else their hypothalamic cells lack a suitable number of receptors for leptin.

Growth Factors

A number of different types of organs and cells produce peptide **growth factors,** which stimulate cell division and mitosis. Some, such as lymphokines, are released into the blood; others diffuse to nearby cells. Growth factors of particular interest are the following:

Granulocyte and macrophage colony-stimulating factor (GM-CSF) is secreted by many different tissues. GM-CSF causes bone marrow stem cells to form either granulocyte or macrophage cells (both are forms of white blood cells, or leukocytes), depending on whether the concentration is low or high.

Platelet-derived growth factor is released from platelets and from many other cell types. It helps in wound healing and causes an increase in the number of fibroblasts, smooth muscle cells, and certain cells of the nervous system.

Epidermal growth factor and *nerve growth factor* stimulate the cells indicated by their names, as well as many others. These growth factors are also important in wound healing.

Tumor angiogenesis factor stimulates the formation of capillary networks and is released by tumor cells. One treatment for cancer is to prevent the activity of this growth factor.

Prostaglandins

Prostaglandins are potent chemical signals produced within cells from arachidonate, a fatty acid. Prostaglandins are not distributed in the blood; instead, they act locally, quite close to where they were produced. In the uterus, prostaglandins cause muscles to contract and may be involved in the pain and discomfort of menstruation. Also, prostaglandins mediate the effects of pyrogens, chemicals that are believed to reset the temperature regulatory center in the brain. For example, aspirin reduces body temperature and controls pain because of its effect on prostaglandins.

Certain prostaglandins reduce gastric secretion and have been used to treat ulcers; others lower blood pressure and have been used to treat hypertension; and still others inhibit platelet aggregation and have been used to prevent thrombosis (the formation of stationary clots in blood vessels). However, different prostaglandins have contrary effects, and it has been very difficult to successfully standardize their use.

Content **CHECK-UP!**

16. From the following list of endocrine glands and their hormones, choose the incorrect pair:

    a. ovaries → androgens   c. kidney → erythropoietin

    b. thymus → thymosin   d. adipose tissue → leptin

17. Which of the following is a local tissue messenger that stimulates nearby cells?

    a. leptin   c. melatonin

    b. prostaglandin   d. thymosin

18. Select the false statement regarding female sex hormones:

    a. They stimulate the growth of the uterus and vagina during puberty.

    b. They are required for breast development.

    c. They are required for monthly menstruation.

    d. Progesterone is necessary for egg maturation and development of secondary sex characteristics.

10.7 Chemical Signals

Chemical signals are molecules that affect the behavior of those cells that have receptor proteins to receive them. For example, a hormone that binds to a receptor protein affects the metabolism of the cell.

Hormones fall into two basic chemical classes. As noted in Table 10.1, most are **peptide hormones,** a category that includes not only those that are peptides but also proteins, glycoproteins, or modified amino acids. The remainder are **steroid hormones,** each having the same four-carbon ring complex, but with different side chains.

The Importance of Chemical Signals

Cells, organs, and even individuals communicate with one another by using chemical signals.

We are most familiar with chemical signals, such as hormones, that are produced by organs some distance from one another in the body. Hormones produced by the anterior pituitary, for example, influence the function of numerous organs throughout the body. Insulin, produced by the pancreas, is transported in blood to muscle, adipose, and other insulin-sensitive cells. The nervous system at times utilizes chemical signals that are produced by an organ distant from the one being affected. For example, the hypothalamus produces releasing hormones that travel in a portal system to the anterior pituitary gland.

Many chemical signals act locally—that is, from cell to cell. Prostaglandins are *local hormones*, and certainly neurotransmitter substances released by one neuron affect a neuron nearby. Growth factors, which fall into this category, are very important regulators of cell division. Some growth factors are being used as medicines to promote the production of blood cells in AIDS and cancer patients. When a tumor develops, cell division occurs even when no detectable stimulatory growth factor has been received. And the tumor produces a growth factor called tumor angiogenesis factor, which promotes the formation of capillary networks to service its cells.

Chemical Signals Between Individuals

Chemical signals that act between individuals are called *pheromones*. Pheromones are well exemplified in other animals, but they may also be effective between people. Humans produce airborne chemicals from a variety of areas, including the scalp, oral cavity, armpits, genital areas, and feet. For example, the armpit secretions of one woman could possibly affect the menstrual cycle of another woman. Women who live in the same household often have menstrual cycles in synchrony. Also, the cycle length becomes more normal when women with irregular cycles are exposed to extracts of male armpit secretions.

10.8 Effects of Aging

Thyroid disorders and diabetes are the most significant endocrine problems affecting health and function as we age. Both hypothyroidism and hyperthyroidism are seen in the elderly. Graves' disease is an autoimmune disease that targets the thyroid, resulting in symptoms of cardiovascular disease, increased body temperature, and fatigue. In addition, a patient may experience weight loss of as much as 20 pounds, depression, and mental confusion. Hypothyroidism (myxedema) may fail to be diagnosed because the symptoms of hair loss, skin changes, and mental deterioration are attributed simply to the process of aging.

The true incidence of IDDM diabetes among the elderly is unknown. Its symptoms can be confused with those of other medical conditions that are present. As in all adults, NIDDM diabetes is associated with being overweight and often can be controlled by proper diet.

The effect of age on the sex organs is discussed in Chapter 17.

10.9 Homeostasis

The endocrine system and the nervous system work together to regulate the organs of the body and thereby maintain homeostasis. It is clear from reviewing the Human Systems Work Together illustration on page 223 that the endocrine system particularly influences the digestive, cardiovascular, and urinary systems in a way that maintains homeostasis.

The endocrine system helps regulate digestion. The digestive system adds nutrients to the blood, and hormones produced by the digestive system influence the gallbladder and pancreas to send their secretions to the digestive tract. Another hormone, gastrin, promotes the digestion of protein by the stomach. Through its influence on the digestive process, the endocrine system promotes the presence of nutrients in the blood.

The endocrine system helps regulate fuel metabolism. Controlling the level of glucose in the blood is the function of insulin and glucagon. Just after eating, insulin encourages the uptake of glucose by cells and the storage of glucose as glycogen in the liver and muscles. In between eating, glucagon stimulates the liver to break down glycogen to glucose so that the blood glucose level stays constant. Adrenaline (epinephrine) from the adrenal medulla also stimulates the liver to release glucose. Glucagon (from the pancreas) and cortisol (from the adrenal cortex) promote the breakdown of protein to amino acids, which can be converted to glucose by the liver. They also promote the metabolism of fatty acids to conserve glucose, a process called **glucose sparing.** Finally, the thyroid hormones thyroxine and triiodothyronine set the body's metabolic rate, and thus are the hormones that ultimately regulate fuel metabolism.

The endocrine system helps regulate blood pressure and volume. ADH produced by the hypothalamus (but secreted by the posterior pituitary) promotes reabsorption of water by the kidneys, especially when we have not been drinking water that day. Aldosterone produced by the adrenal cortex causes the kidneys to reabsorb sodium, and when the level of sodium rises, water is automatically reabsorbed so that blood volume and pressure rise. Regulation by the endocrine system often involves antagonistic hormones; in this case, ANH (atriopeptide) produced by the heart causes sodium excretion.

The endocrine system helps regulate calcium balance. The concentration of calcium (Ca^{2+}) in the blood is critical because this ion is important to nervous conduction, muscle contraction, and the action of hormones. As you know, the bones serve as a reservoir for calcium. When the blood calcium concentration lowers, parathyroid hormone promotes the breakdown of bone and the reabsorption of calcium by the kidneys, and the absorption of calcium by the intestines. Opposing the action of parathyroid hormone, calcitonin secreted by the thyroid brings about the deposit of calcium in

Human Systems Work Together

Integumentary System

Androgens activate sebaceous glands and help regulate hair growth.

Skin provides sensory input that results in the activation of certain endocrine glands.

Skeletal System

Growth hormone regulates bone development; para-thyroid hormone and calcitonin regulate Ca^{2+} content.

Bones provide protection for glands; store Ca^{2+} used as second messenger.

Muscular System

Growth harmone and androgens promote growth of skeletal muscle; epinephrine stimulates heart and constricts blood vessels.

Muscles help protect glands.

Nervous System

Sex hormones affect development of brain.

Hypothalamus is part of endocrine system; nerves innervate glands of secretion.

Cardiovascular System

Epinephrine increases blood pressure; ADH, aldosterone, and atrial natriuretic hormone help regulate blood volume; growth factors control blood cell formation.

Blood vessels transport hormones from glands; blood services glands; heart produces atrial natriuretic hormones.

How the Endocrine System works with other body systems.

- Hypothalamus
- Pineal body
- Pituitary

- Thyroid

- Thymus

- Adrenals
- Pancreas
- Kidney

- Testes (male)

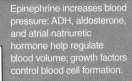

Ovaries (female)

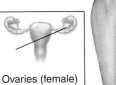

Parathyroids (posterior part of thyroid)

Lymphatic System/Immunity

Thymus is necessary for maturity of T lymphocytes.

Lymphatic vessels pick up excess tissue fluid; immune system protects against infections.

Respiratory System

Epinephrine promotes ventilation by dilating bronchioles; growth factors control production of red blood cells that carry oxygen.

Gas exchange in lungs provides oxygen and rids body of carbon dioxide.

Digestive System

Hormones help control secretion of digestive glands and accessory organs; insulin and glucagon regulate glucose storage in liver.

Stomach and small intestine produce hormones.

Urinary System

ADH, aldosterone, and atrial natriuretic hormone regulate reabsorption of water and Na^+ by kidneys.

Kidneys keep blood values within normal limits so that transport of hormones continues.

Reproductive System

Hypothalamic, pituitary, and sex hormones control sex characteristics and regulate reproductive processes.

Gonads produce sex hormones.

Glucocorticoid Therapy

Glucocorticoids suppress the body's normal reaction to disease: the inflammatory reaction (see Fig. 13.4) and the immune process. Thus, glucocorticoid therapy is useful for treating autoimmune diseases such as rheumatoid arthritis, organ transplant rejection, allergies, and severe asthma. However, glucocorticoids should be used for the minimum time possible because long-term administration of glucocorticoids for therapeutic purposes causes some degree of Cushing's syndrome (see page 214). Glucocorticoid use also predisposes the individual to infection, and long-term use has been linked to increased incidence of cancer. Researchers believe both

effects occur because of impaired immune response. In addition, sudden withdrawal from glucocorticoid therapy causes symptoms of diminished secretory activity by the adrenal cortex. This occurs because glucocorticoids suppress the release of adrenocorticotropic hormone (ACTH) by the anterior pituitary and lead to a decrease in glucocorticoid production by the adrenal cortex. Therefore, withdrawal of glucocorticoids following long-term use must be tapered. During an alternate-day schedule, the dosage is gradually reduced and then finally discontinued as the patient's adrenal cortex resumes activity.

the bones (although this function of calcitonin is more important in growing children than in adults).

The endocrine system helps regulate response to the external environment. In "fight-or-flight" situations, the nervous system stimulates the adrenal medulla to release epinephrine (adrenaline), which has a powerful effect on various organs. This,

too, is important to homeostasis because it allows us to behave in a way that keeps us alive. Any damage due to stress is then repaired by the action of other hormones, including cortisol. As discussed in the Medical Focus on this page, glucocorticoid (e.g., cortisone) therapy is useful for its anti-inflammatory and immunosuppressive effects.

SELECTED NEW TERMS

Basic Key Terms

adrenal cortex (uh-dre'nul kor'teks), p. 213
adrenal gland (uh-dre'nul gland), p. 213
adrenal medulla
    (uh-dre'nul mĕ-dūl'uh), p. 213
adrenocorticotropic hormone
    (uh-dre'no-kor"ti-kō-trōp'ik hor'mōn), p. 208
aldosterone (al"dos'ter-ōn), p. 214
anabolic steroid (an"uh-bol'ik stĕ'royd), p. 219
androgen (an'dro-jen), p. 219
anterior pituitary (an-tēr'ē-or pĭ-tu'ĭ-tār"e), p. 208
antidiuretic hormone (an"tĭ-dī"yū-ret'ik
    hor'mōn), p. 208
atrial natriuretic hormone
    (ā'trē-al nā"trē-yū-ret'ik hor'mōn), p. 214
atriopeptide (ā-tre-yo-pĕp-tīd), p. 214
calcitonin (kal"sĭ-to'nin), p. 212
circadian rhythm (ser"ka'de-an rĭ'thm), p. 219
corticotropin (kor'tĭ-ko-troh-pĭn), p. 213
cortisol (kor'tĭ-sol), p. 214
cyclic AMP (sik'lik AMP), p. 205
endocrine gland (en'do-krin gland), p. 205
epinephrine (ep"ĭ-nef'rin), p. 213
estrogen (es'tro-jen), p. 219
glucagon (glu'kuh-gon), p. 216
glucocorticoid (glu'ko-kor'tĭ-koyd), p. 214
gonad (go'nad), p. 219

gonadotropic hormone
    (go"nad-o-trōp'ik hor'mōn), p. 208
growth factor (grōth fak'tor), p. 221
growth hormone (grōth hor'mōn), p. 210
hormone (hor'mōn), p. 205
hypothalamic-inhibiting hormone (hi"po-thĕ-
    lam'ik-in-hib'it-ing hor'mōn), p. 208
hypothalamic-releasing hormone (hi"po-thĕ-
    lam'ik-re-lēs'-ing hor'mōn), p. 208
hypothalamus (hī'pō-thāl'ə-məs), p. 208
insulin (in'suh-lin), p. 216
leptin (lep'tin), p. 214
melatonin (mel"uh-to'nin), p. 219
mineralocorticoids
    (min"er-al-o-kor'tĭ-koyds), p. 214
norepinephrine (nôr'ĕp-ə-nĕf'rĭn), p. 213
ovary (o'var-e), p. 219
oxytocin (ok"sī-to'sin), p. 211
pancreas (pan'kre-us), p. 216
pancreatic islets (of Langerhans) (pan"kre-at'ik
    ī'lets ov lahng'er-hanz), p. 216
parathyroid gland
    (pār"uh-thi'royd gland), p. 212
parathyroid hormone (pār"uh-thi'royd
    hor'mōn), p. 212
peptide hormone (pep'tid hor'mōn), p. 221
pineal gland (pin'e-ul gland), p. 219
pituitary gland (pĭ-tu'ĭ-tār"ē gland), p. 208

posterior pituitary
    (pōs-tēr'e-or pĭ-tū'-ĭ-tār"ē), p. 208
progesterone (pro-jes'ter-ōn), p. 219
prolactin (pro-lak'tin), p. 210
prostaglandins (pros"tuh-glan'dinz), p. 221
renin (re'nin), p. 214
steroid hormone (stēr'oyd hor'mōn), p. 221
testis (tes'tis), p. 219
testosterone (tes-tos'tĕ-rōn), p. 219
thymosin (thi'mo-sin), p. 219
thymus gland (thi'mus gland), p. 219
thyroid gland (thi'royd gland), p. 211
thyroid-stimulating hormone
    (thi'royd stim'yū-lāt-ing hor'mōn), p. 208
thyrotropin (thī'ro-trō-'pĭn), p. 207
thyroxine (thī-rok'sin), p. 211

Clinical Key Terms

acidosis (ăs'ĭ-dō'sĭs), p. 217
acromegaly (ak"ro-meg'uh-le), p. 210
Addison disease (ă'dĭ-son dĭ-zēz'), p. 214
congenital hypothyroidism (kon-gen'i-tul
    hi"po-thi'-roy-dizm), p. 211
Cushing syndrome (koosh'ing sin'drōm), p. 214
diabetes insipidus
    (dī"uh-be'tēz in-sip'ĭ-dus), p. 208
exophthalmic goiter
    (ek"sof-thal'mik goy'ter), p. 212

glycosuria (gli′ko-sūr′e-uh), p. 216
Graves′ disease (grāvz dĭ-zēz′), p. 212
hyperglycemia (hi″per-gli-se′me-uh), p. 216
hypoglycemia (hi″po-gli-se′me-uh), p. 217
insulin-dependent diabetes mellitus (in′sul-in-de-pen′dent di″uh-be′tēz mē-li′tus), p. 217

ketonuria (ke″to-nū′re-uh), p. 216
myxedema (mik″sě-de′muh), p. 211
noninsulin-dependent diabetes (non′in′sūl-in-de-pen′dent di″uh-bē′tēz), p. 217
pituitary dwarfism (pĭ-tū′ĭ-tār″ē dwarf′-izm), p. 210

polydipsia (pol″e-dip′se-uh), p. 216
polyphagia (pol″e-fa-je-uh), p. 216
polyuria (pol″e-yū-re-uh), p. 216
simple (endemic) goiter (sim′pl ĕn-dĕm′ĭk goy′ter), p. 211
tetany (tet′uh-ne), p. 213

SUMMARY

10.1 Endocrine Glands

A. Endocrine glands secrete hormones into the bloodstream, and from there they are distributed to target organs or tissues.

B. Hormones are either peptides or steroids. Reception of a peptide hormone at the plasma membrane activates an enzyme cascade inside the cell. Steroid hormones combine with a receptor in the cell, and the complex attaches to and activates DNA. Protein synthesis follows. The major endocrine glands and hormones are listed in Table 10.1. Neural mechanisms, hormonal mechanisms, and/or negative feedback control the effects of hormones.

10.2 Hypothalamus and Pituitary Gland

A. Neurosecretory cells in the hypothalamus produce antidiuretic hormone (ADH) and oxytocin, which are stored in axon endings in the posterior pituitary until they are released.

B. The hypothalamus produces hypothalamic-releasing and hypothalamic-inhibiting hormones, which pass to the anterior pituitary by way of a portal system. The anterior pituitary produces at least six types of hormones, and some of these stimulate other hormonal glands to secrete hormones.

10.3 Thyroid and Parathyroid Glands

The thyroid gland requires iodine to produce triiodothyronine (T_3) and thyroxine (T_4), which increase the metabolic rate. If iodine is available in limited quantities, a simple goiter develops; if the thyroid is overactive, an exophthalmic goiter develops. The thyroid gland also produces calcitonin, which helps lower the blood calcium level. The parathyroid glands secrete parathyroid hormone, which raises the blood calcium and decreases the blood phosphate levels.

10.4 Adrenal Glands

The adrenal glands respond to stress: Immediately, the adrenal medulla secretes epinephrine and norepinephrine, which bring about responses we associate with emergency situations. On a long-term basis, the adrenal cortex produces the glucocorticoids (e.g., cortisol) and the mineralocorticoids (e.g., aldosterone). Cortisol stimulates hydrolysis of proteins to amino acids that are converted to glucose; in this way, it raises the blood glucose level. Aldosterone causes the kidneys to reabsorb sodium ions (Na^+) and to excrete potassium ions (K^+). Addison's disease develops when the adrenal cortex is underactive, and Cushing's syndrome develops when the adrenal cortex is overactive.

10.5 Pancreas

The pancreatic islets secrete insulin, which lowers the blood glucose level, and glucagon, which has the opposite effect. The most common illness caused by hormonal imbalance is diabetes mellitus, which is due to the failure of the pancreas to produce insulin and/or the failure of the cells to take it up.

10.6 Other Endocrine Glands

A. The gonads produce the sex hormones. The thymus secretes thymosins, which stimulate T-lymphocyte production and maturation. The pineal gland produces melatonin, which may be involved in circadian rhythms and the development of the reproductive organs.

B. Tissues also produce hormones. Adipose tissue produces leptin, which acts on the hypothalamus, and various tissues produce growth factors. Prostaglandins are produced and act locally.

10.7 Chemical Signals

In the human body, some chemical signals, such as traditional endocrine hormones and secretions of neurosecretory cells, act at a distance. Others, such as prostaglandins, growth factors, and neurotransmitters, act locally. Whether humans have pheromones is under study.

10.8 Effects of Aging

Two concerns often seen in the elderly are thyroid malfunctioning and diabetes mellitus.

10.9 Homeostasis

Hormones particularly help maintain homeostasis in several ways: Hormones help maintain the level of nutrients (e.g., amino acids and glucose in blood); help maintain blood volume and pressure by regulating the sodium content of the blood; help maintain the blood calcium level; help regulate fuel metabolism; and help regulate our response to the external environment.

STUDY QUESTIONS

1. Explain how peptide hormones and steroid hormones affect the metabolism of the cell. (p. 205)

2. Contrast hormonal and neural signals, and show that there is an overlap between the mode of operation of the nervous system and that of the endocrine system. (p. 207)

3. Explain the relationship of the hypothalamus to the posterior pituitary gland and to the anterior pituitary gland. List the hormones secreted by the posterior and anterior pituitary glands. (p. 208)
4. Give an example of the negative feedback relationship among the hypothalamus, the anterior pituitary, and other endocrine glands. (p. 208)
5. Discuss the effect of growth hormone on the body and the result of having too much or too little growth hormone when a young person is growing. What is the result if the anterior pituitary produces growth hormone in an adult? (p. 210)

6. What types of goiters are associated with a malfunctioning thyroid? Explain each type. (p. 211)
7. How do the thyroid and the parathyroid work together to control the blood calcium level? (p. 212)
8. How do the adrenal glands respond to stress? What hormones are secreted by the adrenal medulla, and what effects do these hormones have? (pp. 213–214)
9. Name the most significant glucocorticoid and mineralocorticoid, and discuss their functions. Explain the symptoms of Addison's disease and Cushing's syndrome. (p. 214)

10. Draw a diagram to explain how insulin and glucagon maintain the blood glucose level. Use your diagram to explain the major symptoms of type I diabetes mellitus. (pp. 216–217)
11. Name the other endocrine glands discussed in this chapter, and discuss the functions of the hormones they secrete. (p. 219)
12. What are leptin, growth factors, and prostaglandins? How do these substances act? (p. 221)
13. Discuss five ways the endocrine system helps maintain homeotasis. (pp. 222, 224)

LEARNING OBJECTIVE QUESTIONS

Fill in the blanks.
1. Generally, hormone production is self-regulated by a _____ mechanism.
2. The hypothalamus _____ the hormones _____ and _____, released by the posterior pituitary.
3. The _____ secreted by the hypothalamus control the anterior pituitary.
4. Growth hormone is produced by the _____ pituitary.
5. Simple goiter occurs when the thyroid is producing _____ (too much or too little) _____.

6. Parathyroid hormone increases the level of _____ in the blood.
7. Adrenocorticotropic hormone (ACTH), produced by the anterior pituitary, stimulates the _____ of the adrenal glands.
8. An overproductive adrenal cortex results in the condition called _____.
9. Type I diabetes mellitus is due to a malfunctioning _____, but type II diabetes is due to malfunctioning _____.
10. Prostaglandins are not carried in the _____ as are hormones secreted by the endocrine glands.

11. Whereas _____ hormones are lipid soluble and bind to receptor proteins within the cytoplasm of target cells, _____ hormones bind to membrane-bound receptors, thereby activating second messengers.
12. Whereas the adrenal _____ is under the control of the autonomic nervous system, the adrenal _____ secretes its hormones in response to _____ from the anterior pituitary gland.

MEDICAL TERMINOLOGY EXERCISE

Consult Appendix B for help in pronouncing and analyzing the meaning of the terms that follow.

1. antidiuretic (an"tĭ-di"yū-ret'ik)
2. hypophysectomy (hi-pof"ĭ-sek'to-me)
3. gonadotropic (go"nad-o-trōp'ik)
4. hypokalemia (hi"po-kal"e'me-uh)

5. lactogenic (lak"to-jen'ik)
6. adrenopathy (ad"ren-op'uh-the)
7. adenomalacia (ad"ĕ-no-muh-la'she-uh)
8. parathyroidectomy (pār"uh-thi"roy-dek'to-me)
9. polydipsia (pol"e-dip'se-uh)
10. dyspituitarism (dis-pĭ-tu'ĭ-ter'izm)

11. ketoacidosis (ke'to-as'ĭ-do'sis)
12. thyroiditis (thi-roy-di'tis)
13. glucosuria (glu-co-su're-uh)
14. microsomia (mi'kro-so'me-uh)
15. androgenic alopecia (an'dro-jen'ik al-o-pe'she-uh)

WEBSITE LINK

Visit this text's website at http://www.mhhe.com/maderap6 for additional quizzes, interactive learning exercises, and other study tools.

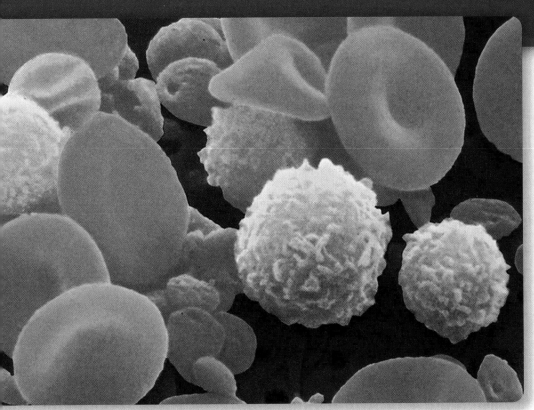

PART IV

CHAPTER

11

Blood

The formed elements of blood—red blood cells, white blood cells, and platelets—are shown in this colorized scanning electron micrograph.

Chapter Outline & Learning Objectives

After you have studied this chapter, you should be able to:

11.1 The Composition and Functions of Blood (p. 228)

- Describe, in general, the composition of blood.
- Divide the functions of blood into three categories, and discuss each category.
- Describe the composition of plasma and the specific functions of the plasma proteins.

11.2 The Blood Cells (p. 230)

- Explain the hematopoietic role of stem cells in the red bone marrow.
- Describe the structure, function, and life cycle of red blood cells and white blood cells.

11.3 Platelets and Hemostasis (p. 234)

- Describe the structure, function, and life cycle of platelets.

- Describe the three events of hemostasis and the reactions necessary to coagulation.
- Discuss disorders of hemostasis.

11.4 Capillary Exchange (p. 236)

- Describe capillary exchange within the tissues.

11.5 Blood Typing and Transfusions (p. 239)

- Explain the ABO and Rh systems of blood typing.
- Explain agglutination and its relationship to transfusions.

11.6 Effects of Aging (p. 241)

- Name the blood disorders that are commonly seen as we age.

Visual Focus

Hematopoiesis (p. 231)

Medical Focus

Avoiding Transfusion: Autotransfusion and Blood Substitutes (p. 233)
Abnormal Red and White Blood Cell Counts (p. 235)

What's New

Zeolite for Hemostasis (p. 237)

Focus on Forensics

Blood at the Crime Scene (p. 242)

227

11.1 The Composition and Functions of Blood

When a blood sample is first prevented from clotting and then spun in a centrifuge tube, it separates into two layers (Fig. 11.1). The lower layer consists first of the white blood cells and the blood platelets, which settle on top in a thin, shiny layer called the **buffy coat.** Below the buffy coat are the red blood cells, which are heavier cells due to their iron-containing hemoglobin. Collectively, these are the **formed elements,** which make up about 45% of the total volume of whole blood. The percentage of blood attributed to formed elements is called the **hematocrit.** The upper layer is **plasma,** a light-yellow fluid that contains a variety of inorganic and organic molecules dissolved or suspended in water. Plasma accounts for about 55% of the total volume of whole blood.

Functions of Blood

The functions of blood fall into three categories: transport, defense, and regulation.

Transport

Blood moves from the heart to all the various organs, where exchange with tissues takes place across thin capillary walls. Blood picks up oxygen from the lungs and nutrients from the digestive tract and transports these to the tissues. It also picks up and transports cellular wastes, including carbon dioxide, away from the tissues to exchange surfaces, such as the lungs and kidneys. We will see that capillary exchanges keep the composition of tissue fluid within normal limits.

Various organs and tissues secrete hormones and other messenger chemicals into the blood, and blood transports these to other organs and tissues, where they serve as signals that influence cellular metabolism.

Defense

Blood defends the body against invasion by **pathogens** (microscopic infectious agents, such as bacteria and viruses). Blood also removes dead and dying cells, clearing room for growth of healthy cells. Mutated cells—which could potentially develop into cancer—are also destroyed by the blood. Certain white blood cells are capable of engulfing and destroying pathogens or cancer cells; others produce and secrete antibodies into the blood. Antibodies incapacitate pathogens, making them subject to destruction by other white blood cells. This function of the blood will be discussed completely in Chapter 13.

When an injury occurs, blood forms a clot, and this prevents blood loss. Blood clotting involves platelets and the plasma proteins prothrombin and fibrinogen. Without blood clotting, we could bleed to death even from a small cut.

Regulation

Blood helps regulate body temperature by picking up heat, mostly from active muscles, and transporting it about the body. If the blood is too warm, the heat dissipates to the environment from dilated blood vessels in the skin.

The salts and plasma proteins in blood act to keep the osmotic pressure in the normal range. Recall that osmotic pressure causes diffusion of water (review section 3.2). Thus, the osmotic pressure of blood causes diffusion of water into the blood. In this way, blood plays a role in maintaining its own water-salt balance.

Because blood contains buffers, it also helps regulate body pH and keeps it relatively constant.

Plasma

Plasma is the liquid portion of blood, and about 92% of plasma is water. The remaining 8% of plasma is composed of various salts (ions) and organic molecules (Table 11.1). The salts, which are simply dissolved in plasma, help maintain the osmotic pressure and pH of the blood. Small organic molecules such as glucose, amino acids, and urea are also dissolved in plasma. Glucose and amino acids are nutrients for cells; urea is a nitrogenous waste product on its way to the kidneys for excretion. The large organic molecules in plasma include hormones and the plasma proteins.

The Plasma Proteins

Three major types of plasma proteins are the albumins, the globulins, and fibrinogen. Most plasma proteins are made in the liver. One exception is the antibodies produced by B lymphocytes, which function in immunity. Certain hormones are also plasma proteins made by various glands.

The plasma proteins have many functions that help maintain homeostasis. They are able to take up and release hydrogen ions; therefore, the plasma proteins help buffer the blood and keep its pH around 7.40. **Osmotic pressure** is the force caused by a difference in solute concentration on either side of a membrane. As mentioned previously, plasma proteins, particularly the **albumins,** contribute to the osmotic pressure.

There are three types of **globulins,** designated alpha, beta, and gamma globulins. The alpha and beta globulins, produced

TABLE 11.1 Blood Plasma Solutes

Plasma proteins	Albumin, globulins, fibrinogen
Inorganic ions (salts)	Na^+, Ca^{2+}, K^+, Mg^{2+}, Cl^-, HCO_3^-, HPO_4^{2-}, SO_4^{2-}
Gases	O_2, CO_2
Organic nutrients	Glucose, fats, phospholipids, amino acids, etc.
Nitrogenous waste products	Urea, ammonia, uric acid
Regulatory substances	Hormones, enzymes, other messenger chemicals

FORMED ELEMENTS	Function and Description	Source
Red Blood Cells (erythrocytes) 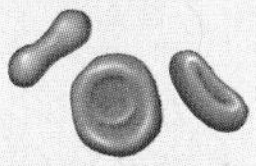 4 million–6 million per mm³ blood	Transport O_2 and help transport CO_2. 7–8 μm in diameter; bright-red to dark-purple biconcave disks without nuclei.	Red bone marrow
White Blood Cells (leukocytes) 5,000–11,000 per mm³ blood	Fight infection. Remove dead/dying cells. Destroy cancer cells.	Red bone marrow
Granular leukocytes		
• Neutrophils 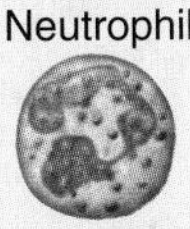 40–70%	Phagocytize pathogens. 10–14 μm in diameter; spherical cells with multilobed nuclei; fine, lilac granules in cytoplasm if stained.	
• Eosinophils 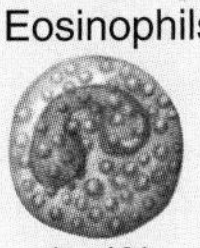 1–4%	Phagocytize antigen-antibody complexes and allergens. 10–14 μm in diameter; spherical cells with bilobed nuclei; coarse, deep-red, uniformly sized granules in cytoplasm if stained.	
• Basophils 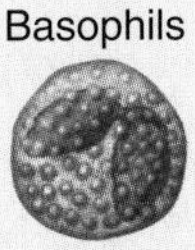 0–1%	Release histamine and heparin, which promote blood flow to injured tissues. 10–12 μm in diameter; spherical cells with lobed nuclei; large, irregularly shaped, deep-blue granules in cytoplasm if stained.	
Agranular leukocytes		
• Lymphocytes  20–45%	Responsible for specific immunity. 5–17 μm in diameter (average 9–10 μm); spherical cells with large, round nuclei.	
• Monocytes 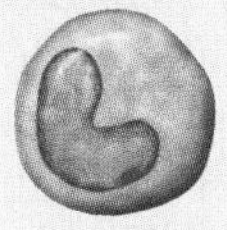 4–8%	Become macrophages that phagocytize pathogens and cellular debris. 10–24 μm in diameter; large, spherical cells with kidney-shaped, round, or lobed nuclei.	
Platelets  150,000–300,000 per mm³ blood	Aid hemostasis. 2–4 μm in diameter; disk-shaped cell fragments with no nuclei; purple granules in cytoplasm.	Red bone marrow

PLASMA	Function	Source
Water (90–92% of plasma)	Maintains blood volume; transports molecules.	Absorbed from intestine
Plasma proteins (7–8% of plasma)	Maintain blood osmotic pressure and pH.	Liver
Albumins	Maintain blood volume and pressure, transport.	Select white blood cells
Globulins	Transport; fight infection.	
Fibrinogen	Coagulation.	
Salts (less than 1% of plasma)	Maintain blood osmotic pressure and pH; aid metabolism.	Absorbed from intestine
Gases		
Oxygen Carbon dioxide	Cellular respiration End product of metabolism.	Lungs Tissues
Nutrients Lipids Glucose Amino acids	Food for cells.	Absorbed from intestine
Nitrogenous wastes Uric acid Urea	Excretion by kidneys.	Liver
Other Hormones, vitamins, etc.	Aid metabolism.	Varied

Plasma 55%

Buffy coat

Formed elements 45%

Figure 11.1 Composition of blood. When a blood sample is prevented from clotting and spun in a centrifuge tube, it forms two layers. The lucent, yellow top layer is plasma, the liquid portion of blood. The formed elements are in the bottom layer. This table describes these components in detail.

by the liver, bind to metal ions and to fat-soluble vitamins. They also bind to lipids, forming the **lipoproteins.** Antibodies, proteins made by white blood cells which help fight infections, are gamma globulins.

Both albumins and globulins combine with and transport large organic molecules. For example, albumin transports the molecule bilirubin, a breakdown product of hemoglobin. Lipoproteins, whose protein portion is a globulin, transport cholesterol.

The other plasma proteins, fibrinogen and prothrombin, are necessary to coagulation (blood clotting), which is discussed on page 236.

Content **CHECK-UP!**

1. Which of the following statements about hematocrit is false?
   a. Typically, 55% of a blood sample is composed of cells.
   b. The buffy coat consists of white blood cells and platelets.
   c. Plasma is a water solution containing dissolved organic and inorganic substances.
   d. Approximately 92% of plasma is water.

2. Which of the following are defense mechanisms of the blood?
   a. Blood cells can engulf and destroy bacteria and viruses.
   b. Antibodies target pathogens for destruction.
   c. The blood clotting mechanism prevents blood loss.
   d. All of the above are defense mechanisms of the blood.

3. Which of the following is not a plasma protein produced by the liver?
   a. albumin   c. gamma globulin antibodies
   b. fibrinogen   d. prothrombin

11.2 The Blood Cells

The formed elements consist of blood cells and platelets. (Platelets are discussed in detail on page 234.) In the adult, the formed elements are produced continuously in the red bone marrow of the skull, ribs, and vertebrae, the iliac crests, and the ends of long bones.

The process by which formed elements are made is called **hematopoiesis** (Fig. 11.2). **Multipotent stem cells** are the red bone marrow cells that mature into all the various types of blood cells. At the top of Figure 11.2 is a multipotent stem cell that first replicates itself by mitosis. Each cell then *differentiates*, producing two other types of stem cells: myeloid and lymphatic stem cells. (When a stem cell differentiates, it commits itself to follow a single pathway in development.) The myeloid stem cell further differentiates to give rise to the cells that become red blood cells, granular leukocytes, monocytes, and megakaryocytes, the parent cell for platelets. The lymphatic stem cell differentiates to produce the lymphocytes.

Many scientists are very interested in developing ways to use blood stem cells, as well as stem cells from other adult tissues, to regenerate the body's tissues in the laboratory.

Red Blood Cells

Red blood cells (RBCs, or erythrocytes) are small, biconcave disks that lack a nucleus when mature. They occur in great quantity; there are 4 to 6 million red blood cells per mm^3 of whole blood.

Red blood cells transport oxygen, and each contains about 200 million molecules of **hemoglobin,** the respiratory pigment. If this much hemoglobin were suspended within the plasma rather than enclosed within the cells, blood would be so viscous that the heart would have difficulty pumping it.

Hemoglobin

In a molecule of hemoglobin, each of four polypeptide chains making up globin has an iron-containing heme group in the center. Oxygen combines loosely with iron when hemoglobin is oxygenated:

$$\text{Hb} + \text{O}_2 \underset{\text{tissues}}{\overset{\text{lungs}}{\rightleftharpoons}} \text{HbO}_2$$

In this equation, the hemoglobin on the right, which is combined with oxygen, is called oxyhemoglobin. Oxyhemoglobin forms in lung capillaries, and has a bright red color. The hemoglobin on the left, which is not combined with oxygen, is called deoxyhemoglobin. Deoxyhemoglobin forms in tissue capillaries, and has a dark maroon color.

Hemoglobin is remarkably adapted to its function of picking up oxygen in lung capillaries and releasing it in the tissues. As discussed in the Medical Focus on page 233, hemoglobin alone can be used as a blood substitute. The higher concentration of oxygen, plus the slightly cooler temperature and slightly higher pH within lung capillaries, causes hemoglobin to take up oxygen. By contrast, in tissues the lower concentration of oxygen, plus the slightly warmer temperature and slightly lower pH within tissue capillaries, causes hemoglobin to give up its oxygen.

Production of Red Blood Cells

Erythrocytes are formed from red bone marrow stem cells (see Fig. 11.2): A multipotent stem cell descendant, called a myeloid stem cell, gives rise to erythroblasts, which divide many times. As they mature, erythroblasts gain many molecules of hemoglobin and lose their nucleus and most of their organelles. Possibly because mature red blood cells lack a nucleus, they live only about 120 days. It is estimated that about 2 million red blood cells are destroyed per second; therefore, an equal number must be produced to keep the red blood cell count in balance.

Whenever blood carries a reduced amount of oxygen, more red blood cells must be produced to maintain homeostasis. Lowered blood oxygen occurs when an individual first takes up residence at a high altitude, or after a hemorrhage when a person loses red blood cells, or in a patient with impaired lung function. These are stimuli for the kidneys to accelerate their release of **erythropoietin** (EPO) (Fig. 11.3).

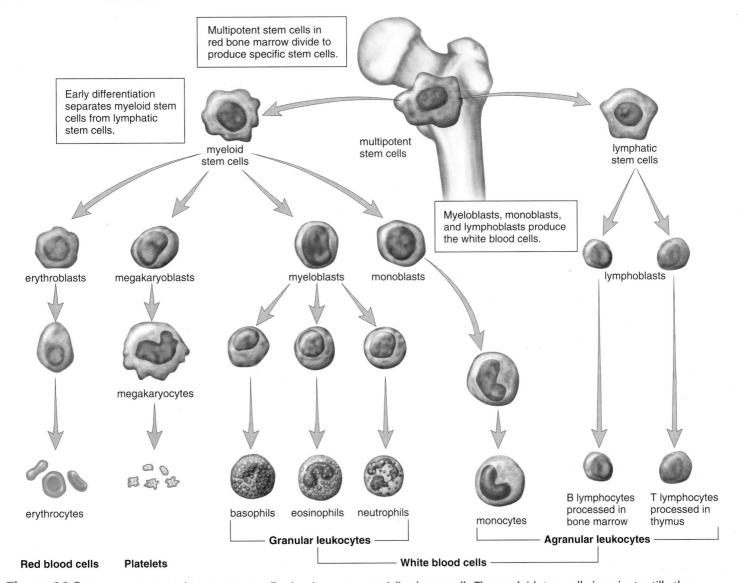

Multipotent stem cells in red bone marrow divide to produce specific stem cells.

Early differentiation separates myeloid stem cells from lymphatic stem cells.

myeloid stem cells

multipotent stem cells

lymphatic stem cells

Myeloblasts, monoblasts, and lymphoblasts produce the white blood cells.

erythroblasts

megakaryoblasts

myeloblasts

monoblasts

lymphoblasts

megakaryocytes

erythrocytes

basophils   eosinophils   neutrophils

monocytes

B lymphocytes processed in bone marrow

T lymphocytes processed in thymus

Red blood cells   **Platelets**

Granular leukocytes

Agranular leukocytes

White blood cells

Figure 11.2 Hematopoiesis. Multipotent stem cells give rise to two specialized stem cells. The myeloid stem cell gives rise to still other cells, which become red blood cells, platelets, and all the whole blood cells except lymphocytes. The lymphatic stem cell gives rise to lymphoblasts, which become lymphocytes.

This hormone stimulates mitosis of stem cells and speeds up the maturation of red blood cells. The liver and other tissues also produce erythropoietin. Erythropoietin, now mass-produced through biotechnology, is often prescribed for cancer patients who are anemic (low red blood cell count) due to chemotherapy or radiation therapy. Unfortunately, erythropoietin is sometimes abused by athletes in order to raise their red blood cell counts and thereby increase the oxygen-carrying capacity of their blood.

Destruction of Red Blood Cells

With age, red blood cells are destroyed in the liver and spleen, where they are engulfed by macrophages. When red blood cells are broken down, hemoglobin is released. The globin portion of the hemoglobin is broken down into its component amino acids, which are recycled by the body. The iron is recovered and returned to the bone marrow for reuse. The heme portion of the molecule undergoes chemical degradation and is excreted as **bile pigments** by the liver into the bile. These bile pigments

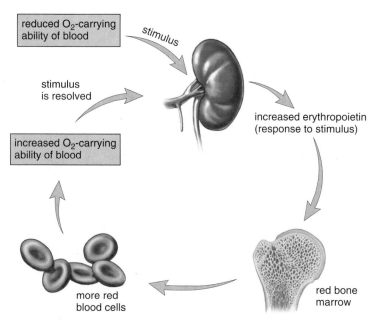

Figure 11.3 Action of erythropoietin. The kidneys release increased amounts of erythropoietin whenever the oxygen capacity of the blood is reduced. Erythropoietin stimulates the red bone marrow to speed up its production of red blood cells, which carry oxygen. Once the oxygen-carrying capacity of the blood is sufficient to support normal cellular activity, the kidneys cut back on their production of erythropoietin.

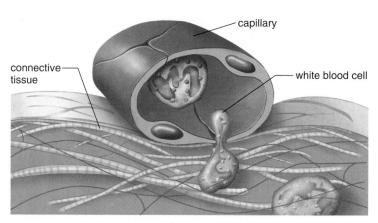

Figure 11.4 Mobility of white blood cells. White blood cells can squeeze between the cells of a capillary wall and enter the tissues of the body.

are bilirubin and biliverdin, which contribute to the color of feces. Chemical breakdown of heme is also what causes a bruise on the skin to change color from red/purple to green to yellow.

Abnormal Red Blood Cell Counts

As discussed in the Medical Focus on page 235, anemia is an illness in which the patient has a tired, run-down feeling. The cells are not getting enough oxygen due to a reduction in the amount of functional hemoglobin or the number of red blood cells. **Hemolysis** (bursting of red blood cells) can also cause anemia.

White Blood Cells

White blood cells (**WBCs,** or **leukocytes**) differ from red blood cells in that they are usually larger, have a nucleus, lack hemoglobin, and are translucent unless stained. White blood cells are not as numerous as red blood cells; there are only 5,000–11,000 per mm^3 of blood. White blood cells fight infection, destroy dead or dying body cells, and recognize and kill cancerous cells. Thus, they are important contributors to homeostasis. This function of white blood cells is discussed at greater length in Chapter 13, which concerns immunity.

White blood cells are derived from stem cells in the red bone marrow, and they, too, undergo several maturation stages (see Fig. 11.2). Each type of white blood cell is apparently capable of producing a specific growth factor that circulates back to the bone marrow to stimulate its own production.

Red blood cells are confined to the blood, but white blood cells are able to squeeze through pores in the capillary wall (Fig. 11.4). Therefore, they are found in tissue fluid and lymph (the fluid within lymphatic vessels) and in lymphatic organs. When an infection is present, white blood cells greatly increase in number. Many white blood cells live only a few days—they probably die while fighting pathogens. Others live months or even years.

Types of White Blood Cells

White blood cells are classified into the **granular leukocytes** and the **agranular leukocytes.** Both types of cells have granules in the cytoplasm surrounding the nucleus, but the granules are more visible upon staining in granular leukocytes. (The white cells in Figures 11.1 and 11.2 have been stained with a dye called Wright stain.) The granules contain various enzymes and proteins, which help white blood cells defend the body. There are four types of granular leukocytes and two types of agranular leukocytes. They differ somewhat by the size of the cell and the shape of the nucleus (see Fig. 11.1), and they also differ in their functions.

Granular Leukocytes Neutrophils (see Fig. 13.3*a*) are the most abundant of the white blood cells. They have a multi-lobed nucleus joined by nuclear threads; therefore, they are also called polymorphonuclear ("many-shaped nuclei"). Some of their granules take up acid stain, and some take up basic stain (creating an overall lilac color). Neutrophils are the first type of white blood cell to respond to an infection, and they engulf pathogens during phagocytosis.

Eosinophils (see Fig. 13.3*b*) have a bilobed nucleus, and their large, abundant granules take up the dye called eosin and become a red color. (This accounts for their name, eosinophil.) Among several functions, they increase in number in the event of a parasitic worm infection. Eosinophils also lessen an allergic reaction by phagocytizing antigen-antibody complexes involved in an allergic attack.

Basophils (see Fig. 13.3*c*) have a U-shaped or lobed nucleus. Their granules take up the basic stain and become dark blue in color. (This accounts for their name, basophil.) In the

Avoiding Transfusion: Autotransfusion and Blood Substitutes

In the emergency room (ER) setting, it's a problem you and your co-workers will face every day. Your patient may have survived a serious automobile accident, or perhaps he was involved in a shooting. A young woman may have hemorrhaged following the unexpected early delivery of her baby. Or maybe your patient is a young acute lymphoblastic leukemia sufferer (see page 235), whose hematocrit (see page 228) has dropped dangerously low because of chemotherapy. These patients all share a common need—an immediate blood transfusion to save their lives. Without transfusion, blood loss will cause tissue cells to die from lack of oxygen. For this reason, emergency room caregivers often refer to the "golden hour" for treatment of patients. Patients who receive the best possible care, including blood transfusions, within an hour of admission to the ER have the best chance of surviving and recovering.

If the emergency room is a major trauma center in a large city hospital, donor blood for transfusion is usually available. The correct blood can be matched to the patient's blood type. If there isn't time to match donor and recipient blood, the ready supply of O-negative blood can theoretically be donated to anyone. But what if the transfusion is needed in a remote area, such as a wartime field hospital or the accident scene on an isolated stretch of highway? Military medics and EMT personnel often can't store and transport whole blood. What if the hospital is in a rural area? Small regional hospitals face regular shortages of donor blood for transfusion.

Even when the blood is properly matched, receiving a transfusion always carries a small but significant risk. Because blood is a tissue, transfusion is in effect a "tissue transplant." If the patient's immune system detects that the proteins on the red blood cell membrane are foreign, the transfused cells will be rejected (see Fig. 11.9). This process, called a transfusion reaction, can be fatal. Donor blood may be infected with viruses, including HIV (which causes AIDS) and hepatitis B and C viruses. Currently, the transfusion recipient also faces the risk of infection with prions (protein infectious particles). Prions, which are smaller than viruses, cause Creutzfeldt-Jacob disease, the human form of mad cow disease (see Medical Focus, Prions: Malicious Proteins p. 33).

Increasingly, physicians are attempting to avoid donor blood transfusion entirely, if at all possible. One technology that can be used for this purpose is intraoperative blood salvage, or *autotransfusion*. This technique uses an apparatus termed a "Cell Saver©", which an emergency room physician or surgeon can use to suction blood from the patient's body. The suctioned blood is immediately mixed with an anticoagulant and then collected in a reservoir. Salvaged blood cells are washed with a saline solution to remove any contaminants. The washed cells are centrifuged and suspended in sterile saline solution. The unit of the patient's own blood is then re-introduced into a vein, just like a traditional transfusion. The procedure is quite fast: To prepare an entire unit of blood to re-infuse into the patient takes only 5 minutes. Autotransfusion with the patient's own blood completely eliminates the possibility of transfusion reactions.

In addition, researchers are currently investigating the use of blood substitutes to solve some of the problems inherent in blood transfusions. The most promising blood substitutes use the hemoglobin molecule as their basic component. Hemoglobin is the oxygen-transporting molecule contained in red blood cells (see page 230 and Fig. 11A). Natural hemoglobin taken out of red blood cells cannot be introduced into the bloodstream. It breaks down immediately into smaller molecules that are toxic, especially to nerve cells, the liver, and the kidneys. However, hemoglobin that is first chemically altered to prevent it from breaking down can be safely transfused. Once in the cardiovascular system, the hemoglobin will transport oxygen in much the same way that it does inside an intact red blood cell. The modified hemoglobin is slowly broken down and eliminated from the body, without harming the patient's liver or kidneys. What's more, adequate supplies of hemoglobin are readily available, and don't rely on human blood donors. One developer uses blood from cattle. Another uses human hemoglobin produced by genetically engineered bacteria (also the source of the human insulin injected by diabetics).

Blood substitutes have additional benefits. Hemoglobin-based blood substitutes are better oxygen transporters than whole blood, although they remain in the patient's body for only a few days. Unlike whole blood, blood substitutes are free of disease-causing contaminants and can be stored for months at room temperature. Moreover, blood substitutes cannot cause a transfusion reaction because they lack the protein membrane of a red blood cell. This makes them the perfect "one-size-fits-all" substance for transfusion, and perhaps the ideal solution for critical-care emergencies. Blood substitutes are currently in widespread clinical trials in South Africa, where the AIDS outbreak has caused a critical shortage of available donors for whole blood. Other clinical trials are under way in the United States and Europe.

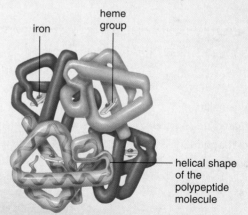

Figure 11A Hemoglobin contains four polypeptide chains (blue). There is an iron-containing heme group in the center of each chain. Oxygen combines loosely with iron when hemoglobin is oxygenated. Oxyhemoglobin is bright red, and deoxyhemoglobin is a dark maroon color.

connective tissues, basophils, as well as a similar type of cells called mast cells, release histamine and heparin. Histamine, which is associated with allergic reactions, dilates blood vessels and causes contraction of smooth muscle. Heparin prevents clotting and promotes blood flow. Natural killer cells (often abbreviated NK cells) are large, granular leukocytes that are similar in structure to lymphocytes. NK cells target and kill virus-infected cells. NK cells are also important defenders against cancer, because they can recognize and destroy cells which have mutated to become cancerous.

Agranular Leukocytes The agranular leukocytes include lymphocytes, which have a spherical nucleus, and monocytes, which have a kidney-shaped nucleus. **Lymphocytes** are responsible for specific immunity to particular pathogens and their toxins (poisonous substances), as well as recognizing and destroying cancer cells. Lymphocytes (see Fig. 13.3*d*) are of two types, B lymphocytes and T lymphocytes. Pathogens have **antigens,** surface molecules that the immune system can recognize as foreign. When an antigen is recognized as foreign, B lymphocytes will form antibodies against it. **Antibodies** are proteins that neutralize antigens. T lymphocytes, on the other hand, directly attack and destroy any cell, such as a pathogen that has foreign antigens. B lymphocytes and T lymphocytes are discussed more fully in Chapter 13.

Monocytes (see Fig. 13.3*e*) are the largest of the white blood cells, and after taking up residence in the tissues, they differentiate into even larger macrophages. **Macrophages** phagocytize pathogens, old cells, and cellular debris. They also stimulate other white blood cells, including lymphocytes, to defend the body.

Abnormal White Cell Counts

Abnormal white blood cell counts are discussed in the Medical Focus on page 235. Because specific white blood cells increase with particular infections, a differential white cell count, also discussed in the Medical Focus, can be quite helpful in diagnosing the cause of a particular illness.

Content **CHECK-UP!**

4. Which of the following is not a stimulus for the release of erythropoietin (EPO)?

   a. hyperventilation   c. blood loss

   b. decreased lung function   d. decreased oxygen in the blood

5. Which of the following is/are bile pigments?

   a. hemoglobin   d. b and c

   b. biliverdin   e. all of the above

   c. bilirubin

6. Granular white blood cells include all of the following except:

   a. eosinophils.   c. neutrophils.

   b. macrophages.   d. basophils.

11.3 Platelets and Hemostasis

Platelets (thrombocytes) are formed elements necessary to the process of **hemostasis,** the cessation of bleeding.

Platelets

Platelets result from fragmentation of certain large cells, called **megakaryocytes,** that develop in red bone marrow. It is important to note that although platelets are called thrombocytes, they are *not* cellular; rather, they are cell *fragments*. Platelets are produced at a rate of 200 billion per day, and the blood contains 150,000–300,000 per mm^3. Because platelets have no nucleus, they last at most ten days, assuming they are not used sooner than that in hemostasis.

Hemostasis

Hemostasis is divided into three events: vascular spasm, platelet plug formation, and coagulation (Fig. 11.5).

    Vascular spasm, the constriction of the smooth muscle layer in a broken blood vessel, is the immediate response to blood vessel injury. Platelets release serotonin, a chemical that prolongs smooth muscle contraction.

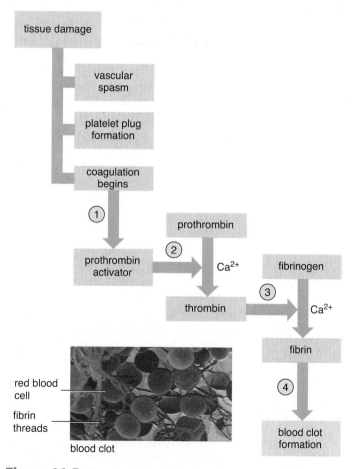

Figure 11.5 Hemostasis requires three events: vascular spasm, platelet plug formation, and coagulation. Coagulation is further broken down into four steps.

Abnormal Red and White Blood Cell Counts

Normal erythrocyte count should be in the range of 4–6 million cells per cubic millimeter of blood (about the size of one large drop). Various disease processes can affect erythrocyte count. **Polycythemia** is a disorder in which an excessive number of red blood cells makes the blood so thick that it is unable to flow properly. An increased risk of clot formation is also associated with this condition.

Normally, the blood hemoglobin level is 12–17 grams per 100 milliliters. In **anemia,** either the number of red cells is insufficient, and/or the cells do not have enough normal hemoglobin. Anemia can be classified in one of several categories. The first category, **hemolytic anemia,** occurs because the rate of red blood cell destruction increases (hemolysis is the rupturing of red blood cells). **Hemolytic disease of the newborn,** discussed at the end of this chapter (see pages 240–241) is a type of anemia.

Sickle-cell disease is a hereditary hemolytic anemia in which the individual has fragile, sickle-shaped red blood cells (Fig. 11B). Such cells tend to rupture easily as they pass through the narrow capillaries, leading to the symptoms of anemia. Sickle-cell disease is most common among people whose ancestors came from areas bordering the equator. Sickle-shaped cells protect against malaria, a disease spread by mosquitoes. Malaria is epidemic in equatorial Africa, India, and Asia. The parasite that causes malaria cannot infect sickle-shaped red blood cells. In the United States, sickle-cell disease is most common in African Americans whose ancestors were from equatorial Africa.

Dietary anemias occur because the patient's diet lacks substances needed for red blood cell development. In **iron deficiency anemia,** a common type of anemia, the person's diet does not contain enough iron, or there is excessive iron loss from the body. The hemoglobin count is low, red blood cells are small

and pale in color, and the individual feels tired and run-down. Including iron supplements in the diet can help prevent this type of anemia. **Pernicious anemia** is another form of dietary anemia. The digestive tract is unable to absorb enough vitamin B_{12}, which is essential to the proper formation of red cells. Without vitamin B_{12}, large numbers of immature red cells tend to accumulate in the bone marrow. A special diet, vitamin supplements, and/or injections of vitamin B_{12} are effective treatments.

In **aplastic anemia,** the red bone marrow has been damaged due to radiation or chemicals, and red blood cells, white blood cells, and platelets are all deficient. Bone marrow transplant is one option to treat this condition.

Hemorrhagic anemia is decreased red blood cell count following a hemorrhage. Transfusions may be administered to increase red blood cell count, although use of transfusion therapy is becoming increasingly less common due to its potential side effects (see the Medical Focus on page 233).

Certain viral illnesses, such as influenza, measles, and mumps, cause the white blood cell count to decrease. **Leukopenia** is a total white blood cell count below 5,000 per cubic millimeter. Other illnesses, including appendicitis and bacterial infections, cause the white blood cell count to increase dramatically. **Leukocytosis** is a white blood cell count above 10,000 per cubic millimeter.

Illness often causes an increase in a particular type of white blood cell. For this reason, a **differential white blood cell count,** involving the microscopic examination of a blood sample and the counting of each type of white blood cell to a total of 100 cells, may be done as part of the diagnostic procedure. For example, the characteristic finding in the viral disease **mononucleosis** is a great number of lymphocytes that are larger than mature lymphocytes and that stain more darkly. This condition takes its name from the fact that lymphocytes are mononuclear.

Leukemia is a form of cancer characterized by uncontrolled production of abnormal white blood cells. These cells accumulate in the bone marrow, lymph nodes, spleen, and liver so that these organs are unable to function properly. **Acute lymphoblastic leukemia (ALL),** which represents over 80% of the acute leukemias in children, also occurs in adults. Chemotherapy is used to destroy abnormal cells and restore normal blood cell production. Intraspinal injection of drugs and craniospinal irradiation are measures that prevent leukemic cells from infiltrating the central nervous system. In general, the prognosis is more favorable for children between the ages of 2 and 10 years than for either older or younger patients. The prognosis is somewhat better in females because leukemia recurs in the testes of 8–16% of males. Remission occurs in 78% of adult patients after chemotherapy, and the median period of remission is 20 months. With chemotherapy, 50–60% of children survive past five years, and of those among this group who do not have a relapse, 85% are considered cured.

Figure 11B Sickle-shaped red blood cells, as seen by a scanning electron microscope.

Platelet plug formation is the next event in hemostasis. Platelets don't normally adhere to damaged blood vessel walls, but when the lining of a blood vessel breaks, connective tissue, including collagen fibers, is exposed. Platelets adhere to collagen fibers and release a number of substances, including one that promotes platelet aggregation so that a so-called *platelet plug* forms. As a part of normal activities, small blood vessels often break, and a platelet plug is usually sufficient to stop the bleeding.

Coagulation, also called blood clotting, is the last event to bring about hemostasis. As you will see, two plasma proteins, called **fibrinogen** and **prothrombin,** participate in blood clotting. Vitamin K, found in green vegetables and also formed by intestinal bacteria, is necessary for the production of prothrombin. If, by chance, vitamin K is missing from the diet, **hemorrhagic bleeding** disorders develop.

Coagulation

Coagulation requires many protein clotting factors which are constantly present in the blood in an inactive state. Most of these clotting factors are produced by the liver, although several are released "on demand" by platelets if tissues or blood vessels are damaged. There are two mechanisms for activation of clotting. The **intrinsic mechanism,** so-named thus because all clotting factors are *intrinsic* to the blood, is the slower pathway. The intrinsic mechanism explains why blood will coagulate if placed in a test tube that has not been treated with an anticoagulant. The **extrinsic mechanism** is activated when damaged tissues release a substance called **tissue thromboplastin** (thus, its activation is *extrinsic* to blood itself). Tissue thromboplastin will interact with platelets, other clotting factors, and calcium ions (Ca^{2+}).

In the body, both clotting pathways are activated simultaneously; their common end point is production of prothrombin activator. Figure 11.5 breaks down the subsequent clotting process into four steps: ① **prothrombin activator** is formed. ② Prothrombin activator then converts prothrombin to **thrombin.** ③ Thrombin, in turn, severs two short amino acid chains from each fibrinogen molecule, and these activated fragments join end-to-end, forming long threads of **fibrin.** ④ Fibrin threads wind around the platelet plug in the damaged area of the blood vessel and provide the framework for the clot. Red blood cells also are trapped within the fibrin threads. These cells make a clot appear red. Once clotting is initiated, the clotting process will be self-limiting and confined to just the area of injury. This important safeguard prevents the clot from becoming too large, possibly blocking blood flow to other body areas. Limiting the clotting process to only the injured area also prevents excessive consumption of the available clotting factors in the blood.

Clot retraction follows, and the clot gets smaller as platelets contract. A fluid called **serum** (plasma minus fibrinogen and prothrombin) is squeezed from the clot. A fibrin clot is present only temporarily. As soon as blood vessel repair is initiated, an enzyme called *plasmin* destroys the fibrin network and restores the fluidity of the plasma.

Naturally, it is vital to activate clotting when tissues have been injured. However, it is just as important that clotting *not* be activated if there is no injury. Normally, the smooth endothelial lining of an intact blood vessel prevents clots from forming in the blood vessel. Anticoagulants, such as heparin produced by basophils and mast cells, also help to prevent accidental clotting.

Disorders of Hemostasis

Both impaired clotting and excessive clotting can be life-threatening or fatal. **Thrombocytopenia,** a low platelet count, is one cause of impaired clotting. It occurs when one's own antibodies attack platelets, or if megakaryocytes in red bone marrow (parent cells that fragment to form platelets) are destroyed. The **hemophilias** are inherited clotting disorders caused by deficiencies of clotting factors. (The most severe form, hemophilia A, is due to the lack of clotting factor VIII, a step in the intrinsic clotting pathway.) In these clotting impairments, the slightest bump can cause bleeding into the tissues. Bleeding into a joint causes cartilage degeneration, and resorption of underlying bone can follow. Bleeding into muscles can lead to nerve damage and muscular atrophy. The most frequent cause of death is bleeding into the brain with accompanying neurological damage.

Undesirable, excessive clotting may block blood flow to tissues. Despite the presence of anticoagulants in the blood, sometimes a clot forms in an unbroken blood vessel. Such a clot is called a **thrombus** if it remains stationary. If the clot dislodges and travels in the blood, it is called an **embolus.** If this embolus blocks a blood vessel, it is a **thromboembolism.** Pulmonary thromboembolism (blockage of the pulmonary arteries by a clot) will seriously inhibit the oxygenation of the blood; cerebral thromboembolism will cause a **cerebrovascular accident** or **stroke.** Coronary thrombosis or thromboembolism causes a heart attack, as discussed in Chapter 12.

Content **CHECK-UP!**

7. The parent cell that fragments to form thrombocytes is called a:
   a. stem cell.
   c. megalocyte.
   b. monocyte.
   d. megakaryocyte.

8. Choose the correct sequence for hemostasis (first step to last):
   a. coagulation → vascular spasm → platelet plug formation
   b. vascular spasm → coagulation → platelet plug formation
   c. vascular spasm → platelet plug formation → coagulation
   d. coagulation → platelet plug formation → vascular spasm

9. Which vitamin is necessary to form prothrombin?
   a. B_{12}
   c. C
   b. folic acid
   d. K

11.4 Capillary Exchange

The pumping of the heart sends blood by way of arteries to smaller arterioles, to the capillaries where exchange takes place across thin capillary walls (Fig. 11.6). Blood that has passed through the capillaries returns to the heart via veins. Capillary walls are largely composed of one layer of epithelial cells connected by tight junctions. Capillaries are extremely numerous. The body most likely contains a billion capillaries, and their total surface area is estimated at 6,300 square meters. Therefore, most cells of the body are near a capillary.

In the tissues of the body, metabolically active cells require oxygen and nutrients and give off wastes, including carbon dioxide. During capillary exchange between tissue capillaries and body cells, oxygen and nutrients leave a capillary. Cellular wastes, including carbon dioxide, enter a capillary. For this reason, systemic arterial blood contains more oxygen and nutri-

ents than does systemic venous blood, and venous blood contains more wastes than does arterial blood. In pulmonary capillaries—the capillaries supplying the lung—the exchange is reversed: Oxygen enters the blood and carbon dioxide leaves.

The internal environment of the body consists of blood and tissue fluid. **Tissue fluid** is simply the fluid that surrounds the cells of the body. In other words, substances that leave a capillary pass through tissue fluid before entering the body's cells, and substances that leave the body's cells pass through tissue fluid before entering a capillary. The composition of tissue fluid stays relatively constant because of capillary exchange. Tissue fluid is a water-based solution that

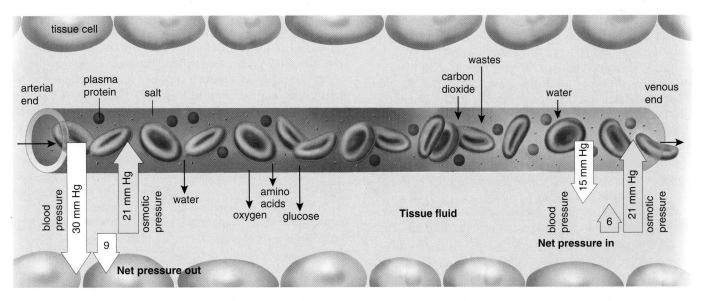

Figure 11.6 Capillary exchange. At the arterial end of a capillary, blood pressure is higher than osmotic pressure; therefore, water tends to leave the bloodstream. In the midsection of a capillary, small molecules follow their concentration gradients: Oxygen and nutrients leave the capillary, while wastes, including carbon dioxide, enter the capillary. At the venous end of a capillary, osmotic pressure is higher than blood pressure; therefore, water tends to enter the bloodstream.

contains sodium chloride, other electrolytes, and scant protein. Any excess tissue fluid is collected by lymphatic capillaries, which are always found near blood capillaries.

Blood Capillaries

Water and other small molecules can cross through the cells of a capillary wall or through tiny clefts that occur between the cells. Large molecules in plasma, such as the plasma proteins, are too large to pass through capillary walls.

Three processes influence capillary exchange—blood pressure, diffusion, and osmotic pressure:

Blood pressure, which is created by the pumping of the heart, pushes the blood through the capillary. Blood pressure also pushes blood against a vessel's (e.g., capillary) walls.

Diffusion, as you know, is simply the movement of substances from the area of higher concentration to the area of lower concentration.

Osmotic pressure is a force caused by a difference in solute concentration on either side of a membrane.

To understand osmotic pressure, consider that water will diffuse across a membrane toward the side that has the greater concentration of solutes, and the accumulation of this water results in a pressure. The presence of the plasma proteins, and also salts to some degree, means that blood has a greater osmotic pressure than does tissue fluid. Therefore, the osmotic pressure of blood pulls water into and retains water inside a capillary.

Notice in Figure 11.6 that a capillary has an arterial end (contains arterial blood) and a venous end (contains venous blood). In between, a capillary has a midsection. We will now consider the exchange of molecules across capillary walls at each of these locations.

Arterial End of Capillary

When arterial blood enters tissue capillaries, it is bright red because the hemoglobin in red blood cells is carrying oxygen. Blood is also rich in nutrients, which are dissolved in plasma.

At the arterial end of a capillary, blood pressure, an outward force, is higher than osmotic pressure, an inward force. Pressure is measured in terms of mm Hg (mercury). In this case, blood pressure is 30 mm Hg, and osmotic pressure is 21 mm Hg. Because blood pressure is higher than osmotic pressure at the arterial end of a capillary, water and other small molecules (e.g., glucose and amino acids) filter out of a capillary at its arterial end.

Red blood cells and a large proportion of the plasma proteins generally remain in a capillary because they are too large to pass through its wall. The exit of water and other small molecules from a capillary creates tissue fluid. Therefore, tissue fluid consists of all the components of plasma, except that it contains fewer plasma proteins.

Midsection of Capillary

Diffusion takes place along the length of the capillary, as small molecules follow their concentration gradient by mov-

ing from the area of higher to the area of lower concentration. In the tissues, the area of higher concentration of oxygen and nutrients is always blood, because after these molecules have passed into tissue fluid, they are taken up and metabolized by cells. The cells use oxygen and glucose in the process of cellular respiration, and they use amino acids for protein synthesis.

As a result of metabolism, tissue cells give off carbon dioxide and other wastes. Because tissue fluid is always the area of greater concentration for waste materials, they diffuse from tissue fluid into a capillary.

Venous End of Capillary

At the venous end of the capillary, blood pressure is much reduced to only about 15 mm Hg, as shown in Figure 11.6. Blood pressure is reduced at the venous end because capillaries have a greater cross-sectional area at their venous end than their arterial end. However, there is no reduction in osmotic pressure, which remains at 21 mm Hg and is now higher than blood pressure. Therefore, water tends to diffuse into a capillary at the venous end. As water enters a capillary, it brings with it additional waste molecules. Blood that leaves the capillaries to drain into veins is deep maroon in color because red blood cells now contain reduced hemoglobin—hemoglobin that has given up its oxygen and taken on hydrogen ions.

In the end, about 85% of the water that left a capillary at the arterial end returns to it at the venous end. Therefore, retrieving fluid by means of osmotic pressure is not completely effective. The body has an auxiliary means of collecting tissue fluid; any excess usually enters lymphatic capillaries.

Lymphatic Capillaries

Lymphatic vessels are a one-way system of vessels. Notice that lymphatic capillaries have blind ends that lie near blood capillaries (Fig. 11.7). Lymphatic vessels have a structure similar to that of cardiovascular veins, except that their walls are thinner and they have more valves. The valves prevent the backward flow of lymph as lymph flows toward the thoracic cavity. Lymphatic capillaries join to form larger vessels that merge into the lymphatic ducts (see Fig. 13.1). Lymphatic ducts empty into the area of the right and left subclavian veins within the thoracic cavity.

Lymph, the fluid carried by lymphatic vessels, has the same composition as tissue fluid. Why? Because lymphatic capillaries absorb excess tissue fluid at the blood capillaries. The lymphatic system contributes to homeostasis in several ways. One way is to maintain normal blood volume and pressure by returning excess tissue fluid to the blood.

Edema

Edema is localized swelling that occurs when tissue fluid accumulates. Edema can be caused by several factors: an increase in capillary permeability; a decrease in the uptake of water at the venous end of blood capillaries due to a decrease in plasma proteins; an increase in venous pressure; or insufficient uptake

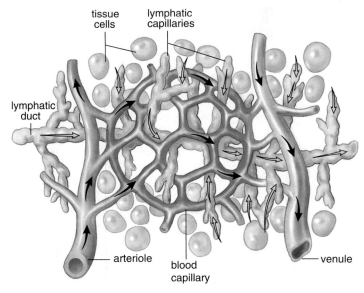

tissue cells

lymphatic capillaries

lymphatic duct

arteriole

blood capillary

venule

Figure 11.7 Lymphatic capillaries. Lymphatic capillaries lie near blood capillaries. The black arrows show the flow of blood. The yellow arrows show that lymph is formed when lymphatic capillaries take up excess tissue fluid.

of tissue fluid by the lymphatic capillaries. Another cause of edema is blocked lymphatic vessels. Surgery to remove a cancer patient's affected lymph nodes can block lymphatic drainage, resulting in a condition called **lymphedema.** One dramatic cause of lymphedema occurs after blockage of lymph nodes by a parasitic worm. An affected leg can become so large that the disease is called **elephantiasis.**

Content CHECK-UP!

10. Blood pressure is greatest at:

  a. the venous end of a capillary bed.

  b. the arterial end of a capillary bed.

  c. in the center of a capillary bed.

  d. none; it is equal throughout the capillary bed.

11. Which of the following statements about capillary exchange is false?

  a. Blood pressure pushes fluid out of the capillary to create tissue fluid.

  b. Osmotic pressure is higher than blood pressure at the venous end of the capillary.

  c. Excess tissue fluid drains into lymphatic vessels in the capillary bed.

  d. Tissue fluid has a greater osmotic pressure than blood.

12. Causes for edema include:

  a. blockage of lymphatic capillaries or lymph nodes.

  b. increase in venous pressure.

  c. increase in permeability of capillaries.

  d. all of the above.

11.5   Blood Typing and Transfusions

A **blood transfusion** is the transfer of blood from one individual into the blood of another. In order for transfusions to be safely done, it is necessary for blood to be typed so that **agglutination** (clumping of red blood cells) does not occur. Blood typing usually involves determining the ABO blood group and whether the individual is Rh^- (negative) or Rh^+ (positive).

ABO Blood Groups

ABO blood typing is based on the presence or absence of two possible antigens, called type A antigen and type B antigen, on the surface of red blood cells. Whether these antigens are present or not depends on the particular inheritance of the individual.

A person with type A antigen on the surface of the red blood cells has type A blood; one with type B blood has type B antigen on the surface of the red blood cells. What antigens would be present on the surface of red blood cells if the person has type AB blood or type O blood? Notice in Figure 11.8 that a person with type AB blood has both antigens. A person with type O blood has no AB antigens on the surface of the red blood cell, though other protein antigens will be present.

It so happens that an individual with type A blood has anti-B antibodies in the plasma; a person with type B blood has anti-A antibodies in the plasma; and a person with type O blood has both antibodies in the plasma (Fig. 11.8). These antibodies are not present at birth, but they appear over the course of several months after birth.

Blood compatibility is very important when transfusions are done. The antibodies in the plasma must not combine with the antigens on the surface of the red blood cells, or else agglutination occurs. With agglutination, anti-A antibodies have combined with type A antigens, or anti-B antibodies have combined with type B antigens, or both types of binding have occurred. Therefore, agglutination is expected if the donor has type A blood and the recipient has type B blood (Fig. 11.9). What about other combinations of blood types? Try out all other possible donors and recipients to see if agglutination will occur. Type O blood is sometimes called the *universal donor* because it has no antigens on the red blood cells, and type AB blood is sometimes called the *universal recipient* because this blood type has no antibodies in the plasma. In practice, however, there are other possible blood groups, aside from ABO blood groups, so it is necessary to physically put the donor's blood on a slide with the recipient's blood and observe whether the blood types match (no agglutination occurs) before blood can be safely given from one person to another. This is called cross-matching blood.

As explained in the Medical Focus on page 233, the use of autotransfusion technology and blood substitutes does away with the problems of matching blood types.

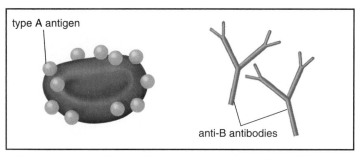

Type A blood. Red blood cells have type A surface antigens. Plasma has anti-B antibodies.

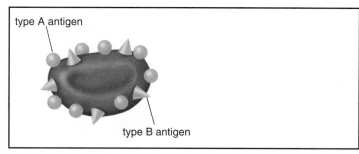

Type AB blood. Red blood cells have type A and type B surface antigens. Plasma has neither anti-A nor anti-B antibodies.

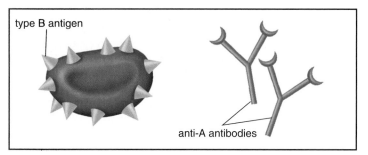

Type B blood. Red blood cells have type B surface antigens. Plasma has anti-A antibodies.

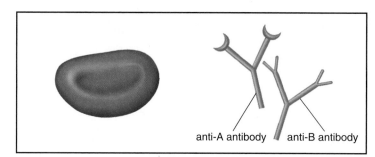

Type O blood. Red blood cells have neither type A nor type B surface antigens. Plasma has both anti-A and anti-B antibodies.

Figure 11.8 Types of blood. In the ABO system, blood type depends on the presence or absence of antigens A and B on the surface of red blood cells. In these drawings, A and B antigens are represented by different shapes on the red blood cells. The possible anti-A and anti-B antibodies in the plasma are shown for each blood type. Notice that an anti-B antibody cannot bind to an A antigen, and vice versa.

Rh Blood Groups

The designation of blood type usually also includes whether the person has or does not have the Rh factor on the red blood cell. Rh⁻ individuals normally do not have antibodies to the Rh factor, but they make them when exposed to the Rh factor.

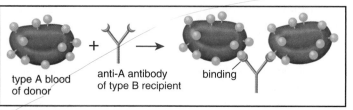

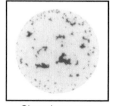

a. No agglutination — No clumping seen. Successful blood type match.

b. Agglutination — Clumping seen. Hemolysis occurs. Unsuccessful blood type match.

Figure 11.9 Blood transfusions. No agglutination **(a)** versus agglutination **(b)** is determined by whether antibodies are present that can combine with antigens.

If a mother is Rh⁻ and the father is Rh⁺, the fetus conceived can be Rh⁺. The Rh⁺ red blood cells may begin leaking across the placenta into the mother's cardiovascular system (Fig. 11.10), as placental tissues normally break down before and at birth. The presence of these Rh⁺ antigens causes the mother to produce anti-Rh antibodies. In a subsequent pregnancy with another Rh⁺ baby, the anti-Rh antibodies may cross the placenta and destroy the child's red blood cells. This is called *hemolytic disease of the newborn (HDN)* because hemolysis continues after the baby is born. Due to red blood cell destruction the baby can be severely anemic at birth. Excess bilirubin in the blood can lead to brain damage and mental retardation or even death.

The Rh problem is prevented by giving Rh⁻ women an Rh immunoglobulin injection no later than 72 hours after giving birth to an Rh⁺ child. This injection, called Rho-Gam, contains anti-Rh antibodies that attack any of the baby's red blood cells in the mother's blood before these cells can stimulate her immune system to

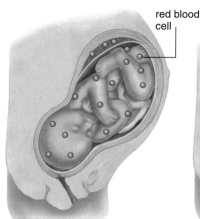

red blood
cell

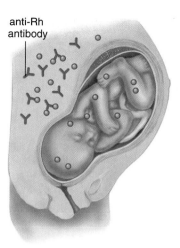

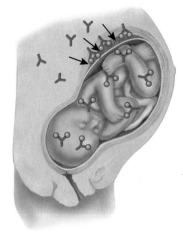

anti-Rh
antibody

Child is Rh positive;
mother is Rh negative.

Red blood cells leak
across placenta.

Mother makes anti-Rh
antibodies.

Antibodies attack Rh-positive
red blood cells in child.

Figure 11.10 Hemolytic disease of the newborn. Due to a pregnancy in which the child is Rh positive, an Rh-negative mother can begin to produce antibodies against Rh-positive red blood cells. In another a subsequent pregnancy, these antibodies can cross the placenta and cause hemolysis of an Rh-positive child's red blood cells.

produce her own antibodies. This injection is not beneficial if the woman has already begun to produce antibodies; therefore, the timing of the injection is most important.

Content CHECK-UP!

13. Which of the following correctly describes type AB blood?

   a. A antigen and B antigen on the red cell membrane, anti-A and anti-B antibodies in plasma

   b. A antigen and B antigen on the red cell membrane, no antibodies in plasma

   c. A antigen on the red cell membrane, anti-B antibodies in plasma

   d. B antigen on the red cell membrane, anti-A antibodies in plasma

14. Which of the following transfusions would most likely be safe to administer?

   a. type A blood to a type B blood recipient

   b. type AB blood to a type B blood recipient

   c. type B blood to a type AB recipient

   d. type AB blood to a type O recipient

15. Hemolytic disease of the newborn occurs in which situation?

   a. Rh$^+$ mother, Rh$^-$ father, Rh$^+$ fetus

   b. Rh$^-$ mother, Rh$^-$ father, Rh$^+$ fetus

   c. Rh$^-$ mother, Rh$^+$ father, Rh$^+$ fetus

   d. Rh$^+$ mother, Rh$^-$ father, Rh$^-$ fetus

11.6 Effects of Aging

Anemias, leukemias, and clotting disorders increase in frequency with age. As with other disorders, good health habits can help prevent these conditions from appearing.

Iron deficiency anemia most frequently results from an iron-deficient diet, but it can also result from a gradual and sometimes undetectable loss of blood. It is vital that the bleeding source be discovered as soon as possible. For example, iron deficiency anemia is often a first symptom of colon cancer in its early stage. The blood is often not visible in a person's stool because it has oxidized and is concealed in the stool. An *occult blood test* (test for oxidized blood present in the stool) can detect this source of blood loss.

Pernicious anemia most often signals that the digestive tract is unable to absorb enough vitamin B_{12}, but it can also be due to a diet deficient in B_{12}. Diets that completely exclude animal-derived protein sources (called *vegan* diets), are often B_{12} deficient. A vitamin supplement containing B_{12} will prevent pernicious anemia in vegan dieters.

Leukemia is a form of cancer that generally increases in frequency with age because of both intrinsic (genetic) and extrinsic (environmental) reasons.

Thromboembolism, a clotting disorder, may be associated with the progressive development of atherosclerosis in an elderly person. When arteries develop plaque (see Fig. 12B, p. 253), thromboembolism often follows. For many people, atherosclerosis can be controlled by diet and exercise, as discussed in the Chapter 12 Medical Focus, "Preventing Cardiovascular Disease."

In cases of violent crime such as homicide, assault, or sexual assault, the presence of blood at the crime scene is often the most important evidence. Blood evidence can tell an entire story about the crime—how a victim was attacked, whether the victim fought back or tried to escape, the type of weapon used, and perhaps even the identity of the attacker. Extensive research on the chemistry and properties of blood evidence, along with detailed examination of the crime scene, can enable a *forensic serologist* to help solve violent crime cases. (Forensic serologists are scientists trained in the examination of blood and body fluid at a crime scene.)

Upon arrival at the scene, investigators must first carry out a careful visual inspection for the presence of blood. Because blood hemoglobin is a protein molecule (globin) containing an iron pigment (heme), its stains are very difficult to remove completely from contaminated areas. As a result, a criminal who tries to clean up blood will almost always leave traces behind. Investigators sample between floorboards, under baseboards, in carpet padding, and on all surface areas. Tests can be done to confirm that a stain is due to human blood and not some other reddish-brown chemical. One test involves spraying the stain with luminol, a chemical that binds with blood and then glows in the dark. The Takayama and Teichmann tests cause crystals to form from hemoglobin, confirming the presence of blood. Furthermore, by testing with anti-human antibodies (protein molecules from rabbits), researchers can prove that a sample is in fact human in origin.

Careful documentation and analysis of the location and shape of any drops, spatters, or smears follows next. This type of evidence is carefully recorded and can also help investigators to piece together the events of the crime—the position of the suspect and the victim, whether a struggle occurred, if the victim moved or was dragged, etc. For example, research has shown that the shape of blood drops can yield useful information about a crime. If sprayed from a short distance, a blood drop will form a perfect circle on a wall or floor; however, blood that travels a greater distance will form a "sunburst" pattern. Droplet shape also yields information about the speed of travel from a person's body. When a droplet travels slowly from the body to an object (for example, a wall), it will form a circular drop. Blood that travels at a fast speed from the body to the object will form an elongated, "teardrop" shape. Blood flying in a trail of droplets from a knife or a club will leave a trail of spatters called a *castoff pattern*. Smeared blood may indicate that the body has been dragged, or that the victim attempted to escape.

If the suspect was wounded in a struggle, the suspect's blood may be present at the crime scene along with that of the victim. Blood chemistry is as individual as a fingerprint, and can positively confirm the identity of the suspect. ABO and Rh blood typing, studies of unique blood proteins, and DNA analysis (see pages 239–240) are carried out on all blood samples. Other body fluids, if present, may also need to be tested because approximately 80% of the human population are *secretors*. Secretors are individuals whose blood proteins and enzymes are present in saliva, tears, urine, semen, and skin, as well as in their blood. The data from these tests identify the source of blood samples, which can be statistically narrowed down to one person in several billion. Thus, for all practical purposes, a positive match against a suspect's blood can be made. Evidence such as this can prove that a suspect was at the crime scene and result in a conviction. Just as important, blood analysis may serve to acquit an innocent person.

SELECTED NEW TERMS

Basic Key Terms

agglutination (uh-glu"tǐ-na'shun), p. 239

agranular leukocyte
    (a-gran'yū-ler lu'ko-sīt), p. 232

albumin (al-byū'min), p. 228

antibody (an'tǐ-bod"e), p. 234

antigen (an'tǐ-jen), p. 234

basophil (ba'so-fil), p. 232

bile pigments (bīl pig'-měnts), p. 231

cerebrovascular accident (sěr'ə-brō-văs'kyl
    -ər ăk'sǐ-dənt), p. 236

coagulation (ko-ag"yū-la'shun), p. 236

edema (ě-de'muh), p. 238

eosinophil (e"o-sin'o-fil), p. 232

erythrocyte (ě-rith"ro-sīt'), p. 230

erythropoietin (ě-rith"ro-poy'ě-tin), p. 230

fibrin (fi'brin), p. 236

fibrinogen (fi-brin'o-jen), p. 236

formed element (formd el'ě-ment), p. 228

globulin (glob'u-lin), p. 228

granular leukocyte (gran'u-ler lu'ko-sīt), p. 232

hematocrit (he-mat'o-krit), p. 228

hematopoiesis (hě'mə-tō-poi-ē'sǐs), p. 230

hemoglobin (he"mo-glo'bin), p. 230

hemolysis (he-mol'ǐ-sis), p. 232

hemostasis (hě'mə-stā'sǐs), p. 234

leukocytes (loo'kə-sīt's), p. 232

lipoproteins (lǐp'ō-prō'tēn's), p. 230

lymph (limf), p. 238

lymphatic vessel (lim-fat'ik ves'l), p. 238

lymphocyte (lim'fo-sīt), p. 234

macrophage, (mak'rō-fāj), p. 234

megakaryocyte(meg'uh-kār'e-o-sīt), p. 234

monocyte (mon'o-sīt), p. 234

multipotent stem cell (mǔl'tǐ-pōt'nt stěm sěl),
    p. 230

neutrophil (nu'tro-fil), p. 232

osmotic pressure (oz-mot'ik presh'ur), p. 228

pathogen (path'o-jen), p. 228

Clinical Key Terms

SUMMARY

11.1 The Composition and Functions of Blood

A. Blood, which is composed of formed elements and plasma, has several functions. It transports hormones, oxygen, and nutrients to the cells and carbon dioxide and other wastes away from the cells. It fights infections. It regulates body temperature, and keeps the pH of body fluids within normal limits. All of these functions help maintain homeostasis.

B. Small organic molecules such as glucose and amino acids are dissolved in plasma and serve as nutrients for cells; urea is a waste product. Large organic molecules include the plasma proteins.

C. Plasma is mostly water (92%) and the plasma proteins (8%). The plasma proteins, most of which are produced by the liver, occur in three categories: albumins, globulins, and fibrinogen. The plasma proteins maintain osmotic pressure, help regulate pH, and transport molecules. Some plasma proteins have specific functions: The gamma globulins, which are antibodies produced by B lymphocytes, function in immunity; fibrinogen and prothrombin are necessary to blood clotting.

11.2 The Blood Cells

All blood cells, including red blood cells, are produced within red bone marrow from stem cells, which are ever capable of dividing and producing new cells.

A. Red blood cells are small, biconcave disks that lack a nucleus. They contain hemoglobin, the respiratory pigment, which combines with oxygen and transports it to the tissues. Red blood cells live about 120 days and are destroyed in the liver and spleen when they are old or abnormal. The production of red blood cells is controlled by the oxygen concentration of the blood. When the oxygen concentration decreases, the kidneys increase their production of erythropoietin, and more red blood cells are produced.

B. White blood cells are larger than red blood cells, have a nucleus, and are translucent unless stained. Like red blood cells, they are produced in the red bone marrow. White blood cells are divided into the granular leukocytes and the agranular leukocytes. The granular leukocytes have conspicuous granules; in eosinophils, granules are red when stained with eosin, and in basophils, granules are blue when stained with a basic dye. In neutrophils, some of the granules take up eosin, and others take up the basic dye, giving them a lilac color. Neutrophils are the most plentiful of the white blood cells, and they are able to phagocytize pathogens. Many neutrophils die within a few days when they are fighting an infection. The agranulocytes include the lymphocytes and the monocytes, which function in specific immunity. On occasion, the monocytes become large phagocytic cells of great significance. They engulf worn-out red blood cells and pathogens at a ferocious rate.

11.3 Platelets and Hemostasis

A. The extremely plentiful platelets result from fragmentation of megakaryocytes.

B. The three events of hemostasis are vascular spasm, platelet plug formation, and coagulation. Extrinsic and intrinsic pathways that activate the clotting mechanism both cause formation of prothrombin activator, which breaks down prothrombin to thrombin. Thrombin changes fibrinogen to fibrin threads, entrapping cells. The fluid that escapes from a clot is called serum and consists of plasma minus fibrinogen and prothrombin.

11.4 Capillary Exchange

A. At the arterial end of a cardiovascular capillary, blood pressure is greater than osmotic pressure; therefore, water leaves the capillary. In the midsection, oxygen and nutrients diffuse out of the capillary, while carbon dioxide and other wastes diffuse into the capillary. (In a pulmonary capillary, gas exchange is reversed as CO_2 leaves capillary blood and oxygen enters.) At the venous end, osmotic pressure created by the presence of proteins exceeds blood pressure, causing water to enter the capillary.

B. Retrieving fluid by means of osmotic pressure is not completely effective. There is always some fluid that is not picked up at the venous end of the cardiovascular capillary. This excess tissue fluid enters the lymphatic capillaries. Lymph is tissue fluid contained within lymphatic vessels. The lymphatic system is a one-way system, and lymph is returned to the blood by way of a cardiovascular vein.

11.5 **Blood Typing and Transfusions**
A. ABO typing is the most common blood typing system used. Type A, type B, both type A and B, or no antigens can be on the surface of red blood cells. In the plasma, there are two possible antibodies: anti-A or anti-B. If the corresponding antigen and antibody are put together during a transfusion, agglutination occurs. Therefore, it is necessary to determine an individual's blood type before a transfusion is given.

B. Another important antigen is the Rh antigen. This particular antigen must also be considered in transfusing blood, and it is important during pregnancy because an Rh$^-$ mother may form antibodies to the Rh antigen when giving birth to an Rh$^+$ child. These antibodies can cross the placenta and destroy the red blood cells of any subsequent Rh$^+$ child.

11.6 **Effects of Aging**
As we age, anemias, leukemias, and clotting disorders increase in frequency.

STUDY QUESTIONS

1. Name the two main components of blood, and describe the functions of blood. (p. 228)
2. List and discuss the major components of plasma. Name several plasma proteins, and give a function for each. (p. 228)
3. What is hemoglobin, and how does it function? (p. 230)
4. Describe the life cycle of red blood cells, and tell how the production of red blood cells is regulated. (pp. 230–231)
5. Name the five types of white blood cells; describe the structure and give a function for each type. (pp. 232, 234)
6. Name the steps that take place when blood clots. Which substances are present in blood at all times, and which appear during the clotting process? (pp. 234, 236)
7. What forces operate to facilitate exchange of molecules across the capillary wall? (pp. 236–239)
8. What are the four ABO blood types? For each, state the antigen(s) on the red blood cells and the antibody(ies) in the plasma. (p. 239)
9. Explain why a person with type O blood cannot receive a transfusion of type A blood. (p. 239)
10. Problems can arise if the mother is which Rh type and the father is which Rh type? Explain why this is so. (p. 240)

LEARNING OBJECTIVE QUESTIONS

I. Fill in the blanks.
1. The liquid part of blood is called _____ .
2. Red blood cells carry _____, and white blood cells _____.
3. Hemoglobin that is carrying oxygen is called _____.
4. Human red blood cells lack a _____ and only live about _____ days.
5. The most common granular leukocyte is the _____, a phagocytic white blood cell.
6. B lymphocytes produce _____, and T lymphocytes attack and _____ pathogens.
7. At a capillary, _____, _____, and _____ leave the arterial end, and _____ and _____ enter the venous end.
8. When a blood clot occurs, fibrinogen has been converted to _____ threads.
9. AB blood has the antigens _____ and _____ on the red blood cells and _____ of these antibodies in the plasma.
10. Hemolytic disease of the newborn can occur when the mother is _____ and the father is _____.

II. Match the terms in the key to the descriptions in questions 11–14.
Key:
 a. hematocrit
 b. red blood cell count
 c. white blood cell count
 d. hemoglobin

11. 5,000 to 11,000 per cubic millimeter
12. 4 to 6 million per cubic millimeter in males
13. Just under 45% of blood volume
14. 200 million molecules in one red blood cell

15. Label the following diagram.

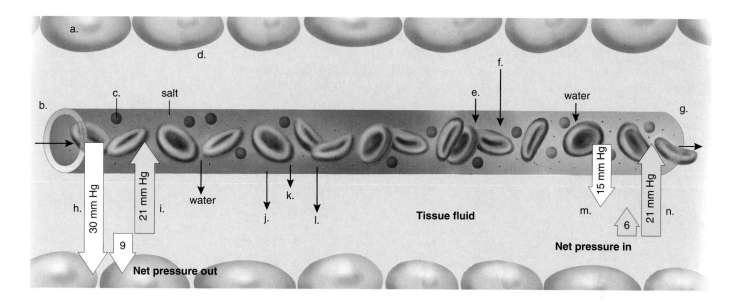

MEDICAL TERMINOLOGY EXERCISE

Consult Appendix B for help in pronouncing and analyzing the meaning of the terms that follow.

1. hematemesis (hem″uh-tem′ĕ-sis)
2. erythrocytometry (ĕ-rith″ro-si-tom′ĕ-tre)
3. leukocytogenesis (lu″ko-si″to-jen′ĕ-sis)
4. hemophobia (he″mo-fo′be-uh)

5. afibrinogenemia (uh-fi″brin-o-jĕ-ne′me-uh)
6. lymphosarcoma (lim″fo-sar-ko′muh)
7. phagocytosis (fag″o-si-to′sis)
8. phlebotomy (flĕ-bot′o-me)
9. hemocytoblast (he′mo-si′to-blast)
10. megaloblastic anemia (meg′uh-lo-blas′tik uh-ne′me-uh)

11. microcytic hypochromic anemia (mi′kro-sit′ik hi″po-kro′mik uh-ne′me-uh)
12. hematology (he′muh-tol′o-je)
13. lymphedema (limf′uh-de′muh)
14. antithrombin (an″te-throm″bin)

WEBSITE LINK

Visit this text's website at http://www.mhhe.com/maderap6 for additional quizzes, interactive learning exercises, and other study tools.

Heart valves control blood flow through the heart. Fibrous chordae tendinae attach the valves to the muscular heart wall.

The Cardiovascular System

Chapter Outline & Learning Objectives

After you have studied this chapter, you should be able to:

Chapter 11 described how oxygen and nutrients are exchanged for carbon dioxide and other waste molecules at tissue capillaries (see Fig. 11.6). We emphasized that cells are dependent on the functioning of capillaries for this purpose. In this chapter, we will study how blood is moved to and from tissue (systemic) capillaries and also to and from lung (pulmonary) capillaries where oxygen enters and carbon dioxide exits the blood.

The cardiovascular system consists of two components: (1) the heart, which pumps blood so that it flows to tissue capillaries and lung capillaries, and (2) the blood vessels through which the blood flows. As you can see in Figure 12.1, the cardiovascular system is divided into two functional systems. The right side of the heart and its blood vessels form the **pulmonary circuit,** which pumps blood to the lungs. The left side of the heart and its vessels form the **systemic circuit,** which supplies blood to the entire body.

In this chapter, we will first study the anatomy and physiology of the heart and of the blood vessels. Then, we will take a look at various branches of the circulatory system. A crucial function of circulation is to connect the body's trillions of cells to the organs of exchange: lungs, small intestine, and kidney. In the lungs, oxygen enters and carbon dioxide exits the blood. The small intestine absorbs nutrient molecules into the blood, and the kidneys allow metabolic wastes to exit the blood. The circulatory system assures homeostasis by carrying nutrients and metabolites to and from these organs.

12.1 Anatomy of the Heart

The heart is located in the thoracic cavity between the lungs within the mediastinum. It is a hollow, cone-shaped, muscular organ. To approximate the size of your heart, make a fist and then clasp the fist with your opposite hand. Figure 12.1 shows that the base (the widest part) of the heart is superior to its apex (the pointed tip), which rests on the diaphragm. Also, the heart is on a slant; the base is directed toward the right shoulder, and the apex points to the left hip. The base is deep to the second rib, and the apex is at the level of the fifth intercostal space.

As the heart pumps the blood through the pulmonary and systemic vessels, it performs these functions:

1. keeps O_2-poor blood separate from O_2-rich blood;
2. keeps the blood flowing in one direction—blood flows away from and then back to the heart in each circuit;
3. creates blood pressure, which moves the blood through the circuits;
4. regulates the blood supply based on the current needs of the body.

The Wall and Coverings of the Heart

In Chapter 4, we mentioned that the heart is enclosed by a two-layered serous membrane called the **pericardium.** One layer, the **visceral** (meaning "organ") **pericardium,** is

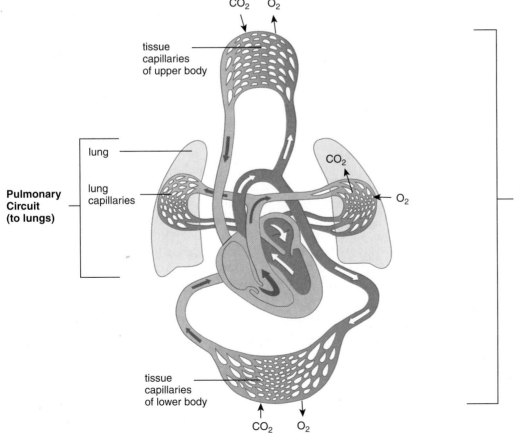

Figure 12.1 Cardiovascular system. The right side of the heart pumps blood through vessels of the pulmonary circuit. The left side of the heart pumps blood through vessels of the systemic circuit. Gas exchange occurs as blood passes through lung (pulmonary) capillaries. Gas exchange and nutrient-for-waste exchange occur as blood passes through tissue (systemic) capillaries. In this illustration, red vessels carry O_2-rich blood, and blue vessels carry O_2-poor blood.

CO₂   O₂

tissue capillaries of upper body

lung

lung capillaries

Pulmonary Circuit (to lungs)

CO₂

O₂

Systemic Circuit (to body)

tissue capillaries of lower body

CO₂   O₂

considered part of the heart wall; it forms the **epicardium,** the outer surface of the heart. The **myocardium** is the thickest part of the heart wall and is made up of cardiac muscle (see Fig. 4.15). When cardiac muscle fibers contract, the heart beats. The inner **endocardium** is composed of simple squamous epithelium. Endothelium not only lines the heart but it also continues into and lines the blood vessels. Its smooth nature helps prevent blood from clotting unnecessarily.

The pericardial cavity develops when the visceral pericardium doubles back to become the parietal (meaning "wall") pericardium, the other serous layer. The two serous membranes (epicardium and parietal pericardium) secrete pericardial fluid (a fluid similar to plasma). The pericardial fluid reduces friction as the heart beats. The parietal pericardium is fused to the outermost fibrous pericardium (Fig. 12.2). The fibrous pericardium is a layer of fibrous connective tissue that adheres to the great blood vessels at the heart's base and anchors the heart to the wall of the mediastinum. The coverings of the heart protect the heart, confine it to its location, and prevent it from overfilling, while still allowing the heart to contract and carry out its function of pumping the blood.

A layer of the heart can become inflamed due to infection, cancer, injury, or a complication of surgery. The suffix "itis" added to the name of a heart condition tells which layer is affected. For example, pericarditis refers to inflammation of the pericardium, and endocarditis refers to inflammation of the endocardium.

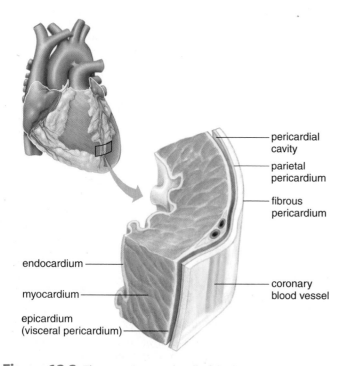

pericardial cavity

parietal pericardium

fibrous pericardium

endocardium

myocardium

epicardium (visceral pericardium)

coronary blood vessel

Figure 12.2 The coverings and wall of the heart. The heart wall has three layers, from deep to superficial: endocardium, myocardium, and epicardium.

Chambers of the Heart

The heart has four hollow chambers: two superior **atria** (sing., **atrium**) and two inferior **ventricles** (Fig. 12.3). Each atrium has a wrinkled anterior pouch called an auricle. Internally, the atria are separated by the **interatrial septum,** and the ventricles are separated by the **interventricular septum.** Therefore, the heart's pulmonary circuit (its right side) is completely separated from its systemic circuit (the left side) by the septum.

The thickness of each chamber's myocardium is suited to its function. The atria have thin walls and each pumps blood into the ventricle below. The ventricles are thicker and they pump blood into blood vessels that travel to other parts of the body. The thinner myocardium of the right ventricle is suited for pumping blood to the lungs, which are nearby in the thoracic cavity. The left ventricle has a thicker wall than the right ventricle. Thicker myocardium enables the left ventricle to pump its blood to all other parts of the body.

Right Atrium

At its posterior wall, the **right atrium** receives O_2-poor blood from three veins: the *superior vena cava, the coronary sinus,* and the *inferior vena cava.* Venous blood passes from the right atrium into the right ventricle through an **atrioventricular (AV) valve.** This valve, like the other heart valves, directs the flow of blood and prevents any backflow. The AV valve on the right side of the heart is specifically called the **tricuspid valve** because it has three *cusps,* or flaps.

Right Ventricle

In the **right ventricle**, the cusps of the tricuspid valve are connected to fibrous cords, called the **chordae tendineae** (meaning "heart strings"). The chordae tendineae in turn are connected to the **papillary muscles,** which are conical extensions of the myocardium.

Blood from the right ventricle passes through a **semilunar valve** into the *pulmonary trunk.* Semilunar valves are so called because their cusps are thought to resemble half-moons. This particular semilunar valve, called the **pulmonary semilunar valve,** prevents blood from flowing back into the right ventricle.

Note in Figure 12.3 that the pulmonary trunk divides into the left and right pulmonary arteries. For help in remembering how blood flows through the heart, trace the path of O_2-poor blood from the vena cava to the pulmonary arteries that take blood to the lungs (see Figs. 12.1 and 12.3*b*).

Left Atrium

At its posterior wall, the **left atrium** receives O_2-rich blood from four *pulmonary veins.* Two pulmonary veins come from each lung. Blood passes from the left atrium into the left ventricle through an AV valve. The AV valve on the left side is specifically called the **bicuspid valve** because it has two cusps. (In the United States, the bicuspid valve is more commonly

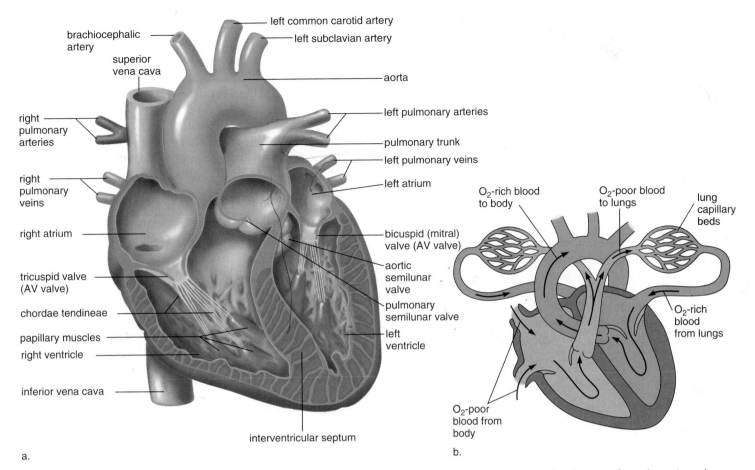

brachiocephalic artery
left common carotid artery
left subclavian artery
superior vena cava
aorta
right pulmonary arteries
left pulmonary arteries
pulmonary trunk
left pulmonary veins
right pulmonary veins
left atrium
right atrium
bicuspid (mitral) valve (AV valve)
aortic semilunar valve
tricuspid valve (AV valve)
pulmonary semilunar valve
chordae tendineae
left ventricle
papillary muscles
right ventricle
inferior vena cava
interventricular septum

a.

O_2-rich blood to body
O_2-poor blood to lungs
lung capillary beds
O_2-rich blood from lungs
O_2-poor blood from body

b.

Figure 12.3 Internal heart anatomy. **a.** The heart has four valves. The two atrioventricular valves allow blood to pass from the atria to the ventricles, and the two semilunar valves allow blood to pass out of the heart. **b.** A diagrammatic representation of the heart allows you to trace the path of the blood through the heart.

referred to as the **mitral valve,** so called because the valve is similar in shape to a bishop's hat, or mitre.)

Left Ventricle

The **left ventricle** forms the apex of the heart. The papillary muscles in the left ventricle are quite large, and the chordae tendineae attached to the AV valve are thicker and stronger than those in the right ventricle. As mentioned, the AV valve on the left side is also called the bicuspid (or mitral) valve.

Blood passes from the left ventricle through a semilunar valve into the aorta. This semilunar valve is appropriately called the **aortic semilunar valve.** The semilunar cusps of this valve are larger and thicker than those of the pulmonary semilunar valve.

Just beyond the aortic semilunar valve lie the first branches from the aorta. These are the **coronary arteries**—blood vessels that lie on and nourish the heart itself. The rest of the blood stays in the aorta, which continues as the arch of the aorta and then the descending aorta.

To make sure you understand this discussion, trace the path of O_2-rich blood through the heart, from the pulmonary veins to the aorta (see Figs. 12.1 and 12.3b).

Operation of the Heart Valves and Heart Sounds

Let's take a look at how the valves of the heart operate to direct a one-way flow of blood from the atria to the ventricles to the arteries. The AV valves (tricuspid and mitral valve) are normally open. When a ventricle contracts, however, the increasing pressure of the blood inside the ventricle forces the cusps of the AV valve to slam shut. The force of the blood can be compared to a strong wind that can blow a door (the valve cusps) shut. However, when the ventricle contracts, the papillary muscles also contract, causing the chordae tendinae to tighten and pull on the valve. Thus, AV valves in the normal heart are prevented from inverting back up into the atrium.

The semilunar valves (pulmonary and aortic) are normally closed. However, the contraction of the ventricles pushes blood at high pressure against the valve cusps, forcing the valves open. Then, when the ventricle relaxes, the blood in the artery pushes backward, closing the valve once again.

A heartbeat produces the familiar "LUB-DUP" sounds as the chambers contract and the valves close. The first heart sound, "lub," is heard when the ventricles begin to contract and the atrioventricular (tricuspid and mitral) valves close. This sound lasts the longest and has a lower pitch. The second heart sound, "dup," is heard when the relaxation of the ventricles allows the semilunar (pulmonary and aortic) valves to close.

Like mechanical valves, the heart valves are sometimes leaky. When valves don't close properly, there is a backflow of blood. Heart murmurs, which are clicking or swishing sounds commonly heard after the first heart sound ("lub"), are often due to ineffective, leaky valves. These leaky valves allow blood to pass back into the atria after the atrioventricular valves have closed, or back into the ventricles after the semilunar valves have closed. Most often, the valves of the systemic circulation—bicuspid (mitral) valve and the aortic semilunar valve—become leaky. A trained physician or health professional can diagnose heart murmurs from their sound and timing. A person may be born with a valve deformity (called a congenital valve defect), or the valve can be damaged by infection. When a person has rheumatic fever, for example, a bacterial infection that began in the throat has spread throughout the body. The bacteria attack connective tissue in the heart valves as well as other organs. Defective heart valves can be repaired surgically, or replaced with a synthetic valve or one taken from a pig's heart.

Coronary Circulation

Cardiac muscle fibers and the other types of cells in the wall of the heart are not nourished by the blood in the chambers; diffusion of oxygen and nutrients from this blood to all the cells that make up the heart would be too slow. Instead, these cells receive nutrients and rid themselves of wastes at capillaries embedded in the heart wall.

As mentioned previously, the right and left coronary arteries branch from the aorta just beyond the aortic semilunar valve (Fig. 12.4). Each of these arteries branches and then rebranch until the heart is encircled by small arterial blood vessels. Some of these join so that there are several routes to reach any particular capillary bed in the heart. Alternate routes are helpful if an obstruction should occur along the path of blood reaching cardiac muscle cells. The Medical Focus on p. 252 describes the cause, and some treatment options, for obstruction of the coronary arteries.

After blood has passed through cardiac capillaries, it is taken up by vessels that join to form veins. The coronary veins are specifically called **cardiac veins**. The cardiac veins enter a coronary sinus, which is essentially a thin-walled vein. The coronary sinus enters the right ventricle.

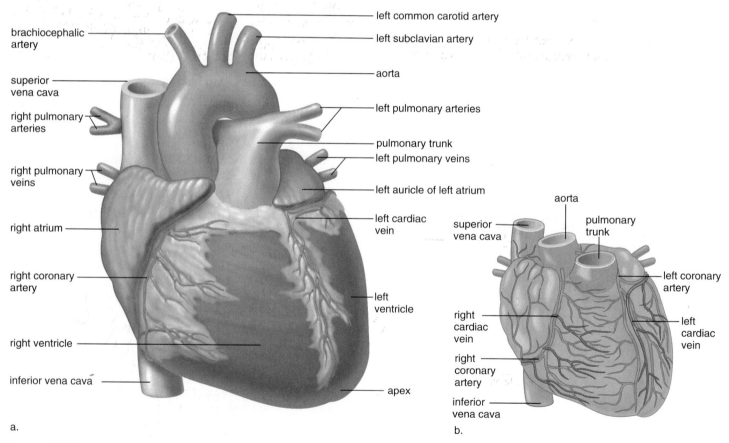

Figure 12.4 Anterior view of exterior heart anatomy. **a.** The great vessels (venae cavae, pulmonary trunk, pulmonary arteries, and aorta) are attached to the base of the heart. The right ventricle forms most of the anterior surface of the heart, and the left ventricle forms most of the posterior surface. **b.** The coronary arteries and cardiac veins pervade cardiac muscle. The coronary arteries bring oxygen and nutrients to cardiac cells, which derive little from the blood coursing through the heart.

12.2 Physiology of the Heart

The physiology of the heart pertains to its pumping action—that is, the heartbeat. It is estimated that the heart beats two-and-a-half billion times in a lifetime, continuously recycling some 5 liters (L) of blood to keep us alive. In this section, we will consider what causes the heartbeat, what it consists of, and its consequences.

Conduction System of the Heart

The **conduction system of the heart** is a route of specialized cardiac muscle fibers that initiate and stimulate contraction of the atria and ventricles. The conduction system is said to be *intrinsic*, meaning that the heart beats automatically without the need for external nervous stimulation. The conduction system coordinates the contraction of the atria and ventricles so that the heart is an effective pump. Without this conduction system, the atria and ventricles would contract at different rates.

Nodal Tissue

The heartbeat is controlled by nodal tissue, which has both muscular and nervous characteristics. This unique type of cardiac muscle is located in two regions of the heart: The **SA (sinoatrial) node** is located in the upper posterior wall of the right atrium. The **AV (atrioventricular) node** is located in the base of the right atrium very near the interatrial septum (Fig. 12.5).

The SA node initiates the heartbeat and automatically sends out an excitation impulse every 0.85 second. The SA node normally functions as the pacemaker for the entire heart because its intrinsic rate is the fastest in the system. From the SA node, impulses spread out over the atria, causing them to contract.

When the impulses reach the AV node, there is a slight delay that allows the atria to finish their contraction before the ventricles begin their contraction. The signal for the ventricles to contract travels from the AV node through the two branches of the **atrioventricular bundle (AV bundle)** before reaching the numerous and smaller **Purkinje fibers.** The AV bundle,

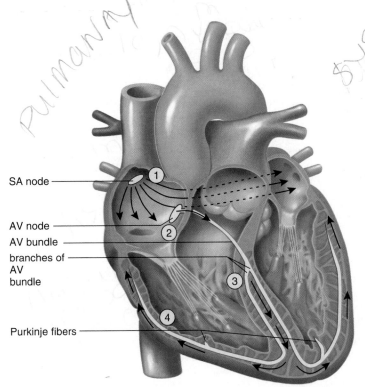

1. Stimulus originates in the SA node and travels across the walls of the atria, causing them to contract.

2. Stimulus arrives at the AV node and travels along the AV bundle.

3. Stimulus descends to the apex of the heart through the bundle branches.

4. After stimulus reaches the Purkinje fibers, the ventricles contract.

SA node

AV node

AV bundle

branches of AV bundle

Purkinje fibers

Figure 12.5 Conduction system of the heart. ① The SA node sends out a stimulus, which causes the atria to contract. ② When this stimulus reaches the AV node, it passes through the atrial wall to the AV bundle. ③ Impulses pass down the two branches of the atrioventricular bundle to the Purkinje fibers, and ④ thereafter, the ventricles contract.

Arteriosclerosis, Atherosclerosis, and Coronary Artery Disease

The number-one killer of men and women in North America and Western Europe is arteriosclerosis. It is the leading cause of heart attack and stroke. **Arteriosclerosis** is the generalized term for abnormal thickening and hardening of the walls of the arteries over time. The most common form of arteriosclerosis is **atherosclerosis.** Scientists agree that atherosclerosis begins with injury to the arterial wall. Research has suggested several possible causes for arteries to be injured: smoking, high blood pressure (called **hypertension**), and elevated levels of blood lipids, cholesterol, and homocysteine (a by-product of protein metabolism). Diabetics (especially those with type II or non-insulin dependent diabetes mellitus) are at increased risk for atherosclerosis, probably because their disease causes high levels of blood lipids and cholesterol.

Research also indicates that low-level bacterial or viral infection that spreads to the blood may cause the injury that starts the process of atherosclerosis. This infection may originate with gum disease, or it can be caused by a bacterium called *Helicobacter pylori* (the microbe that also causes stomach ulcers). Antibodies specific to these microbes are found in people with atherosclerosis. In addition, a protein in the blood called *C*-reactive protein, or CRP, is an important piece of evidence suggesting that atherosclerosis may be caused by infection. For example, CRP levels rise in your blood if you suffer from a cold or are recovering from a wound. High blood levels of CRP in an otherwise healthy person imply that the arteries have been damaged. Indeed, recent studies show that people with the highest blood levels of CRP have double the risk of heart attack.

Once the arterial wall is injured, the body's defense mechanisms respond. White blood cells called **macrophages** invade the injured area, where they will stick to the endothelial wall. These macrophages ingest low-density lipoprotein, or LDL (physicians often call this lipid "bad cholesterol"). Macrophages filled with LDL are called *foam cells* (because mixing fat with the cell's watery cytoplasm creates a foamy appearance). A collection of foam cells creates a **fatty streak.** Sadly, post-mortem studies on the arteries of young people have shown that these fatty streaks begin to develop during the early teenage years.

Over time, smooth muscle cells from the artery's muscular layer migrate to cover the fatty streak. Finally, fibroblasts and scar tissue cover the smooth muscle cells. Calcium ions invade the tissue, causing it to harden into an atherosclerotic **plaque.** Atherosclerotic plaques can grow so large as to completely block blood flow through an artery, causing the tissue supplied by the artery to die. Because the surface of a plaque is very rough compared to the normally smooth blood vessel lining, the plaque may also trigger the clotting mechanism and cause the formation of a stationary blood clot, or **thrombus,** inside a blood vessel. Thrombi may also form if the surface of the plaque ulcerates (cracks open and bleeds). As mentioned in Chapter 11, **thromboembolism** is present when a blood clot breaks away from its place of origin and is carried to a new location.

Coronary artery disease is the term for atherosclerosis of the coronary arteries (Fig. 12A). If the coronary artery is partially occluded (blocked) by atherosclerosis, the individual may suffer from **ischemic heart disease.** Although enough oxygen may normally reach the resting heart, the person experiences insufficiency during exercise or stress. This may lead to **angina pectoris,** chest pain that is often accompanied by a radiating pain in the left arm. Angina pain is a warning sign that blood flow to the heart is reduced, and this warning must not be ignored. Should the coronary artery become completely blocked due to atherosclerotic plaque or thrombus, a portion of the heart dies due to lack of oxygen. Dead tissue is called an **infarct,** and therefore, a heart attack is termed a **myocardial infarction.**

Surgical Repair

Two surgical procedures can re-open occluded coronary arteries. In balloon angioplasty, a plastic tube is threaded into an artery of an arm or leg and is guided through a major blood vessel toward the heart. Once the tube reaches a blockage, a balloon attached to the end of the tube can be inflated to break up a blood clot or to open up a vessel clogged with plaque (Fig. 12B). In some cases, a small metal-mesh cylinder called a vascular **stent** is inserted into a blood vessel during balloon angioplasty. The stent holds the vessels open and decreases the risk of future occlusion. Stent devices currently in use have built-in medication, which is slowly released in the artery. These medications prevent the formation of blood clots and additional scar tissue, thus helping to keep the stent open and blood flowing.

In a **coronary bypass operation,** a portion of a blood vessel from another part of the body (usually one of the mammary arteries from the chest), is sutured from the aorta to the coronary artery, past the point of obstruction. This procedure allows blood to flow normally again from the aorta to the heart muscle. Figure 12C shows a triple bypass in which three blood vessels have been used to allow blood to flow freely from the aorta to cardiac muscle by way of the coronary artery.

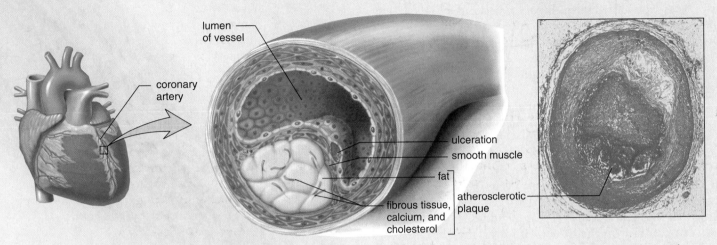

Figure 12A Coronary artery disease. Atherosclerosis begins with injury to an artery; a fatty streak develops at the site and smooth muscle grows over the lesion. Fibrous tissue and ionic calcium enlarge and stiffen the atherosclerotic plaque. When plaque is present in a coronary artery, restricted blood flow may result in a heart attack.

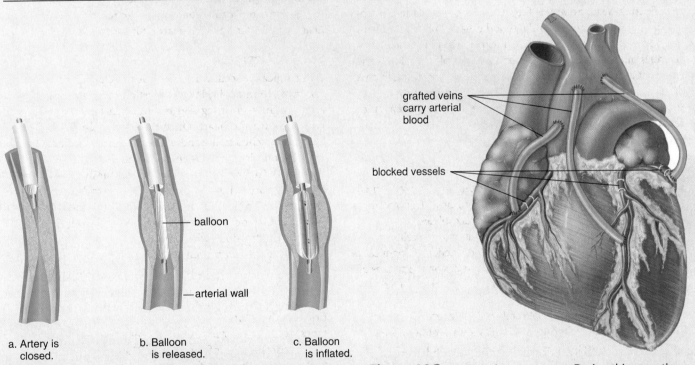

a. Artery is closed.

b. Balloon is released.

c. Balloon is inflated.

Figure 12B Balloon angioplasty. As described in the text, a balloon inserted in an artery can be inflated to open up a clogged coronary blood vessel.

Figure 12C Coronary bypass surgery. During this operation, the surgeon grafts segments of another vessel between the aorta and the coronary vessels, bypassing areas of blockage.

its branches, and the Purkinje fibers consist of specialized cardiac muscle fibers that efficiently spread an electrical signal throughout the ventricles. Recall from Chapter 4 (p. 73) that cardiac muscle cells are connected to each other by specialized **gap junctions** called intercalated discs. These specialized intercellular connections allow electrical current to flow from cell to cell. Once stimulated electrically, the ventricular muscle contracts purposefully to pump blood.

The SA node **pacemaker** usually keeps the heartbeat regular. If the SA node fails to work properly, the ventricles still beat due to impulses generated by the AV node. But the beat is slower (40 to 60 beats per minute). To correct this condition, it is possible to implant an artificial pacemaker, which automatically gives an electrical stimulus to the heart every 0.85 second. Should the AV node be damaged, the ventricles still beat because all cardiac muscle cells can contract on their own. However, the beat is so slow that the condition is called a **heart block.**

An area other than the SA node can become the pacemaker when it develops a rate of contraction that is faster than the SA node. This site, called an **ectopic pacemaker,** may cause an extra beat, if it operates only occasionally, or it can even pace the heart for a while. Caffeine and nicotine are two substances that can stimulate an ectopic pacemaker.

Time	Atria	Ventricles	Atrioventricular Valves (tricuspid, bicuspid)	Semilunar Valves (pulmonary, aortic)
0.15 sec	Systole	Diastole	open	closed
0.30 sec	Diastole	Systole	closed	open
0.40 sec	Diastole	Diastole	open	closed

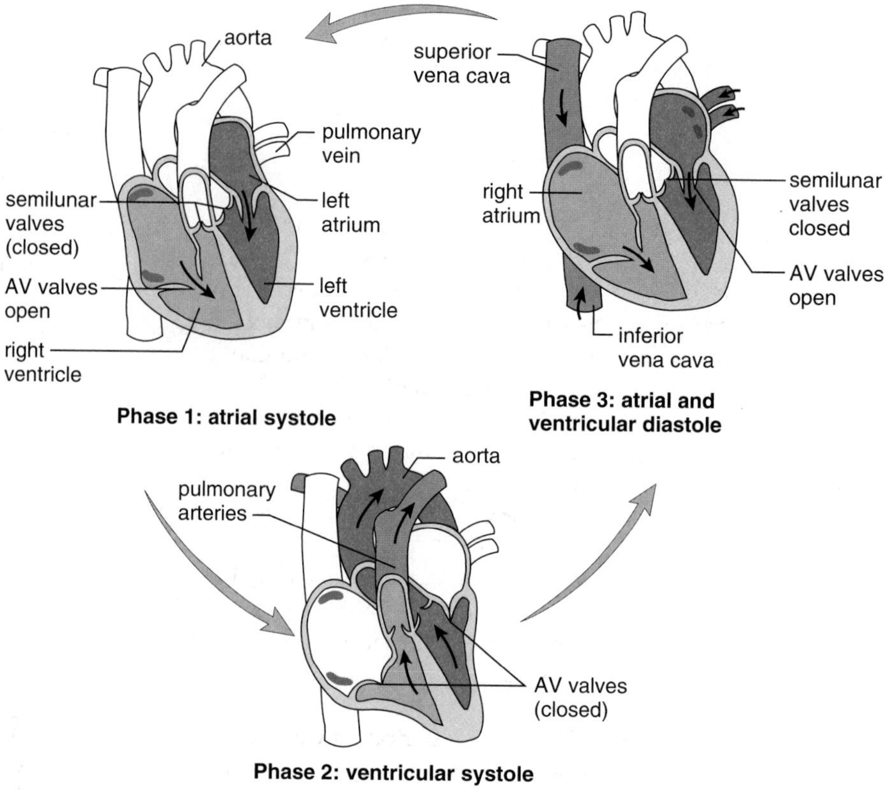

Figure 12.6 Stages in the cardiac cycle. Phase 1: atrial systole. Phase 2: ventricular systole. Phase 3: atrial and ventricular diastole.

Electrocardiogram

With the contraction of any muscle, including the myocardium, electrolyte changes occur that can be detected by electrical recording devices. These changes occur as a muscle action potential sweeps over the cardiac muscle fibers. The resulting record, called an electrocardiogram, helps a physician detect and possibly diagnose the cause of an irregular heartbeat. There are many types of irregular heartbeats, called **arrhythmias.** The Medical Focus page 255 discusses the electrocardiogram and some types of arrhythmias.

Cardiac Cycle

A **cardiac cycle** includes all the events that occur during one heartbeat. On average, the heart beats about 70 times a minute, although a normal adult heart rate can vary from 60 to 100 beats per minute. After tracing the path of blood through the heart, it might seem that the right and left sides of the heart beat independently of one another, but actually, they contract together. First the two atria contract simultaneously; then the two ventricles contract together. The term **systole** refers to contraction of heart muscle, and the term **diastole** refers to relaxation of heart muscle.

During the cardiac cycle, *atrial systole* is followed by *ventricular systole.*

As shown in Figure 12.6, the three phases of the cardiac cycle are:

Phase 1: Atrial Systole. Time = 0.15 sec. During this phase, both atria are in systole (contracted), while the ventricles are in diastole (relaxed). Rising blood pressure in the atria forces the blood to enter the two ventricles through the AV valves. At this time, both atrioventricular valves are open, and the semilunar valves are closed. Atrial systole ends when the atrioventricular valves (tricuspid and bicuspid/mitral) slam shut. Closure of the AV valves is caused by the rising pressure of blood filling the ventricle. Remember that closure of the AV valves causes the first heart sound, "lub" (page 249).

Phase 2: Ventricular Systole. Time = 0.30 sec. During this phase, both ventricles are in systole (contracted), while the atria are in diastole (relaxed). Rising blood pressure in the ventricles forces the semilunar valves (aortic and pulmonary) to open. Blood in the right ventricle exits

MEDICAL FOCUS

The Electrocardiogram

A graph that records the electrical activity of the myocardium during a cardiac cycle is called an **electrocardiogram,** or **ECG.** * An ECG is obtained by placing several electrodes on the patient's skin, then wiring the electrodes to a voltmeter (an instrument for measuring voltage). As the heart's chambers contract and then relax, the change in polarity is measured in millivolts.

An ECG consists of a set of waves: the P wave, a QRS complex, and a T wave (Figure 12D). The P wave represents depolarization of the atria as an impulse started by the SA node travels throughout the atria. The P wave signals that the atria are going to be in systole and that the atrial myocardium is about to contract. The QRS complex represents depolarization of the ventricles following excitation of the Purkinje fibers. It signals that the ventricles are going to be in systole and that the ventricular myocardium is about to contract. The QRS complex shows greater voltage changes than the P wave because the ventricles have more muscle mass than the atria. The T wave represents repolarization of the ventricles. It signals that the ventricles are going to be in diastole and that the ventricular myocardium is about to relax. Atrial diastole does not show up on an ECG as an independent event because the voltage changes are masked by the QRS complex.

An ECG records the duration of electrical activity and therefore can be used to detect arrhythmia, an irregular or abnormal heartbeat. A rate of fewer than 60 heartbeats per minute is called **bradycardia,** and more than 100 heartbeats per minute is called **tachycardia.** Another type of arrhythmia is **fibrillation,** in which the heart beats rapidly but the contractions are uncoordinated. The heart can sometimes be defibrillated by briefly applying a strong electrical current to the chest.

*Also known as EKG (German, ElectroKardioGramm)

It is important to understand that an ECG only supplies information about the heart's electrical activity. To be used in diagnosis, an ECG must be coupled with other information, including X rays, studies of blood flow, and a detailed history from the patient.

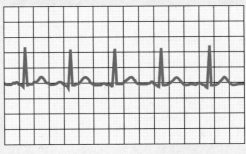

a.

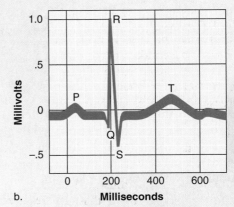

b.

Figure 12D Electrocardiogram. **a.** A portion of an electrocardiogram. **b.** An enlarged normal cycle.

---

through the pulmonary artery trunk to the right and left pulmonary arteries. Simultaneously, blood in the left ventricle exits into the aorta. During ventricular systole, both semilunar valves are open, and the atrioventricular valves are closed. Ventricular systole ends as the ventricles complete their pumping job; recall that backflow of blood in the pulmonary artery and aorta forces the semilunar valves to slam shut once more (page 250). Closure of the semilunar valves causes the second heart sound "dup."

Phase 3: Atrial and Ventricular Diastole. Time = 0.40 sec. During this period, both atria and both ventricles are in diastole (relaxed). At this point, pressure in all the heart chambers is low. Blood returning to the heart from the superior and inferior venae cavae and the pulmonary

veins fills the right and left atria and flows passively into the ventricles. At this time, both atrioventricular valves are open and the semilunar valves are closed.

Cardiac Output

Cardiac output (CO) is the volume of blood pumped out of a ventricle in one minute. (The same amount of blood is pumped out of each ventricle in one minute.) Cardiac output is dependent on two factors:

- *heart rate (HR)* = beats per minute
- *stroke volume (SV)* = amount of blood pumped by a ventricle each time it contracts

Thus, cardiac output = HR × SV.

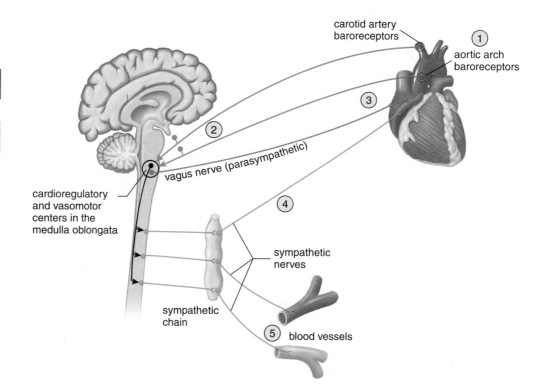

Regulation of heart rate:

① Baroreceptors in the aortic arch and carotid arteries monitor blood pressure.

② Nerve impulses from the baroreceptors signal the cardioregulatory center.

③ Increased parasympathetic impulses decrease heart rate.

④ Increased sympathetic impulses increase heart rate.

Regulation of blood pressure:

⑤ Increased sympathetic impulses cause blood vessels to constrict.

Figure 12.7 The cardioregulatory center regulates the heart rate and the vasomotor center regulates constriction of blood vessels, according to input received from baroreceptors in the carotid artery and aortic arch.

The CO of an average human is 5,250 ml (or 5.25 L) per minute, which equates to about the total volume of blood in the human body. Each minute, the right ventricle pumps about 5.25 L through the pulmonary circuit, while the left ventricle pumps about 5.25 L through the systemic circuit. And this is only the resting cardiac output!

Cardiac output can vary because stroke volume and heart rate can vary, as discussed next. In this way, the heart regulates the blood supply, dependent on the body's needs.

Heart Rate

A **cardioregulatory center** in the medulla oblongata of the brain can alter the heart rate by way of the autonomic nervous system (Fig. 12.7). Parasympathetic motor impulses conducted by the vagus nerve cause the heart rate to slow, and sympathetic motor impulses conducted by sympathetic motor fibers cause the heart rate to increase.

The cardioregulatory center receives sensory input from receptors within the cardiovascular system. For example, *baroreceptors* are present in the aorta just after it leaves the heart, and also in the carotid arteries, which take blood from the aorta to the brain. If blood pressure falls, as it sometimes does when we stand up quickly, the baroreceptors signal the cardioregulatory center. Thereafter, sympathetic motor impulses to the heart cause the heart rate to increase. Once blood pressure begins to rise above normal, nerve impulses from the cardioregulatory center cause the heart rate to decrease. Such reflexes help control cardiac

output and, therefore, blood pressure, as discussed in section 12.4.

The cardioregulatory center is under the influence of the cerebrum and the hypothalamus. Therefore, when we feel anxious, the sympathetic motor nerves are activated. In addition, the adrenal medulla releases the hormones norepinephrine and epinephrine. The result is an increase in heart rate. On the other hand, activities such as yoga and meditation lead to activation of the vagus nerve, which slows the heart rate.

Other factors affect the heart rate as well. For example, a low body temperature slows the rate. Also, the proper electrolyte concentrations are needed to keep the heart rate regular.

Stroke Volume

Stroke volume, which is the amount of blood that leaves a ventricle, depends on the strength of contraction. The degree of contraction depends on the blood electrolyte concentration and the activity of the autonomic system. Two additional factors also influence the strength of contraction.

Venous Return Venous return is the amount of blood entering the heart by way of the venae cavae (right side of heart) or pulmonary veins (left side of heart). The heart adjusts the strength of its own contraction beat by beat, based upon venous return. This principle is called the Frank-Starling law. The more blood returned to the heart before a given beat, the more strongly the heart will contract with that next beat. Thus, any event that increases the volume of blood entering the heart will increase the stroke volume leaving the heart. For example, exercise increases the strength of cardiac contraction because skeletal muscle contraction squeezes the veins within muscles and increases venous return. The opposite is also true: If venous return decreases, stroke volume decreases for the next beat. A low venous return, as might happen if there is blood loss, decreases the strength of cardiac contraction.

Difference in Blood Pressure The strength of ventricular contraction has to be strong enough to oppose the blood pressure within the attached arteries. If a person has hypertension or atherosclerosis, the opposing arterial pressure may reduce the effectiveness of contraction and the stroke volume.

Content CHECK-UP!

4. The SA node is the pacemaker of the heart and controls heart rate because:
   a. it sits at the top of the heart.
   b. it has the fastest intrinsic rate.
   c. it has the largest myocardial cells.
   d. it has the slowest intrinsic rate.

5. The semilunar valves are open during which phase of the cardiac cycle?
   a. atrial systole
   b. ventricular systole
   c. atrial and ventricular diastole

6. Cardiac output is equal to:
   a. heart rate plus stroke volume.
   b. stroke volume divided by heart rate.
   c. heart rate divided by stroke volume.
   d. heart rate times stroke volume.

12.3 Anatomy of Blood Vessels

Blood vessels are of three types: arteries, capillaries, and veins (Fig. 12.8). These vessels function to:

1. transport blood and its contents (see page 228);
2. carry out exchange of gases in the pulmonary capillaries and exchange of gases plus nutrients for waste at the systemic capillaries (see page 237);
3. regulate blood pressure;
4. direct blood flow to those systemic tissues that most require it at the moment.

Arteries and Arterioles

Arteries (Fig. 12.8*a*) transport blood away from the heart. They have thick, strong walls composed of three layers: (1) The **tunica interna** is an endothelium layer with a basement membrane. (2) The **tunica media** is a thick middle layer of smooth muscle and elastic fibers. (3) The **tunica externa** is an outer connective tissue layer composed principally of elastic and collagen fibers. Arterial walls are sometimes so thick that they are supplied with blood vessels. The radius of an artery allows the blood to flow rapidly and the elasticity of an artery allows it to expand when the heart contracts and recoil when the heart rests. This means that blood continues to flow in an artery even when the heart is in diastole.

Arterioles are small arteries just visible to the naked eye. The middle layer of these vessels has some elastic tissue but is composed mostly of smooth muscle whose fibers encircle the arteriole. If the muscle fibers contract, the lumen (cavity) of the arteriole decreases; if the fibers relax, the lumen of the arteriole enlarges. Whether arterioles are constricted or dilated affects blood distribution and blood pressure. When a muscle is actively contracting, for example, the arterioles in the vicinity dilate and additional blood flows to the muscle. In this way, the needs of the muscle for oxygen and glucose are met. As we shall see, the autonomic nervous system helps control blood pressure by regulating the number of arterioles that are contracted. The greater the number of vessels contracted, the higher the resistance to blood flow, and hence, the higher the blood pressure. The greater the number of vessels dilated, the lower the resistance to blood flow, and hence, the lower the blood pressure.

Capillaries

Arterioles branch into **capillaries** (Fig. 12.8*b*), which are extremely narrow, microscopic blood vessels with a wall composed of only one layer of endothelial cells. *Capillary beds* (networks of many capillaries) are present in all regions of the body; each supplies the needs of neighboring cells. Capillaries are an important part of the cardiovascular system because nutrient and waste molecules are exchanged only across their thin walls. Oxygen and glucose diffuse out of capillaries into the tissue fluid that surrounds cells, and carbon dioxide and other wastes diffuse into the capillaries (see Fig. 11.6). Because capillaries serve the needs of the cells, the heart and other vessels of the cardiovascular system can be considered a means by which blood is conducted to and from the capillaries.

Not all capillary beds are open or in use at the same time. For instance, after a meal, the capillary beds of the digestive tract are usually open, and during muscular exercise, the capillary beds of the skeletal muscles are open.

Most capillary beds have a shunt that allows blood to move directly from an arteriole to a venule (a small vessel leading to a vein) when the capillary bed is closed. Sphincter muscles, called *precapillary sphincters,* encircle the entrance to each capillary (Fig. 12.9). When the capillary sphincters are constricted, the capillary bed is closed, preventing blood from entering the capillaries. Conversely, when the capillary sphincters are relaxed, the capillary bed is open. As would be expected, the larger the number of capillary beds open, the lower the blood pressure in the body.

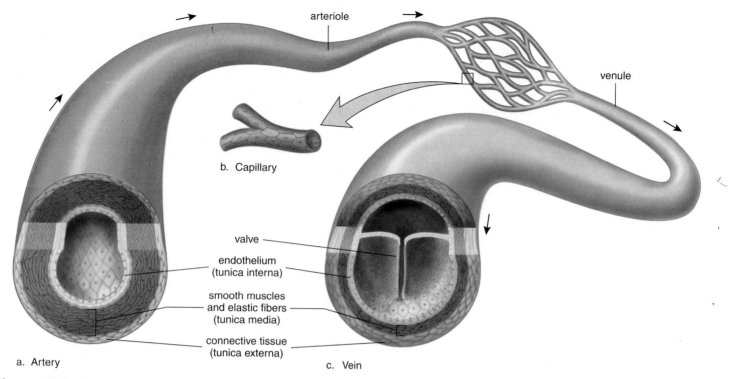

arteriole

venule

b. Capillary

valve
endothelium
(tunica interna)
smooth muscles
and elastic fibers
(tunica media)
connective tissue
(tunica externa)

a. Artery

c. Vein

Figure 12.8 Blood vessels. The walls of arteries and veins have three layers. The tunica interna is an endothelium with a basement membrane; the tunica media is smooth muscle tissue and elastic fibers; the tunica externa is composed of connective tissue. **a.** Arteries have a thicker wall than veins because they have a thicker middle layer than veins. **b.** Capillary walls are one-cell-thick endothelium. **c.** Veins are larger in diameter than arteries, so collectively, veins have a larger holding capacity than arteries.

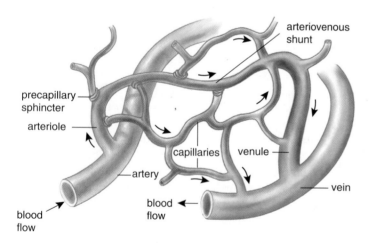

arteriovenous
shunt

precapillary
sphincter

arteriole

capillaries

venule

artery

vein

blood
flow

blood
flow

Figure 12.9 Anatomy of a capillary bed. A capillary bed forms a maze of capillary vessels that lies between an arteriole and a venule. When sphincter muscles are relaxed, the capillary bed is open, and blood flows through the capillaries. When sphincter muscles are contracted, blood flows through a shunt that carries blood directly from an arteriole to a venule. As blood passes through a capillary in the tissues, it gives up its oxygen (O_2). Therefore, blood goes from being O_2-rich in the arteriole (red color) to being O_2-poor in the vein (blue color).

Veins and Venules

Veins and smaller vessels called **venules** return blood from the capillary beds to the heart. The venules first drain the blood from the capillaries and then join together to form a vein. The wall of a vein is much thinner than that of an artery because the middle layer of muscle and elastic fibers is thinner (see Fig. 12.8c). Within some veins, especially the major veins of the arms and legs, **valves** allow blood to flow only toward the heart when they are open and prevent the backward flow of blood when they are closed.

At any given time, more than half of the total blood volume is found in the veins and venules. If blood is lost due to, for example, hemorrhaging, sympathetic nervous stimulation causes the veins to constrict, providing more blood to the rest of the body. In this way, the veins act as a blood reservoir.

Varicose Veins and Phlebitis

Varicose veins are abnormal and irregular dilations in superficial (near the surface) veins, particularly those in the lower legs. Varicose veins in the rectum, however, are commonly called piles, or more properly, **hemorrhoids**. Varicose veins develop when the valves of the veins become weak and ineffective due to backward pressure of the blood.

Phlebitis, or inflammation of a vein, is a more serious condition because thromboembolism can occur. In this instance, the embolus may eventually come to rest in a pulmonary arteriole, blocking circulation through the lungs. This condition, termed **pulmonary embolism,** can be fatal.

Content CHECK-UP!

7. The order for the three layers of a blood vessel, from superficial to deep, is:
   a. tunica externa—tunica intima—tunica media
   b. tunica media—tunica externa—tunica intima
   c. tunica externa—tunica media—tunica intima

8. Which of the following statements comparing arteries and veins is incorrect?
   a. Arteries have a thicker wall than veins.
   b. Veins have valves but arteries do not.
   c. Veins return blood to the heart; arteries carry blood away from the heart.
   d. Arteries have three tissue layers in their walls and veins do not.

9. Select one incorrect statement about capillaries:
   a. Their walls are composed of three tissue layers.
   b. Oxygen and nutrients can diffuse through their walls to supply tissue cells.
   c. A network of many capillaries is called a capillary bed.
   d. Not all capillary beds are open or in use at the same time.

12.4 Physiology of Circulation

Circulation is the movement of blood through blood vessels, from the heart to the body and then back to the heart. In this section, we discuss various factors affecting circulation.

Velocity of Blood Flow

The velocity of blood flow is slowest in the capillaries. What might account for this? Consider that the aorta branches into the other arteries, and these in turn branch into the arterioles, and so forth until blood finally flows into the capillaries. Each time an artery branches, the total cross-sectional area of the blood vessels increases, reaching the maximum cross-sectional area in the capillaries (Fig. 12.10). The slow rate of blood flow in the capillaries is beneficial because it allows time for the exchange of gases in pulmonary capillaries and for the exchange of gases and nutrients for wastes in systemic capillaries (see Fig. 11.6).

Conversely, blood flow increases as venules combine to form veins, and velocity is faster in the venae cavae than in the smaller veins. The cross-sectional area of the two venae cavae is more than twice that of the aorta, and the velocity of the blood returning to the heart remains low compared to the blood leaving the heart. In a resting individual, it takes only a minute for a drop of blood to go from the heart to the foot and back again to the heart! Blood pressure causes blood flow because blood always flows from a higher to a lower pressure difference.

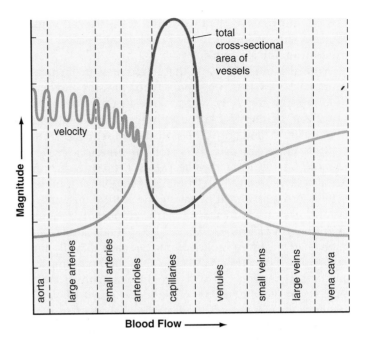

Figure 12.10 Velocity of blood flow changes throughout the systemic circuit. Velocity changes according to the total cross-sectional area of vessels.

Blood Pressure

Blood pressure is the force of blood against the walls of blood vessels. You would expect arterial blood pressure to be highest in the aorta. Why? Because the pumping action of the powerful, thick-muscled left ventricle forces blood into the aorta. Further, Figure 12.11 shows that systemic blood pressure

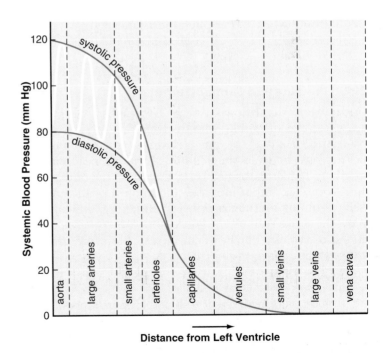

Figure 12.11 Blood pressure changes throughout the systemic circuit. Blood pressure decreases with distance from the left ventricle.

decreases progressively with distance from the left ventricle. Blood pressure is lowest in the venae cavae because they are farthest from the left ventricle.

Note also in Figure 12.11 that blood pressure fluctuates in the arterial system between systolic blood pressure and diastolic blood pressure. Certainly, we can correlate this with the action of the heart. During systole, the left ventricle is pumping blood out of the heart, and during diastole the left ventricle is resting.

More important than the systolic and diastolic pressure is the *mean arterial blood pressure (MABP)*, the pressure in the arterial system averaged over time. It is important to note that MABP is not determined by taking the average of systolic and diastolic pressures. Rather, MABP is the product of *cardiac output (CO)* times *peripheral resistance*. (Recall that cardiac output equals heart rate times stroke volume; see page 255.) To put it as a simple math equation, MABP = CO × PR. According to this equation, increasing CO will also increase MABP. In other words, the greater the amount of blood leaving the left ventricle, the greater the pressure of blood against the wall of an artery.

Another factor that determines blood pressure is peripheral resistance, which is the resistance to flow between blood and the walls of a blood vessel. All things being equal, the smaller the blood vessel diameter, the greater the resistance and the higher the blood pressure. As an analogy, imagine a skinny 1-inch-diameter garden hose (high resistance) compared to a fire-fighter's 12-inch canvas hose (low resistance). Similarly, total blood vessel length increases blood pressure because a longer vessel offers greater resistance. For this reason, an obese person is apt to have high blood pressure because about 200 miles of additional blood vessels develop for each extra pound of fat.

Let's summarize our discussion so far. The two factors that affect blood pressure are:

Cardiac output	Peripheral resistance
Heart rate	Arterial diameter and length
Stroke volume	

Blood Pressure and Cardiac Output

Our previous discussion on page 255 emphasized that the heart rate and the stroke volume determine cardiac output. We learned that the heart rate is intrinsic but is under extrinsic (nervous and endocrine) control. Therefore, heart rate can speed up. The faster the heart rate, the greater the cardiac output. As cardiac output increases, blood pressure increases as well (assuming constant peripheral resistance). Similarly, the larger the stroke volume, the greater the blood pressure. However, stroke volume and heart rate increase blood pressure *only* if the venous return is adequate.

Venous Return Venous return depends on three factors:

1. a blood pressure difference—blood pressure is about 16 mm Hg in venules versus 0 mm Hg in the right atrium;
2. the skeletal muscle pump and the respiratory pump, both of which are effective because of the presence of valves in veins;
3. total blood volume in the cardiovascular system.

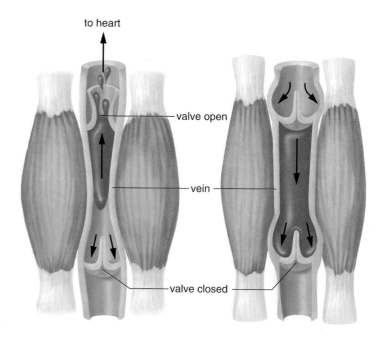

a. Contracted skeletal muscles   b. Relaxed skeletal muscles

Figure 12.12 Skeletal muscle pump. **a.** When skeletal muscles contract and compress a vein, blood is squeezed past a valve. **b.** When muscles relax, the backward flow of blood closes the valve.

The **skeletal muscle pump** works like this: When skeletal muscles contract, they compress the weak walls of the veins. This causes the blood to move past a valve (Fig. 12.12). Once past the valve, backward pressure of blood closes the valve and prevents its return. Blood in veins will always return to the heart.

As you might suspect, gravity can assist the return of venous blood from the head to the heart but not the return of blood from the extremities and trunk to the heart. The importance of the skeletal muscle pump in maintaining CO and blood pressure can be demonstrated by forcing a person to stand rigidly still for a number of hours. Frequently, the person faints because blood collects in the limbs, limiting venous return to the heart. As venous return decreases, cardiac output decreases as well. Blood pressure falls, robbing the brain of oxygen. In this case, fainting is beneficial. The horizontal body position caused by the faint aids in getting blood to the brain.

The **respiratory pump** works like this: When inhalation occurs, thoracic pressure falls and abdominal pressure rises as the chest expands. This aids in the flow of venous blood back to the heart because blood flows from areas of higher pressure (in the abdominal cavity) to areas of lower pressure (in the thoracic cavity). During exhalation, the pressure reverses, but the valves in the veins prevent backward flow.

As stated, the amount of venous return also depends on the total blood volume in the cardiovascular system. As you know, this volume in the pulmonary circuit and the systemic circuit is 5 L. If this amount of blood decreases, say due to hemorrhaging, blood pressure falls. On the other hand, if blood volume increases (due to water retention, for example), blood pressure rises.

Blood Pressure and Peripheral Resistance

The nervous system and the endocrine system both affect peripheral resistance.

Neural Regulation of Peripheral Resistance A **vasomotor center** in the medulla oblongata controls vasoconstriction. This center is under the control of the cardioregulatory center. As mentioned on page 256, if blood pressure falls, baroreceptors in the blood vessels signal the cardioregulatory center. The cardioregulatory center will activate its vasomotor center. The vasomotor center then stimulates sympathetic nerve fibers, which cause the heart rate to increase *and* the arterioles to constrict. Increasing heart rate increases cardiac output; constricting the arterioles increases peripheral resistance. The result is a rise in blood pressure. What factors lead to a reduction in blood pressure? If blood pressure rises above normal, the baroreceptors signal the cardioregulatory center in the medulla oblongata. Subsequently, the heart rate decreases *and* the arterioles dilate.

Nervous control of blood vessels also causes blood to be shunted from one area of the body to another. During exercise, arteries in the viscera and skin are more constricted than those in the muscles. Therefore, blood flow to the muscles increases. Also, dilation of the precapillary sphincters in muscles means that blood will flow to the muscles and not to the viscera.

Hormonal Regulation of Peripheral Resistance Certain hormones cause blood pressure to rise. Epinephrine and norepinephrine, the hormones from the adrenal medulla, increase the heart rate. As heart rate increases, so too does cardiac output and blood pressure. Epinephrine and norepinephrine also constrict arterioles in the capillary beds supplying the skin, abdominal viscera, and kidneys. Arteriolar vasoconstriction increases blood pressure by increasing peripheral resistance.

When the blood volume and blood sodium level are low, the kidneys secrete the enzyme **renin.** Renin converts the plasma protein angiotensinogen to angiotensin I, which is changed to angiotensin II by a converting enzyme found in the lungs. Angiotensin II stimulates the adrenal cortex to release aldosterone. The effect of this system, called the renin-angiotensin-aldosterone system, is to raise the blood volume and pressure in two ways. First, angiotensin II constricts the arterioles directly. Second, aldosterone causes the kidneys to reabsorb sodium. When the blood sodium level rises, water is reabsorbed, and blood volume and pressure are maintained.

Two other hormones play a role in the homeostatic maintenance of blood volume and blood pressure. As discussed in Chapter 10, antidiuretic hormone (ADH) helps increase blood volume by causing the kidneys to reabsorb water. ADH also causes vasoconstriction of smooth muscle in arteries and veins throughout the body. Thus, ADH boosts blood pressure by simultaneously increasing blood volume (which increases cardiac output) *and* by causing vasoconstriction (which increases peripheral resistance).

The hormonal mechanism for decreasing blood pressure involves an endocrine hormone secreted by the heart. When

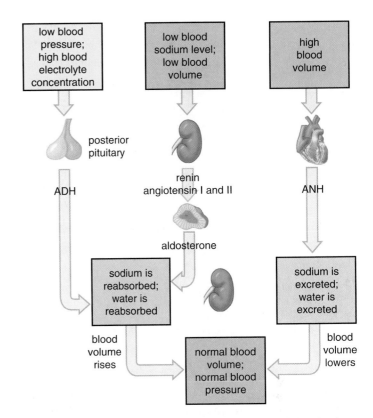

Figure 12.13 Blood volume maintenance. Normal blood volume is maintained by ADH (antidiuretic hormone) and aldosterone, whose actions raise blood volume, and by ANH (atrial natriuretic hormone), whose actions lower blood volume.

the atria of the heart are stretched due to increased blood volume, cardiac cells release a hormone called **atrial natriuretic hormone (ANH),** which inhibits renin secretion by the kidneys and aldosterone secretion by the adrenal cortex. The effect of ANH, therefore, is to cause sodium excretion—that is, *natriuresis* (a term that means "salty urination"). When sodium is excreted, so is water, and therefore blood volume and blood pressure decrease (Fig. 12.13).

Evaluating Circulation

Taking a patient's pulse and blood pressure are two ways to evaluate circulation.

Pulse

The surge of blood entering the arteries causes their elastic walls to stretch, but then they almost immediately recoil. This alternating expansion and recoil of an arterial wall can be felt as a **pulse** in any artery that runs close to the body's surface. These superficial arteries are called pulse points (Fig. 12.14). It is customary to feel the pulse by placing several fingers on the radial artery, which lies near the outer border of the palm side of a wrist. The common carotid artery, on either side of the trachea in the neck, is another accessible location for feeling the pulse. Normally, the pulse rate indicates the rate of the

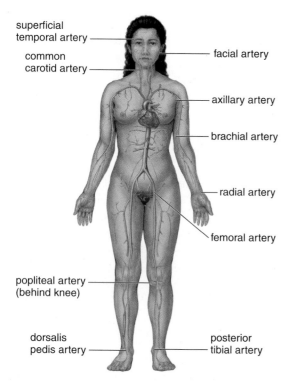

Figure 12.14 The pulse rate. Pulse points are the locations where the pulse can be taken. Each pulse point is named after the appropriate artery.

heartbeat because the arterial walls pulse whenever the left ventricle contracts. The pulse is usually 70 times per minute, but can vary between 60 and 80 times per minute.

Blood Pressure

Blood pressure is usually measured in the brachial artery with a sphygmomanometer, an instrument that records changes in terms of millimeters (mm) of mercury (Fig. 12.15). A blood pressure cuff connected to the sphygmomanometer is wrapped around the patient's arm, and a stethoscope is placed over the brachial artery. The blood pressure cuff is inflated until no blood flows through it; therefore, no sounds can be heard through the stethoscope. The cuff pressure is then gradually lowered. As soon as the cuff pressure declines below systolic pressure, blood flows through the brachial artery each time the left ventricle contracts. The blood flow is turbulent below the cuff. This turbulence produces vibrations in the blood and surrounding tissues that can be heard through the stethoscope. These sounds are called *Korotkoff sounds,* and the cuff pressure at which the Korotkoff sounds are heard the first time is the systolic pressure. As the pressure in the cuff is lowered still more, the Korotkoff sounds change tone and loudness. When the cuff pressure no longer constricts the brachial artery, no sound is heard. The cuff pressure at which the Korotkoff sounds disappear is the diastolic pressure.

Normal resting blood pressure for a young adult is 120/80. The higher number is the *systolic pressure,* the pressure recorded in an artery when the left ventricle contracts. The lower number is the *diastolic pressure,* the pressure recorded in an artery when the left ventricle relaxes.

It is estimated that about 20% of all Americans suffer from hypertension, which is high blood pressure. Hypertension is present when the systolic blood pressure is 140 or greater, or the diastolic blood pressure is 90 or greater. While both systolic and diastolic pressures are considered important, the diastolic pressure is emphasized when medical treatment is being considered.

Hypertension is sometimes called a silent killer because it may not be detected until a stroke or heart attack occurs. It has long been thought that a certain genetic makeup might account for the development of hypertension. Now researchers have discovered two genes that may be involved in some individuals. One gene codes for angiotensinogen, the plasma protein mentioned previously (see page 261). Angiotensinogen is converted to a powerful vasoconstrictor in part by the product of the second gene. Persons with hypertension due to overactivity of these genes might one day be cured by gene therapy.

At present, however, the best safeguard against developing hypertension is to have regular blood pressure checks and to adopt a lifestyle that lowers the risk of hypertension as described in the Medical Focus on page 270.

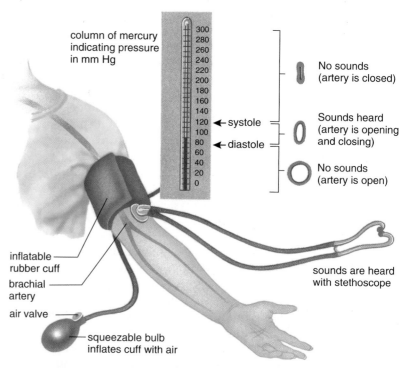

Figure 12.15 Use of a sphygmomanometer. The technician inflates the cuff with air, gradually reduces the pressure, and listens with a stethoscope for the sounds that indicate blood is moving past the cuff in an artery. This is systolic blood pressure. The pressure in the cuff is further reduced until no sound is heard, indicating that blood is flowing freely through the artery. This is diastolic pressure.

10. Blood flow is slowest through which blood vessels?

    a. arterioles   c. capillaries

    b. veins   d. venules

11. Mean arterial blood pressure is equal to:

    a. cardiac output times peripheral resistance.

    b. cardiac output plus peripheral resistance.

    c. cardiac output/peripheral resistance.

    d. cardiac output minus peripheral resistance.

12. Which of the following will not increase mean arterial blood pressure?

    a. increased venous return to the heart

    b. increased constriction of arteries and arterioles throughout the body

    c. release of epinephrine from the adrenal medulla

    d. decrease in heart rate

12.5 Circulatory Routes

Blood vessels belong to either the pulmonary circuit or the systemic circuit. The path of blood through the pulmonary circuit can be traced as follows: Blood from all regions of the body first collects in the right atrium and then passes into the right ventricle, which pumps it into the pulmonary trunk.

The pulmonary trunk divides into the **pulmonary arteries,** which in turn divide into the arterioles of the lungs. The arterioles then take blood to the pulmonary capillaries where carbon dioxide and oxygen are exchanged. The blood then enters the pulmonary venules and flows through the **pulmonary veins** back to the left atrium. Because the blood in the pulmonary arteries is O_2-poor but the blood in the pulmonary veins is O_2-rich, it is not correct to say that all arteries carry blood that is high in oxygen and that all veins carry blood that is low in oxygen. In fact, just the reverse is true in the pulmonary circuit.

Congestive Heart Failure

In **congestive heart failure,** a damaged left side of the heart fails to pump adequate blood, and blood backs up in the pulmonary circuit. Therefore, pulmonary blood vessels have become *congested.* The congested vessels leak fluid into tissue spaces, causing pulmonary edema. The result is shortness of breath, fatigue, and a constant cough with pink, frothy sputum. Treatment consists of the three Ds: diuretics (to increase urinary output), digoxin (to increase the heart's contractile force), and dilators (to relax the blood vessels). If necessary, a heart transplant is done.

The systemic circuit includes all of the other arteries and veins of the body. The largest artery in the systemic circuit is the aorta, and the largest veins are the **superior vena cava** and **inferior vena cava.** The superior vena cava collects blood from the head, chest, and arms, and the inferior vena cava collects blood from the lower body regions. Both venae cavae enter the right atrium. The aorta and venae cavae are the major pathways for blood in the systemic system.

The path of systemic blood to any organ in the body begins in the left ventricle, which pumps blood into the **aorta.** Branches from the aorta go to the major body regions and organs. Tracing the path of blood to any organ in the body requires mentioning only the aorta, the proper branch of the aorta, the organ, and the returning vein to the vena cava. In many instances, the artery and vein that serve the same organ have the same name. For example, the path of blood to and from the kidneys is: left ventricle; aorta; renal artery; arterioles, capillaries, venules; renal vein; inferior vena cava; right atrium. In the systemic circuit, unlike the pulmonary circuit, arteries contain O_2-rich blood and appear bright red, while veins contain O_2-poor blood and appear dark maroon.

The Major Systemic Arteries

After the aorta leaves the heart, it divides into the *ascending aorta,* the *aortic arch,* and the *descending aorta* (Fig. 12.16). The left and right coronary arteries, which supply blood to the heart, branch off the ascending aorta (Table. 12.1).

Three major arteries branch off the aortic arch: the **brachiocephalic artery,** the **left common carotid artery,** and the **left subclavian artery.** The brachiocephalic artery divides into the **right common carotid** and the **right subclavian arteries.** These blood vessels serve the head (right and left common carotids) and arms (right and left subclavians).

The descending aorta is divided into the *thoracic aorta,* which branches off to the organs within the thoracic cavity, and the *abdominal aorta,* which branches off to the organs in the abdominal cavity. See the What's New box on p. 265.

The descending aorta ends when it divides into the **common iliac arteries** that branch into the **internal iliac artery** and the **external iliac artery.** The internal iliac artery serves the pelvic organs, and the external iliac artery serves the legs. These and other arteries are shown in Figure 12.16.

The Major Systemic Veins

Figure 12.17 shows the major veins of the body. The **external** and **internal jugular veins** drain blood from the brain, head, and neck. An external jugular vein enters a **subclavian vein** that, along with an internal jugular vein, enters a **brachiocephalic vein.** Right and left brachiocephalic veins merge, giving rise to the superior vena cava.

In the abdominal cavity, as discussed in more detail later, the **hepatic portal vein** receives blood from the abdominal viscera and enters the liver. Emerging from the liver, the **hepatic veins** enter the inferior vena cava.

In the pelvic region, veins from the various organs enter the **internal iliac** veins, while the veins from the legs enter

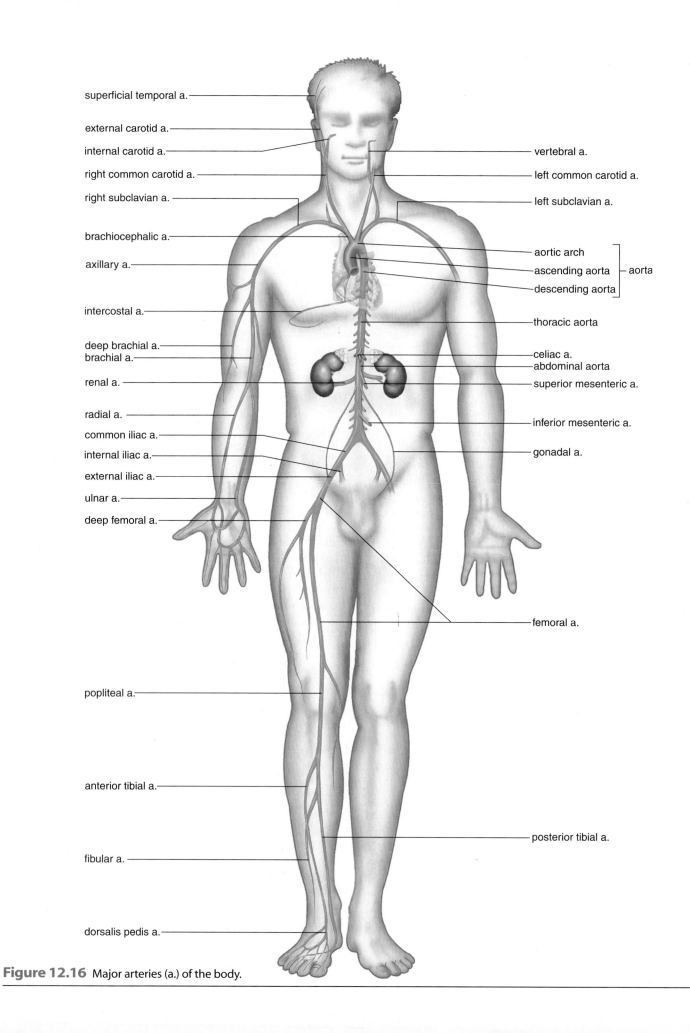

superficial temporal a.

external carotid a.

internal carotid a.

right common carotid a.

right subclavian a.

brachiocephalic a.

axillary a.

intercostal a.

deep brachial a.
brachial a.

renal a.

radial a.

common iliac a.

internal iliac a.

external iliac a.

ulnar a.

deep femoral a.

popliteal a.

anterior tibial a.

fibular a.

dorsalis pedis a.

vertebral a.

left common carotid a.

left subclavian a.

aortic arch

ascending aorta

descending aorta

aorta

thoracic aorta

celiac a.
abdominal aorta

superior mesenteric a.

inferior mesenteric a.

gonadal a.

femoral a.

posterior tibial a.

Figure 12.16 Major arteries (a.) of the body.

Aneurysm Sensors

Years of untreated high blood pressure, or hypertension, will take its toll on arteries throughout the body. Hypertension is a major risk factor in the development of atherosclerosis—the process of inflammation that causes arteries to narrow and lose their flexibility (see the Medical Focus on page 252). Hypertension and subsequent atherosclerosis can also cause the formation of **aneurysms**—weakened areas in the artery wall that balloon outward. If an aneurysm ruptures, it will likely cause a fatal hemorrhage. One area that is particularly vulnerable is the abdominal aorta (Fig. 12E*a*).

Treatment for an abdominal aortic aneurysm (AAA) involves placement of a stent graft, a flexible fabric and metal sleeve, inside the weakened artery. Stents are inserted through small incisions in the patient's femoral artery and directed into the abdominal aorta. Once in place, the stent reinforces the artery and prevents it from tearing (Fig. 12E*b*).

After stent placement, patients must undergo yearly CT scans of the abdomen to ensure that the stent has not shifted position. However, this technique is costly and requires injection of contrast dye that can be toxic to the kidneys. CT scans also do not detect small tears in the artery, which sometimes lead to total failure of the stent.

A new, implantable, wireless aneurysm sensor now allows the stent to be monitored continuously. The sensor is about the size of a quarter and is implanted alongside the stent. To obtain information from the sensor, the physician uses an external wand device that is held over the patient's abdomen. Unlike CT scans, the sensor can detect small pressure increases around the stent, which might indicate failure of the stent. Early detection of small tears around the stent provides the time needed for a successful surgical repair.

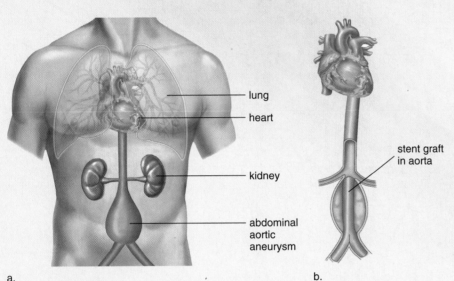

lung
heart
kidney
abdominal aortic aneurysm

stent graft in aorta

a.

b.

Figure 12E Abdominal aortic aneurysm **(a)** before and **(b)** after stent placement.

TABLE 12.1 The Aorta and Its Principal Branches

Portion of Aorta	Major Branch	Regions Supplied
Ascending aorta	Left and right coronary arteries	Heart
Aortic arch	Brachiocephalic artery	
	Right common carotid	Right side of head
	Right subclavian	Right arm
	Left common carotid artery	Left side of head
	Left subclavian artery	Left arm
Descending aorta		
Thoracic aorta	Intercostal artery	Thoracic wall
Abdominal aorta	Celiac artery	Stomach, spleen and liver
	Superior mesenteric artery	Small and large intestines (ascending and transverse colons)
	Renal artery	Kidney
	Gonadal artery	Ovary or testis
	Inferior mesenteric artery	Lower digestive system (transverse and descending colons, and rectum)
	Common iliac artery	Pelvic organs and legs

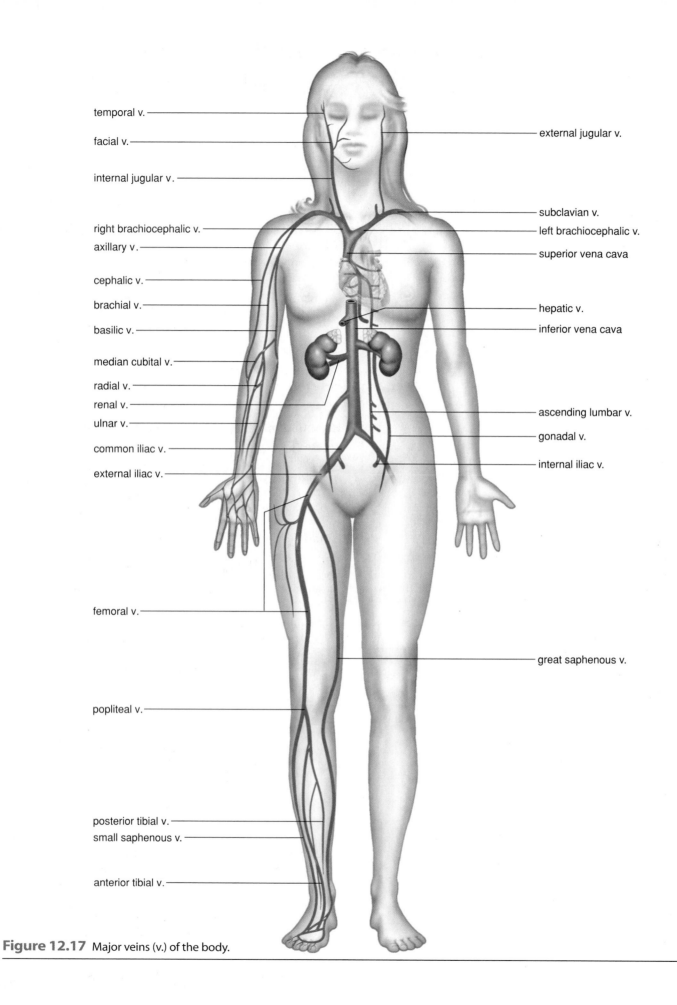

temporal v.

facial v.

internal jugular v.

right brachiocephalic v.

axillary v.

cephalic v.

brachial v.

basilic v.

median cubital v.

radial v.

renal v.

ulnar v.

common iliac v.

external iliac v.

femoral v.

popliteal v.

posterior tibial v.

small saphenous v.

anterior tibial v.

external jugular v.

subclavian v.

left brachiocephalic v.

superior vena cava

hepatic v.

inferior vena cava

ascending lumbar v.

gonadal v.

internal iliac v.

great saphenous v.

Figure 12.17 Major veins (v.) of the body.

TABLE 12.2 Principal Veins That Join the Venae Cavae		
Vein	**Region Drained**	**Vena Cava**
Right and left brachiocephalic veins	Head, neck and upper extremities	Form superior vena cava
Right and left common iliac veins	Lower extremities	Form inferior vena cava
Right and left renal veins	Kidneys	Enters inferior vena cava
Right and left hepatic veins	Liver, digestive tract and spleen	Enters inferior vena cava

the **external iliac veins.** The internal and external iliac veins become the **common iliac veins** that merge, forming the inferior vena cava. Table 12.2 lists the principal veins that enter the venae cavae.

Special Systemic Circulations

Hepatic Portal System

The **hepatic portal system** (Fig. 12.18) carries blood from the stomach, intestines, and other organs to the liver. The term portal system is used to describe the following unique pattern of circulation:

$$capillaries \rightarrow vein \rightarrow capillaries \rightarrow vein$$

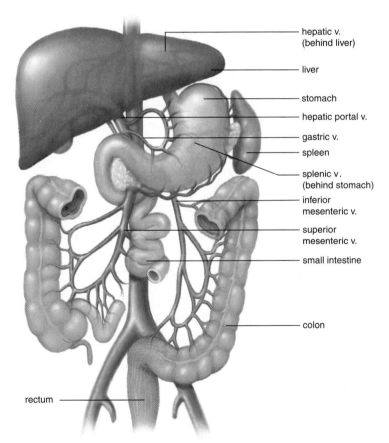

hepatic v. (behind liver)

liver

stomach

hepatic portal v.

gastric v.

spleen

splenic v. (behind stomach)

inferior mesenteric v.

superior mesenteric v.

small intestine

colon

rectum

Figure 12.18 Hepatic portal system. This system provides venous drainage of the digestive organs and takes venous blood to the liver. (v. = vein.)

Capillaries of the digestive tract drain into the superior mesenteric vein and the splenic vein, which join to form the hepatic portal vein. The gastric veins empty directly into the hepatic portal vein. The hepatic portal vein carries blood to capillaries in the liver. The hepatic capillaries allow nutrients and wastes to diffuse into liver cells for further processing. Then, hepatic capillaries join to form venules that enter a hepatic vein. The hepatic veins enter the inferior vena cava.

In addition to receiving venous blood from the intestine, the liver also receives arterial blood via the hepatic artery. The hepatic artery is not a part of the hepatic portal system.

Hypothalamus-Hypophyseal Portal System

The body has other portal systems. For example, the vascular link between the hypothalamus and the anterior pituitary through which the hypothalamus sends hypothalamic-releasing hormones to the anterior pituitary is a portal system.

Blood Supply to the Brain

The brain is supplied with O_2-rich blood by the anterior and posterior cerebral arteries and the carotid arteries. These arteries give off branches. that join to form the **cerebral arterial circle** (circle of Willis), a vascular route in the region of the pituitary gland (Fig. 12.19). Because the blood vessels form a circle, alternate routes are available for bringing arterial blood to the brain and thus supplying the brain with oxygen. The presence of the cerebral arterial circle also equalizes blood pressure in the brain's blood supply.

Fetal Circulation

As Figure 12.20 shows, the fetus has four circulatory features that are not present in adult circulation:

1. The **foramen ovale,** or *oval window,* is an opening between the two atria. This window is covered by a flap of tissue that acts as a valve.
2. The **ductus arteriosus,** or *arterial duct,* is a connection between the pulmonary artery and the aorta.
3. The **umbilical arteries** and **vein** are vessels that travel to and from the placenta, leaving waste and receiving nutrients.
4. The **ductus venosus,** or *venous duct,* is a connection between the umbilical vein and the inferior vena cava.

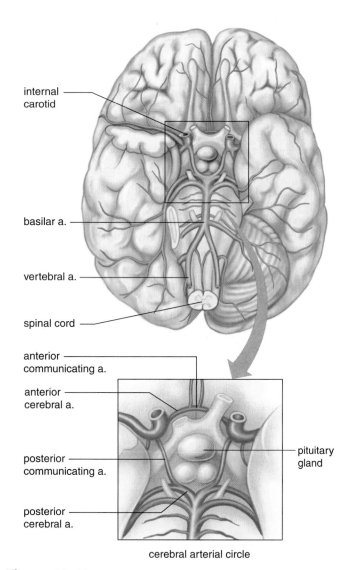

internal
carotid

basilar a.

vertebral a.

spinal cord

anterior
communicating a.

anterior
cerebral a.

posterior
communicating a.

posterior
cerebral a.

pituitary
gland

cerebral arterial circle

Figure 12.19 Cerebral arterial circle. The arteries that supply blood to the brain form the cerebral arterial circle (circle of Willis). (a. = artery.)

All of these features can be related to the fact that the fetus does not use its lungs for gas exchange, since it receives oxygen and nutrients from the mother's blood at the placenta. During development, the lungs receive only enough blood to supply their developmental need for oxygen and nutrients.

The path of blood in the fetus can be traced, beginning from the right atrium (Fig. 12.20). Most of the blood that enters the right atrium passes directly into the left atrium by way of the foramen ovale because the blood pressure in the right atrium is somewhat greater than that in the left atrium. The rest of the fetal blood entering the right atrium passes into the right ventricle and out through the pulmonary trunk. However, because of the ductus arteriosus, most pulmonary trunk blood passes directly into the aortic arch. Notice that, whatever route blood takes, most of it reaches the aortic arch instead of the pulmonary circuit vessels.

MEDICAL FOCUS

Stroke: Beware of "Brain Attack"!

Heart attack, or myocardial infarction, and **cerebrovascular accident,** or stroke, are two disease processes with a lot in common. Both are leading causes of death and disability in North America. Both most often result from atherosclerosis, the hardening and narrowing of arteries that results from injury and inflammation (see the Medical Focus on page 252). Just as a myocardial infarction causes death to heart muscle, a cerebrovascular accident causes death of nerve cells. Thus, a stroke could be termed a "brain attack."

An atherosclerotic plaque formed inside an artery often triggers the clotting cascade when blood clotting factors contact the roughened surface of the plaque, or if the plaque itself cracks open and bleeds. The subsequent clot causes a **thrombotic stroke** if it blocks arteries supplying the brain. A clot that forms elsewhere in the body, but breaks off to lodge in a cerebral vessel, will cause an **embolic stroke. Hemorrhagic stroke** occurs when intracranial blood vessels burst; this form of stroke is much less common (approximately 10% of strokes occur due to hemorrhage). Again, atherosclerosis and hypertension cause most hemorrhagic strokes because both can weaken the wall of an artery to the point that an aneurysm develops. An **aneurysm** is a weakened area in the arterial wall; rupture of the aneurysm causes the hemorrhage. Very rarely, an aneurysm results from congenital malformation of the cranial blood vessels during fetal development.

Just as a patient with angina pectoris is warned by his chest pain that blood flow through his coronary arteries is reduced, a potential stroke victim may be warned by a **transient ischemic attack (TIA).** Numbness or tingling in the hands or face, difficulty in speaking or understanding speech, and visual disturbances are all symptoms of a transient ischemic attack to the brain. Prompt medical attention during a TIA may prevent an actual stroke from occurring.

Heart attack and stroke can be prevented by minimizing the risk factors they also share in common: smoking; hypertension; elevated blood lipid, cholesterol, and homocysteine levels; and uncontrolled diabetes mellitus. See the Medical Focus on page 270 for a complete discussion on preventing cardiovascular disease.

Blood within the aorta travels to the various branches, including the iliac arteries, which connect to the umbilical arteries leading to the placenta. Exchange between maternal and fetal blood takes place at the placenta. Blood in the umbilical arteries is O_2-poor, but blood in the umbilical vein, which travels from the placenta, is O_2-rich. The umbilical vein enters the ductus venosus, which passes directly through the

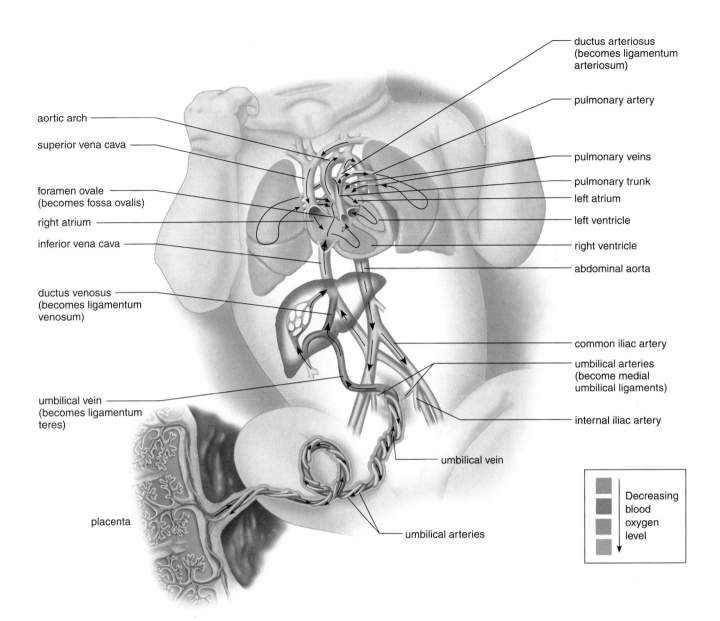

Figure 12.20 Fetal circulation. Arrows indicate the direction of blood flow. The lungs are not functional in the fetus. The blood passes directly from the right atrium to the left atrium via the foramen ovale or from the right ventricle to the aorta via the pulmonary trunk and ductus arteriosus. The umbilical arteries take fetal blood to the placenta where exchange of molecules between fetal and maternal blood takes place. Oxygen and nutrient molecules diffuse into the fetal blood, and carbon dioxide and urea diffuse from the fetal blood. The umbilical vein returns blood from the placenta to the fetus.

liver. The ductus venosus then joins with the inferior vena cava, a vessel that contains O_2-poor blood. The vena cava returns this mixture to the right atrium.

Changes at Birth Sectioning and tying the umbilical cord permanently separates the newborn from the placenta. The first breath inflates the lungs and oxygen enters the blood at the lungs instead of the placenta. O_2-rich blood returning from the lungs to the left side of the heart usually causes a flap on the left side of the interatrial septum to close the foramen ovale. What remains is a depression called the *fossa ovalis.* Incomplete closure occurs in nearly one out of four individuals, but even so, blood rarely passes from the right

atrium to the left atrium because either the opening is small or it closes when the atria contract. In a small number of cases, the passage of O_2-poor blood from the right side to the left side of the heart is sufficient to cause **cyanosis,** a bluish cast to the skin. This condition can now be corrected by open-heart surgery.

The fetal blood vessels and shunts constrict and become fibrous connective tissue called ligamentums in all cases except the distal portions of the umbilical arteries, which become the medial umbilical ligaments. Regardless, these structures run between internal organs. For example, the ligamentum teres (which is the remnant of the umbilical vein) attaches the umbilicus to the liver.

Preventing Cardiovascular Disease

All of us can take steps to prevent cardiovascular disease, the most frequent cause of death in the United States. Genetic factors that predispose an individual to cardiovascular disease include family history of heart attack under age 55, male gender, and ethnicity (African Americans are at greater risk). However, people with one or more of these risk factors need not despair. It only means that they need to pay particular attention to the following guidelines for a heart-healthy lifestyle.

The Don'ts

Smoking

Hypertension is recognized as a major contributor to cardiovascular disease. When a person smokes, the drug nicotine, present in cigarette smoke, enters the bloodstream. Nicotine causes the arterioles to constrict and the blood pressure to rise. Restricted blood flow and cold hands are associated with smoking in most people. Cigarette smoke also contains carbon monoxide, and hemoglobin combines preferentially and nonreversibly with carbon monoxide. Therefore, the presence of carbon monoxide lowers the oxygen-carrying capacity of the blood, and the heart must pump harder to propel the blood through the lungs. Smoking also damages the arterial wall and accelerates the formation of atherosclerosis and plaque.

Drug Abuse

Stimulants, such as cocaine and amphetamines, can cause an irregular heartbeat and lead to heart attacks in people who are using drugs even for the first time. Intravenous drug use may also result in a cerebral blood clot and stroke.

Too much alcohol can destroy just about every organ in the body, the heart included. But investigators have discovered that people who take an occasional drink have a 20% lower risk of heart disease than do teetotalers. Two to four drinks a week is the recommended limit for men; one to three drinks is the recommendation for women.

Weight Gain

Hypertension also occurs more often in persons who are more than 20% above the recommended weight for their height. Because more tissue requires servicing, the heart must send extra blood out under greater pressure in those who are overweight. It may be very difficult to lose weight once it is gained, and therefore weight control should be a lifelong endeavor. Even a slight decrease in weight can bring a reduction in hypertension. A 4.5-kilogram weight loss doubles the chance that blood pressure can be normalized without drugs.

The Do's

Healthy Diet

It was once thought that a low-salt diet could protect against cardiovascular disease, and that still may be true in certain persons. Theoretically, hypertension occurs because the more salty the blood, the greater the osmotic pressure and the higher the water content. However, in recent years, the emphasis has switched to a diet low in saturated fats and cholesterol as being protective

---

Content CHECK-UP!

13. The superior and inferior vena cavae return their blood to the:

   a. right atrium.
   c. pulmonary artery.
   b. left atrium.
   d. pulmonary veins.

14. In an adult heart, the oxygen content is highest in blood found in the:

   a. pulmonary arteries.
   c. pulmonary veins.
   b. pulmonary trunk.
   d. right ventricle.

15. From the following list of pairs of major arteries and the smaller arteries that branch from them, choose the incorrect pair:

   a. brachiocephalic—right common carotid
   b. abdominal aorta—intercostal artery
   c. aortic arch—left subclavian
   d. abdominal aorta—inferior mesenteric artery

12.6 Effects of Aging

The heart generally grows larger with age, primarily because of fat deposition in the epicardium and myocardium. In many middle-aged people, the heart is covered by a layer of fat, and the number of collagenous fibers in the endocardium increases. With age, the valves, particularly the aortic semilunar valve, become thicker and more rigid.

As a person ages, the myocardium loses some of its contractile power and some of its ability to relax. The resting heart rate decreases throughout life, and the maximum possible rate during exercise also decreases. With age, the contractions become less forceful; the heart loses about 1% of its reserve pumping capacity each year after age 30.

In the elderly, arterial walls tend to thicken with plaque and become inelastic, signaling that atherosclerosis and arteriosclerosis are present. The chances of coronary thrombosis and heart attack increase with age. Increased blood pressure was once be-against cardiovascular disease. Cholesterol is ferried in the blood by two types of plasma lipoproteins, called LDL (low-density lipoprotein) and HDL (high-density lipoprotein). LDL (often referred to as "bad cholesterol" or "bad lipoprotein" by physicians) takes cholesterol from the liver to the tissues. Recall that LDL is oxidized by macrophages to form foam cells, and later, a fatty streak that begins an atherosclerotic plaque. HDL ("good" cholesterol or lipoprotein) transports cholesterol out of the tissues to the liver, where cholesterol is metabolized. When the LDL level in the blood is abnormally high or the HDL level is abnormally low, cholesterol accumulates in artery walls.

It is recommended that everyone know his or her blood cholesterol level. Individuals with a high blood cholesterol level (240 mg/100 ml) should be further tested to determine their LDL cholesterol level. Blood tests can also determine high levels of homocysteine and C-reactive protein, two important indicators that atherosclerosis is occurring. Levels of these two markers, along with LDL cholesterol level, must be considered together with other cardiovascular risk factors such as age, family history, general health, and whether the patient smokes. A person with a moderate risk of cardiovascular disease may have success with dietary therapy to lower LDL. Cholesterol-lowering drugs are reserved for high-risk patients.

Evidence is mounting to suggest a role for antioxidant vitamins (A, E, and C) in the prevention of cardiovascular disease. Antioxidants protect the body from free radicals that may damage HDL cholesterol through oxidation or damage the lining of an artery, leading to a blood clot that can block the vessel. Nutritionists believe that consuming at least five servings of fruit and vegetables a day may protect against cardiovascular disease.

Proper Dental Care

Periodontal disease, the inflammation of the gums caused by poor dental hygiene, has been suggested as a cause for atherosclerosis. Scientists suspect that people with tooth decay and gum disease may have a low-level bacterial infection in the blood—not severe enough to cause illness, but enough to injure the endothelial lining and start the formation of atherosclerotic plaques. Proper care of the teeth and gums, along with regular visits to the dentist, just might prevent cardiovascular disease.

Exercise

People who exercise are less apt to have cardiovascular disease. One study found that moderately active men who spent an average of 48 minutes a day on a leisure-time activity such as gardening, bowling, or dancing had one-third fewer heart attacks than their peers who spent an average of only 16 minutes each day on such activities. Exercise helps keep weight under control, may help minimize stress, and reduces hypertension. The heart beats faster when exercising, but exercise slowly increases the heart's capacity. This means that the heart can beat more slowly when we are at rest and still do the same amount of work. One physician recommends that his cardiovascular patients walk for one hour, three times a week. In addition, they are to practice meditation and yoga-like stretching and breathing exercises to reduce stress.

lieved to be inevitable with age, but now hypertension is known to result from other conditions, such as kidney disease and atherosclerosis. The Medical Focus on these pages describes how diet and exercise in particular can help prevent atherosclerosis.

The occurrence of varicose veins increases with age, particularly in people who are required to stand for long periods. Thromboembolism as a result of varicose veins can lead to death if a blood clot settles in a major branch of a pulmonary artery. (This disorder is called pulmonary embolism.)

12.7 Homeostasis

Homeostasis is possible only if the cardiovascular system delivers oxygen and nutrients to and takes metabolic wastes from the tissue fluid surrounding the cells. Human Systems Work Together on page 272 tells how the cardiovascular system works with other systems of the body to maintain homeostasis.

Maintaining Blood Composition, pH, and Temperature

The composition of the blood is maintained by the other systems of the body. Growth factors regulate the manufacture of formed elements in the red bone marrow, which is a lymphatic organ. In this way, the skeletal system contributes to the cardiovascular system. Red blood cells assist the respiratory system by carrying oxygen, and the immune system could not function without the ability of white blood cells to fight infection.

The digestive system absorbs nutrients into the blood, and the lungs and kidneys remove metabolic wastes from blood. One of the most important functions of the kidneys is to maintain the pH of the blood within normal limits. The liver, of course, is a key regulator of blood components by producing plasma proteins, storing glucose until it is needed, transforming ammonia into urea, and changing other poisons into molecules that are also excreted.

Human Systems Work Together

Integumentary System

Blood vessels deliver nutrients and oxygen to skin; carry away wastes; blood clots if skin is broken.

Skin prevents water loss; helps regulate body temperature; protects blood vessels.

Skeletal System

Blood vessels deliver nutrients and oxygen to bones; carry away wastes.

Rib cage protects heart; red bone marrow produces blood cells; bones store Ca^{2+} for blood clotting.

Muscular System

Blood vessels deliver nutrients and oxygen to muscles; carry away wastes.

Muscle contraction keeps blood moving in heart and blood vessels.

Nervous System

Blood vessels deliver nutrients and oxygen to neurons; carry away wastes.

Brain controls nerves that regulate the heart and dilation of blood vessels.

Endocrine System

Blood vessels transport hormones from glands; blood services glands; heart produces atrial natriuretic hormone.

Epinephrine increases blood pressure; ADH, aldosterone, and atrial natriuretic hormone factors help regulate blood volume; growth factors control blood cell formation.

How the Cardiovascular System works with other body systems.

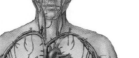

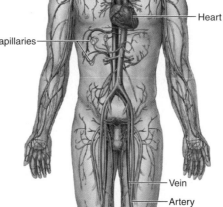
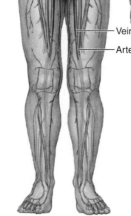

Heart

Capillaries

Vein

Artery

Reproductive System

Blood vessels transport sex hormones; vasodilation causes genitals to become erect; blood services reproductive organs.

Sex hormones influence cardiovascular health; sexual activities stimulate cardiovascular system.

Lymphatic System/Immunity

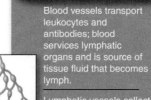

Blood vessels transport leukocytes and antibodies; blood services lymphatic organs and is source of tissue fluid that becomes lymph.

Lymphatic vessels collect excess tissue fluid and return it to blood vessels; lymphatic organs store lymphocytes; lymph nodes filter lymph, and the spleen filters blood.

Respiratory System

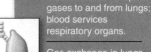

Blood vessels transport gases to and from lungs; blood services respiratory organs.

Gas exchange in lungs rids body of carbon dioxide, helping to regulate the pH of blood; breathing aids venous return.

Digestive System

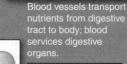

Blood vessels transport nutrients from digestive tract to body; blood services digestive organs.

Digestive tract provides nutrients for plasma protein formation and blood cell formation; liver detoxifies blood, makes plasma proteins, destroys old red blood cells.

Urinary System

Blood vessels deliver wastes to be excreted; blood pressure aids kidney function; blood services urinary organs.

Kidneys filter blood and excrete wastes; maintain blood volume, pressure, and pH; produce renin and erythropoietin.

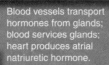

The blood distributes heat created by muscle contraction to the rest of the body. Blood vessels in the skin dilate when body temperature rises and constrict when heat needs to be conserved. In this way, the integumentary system plays a key role in regulating body temperature.

Maintaining Blood Pressure

The pumping of the heart is critical to creating the blood pressure that moves blood to the lungs, where oxygen is exchanged for carbon dioxide, and to the tissues, where gas exchange and nutrient-for-waste exchange take place. Only then is the brain able to think, the lungs to breathe, and the muscles to move. The importance of the heart to survival can be seen in the speed with which it develops during prenatal life. Long before other major organs, the heart and its vessels have taken shape and are ready to function.

The body has multiple ways to maintain blood pressure. Sensory receptors within the aortic arch signal regulatory centers in the brain when blood pressure falls. This center subsequently increases heartbeat and constricts blood vessels. Thereafter, blood pressure is restored. The lymphatic system collects excess tissue fluid at blood capillaries and returns it to cardiovascular veins in the thoracic cavity. In this way, the lymphatic system makes an important contribution to regulating blood volume and pressure.

The endocrine system assists the nervous system in maintaining homeostasis, so it is not surprising that hormones are also involved in regulating blood pressure. Epinephrine and norepinephrine bring about the constriction of arterioles. Other hormones, such as aldosterone, ADH, and ANH, regulate urine excretion. After all, if water is retained, blood volume and pressure will rise, and if water is excreted, blood volume and pressure will drop. In fact, some drugs prescribed for hypertension increase the amount of urine excreted.

Venous return from the capillaries to the heart is assisted by two other systems of the body: the muscular and respiratory systems. Skeletal muscle contraction pushes blood past the valves in the veins, and breathing movements encourage the flow of blood toward the heart in the thoracic cavity. Without smooth muscle, the walls of arterioles would not be able to constrict and in this way help raise blood pressure.

Platelets are necessary to blood clotting, which prevents the loss of blood and the loss of pressure. Clots, however, are not enough to stop massive blood loss. An individual who loses more than 10% of his or her blood will suffer a sudden drop in blood pressure and they usually go into shock. The decreased pressure triggers the body's last defense: A powerful wave of sympathetic impulses constricts the veins and arterioles throughout the body to slow the drop in blood pressure. Heart rate soars as high as 200 beats a minute to maintain blood flow, especially to the brain and heart itself. Because of this reflex, you can lose as much as 40% of your total blood volume and still live.

SELECTED NEW TERMS

Basic Key Terms

aorta (a-or'tuh), p. 263

arteriole (ar-te're-ōl), p. 257

artery (ar'ter-e), p. 257

atrioventricular (AV) node (a"tre-o-ven-trik'yū-ler nōd), p. 251

atrioventricular valve (a"tre-o-ven-trik'yū-ler valv), p. 248

atrium (a'tre-um), p. 248

bicuspid valve (bi-kus'pid valv), p. 248

capillary (kap'ĭ-lār"e), p. 257

cardioregulatory center (kar"de-o-reg'yū-luh-tor-e sen'ter), p. 256

cerebral arterial circle (sĕr'ĕ-bral ar-te're'al ser'kl), p. 267

chordae tendinae (kor'de ten'dĭ-nēa), p. 248

coronary artery (kor'ŏ-na-re ar'ter-e), p. 249

diastole (di-as'to-le), p. 254

ductus arteriosus (duk'tus ar-tēr-e-o'sus), p. 267

ductus venosus (duk'tus vĕ-no'sus), p. 267

endocardium (en"do-kar'de-um), p. 248

foramen ovale (fo-ra'men o-vah'le), p. 267

hepatic portal system (hĕ-pat'ik por'tal sis'tem), p. 267

inferior vena cava (in-fēr'e-or ve'nuh ka'vuh), p. 263

interatrial septum (in"ter-a'tre-al sep'tum), p. 248

interventricular septum (in"ter-ven-trik'yū-ler sep'tum), p. 248

mitral valve (mi'trəl vălv), p. 249

myocardium (mi"o-kar'de-um), p. 248

pacemaker (pās'mā'kər), p. 254

papillary muscles (păp'ə-lĕr'ē mŭs'əl), p. 248

pulmonary artery (pul'mo-nĕr"e ar'ter-e), p. 263

pulmonary circuit (pul'mo-nĕr"e ser"kyū-la'shun), p. 247

pulmonary vein (pul'mo-nĕr"e vān), p. 263

pulse (puls), p. 261

Purkinje fiber (per-kin'je fi'ber), p. 251

renin (rē'nĭn), p. 261

semilunar valve (sem"e-lu'ner valv), p. 248

sinoatrial (SA) node (si"no-a'tre-ul nōd), p. 251

superior vena cava (su-pēr'e-or ve'nuh ka'vuh), p. 263

systemic circuit (sis-tem'ik ser"kut), p. 247

systole (sis'to-le), p. 254

tricuspid valve (tri-kus'pid valv), p. 248

tunica externa (tōō'nĭ-kə ĕk'stûrnə), p. 257

tunica interna (tōō'nĭ-kə ĭn-tûr'nə), p. 257

tunica media (tōō'nĭ-kə mē'dē-ə), p. 257

umbilical artery and vein (um-bil'ĭ-kl ar'ter-e and vān), p. 267

vasomotor center (vă'zō-mō'tər sĕn'tər), p. 261

vein (vān), p. 258

venule (ven'ūl), p. 258

ventricles (vĕn'trĭ-kəlz), p. 248

viseral pericardium, p. 257

Clinical Key Terms

aneurysm (an'yer'ĭzm), p. 265

angina pectoris (an-ji'nuh pek'to-ris), p. 252

arrhythmias (ə-rĭth'mē-əz), p. 254

arteriosclerosis (ar-te"re-o-sklĕ-ro'sis), p. 252

atherosclerosis (ath"er-o"sklĕ-ro'sis), p. 252

bradycardia (brad"e-kar'de-uh), p. 255

cerebrovascular accident (sĕr"e-bro-vas'kyū-ler ak'si-dent), p. 268

congestive heart failure (kon-jes'tiv hart fāl'yer), p. 263

coronary artery disease (kôr'ə-nĕr'ē är'tə-rē dī-zēz), p. 252

coronary bypass operation (kor'ŏ-na-re bi'pas op-er-a'shun), p. 252

cyanosis (si"uh-no'sis), p. 269

ectopic pacemaker (ek-top'ik pās'ma-ker), p. 254

electrocardiogram (e-lek"tro-kar'de-o-gram"), p. 255

fibrillation (fĭ″brĭ-la′shun), p. 255
heart block (hart blok), p. 254
hemorrhoid (hem′royd), p. 258
hypertension (hī′pər-tĕn′shən), p. 252
ischemic heart disease (is-kem′ik hart dĭ-zēz′), p. 252

myocardial infarction (mi″o-kar′de-ul in-fark′shun), p. 252
phlebitis (flĭ-bi′tus), p. 258
plaque (plak), p. 252
pulmonary embolism (pul′mo-nĕr″e em′bo-lizm), p. 259
stent (stĕnt), p. 252

tachycardia (tak′ĭ kar′de-uh), p. 255
thromboembolism (thrŏm′bō-em′bə-lĭz′əm), p. 252
transient ischemic attack (TIA) (trăn-zē-ənt is-kem′ik′ ə-tăk′), p. 268
varicose vein (văr′ĭ-kōs vān), p. 258

SUMMARY

12.1 Anatomy of the Heart

A. The heart keeps O_2-poor blood separate from O_2-rich blood and blood flowing in one direction. It creates blood pressure and regulates the supply of blood to meet current needs.

B. The heart is covered by the pericardium. The visceral pericardium is equal to the epicardium of the heart wall. Myocardium is cardiac muscle, and endocardium is its lining.

C. The heart has a right and left side and four chambers, consisting of two atria and two ventricles. The heart valves are the tricuspid valve, the pulmonary semilunar valve, the bicuspid valve, and the aortic semilunar valve. The first heart sound, "lub," is caused by AV valve closure. The second heart sound, "dup," occurs when the semilunar valves close.

D. The right side of the heart pumps blood to the lungs (pulmonary circuit), and the left side pumps blood to the tissues (systemic circuit). The myocardium is serviced by blood in the coronary circuit.

12.2 Physiology of the Heart

A. The conduction system of the heart includes the SA node, the AV node, the AV bundle, the bundle branches, and the Purkinje fibers. The SA node causes the atria to contract. The AV node and the rest of the conduction system cause the ventricles to contract.

B. The heartbeat (cardiac cycle) is divided into three phases: (1) In atrial systole, the atria contract; (2) in ventricular systole, the ventricles contract; and (3) in atrial and ventricular diastole, both the atria and the ventricles rest.

C. The cardiac output (amount of blood discharged by the heart in one minute) is the product of stroke volume and heart rate. The heart rate is regulated largely by the cardioregulatory center and the autonomic nervous system.

12.3 Anatomy of Blood Vessels

A. Blood vessels transport blood; carry out exchange in pulmonary capillaries and systemic capillaries; regulate blood pressure; and direct blood flow.

B. Arteries and arterioles carry blood away from the heart; veins and venules carry blood to the heart; and capillaries join arterioles to venules.

12.4 Physiology of Circulation

A. Velocity of blood flow varies according to total cross-sectional area; therefore, blood flow is slowest in the capillaries.

B. Blood pressure decreases with distance from the left ventricle. Cardiac output (CO) and resistance to flow determine blood pressure. Venous return affects CO. The skeletal muscle pump and the respiratory pump assist venous return. A vasomotor center regulates peripheral resistance. Neural regulation of peripheral resistance is via a vasomotor center in the medulla that is under the control of the cardioregulatory center. Several different hormones regulate blood pressure through their influence over kidney reabsorption of water.

C. To evaluate a person's circulation, it is customary to take the pulse and blood pressure.

12.5 Circulatory Routes

A. The pulmonary arteries transport O_2-poor blood to the pulmonary capillaries, and the pulmonary veins return O_2-rich blood to the heart. Congestive heart failure occurs if blood backs up in the pulmonary vessels. In the systemic circuit, blood travels from the left ventricle to the aorta, systemic arteries, arterioles, and capillaries, and then from the capillaries to the venules and veins to the right atrium of the heart. The systemic circuit serves the body proper.

B. The hepatic portal system carries blood from the stomach and intestines to the liver.

C. Circulation to the brain includes the cerebral arterial circle, which protects all regions of the brain from reduced blood supply.

D. Fetal circulation includes four unique features: (1) the foramen ovale, (2) the ductus arteriosus, (3) the umbilical arteries and vein, and (4) the ductus venosus. These features are necessary because the fetus does not use its lungs for gas exchange.

12.6 Effects of Aging

As we age, the cardiovascular system is more apt to suffer from all the disorders discussed in this chapter.

12.7 Homeostasis

The cardiovascular system is essential to homeostasis because it functions to assure exchange at the pulmonary capillaries and the systemic capillaries. There are many examples of the interaction of the cardiovascular system with other systems. For example, the endocrine system is dependent on the cardiovascular system to transport its hormones; and hormones help maintain blood pressure. Blood vessels deliver wastes to the kidneys and the kidneys help maintain blood pressure. The respiratory system is dependent on the cardiovascular system to transport gases to and from cells, and the respiratory system assists venous return.

STUDY QUESTIONS

1. State the location and functions of the heart. (p. 247)
2. Describe the wall and coverings of the heart. (pp. 247–248)
3. Name the chambers and valves of the heart. Trace the path of blood through the heart. (pp. 248–249)
4. Describe the coronary circuit, and discuss several coronary circuit disorders. (pp. 250, 252)
5. Describe the conduction system of the heart and an electrocardiogram. (pp. 251, 254, 255)
6. Describe the cardiac cycle (using the terms systole and diastole), and explain the heart sounds. (p. 254)
7. What is cardiac output (CO)? What two factors determine CO? How are these factors regulated? (pp. 255–256)
8. What types of blood vessels are in the body? Discuss their structure and function. (pp. 257–258)
9. What factors determine velocity of blood flow? Blood pressure? In what vessel is blood pressure highest? Lowest? (p. 259)
10. What mechanisms assist venous return to the heart? Discuss nervous and hormonal control of blood pressure. (pp. 260–261)
11. What is pulse? How do you take a person's pulse? How do you take a person's blood pressure? What does a blood pressure of 120/80 mean? (pp. 261–262)
12. What are hypertension, stroke, aneurysm, and congestive heart failure? (pp. 263, 268)
13. Trace the path of blood from the superior mesenteric artery back to the aorta, indicating which of the vessels are in the systemic circuit and which are in the pulmonary circuit. (pp. 263–267)
14. Give examples to show that the cardiovascular system functions to maintain homeostasis and that interactions with other systems help it and the other systems maintain homeostasis. (p. 272)

LEARNING OBJECTIVE QUESTIONS

Fill in the blanks.

1. When the left ventricle contracts, blood enters the _____.
2. The right side of the heart pumps blood to the _____.
3. The _____ node is known as the pacemaker.
4. Arteries are blood vessels that take blood _____ the heart.
5. The blood vessels that serve the heart are the _____ arteries and veins.
6. The major blood vessels taking blood to and from the arms are the _____ arteries and veins. Those taking blood to and from the legs are the _____ arteries and veins.
7. Blood vessels to the brain end in a circular path known as the _____.
8. The human body contains a hepatic portal system that takes blood from the _____ to the _____.
9. The force of blood against the walls of a vessel is termed _____.
10. Blood moves in arteries due to _____ and in veins movement is assisted by _____.
11. The blood pressure recorded when the left ventricle contracts is called the _____ pressure, and the pressure recorded when the left ventricle relaxes is called the _____ pressure.
12. The two factors that affect blood pressure are _____ and _____.
13. In the fetus, the opening between the two atria is called the _____ and the connection between the pulmonary artery and the aorta is called the _____.
14. The valve between the left atrium and left ventricle is the _____, or mitral, valve.

MEDICAL TERMINOLOGY EXERCISE

Consult Appendix B for help in pronouncing and analyzing the meaning of the terms that follow.

1. cryocardioplegia (kri-o-kar″de-o-ple′je-uh)
2. echocardiography (ek″o-kar″de-og′ruh-fe)
3. percutaneous transluminal coronary angioplasty (per″kyŭ-ta′ne-us trans″lu′ mĭ-nal kor′ō-nā-re an′je-o-plas″te)
4. vasoconstriction (vas″o-kon-strik′shun)
5. valvuloplasty (val′vu-lo-plas″te)
6. antihypertensive (an″tĭ-hi″per-ten′siv)
7. arrhythmia (uh-rith′me-uh)
8. thromboendarterectomy (throm″bo-end″ar-ter-ek′to-me)
9. cardiovalvulitis (kar′de-o-val-yŭ-li′tis)
10. vasospasm (va′-so-spazm)
11. pericardiocentesis (pĕr-ĭ-kar′de-o-sen-te′sis)
12. ventriculotomy (ven-trik-yŭ-lot′o-me)
13. phlebectasia (fleb-ek-ta′ze-uh)
14. myocardiorrhaphy (mi′o-kar-de-or′uh-fe)

WEBSITE LINK

Visit this text's website at http://www.mhhe.com/maderap6 for additional quizzes, interactive learning exercises, and other study tools.

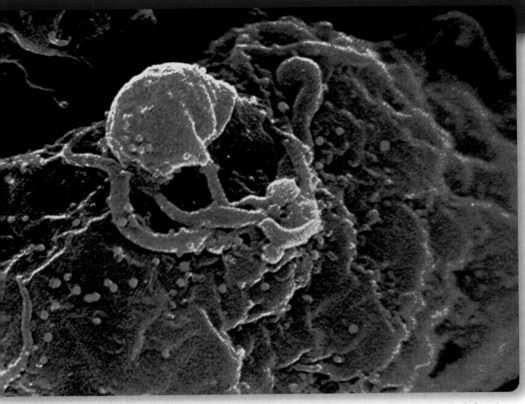

A scanning election micrograph captures the last moments of a white blood cell, as the HIV virus buds from its surface. HIV is the virus which causes AIDS, or acquired immune deficiency syndrome.

Chapter Outline & Learning Objectives

After you have studied this chapter, you should be able to:

13.1 Lymphatic System

The **lymphatic system** consists of lymphatic vessels and the lymphatic organs. This system, which is closely associated with the cardiovascular system, has three main functions that contribute to homeostasis:

1. **Fluid balance.** The lymphatic system takes up excess tissue fluid and returns it to the bloodstream. Recall that lymphatic capillaries lie very near blood capillaries, and they serve as an auxiliary way to take up fluid that has exited the blood capillaries (see Fig. 11.7).
2. **Fat absorption.** The lymphatic system absorbs fats from the digestive tract and transports them to the bloodstream. Special lymphatic capillaries called **lacteals** are located in the intestinal villi (see Fig. 15.7). This function ensures the absorption of dietary lipids as well as lipid-soluble vitamins.
3. **Defense.** The lymphatic system helps defend the body against disease. This function is carried out by the white blood cells present in lymphatic vessels and lymphatic organs, as well as those present in the blood and scattered between tissue cells. In addition to destroying foreign invaders, white blood cells are important in the destruction of dead and dying tissue cells. White blood cells also defend against body cells that have mutated to become cancerous.

Lymphatic Vessels

Lymphatic vessels form a one-way system that begins with lymphatic capillaries. Most regions of the body are richly supplied with lymphatic capillaries. These are tiny, closed-ended vessels whose walls consist of simple squamous epithelium (Fig. 13.1). Lymphatic capillaries take up excess tissue fluid. Tissue fluid is mostly water, but it also contains solutes (e.g., nutrients, electrolytes, and oxygen) derived from plasma and cellular products (i.e., hormones, enzymes, and wastes) secreted by cells. These all become **lymph,** the fluid inside lymphatic vessels.

The lymphatic capillaries join to form lymphatic vessels that merge before entering one of two ducts: the thoracic duct or the right lymphatic duct. The larger, thoracic duct returns lymph collected from the body below the thorax and the left arm and left side of the head and neck into the left subclavian vein. The right lymphatic duct returns lymph from the right arm and right side of the head and neck into the right subclavian vein.

The construction of the larger lymphatic vessels is similar to that of cardiovascular veins, including the presence of valves. The movement of lymph within lymphatic capillaries is largely dependent upon skeletal muscle contraction. Lymph forced through lymphatic vessels as a result of muscular compression is prevented from flowing backward

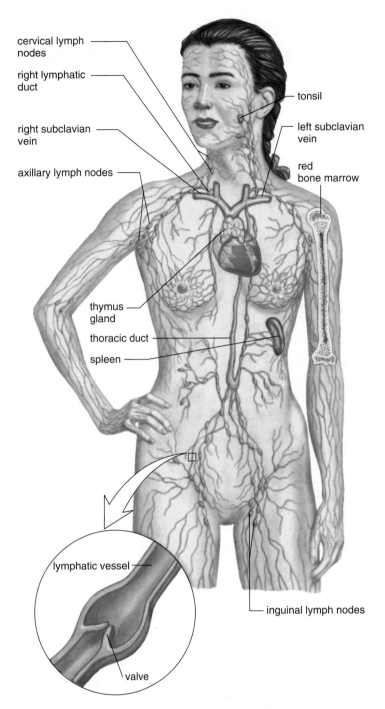

Figure 13.1 The lymphatic system. Lymphatic vessels drain excess fluid from the tissues and return it to the cardiovascular system. The enlargement shows that lymphatic vessels, like cardiovascular veins, have valves to prevent backward flow. The tonsils, spleen, thymus gland, and red bone marrow are among those lymphatic organs that assist immunity.

by one-way valves (see Fig. 13.1). Failure of the lymphatic system to properly collect and return excessive tissue fluid to the bloodstream results in **edema,** as described in the Medical Focus on page 281.

Bone marrow is the source of the stem cells that give rise to all of the formed elements of the blood: erythrocytes, leukocytes, and megakaryocytes (parent cells that fragment to form platelets or thrombocytes). In *aplastic anemia*, bone marrow is destroyed, typically by toxins or viruses. A bone marrow transplant is essential. Cancer patients may also require a bone marrow transplant if the cancer destroys bone marrow, as in leukemia. High doses of chemotherapy and/or radiation therapy used to fight cancer will kill off cancerous cells and improve the chances of curing the disease. However, chemotherapy and radiation may also destroy bone marrow. Without healthy bone marrow, the patient will die of infection, anemia, or hemorrhage.

In *autologous* marrow transplants, the marrow is removed from the patient before cancer therapy begins; the marrow is stored alive, and then it is returned to the patient. In *allogenic* marrow transplants, the marrow is donated by someone else. As with any other allogenic transplant, bone marrow transplants require careful matching of donor and recipient tissue and administration of drugs to suppress the immune system and avoid transplant rejection.

Bone marrow for a transplant is obtained surgically using general anesthesia, though it is usually an outpatient procedure requiring only a short hospital stay for the donor. With the donor lying on his or her stomach or side, a large needle is positioned perpendicular to the pelvis and pushed into the bone, using a screwing motion. When the needle is deep enough in the bone to be anchored, a syringe is attached in order to remove a sample of bone marrow. Then, to perform the transplant, the marrow is injected into the recipient's bloodstream. The bone marrow stem cells are expected to migrate to the recipient's marrow and produce new formed elements.

If available, umbilical cord blood can also be used for transplantation. The immature cells found in cord blood are easier to match between nonrelated people than are bone marrow cells. When cord blood is used, there is also a far less chance of the recipient rejecting the transplant.

Content CHECK-UP!

1. Which of the following statements about the lymphatic system is false?
   a. It takes up excess tissue fluid and returns it to the circulation.
   b. It houses white blood cells.
   c. Lymphatic capillaries are composed of stratified squamous epithelium.
   d. Lacteals in the intestines absorb dietary lipids.

2. Imagine that lymph is draining from your right arm to return to the heart. What is the correct pathway for it to take?
   a. lymphatic capillaries → larger lymphatic vessels → right lymphatic duct → right subclavian vein → superior vena cava
   b. lymphatic capillaries → larger lymphatic vessels → thoracic duct → right subclavian vein → superior vena cava
   c. lymphatic capillaries → larger lymphatic vessels → thoracic duct → left subclavian vein → superior vena cava
   d. lymphatic capillaries → larger lymphatic vessels → cervical duct → left subclavian vein → superior vena cava

3. Large lymphatic vessels are similar in structure to:
   a. capillaries in the cardiovascular system.
   b. arteries in the cardiovascular system.
   c. veins in the cardiovascular system.

13.2 Organs, Tissues, and Cells of the Immune System

The **immune system,** which plays an important role in keeping us healthy, consists of a network of lymphatic organs, tissues, and cells as well as products of these cells, including antibodies and regulatory agents. **Immunity** is the ability to react to antigens so that the body remains free of disease. Disease results from a failure of homeostasis. It can be due to infection by foreign microbes, and/or to the failure of the immune system to function properly.

Primary Lymphatic Organs

Lymphatic (lymphoid) organs contain large numbers of lymphocytes, the type of white blood cell that plays a pivotal role in immunity. The *primary lymphatic organs* are the red bone marrow and the thymus gland (Fig. 13.2, *left*). Lymphocytes originate and/or mature in these organs. The Medical Focus on this page discusses bone marrow transplant, a life saving therapy used when bone marrow is destroyed.

Red Bone Marrow

Red bone marrow is the site of stem cells that are ever capable of dividing and producing blood cells. Some of these cells become the various types of white blood cells:

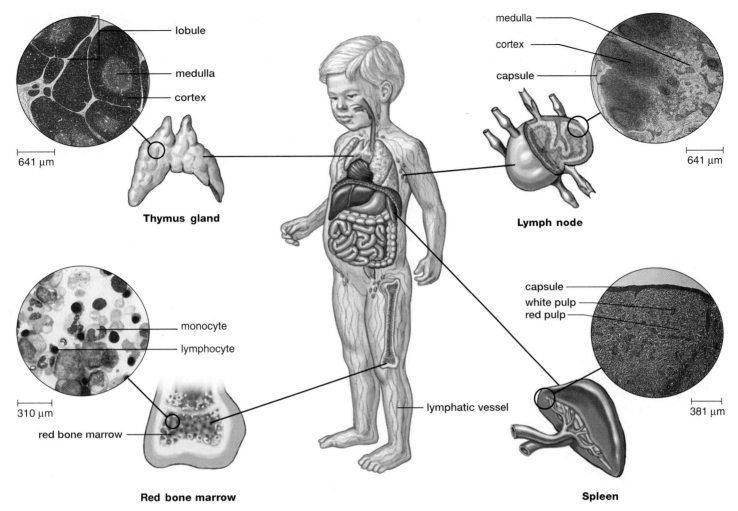

Figure 13.2 The lymphatic organs. *Left:* The red bone marrow and thymus gland are the primary lymphatic organs. *Right:* Lymph nodes and the spleen, as well as other lymphatic organs such as the tonsils, are secondary lymphatic organs.

neutrophils, eosinophils, basophils, lymphocytes, and monocytes (Fig. 13.3).

In a child, most bones have red bone marrow, but in an adult it is limited to the sternum, vertebrae, ribs, the skull, part of the pelvic girdle, and the proximal heads of the humerus and femur.

The red bone marrow consists of a network of reticular tissue fibers, which support the stem cells and their progeny. They are packed around thin-walled sinuses filled with venous blood. Differentiated blood cells enter the bloodstream at these sinuses.

Lymphocytes differentiate into the B lymphocytes and the T lymphocytes. Bone marrow is not only the source of B lymphocytes, but also the place where B lymphocytes mature. T lymphocytes mature in the thymus.

Thymus Gland

The soft, bilobed **thymus gland** is located in the thoracic cavity between the trachea and the sternum superior to the heart. The thymus varies in size, but it is largest in children and shrinks as we get older. Connective tissue divides the thymus into lobules, which are filled with lymphocytes. The thymus gland produces thymic hormones, such as thymosin, that are thought to aid in the maturation of T lymphocytes. Thymosin may also have other functions in immunity.

Immature T lymphocytes migrate from the bone marrow through the bloodstream to the thymus, where they mature. Only about 5% of these cells ever leave the thymus. These T lymphocytes have survived a critical test: If any show the ability to react with "self" cells (i.e., one's own tissue cells), they die. If they have potential to attack a foreign cell, they leave the thymus.

The thymus is absolutely critical to immunity; without a thymus, the body does not reject foreign tissues, blood lymphocyte levels are drastically reduced, and the body's response to most antigens is poor or absent.

Secondary Lymphatic Organs

The secondary lymphatic organs are the spleen, the lymph nodes, and other organs, such as the tonsils, Peyer patches, and the appendix. All the secondary organs are places where

a. Neutrophil
40–70%
Phagocytizes
primarily bacteria

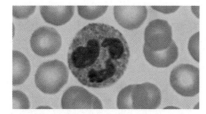

b. Eosinophil
1–4%
Phagocytizes and
destroys antigen-antibody
complexes and parasitic
organisms

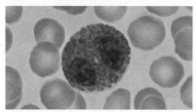

c. Basophil
0–1%
Releases histamine and
other chemicals when
stimulated

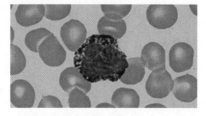

d. Lymphocyte
20–45%
B type produces
antibodies in blood and
lymph; T type kills virus-
containing cells and
cancer cells

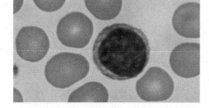

e. Monocyte
4–8%
Becomes macrophage—
phagocytizes bacteria
and viruses

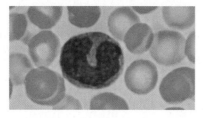

Figure 13.3 The five types of white blood cells. These cell types differ according to structure and function. The frequency of each type of cell is given as a percentage of the total found in circulating blood.

lymphocytes encounter and bind with antigens, after which they proliferate and become actively engaged cells.

Spleen

The **spleen,** the largest lymphatic organ, is located in the upper left region of the abdominal cavity posterior to the stomach. Connective tissue divides the spleen into partial compartments, each of which contains tissue known as white pulp and red pulp (see Fig. 13.2). The white pulp contains a concentration of lymphocytes; the red pulp, which surrounds venous sinuses, is involved in filtering the blood. Blood entering the spleen must pass through the sinuses before exiting. Lymphocytes and macrophages react to pathogens, and macrophages engulf debris and also remove any old, worn-out red blood cells.

The spleen's outer capsule is relatively thin, and an infection or a blow can cause the spleen to burst. Although the spleen's functions are replaced by other organs, a person without a spleen is often slightly more susceptible to infections and may have to receive antibiotic therapy indefinitely.

Lymph Nodes

Lymph nodes, which are small, ovoid structures, occur along lymphatic vessels. Connective tissue forms the capsule of a lymph node and also divides the organ into compartments (see Fig. 13.2). Each compartment contains a nodule packed with B lymphocytes and a sinus that increases in size toward the center of the node. As lymph courses through the sinuses, it is filtered by macrophages (phagocytic white blood cells), which engulf pathogens and debris. T lymphocytes, also present in sinuses, fight infections and attack cancer cells.

Each portion of the anterior cavity (see Fig 1.5) contains superficial and deep lymph nodes, named for their location. For example, inguinal nodes are in the groin, and axillary nodes are in the armpits. Physicians often examine for the presence of swollen, tender lymph nodes as evidence that the body is fighting an infection. This is a noninvasive, preliminary way to help make such a diagnosis.

Lymphatic Nodules

Lymphatic nodules are concentrations of lymphatic tissue not surrounded by a capsule. The **tonsils** are lymphatic nodules located in the posterior pharynx (see Fig. 14.2). A single **pharyngeal tonsil** (also referred to as adenoids) is in the nasopharynx, whereas **lingual tonsils** sit at the base of the tongue. **Palatine tonsils** are visible in the posterior oral cavity. The tonsils perform the same functions as lymph nodes, but because of their location, they are the first to encounter pathogens and antigens that enter the body by way of the nose and mouth.

Peyer patches are located in the intestinal wall and the walls of the **appendix.** These structures encounter pathogens that enter the body by way of the intestinal tract.

Content **CHECK-UP!**

4. Which lymphatic organ functions as a blood reservoir and removes dead and dying red blood cells?

   a. thymus   c. appendix

   b. Peyer patches   d. spleen

5. Which of the following is not a function of the thymus gland?

   a. provides place where T lymphocytes mature

   b. produces and secretes thymic hormones

   c. provides a place where B lymphocytes mature

   d. vital to normal development of the body's immune response

6. Which of the following structures has a connective tissue capsule?

   a. axillary lymph nodes   c. Peyer patches

   b. palatine tonsils   d. adenoids (pharyngeal tonsil)

The Lymphatic System and Illness

As the "next-door neighbor" to the cardiovascular system, the parallel lymphatic system defends against infectious disease and cancer, and regulates fluid balance in the tissues. Failure of the lymphatic system to maintain homeostatic balance in these critical areas results in disease.

The internal structure of a lymph node is designed to filter out and destroy any foreign material, including microbes and cancer cells, from lymph. An infection that causes swelling and tenderness of nearby lymph nodes is called **lymphadenitis.** If the infection is not contained, **lymphangitis,** an infection of the lymphatic vessels, may result. Red streaks can be seen through the skin, indicating that the infection may spread to the bloodstream.

Like microbes, cancer cells sometimes enter lymphatic vessels and move undetected to other regions of the body, where they produce secondary tumors. In this way, the lymphatic system sometimes assists metastasis, the spread of cancer far from its place of origin. Because of this potential for metastasis, regional lymph nodes are usually removed for examination whenever surgery is used to diagnose or treat cancer. The presence or absence of tumor cells in the nodes can be used to determine how far the cancer has spread. Lymph node *biopsy* (the microscopic study of the tissue) also aids in making decisions concerning additional treatment, such as radiation or chemotherapy.

Failure of the lymphatic vessels to properly remove tissue fluid results in a condition called edema. Edema forms either because too much tissue fluid is made, and/or not enough tissue fluid is drained away. Excessive tissue fluid can compress the blood vessels within a tissue, preventing the flow of blood, oxygen and nutrients to tissue cells. Thus, edema can lead to tissue damage and eventual death, illustrating the importance of tissue fluid collection by the lymphatic system.

Blockage of the lymphatic vessels prevents proper tissue fluid drainage, resulting in edema. As mentioned previously, surgical removal of lymph nodes is a common treatment for cancer. However, node removal often blocks lymphatic drainage, producing a painful and debilitating **lymphedema** for the patient. A dramatic example of edema caused by lymphatic obstruction occurs when a parasitic roundworm clogs the lymphatic vessels. Tremendous swelling of the arm, leg, or external genitals results, causing a condition called **elephantiasis.**

Edema can also be due to a low osmotic pressure of the blood, as when plasma proteins are excreted by the kidneys instead of being retained in the blood. Extra tissue fluid forms and lymphatic vessels may not be able to absorb it all. **Pulmonary edema** is a life-threatening condition associated with congestive heart failure. When the heart muscle is damaged and unable to effectively empty the heart, blood backs up in the pulmonary circulation. Pulmonary capillary blood pressure increases, which leads to excess fluid in lung tissue. Congested lung tissue cannot properly exchange oxygen and carbon dioxide with the blood, and the patient may suffocate.

Cancer of lymphatic tissue is called **lymphoma.** In **Hodgkin disease,** billions of lymphoma cells create swollen lymph nodes throughout the body. The lymphoma cells can migrate and grow in the spleen, liver and bone marrow. The prognosis is good, however, if Hodgkin disease is diagnosed early.

13.3 Nonspecific and Specific Defenses

Immunity includes nonspecific defenses and specific defenses. The five types of nonspecific defenses—barriers to entry, the inflammatory reaction, nonspecific phagocytic white blood cells, natural killer cells, and protective proteins—are effective against many types of infectious agents. Specific defenses are effective against a particular infectious agent.

Nonspecific Defenses

Barriers to Entry

The body has built-in barriers, both physical and chemical, that help to prevent infection by microbes. The intact skin is generally a very effective physical barrier that prevents infection. Mucous membranes lining the respiratory, digestive, reproductive, and urinary tracts are also physical barriers to entry by pathogens. (However, it must be noted that their effectiveness is limited, especially in the oral and nasal cavities, because most cold viruses enter the body by crossing these membranes.) The ciliated cells that line the upper respiratory tract sweep mucus and trapped particles up into the throat, where they can be swallowed or expectorated (spit out).

Chemical barriers to infection include the secretions of sebaceous (oil) glands in the skin, which contain chemicals that weaken or kill certain bacteria on the skin. Perspiration, saliva, and tears contain an antibacterial enzyme called *lysozyme*. Saliva also helps to wash microbes off the teeth and tongue. Similarly, as urine is voided from the body, it flushes bacteria from the urinary tract. The acidic pH of the stomach inhibits bacterial growth or kills many types of bacteria. Finally, a significant chemical barrier to infection is created by the *normal flora*—microbes that normally reside in the mouth, intestine, and other areas. By using up available nutrients and releasing their own waste, these normal germs prevent potential pathogens from taking up residence. For this reason, abusing antibiotics can make a person susceptible to pathogenic infection by killing off the normal flora.

Inflammatory Reaction

Whenever tissue is damaged by physical or chemical agents or by pathogens, a series of events occurs that is known as the **inflammatory reaction.** Figure 13.4 illustrates the participants in the inflammatory reaction. **Mast cells,** which occur in tissues, resemble basophils, one of the types of white cells found in the blood.

The inflamed area has four outward signs: redness, heat, swelling, and pain. All of these signs are due to capillary changes in the damaged area, and all serve to protect the body. Chemical mediators, such as **histamine,** released by damaged tissue cells and mast cells, cause the capillaries to dilate and become more permeable. Excess blood flow due to enlarged capillaries causes the skin to redden and become warm. Increased temperature in an inflamed area tends to inhibit growth of some pathogens, and increased blood flow brings more defensive white blood cells to the area. Increased permeability of capillaries allows fluids and proteins, including blood clotting factors, to escape into the tissues. Swelling and clot formation in the injured area will result, helping to "wall off" the area from the rest of the body. Thus, if pathogens have entered the body during injury, they will not be able to cause a system-wide infection. Increased pressure in the swollen area, as well as the inflammatory chemicals themselves, stimulate free nerve endings to cause the sensation of pain. Although we can all agree that pain is a big nuisance, it is also a warning that tissue has been injured and corrective measures need to be taken.

Migration of phagocytes, namely neutrophils and monocytes, also occurs during the inflammatory reaction. Neutrophils and monocytes are amoeboid and can change shape to squeeze through capillary walls and enter tissue fluid. After monocytes appear on the scene, they differentiate into **macrophages,** large phagocytic cells that are able to devour as many as a hundred pathogens and still survive. Some tissues, particularly connective tissue, have resident macrophages, which routinely act as scavengers, devouring old blood cells, bits of dead tissue, and other debris. Macrophages also release colony-stimulating factors, which pass by way of blood to the red bone marrow, where the factors stimulate the production and the release of white blood cells, primarily neutrophils. During the process of phagocytosis, endocytic vesicles form when neutrophils and macrophages engulf pathogens (see section 3.2, page 50). When the vesicle combines with a lysosome, a cellular organelle, the pathogen is destroyed by hydrolytic enzymes. As the infection is being overcome, some phagocytes die. These—along with dead tissue cells, dead bacteria, and living white blood cells—form **pus,** a whitish material. The presence of pus indicates that the body is destroying dead tissue cells, and/or trying to overcome an infection.

Sometimes an inflammation persists, and the result is chronic inflammation that is often treated by administering anti-inflammatory agents such as aspirin, ibuprofen, or cortisone. These medications act against the chemical mediators released by the white blood cells in the damaged area and also decrease the phagocytic response by white blood cells.

Although an inflammatory reaction is a nonspecific defense mechanism that occurs whenever tissue is injured, it can be accompanied by a specific, directed response to the injury. As infectious microbes are destroyed by neutrophils and macrophages, bits and pieces of the dead cells can cross the barrier created by inflammation. These fragments, called **antigens,** can travel along with the released chemical mediators through the tissue fluid and lymph to the lymph nodes. Now lymphocytes mount a specific defense to the infection, as described on page 284.

Natural Killer Cells

Natural killer (NK) cells kill virus-infected cells and tumor cells by cell-to-cell contact. They are large, granular leukocytes, similar in structure to lymphocytes, but with no specificity and no memory. Their number is not increased by prior exposure to any kind of cell.

Protective Proteins

The **complement system,** often simply called complement, is composed of a number of blood plasma proteins designated by the letter C and a subscript. A limited amount of activated complement protein is needed because a cascade effect occurs: Each activated protein in a series is capable of activating many other proteins.

The complement proteins are activated when pathogens enter the body. The proteins "complement" certain immune responses, which accounts for their name. For example, they are involved in and amplify the inflammatory response because complement proteins attract phagocytes to the scene. Some complement proteins bind to the surface of pathogens already coated with antibodies (specific antibacterial protein molecules), which ensures that the pathogens will be phagocytized by a neutrophil or macrophage.

Certain other complement proteins join to form a membrane attack complex that punches holes in the walls and plasma membranes of bacteria. Fluids and salts then enter the bacterial cell to the point that it bursts.

Interferon is a protein produced by virus-infected cells. Interferon binds to receptors of noninfected cells, causing them to prepare for possible attack by producing substances that interfere with viral replication. Interferon is specific to the species; therefore, only human interferon can be used in humans.

Specific Defenses

Specific defenses respond to antigens, which are molecules the immune system recognizes to be foreign. Antigens are typically large molecules like proteins. Fragments of bacteria, viruses, molds, or parasites can all be antigenic; further, abnormal cell membrane proteins produced by cancer cells may also be antigens. Because we do not ordinarily become immune to our own normal cells, it is said that the immune system is able to distinguish "self" from "nonself." Lymphocytes are capable of recognizing an antigen because they have antigen receptors—plasma membrane receptor proteins that combine with a specific antigen.

Immunity usually lasts for some time. For example, once a child has been properly immunized against measles virus,

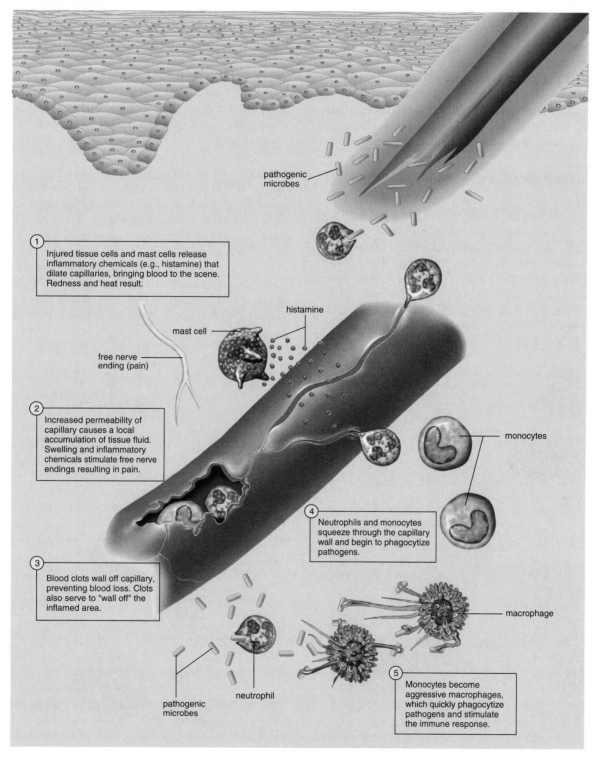

① Injured tissue cells and mast cells release inflammatory chemicals (e.g., histamine) that dilate capillaries, bringing blood to the scene. Redness and heat result.

histamine

mast cell

free nerve ending (pain)

② Increased permeability of capillary causes a local accumulation of tissue fluid. Swelling and inflammatory chemicals stimulate free nerve endings resulting in pain.

③ Blood clots wall off capillary, preventing blood loss. Clots also serve to "wall off" the inflamed area.

pathogenic microbes

monocytes

④ Neutrophils and monocytes squeeze through the capillary wall and begin to phagocytize pathogens.

macrophage

⑤ Monocytes become aggressive macrophages, which quickly phagocytize pathogens and stimulate the immune response.

neutrophil

pathogenic microbes

Figure 13.4 Inflammatory reaction. Mast cells, which are related to basophils, a type of white blood cell, are involved in the inflammatory reaction. When tissue cells are injured, mast cells release substances such as histamine. Histamine dilates blood vessels and increases their permeability so that tissue fluid leaks from the vessel. Swelling and inflammatory chemicals stimulate pain receptors (free nerve endings). Neutrophils and monocytes (which become macrophages) squeeze through the capillary wall. These white blood cells begin to phagocytize pathogens (e.g., disease-causing viruses and bacteria), especially those combined with antibodies. Blood clotting seals off the capillary, preventing blood loss and walling off the inflamed area.

the immune system confers a lifelong protection (see the Medical Focus on page 290).

Immunity is primarily the result of the action of the **B lymphocytes** and the **T lymphocytes.** B lymphocytes mature in the *b*one marrow,[1] and T lymphocytes mature in the *t*hymus gland. B lymphocytes, also called B cells, give rise to plasma cells, which produce antibodies. **Antibodies** are proteins shaped like the antigen receptor and capable of combining with and neutralizing a specific antigen. These antibodies are secreted into the blood, lymph, and other body fluids. In contrast, T lymphocytes, also called T cells, do not produce antibodies. Instead, certain T cells directly attack cells that bear nonself proteins. Other T cells regulate the immune response.

B Cells and Antibody-Mediated Immunity

When a B cell encounters a specific antigen, it is activated to divide many times. Most of the resulting cells are plasma cells. A **plasma cell** is a mature B cell that mass-produces antibodies against a specific antigen. The **clonal selection theory** states that the antigen *selects* one particular lymphocyte to reproduce multiple copies of itself. The resulting group of identical cells is called a *clone*. The majority of the B lymphocytes in the clone

will mature to form plasma cells, all bearing the same type of antigen receptor. Notice in Figure 13.5 that different types of antigen receptors are represented by color. The B cell with blue receptors undergoes clonal expansion because a specific antigen (red dots) is present and binds to its receptors. B cells are stimulated to divide and become plasma cells by helper T-cell secretions called cytokines, as discussed later in this section. Some members of the clone become memory cells, which are the means by which long-term immunity is possible. If the same antigen enters the system again, **memory B cells** quickly divide and give rise to more lymphocytes capable of quickly producing antibodies. A second exposure to the same antigen produces a stronger, faster immune response.

Once the threat of an infection has passed, the development of new plasma cells ceases, and those present undergo apoptosis. **Apoptosis** is a process of programmed cell death involving a cascade of specific cellular events leading to the death and destruction of the cell. It is important that these cells die once they are no longer needed. Otherwise they could mistakenly destroy body cells in an **autoimmune response.**

Defense by B cells is called **antibody-mediated immunity** because the various types of B cells produce antibodies. It is also called **humoral immunity** because these antibodies are present in blood and lymph. A humor is any fluid normally occurring in the body.

---

[1] Historically, the B stands for bursa of Fabricius, an organ in the chicken where these cells were first identified.

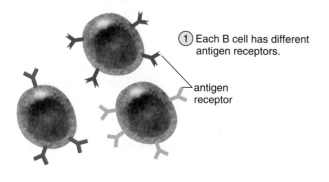

① Each B cell has different antigen receptors.

antigen receptor

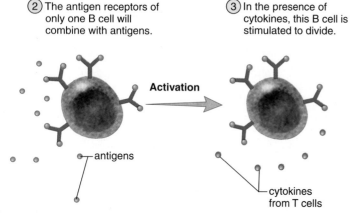

② The antigen receptors of only one B cell will combine with antigens.

Activation

③ In the presence of cytokines, this B cell is stimulated to divide.

antigens

cytokines from T cells

Figure 13.5 Clonal selection theory as it applies to B cells.

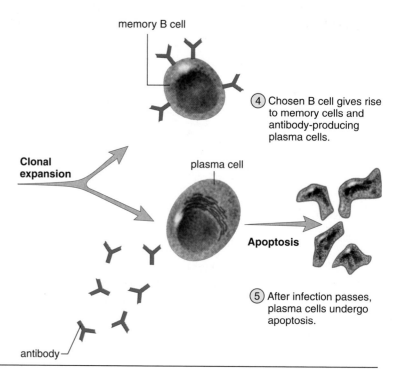

memory B cell

④ Chosen B cell gives rise to memory cells and antibody-producing plasma cells.

Clonal expansion

plasma cell

Apoptosis

⑤ After infection passes, plasma cells undergo apoptosis.

antibody

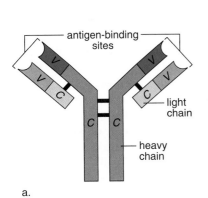

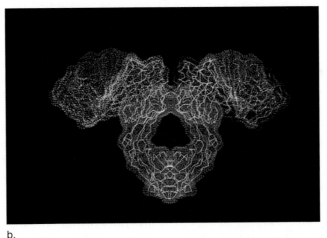

Figure 13.6 Structure of a single antibody molecule (monomer). **a.** Monomers contain two heavy (long) polypeptide chains and two light (short) chains arranged so that there are two variable regions, where a particular antigen is capable of binding with an antibody (V = variable region, C = constant region). **b.** Computer model of an antibody molecule. The antigen combines with the two side branches.

Structure and Function of Antibodies

The basic unit that composes antibody molecules is a Y-shaped protein molecule with two arms. Each arm has a "heavy" (long) polypeptide chain and a "light" (short) polypeptide chain (Fig. 13.6). These chains have constant regions, located at the trunk of the Y, where the sequence of amino acids is set. The class of antibody for each individual molecule is determined by the structure of the antibody's constant region. The variable regions form an antigen-binding site, and their shape is specific to a particular antigen. The antigen combines with the antibody at the antigen-binding site in a lock-and-key manner. Antibodies may consist of single Y-shaped molecules called *monomers,* or they can be paired together in a molecule called a *dimer.* Very large antibodies belonging to the M class are *pentamers*—clusters of five Y-shaped molecules linked together.

The antigen-antibody reaction can take several forms. Antibodies react with viruses and toxins (the poisons made by bacteria) by coating them completely—a process called *neutralization.* Often the reaction produces *an immune complex*—a clump of antigens combined with antibodies. Clustering antigens into an immune complex marks the antigens for destruction. For example, an antigen-antibody complex may be engulfed by neutrophils or macrophages, or it may activate

complement. Complement makes pathogens more susceptible to phagocytosis, as discussed previously.

Classes of Antibodies

There are five different classes of circulating antibody proteins, or immunoglobulins (Igs) (Table 13.1). IgG antibodies are the major type in blood, and lesser amounts are also found in lymph and tissue fluid. IgG antibodies bind to pathogens and their toxins. IgG antibodies can cross the placenta from a mother to her fetus, so that the newborn has a temporary, partial immune response. As mentioned previously, IgM antibodies are pentamers. These antibodies are "first responders": the first antibodies produced by a newborn's body. Subsequently, IgM antibodies are the first to appear in blood soon after an infection begins and the first to disappear before the infection is over. They are good activators of the complement system. IgA antibodies are monomers or dimers containing two Y-shaped structures. They are the main type of antibody found in body secretions: saliva, tears, mucus, and breast milk. IgA molecules bind to pathogens and prevent them from reaching the bloodstream. The main function of IgD molecules seems to be to serve as antigen receptors on immature B cells. IgE antibodies are responsible for

TABLE 13.1 Antibodies		
Classes	**Presence**	**Function**
IgG	Main antibody type in circulation; crosses the placenta from mother to fetus	Binds to pathogens, activates complement proteins, and enhances phagocytosis
IgM	Antibody type found in circulation; largest antibody; first antibody formed by a newborn; first antibody formed with any new infection	Activates complement proteins; clumps cells
IgA	Main antibody type in secretions such as saliva and milk	Prevents pathogens from attaching to epithelial cells in digestive and respiratory tract
IgD	Antibody type found on surface of immature B cells	Presence signifies readiness of B cell
IgE	Antibody type found as antigen receptors on basophils in blood and on mast cells in tissues	Responsible for immediate allergic response and protection against certain parasitic infections

prevention of parasitic infections, but they can also cause immediate allergic responses.

T Cells and Cell-Mediated Immunity

When T cells leave the thymus, they have unique antigen receptors just as B cells do. Unlike B cells, however, T cells are unable to recognize an antigen present in lymph, blood, or the tissues without help. The antigen must be presented to them by an **antigen-presenting cell (APC)**, which is a specialized form of macrophage cell. When an APC presents a viral or cancer cell antigen to the T cell, the antigen is first linked to a major histocompatibility complex (MHC) protein in the APC cell's plasma membrane.

Human MHC proteins are called **HLA (human leukocyte antigens)**. These proteins are found on all of the body's cells. More than 50 different proteins have been identified, and each individual human being has a unique combination—no two sets are exactly alike. An exception is the set belonging to monozygotic, or identical twins. Because identical twins arise from division of a single zygote, their HLA proteins are identical. Because HLA antigens mark the cell as belonging to a particular individual, they are *self* proteins. The importance of self proteins in plasma membranes was first recognized when it was discovered that they contribute to the specificity of tissues and make it difficult to transplant tissue from one human to another. Comparison studies of HLA antigens must always be carried out before a transplant is attempted. The more

of the 50-plus proteins that the donor and recipient share in common, the better the tissue match. In other words, when the donor and the recipient are histo (tissue)-compatible, a transplant is more likely to be successful.

When an antigen-presenting cell links a foreign antigen to the self protein on its plasma membrane, it carries out an important safeguard for the rest of the body. Now the T cell to be activated can compare the antigen and self protein side by side. The activated T cell and all of the daughter cells it will form can recognize "foreign" from "self," and go on to destroy cells carrying foreign antigens while leaving normal body cells unharmed.

Figure 13.7 shows a macrophage presenting an antigen, represented by a red circle, to a particular T cell. This T cell has the type of antigen receptor that will combine with this specific antigen. In the figure, the different types of antigen receptors are represented by color. Presentation of the antigen leads to activation of the T cell. An activated T cell produces cytokines and undergoes clonal expansion. **Cytokines** are signaling chemicals that stimulate various immune cells (e.g., macrophages, B cells, and other T cells) to perform their functions. Many copies of the activated T cell are produced during clonal expansion. They destroy any cell, such as a virus-infected cell or a cancer cell, that displays the antigen presented earlier.

As the illness disappears, the immune reaction wanes, and fewer cytokines are produced. Now, the activated T cells become susceptible to apoptosis. As mentioned previously, apoptosis is programmed cell death that contributes to homeostasis by regulating the number of cells present in an organ, or

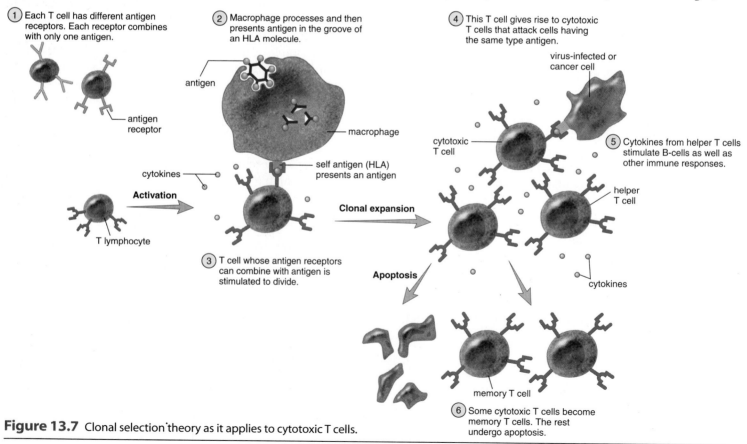

Figure 13.7 Clonal selection theory as it applies to cytotoxic T cells.

in this case, in the immune system. When apoptosis does not occur as it should, the potential exists for self-destruction called an (autoimmune response). Occasionally, T-cell cancers (i.e., lymphomas and leukemias) can result.

Apoptosis also occurs in the thymus as T cells are maturing. Any T cell that has the potential to destroy the body's own cells undergoes suicide.

Types of T Cells

The two main types of T cells are cytotoxic T cells and helper T cells. **Cytotoxic T (T_c) cells**, also called CD_8 leukocytes, can bring about the destruction of antigen-bearing cells, such as virus-infected or cancer cells. Cancer cells also have nonself proteins.

Cytotoxic T cells have storage vacuoles containing perforin molecules. **Perforin** molecules punch holes into a plasma membrane, forming a pore that allows water and salts to enter. The cell then swells and eventually bursts. Cytotoxic T cells are responsible for so-called **cell-mediated immunity** (Fig. 13.8).

Helper T (T_h) cells, also called CD_4 leukocytes, regulate immunity by secreting cytokines, the chemicals that enhance the response of other immune cells. Because HIV, the virus that causes AIDS, infects helper T cells and certain other cells of the immune system, it inactivates the immune response. AIDS is described in the Medical Focus, p. 288.

Notice in Figure 13.7 that a few of the clonally expanded T cells are **memory T cells.** They remain in the body and can jump-start an immune reaction to an antigen previously present in the body.

Cytokines in Cancer Chemotherapy

Whenever cancer develops, it is possible that cytotoxic T cells have not been activated. With this possibility in mind, cytokines have been used as immunotherapeutic drugs to enhance the ability of T cells to fight cancer. Interferon, discussed on page 282, and also **interleukins,** which are cytokines produced by various white blood cells, are also being administered for this purpose.

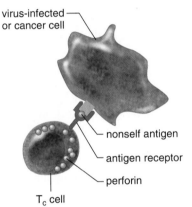

virus-infected or cancer cell

nonself antigen

antigen receptor

perforin

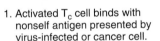

T_c cell

1. Activated T_c cell binds with nonself antigen presented by virus-infected or cancer cell.

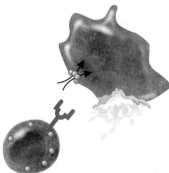

pore

2. T_c cell discharges perforin molecules, which combine to form pores in target cell's plasma membrane.

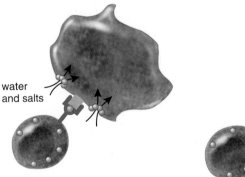

water and salts

3. Water and salts enter virus-infected or cancer cell.

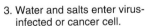

4. The target cell bursts.

a. Cytotoxic (T_c) cell attacks a target cell.

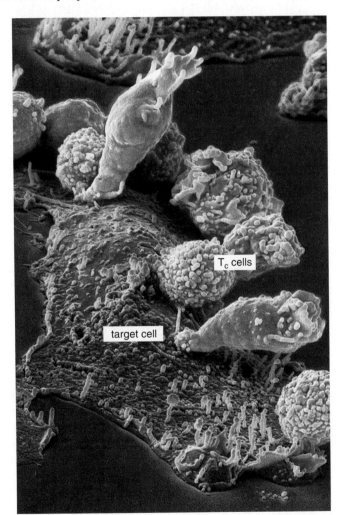

T_c cells

target cell

b. Scanning electron micrograph

1 μm

Figure 13.8 Cell-mediated immunity. **a.** How a cytotoxic T (T_c) cell destroys a virus-infected or cancer cell. **b.** The scanning electron micrograph shows T_c cells attacking and destroying a cancer cell (target cell).

AIDS Epidemic

Acquired immunodeficiency syndrome (AIDS) is caused by a group of related retroviruses known as HIV (human immunodeficiency viruses). In the United States, AIDS is usually caused by HIV-1, which enters a host by attaching itself to a cell membrane receptor called a CD4 receptor. HIV-1 infects helper T cells, the type of lymphocyte that stimulates B cells to produce antibodies and cytotoxic T cells to destroy virus-infected cells. Macrophages, which present antigens to helper T cells and thereby stimulate them, are also under attack. Other cells that have CD_4 receptors can also be infected and later destroyed.

HIV is a **retrovirus,** meaning that its genetic material consists of RNA instead of DNA. Once inside the host cell, HIV uses a special enzyme called *reverse transcriptase* to make a DNA copy (called cDNA) of its genetic material. Now cDNA integrates into a host chromosome, where it directs the production of more viral RNA. Each strand of viral RNA brings about synthesis of an outer protein coat called a capsid. The viral enzyme protease is necessary to the formation of capsids. Capsids assemble with RNA strands to form viruses, which bud from the host cell and spread to other cells.

Transmission of AIDS

HIV infection spreads when infected cells in body secretions, such as semen, and in blood are passed to another individual. To date, as many as 100 million people worldwide may have contracted HIV, and almost 31 million have died. A new infection is believed to occur every 15 seconds, the majority in heterosexuals. HIV infections are not distributed equally throughout the world. Most infected people live in Africa (66%) where the infection first began, but new infections are now occurring at the fastest rate in Southeast Asia and the Indian subcontinent.

HIV is transmitted by sexual contact with an infected person, including vaginal or rectal intercourse and oral/genital contact. Also, needle-sharing among intravenous drug users is high-risk behavior. Babies born to HIV-infected women may become infected before or during birth, or through breast-feeding after birth.

HIV first spread through the homosexual community, and male-to-male sexual contact still accounts for the largest percentage of new AIDS cases in the United States. But the largest increases in HIV infections are occurring through heterosexual contact or by intravenous drug use. Now, women account for 20% of all newly diagnosed cases of AIDS. The rise in the incidence of AIDS among women of reproductive age is paralleled by a rise in the incidence of AIDS in children younger than 13.

Phases of an HIV Infection

The Centers for Disease Control and Prevention recognize three stages of an HIV-1 infection, called categories A, B, and C. During the category A stage, the helper T-lymphocyte count is 500 per mm^3 or greater (Fig. 13A). For a period of time after the initial infection with HIV, people don't usually have any symptoms at all. A few (1–2%) do have mononucleosis-like symptoms that may include fever, chills, aches, swollen lymph nodes, and an itchy rash. These symptoms disappear, however, and no other symptoms appear for quite some time. Although there are no symptoms, the person is highly infectious. Despite the presence of a large number of viruses in the plasma, the HIV blood test is not yet positive because it tests for the presence of antibodies and not for the presence of HIV itself. This means that HIV can still be transmitted before the HIV blood test is positive.

Several months to several years after a nontreated infection, the individual will probably progress to category B, in which the helper T-lymphocyte count is 200 to 499 per mm^3. During this stage, the patient may experience swollen lymph nodes in the neck, armpits, or groin that persist for three months or more. Other symptoms that indicate category B are severe fatigue not related to exercise or drug use; unexplained persistent or recurrent fevers, often with night sweats; persistent cough not associated with smoking, a cold, or the flu; and persistent diarrhea.

The development of non-life-threatening but recurrent infections is a signal that the disease is progressing. One possible infection is thrush, a fungal infection that is identified by the presence of white spots and ulcers on the tongue and inside the mouth. The fungus may also spread to the vagina, resulting in a chronic infection there. Another frequent infection is herpes simplex, with painful and persistent sores on the skin surrounding the anus, the genital area, and/or the mouth.

Previously, the majority of infected persons proceeded to category C, in which the helper T-lymphocyte count is below 200 per mm^3 and the lymph nodes degenerate. The patient is now suffering from AIDS, characterized by severe weight loss and weakness due to persistent diarrhea and coughing, and will most likely contract an opportunistic infection. An **opportunistic infection** is one that only has the opportunity to occur because the immune system is severely weakened. Such infections are extremely rare in healthy

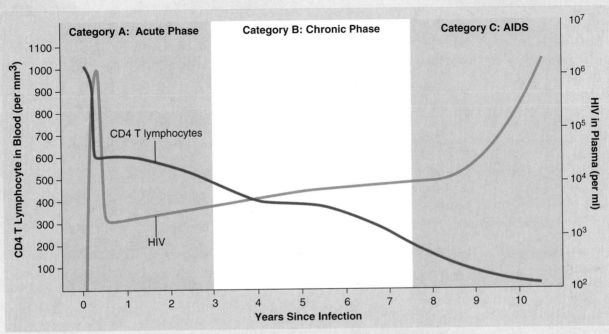

Figure 13A Stages of an HIV infection. In category A individuals, the number of HIV in plasma rises upon infection and then falls. The number of CD4 T lymphocytes falls, but stays above 400 per mm³. In category B individuals, the number of HIV in plasma is slowly rising, and the number of T lymphocytes is decreasing. In category C individuals, the number of HIV in plasma rises dramatically as the number of T lymphocytes falls below 200 per mm³.

individuals. Persons with AIDS die from one or more opportunistic diseases, such as *Pneumocystis carinii* pneumonia, *Mycobacterium tuberculosis*, toxoplasmic encephalitis, Kaposi's sarcoma, or invasive cervical cancer. This last condition has been added to the list because the incidence of AIDS has now increased in women.

Treatment for AIDS

Therapy usually consists of combining two drugs that inhibit reverse transcriptase with another that inhibits protease, an enzyme needed for formation of a viral capsid. This multidrug therapy, when taken according to the manner prescribed, usually seems to prevent mutation of the virus to a resistant strain. The sooner drug therapy begins after infection, the better the chances that the immune system will not be destroyed by HIV. Also, medication must be continued indefinitely. The medications all have serious side effects as well, including bone and joint destruction. Unfortunately, an

HIV strain resistant to all known drugs has been reported, and persons who become infected with this strain have no drug therapy available to them.

The likelihood of transmission from mother to child at birth can be lessened if the mother takes an inhibitor of reverse transcriptase called AZT and if the child is delivered by cesarean section. The mother also must not breast-feed her child because the virus is transmissible in breast milk.

Many investigators are working on a vaccine for AIDS. Some are trying to develop a vaccine in the traditional way. Others are working on subunit vaccines that utilize just a single HIV protein as the vaccine. So far, no method has resulted in sufficient antibodies to keep an infection at bay. After many clinical trials, none too successful, most investigators now agree that a combination of various vaccines may be the best strategy to bring about a response in both B lymphocytes and cytotoxic T cells.

Immunization: The Great Protector

Immunization protects children and adults from diseases. The success of immunization is witnessed by the fact that the smallpox vaccination is no longer required because the disease has been eradicated. However, parents today often fail to get their children immunized because they do not realize the importance of immunizations, fear potential side effects (which occur rarely), or cannot bear the expense. Newspaper accounts of an outbreak of measles at a U.S. college or hospital, therefore, are not uncommon because many adults were not immunized as children.

Figure 13B shows a recommended immunization schedule for children. The United States is now committed to the goal of immunizing all children against the common types of childhood diseases listed. Diphtheria, whooping cough, *Haemophilus influenzae*, and *Pneumococcus* infection are all life-threatening respiratory diseases. Tetanus is characterized by muscular rigidity, including a locked jaw. These extremely serious infections are all caused by bacteria; the rest of the diseases listed are caused by viruses. *Varicella* virus causes chickenpox. Polio is a type of paralysis; measles and rubella, sometimes called German measles, are characterized by high fever and skin rashes; and mumps is characterized by fever and enlarged parotid and other salivary glands.

Hepatitis B virus (HBV) is a blood-borne pathogen that is spread in the United States mainly by sexual contact and intravenous drug use. Health-care workers who are exposed to blood or blood products are also at risk, and maternal-neonatal transmission is a possibility as well. Recovery from an initial bout of hepatitis (inflammation of the liver) can lead to chronic hepatitis and then cancer of the liver. Fortunately, a hepatitis B vaccine is now available. It is rare for a child to come into contact with hepatitis B virus. However, physicians feel that the potential for liver damage from exposure to the virus warrants vaccination early in life. Infants now receive their first dose of vaccine before leaving the hospital, with booster immunizations at approximately 3 months and 12 months of age.

Cervical cancer has recently been linked to the occurrence of genital warts, a sexually transmitted disease caused by human papillomavirus. Therefore, a new vaccine for papillomavirus type 16, the most frequent cause of genital warts, has been developed. The vaccine is not currently recommended by the Centers for Disease Control, but it may be administered in the future. Perhaps more vaccines for sexually transmitted diseases will one day become available.

Even though bacterial infections (e.g., tetanus) can be cured by antibiotic therapy, it is better to be immunized. Some patients are allergic to antibiotics, and their reaction to them can be fatal. In addition, antibiotics not only kill off disease-causing bacteria, but they also reduce the number of beneficial bacteria in the intestinal tract and elsewhere. These beneficial bacteria may have checked the spread of pathogens that now are free to multiply and to invade the body. This is why antibiotic therapy

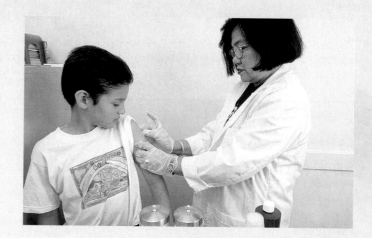

Content CHECK-UP!

7. Which type of cell produces the chemical mediators that cause the four outward signs of an inflammatory response?

a. macrophage

b. mast cell

c. neutrophil

d. natural killer cells

8. Which sign of inflammation occurs when tissue capillaries become more permeable and leak protein and tissue fluid into the tissue spaces?

a. swelling

b. redness

c. heat

d. formation of pus

9. Which protein produced by virus-infected cells prevents other cells from being infected?

a. complement

b. immunoglobulin

c. cytokine

d. interferon

is often followed by a secondary infection, such as a vaginal yeast infection in women.

Abuse of antibiotic therapy also leads to resistant bacterial strains that are difficult to cure, even with antibiotics. For example, a once-harmless skin bacterium called *Staphylococcus aureus* now causes thousands of life-threatening infections and deaths every year. Post-surgical patients are particularly at risk, but anyone who is hospitalized can become infected. After decades of antibiotic exposure, this microbe (abbreviated MRSA, for *m*ethicillin-*r*esistant *S*taphylococcus *a*ureus) and others have become resistant to all but the strongest antibiotics.

Therefore, everyone should avail themselves of appropriate vaccinations. Preventing a disease by becoming actively immune to it is preferable to becoming ill and needing antibiotic therapy to be cured.

	Birth	1 month	2 mos.	4 mos.	6 mos.	12 mos.	15 mos.	18 mos.	24 mos.	4-6 yrs.	11-12 yrs.	13-18 yrs.
Hepatitis B	1st Dose	Booster 1			Booster 2				Catch Up*			Catch Up*
Diphtheria, Pertussis, Tetanus			1st Dose	2nd Dose	3rd Dose		Booster 1			Booster 2	Tetanus, Diphtheria booster	Catch Up*
Haemophilus influenzae			1st Dose	2nd Dose	3rd Dose**	Final Dose						
Pneumococcal			1st Dose	2nd Dose	3rd Dose	Final Dose			Catch Up*			
Polio			1st Dose	2nd Dose	Final Dose	Catch Up*				Booster ***		
Measles, Mumps, Rubella						1st Dose				2nd Dose	Catch Up*	Catch Up*
Varicella (chicken pox)						1st Dose	2nd Dose	Final Dose	Catch Up*			Catch Up*

*for children or adolescents not previously immunized as infants
**optional, not necessary if doses 1 and 2 have been administered previously
***optional, not necessary if three doses administered previously

Figure 13B Suggested immunization schedule for infants, young children, and adolescents. Children who are immune-compromised may need additional immunizations. Always consult with your health care provider for up-to-date information on recommended immunizations for your area.

Source: From Centers for Disease Control and Prevention. Recommended childhood and adolescent immunization schedule—United States, 2005. MMWR 2005:53.

13.4 Induced Immunity

Immunity occurs naturally through infection or is brought about artificially (induced) by medical intervention. The two types of induced immunity are active and passive. In **active immunity,** the individual alone produces antibodies against an antigen; in **passive immunity,** the individual receives prepared antibodies from another person.

Active Immunity

Active immunity often develops naturally after a person is infected with a pathogen. Once recovery from illness is complete, the person will have a *natural* active immunity from the antibodies used to fight off the infection. However, active immunity can also be induced while a person is well, so that future infection will not take place. To prevent infections, people are immunized artificially against them. Successful immunization provides *artificial* active immunity, because the individual produces his or her own antibodies against the antigen given. The United States is committed to immunizing all children against the common types of childhood disease, as discussed in the Medical Focus on pages 290–291.

Immunization involves the use of **vaccines,** substances that contain an antigen to which the immune system responds. Traditionally, vaccines are the pathogens themselves, or their products, that have been treated so they are no longer virulent (able to cause disease). Today, it is possible to genetically engineer bacteria to mass-produce a protein from pathogens, and this protein can be used as a vaccine. This method has now

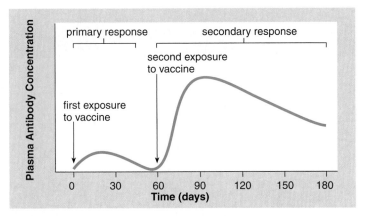

Figure 13.9 During immunization, the primary response, after the first exposure to a vaccine, is minimal, but the secondary response, which may occur after the second exposure, shows a dramatic rise in the amount of antibody present in plasma.

produced a vaccine against hepatitis B, a viral-induced disease, and is being used to prepare a vaccine against malaria, a protozoan-induced disease.

After a vaccine is given, it is possible to follow an immune response by determining the amount of antibody present in a sample of plasma. This is called the **antibody titer.** After the first exposure to a vaccine, a primary response occurs. For a period of several days, no antibodies are present; then the titer rises slowly, levels off, and gradually declines as the antibodies bind to the antigen or simply break down (Fig. 13.9). After a second exposure to the vaccine, a secondary response is expected. The titer rises rapidly to a level much greater than before; then it slowly declines. The second exposure is called a "booster" because it boosts the antibody titer to a high level. The high antibody titer now is expected to help prevent disease symptoms even if the individual is exposed to the disease-causing antigen.

Active immunity is dependent upon the presence of memory B cells and memory T cells that are capable of responding to lower doses of antigen. Active immunity is usually long-lasting, although a booster may be required every few years.

Passive Immunity

Passive immunity occurs when an individual is given prepared antibodies (immunoglobulins) to combat a disease. Since these antibodies are not produced by the individual's plasma cells, passive immunity is temporary. For example, newborn infants are passively immune to some diseases because antibodies have crossed the placenta from the mother's blood. This type of immunity is called natural passive immunity. These antibodies soon disappear, however, so that within a few months, infants become more susceptible to infections. Breast-feeding prolongs the natural passive immunity an infant receives from the mother because antibodies are present in the mother's milk (Fig. 13.10).

Even though passive immunity does not last, it is sometimes used to prevent illness in a patient who has been unexpectedly exposed to an infectious disease. Usually, the patient receives a **gamma globulin** injection (serum that contains IgG class anti-

Figure 13.10 Passive immunity. Breast-feeding is believed to prolong the passive immunity an infant receives from the mother because antibodies are present in the mother's milk.

bodies), perhaps taken from individuals who have recovered from the illness. For example, a health care worker who suffers an accidental needle stick may come into contact with the blood from a patient infected with hepatitis B virus. Immediate treatment with an antiviral gammaglobulin injection (along with simultaneous vaccination against the virus) can typically prevent the virus from causing infection in the health care worker. The gammaglobulin injection imparts an artificial *passive* immunity. The vaccination will cause the worker to develop her own antibodies, thus creating artificial *active* immunity as well.

Monoclonal Antibodies

Every plasma cell derived from the same B cell secretes antibodies against one specific antigen. These are **monoclonal antibodies** because all of them are the same type and because they are produced by plasma cells derived from the same B cell. One method of producing monoclonal antibodies *in vitro* (outside the body in a laboratory) is depicted in Figure 13.11. B lymphocytes are removed from an animal (today, usually mice are used) and are exposed to a particular antigen. The resulting plasma cells are fused with myeloma cells (malignant plasma cells that live and divide indefinitely). The fused cells are called hybridomas—*hybrid-* because they result from the fusion of two different cells, and *-oma* because one of the cells is a cancer cell. It is important to note that the antibodies from these cells *cannot* cause cancer, however.

At present, monoclonal antibodies are being used for quick and certain diagnosis of various conditions. For example,

Avian Flu—Potential Pandemic?

"Over there, over there, Send the word, send the word over there—
That the Yanks are coming, The Yanks are coming . . . And we
won't come back till it's over, Over there!"
—"Over There," George M. Cohan

In April 1918, American soldiers who proudly marched off to World War I battlefields "over there" (referring to Europe, in the words of the popular patriotic ballad of the time) took an un-wanted hitchhiker with them. A particularly virulent strain of in-fluenza A, nicknamed the "Spanish flu" because of the huge numbers of victims in that country, was carried to Europe. It quickly became a worldwide epidemic, or **pandemic,** causing an estimated 50 million deaths. Spanish flu caused twice as many deaths among the soldiers as those that resulted from combat. Unlike previous strains of influenza, this virus was capable of striking down young, otherwise healthy people as well as the el-derly and infirm. This particularly savage virus is once again a concern because of characteristics it shares with a modern-day virus: the avian influenza, or "bird flu" virus.

Influenza viruses, like all viruses, have two basic parts. The core of the virus is its genetic material, which can be either DNA or RNA. Influenza A viruses are RNA viruses. Their genetic mate-rial is covered by a protective protein coat called a **capsid.** An ad-ditional third piece, called the lipid envelope, contains protein spikes that allow the virus to attach only to its own specific kind of host cell. The two types of protein spikes, abbreviated "H" (for hemagglutinin) and "N" (for neuraminidase) allow each type of influenza virus to be categorized. There are 16 known types of H proteins, and 9 known types of N proteins. Thus, many different combinations are possible, meaning there are many different ways for viruses to attach to different kinds of host cells. Each different combination of H and N represents a new form of flu virus. Spanish flu virus is denoted "H1N1," and bird flu, avian influenza A, is denoted "H5N1." The ever-changing nature of influenza viruses explains why each year a new—and different—flu shot is needed.

Avian flu is highly contagious among birds and causes the death of entire flocks of wild and domesticated birds in a very short period. Yet, if avian influenza, and other animal viruses, were limited to infecting birds and/or other animals, avian and other animal influenzas would be largely just a vet-erinary problem. However, the genetic material of influenza virus is well known for its high mutation rate. Some of these mutations affect the structure of the protein spikes, so that a virus that previously could only infect a particular animal species can "jump species" to infect humans. In the case of

H5N1, the jump has already occurred. More than 130 cases of H5N1 infection in humans were reported between January 2004 through January 2006 in Southeast Asia and China. Most infec-tions occurred when the victims came into contact with infected birds, but a troubling few seem to have been due to human-to-human transmission. Scientists studying H1N1, the deadly Spanish flu virus of 1918, have noted that it probably also origi-nated as a "bird flu" virus that mutated to become transmissible among human beings.

Up until now, H5N1 viruses have not usually infected hu-man beings, so the recent human infection cases have public health officials especially concerned. Because the human popu-lation has not had ongoing exposure to this virus, little or no nat-urally immunity is thought to exist among humans. During re-cent outbreaks, avian flu has killed more than half of those infected with the virus, and most of its victims were previously healthy children and young adults—a fact eerily reminiscent of the 1918 Spanish flu pandemic.

Researchers and public health officials are trying to be pre-pared, just in case the virus achieves widespread human infection. Fortunately, thanks to the advance of modern technology, we will have a few more weapons in our antiviral arsenal than our grand-parents or great-grandparents had back in 1918. Vaccinations, drugs that prevent viral replication, antibiotics for secondary in-fections such as pneumonia, as well as communication tools like radio, television, and the Internet—all will hopefully create a dif-ferent outcome if a pandemic threatens the world again.

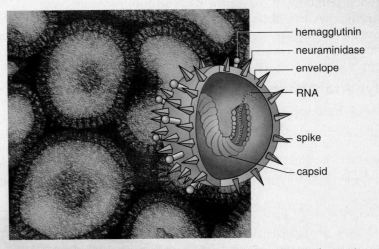

Figure 13C Influenza virus is an RNA virus, with protein spikes on its outer envelope that make it specific to one kind of host cell.

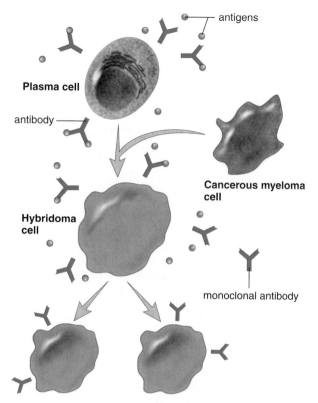

antigens

Plasma cell

antibody

Cancerous myeloma cell

Hybridoma cell

monoclonal antibody

Figure 13.11 Production of monoclonal antibodies. Plasma cells of the same type (derived from immunized mice) are fused with myeloma (cancerous) cells, producing hybridoma cells that are "immortal." Hybridoma cells divide and continue to produce the same type of antibody, called monoclonal antibodies.

a particular hormone is present in the urine of a pregnant woman. A monoclonal antibody can be used to detect this hormone; if it is present, the woman knows she is pregnant. Monoclonal antibodies are also used to identify infections. And because they can distinguish between cancerous and normal tissue cells, they are used to carry radioactive isotopes or toxic drugs to tumors, which can then be selectively destroyed.

Hypersensitivity Reactions

The immune system usually protects us from disease because it can distinguish self from nonself. Sometimes, however, it responds in a manner that harms the body, as when individuals develop allergies, suffer tissue rejection, or have an autoimmune disease. These are all forms of hypersensitivity reactions.

Allergies

Allergies are hypersensitivities to substances such as pollen or animal hair that ordinarily would do no harm to the body. The response to these antigens, called **allergens,** usually includes some degree of tissue damage. There are four types of allergic responses, but we will consider only two of them: IgE-mediated allergic response and T-cell-mediated allergic response.

IgE-Mediated Allergic Response An IgE-mediated allergic **response** is often referred to as an *immediate allergic response* because it can occur within seconds of contact with an antigen. The response is caused by antibodies known as IgE (see Table 13.1). IgE antibodies are attached to the plasma membrane of mast cells in the tissues and also to basophils in the blood. When an allergen attaches to the IgE antibodies on these cells, mast cells release histamine and other substances that bring about the allergic symptoms. When pollen is an allergen, histamine stimulates the mucosal membranes of the nose and eyes to release fluid, causing the runny nose and watery eyes typical of **hay fever.** If a person has **asthma,** the airways leading to the lungs constrict, resulting in difficult breathing accompanied by wheezing. When food contains an allergen, nausea, vomiting , and diarrhea result.

Anaphylactic shock is an Ig-E mediated allergic response that occurs because the allergen has entered the bloodstream. Bee stings and penicillin shots are known to cause this reaction because both inject the allergen into the blood. Anaphylactic shock is characterized by a sudden and life-threatening drop in blood pressure due to increased blood vessel dilation and permeability of the capillaries, caused by histamine. Taking epinephrine can delay this reaction until medical help is available.

Allergy shots sometimes prevent the onset of Ig-E mediated allergic responses. It has been suggested that injections of the allergen may cause the body to build up high quantities of IgG antibodies, and these combine with allergens received from the environment before they have a chance to reach the IgE antibodies located in the membrane of mast cells and basophils.

T-Cell Mediated Allergic Response A T-cell-mediated **allergic response** is initiated by memory T-cells at the site of allergen contact in the body. T-cell mediated responses are sometimes referred to as *delayed* allergic responses, because they develop more slowly than Ig-E mediated (*immediate*) responses. The allergic response is regulated by the cytokines secreted by both T cells and macrophages.

A classic example of a T-cell mediated allergic response is the skin test for tuberculosis (TB). When the test result is positive, the tissue where the antigen was injected becomes red and hardened. This shows that there was prior exposure to tubercle bacilli, the cause of TB. Contact dermatitis, which occurs when a person is allergic to poison ivy, jewelry, cosmetics, and many other substances that touch the skin, is also an example of a T-cell mediated allergic response.

Tissue Rejection

Certain organs, such as skin, the heart, and the kidneys, could be transplanted easily from one person to another if the body did not attempt to reject them. Rejection of transplanted tissue results because the recipient's immune system recognizes that the transplanted tissue is not "self." Cytotoxic T cells respond by causing disintegration of the transplanted tissue.

Organ rejection can be controlled by carefully selecting the organ to be transplanted and administering **immuno-suppressive** drugs. It is best if the transplanted organ has the same type of HLA antigens as those of the recipient, because T_c cells recognize foreign HLA antigens. Two well-known immunosuppressive drugs, cyclosporine and tacrolimus, both act by inhibiting the response of T cells to cytokines.

Researchers hope that tissue engineering, including the production of organs that lack antigens or that can be protected in some way from the immune system, will one day do away with the problem of rejection.

Autoimmune Disease

When a person has an **autoimmune disease,** cytotoxic T cells or antibodies mistakenly attack the body's own cells as if they bear foreign antigens. Exactly what causes autoimmune diseases is not known. However, sometimes they occur after an individual has recovered from an infection.

Virtually any tissue of the body can be a target for autoimmune disease. In **myasthenia gravis,** the synaptic junctions between motor nerves and skeletal muscles are destroyed. Neuromuscular junctions do not work properly and muscular weakness results. In **multiple sclerosis (MS),** the myelin sheath of nerve fibers is attacked. This can cause a wide variety of neurological and neuromuscular symptoms: muscle weakness, paralysis, blurred vision, dizziness, and deafness. A person with **systemic lupus erythematosus (SLE)** has an extremely serious disease characterized by antibodies directed against multiple self antigens, including her own DNA molecules. Symptoms can include facial rashes, arthritis, anemia, and kidney damage. In **rheumatoid arthritis,** the joints are affected. Researchers suggest that heart damage following rheumatic fever, and type I diabetes are also autoimmune illnesses. As yet, there are no cures for autoimmune diseases, but they can be managed with medication.

Immune Deficiency

When a person has an immune deficiency, the immune system is unable to protect the body against disease. AIDS (see the Medical Focus on page 288) is an example of an acquired immune deficiency. As a result of a weakened immune system, AIDS patients show a greater susceptibility to a variety of diseases, and also have a higher risk of cancer. Immune deficiency may also be genetic (that is, inherited from one's parents) or congenital (due to failure of lymphatic tissue to develop). For example, in some infants the thymus gland fails to develop, producing a child with severe immune deficiency. Infrequently, a child may be born with an impaired B- or T-cell system caused by a defect in lymphocyte development. In **severe combined immunodeficiency disease (SCID),** a genetic disorder, both antibody and cell-mediated immunity are lacking or inadequate. Without treatment, even common infections can be fatal. Gene therapy has been successful in SCID patients (see page 434).

13.5 Effects of Aging

With advancing age, people become more susceptible to all types of infections and disorders because the immune system exhibits lower levels of function. One reason is that the thymus gland degenerates. Having reached its maximum size in early childhood, it begins to shrink after puberty and has virtually disappeared by old age. As the gland decreases in size, so does the number of T cells. The T cells remaining do not respond to foreign antigens; therefore, the chance of having cancer increases with age.

Among the elderly, the B cells sometimes fail to form clones. Or, when they do form clones, the antibodies released may not function well. Therefore, infections are more common among the elderly. Also, the antibodies are more likely to attack the body's own tissues, increasing the incidence of autoimmune diseases.

The response of elderly individuals to vaccines is decreased. However, considering that their overall level of immune response is low, it is better that these people be vaccinated than not. For this reason, elderly individuals are encouraged to get an influenza (flu) vaccination each year.

13.6 Homeostasis

The three functions of the lymphatic system listed on page 277 assist homeostasis. The lymphatic system helps the digestive system by absorbing fats. In the process of absorbing dietary fats, lacteals also absorb fat-soluble vitamins. The lymphatic system assists the cardiovascular system by absorbing excess tissue fluid. The lymphatic vessels return excess tissue fluid as lymph to cardiovascular veins in the thorax. Without

Human Systems Work Together

LYMPHATIC SYSTEM

Integumentary System

Lymphatic vessels pick up excess tissue fluid; immune system protects against skin infections.

Skin serves as a barrier to pathogen invasion; Langerhans cells phagocytize pathogens; protects lymphatic vessels.

Skeletal System

Lymphatic vessels pick up excess tissue fluid; immune system protects against infections.

Red bone marrow produces leukocytes involved in immunity.

Muscular System

Lymphatic vessels pick up excess tissue fluid; immune system protects against infections.

Skeletal muscle contraction moves lymph; physical exercise enhances immunity.

Nervous System

Lymphatic vessels pick up excess tissue fluid; immune system protects against infections of nerves.

Microglia engulf and destroy pathogens.

Endocrine System

Lymphatic vessels pick up excess tissue fluid; immune system protects against infections.

Thymus is necessary to maturity of T lymphocytes.

How the Lymphatic System works with other body systems.

tonsils

cervical lymph nodes

thymus

axillary lymph nodes

thoracic duct

spleen

inguinal lymph nodes

popliteal lymph node

lymphatic vessel

Cardiovascular System

Lymphatic organs produce and store formed elements; lymphatic vessels transport leukocytes and return tissue fluid to blood vessels; spleen serves as blood reservoir, filters blood.

Blood vessels transport leukocytes and antibodies; blood services lymphatic organs and is source of tissue fluid that becomes lymph.

Respiratory System

Lymphatic vessels pick up excess tissue fluid; immune system protects against respiratory tract and lung infections.

Tonsils and adenoids occur along respiratory tract; breathing aids lymph flow.

Digestive System

Lacteals absorb fats; Peyer patches prevent invasion of pathogens; appendix contains lymphatic tissue.

Digestive tract provides nutrients for lymphatic organs; stomach acidity prevents pathogen invasion of body.

Urinary System

Lymphatic system picks up excess tissue fluid, helping to maintain blood pressure for kidneys to function; immune system protects against infections.

Kidneys control volume of body fluids, including lymph.

Reproductive System

Female immune system does not attack sperm or fetus, even though they are foreign to the body.

Sex hormones influence immune functioning; acidity of vagina helps prevent pathogen invasion of body; milk passes antibodies to newborn.

this assistance, it would be more difficult for the body to maintain the blood volume and pressure needed for capillary exchange.

The lymphatic organs, along with the immune system, protect us from infectious diseases. Nonspecific ways of protecting the body from disease precede specific immunity. The skin and the mucous membranes of the respiratory tract, the digestive tract, and the reproductive and urinary systems all resist invasion by viruses and bacteria. If a pathogen should enter the body, the infection is localized as much as possible. During the inflammatory reaction, the phagocytic white blood cells immediately rush to the scene and engulf as many pathogens as possible. Macrophages are especially good at devouring viruses and bacteria by phagocytosis. If the infection cannot be confined and pathogens do enter the blood, complement is a series of proteins that work in diverse ways to keep the blood free of disease-causing organisms and their toxins.

Not surprisingly, specific defenses are dependent upon blood cells; the lymphocytes and macrophages play central roles. B and T cells have antigen receptors and can distinguish self from nonself. The binding of the antigen selects which specific B or T cells will undergo clonal expansion. B cells are capable of recognizing an antigen directly, but T cells must have the antigen displayed by an APC in the groove of an HLA antigen. Plasma cells (mature B cells) produce antibodies, but T cells kill virus-infected and cancer cells outright.

The lymphatic organs play a central role in immunity. White blood cells are made in the red bone marrow where B cells also mature. T cells mature in the thymus. The spleen filters the blood directly. Clonal expansion of lymphocytes occurs in the lymph nodes, which also filter the lymph.

A strong connection exists between the immune, nervous and endocrine systems. Lymphocytes have receptor proteins for a wide variety of hormones and the thymus gland produces hormones that influence the immune response. Cytokines produced by T lymphocytes help the body recover from disease by affecting the brain's temperature control center. The high body temperature of a fever is thought to create an unfavorable environment for the foreign invaders. Also, cytokines bring about a feeling of sluggishness, sleepiness, and loss of appetite. These behaviors tend to make us take care of ourselves until we feel better. A close connection between the immune and endocrine systems is illustrated by the ability of cortisone to decrease the inflammatory reaction in the joints.

SELECTED NEW TERMS

Basic Key Terms

active immunity (ak′tiv ĭ-myū′ nĭ-te), p. 291

antibodies (ăn′tĭ-bŏd′ēz), p. 284

antibody-mediated immunity (an″tĭ-bod″ e-me′de-āt-ed ĭ-myū′nĭ-te), p. 284

antigens (ăn′tĭ-jən), p. 282

antigen-presenting cell (ăn′tĭ-jən prĭ-zĕnt-ing sĕl), p. 286

apoptosis (ăp′əp-tōsĭs), p. 284

appendix (ă-pen′-diks), p. 280

autoimmune response (ô′tō-ĭmyo͞on rĭ-spŏns), p. 284

B lymphocytes (B lim′fo-sītz), p. 284

cell-mediated immunity (sel me″de-āt′ed ĭ-myū′nĭ-te), p. 287

complement system (kom′plĕ-ment sis′tem), p. 282

cytokines (si′to-kīnz), p. 286

cytotoxic T cells (si′to-tok′sik T selz), p. 287

helper T cells (help′er T selz), p. 287

histamine (his′tuh-min), p. 282

HLA (human leukocyte) antigens (hyū′ mun lu′ko-sīt an′tĭ-jenz), p. 286

immunity (ĭ-myū-nĭ-te), p. 278

inflammatory reaction (in-flam′uh-to″re re-ak′shun), p. 282

lacteal (lak′tē-ăl), p. 277

lymph (lĭmf), p. 277

lymphatic organ (lim-fat′ik or′gan), p. 278

lymphatic system (lim-fat′ik sis′tem), p. 277

lymphatic vessels (lĭm-făt′ĭk vĕs′əl), p. 277

lymph node (limf nōd), p. 280

lymphoma (lĭm-fō′mə), p. 281

macrophages (măk′rə-fāj-ez), p. 282

mast cells (măst sĕlz), p. 282

memory B cells (mem′o-re B selz), p. 284

monoclonal antibodies (mon″o-klōn′al an″tĭ-bod′ēz), p. 292

natural killer cells (nat′u-ral kil′er selz), p. 282

passive immunity (păs′-iv ĭ-myū′ nĭ-te), p. 291

perforin (pĕr′for-ən), p. 287

Peyer patches (pi′er pach″ēz), p. 280

plasma cells (plaz′muh selz), p. 284

pus (pus), p. 282

red bone marrow (rĕd bōn măr′ō), p. 278

retrovirus (rĕt′rō-vī′rəs), p. 288

spleen (splēn), p. 280

thymus gland (thī′məs glănd), p. 279

T lymphocytes (T lim′fo-sītz), p. 284

tonsils (ton′silz), p. 280

Clinical Key Terms

AIDS (acquired immunodeficiency syndrome) (uh-kwīr′d ĭ-myū″no-dĭ-fī′shun-se sin′drōm), p. 288

allergens (al′er-jenz), p. 294

allergies (al′er-jēz), p. 294

anaphylactic shock (an″uh-fī-lak′tik shok), p. 294

antibody titer (an″tĭ-bod″e ti′ter), p. 292

asthma (az′muh), p. 294

autoimmune disease (aw″to-ĭ-myūn′ dĭ-zez′), p. 295

edema (ĕ-de′muh), p. 277

elephantiasis (ĕ″luh-fun-ti′uh-sis), p. 281

gamma globulin (găm-uh glŏb′yoo-lin), p. 292

hay fever (hā fē′vĕr), p. 294

Hodgkin disease (hoj″kin dĭ-zĕz′), p. 281

IgE-mediated allergic response (i-jē-mĕ′dē-at′ĕd ə-lûr′jik rĭ-spŏns′), p. 294

immunization (ĭ-myū-nĭ-za′shun), p. 291

immunosuppressive drugs (ĭ-myū″ no-sŭ-pres′iv drugz), p. 295

interferon (in″ter-fēr′on), p. 282

interleukin, (in-ter-loo′kin), p. 287

lymphadenitis (lim-fad″ĕ-ni′tis), p. 281

lymphangitis (lim″fan-jī′tis), p. 281

lymphedema (lim″fe-de′mah), p. 281

SUMMARY

13.1 Lymphatic System

A. The lymphatic system consists of lymphatic vessels and lymphatic organs. The lymphatic vessels return excess tissue fluid to the bloodstream, absorb fats at intestinal villi and help the immune system defend the body against disease.

B. Lymphatic capillaries have thin walls. Larger vessels are structured the same as cardiovascular veins, with valves that prevent backward flow.

13.2 Organs, Tissues, and Cells of the Immune System

Lymphocytes are produced and accumulate in the lymphatic organs (primary organs: red bone marrow, thymus gland; secondary organs: lymph nodes, spleen, and other lymphatic tissues). Lymph is cleansed of pathogens and/or their toxins in lymph nodes and blood is cleansed of pathogens in the spleen. T lymphocytes mature in the thymus, while B lymphocytes mature in the red bone marrow where all blood cells are produced. White blood cells are necessary for nonspecific and specific defenses.

13.3 Nonspecific and Specific Defenses

A. Immunity involves nonspecific and specific defenses. Nonspecific defenses include barriers to entry, the inflammatory reaction, nonspecific white blood cells, natural killer cells, and protective proteins.

B. Specific defenses require B lymphocytes and T lymphocytes, also called B cells and T cells. B cells undergo clonal selection with production of plasma cells and memory B cells after their antigen receptors combine with a specific antigen. Plasma cells secrete antibodies and eventually undergo apoptosis. Plasma cells are responsible for antibody-mediated immunity. Antibodies are Y-shaped molecules with two binding sites for a specific antigen. Memory B cells remain in the body and produce antibodies if the same antigen enters the body at a later date.

C. T cells are responsible for cell-mediated immunity. The two main types of T cells are cytotoxic T cells and helper T cells. Cytotoxic T cells kill virus-infected or cancer cells on contact because they bear a nonself antigen. Helper T cells produce cytokines and stimulate other immune cells. Like B cells, each T cell bears antigen receptors. However, for a T cell to recognize an antigen, the antigen must be presented by an antigen-presenting cell (APC), usually a macrophage, in the groove of an HLA (human leukocyte-associated antigen). Thereafter, the activated T cell undergoes clonal expansion until the illness has been stemmed. Then most of the activated T cells undergo apoptosis. A few cells remain, however, as memory T cells. Cytokines, including interferon and interleukins, are used in an attempt to promote the body's ability to recover from cancer and to treat AIDS.

13.4 Induced Immunity

A. Immunity can be induced in various ways. Vaccines are available to induce long-lasting, active immunity. Antibodies sometimes are available to provide an individual with temporary, passive immunity. Monoclonal antibodies are produced in the laboratory and used for diagnosis and treatment purposes.

B. Allergic responses occur when the immune system reacts vigorously to substances not normally recognized as foreign. IgE-mediated or immediate allergic responses, usually consisting of coldlike symptoms, are due to the activity of antibodies. T cell mediated allergic responses, such as contact dermatitis, are due to the activity of T cells.

13.5 Effects of Aging

The thymus gets smaller as we age, and fewer antibodies are produced. The elderly are at great risk of infections, cancer and autoimmune diseases.

13.6 Homeostasis

The lymphatic system assists the cardiovascular system by returning excess tissue fluid to the bloodstream. It assists the digestive system by absorbing fats from the intestinal tract, and it assists the immune system through the functioning of its lymphatic organs.

STUDY QUESTIONS

1. What is the lymphatic system, and what are its three functions? (p. 277)
2. Describe the structure and the function of red bone marrow, the thymus, the spleen, lymph nodes and the tonsils. (pp. 278–280)
3. What are the body's nonspecific defense mechanisms? (p. 281)
4. Describe the inflammatory reaction and give a role for each type of cell and molecule that participates in the reaction. (p. 282)

5. What is the clonal selection theory as it applies to B cells? B cells are responsible for which type of immunity? (p. 284)
6. Describe the structure of an antibody, and define the terms variable regions and constant regions. (p. 285)
7. Describe the clonal selection theory as it applies to T cells. (p. 286)
8. Name the two main types of T cells and state their functions. (p. 287)
9. What are cytokines and how are they used in immunotherapy? (pp. 286–287)
10. How is active immunity artificially achieved? How is passive immunity achieved? (pp. 291–292)
11. How are monoclonal antibodies produced, and what are their applications? (p. 292)
12. Discuss allergies, tissue rejection and autoimmune diseases as they relate to the immune system. (pp. 294–295)
13. How do the lymphatic and immune systems help maintain homeostasis? (pp. 295, 297)
14. How does the skeletal system assist the immune system in maintaining homeostasis? (p. 296)

LEARNING OBJECTIVE QUESTIONS

Fill in the blanks.
1. Lymphatic vessels contain _____, which close, preventing lymph from flowing backward.
2. _____ and _____ are two types of white blood cells produced and stored in the lymphatic organs.
3. Lymph nodes cleanse the _____, while the spleen cleanses the _____.
4. _____ and _____ are phagocytic white blood cells.
5. T lymphocytes have matured in the _____.
6. A stimulated B cell divides and differentiates into antibody-secreting _____ cells and also into _____ cells that are ready to produce the same type of antibody at a later time.
7. B cells are responsible for _____-mediated immunity.
8. Cytotoxic T cells are responsible for _____-mediated immunity.
9. Immunization with _____ brings about active immunity.
10. Allergic reactions are associated with the release of _____ from mast cells.
11. Whereas _____ immunity occurs when an individual is given antibodies to combat a disease, _____ immunity occurs when an individual develops the ability to produce antibodies against a specific antigen.
12. Barriers to entry, protective proteins, and the inflammatory reaction are all examples of _____ defenses.
13. Proteins that function to form holes in bacterial cell walls comprise the _____ system.

MEDICAL TERMINOLOGY EXERCISE

Consult Appendix B for help in pronouncing and analyzing the meaning of the terms that follow.

1. metastasis (mĕ-tas'tuh-sis)
2. allergist (al'er-jist)
3. immunosuppressant (i-myū"no-sŭ-pres'ant)
4. immunotherapy (i-myū"no-thĕr'uh-pe)
5. splenorrhagia (sple"no-ra'je-uh)
6. lymphadenopathy (lim-fad"ĕ-nop'uh-the)
7. lymphangiography (lim-fan"je-og'ruh-fe)
8. eosinophilia (e'oh-sin'o-fil'e-uh)
9. thymectomy (thi-mek'to-me)
10. lymphopenia (limf'o-pe'ne-uh)
11. agammaglobulinemia (ā-gam'uh-glob'yū-lĭ-ne'me-uh)
12. pyemia (pi-ē'me-uh)
13. tonsillectomy (tŏn'sə-lēk'tə-mē)
14. hypersensitivity (hi'per-sen-sĭ-tiv'ĭ-te)

WEBSITE LINK

Visit this text's website at http://www.mhhe.com/maderap6 for additional quizzes, interactive learning exercises, and other study tools.

The cilia of cells lining the bronchial wall help keep the lungs clean by moving trapped particles.

Chapter Outline & Learning Objectives
After you have studied this chapter, you should be able to:

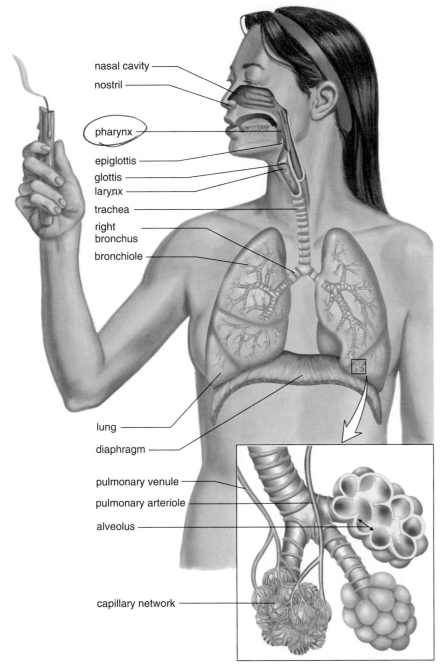

nasal cavity

nostril

pharynx

epiglottis

glottis

larynx

trachea

right bronchus

bronchiole

lung

diaphragm

pulmonary venule

pulmonary arteriole

alveolus

capillary network

Figure 14.1 The respiratory tract extends from the nasal cavities to the lungs, which are composed of air sacs called alveoli. Gas exchange occurs between the air in the alveoli and the blood within a capillary network that surrounds the alveoli. Notice in the blow-up that the pulmonary arteriole is colored blue—it carries O_2-poor blood away from the heart to the alveoli. Then carbon dioxide leaves the blood, and oxygen enters the blood. The pulmonary venule is colored red—it carries O_2-rich blood from the alveoli toward the heart.

14.1 The Respiratory System

The primary function of the respiratory system is gas exchange—allowing oxygen from the air to enter the blood and carbon dioxide from the blood to exit into the air. During **inspiration,** or inhalation (breathing in), and **expiration,** or

exhalation (breathing out), air is conducted toward or away from the lungs by a series of cavities, tubes, and openings, illustrated in Figure 14.1.

The respiratory system also works with the cardiovascular system to accomplish these four respiratory events:

1. pulmonary ventilation (breathing): the entrance and exit of air into and out of lungs
2. external respiration: the exchange of gases (oxygen and carbon dioxide) between air and blood
3. internal respiration: the exchange of gases between blood and tissue fluid
4. transport of gases to and from the lungs and the tissues.

Cellular respiration, which produces ATP, uses the oxygen and produces the carbon dioxide that makes gas exchange with the environment necessary. Without a continuous supply of ATP, the cells cease to function. The four events listed here allow cellular respiration to continue.

The Respiratory Tract

Table 14.1 traces the path of air from the nose to the lungs. As air moves in along the airways, it is cleansed, warmed, and moistened by the mucous membrane lining of the nasal cavity. Cleansing is accomplished by coarse hairs just inside the nostrils and by cilia and mucus in the nasal cavities and the other airways of the respiratory tract. Nasal hairs, cilia, and mucus act as screening devices for dust, dirt, pollen, fungal spores, etc., in inhaled air. Cilia inside the nasal cavity beat backwards, sweeping the mucous sheet into the throat where it can be swallowed. Lysozyme in the mucus helps to kill bacteria, while lymphocytes populating underlying tissue create an immune response whenever pathogens are accidentally inhaled. In the trachea and other airways, the cilia beat upward, carrying mucus, dust, and other trapped potential contaminants into the pharynx. This important protective mechanism is called the **mucociliary escalator.** Once in the pharynx, the mucous accumulation can be swallowed (thus, destroyed by stomach acid and digestive enzymes) or spit out. Inhaled air is also warmed by heat given off by blood vessels lying close to the surface of the lining of the airways, and it is moistened by the wet surface of the mucous membrane.

Conversely, as air moves out during expiration, it cools and loses its moisture. As the air cools, it deposits its moisture on the lining of the trachea and the nose, and the nose may even drip as a result of this condensation. The air still retains so much moisture, however, that upon expiration on a cold day, it condenses and forms a small cloud.

TABLE 14.1 Path of Air

Structure	Description	Function
UPPER RESPIRATORY TRACT		
Nasal cavities	Hollow spaces in nose	Filter, warm and moisten air
Pharynx	Chamber posterior to oral cavity; lies between nasal cavity and larynx	Connection to surrounding regions
Glottis	Opening into larynx	Passage of air into larynx
Larnyx	Cartilaginous organ that houses the vocal cords; voice box	Sound production
LOWER RESPIRATORY TRACT		
Trachea	Flexible tube that connects larynx with bronchi	Passage of air to bronchi
Bronchi	Paired tubes inferior to the trachea that enter the lungs	Passage of air to lungs
Bronchioles	Branched tubes that lead from bronchi to alveoli	Passage of air to each alveolus
Lungs	Soft, cone-shaped organs that occupy lateral portions of thoracic cavity	Gas exchange Acid-base balance; conversion of angiotensin I to angiotensin II
Alveoli	Air sacs that branch from bronchioles	Passage of gases into and out of pulmonary capillaries
Pulmonary capillaries	Capillaries that cover surface of alveoli	Receive oxygen from alveoli; oxgenation of hemoglobin; deliver CO_2 from tissues to alveoli

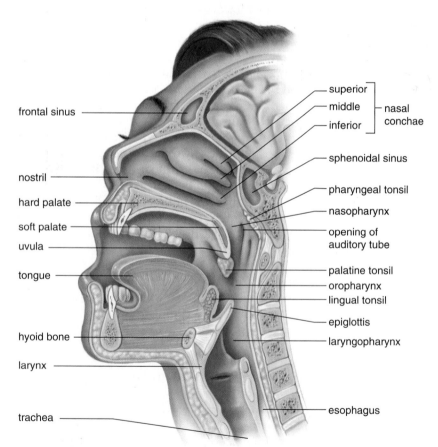

Figure 14.2 The path of air. This drawing shows the path of air from the nasal cavities to the trachea, which is a part of the lower respiratory tract. The other organs are in the upper respiratory tract.

Labels: frontal sinus, nostril, hard palate, soft palate, uvula, tongue, hyoid bone, larynx, trachea; superior, middle, inferior nasal conchae; sphenoidal sinus, pharyngeal tonsil, nasopharynx, opening of auditory tube, palatine tonsil, oropharynx, lingual tonsil, epiglottis, laryngopharynx, esophagus

The Nose

The **nose,** a prominent feature of the face, is the only external portion of the respiratory system. Air enters the nose through external openings called **nostrils.** The nose contains two **nasal cavities,** which are narrow canals separated from one another by a septum composed of bone and cartilage (Fig. 14.2). Mucous membrane lines the nasal cavities. The nasal **conchae** are bony ridges that project laterally into the nasal cavity. They increase the surface area for moistening and warming air during inhalation and for trapping water droplets during exhalation. Odor receptors are on the cilia of cells in the olfactory epithelium, located high in the recesses of the nasal cavities.

The tear (lacrimal) glands drain into the nasal cavities by way of tear ducts. For this reason, crying produces a runny nose. The nasal cavities also communicate with the **paranasal sinuses,** air-filled spaces that reduce the weight of the skull and act as resonating chambers for the voice. If the ducts leading from the sinuses become inflamed, fluid may accumulate, causing a sinus headache. The nasal cavities are separated from the oral cavity by a partition called the palate which has two portions. Anteriorly, the hard palate is supported by the maxilla and palatine bones. Posteriorly, the soft palate is composed of muscle and glandular tissue.

The Pharynx

The **pharynx** is a funnel-shaped passageway that connects the nasal and oral cavities to the larynx. Consequently, the pharynx, commonly referred to as the "throat," has three parts: the nasopharynx, where the nasal cavities open posterior to the soft palate; the oropharynx, where the oral cavity joins the pharynx; and the laryngopharynx, which opens into the larynx. The soft palate has a soft extension called the uvula that can be seen projecting into the oropharynx.

The single pharyngeal tonsil (also called the adenoids) in the posterior nasal cavity helps to defend against infection. Because inhaled air passes directly over this tissue, it is the primary lymphatic tissue defense for breathing. Its action is supported by the paired palatine tonsils at the rear of the oropharynx, as well as the lingual tonsils at the base of the tongue. Being lymphatic tissue, the tonsils contain lymphocytes that protect against invasion of inhaled pathogens. Here, both B cells and T cells are prepared to respond to antigens that may subsequently invade internal tissues and fluids. In this way, the respiratory tract assists the immune system in maintaining homeostasis.

In the pharynx, the air passage and the food passage cross because the larynx, which receives air, is anterior to the esophagus, which receives food. The larynx lies at the top of the trachea. The larynx and trachea are normally open, allowing air to pass, but the esophagus is normally collapsed and opens only when a person swallows.

The Larynx

The **larynx** is a cartilaginous structure that serves as a passageway for air between the pharynx and the trachea. The larynx can be pictured as a triangular box whose apex, the Adam's apple, is located at the anterior of the neck. The larynx is called the voice box because it houses the vocal cords. The **vocal cords** are mucosal folds supported by elastic ligaments, and the slit between the vocal cords is an opening called the **glottis** (Fig. 14.3). When air is expelled past the vocal cords through the glottis, the vocal cords vibrate, producing sound. At the time of puberty, the growth of the larynx and the vocal cords is much more rapid and accentuated in the male than in the female, causing the male to have a more prominent Adam's apple and a deeper voice. The voice "breaks" in the young male due to his inability to control the longer vocal cords. These changes cause the lower pitch of the voice in males.

The high or low pitch of the voice is regulated when speaking and singing by changing the tension on the vocal cords. The greater the tension, as when the glottis becomes more narrow, the higher the pitch. When the vocal cords relax, the glottis is wider, and the pitch is lower (Fig. 14.3b). The loudness, or intensity, of the voice depends upon the amplitude of the vibrations—that is, the degree to which the vocal cords vibrate.

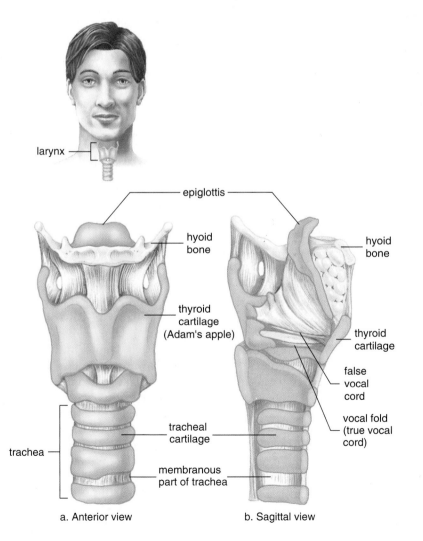

a. Anterior view   b. Sagittal view

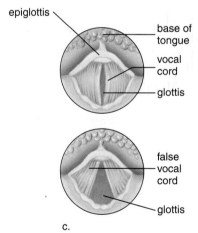

c.

Figure 14.3 Anatomy of the larynx. **a.** The larynx lies posterior to the thyroid cartilage of the neck (commonly called the Adam's apple). **b.** Sagittal view of the larynx, showing location of the vocal cords. **c.** Viewed from above, the vocal cords can be seen to stretch from anterior to posterior across the larynx. When air is forced past the vocal cords, they vibrate, producing sound. The vocal cords are taut when we produce a high-pitched sound (*top*), and they relax as the pitch deepens (*bottom*).

When food is swallowed, the larynx moves upward against the **epiglottis**, a flap of elastic cartilage that prevents food from passing through the glottis into the larynx. You can detect this movement by placing your hand gently on your larynx and swallowing.

The Trachea

The **trachea**, commonly called the windpipe, is a tube connecting the larynx to the primary bronchi. The trachea lies ventral to the esophagus and is held open by C-shaped cartilaginous rings. The open part of the C-shaped rings forms the anterior wall of the esophagus, and this allows the esophagus to expand when swallowing. The mucosa that lines the trachea has a layer of pseudostratified ciliated columnar epithelium. (Pseudostratified—"false layers"—means that while the epithelium appears to be layered, actually each cell touches the basement membrane.) As mentioned previously, the cilia that project from the epithelium keep the lungs clean by sweeping mucus, produced by goblet cells, and debris toward the pharynx (called the mucociliary escalator).

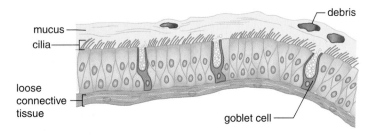

Smoking is known to destroy these cilia, and consequently the soot in cigarette smoke collects in the lungs. Smoking is discussed more fully in the Medical Focus on page 316.

If the trachea is blocked because of illness or the accidental swallowing of a foreign object, it is possible to insert a breathing tube by way of an incision made in the trachea. This tube acts as an artificial air intake and exhaust duct. The operation is called a **tracheostomy.**

The Bronchial Tree

The trachea divides into right and left primary bronchi (sing., **bronchus**), which lead into the right and left lungs (see Fig. 14.1). The primary bronchi then branch into secondary bronchi: one for each lobe of the lung. Thus, there are three secondary bronchi for the right lung, which has three lobes. Two secondary bronchi supply the left lung, which has only two lobes in order to allow room for the heart. Each secondary bronchus then divides into smaller tertiary bronchi. These smaller bronchi are supported by smaller plates of cartilage, in place of the cartilage rings of the trachea. **Bronchioles** are the smallest conducting airways. They lack cartilage support, but possess a ciliated epithelium and a well-developed smooth muscle layer. During an asthma attack, the smooth muscle of the bronchioles contracts, causing bronchiolar constriction and characteristic wheezing. Each bronchiole leads to an elongated space enclosed by a multitude of air

pockets, or sacs, called alveoli (sing., **alveolus**). The components of the bronchial tree beyond the primary bronchi, including the alveoli, compose the lungs.

The Lungs

The **lungs** are paired, cone-shaped organs. Each fills its own pleural cavity inside the thoracic cavity, separated by the mediastinum. Recall that the mediastinum is the central membrane that separates the thoracic cavity, covering the primary bronchi, heart, thymus gland, trachea, and esophagus (see Chapter 1, page 7). The apex is the superior narrow portion of a lung, and the base is the inferior broad portion that curves to fit the dome-shaped diaphragm, the muscle of respiration that separates the thoracic cavity from the abdominal cavity. The lateral surfaces of the lungs follow the contours of the ribs in the thoracic cavity.

Each lobe of the lung is further divided into lobules, and each lobule has a bronchiole supplying many alveoli. Pulmonary arteries travel alongside the bronchi; likewise, pulmonary arterioles parallel the bronchioles. Each pulmonary arteriole then further branches to form **pulmonary capillaries.** Pulmonary capillaries surround and cover each alveolus of the lung. Elastic connective tissue binds the air passages to the blood vessels within each lung; this elastic tissue helps the lungs return to their resting position, or *recoil*, when a person exhales.

Each lung is enclosed by a double layer of serous membrane called the **pleurae** (sing., **pleura**). The visceral pleura adheres to the surface of the lung; the parietal pleura lines the inside of the thoracic cavity. The pleurae produce a lubricating serous fluid that reduces friction and allows the two layers to slide across one another. Serous fluid, a water-based solution, also creates **surface tension:** the tendency for water molecules to cling to each other (due to hydrogen bonding between the molecules) and to form a droplet. Surface tension holds the two pleural layers together, thus holding the lungs open against the chest wall.

The Alveoli

With each inhalation, air passes by way of the bronchial tree to the alveoli. An alveolar sac is made up of simple squamous epithelium surrounded by pulmonary capillaries. Gas exchange occurs between the air in the alveoli and the blood in the capillaries (Fig. 14.4). Oxygen diffuses across the alveolar and capillary walls to enter the bloodstream, while carbon dioxide diffuses from the blood across these walls to enter the alveoli.

Each alveolus is covered with an extremely thin layer of water-based tissue fluid. Gas exchange takes place across moist cellular membranes, and the attractive force created by the fluid's surface tension helps the distended lung tissue to return to its resting position when a person exhales. However, the alveoli must stay open to receive the inhaled air if gas exchange is to occur, and the surface tension of water lining the alveoli is capable of causing them to close up. (As an analogy, rinse a balloon several times with water, then empty the balloon completely. Note how the balloon's sides stick together and

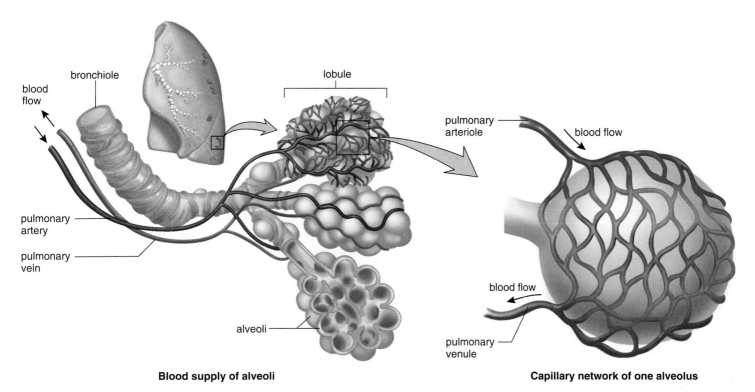

blood flow

bronchiole

lobule

pulmonary arteriole

blood flow

pulmonary artery

pulmonary vein

alveoli

blood flow

pulmonary venule

Blood supply of alveoli

Capillary network of one alveolus

Figure 14.4 Gas exchange in the lungs. The lungs consist of portions of the bronchial tree leading to the alveoli, each of which is surrounded by an extensive capillary network. Notice that the pulmonary artery and arteriole carry O_2-poor blood (colored blue), and the pulmonary vein and venule carry O_2-rich blood (colored red).

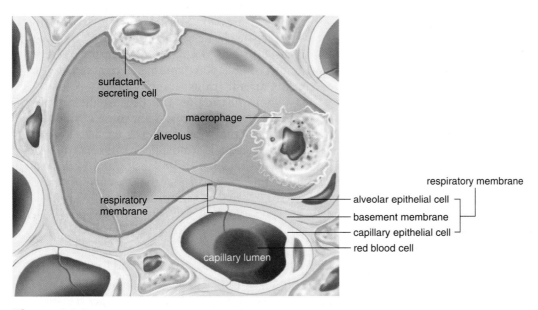

surfactant-secreting cell

macrophage

alveolus

respiratory membrane

respiratory membrane

alveolar epithelial cell

basement membrane

capillary epithelial cell

red blood cell

capillary lumen

Figure 14.5 The respiratory membrane consists of the alveolar wall and the capillary wall.

how hard it is to blow it up.) Normal alveoli are lined with **surfactant**—a film of lipoprotein that lowers surface tension to an acceptable level. Thus, in healthy lungs surface tension is high enough to help the lungs recoil (return to a resting position) yet low enough to prevent the alveoli from collapsing. The lungs collapse in some newborn babies, especially premature infants, who lack this surfactant film. The condition, called **infant respiratory distress syndrome,** is now treatable by surfactant replacement therapy.

Respiratory Membrane Gas exchange occurs very rapidly because of the characteristics of the so-called respiratory membrane (Fig. 14.5). The **respiratory membrane** consists of the juxtaposed alveolar epithelium and the capillary endothelium. Throughout most of the lungs, their basement membranes are fused, meaning that very little tissue fluid separates the two portions[1] of the respiratory membrane and they are indeed one membrane. This membrane is extremely thin—only 0.2–0.6 μm thick.

The total surface area of the respiratory membrane is the same as the area of the alveoli, namely 50–70 m². The blood that enters the many pulmonary capillaries spreads thin. The red blood cells within the capillaries are pressed up against the narrow capillary walls, and little plasma is present. This too facilitates the speed of gas exchange during external respiration. As

[1]The respiratory membrane has two portions but four layers: alveolar epithelium plus its basement membrane and a basement membrane of endothelium plus the capillary endothelium.

discussed in the Medical Focus on page 310, a person with emphysema has a reduced amount of respiratory membrane.

discussed in the Medical Focus on page 310,

Content CHECK-UP!

1. Which structure covers the larynx as swallowing occurs?
   a. epiglottis
   c. uvula
   b. soft palate
   d. hard palate

2. All of the following structures of the respiratory tree have cartilage supporting their walls except:
   a. trachea
   c. tertiary bronchi
   b. secondary bronchi
   d. bronchioles

3. Which membrane structure divides the thoracic cavity into two separate cavities?
   a. parietal pleura
   c. parietal pericardium
   b. mediastinum
   d. visceral pericardium

14.2 Mechanism of Breathing

Ventilation

To understand **ventilation,** the manner in which air enters and exits the lungs, it is helpful to be aware of the following conditions:

1. *The lungs lie within the sealed-off thoracic cavity.* The rib cage, consisting of the ribs joined to the vertebral column posteriorly and to the sternum anteriorly, forms the top and sides of the thoracic cavity. The intercostal muscles lie between the ribs. The diaphragm and connective tissue form the floor of the thoracic cavity.

2. *The lungs adhere to the thoracic wall by way of the pleurae.* The visceral pleura covering the lungs attaches to the parietal pleura covering the chest wall, by utilizing the force of surface tension created by the fluid between them. Any space between the two layers of the pleura is minimal. (As an analogy, wet a clean glass microscope slide with a drop of water, then cover it with a second glass slide. Note the two pieces of glass will slide back and forth, but can't be easily pulled apart.) This surface tension force creates **intrapleural pressure**—the pressure between the pleurae. Intrapleural pressure is less than atmospheric pressure, and this also helps to keep the lungs inflated. The importance of a reduced intrapleural pressure is demonstrated when, by design or accident, atmospheric air enters the intrapleural space and the lung then collapses.

3. *A continuous column of air extends from the pharynx to the alveoli of the lungs.*

Inspiration

Inspiration is the active phase of ventilation because this is the phase in which the diaphragm and the external intercostal muscles contract (Fig. 14.6*a*). In its relaxed state, the diaphragm is dome-shaped; during inspiration, it contracts and becomes a flattened sheet of muscle. Simultaneously, the external intercostal muscles contract, and the rib cage moves upward and outward.

As the thoracic volume increases, the lungs increase in volume as well because the lung adheres to the wall of the thoracic cavity. As lung volume increases, the air pressure within the alveoli (called intrapulmonary pressure) decreases, creating a partial vacuum. In other words, alveolar pressure is now less than atmospheric pressure (air pressure outside the lungs). Because a continuous column of air fills the lungs, air will naturally flow from outside the body into the respiratory passages and into the alveoli. Air flow continues, until intrapulmonary pressure equals atmospheric pressure.

It is important to realize that air comes into the lungs because they have already opened up; air does not force the lungs open. This is why it is sometimes said that *humans breathe by subatmospheric pressure.* The creation of a partial vacuum in the alveoli causes air to enter the lungs. While inspiration is the active phase of breathing, the actual flow of air into the alveoli is passive.

Expiration

Usually, expiration is the passive phase of ventilation, and no muscular effort is required to bring it about. During expiration, the diaphragm and the external intercostal muscles relax. Therefore, the diaphragm resumes its dome shape and the rib cage moves down and in (Fig. 14.6*b*). As the volume of the thoracic cavity decreases, the lungs are free to recoil. Lung recoil occurs due to the elastic tissue built into the lungs' walls, and also due to the slight alveolar surface tension. As thoracic cavity volume and, thus, lung volume, decrease, the air pressure within the alveoli (called intrapulmonary pressure) increases above atmospheric pressure. Because the airways are filled with a continuous column of air, some of that air will naturally flow out of the body. Air flow continues until intrapulmonary pressure equals atmospheric pressure.

What keeps the alveoli from collapsing as they decrease in size during expiration? Recall that the presence of surfactant lowers the surface tension within the alveoli. Also, as the lungs recoil, pressure between the two layers of pleura decreases, and this also helps the alveoli stay open.

Maximum Inspiratory Effort and Forced Expiration

If you recall the last time you exercised vigorously—perhaps running in a race or even just climbing all those stairs to your classroom—you'll probably remember that you were breathing a lot harder than normal during and immediately after that heavy exercise. Maximum inspiratory effort involves the *accessory muscles* of respiration: erector spinae muscles of the back, pectoralis minor (chest), and scalene and sternocleidomastoid muscles of the anterior neck. Their combined efforts can help to increase the size of the thoracic cavity larger than normal, thus allowing maximum expansion of the lungs.

While inspiration is always the active phase of breathing, expiration is usually passive—that is, the diaphragm and

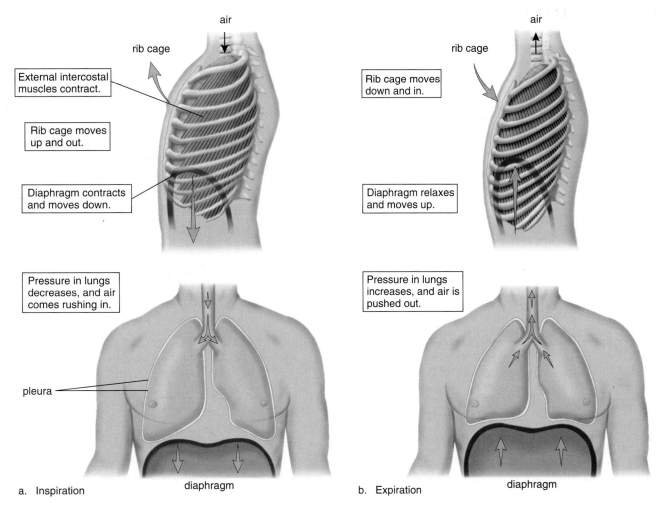

External intercostal muscles contract.

Rib cage moves up and out.

Diaphragm contracts and moves down.

Pressure in lungs decreases, and air comes rushing in.

pleura

a. Inspiration

diaphragm

Rib cage moves down and in.

Diaphragm relaxes and moves up.

Pressure in lungs increases, and air is pushed out.

b. Expiration

diaphragm

Figure 14.6 Inspiration versus expiration. **a.** During inspiration, the thoracic cavity and lungs expand so that intrapulmonary pressure decreases. Now air flows into the lungs. **b.** During expiration, the thoracic wall and lungs recoil, assuming their original positions and pressures. Now air is forced out. The internal intercostal muscles only contract during forceful expiration.

external intercostal muscles are simply allowed to relax, the lungs recoil, and expiration occurs. However, expiration can also be forced. Forced expiration accompanies the maximum inspiratory efforts of heavy exercise. Forced expiration is also necessary to sing, blow air into a trumpet, or blow out birthday candles. Contraction of the internal intercostal muscles can force the rib cage to move downward and inward. Also, when the abdominal wall muscles contract, they push on the viscera, which push against the diaphragm, and the increased pressure in the thoracic cavity helps to expel air.

During breathing, air moves into the lungs from the nose or mouth (called inspiration, or inhalation), and then moves out of the lungs during expiration, or exhalation. A free flow of air from the nose or mouth to the lungs and from the lungs to the nose or mouth is vitally important. Therefore, a technique has been developed that allows physicians to determine if there is a medical problem that prevents the lungs from filling with air upon inspiration and releasing air from the body upon expiration. An instrument called a **spirometer** records the volume of air exchanged during normal breathing and during deep breathing. A spirogram (recording from a spirom-

eter) shows the measurements recorded by a spirometer when a person breathes as directed by a technician (Fig. 14.7).

Respiratory Volumes

Normally when we are relaxed, only a small amount of air moves in and out with each breath. This amount of air, called the **tidal volume,** is only about 500 ml.

It is possible to increase the amount of air inhaled, and therefore the amount exhaled, by deep breathing. The maximum volume of air that can be moved in plus the maximum volume that can be moved out during a single breath is the **vital capacity.** It is called vital capacity because your life depends on breathing, and the more air you can move, the better off you are. A number of different illnesses, discussed in section 14.4, can decrease vital capacity.

Vital capacity varies by how much we can increase inspiration and expiration over the tidal volume amount. We can increase inspiration by using the accessory muscles of respiration (scalenes, sternocleidomastoid, pectoralis minor, etc.) to expand the chest, and also by lowering the diaphragm to the

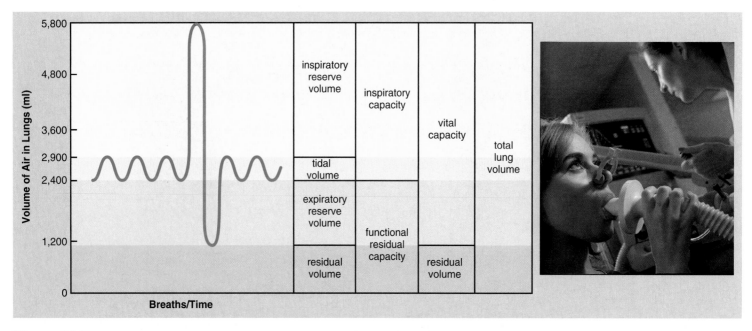

Figure 14.7 Vital capacity. A spirometer measures the amount of air inhaled and exhaled with each breath. During inspiration, the pen moves up, and during expiration, the pen moves down. The volume of one normal breath (tidal volume) multiplied by the number of breaths per minute is called the minute ventilation. A lower-than-normal minute ventilation can be a sign of pulmonary malfunction. Vital capacity (red) is the maximum amount of air a person can exhale after taking the deepest inhalation possible.

maximum extent possible. Forced inspiration usually increases the volume of air beyond the tidal volume by 2,900 ml, and that amount is called the **inspiratory reserve volume.** We can increase the amount of air expired by contracting the abdominal and internal intercostal muscles. This so-called **expiratory reserve volume** is usually about 1,400 ml of air. You can see from Figure 14.6 that vital capacity is the sum of the tidal, inspiratory reserve, and expiratory reserve volumes.

It's a curious fact that some of the inhaled air never reaches the alveoli; instead, it fills the nasal cavities, trachea, bronchi, and bronchioles (see Fig. 14.1). In an average adult, some 70% of the tidal volume does reach the aveoli; but 30% remains in the airways. These passages are not used for gas exchange, and therefore they are said to contain **dead-space** air. To ensure that a large portion of inhaled air reaches the lungs, it is better to breathe slowly and deeply. Also, note in Figure 14.8 that even after a very deep exhalation, some air (about 1,000 ml) remains in the alveoli; this is called the **residual volume.** This air is not as useful for gas exchange because it contains a great deal of CO_2 and has been depleted of oxygen. In some lung diseases, such as emphysema (see the Medical Focus on page 310), the residual volume increases because the individual has difficulty emptying the lungs. This means that the vital capacity is reduced because the lungs have more residual volume.

Control of Ventilation

Normally, adults have a breathing rate of 12 to 20 ventilations per minute. The basic rhythm of ventilation is controlled by a **primary respiratory center** located in the medulla oblongata of the brain.

The primary respiratory center automatically sends out motor nerve signals by way of the phrenic nerve to the diaphragm. Simultaneously, the intercostal nerves stimulate the external intercostal muscles of the rib cage (Fig. 14.6). When these muscles contract, the thoracic volume and lung volume increase, and the person inhales. When the respiratory center stops sending neuronal signals to the diaphragm and the rib cage, the diaphragm relaxes, resuming its dome shape. The rib cage moves down and in. Decreasing thoracic and lung volumes allow the person to exhale.

The primary respiratory center in the medulla oblongata allows the *basic* pattern of inhalation and exhalation. However, if the medulla functions alone (as might occur in a person with a serious head injury, if nerves from higher centers are damaged), respiration is short and gasping. Breathing rhythmically at a normal rate and volume requires nervous input from the **pons,** the brainstem center immediately superior to the medulla. Functioning together, these two brain centers allow normal, quiet breathing: smooth, sustained inspiration, followed by a smooth, sustained expiration.

Although the respiratory center controls the rate and depth of breathing, its activity can be influenced by nervous input and chemical input.

Nervous Input The observation of one's own breathing patterns during day-to-day living makes it evident that higher centers of the brain influence the rate and depth of respiration. Input from the cerebral cortex, limbic system, hypothalamus, and other brain centers account for the fact that respiration rate and depth increase if one is angry, frightened or otherwise upset. Conversely, respiration rate and depth

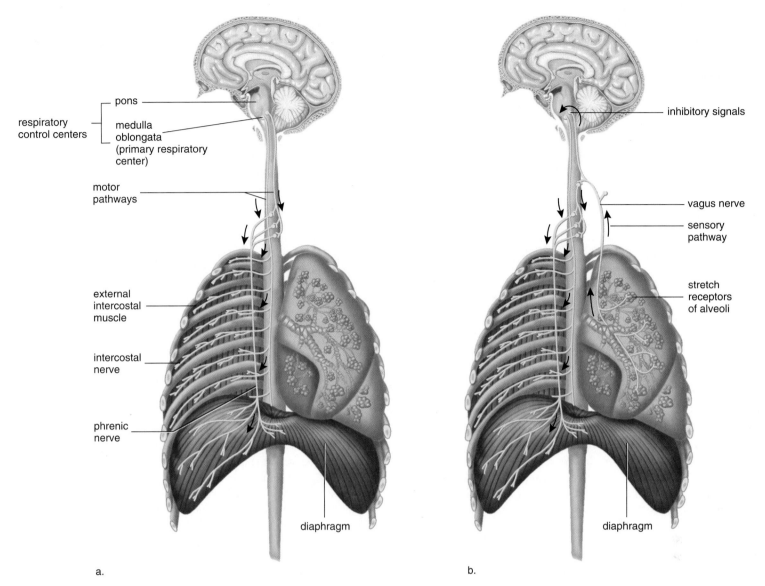

Figure 14.8 Nervous control of breathing. **a.** During inspiration, respiratory control areas of the pons and medulla stimulate the intercostal and phrenic nerves. Intercostal nerves cause the external intercostal muscles to contract. The phrenic nerve stimulates the diaphragm. Expiration occurs rhythmically when respiratory control areas stop nervous stimulation and allow the muscles to relax. **b.** Hering-Breuer reflex. When tidal volume increases above 1.5 liters, alveolar stretch receptors signal the respiratory centers via the vagus nerve and inspiration ceases. This prevents overinflation of the lungs.

decrease in the soundest stages of sleep. Conscious control of respiration allows a person to hold his breath for a time, or to voluntarily hyperventilate. Neural influence over the respiratory centers also helps to account for the increased respiratory rate and depth that occur during exercise.

Nervous control over respiration also helps to protect delicate lung tissue, as illustrated by the Hering-Breuer reflex. During exercise, the depth of inspiration can increase, due to recruitment of muscle fibers in the diaphragm, intercostal muscles, and accessory muscles. Then, stretch receptors in the alveolar walls are stimulated, and they initiate inhibitory nerve impulses that travel from the inflated lungs to the respiratory center. This causes the respiratory center to stop sending out nerve impulses. This reflex helps support rhythmic res-

piratory movements by limiting the extent of inspiration. The reflex also prevents over-expansion of lung tissue.

Chemical Input The respiratory center is directly sensitive to the levels of carbon dioxide (CO_2) and hydrogen ions (H^+). When they rise, due to increased cellular respiration (during exercise, for example), the respiratory center increases the rate and depth of breathing. The center is not affected directly by low oxygen (O_2) levels. However, chemoreceptors in the **carotid bodies,** located in the carotid arteries, and in the **aortic bodies,** located in the aorta, are sensitive to the level of oxygen in the blood. (Do not confuse the carotid and aortic bodies with the carotid and aortic baroreceptors, which monitor blood pressure.) When the concentration of oxygen decreases, these bodies communicate

Emphysema: Medical and Surgical Treatment

"By the time I was diagnosed," the patient complains, "I couldn't walk across this room without stopping to catch my breath. I lost 40 pounds because I didn't have the energy to eat. I had to quit a job I loved, because I just couldn't do the work anymore. Now I'm hooked to this oxygen tank. I know there are risks involved, but I'm willing to take a chance on surgery. I just can't live my life like this."

This patient has emphysema, a degenerative disease most often caused by prolonged cigarette smoking. The respiratory membrane breaks down, and there are fewer (but enlarged) alveoli. Gas exchange is reduced because there is less surface area for the diffusion of respiratory gases into the blood. Air is trapped in the alveoli, and exhaling becomes very difficult. The lungs themselves become larger and less elastic, and the patient develops a barrel-chested appearance. On X ray, the diaphragm (which is normally curved when at rest; see page 307) appears flattened as the lungs increase in size. Exercise—and even normal daily activities—become impossible for the person.

Medical treatment for emphysema is limited to drugs that dilate the bronchi and bronchioles by relaxing the smooth muscle, along with inhaled oxygen to improve oxygenation of the blood.

Two surgical options remain for some patients with end-stage emphysema: lung volume reduction surgery (LVRS) and lung transplantation. In a recently completed nationwide study, it was shown that the best candidate for LVRS is a patient with poor ability to exercise, whose disease is confined to the upper lobes of the lungs. In LVRS, the diseased upper lobes of the lungs are surgically removed. The remaining healthy lung tissue is sealed, usually with a flap of pericardium. Removal of the diseased portion of the lung allows the remaining healthy tissue to expand and fill the thoracic cavity. This improves gas exchange because the thorax is filled with the healthiest possible tissue. The diaphragm can return to its normal curved position at rest, allowing the patient to once again take a deep breath.

Before being approved for this surgery, the patient must undergo preoperative rehabilitation. The first step in the rehabilitation process is to quit smoking—forever. The next step includes drug treatment to widen the bronchial tubes and improve blood circulation. Physical and respiratory therapy enables the patient to tolerate exercise. Similar treatments are continued after the surgery has been completed. Studies have shown that patients who complete the rehabilitation before and after surgery have the best recovery.

LVRS is a complicated and potentially dangerous procedure, and it certainly is not a cure for emphysema. However, the benefits of the surgery can be dramatic for some patients. Improvements in oxygenation of the blood, tolerance for exercise, and overall lung function enable some patients to resume near-normal lives.

Lung transplant is the other surgical option for emphysema sufferers. This option is rarely used, however, due to the scarcity of donor organs. As with all transplant recipients, the patient must remain on lifelong anti-rejection therapy after a lung transplant.

---

with the respiratory center, and the rate and depth of breathing increase. The Medical Focus on page 311 describes some modified ventilation rates that occur due to various circumstances.

Content CHECK-UP!

4. Muscles that are contracted when a person inhales include all of the following except:
   a. internal intercostals.
   c. external intercostals.
   b. diaphragm.
   d. pectoralis minor.

5. Which of the following has to be true for a person to inhale?
   a. Intrapulmonary pressure is greater than atmospheric pressure.
   b. Atmospheric pressure is greater than intrapulmonary pressure.
   c. The abdominal muscles must contract.
   d. a and c

6. Which of the following is the volume of air left over in the lungs after the maximum expiratory effort?
   a. inspiratory reserve volume
   c. residual volume
   b. expiratory reserve volume
   d. alveolar volume

14.3 Gas Exchange and Transport

Gas exchange and transport are critical to homeostasis. As mentioned previously, respiration includes not only the exchange of gases in the lungs, but also the exchange of gases in the tissues. Recall that diffusion is the movement of molecules from the area of higher concentration to the area of lower concentration. The principles of diffusion alone govern whether oxygen (O_2) or carbon dioxide (CO_2) enters or leaves the blood in the lungs and in the tissues.

External Respiration

External respiration is the exchange of gases in the lungs. Specifically, during external respiration, gases are exchanged between the air in the alveoli and the blood in the pulmonary capillaries. Blood that enters the pulmonary capillaries is dark maroon because it is relatively O_2-poor.

Once inspiration has occurred, the alveoli have a higher concentration of O_2 than does blood entering the lungs. Therefore, O_2 diffuses from the alveoli into the blood. The reverse is true of CO_2. The alveoli have a lower concentration of CO_2 than does blood entering the lungs. Therefore, CO_2

MEDICAL FOCUS

Respiratory and Nonrespiratory Patterns

The normal pattern of quiet breathing is termed **eupnea.** A condition called **Cheyne-Stokes respiration** is characterized by alternate periods of **hyperpnea** (deep and labored breathing) and **apnea** (no breathing or shallow breathing). In this condition, the respiratory center is apparently being controlled largely by chemical input so that the breathing rate first increases when CO_2 and H^+ are high and O_2 is low and then decreases when CO_2 and H^+ are low and O_2 is high. Cheyne-Stokes respiration is associated with abnormal environmental conditions (e.g., high altitude) or physiological disorders (e.g., congestive heart failure).

Other factors can also affect respiration. A sudden cold stimulus, such as a plunge into cold water, causes temporary apnea. A sudden, severe pain has the same effect, but prolonged pain triggers the stress syndrome, which causes an increased breathing rate. A rather interesting stimulus is stretching of the anal sphincter muscle, which causes inspiration. Various other patterns of nonrespiratory air movements are of interest.

Prior to a *cough*, the glottis closes. Then the glottis suddenly opens as a blast of air is forced upward from the lower respiratory tract. A *sneeze* is like a cough except that the blast of air is directed into the nasal passages by a depression of the uvula that closes off the pharynx and mouth. When a person *laughs* or cries, air is released in a series of short expirations; therefore, it is necessary to study the facial expression to know if a person is crying or laughing. A *hiccup* occurs when the diaphragm contracts spasmodically while the glottis is closed. Air striking the vocal cords causes the hiccup.

Sighing and *yawning* require a long inspiration followed by a shorter expiration. The sound that accompanies expiration serves as a means of communication. Yawning has the benefit of making a person more alert when drowsiness occurs. The neurophysiology of yawning is complex, and knowledge of its mechanisms is incomplete because it is apparently under the control of various neurotransmitters.

diffuses out of the blood into the alveoli. This CO_2 exits the body during expiration.

Another way to explain gas exchange in the lungs is to consider the partial pressure of the gases involved. Gases exert pressure, and the amount of pressure each gas exerts is its partial pressure, symbolized as P_{O_2} and P_{CO_2}. Alveolar air has a much higher P_{O_2} than does blood. Therefore, O_2 diffuses into the blood from the alveoli. The pressure pattern is the reverse for CO_2. Blood entering the pulmonary capillaries has a higher P_{CO_2} than the air in the alveoli. Therefore, CO_2 diffuses out of the blood into the alveoli.

Internal Respiration

Internal respiration refers to the exchange of gases in the tissues. Specifically, during internal respiration, gases are exchanged between the blood in systemic capillaries and the tissue fluid. Blood that enters the systemic capillaries is a bright red color because the blood is O_2-rich. Tissue fluid, on the other hand, has a low concentration of O_2. Why? Because the cells are continually consuming O_2 during cellular respiration. Therefore, O_2 diffuses from the blood into the tissue fluid. Tissue fluid has a higher concentration of CO_2 than does the blood entering the tissues. Why? Because CO_2 is an end product of cellular respiration. Therefore, CO_2 diffuses from the tissue fluid into the blood. Figure 14.9 summarizes our discussion of gas exchange in the lungs and tissues and shows the differences in O_2 and CO_2 that lead to diffusion of these gases.

Again, we can explain exchange in the tissues by considering the partial pressure of the gases involved. In this case, oxygen diffuses out of the blood into the tissues because the P_{O_2} in tissue fluid is lower than that of the blood. Carbon dioxide diffuses into the blood from the tissues because the P_{CO_2} in tissue fluid is higher than that of the blood.

Gas Transport

The mode of transport of oxygen and carbon dioxide in the blood differs, although red blood cells are involved in transporting both of these gases.

Oxygen Transport

After O_2 enters the blood contained in pulmonary capillaries of the lungs, it enters red blood cells and combines with the iron portion of **hemoglobin,** the pigment in red blood cells. In addition, a small amount of oxygen is transported as a dissolved gas in the watery blood plasma. This dissolved oxygen amounts to only 2–3% of the body's oxygen at any given time because oxygen is not very soluble in water.

Hemoglobin is remarkably suited to the task of transporting oxygen because it can either combine with or release oxygen (depending upon its surroundings). The higher concentration of oxygen in the alveoli, plus the slightly higher pH and slightly cooler temperature, causes hemoglobin to take up oxygen and become **oxyhemoglobin** (HbO_2). The lower concentration of oxygen in the tissues, plus the slightly lower pH and slightly warmer temperature in the tissues, causes hemoglobin to release oxygen and become deoxyhemoglobin (Hb). This equation summarizes our discussion of oxygen transport:

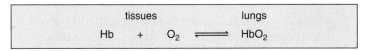

tissues		lungs
Hb $+$ O_2	$\rightleftharpoons$	HbO_2

Carbon Dioxide Transport

Transport of CO_2 to the lungs involves a number of steps. After CO_2 diffuses into the blood at the tissues, it can be transported in one of three ways:

1. Approximately 10% of CO_2 is transported as a dissolved gas in blood plasma and in the cytoplasm of red blood

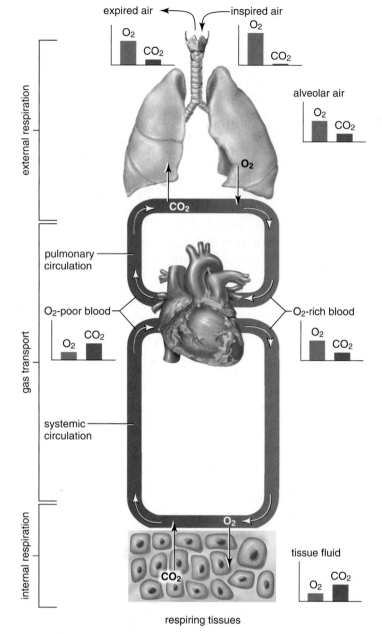

Figure 14.9 External and internal respiration. During external respiration in the lungs, CO_2 leaves the blood and O_2 enters the blood passively by diffusion. During internal respiration in the tissues, O_2 leaves the blood and CO_2 enters the blood passively by diffusion.

cells. Carbon dioxide is much more soluble in water than is oxygen and, thus, three to five times as much CO_2 can be transported as a dissolved gas.

2. Roughly 30% of CO_2 molecules formed are taken up by the protein (globin) portion of hemoglobin, forming a compound called **carbaminohemoglobin.**

3. Most of the CO_2—60%—combines with water, forming carbonic acid (H_2CO_3). The carbonic acid dissociates to hydrogen ions (H^+) and bicarbonate

ions (HCO_3^-). This reaction is assisted by an enzyme, **carbonic anhydrase,** which is present in all body cells but is especially abundant in red blood cells. The release of hydrogen ions from carbonic acid explains why the blood in systemic capillaries has a lower pH than the blood in pulmonary capillaries. However, the difference in pH is slight because the globin portion of hemoglobin combines with excess hydrogen ions and becomes **reduced hemoglobin (HHb).**

Bicarbonate ions are carried in the plasma because they diffuse out of red blood cells and go into the plasma. Most of the carbon dioxide in blood is carried as HCO_3^-, the **bicarbonate ion.** As bicarbonate ions diffuse out of red blood cells, chloride ions (Cl^-) diffuse into them. This so-called **chloride shift** maintains the electrical balance between the plasma and the red blood cells.

In pulmonary capillaries, a reverse reaction occurs. Bicarbonate combines with hydrogen ions to form carbonic acid, which this time splits into CO_2 and H_2O, and the CO_2 diffuses out of the blood into the alveoli. The following equation summarizes the reaction between carbon dioxide and water:

$$CO_2 \ + \ H_2O \ \underset{\text{lungs}}{\overset{\text{tissues}}{\rightleftharpoons}} \ H_2CO_3 \ \underset{\text{lungs}}{\overset{\text{tissues}}{\rightleftharpoons}} \ H^+ \ + \ HCO_3^-$$

Content **CHECK-UP!**

7. Which of the following statements describing internal respiration is incorrect?
   a. Oxygen diffuses from the blood into the tissue fluid.
   b. The partial pressure of oxygen is higher in tissue fluid than in the blood.
   c. Carbon dioxide diffuses from tissue fluid into the blood.
   d. Internal respiration takes place in systemic capillaries.

8. Hemoglobin with carbon dioxide attached is termed:
   a. reduced hemoglobin.   c. oxyhemoglobin.
   b. carbaminohemoglobin.   d. dioxyhemoglobin.

9. Which of the following is not a method of transporting carbon dioxide in the blood?
   a. as bicarbonate ion
   b. as a dissolved gas
   c. attached to the globin portion of hemoglobin
   d. attached to the heme portion of hemoglobin

14.4 Respiration and Health

The respiratory tract is constantly exposed to environmental air. The quality of this air can affect our health. The presence of a disease means that homeostasis is threatened and if the condition is not brought under control, death is inevitable.

Upper Respiratory Tract Infections

The upper respiratory tract consists of the nasal cavities, the pharynx, and the larynx. Upper respiratory infections (URI) can spread from the nasal cavities to the sinuses, middle ears, and larynx. Viral infections sometimes lead to secondary bacterial infections. What we call "strep throat" is a primary bacterial infection caused by *Streptococcus pyogenes* that can lead to a generalized upper respiratory infection and even a systemic (affecting the body as a whole) infection. Although antibiotics have no effect on viral infections, they are successfully used to treat most bacterial infections, including strep throat. The symptoms of strep throat are severe sore throat, high fever, and white patches on a dark red throat.

Sinusitis

Sinusitis is an infection of the cranial sinuses, the cavities within the facial skeleton that drain into the nasal cavities. Only about 1–3% of upper respiratory infections are accompanied by sinusitis. Sinusitis develops when nasal congestion blocks the tiny openings leading to the sinuses. Symptoms include postnasal discharge as well as facial pain that worsens when the patient bends forward. Pain and tenderness usually occur over the lower forehead or over the cheeks. If the latter, toothache is also a complaint. Successful treatment depends on restoring proper drainage of the sinuses. Antibiotic therapy is necessary if the condition results from bacterial infection. Even a hot shower and sleeping upright can be helpful. Otherwise, spray decongestants are preferred over oral antihistamines, which thicken rather than liquefy the material trapped in the sinuses.

Otitis Media

Otitis media is a bacterial infection of the middle ear. The middle ear is not a part of the respiratory tract, but this infection is considered here because it is a complication often seen in children who have a nasal infection. Infection can spread by way of the **auditory (eustachian) tube** that leads from the nasopharynx to the middle ear. Pain is the primary symptom of a middle ear infection. A sense of fullness, hearing loss, vertigo (dizziness), and fever may also be present. Antibiotics almost always bring about a full recovery, and recurrence is probably due to a new infection. Tubes (called tympanostomy tubes) are sometimes placed in the eardrums of children with multiple recurrences to help promote fluid drainage, as well as to prevent the buildup of pressure in the middle ear and the possibility of hearing loss. Normally, the tubes fall out with time.

Tonsillitis

Tonsillitis occurs when the **tonsils,** masses of lymphatic tissue in the pharynx, become inflamed and enlarged. If tonsillitis occurs frequently and enlargement makes breathing difficult, the tonsils can be removed surgically in a **tonsillectomy.** Fewer tonsillectomies are performed today than in the past because we now know that the tonsils remove many of the pathogens that enter the pharynx; therefore, they are a first line of defense against invasion of the body.

Laryngitis

Laryngitis is an inflammation of the larynx, with accompanying hoarseness leading to the inability to talk in an audible voice. It may result from an upper respiratory infection, or simply from overuse of the larynx (if, for example, you've screamed for several hours at a sports event or concert.) So long as the immune system is functioning properly, an upper respiratory infection usually resolves, with rest, over time. Likewise, laryngitis disappears over time as long as the larynx is rested. Persistent hoarseness without the presence of an upper respiratory infection is one of the warning signs of cancer, and therefore should be looked into by a physician.

Lower Respiratory Tract Disorders

Lower respiratory tract disorders include infections, restrictive pulmonary disorders, obstructive pulmonary disorders, and lung cancer.

Lower Respiratory Infections

Acute bronchitis, pneumonia, and tuberculosis are infections of the lower respiratory tract. **Acute bronchitis** is an infection of the primary and secondary bronchi. Usually, it is preceded by a viral URI that has led to a secondary bacterial infection. Most likely, a nonproductive cough has become a deep cough that expectorates mucus and perhaps pus.

Pneumonia is a viral or bacterial infection of the lungs in which the bronchi and alveoli fill with thick fluid (Fig. 14.10). Risk factors for pneumonia include advanced age (it is most common and most often fatal in the elderly), weakened immune system, smoking, and being immobilized. A common scenario for development of pneumonia is an elderly person who is immobilized due to a fractured hip. High fever, productive cough, difficulty in breathing, headache, and chest pain are symptoms of pneumonia. Rather than being a generalized lung infection, pneumonia may be localized in specific lobules of the lungs. Obviously, the more lobules involved, the more serious is the infection. Pneumonia can be caused by a bacterium that is usually held in check but has gained the upper hand due to stress and/or reduced immunity. AIDS patients are subject to a particularly rare form of pneumonia caused by the protozoan *Pneumocystis carinii.* Pneumonia of this type is almost never seen in individuals with a healthy immune system.

Pulmonary tuberculosis is caused by the tubercle bacillus, a type of bacterium. When tubercle bacilli invade the lung tissue, the cells build a protective capsule about the foreigners, isolating them from the rest of the body. This tiny capsule is called a tubercle. If the resistance of the body is high, the imprisoned organisms die, but if the resistance is low, the organisms eventually can be liberated. If a chest X ray detects active tubercles, the individual is put on appropriate antibiotic therapy to ensure the localization of the disease and the eventual destruction of any live bacteria. It is possible to tell if a person has ever been exposed to tuberculosis with a test in which a highly diluted extract of the bacillus is injected into the skin of the patient. A person who has never been in contact with occurred. It has been projected that two million deaths caused by asbestos exposure—mostly in the workplace—will occur in the United States between 1990 and 2020.

Obstructive Pulmonary Disorders

In obstructive pulmonary disorders, air does not flow freely in the airways, and the time it takes to inhale or exhale maximally is greatly increased. Chronic bronchitis and emphysema are referred to as **chronic obstructive pulmonary disorders (COPD)** because they develop slowly, over a long period of time, and are recurrent. Asthma is an obstructive disorder as well. However, asthma is generally an acute illness—that is, one that occurs in intermittent episodes that flare up and disappear quickly.

In **chronic bronchitis,** the airways are inflamed and filled with mucus. A cough that brings up mucus is common. The bronchi have undergone degenerative changes, including the loss of cilia and their normal cleansing action. Under these conditions, an infection is more likely to occur. Smoking cigarettes and cigars is the most frequent cause of chronic bronchitis. Exposure to other pollutants can also cause chronic bronchitis.

Emphysema is a chronic and incurable disorder in which the alveoli are distended and their walls damaged so that the surface area available for gas exchange is reduced. Emphysema is often preceded by chronic bronchitis. Air trapped in the lungs leads to alveolar damage and a noticeable ballooning of the chest. Breakdown in the elastic tissue of the lung diminishes the elastic recoil. Not only are the airways narrowed, but the driving force behind expiration is also reduced. The patient has a very difficult time exhaling, and residual volume increases. The victim often feels a breathless sensation and may have a cough. Because the surface area for gas exchange in the lungs is reduced, less oxygen reaches the heart and the brain. Even so, the heart works furiously to force more blood through the lungs, and an increased workload on the heart can result. Lack of oxygen to the brain can make the person feel depressed, sluggish, and irritable. Before therapy can be effective, the patient must stop

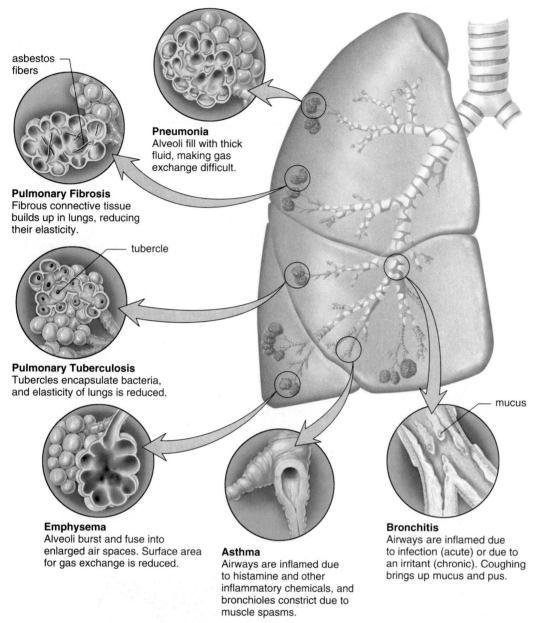

Pulmonary Fibrosis
Fibrous connective tissue builds up in lungs, reducing their elasticity.

asbestos fibers

Pneumonia
Alveoli fill with thick fluid, making gas exchange difficult.

tubercle

Pulmonary Tuberculosis
Tubercles encapsulate bacteria, and elasticity of lungs is reduced.

Emphysema
Alveoli burst and fuse into enlarged air spaces. Surface area for gas exchange is reduced.

Asthma
Airways are inflamed due to histamine and other inflammatory chemicals, and bronchioles constrict due to muscle spasms.

mucus

Bronchitis
Airways are inflamed due to infection (acute) or due to an irritant (chronic). Coughing brings up mucus and pus.

Figure 14.10 Common bronchial and pulmonary diseases. Exposure to infectious pathogens and/or polluted air, including tobacco smoke, causes the diseases and disorders shown here.

the tubercle bacillus shows no reaction, but one who has had or is fighting an infection shows an area of inflammation that peaks in about 48 hours.

Restrictive Pulmonary Disorders

In restrictive pulmonary disorders, vital capacity is reduced because the lungs have lost their elasticity. Inhaling particles such as silica (sand), coal dust, asbestos, clay, cement, flour and fiberglass can lead to **pulmonary fibrosis,** a condition in which fibrous connective tissue builds up in the lungs. The lungs cannot inflate properly and are always tending toward deflation. Breathing asbestos is also associated with the development of cancer. Because asbestos was formerly used widely as a fireproofing and insulating agent, unwarranted exposure has smoking. Then, exercise, drug therapy, supplemental oxygen, and surgery (see the Medical Focus on page 310) may relieve the symptoms and possibly slow the progression of emphysema.

Asthma is a disease of the bronchi and bronchioles that is marked by wheezing, breathlessness, and sometimes a cough and expectoration of mucus. The airways are unusually sensitive to specific irritants, which can include a wide range of allergens such as pollen, animal dander, dust, cigarette smoke, and industrial fumes. Even cold air can be an irritant. When exposed to the irritant, the smooth muscle in the bronchioles undergoes spasms. It now appears that chemical mediators given off by immune cells in the bronchioles cause the spasms. Most asthma patients have some degree of bronchial inflammation that reduces the diameter of the airways and contributes to the seriousness of an attack. Asthma is not curable, but it is treatable. Special inhalers can control the inflammation and hopefully prevent an attack, while other types of inhalers can stop the muscle spasms should an attack occur.

Lung Cancer

Lung cancer used to be more prevalent in men than in women, but recently it has surpassed breast cancer as a cause of death in women. The recent increase in the incidence of lung cancer in women is directly correlated to increased numbers of women who smoke. Autopsies on smokers have revealed the progressive steps by which the most common form of lung cancer develops. The first event appears to be thickening and callusing of the cells lining the primary bronchi. (Callusing occurs whenever cells are exposed to irritants.) Then cilia are lost, making it impossible to prevent dust and dirt from settling in the lungs. Following this, cells with atypical nuclei appear in the callused lining. A tumor consisting of disordered cells with atypical nuclei is considered cancer in situ (at one location) (Fig. 14.11). A final step occurs when some of these cells break loose and penetrate other tissues, a process called **metastasis.** Now the cancer has spread. The original tumor may grow until a bronchus is blocked, cutting off the supply of air to that lung. The entire lung then collapses, the secretions trapped in the lung spaces become infected, and pneumonia or a lung abscess (localized area of pus) results. The only treatment that offers a possibility of cure is to remove a lobe or the whole lung before metastasis has had time to occur. This operation is called **pneumonectomy.** If the cancer has spread, chemotherapy and radiation are also required.

The Medical Focus on pages 316–317 lists the various illnesses, including cancer, that are apt to occur when a person smokes. Current research indicates that passive smoking—exposure to smoke created by others who are smoking—can also cause lung cancer and other illnesses associated with smoking. If a person stops voluntary smoking and avoids passive smoking, and if the body tissues are not already cancerous, they may return to normal over time.

Content CHECK-UP!

10. Obstructive pulmonary diseases include:
    a. emphysema.   c. asthma.
    b. chronic bronchitis.   d. all of the above.

11. Which of the following increases in a person with emphysema?
    a. amount of surface area in the lung for gas exchange
    b. residual volume
    c. amount of elastic tissue in the lungs
    d. amount of oxygen delivered to the brain and heart

12. Antibiotics would be useful medication for all of the following except:
    a. streptococcal infection.   c. bacterial sinusitis.
    b. tuberculosis.   d. viral upper respiratory infection.

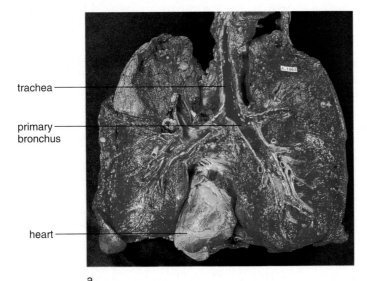

trachea

primary bronchus

heart

a.

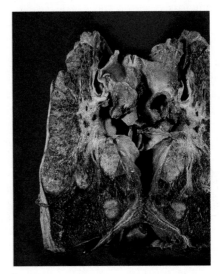

b.

Figure 14.11 Normal lung versus cancerous lung. **a.** Normal lung with heart in place. Note the healthy red color. **b.** Lungs of a heavy smoker. Notice how black the lungs are except where cancerous tumors have formed.

The Most-Often-Asked Questions About Tobacco and Health

Is There a Safe Way To Smoke?

No. All forms of tobacco can cause damage and smoking even a small amount is dangerous. Tobacco is perhaps the only legal product whose advertised and intended use—that is, smoking it—will hurt the body.

Does Smoking Cause Cancer?

Yes, and not only lung cancer. Besides lung cancer, smoking a pipe, cigarettes, or cigars is also a major cause of cancers of the mouth, larynx (voice box) and esophagus. In addition, smoking increases the risk of cancer of the bladder, kidney, pancreas, stomach and uterine cervix.

What are the Chances of Being Cured of Lung Cancer?

Very low; the five-year survival rate is only 13%. However, it is important to remember that lung cancer, or **bronchogenic carcinoma** in technical terms, is very rare (although not unheard of) in nonsmokers. Thus, lung cancer is largely preventable: Don't smoke, and avoid exposure to smoke.

Does Smoking Cause Other Lung Diseases?

Yes. Smoking leads to chronic bronchitis, a disease in which the airways produce excess mucus, forcing the smoker to cough frequently. Smoking is also the major cause of emphysema, a disease that slowly destroys a person's ability to breathe. Chronic bronchitis and pulmonary emphysema are higher in smokers than in nonsmokers.

Why Do Smokers have "Smoker's Cough"?

Remember the mucociliary escalator (page 301)? This is the protective mechanism for the lungs and the airways: a sheet of mucus present in the trachea, bronchi and bronchioles. Normally, constantly beating cilia on the cells lining the airways move the sheet upward. Once at the pharynx, mucus can be swallowed or spit out. Smoking is a double assault on the mucociliary escalator: First, inhaled nasty toxins and particles cling to the sticky mucus. Then, a big dose of nicotine paralyzes the cilia for 30 minutes or more. Long-term smokers will eliminate the delicate, ciliated epithelial cells entirely, replacing them with a sturdier epithelium better able to stand the heat of cigarette smoke. The result: no way to clear accumulated mucus (and the trapped contaminants in it) except to cough it out. The longer a person smokes, the more pronounced and productive the cough becomes.

If You Smoke But Don't Inhale, is There Any Danger?

Yes. Wherever smoke touches living cells, it does harm. So, even if smokers of pipes, cigarettes and cigars don't inhale, they are at an increased risk for lip, mouth and tongue cancer.

Does Smoking Affect the Heart?

Yes. Smoking increases the risk of heart disease, which is the number one killer in the United States. Smoking, high blood pressure, high cholesterol, and lack of exercise are all risk factors for heart disease. Smoking alone doubles the risk of heart disease.

Is There Any Risk for Pregnant Women and Their Babies?

Pregnant women who smoke endanger the health and lives of their unborn babies. When a pregnant woman smokes, she really is smoking for two because the nicotine, carbon monoxide and other dangerous chemicals in smoke enter her bloodstream and then pass into the baby's body. Smoking mothers have more stillbirths and babies of low birthweight than nonsmoking mothers.

Does Smoking Cause Any Special Health Problems for Women?

Yes. Women who smoke and use the birth control pill have an increased risk of stroke and blood clots in the legs. In addition, women who smoke increase their chances of getting cancer of the uterine cervix.

What are Some of the Short-Term Effects of Smoking Cigarettes?

Almost immediately, smoking can make it hard to breathe. Within a short time, it can also worsen asthma and allergies. Only seven seconds after a smoker takes a puff, nicotine reaches the brain, where it produces a morphinelike effect.

14.5 Effects of Aging

Respiratory fitness decreases with age. Maximum breathing capacities decline, while the likelihood of fatigue increases. Inspiration and expiration are not as effective in older persons. With age, weakened intercostal muscles and increased inelasticity of the rib cage combine to reduce the inspiratory reserve volume, while the lungs' inability to recoil reduces the expiratory reserve volume. More residual air is found in the lungs of older people.

Are There Any Other Risks to the Smoker?

Yes, there are many more risks. Smoking is a cause of stroke, which is the third leading cause of death in the United States. Smokers are more likely to have and die from stomach ulcers than nonsmokers. Smokers have a higher incidence of cancer in general. If a person smokes and is exposed to radon or asbestos, the risk for lung cancer increases dramatically.

What are the Dangers of Passive Smoking?

Passive smoking causes lung cancer in healthy nonsmokers. Children whose parents smoke are more likely to suffer from pneumonia or bronchitis in the first two years of life than children who come from smoke-free households. Passive smokers have a 30% greater risk of developing lung cancer than do nonsmokers who live in a smoke-free house.

Are Chewing Tobacco and Snuff Safe Alternatives to Cigarette Smoking?

No, they are not. Many people who use chewing tobacco or snuff believe it can't harm them because there is no smoke. Wrong. Smokeless tobacco contains nicotine, the same addicting drug found in cigarettes and cigars. Although not inhaled through the lungs, the juice from smokeless tobacco is absorbed through the lining of the mouth. There it can cause sores and white patches, which often lead to cancer of the mouth. Snuff dippers actually take in an average of over ten times more cancer-causing substances than cigarette smokers.

OK, OK, enough already! You've convinced me that I should quit. But how? I've tried before. Is there a better way to quit smoking?

The U.S. Office of the Surgeon General has done extensive research on the most effective way to quit smoking. This research has determined that to be successful, a smoker should use the **Five Keys for Quitting:**

1. **Get ready to quit:** Set your date and plan ahead. Review what worked, or didn't, in the past. Throw out all smoking materials: cigarettes, pipes, ashtrays, lighters, etc. Inform your family, friends and co-workers of your intention to quit, and ask them not to smoke around you. Once you quit, don't take another puff from a cigarette.

2. **Get support and encouragement:** See your health care provider for help. Private and group smoking cessation counseling often helps and is available at a low cost or is free. Your local health department will have information about programs in your area. Avoid spending time with people who might undermine your efforts.

3. **Learn new behaviors that aren't associated with smoking:** When you first quit, change your routine completely so that subconscious cues to smoke don't keep appearing. Take a different route to work, eat breakfast in a different place, drink tea instead of coffee. When the urge to smoke recurs, distract yourself with physical or mental activity. Exercise or talk to someone to reduce stress. Each day, plan something fun and pleasurable for yourself.

4. **Use medication correctly:** Never be embarrassed to ask for, and use, anti-smoking medication. Nicotine comes in gum and patches (both sold over the counter) and in prescription inhaler and nasal spray forms. Using these products can help to overcome the cravings for nicotine that occur immediately after quitting. In addition, use of antidepressants (available by prescription) has been shown to significantly increase a patient's chances of quitting for good.

5. **If at first you don't succeed, try, try again!** The first 3–6 months after quitting will be the toughest, and that's when many people go back to smoking. Don't be discouraged if this happens to you. On average, most people try to quit at least twice before finally succeeding. To avoid a relapse, don't drink alcohol and/or hang around with other smokers. Studies have shown that both can lead to failure in the attempt to quit. Weight gain and increased irritability or depression can also be expected. Exercise and eating a healthy diet will minimize both of these side effects.

From *You Can Quit Smoking Consumer Guide*, available online at http://www.surgeongeneral.gov/tobacco/consquits.htm.

With age, gas exchange in the lungs becomes less efficient, not only due to changes in the lungs but also due to changes in the blood capillaries. The respiratory membrane thickens, and the gases cannot diffuse as rapidly as they once did.

In the elderly, the ciliated cells of the trachea are reduced in number, and those remaining are not as effective as they once were. Respiratory diseases, such as those discussed in section 14.4, are more prevalent in older people than in the general public. Pneumonia and other respiratory infections are among the leading causes of death in older persons.

14.6 Homeostasis

The respiratory system contributes to homeostasis in many ways—in particular, by carrying on gas exchange and regulating blood pH.

Gas Exchange

First and foremost, the respiratory system performs gas exchange. Carbon dioxide, a waste molecule given off by cellular respiration, exits the body, and oxygen, a molecule needed for cellular respiration, enters the body at the lungs. Cellular respiration produces ATP, a molecule that allows the body to perform all sorts of work, including muscle contraction and nerve conduction. It is estimated that the brain uses 15–20% of the oxygen taken into the blood. Not surprisingly, a lack of oxygen affects the functioning of the brain before it affects other organs.

Regulation of Blood pH

The respiratory system can alter blood pH by changing blood carbon dioxide levels. In the tissues, carbon dioxide enters the blood and red blood cells where this reaction occurs. The bicarbonate ion (HCO_3^-) diffuses out of the red blood cells to be carried in the plasma:

$$CO_2 + H_2O \xrightleftharpoons[\text{lungs}]{\text{tissues}} H_2CO_3 \xrightleftharpoons[\text{lungs}]{\text{tissues}} H^+ + HCO_3^-$$

This reaction lowers blood pH[2] because free H^+ (hydrogen ions) are formed during the reaction as carbonic acid dissociates. The greater the amount of CO_2 in the blood, the greater the carbonic acid formed, and the more free hydrogen ions are formed. Conversely, when carbon dioxide starts to diffuse out of the blood in the lungs, the reaction occurs in the reverse direction. Carbon dioxide concentration in blood decreases (because it is being exhaled), and carbonic acid concentration decreases, too. Now, the blood pH rises.

What happens to your blood pH if you hypoventilate—that is, breathe at a lower than normal rate? A low blood pH, called **acidosis,** results because carbon dioxide is not being exhaled at the normal rate. Too much carbonic acid is formed from the excess carbon dioxide. Any respiratory condition, such as emphysema, that hinders the passage of carbon dioxide out of the blood also results in acidosis. By contrast, what happens to your blood pH if you hyperventilate—that is, breathe at a higher than normal rate? A high blood pH, called **alkalosis,** results because carbon dioxide is leaving the body at a high rate. This decreases the concentration of carbonic acid and the concentration of free H^+. Severe anxiety can cause a person to hyperventilate.

Control of Blood Pressure

The lungs play a role in the control of blood pressure by assisting in the renin-angiotensin-aldosterone pathway (see p. 261). The kidneys produce the enzyme renin whenever blood pressure falls. Renin activates **angiotensin** I (an inactive plasma protein produced by the liver). As blood that contains angiotensin I flows through the lungs, angiotensin I is further activated to form angiotensin II by angiotensin-converting enzyme (ACE), which is found in endothelial cells that compose the pulmonary capillaries. Once formed, angiotensin II causes arterioles to constrict, increasing blood vessel resistance. Angiotensin II also causes the release of aldosterone from the adrenal gland. In turn, aldosterone causes the kidneys to reabsorb salt and water. Through this mechanism, blood pressure returns to normal and homeostasis is maintained.

Working with Other Systems

The illustration in Human Systems Work Together on page 319 tells how the respiratory system depends on and assists other systems of the body.

The contributions of the respiratory system to homeostasis cannot be overemphasized. The respiratory tract assists defense against pathogens by preventing their entry into the body and by removing them from respiratory surfaces. For example, the cilia that line the trachea sweep impurities toward the throat. The respiratory tract also assists immunity. We now know that the tonsils serve as a location where T cells are presented with antigens before they enter the body as a whole. This action helps the body prepare to respond to an antigen before it enters the bloodstream.

The cardiovascular system transports oxygen from the lungs to the tissues and carbon dioxide from the tissues to the lungs. As mentioned in Chapter 12, expansion of the chest during inspiration causes a reduced pressure that promotes the flow of blood toward the thoracic cavity and the heart. Therefore, the act of breathing assists the return of blood to the heart and the transport of carbon dioxide to the lungs.

The rib cage protects the lungs, and inspiration could not occur without the contraction of external intercostal muscles, which lift the rib cage. The respiratory system is able to respond to the increased gas exchange needed by the body during exercise. Exercise causes the tissues to warm up and the pH to lower; under these conditions, hemoglobin unloads more oxygen than usual. These conditions are also detected by the carotid and aortic bodies, leading to an increase in the ventilation rate.

In the nervous system, the brain stem controls rhythmic ventilation, but it is possible through the cerebral cortex to consciously increase or decrease the rate and depth of the respiratory movements. This is often done while we are talking or singing, for example. The nasal cavities house the sense organs for olfaction. The sensation of smell only occurs after airborne molecules are drawn into the nasal cavity.

---

[2]pH means the negative logarithm of free hydrogen ion concentration, so the smaller the pH number, the greater the number of free hydrogen ions. The pH of arterial blood should range from 7.35 to 7.45. Blood with pH 7.3 (acidosis) has *more* free hydrogen ions than blood with pH 7.4 (normal). See page 25 to review the concept of pH.

Human Systems Work Together

Integumentary System

Gas exchange in lungs provides oxygen to skin and rids body of carbon dioxide from skin.

Skin helps protect respiratory organs and helps regulate body temperature.

Skeletal System

Gas exchange in lungs provides oxygen and rids body of carbon dioxide.

Rib cage protects lungs and assists breathing; bones provide attachment sites for muscles involved in breathing.

Muscular System

Lungs provide oxygen for contracting muscles and rid the body of carbon dioxide from contracting muscles.

Muscle contraction assists breathing; physical exercise increases respiratory capacity.

Nervous System

Lungs provide oxygen for neurons and rid the body of carbon dioxide produced by neurons.

Respiratory centers in brain regulate breathing rate.

Endocrine System

Gas exchange in lungs provides oxygen and rids body of carbon dioxide.

Epinephrine promotes ventilation by dilating bronchioles; growth factors control production of red blood cells that carry oxygen.

How the Respiratory System works with other body systems

nasal cavity
nose
pharynx (throat)
larynx
trachea
bronchi
lungs

Cardiovascular System

Gas exchange in lungs rids body of carbon dioxide, helping to regulate the pH of blood; breathing aids venous return.

Blood vessels transport gases to and from lungs; blood services respiratory organs.

Lymphatic System/Immunity

Tonsils and adenoids occur along respiratory tract; breathing aids lymph flow; lungs carry out gas exchange.

Lymphatic vessels pick up excess tissue fluid; immune system protects against respiratory tract and lung infections.

Digestive System

Gas exchange in lungs provides oxygen to the digestive tract and excretes carbon dioxide from the digestive tract.

Breathing is possible through the mouth because digestive tract and respiratory tract share the pharynx.

Urinary System

Lungs excrete carbon dioxide, provide oxygen, and convert angiotensin I to angiotensin II, leading to kidney regulation.

Kidneys compensate for water lost through respiratory tract; work with lungs to maintain blood pH.

Reproductive System

Gas exchange increases during sexual activity.

Sexual activity increases breathing; pregnancy causes breathing rate and vital capacity to increase.

SELECTED NEW TERMS

Basic Key Terms

alveolus (al-ve'o-lus), p. 304

angiotensin (an-jē-o-ten'sin), p. 318

aortic body (a-or'tik bod'e), p. 309

auditory (eustachian) tube (ô'dĭ-tôr'ē (yōō-stā'shən) tōōb), p. 313

bicarbonate ion (bi-kar'bo-nāt i'on), p. 312

bronchiole (brong'ke-ōl), p. 304

bronchus (brong'kus), p. 304

carbaminohemoglobin (kar-buh-me"no-he"mo-glo'bin), p. 312

carbonic anhydrase (kär-bŏn'ĭk ăn-hī'drəs), p. 312

carotid body (kar-ah'tid bod'e), p. 309

chloride shift (klōr'ĭd shĭft), p. 312

concha (kong'kuh), p. 302

epiglottis (ĕ"pĭ-glot'is), p. 304

expiration (eks"pĭ-ra'shun), p. 301

expiratory reserve volume (ek-spi'ruh-to-re re-zerv' vol'yūm), p. 308

external respiration (eks-ter'nal res"pĭ-ra'shun), p. 310

glottis (glot'is), p. 303

hemoglobin (hĕ'mə-glō'bĭn), p. 311

inspiration (in"spĭ-ra'shun), p. 301

inspiratory reserve volume (in-spi'ruh-to-re re-zerv' vol'yūm), p. 308

internal respiration (in-ter'nal res"pĭ-ra'shun), p. 311

intrapleural pressure (in'trah-ploor'al prĕsh'ər), p. 306

larynx (lār'inks), p. 303

lungs (lungz), p. 304

mucociliary escalator (mū-kō-sil'ē-ă-rē), p. 301

nasal cavities (na'zal kav'ĭ-tēz), p. 302

nose (nōz), p. 302

nostrils (nŏs'trĭls), p. 302

oxyhemoglobin (ok"se-he"mo-glo'bin), p. 311

paranasal sinus (pār-uh-na'zul si'nus), p. 302

pharynx (făr'inks), p. 302

pleurae (pleura) (plŏŏr'-ā), p. 304

pons (pŏnz), p. 308

primary respiratory center (prīmĕr'ē res'pĭ-ruh-tor-e sen'ter), p. 308

pulmonary capillaries (pŭl'-mōh-nār-ee kap'ĭl-lār-ee), p. 304

reduced hemoglobin (re-dūs'd he'mo-glo-bin), p. 312

residual volume (re-zid'yŭ-ul vol'yūm), p. 308

spirometer (spī-rŏm'-ĕ-ter), p. 307

surface tension (ser'fus ten'shun), p. 304

surfactant (sur-fak'tant), p. 305

tidal volume (ti'dul vol'yūm), p. 307

tonsils (tŏn'səlz), p. 313

trachea (tra'ke-uh), p. 304

ventilation (ven"tĭ-la'shun), p. 306

vital capacity (vi'tal kuh-pas'ĭ-te), p. 307

vocal cord (vo'kul kord), p. 303

Clinical Key Terms

acute bronchitis (uh-kyūt brong-ki'tis), p. 313

apnea (ap'ne-uh), p. 311

asthma (ăz'mə), p. 315

bronchogenic carcinoma (brong-ko-jen'ik kär'sə-nō'mə), p. 316

Cheyne-Stokes respiration (shān-stōks res"pĭ-ra'shun), p. 311

chronic bronchitis (kron'ik brong-ki'tis), p. 314

chronic obstructive pulmonary disorders (krŏn'ĭk əb-strŭkt'ive dĭs-or'dərs), p. 314

emphysema (em"fĭ-se'muh), p. 314

eupnea (yūp-ne'uh), p. 311

hyperpnea (hi"per-ne'uh), p. 311

infant respiratory distress syndrome (in'funt res'pĭ-ruh-tor-e dis-tres' sin'drōm), p. 305

laryngitis (lār-in-ji'tis), p. 313

lung cancer (lung kan'ser), p. 315

metastasis (mĕ-tas'-tă-sis), p. 315

otitis media (o-ti'tis me'-de-uh), p. 313

pneumonectomy (nu-mah-nek'tuh-me), p. 315

pneumonia (nu-mo'ne-uh), p. 313

pulmonary fibrosis (pul'mo-nēr"e fi-bro'sis), p. 314

pulmonary tuberculosis (pul'mo-nēr"e tū-ber"kyū-lo'sis), p. 313

sinusitis (si-nŭ-si'tis), p. 313

tonsillectomy (ton'sĭ-lek'to-me), p. 313

tonsillitis (ton-sil-i'tis), p. 313

tracheostomy (tra"ke-ahs'to-me), p. 304

SUMMARY

14.1 The Respiratory System

A. The nasal cavities, which filter, warm and humidify incoming air, open into the pharynx.

B. The food and air passages cross in the pharynx, which conducts air to the larynx and food to the esophagus.

C. The larynx is the voice box. It houses the vocal cords. The glottis, a slit between the vocal cords, is covered by the epiglottis when food is being swallowed.

D. The trachea and the primary bronchi are held open by cartilaginous rings. The rings gradually disappear as the primary bronchi branch into bronchioles, which enter the alveoli. The lungs are composed of the air tubes and alveoli beyond the primary bronchi and the alveoli.

E. The respiratory membrane is the juxtaposed alveolar wall and the capillary wall. The large surface area and thinness of the respiratory membrane allows rapid gas exchange.

14.2 Mechanism of Breathing

A. Ventilation is the movement of gases into and out of the lungs. The pleural membranes, which enclose the lungs, attach them to the thoracic wall. During inspiration, the thoracic cavity increases in size as the diaphragm lowers and the rib cage moves upward and outward. Therefore, the lungs expand, creating a partial vacuum, which causes air to rush in. During expiration, the diaphragm relaxes and resumes its dome shape. As the rib cage and lungs recoil, air is pushed out of the lungs. Expiration can be forceful when the internal intercostal muscles contract, causing the rib cage to move farther downward and inward. Also, contraction of the abdominal wall pushes the viscera against the diaphragm, and the increased pressure expels air.

B. Tidal volume is the amount of air inhaled and exhaled with each breath. Vital capacity is the total volume of air that can be moved in and out of the lungs during a single breath. Some air remains in the lungs after expiration. This is called the residual volume. Passages within the airways are called dead space because no gas exchange takes place in the airways.

C. The primary respiratory center in the medulla oblongata is assisted by the pons. It rhythmically controls the ventilation rate. The respiratory center increases the rate when CO_2 and H^+ levels increase, as detected by chemoreceptors in the respiratory center or the carotid and aortic bodies. The latter also detects a low O_2 level and stimulates the respiratory center, which then increases the ventilation rate.

14.3 Gas Exchange and Transport
A. Diffusion accounts for the movement of gases during external respiration and internal respiration. External respiration occurs when CO_2 moves from the area of higher concentration in the blood to lower concentration in the alveoli. O_2 moves from a higher concentration in the alveoli to a lower concentration in the blood. Internal respiration occurs when O_2 moves from a

higher concentration in the blood to a lower concentration in the tissue fluid, and CO_2 moves from a higher concentration in the tissue fluid to a lower concentration in the blood. The occurrence of cellular respiration always causes gas transport.

B. Oxygen is transported to the tissues in combination with hemoglobin as oxyhemoglobin (HbO_2). Carbon dioxide is mainly carried to the lungs within the plasma as the bicarbonate ion (HCO_3^-). Hemoglobin combines with hydrogen ions and becomes reduced (HHb) and this helps maintain the pH level of the blood within normal limits.

14.4 Respiration and Health
A. A number of illnesses are associated with the respiratory tract. These disorders can be divided into those that affect the upper respiratory tract and those that affect the lower respiratory tract. Infections of the nasal cavities, sinuses, throat, tonsils, and larynx are all well known. In addition, infections can spread from the nasopharynx to the ears.

B. The lower respiratory tract is subject to infections such as acute bronchitis, pneumonia, and pulmonary tuberculosis. In restrictive pulmonary disorders, exemplified by pulmonary fibrosis, the lungs lose their elasticity. In obstructive pulmonary disorders,

exemplified by chronic bronchitis, emphysema, and asthma, the bronchi (and bronchioles) do not effectively conduct air to and from the lungs. Smoking, which is associated with chronic bronchitis and emphysema, can eventually lead to lung cancer.

14.5 Effects of Aging
All aspects of respiration decline with age. The elderly often die from pulmonary infections.

14.6 Homeostasis
A. The respiratory system carries on two main functions: (1) gas exchange, which is essential to the process of cellular respiration, and (2) maintenance of blood pH. The respiratory system helps maintain blood pressure by converting angiotensin I to angiotensin II.

B. The respiratory system works with other systems of the body. The cardiovascular system transports gases, and breathing helps systemic venous blood return to the heart. The respiratory tract assists defense against pathogens by keeping the tract clean of debris. Also, the tonsils are lymphatic tissue where antigens are presented to T cells. The nervous system maintains rhythmic ventilation, and the sensory organs for olfaction are located in the nasal cavities. The respiratory center responds to the increased gas exchange needs of the muscular system when we exercise.

STUDY QUESTIONS

1. Name and explain the four events that comprise respiration. (p. 301)
2. What is the path of air from the nose to the lungs? Describe the structure and state the function of all the organs mentioned. (pp. 301–305)
3. What is the respiratory membrane, and how does its structure promote rapid gas exchange? (p. 305)
4. What three conditions should be borne in mind in order to understand ventilation? (p. 306)

5. Explain the volume and pressure changes necessary to inspiration and expiration. How is ventilation controlled? (pp. 306–307)
6. What is the difference between tidal volume and vital capacity? Of the air we inhale, some is not used for gas exchange. Why not? (pp. 307–308)
7. Contrast external respiration with internal respiration, and explain why there is a flow of gases during each of these. (pp. 310–311)

8. How are oxygen and carbon dioxide transported in the blood? What role does hemoglobin play in the transport of CO_2? (pp. 311–312)
9. Name and describe several upper and several lower respiratory tract disorders (other than cancer). If appropriate, explain why breathing is difficult with these conditions. (pp. 313–314)
10. List the steps by which lung cancer develops. (p. 315)

LEARNING OBJECTIVE QUESTIONS

Fill in the blanks.

1. In tracing the path of air, the _____ immediately follows the pharynx.
2. The lungs contain air sacs called _____.
3. The breathing rate is primarily regulated by the amount of _____ in the blood.
4. Air enters the lungs after they have _____.
5. Gas exchange is dependent on the physical process of _____.
6. During external respiration, oxygen _____ the blood.
7. During internal respiration, carbon dioxide _____ the blood.
8. Carbon dioxide is carried in the blood as _____ ions.
9. The most likely cause of emphysema and chronic bronchitis is _____.
10. Most cases of lung cancer actually begin in the _____.
11. The amount of air moved in and out of the respiratory system with each normal breath is called the _____.
12. The total amount of air that can be moved in and out of the lungs during a single breath is called the _____.
13. The _____ closes the opening into the larynx during swallowing.

14. The respiratory membrane consists of the walls of the _____ and _____.

15. Label the following diagram of the respiratory tract.

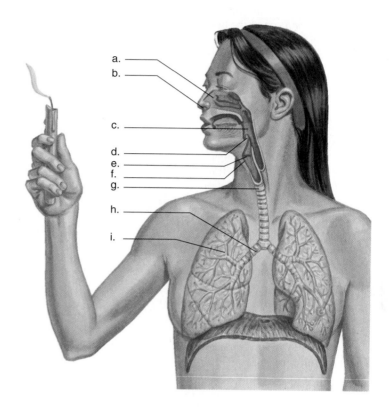

a. _____
b. _____
c. _____
d. _____
e. _____
f. _____
g. _____
h. _____
i. _____

MEDICAL TERMINOLOGY EXERCISE

Consult Appendix B for help in pronouncing and analyzing the meaning of the terms that follow.

1. eupnea (yūp-ne'uh)
2. nasopharyngitis (na"zo-făr"in-ji'tis)
3. tracheostomy (tra"ke-ahs'to-me)
4. pneumonomelanosis (nu-mo"no-mel"uh-no'sis)
5. pleuropericarditis (pler"o-pĕr"ĭ-kar"di'tis)
6. bronchoscopy (brong-kos'kuh-pe)
7. dyspnea (disp'ne-uh)
8. laryngospasm (luh-ring'go-spazm)
9. hemothorax (he"mo-tho'raks)
10. otorhinolaryngology (o"to-ri"no-lăr"in-gol'o-je)
11. hypoxemia (hi"pok-se'me-uh)
12. pulmonectomy (pul-mo-nek'to-me)
13. hypercapnea (hi-per-kap'ne-uh)
14. spirometry (spy-rom'uh-tre)
15. thoracentesis (thor'uh-sen-te'sis)

WEBSITE LINK

Visit this text's website at <u>http://www.mhhe.com/maderap6</u> for additional quizzes, interactive learning exercises, and other study tools.

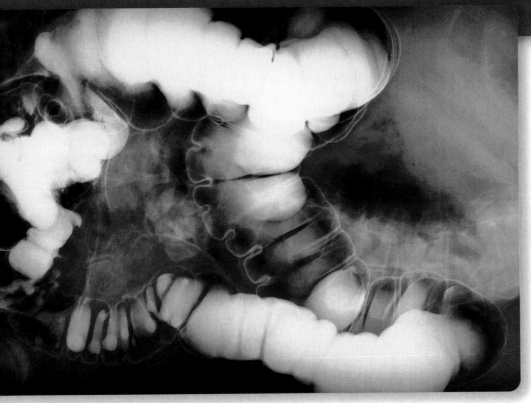

Xray of human intestine. To illuminate soft tissues such as the intestine on an Xray, the patient must drink an opaque dye.

CHAPTER 15

The Digestive System

Chapter Outline & Learning Objectives

After you have studied this chapter, you should be able to:

15.1 Anatomy of the Digestive System

The organs of the digestive system are located within a tube called the **alimentary canal,** or **gastrointestinal tract.** The tube begins with the mouth and ends with the anus (Fig. 15.1). Although the term digestion, strictly speaking, means the breakdown of food by enzymatic action, we will expand the term to include both the physical and chemical processes that reduce food to small, soluble molecules that can be absorbed into the bloodstream.

The functions of the digestive system are to:

1. ingest the food;
2. break food down into small molecules that can cross plasma membranes;
3. absorb these nutrient molecules;
4. eliminate nondigestible wastes.

The Mouth

The **mouth,** also termed the oral cavity, receives food and begins physical and chemical digestion. The oral cavity is bounded externally by the lips and cheeks. The space between the teeth, the lips and cheeks is the **vestibule.** The structure of human teeth is discussed in the Medical Focus on p. 325.

The tongue is composed of skeletal muscle whose contraction changes the shape of the tongue. Muscles exterior to the tongue cause it to move about. Rough projections on the tongue, called papillae, help it handle food and also contain the sensory receptors called taste buds. A fold of mucous membrane, called the lingual frenulum, on the underside of the tongue attaches it to the floor of the mouth. If the frenulum is too short, the individual cannot speak clearly and is said to be tongue-tied. Posteriorly, the tongue is anchored to the hyoid bone.

The mouth has a roof that separates it from the nasal cavities. The roof has two parts: an anterior (toward the front) **hard palate** and a posterior (toward the back) **soft palate** (see Fig. 15.2). The hard palate is formed by the maxilla and palatine bones; the soft palate is formed by muscle and glandular tissue. The soft palate ends in a finger-shaped projection called the **uvula.**

Three pairs of **salivary glands** send saliva by way of ducts to the mouth. The parotid glands lie anterior and somewhat inferior to the ears between the cheek and the masseter muscle. They have ducts that open on the inner surface of the cheek at the location of the second upper molar. The parotid glands swell when a person has the mumps, a disease caused by a viral infection. The sublingual glands are located beneath the tongue, and the submandibular glands are in the floor of the mouth on the inside surface of the lower jaw. The ducts from the sublingual and submandibular glands open under the tongue. You can locate the openings for the salivary glands if you use your tongue to feel for small flaps on the inside of your cheek and under your tongue. Saliva is a solution of mucus and water. It also contains bicarbonate and an enzyme called **salivary amylase,** which begins the process of digesting carbohydrate. Saliva moistens food and prepares it for swallowing. Saliva contains antibacterial lysozyme, as well as secretory antibodies, which help to protect the body. By constantly bathing the teeth, tongue, and oral mucous membrane, saliva removes microbes. Swallowed microbes can be destroyed by stomach acid and enzymes.

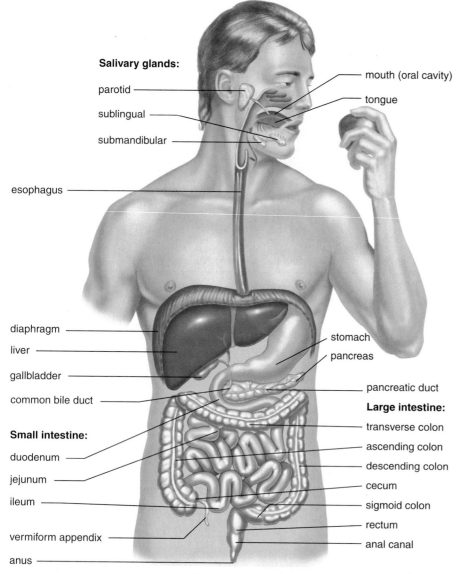

Figure 15.1 Digestive system. Trace the path of food from the mouth to the anus. The large intestine consists of the cecum, the colon (composed of the ascending, transverse, descending, and sigmoid colon), the rectum, and the anal canal. Note also the location of the accessory organs of digestion: the pancreas, the liver, and the gallbladder.

Salivary glands:
parotid
sublingual
submandibular

esophagus

diaphragm
liver
gallbladder
common bile duct

Small intestine:
duodenum
jejunum
ileum
vermiform appendix
anus

mouth (oral cavity)
tongue

stomach
pancreas

pancreatic duct

Large intestine:
transverse colon
ascending colon
descending colon
cecum
sigmoid colon
rectum
anal canal

Human Teeth

During the first two years of life, the 20 deciduous or baby teeth appear. Eventually, the deciduous teeth are replaced by the adult teeth. Normally, adults have 32 teeth (Fig. 15A*a*). The maxilla (upper jaw bone) and the mandible (lower jaw bone) each contain teeth of four different types. Anteriorly, four chisel-shaped **incisors** function for biting. Flanking them on either side are two pointed canine teeth, or **cuspids,** which help tear food. Laterally, four flat premolar teeth, or **bicuspids,** are used for grinding food. The most posterior are the six **molar** teeth, designed for crushing and grinding food. The last molars, called the wisdom teeth, may fail to come in, or if they do, they may grow in crooked and be useless. Frequently, wisdom teeth are extracted.

Each tooth (Fig. 15A*b*) has a crown and a root. The crown has a layer of enamel, an extremely hard outer covering of calcium compounds; dentin, a thick layer of bonelike material; and an inner pulp, which contains the nerves and blood vessels. Dentin and pulp are also in the root. **Caries,** tooth decay commonly called cavities, occur when bacteria within the mouth break down sugar and give off acids that corrode the teeth. Once these acids dissolve the enamel and dentin, the pulp is compromised, triggering a toothache. Fluoride treatments, particularly in children, can make the enamel stronger and more resistant to decay.

Gum disease is more likely as we age. One example is inflammation of the gums, called **gingivitis,** that may spread to the periodontal membrane lining the tooth socket (Fig. 15A*b*). When this occurs, the individual develops **periodontitis,** characterized by loss of bone and loosening of the teeth. Brushing and flossing the teeth after every meal cleans the teeth and stimulates the gums, preventing these conditions. Care should be taken to brush away from the gums to prevent gum recession.

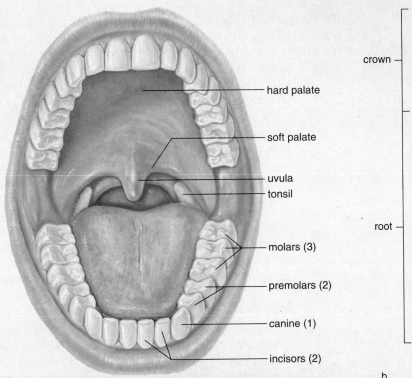

a.

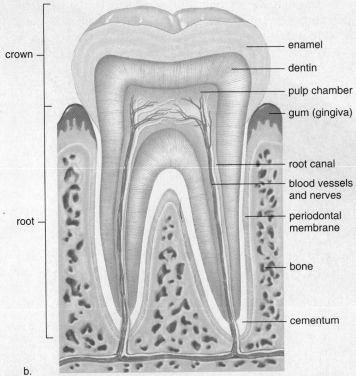

b.

Figure 15A a. The chisel-shaped incisors bite; the pointed canines tear; the fairly flat premolars grind; and the flattened molars crush food. The last molar, called a wisdom tooth, may fail to erupt, or if it does, it is sometimes crooked and useless. Often dentists recommend the extraction of the wisdom teeth. **b.** Longitudinal section of a tooth. The crown is the portion that projects above the gum line and can be replaced by a dentist if damaged. When a "root canal" is done, the nerves are removed. When the periodontal membrane is inflamed, the tooth can loosen.

TABLE 15.1 Path of Food in the Digestive Process

Organ	Function of Organ	Anatomic Features	Function of Anatomic Features
Oral cavity	Mechanical breakdown of food; swallowing; starch digestion	Teeth Tongue	Biting, tearing, chewing food Forms bolus; swallowing
Pharynx	Transport of food to esophagus	Soft palate and uvula Epiglottis	Prevents food from entering nasal cavity Prevents food from entering glottis
Esophagus	Transport of food to stomach	Four-layer construction	Muscularis layer carries out peristalsis
Stomach	Food storage; antibacterial; begins protein digestion; intrinsic factor production; slow release of chyme to small intestine	Rugae Parietal cells of gastric glands Chief cells of gastric glands Parietal cells of gastric glands Pyloric sphincter	Expand as stomach fills Produce hydrochloric acid Produce pepsin Produce intrinsic factor Controls release of chyme
Small intestine	Completes digestion of food; absorption of nutrients	Brush border enzymes Intestinal villi; microvilli	Complete digestive process Absorb nutrients
Large intestine	Absorption of water; storage of indigestible remains; elimination of fecal material	Epithelial cells Haustra Teniae coli and muscularis	Water absorption by osmosis Pouch-like structures store material Move feces toward anus

The Pharynx

Table 15.1 traces the path of food. From the mouth, food passes through the pharynx and esophagus to the stomach, small intestine, and large intestine. The food passage and the air passage cross in the **pharynx** because the trachea is anterior to the esophagus, a long muscular tube that takes food to the stomach (Fig. 15.2).

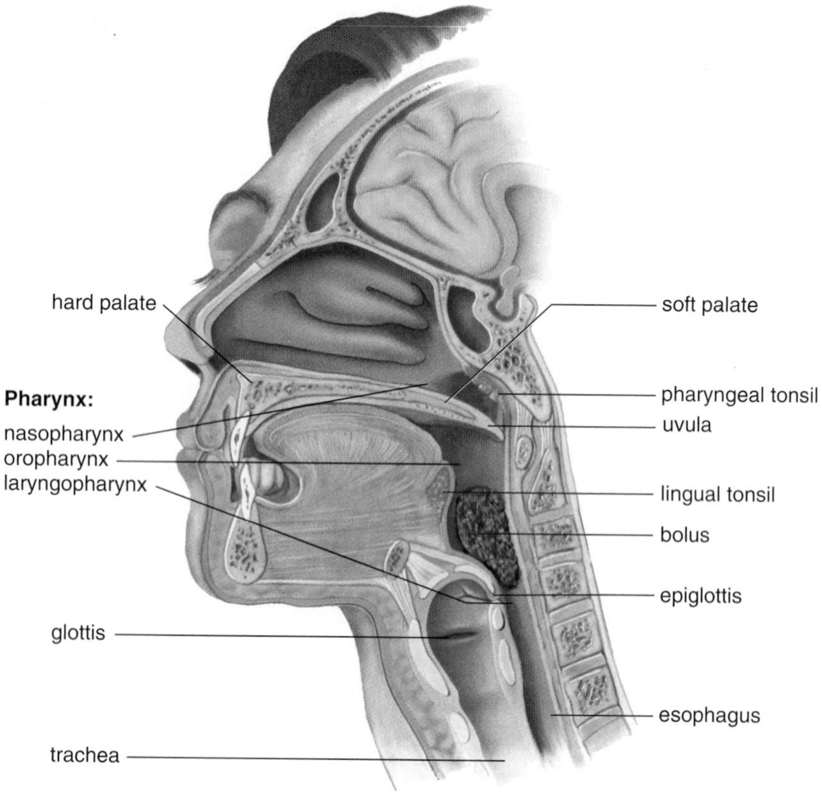

The tonsils are embedded in the mucous membrane of the tongue and pharynx. The **palatine tonsils** are on either side of the tongue close to the soft palate, and the **lingual tonsils** sit at the base of the tongue. Together, they help to protect the body from infection caused by ingested and/or inhaled microbes. The single **pharyngeal tonsil** (sometimes called the **adenoids**) sits in the posterior nasopharynx. As mentioned in Chapter 14, this tonsillar tissue defends against microbes in inhaled air. A person who has inflamed tonsils has tonsillitis. If the tonsillitis keeps recurring, the tonsils may be surgically removed (called a tonsillectomy).

The pharynx has three parts: (1) The **nasopharynx**, posterior to the nasal cavity, serves as a passageway for air; (2) the **oropharynx**, posterior to the soft palate, is a passageway for both air and food; and (3) the **laryngopharynx**, just superior to the esophagus, is a passageway for food entering the esophagus.

Swallowing

During swallowing, food normally enters the esophagus because other possible avenues are blocked. Swallowing has a voluntary phase—from day-to-day living, you know that you can swallow voluntarily (try it). However, once food or drink is pushed back to the oropharynx, swallowing then becomes a **reflex action** performed automatically (without our willing it). When we swallow, the soft palate moves back to close off the **nasopharynx**, and the trachea moves up under the **epiglottis** so that food is less likely to enter it. (We do not breathe when we swallow.) The tongue presses against the soft palate, sealing off the oral cavity, and the esophagus opens to receive a food **bolus** (Fig. 15.2).

hard palate
soft palate
Pharynx:
nasopharynx
oropharynx
laryngopharynx
pharyngeal tonsil
uvula
lingual tonsil
bolus
epiglottis
glottis
esophagus
trachea

Figure 15.2 Swallowing. When food is swallowed, the soft palate closes off the nasopharynx, and the epiglottis covers the glottis, forcing the bolus to pass down the esophagus. Therefore, a person does not breathe while swallowing.

A bolus is the technical term for a morsel of chewed and swallowed food or drink.

Unfortunately, we have all had the unpleasant experience of having food "go the wrong way." The wrong way may be either into the nasal cavities or into the trachea. If it is the latter, coughing will most likely force the food up out of the trachea and into the pharynx again.

The Esophagus

The **esophagus** is a muscular tube that passes from the pharynx through the thoracic cavity and diaphragm into the abdominal cavity where it joins the stomach. The esophagus is ordinarily collapsed but it opens and receives the bolus when swallowing occurs.

A rhythmic contraction called **peristalsis** pushes the food along the alimentary canal. In peristalsis, short segments of smooth muscle built into the wall of the digestive tract alternately contract and then relax. Peristalsis constantly moves food forward through the digestive tract. (As an analogy, think of squeezing a toothpaste tube, starting at the bottom, then continuing upward along the entire length of the tube. If done correctly, the toothpaste moves forward until the tube is empty.) Peristalsis begins in the esophagus and continues in all the organs of the alimentary canal. Occasionally, peristalsis begins even though there is no food in the esophagus. This produces the sensation of a lump in the throat.

The esophagus plays no role in the chemical digestion of food. Its sole purpose is to transport the food bolus from the mouth to the stomach. **Sphincters** are muscles that encircle tubes and act as valves; tubes close when sphincters contract, and they open when sphincters relax. The entrance of the esophagus to the stomach is marked by a constriction, often called the *esophageal sphincter,* although the muscle is not as developed as in a true sphincter. Relaxation of the sphincter allows the bolus to pass into the stomach, while contraction prevents the acidic contents of the stomach from backing up into the esophagus.

Heartburn, which feels like a burning pain rising up into the throat, occurs during reflux when some of the stomach contents escape into the esophagus. When vomiting occurs, a contraction of the abdominal muscles and diaphragm propels the contents of the stomach upward through the esophagus.

The Wall of the Digestive Tract

The entire digestive tract, from the esophagus to the rectum, is a continuous tube composed of four layers (Fig. 15.3). From deepest to most superficial, the layers are:

Mucosa (mucous membrane layer) A layer of epithelium supported by connective tissue and smooth muscle lines

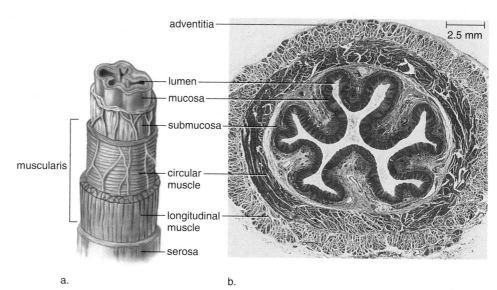

Figure 15.3 Wall of the alimentary canal. **a.** Several different types of tissues are found in the wall of the alimentary canal. Note the placement of circular muscle inside longitudinal muscle. **b.** Micrograph of the wall of the esophagus.

the **lumen** (central cavity). This layer contains glandular epithelial cells that secrete digestive enzymes and goblet cells that secrete mucus.

Submucosa (submucosal layer) A broad band of loose connective tissue that contains blood vessels, lymphatic vessels, and nerves lies beneath the mucosa. The submucosa joins the mucosa to the muscularis layer. Lymph nodules, called Peyer patches, are scattered throughout the submucosa, of the small intestine. Like the tonsils, they help protect us from disease.

Muscularis (smooth muscle layer) Two layers of smooth muscle make up this section. The inner, circular layer encircles the gut; the outer, longitudinal layer lies in the same direction as the gut. The stomach also has oblique muscles.

Serosa (serous membrane layer) Most of the alimentary canal has a serosa, a very thin, outermost layer of squamous epithelium supported by connective tissue. The serosa secretes a serous fluid that keeps the outer surface of the intestines moist so that the organs of the abdominal cavity slide against one another. The esophagus has an outer layer composed only of loose connective tissue called the *adventitia.*

The structure of the digestive tract (a continuous tube from mouth to anus) allows study by a fascinating new device. The What's New reading on p. 332 describes a miniature camera, which passes through the digestive system while photographing the internal organs.

The Stomach

The **stomach** (Fig. 15.4) is a thick-walled, J-shaped organ that lies on the left side of the abdominal cavity inferior to the

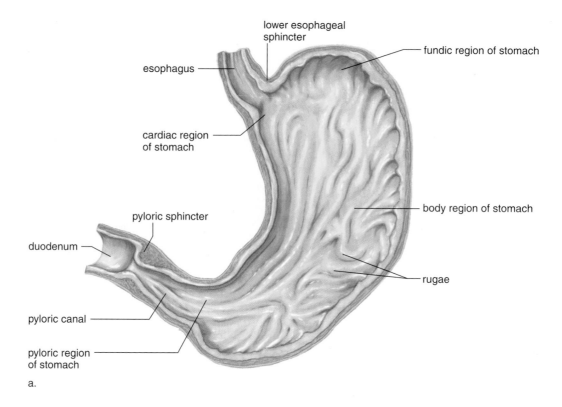

a.

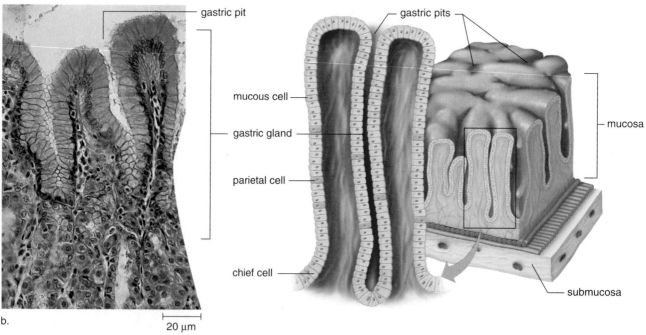

b.

20 μm

Figure 15.4 Anatomy and histology of the stomach. **a.** The stomach has a thick wall with deep folds, rugae, that allow it to expand and fill with food. **b.** The lining of the stomach has gastric glands, which secrete mucus and a gastric juice active in protein digestion.

diaphragm and posterior to the liver. The stomach is continuous with the esophagus above and the duodenum of the small intestine below.

The length of the stomach remains at about 25 cm (10 in.) regardless of the amount of food it holds, but the diameter varies, depending on how full it is. As the stomach expands, deep folds in its wall, called **rugae**, stretch out and gradually disappear. When full, the stomach can hold about 4 liters (1 gallon). The stomach receives food from the esophagus, stores food, liquifies food by mixing food with its juices (thereby starting the digestion of proteins), and moves food into the small intestine.

Regions of the Stomach

The stomach has four regions. The cardiac region, which is near the heart, surrounds the lower esophageal sphincter where food enters the stomach. The fundic region, which holds food temporarily, is an expanded portion superior to the cardiac region. The body region, which comes next, is the main part. The pyloric region narrows to become the pyloric canal leading to the pyloric sphincter. Food passes through this sphincter and enters the duodenum, the first part of the small intestine.

Digestive Functions of the Stomach

The stomach acts on food both chemically and physically. Its wall contains three muscle layers: One layer is longitudinal, another is circular, and the third is obliquely arranged. This muscular wall not only moves the food along, but it also churns, mixing the food with gastric juice and breaking it down to small pieces.

The term *gastric* always refers to the stomach. The columnar epithelial lining of the stomach has millions of *gastric pits,* which lead into **gastric glands;** (Fig. 15.4*a,b*). Gastric glands contain four types of secretory cells: *chief cells, parietal cells, enteroendocrine cells,* and *mucous cells.* The gastric glands produce **gastric juice,** a watery solution that contains pepsinogen, hydrochloric acid (HCl), intrinsic factor, and mucus. Chief cells secrete pepsinogen, which becomes the protein-digesting enzyme **pepsin** when exposed to hydrochloric acid. Parietal cells produce the hydrochloric acid. The HCl causes the stomach to have a high acidity with a pH of about 2; this is beneficial because it kills most of the bacteria present in food. Although HCl does not digest food, it does break down the connective tissue of meat and activates pepsin.

Parietal cells also produce **intrinsic factor,** a glycoprotein that binds to vitamin B_{12}, preventing this vitamin from being destroyed in the harsh, acidic environment of the stomach. If the stomach fails to produce intrinsic factor, or if the diet is deficient in vitamin B_{12}, a serious disorder called *pernicious anemia* will result. Lacking the vitamin, red blood cells will fail to develop.

Enteroendocrine cells of the gastric glands produce the hormone **gastrin.** This hormone enters stomach blood vessels and is circulated throughout the stomach. Gastrin regulates muscular contraction and secretion by the stomach.

The wall of the stomach is protected by the thick layer of mucus secreted by the *mucous cells.* If, by chance, HCl penetrates this mucus, the wall can begin to break down, and an ulcer results. An **ulcer** is an open sore in the wall caused by the gradual disintegration of tissue. It now appears that most ulcers are due to a bacterial infection *(Helicobacter pylori)* that impairs the ability of mucous cells to produce protective mucus.

Alcohol and water are absorbed in the stomach, but food substances are not. Normally, the stomach empties in about 2–6 hours. When food leaves the stomach, it is a thick, soupy liquid called **chyme.** Chyme enters the small intestine in

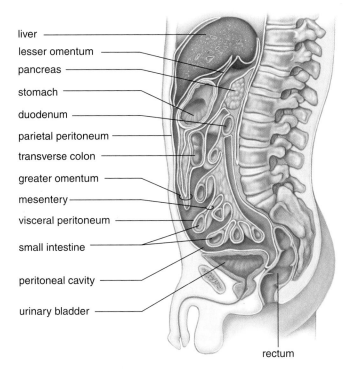

Figure 15.5 Mesentery formed by two layers of the peritoneal membrane supports the abdominal viscera. Deep folds of the peritoneal membrane, called the greater omentum, cover these organs anteriorly.

squirts by way of the pyloric sphincter, which acts like a valve, repeatedly opening and closing.

Peritoneum

The abdominal wall and the organs of the abdomen are covered by **peritoneum,** a serous membrane (Fig. 15.5). The portion of the peritoneum that lines the wall is called the parietal peritoneum. The portion that covers the organs is called the visceral peritoneum. In between the organs, the visceral peritoneum is a double-layered **mesentery** that supports the visceral organs, including the blood vessels, nerves, and lymphatic vessels.

Some portions of the mesentery have specific names. The **lesser omentum** is mesentery that runs between the stomach and the liver. The **greater omentum** is indeed "greater." It hangs down in front of the intestines like a large, double-layered apron. The greater omentum has several functions: It contains fat that cushions and insulates the abdominal cavity; it contains macrophages that can take up and rid the body of pathogens; and it can wall off portions of the alimentary wall that may be infected, keeping the infection from spreading to other parts of the peritoneal cavity.

The Small Intestine

The **small intestine** extends from the pyloric valve of the stomach to the *ileocecal valve* where it joins the large intestine. It is named for its small diameter (compared to that of the large intestine), but perhaps it should be called the long intestine. The small intestine takes up a large portion of the abdominal cavity, averaging about 6 m (18 ft) in length.

All the contents of food—fats, proteins, and carbohydrates—are digested in the small intestine to molecules that can be absorbed. To this end, the small intestine receives secretions from the pancreas and liver and produces intestinal juices. Absorption of nutrients for the body's cells, such as amino acids and sugars, occurs in the small intestine. It also transports nondigestible remains to the large intestine.

Regions of the Small Intestine

The small intestine has the following regions (Fig. 15.6):

Duodenum The first 25 cm (10 in.) contain distinctive glands that secrete mucus and also receive the pancreatic secretions and the bile from the liver through a common duct. Folds and villi (Fig. 15.6) are more numerous at the end than at the beginning.

Jejunum The next 1 m (3 ft) contains folds and villi, more at the beginning than at the end.

Ileum The last 2 m (6–7 ft) contain fewer folds and villi than the jejunum. The ileum wall contains Peyer patches, aggregates of lymph nodules mentioned in Chapter 13.

Wall of the Small Intestine

It has been suggested that the surface area of the small intestine is approximately that of a tennis court. Three features

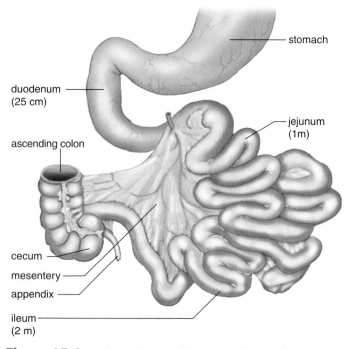

Figure 15.6 Regions of the small intestine. The duodenum is attached to the stomach. The jejunum leads to the ileum, which is attached to the large intestine.

contribute to increasing its surface area: circular folds, villi, and microvilli (Fig. 15.7). The **circular folds** are permanent transverse folds involving the mucosa and submucosa of the small intestine. The **villi** (sing., villus) are fingerlike projections of the mucosa into the lumen of the small intestine. The villi are so numerous and closely packed that they give the wall a velvet-like appearance. A villus has an outer layer of columnar epithelial cells, and each of these cells has thousands of microscopic extensions called **microvilli**. Collectively, in electron micrographs, microvilli give the villi a fuzzy border known as a "brush border" (Fig. 15.7d). Because the microvilli bear the intestinal enzymes, these enzymes are called brush-border enzymes.

Functions of the Small Intestine The digestive process is brought to completion in the small intestine. Ducts from the gallbladder and pancreas join to form one duct that enters the duodenum (Fig. 15.8). The small intestine receives bile from the gallbladder and pancreatic juice from the pancreas via this duct. **Bile,** which is produced by the liver but stored in the gallbladder, emulsifies fat—emulsification causes fat droplets to disperse in water. The intestine has a slightly basic pH because pancreatic juice contains sodium bicarbonate ($NaHCO_3$), which neutralizes the acidic chyme from the stomach. The enzymes in pancreatic juice and the enzymes produced by the intestinal wall complete the process of food digestion.

The other primary function of the small intestine is *absorption of nutrients*. The tremendous increase in surface area created by the circular folds, villi, and microvilli makes this an efficient process—the greater the surface area, the greater is the volume of intake in a given unit of time. Also, a villus contains a generous supply of blood capillaries and a small lymphatic capillary, called a **lacteal** (Fig. 15.7c,d). The lymphatic system, as you know, is an adjunct to the cardiovascular system; its vessels carry a fluid called lymph to the cardiovascular veins. Sugars (digested in part from carbohydrates) and amino acids (digested from protein) enter the blood capillaries of a villus. Glycerol and fatty acids (digested from fats) enter the epithelial cells of the villi, and within these cells are joined and packaged as lipoprotein droplets, which enter a lacteal. Thus, lipids first enter the lymphatic capillaries and are ultimately transported in the lymph to the lymphatic ducts. Recall that the lymphatic ducts return lymph to the bloodstream by way of the subclavian veins (p. 227). After nutrients are absorbed into the blood, they are eventually carried to all the cells of the body by the bloodstream.

As we noted previously, a third function of the small intestine is movement of nondigested remains to the large intestine. The wall of the small intestine has two types of movements: segmentation and peristalsis. Segmentation refers to localized contractions and constrictions that mix the chyme thoroughly with digestive juices. Segmentation also encourages absorption of nutrients into the bloodstream or lymph. Peristalsis then moves nondigested remains toward the large intestine.

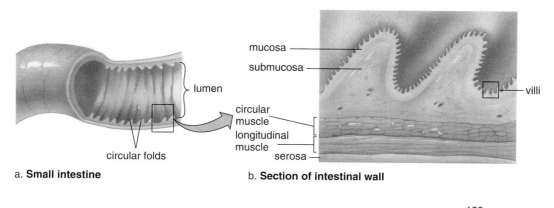

a. **Small intestine**

b. **Section of intestinal wall**

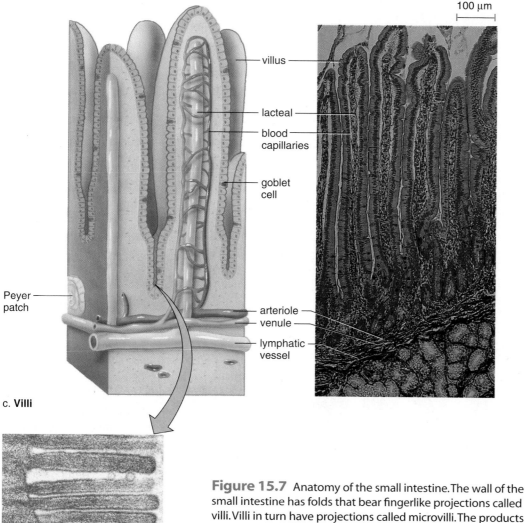

100 μm

c. **Villi**

d. **Electron micrograph of microvilli**

Figure 15.7 Anatomy of the small intestine. The wall of the small intestine has folds that bear fingerlike projections called villi. Villi in turn have projections called microvilli. The products of digestion are absorbed by microvilli and they enter the blood capillaries and the lacteals of the villi.

matically stimulate gastric secretion. Filling the stomach during the first part of digestion causes stimulation of a built-in gastric nerve network. Gastric contraction and secretion by gastric glands are controlled internally by this network. A similar nerve network causes the small intestine to begin its own contractions, as food from the stomach empties into it. Both the stomach and the duodenum also function as endocrine glands, producing their own set of hormones. Recall that a hormone is a substance produced by one set of cells that affects a different set of cells, the so-called target cells. Hormones are transported by the bloodstream to the target cells.

For example, when a person has eaten a meal particularly rich in protein, the enteroendocrine cells of the stomach produce the hormone gastrin. Gastrin enters the bloodstream, and soon the stomach is churning, and the secretory activity of gastric glands is increasing. A hormone produced by the duodenal wall, GIP (gastric inhibitory peptide), works opposite to gastrin: It inhibits gastric gland secretion.

Cells of the duodenal wall produce two other hormones that are of particular interest—**secretin** and CCK (**cholecystokinin**). Acid, especially hydrochloric acid (HCl) present in chyme, stimulates the release of secretin, while partially digested protein and fat stimulate the release of CCK. Secretin stimulates secretion of bicarbonate solution by the pancreas, and cholecystokinin stimulates pancreatic enzyme secretion. Thus, when these hormones enter the bloodstream, the pancreas increases its output of pancreatic juice. Pancreatic juice buffers the acidic chyme entering from the stomach and helps to digest food. At the same time, the liver increases its production of bile, in response to secretin. The gallbladder contracts to release stored bile, in response to CCK. Figure 15.8 summarizes the actions of gastrin, secretin, and CCK.

Regulation of Contraction and Secretion in the Digestive Tract

Muscular contraction and the secretion of digestive juices in the digestive tract are controlled by the nervous system, as well as by hormones. When you look at or smell food, autonomic nerves belonging to the parasympathetic nervous system auto-

"I really just don't even want to eat anymore. My gut just hurts so badly, and then I can't even control the diarrhea. I moved out of the dorm last year because I was so embarrassed to always be in the bathroom." The patient, a 20-year-old college student, is pale and underweight. A complete blood count reveals that she is anemic and her white blood cell count is slightly elevated. She identifies the lower right abdominal quadrant as the source of her pain. The pain has persisted, off and on, for more than six months.

To diagnose problems involving the digestive system, gastroenterologists (physicians who specialize in the digestive tract) have previously used several techniques: endoscopy, colonoscopy, and barium X ray. In endoscopy, a long, flexible tube equipped with a video camera is inserted into the esophagus to examine the upper GI tract: the esophagus, stomach, and upper small intestine. Colonoscopy is a similar technique, except that the video camera tube is inserted through the anus to examine the rectum and colon. Neither technique effectively examines the center portion in between: the ileum, the final section of the small intestine. To illuminate the ileum, patients drink a barium solution that coats the entire GI tract and makes the digestive organs visible, after which a series of X rays can be taken. In addition to being uncomfortable for the patient, the aforementioned techniques occasionally missed problems in the ileum as well.

A new technique involves placing a video camera directly into the digestive tract, in the form of a camera pill (Fig. 15B). The patient swallows the camera pill, which is about the size of a large vitamin. The patient wears a belt pack that receives images transmitted by the camera for the next eight to twelve hours, as the pill passes through the patient's body. Once the camera has been passed, physicians can study the video images from the belt pack receiver. In clinical studies, these video images have been extremely useful in diagnosis of disorders affecting the ileum: intestinal tumors, celiac disease (an allergic condition affecting the small intestine), and chronic inflammatory ileitis, or Crohn's disease.

In the case of the student mentioned at the beginning of this feature, video images taken by the camera pill reveal that she has Crohn's disease. Crohn's disease is thought to be an autoimmune disorder caused by a misdirected immune response, possibly triggered by a viral infection. There is no cure for Crohn's disease, but its symptoms can be controlled with medication. A variety of anti-inflammatory medications can relieve its symptoms. Anti-diarrheal medications help to control diarrhea and nutritional supplements help patients regain lost weight. Over time, Crohn's disease symptoms usually resolve, but sufferers must learn to expect flare-ups.

The Crohn's and Colitis Foundation of America is a support network for people with Crohn's disease and similar disorders. It offers tips for managing this chronic illness. Visit the foundation's website at www.ccfa.org.

Actual size

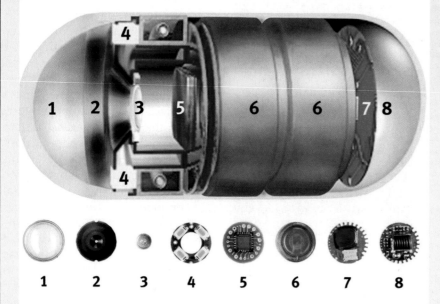

INSIDE THE Pill cam

1. Optical dome
2. Lens holder
3. Lens
4. Illuminating LEDs (Light Emitting Diode)
5. CMOS (Complementary Metal Oxide Semiconductor) imager
6. Battery
7. ASIC (Application Specific Integrated Circuit) transmitter
8. Antenna

Figure 15B The camera pill is about the size of a large vitamin. A belt pack receives video images, which are later viewed by a physician.

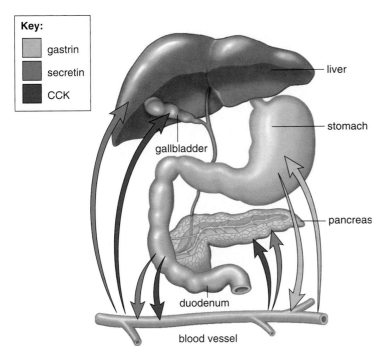

Key:
- gastrin
- secretin
- CCK

liver

stomach

gallbladder

pancreas

duodenum

blood vessel

Figure 15.8 Hormonal control of digestive gland secretions. Gastrin (blue), produced by the lower part of the stomach, enters the bloodstream and thereafter stimulates the upper part of the stomach to produce more gastric juice. Secretin (green) and CCK (purple), produced by the duodenal wall, stimulate the pancreas to secrete its juice, the liver to produce bile, and the gallbladder to release stored bile.

The Large Intestine

The **large intestine,** which includes the cecum, the colon, the rectum, and the anal canal (Fig. 15.9), is larger in diameter than the small intestine (6.5 cm compared to 2.5 cm), but it is shorter in length (see Fig. 15.1). The large intestine absorbs water, salts, and some vitamins. It also stores indigestible material until it is eliminated at the anus.

The **cecum,** which lies below the junction with the small intestine, is the blind end of the large intestine. The cecum has a small projection called the **vermiform appendix** (*vermiform* means wormlike). In humans, the appendix also may play a role in fighting infections. This organ is subject to inflammation, a condition called appendicitis. If inflamed, the appendix should be removed before the fluid content rises to the point that the appendix bursts. A ruptured (burst) appendix will cause **peritonitis,** a generalized infection of the lining of the abdominal cavity. Peritonitis can lead to death.

The **colon** has four portions: the **ascending colon,** which goes up the right side of the body to the level of the liver; the **transverse colon,** which crosses the abdominal cavity just below the liver and the stomach; the **descending colon,** which passes down the left side of the body; and the **sigmoid colon,** which enters the **rectum,** the last 20 cm of the large intestine. The colon is characterized by two distinct anatomic features:

pouches called haustra along its length, and a band of muscle called taenia coli on its surface (see Fig. 15.9). Haustra expand to store material in the colon. The taenia coli and the muscularis layer of the colon cause peristaltic movement called a mass movement several times daily. For most people, this mass movement fills the rectum and is accompanied by the urge to defecate. The rectum opens at the **anus,** where *defecation,* the expulsion of feces, occurs. When feces are forced into the rectum by peristalsis, a defecation reflex occurs. The stretching of the rectal wall initiates nerve impulses to the spinal cord, and shortly thereafter the rectal muscles contract and the anal sphincters relax (see Fig. 15C). Ridding the body of indigestible remains is another way the digestive system helps maintain homeostasis.

Feces are three-quarters water and one-quarter solids. Bacteria, **fiber** (indigestible remains), and other indigestible materials are in the solid portion. Bacterial action on indigestible materials causes the odor of feces and also accounts for the presence of gas. A breakdown product of bilirubin (see page 338) and the presence of oxidized iron cause the brown color of feces.

For many years, it was believed that facultative bacteria (bacteria that can live with or without oxygen), such as *Escherichia coli,* were the major inhabitants of the colon, but new culture methods show that over 99% of the colon bacteria are obligate anaerobes (bacteria that die in the presence of oxygen). Not only do the bacteria break down indigestible material, but they also produce B-complex vitamins and most of the vitamin K needed by our bodies. In this way, they perform a service for us.

Although most intestinal bacteria are harmless, many cause disease. To ensure water quality, public health departments constantly monitor water supplies for their coliform (nonpathogenic intestinal) bacterial count. A high count indicates that a significant amount of feces has entered the water. The more feces present, the greater is the possibility that disease-causing bacteria are also present.

Diarrhea and Constipation

Two common everyday complaints associated with the large intestine are **diarrhea** and **constipation.** The major causes of diarrhea are infection of the lower intestinal tract and nervous stimulation. In the case of infection, such as food poisoning caused by eating contaminated food, the intestinal wall becomes irritated and peristalsis increases. Water is not absorbed and the diarrhea that results rids the body of the infectious organisms. In nervous diarrhea, the nervous system stimulates the intestinal wall excessively and diarrhea results. Prolonged diarrhea can lead to dehydration because of water loss. In severe cases, disturbances in the heart's contraction may occur due to an imbalance of salts in the blood.

When a person is constipated the feces are dry and hard. The Medical Focus on page 335 discusses the causes of constipation and how it can be prevented. Chronic constipation is associated with the development of **hemorrhoids,** enlarged and inflamed blood vessels at the anus.

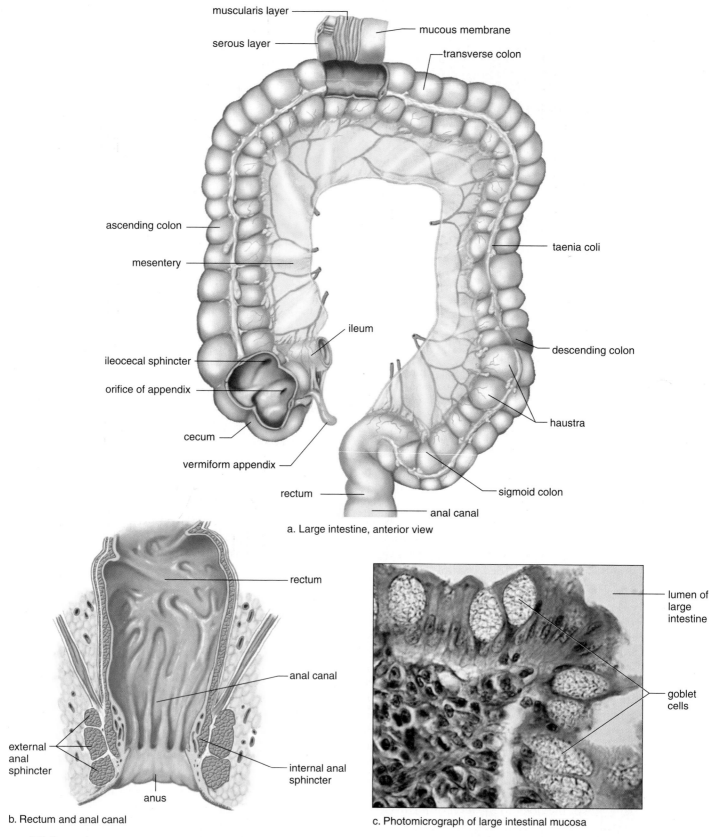

muscularis layer

serous layer

mucous membrane

transverse colon

ascending colon

mesentery

taenia coli

ileum

ileocecal sphincter

descending colon

orifice of appendix

cecum

haustra

vermiform appendix

rectum

sigmoid colon

anal canal

a. Large intestine, anterior view

rectum

anal canal

external anal sphincter

internal anal sphincter

anus

b. Rectum and anal canal

lumen of large intestine

goblet cells

c. Photomicrograph of large intestinal mucosa

Figure 15.9 The large intestine. **a.** The colon has four regions: the ascending colon, the transverse colon, the descending colon, and the sigmoid colon. **b.** The rectum and anal canal are at the distal end of the alimentary canal. **c.** The intestinal mucosa has many goblet cells.

Constipation

The colon of the large intestine has four regions: the ascending colon, the transverse colon, the descending colon, and the sigmoid colon (see Fig. 15.9a). Water is removed from the non-digestible intestinal-contents entering the ascending colon from the small intestine. At this point, bacteria begin their action; they use cellulose as an energy source as they produce fatty acids and vitamins that can also be used by their host. They also release hydrogen gas and sulfur-containing compounds that contribute to human flatulence (gas). Feces, which consist of nondigested intestinal contents, bacteria, and sloughed-off intestinal cells, begin to form in the transverse colon. From there, they are propelled down the descending colon toward the rectum by periodic, firm contractions called peristalsis. When sufficient feces are in the rectum (130–200 grams), a defecatory urge is felt. The involuntary defecation reflex contracts the rectal muscles and relaxes the internal anal sphincter, a ring of muscle that closes off the rectum (Fig. 15C). Then, feces move toward the anus. A pushing motion, along with relaxation of the external anal sphincter, propels feces from the body. Since these activities are under voluntary control, it is possible to control defecation.

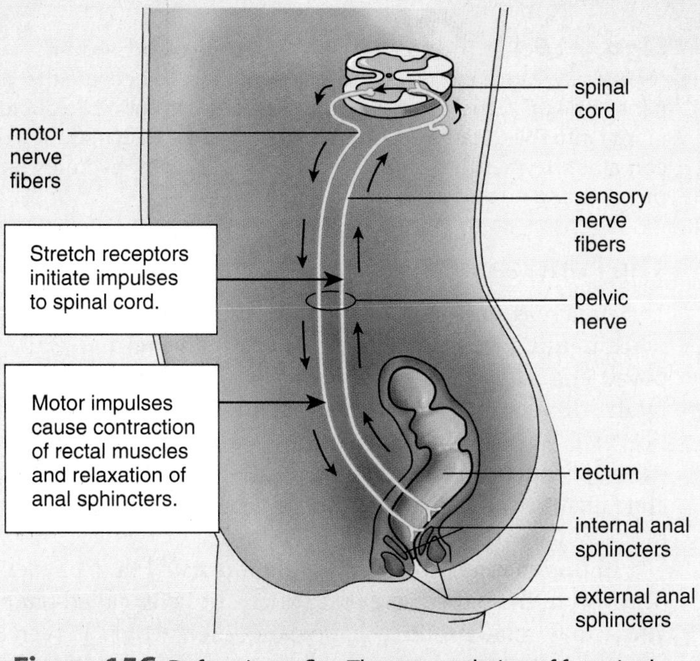

Defecation normally occurs from three times a week to three times a day; therefore, some variation in occurrence is nothing to worry about. However, if the frequency of defecation declines and if defecation becomes difficult, constipation is present. If constipation is a continuing problem, a physician can help record the movement of materials through the large intestine via several tests. The patient swallows about 20 small markers that will show up on an X ray. At intervals during the following week, X rays are taken, and the number and locations of the markers are noted. If muscle contraction of the intestinal wall is insufficient, the markers move slowly along their course. Injured nerves, certain drugs, dehydration, and prolonged overuse of stimulatory laxatives can bring about this difficulty. Some or all of these problems frequently occur in the elderly. On the other hand, markers may move normally at first and then slow down considerably in the descending colon and rectum. Habitual disregard of the defecatory urge may have caused this problem, or a cancerous polyp might be obstructing normal movement. If the former is the case, it is possible to retrain the rectum to work properly. Sitting on the toilet for about 20 minutes each morning can encourage a return of the reflexes that have disappeared, but straining is not recommended.

Temporary constipation due to traveling, pregnancy, or medication can sometimes be relieved by increasing dietary fiber, drinking plenty of water, and getting moderate amounts of exercise. The use of oral laxatives (agents that aid emptying of the intestine) is a last resort. Bulk-forming laxatives, such as those that contain bran, psyllium, and methyl cellulose, are considered best because they promote the defecation reflex. Laxatives that contain osmotic agents, such as carbohydrates or salts (lactulose, milk of magnesia, or Epsom salts), cause water to move into rather than out of the colon. Stool softeners (mineral oil or those that contain docusate) should be used sparingly. Mineral oil reduces the absorption of fat-soluble vitamins, and docusate can cause liver damage. Laxatives that contain chemical stimulants (such as phenolphthalein in Ex-Lax and Feen-A-Mint) can damage the defecation reflex and lead to a dependence on their use. Aside from laxatives, rectal suppositories are sometimes helpful in providing lubrication and stimulating the defecation reflex. Enemas introduce water into the colon and, therefore, also help stimulate defecation.

motor nerve fibers

Stretch receptors initiate impulses to spinal cord.

Motor impulses cause contraction of rectal muscles and relaxation of anal sphincters.

spinal cord

sensory nerve fibers

pelvic nerve

rectum

internal anal sphincters

external anal sphincters

Figure 15C Defecation reflex. The accumulation of feces in the rectum causes it to stretch, which initiates a reflex action resulting in rectal contraction.

Other Disorders of the Large Intestine

The colon is subject to the development of **polyps,** small growths arising from the epithelial lining. Polyps, whether benign or cancerous, can be removed surgically along with a portion of the colon if necessary. If colon cancer is detected while still confined to a polyp, the expected outcome is a complete cure. If the last portion of the rectum and the anal canal must be removed, then the intestine is sometimes attached to the abdominal wall through a procedure known as a **colostomy**, and the digestive remains are collected in a plastic bag fastened around the opening. Recently, the use of metal staples has permitted surgeons to join the colon to a piece of rectum that formerly was considered too short.

Some investigators believe that dietary fat increases the likelihood of colon cancer because dietary fat causes an increase in bile secretion. Researchers believe that intestinal bacteria convert bile salts to substances that promote the development of cancer. On the other hand, fiber in the diet seems to inhibit the development of colon cancer. Dietary fiber absorbs water and adds bulk, thereby diluting the concentration of bile salts and facilitating the movement of substances through the intestine. Regular elimination reduces the time that the colon wall is exposed to any cancer-promoting agents in feces.

Diverticulosis is characterized by the presence of diverticula, or saclike pouches, in the colon. Ordinarily, these pouches cause no problems. But about 15% of people with diverticulosis develop an inflammation known as diverticulitis. The symptoms of diverticulitis are similar to those of appendicitis—cramps or steady pain with local tenderness. Fever, loss of appetite, nausea, and vomiting may also occur. Today, high-fiber diets are recommended to prevent the development of these conditions and of cancer of the colon.

Content CHECK-UP!

1. The correct order of layering in the digestive tract wall, from deep to superficial, is:

    a. mucosa, muscularis, submucosa, serosa.

    b. serosa, mucosa, submucosa, muscularis.

    c. mucosa, submucosa, muscularis, serosa.

    d. submucosa, mucosa, muscularis, serosa.

2. Which hormone is produced by the stomach and controls stomach contraction and secretion?

    a. pepsin   c. secretin

    b. gastrin   d. cholecystokinin (CCK)

3. Which nutrient is absorbed by lymphatic lacteals of the small intestine?

    a. amino acids from protein digestion

    b. fatty acids and glycerol from fat digestion

    c. sugars

15.2 Accessory Organs of Digestion

The salivary glands and even the teeth are accessory organs of digestion that were discussed earlier (see pages 324–325). The pancreas, liver, and gallbladder are also accessory digestive organs. Figure 15.10 shows how the pancreatic duct from the pancreas and the common bile duct from the liver and gallbladder join before entering the duodenum.

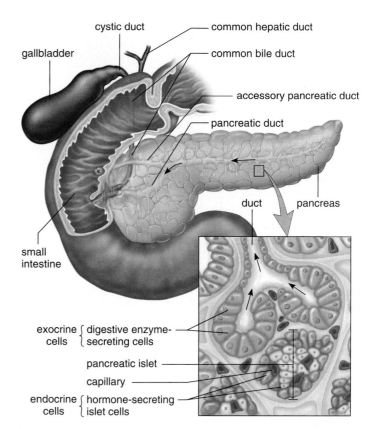

Figure 15.10 The pancreas is an exocrine gland when it secretes digestive enzymes into tubes that join to become the pancreatic duct. The pancreatic duct and the common bile duct empty into the duodenum of the small intestine. Pancreatic juice contains enzymes that digest all types of food: carbohydrates, fats, proteins, and nucleic acids.

The Pancreas

The **pancreas** lies deep in the abdominal cavity, behind the peritoneum, resting on the posterior abdominal wall. Its broad end, called the head, more than fills the loop formed by the duodenum, and its tail extends in the opposite direction (Fig. 15.10). The pancreas has both an endocrine and an exocrine function. Pancreatic islets (islets of Langerhans) secrete insulin and glucagon, hormones that help keep the blood glucose level within normal limits.

In this chapter, however, we are interested in the exocrine function of the pancreas. Most pancreatic cells, called pancreatic acinar cells, produce pancreatic juice, which is secreted into tiny tubes that unite, forming ever-larger ones. Finally, a pancreatic duct extends the length of the pancreas to the duodenum.

Pancreatic Juice

Pancreatic juice contains sodium bicarbonate ($NaHCO_3$) and digestive enzymes for all types of food. Sodium bicarbonate neutralizes the acidic chyme from the stomach. This is important because pancreatic enzymes require a slightly basic pH to function optimally. **Pancreatic amylase** digests starch, and lipase digests fat. There are three protein-digesting enzymes in pancreatic juice: trypsin, chymotrypsin, and carboxypeptidase. Pancreatic juice also contains two nucleases, which are enzymes that break down nucleic acid molecules into nucleotides.

In *cystic fibrosis,* a thick mucus blocks the pancreatic duct, and the patient must take supplemental pancreatic enzymes by mouth for proper digestion to occur.

The Liver

The **liver,** which is the largest organ in the body, lies mainly in the upper right section of the abdominal cavity, just inferior to the diaphragm (see Fig. 15.1).

Liver Structure

The liver has two main lobes, the right lobe and the smaller left lobe, separated by a ligament. Each lobe is divided into many hepatic lobules that serve as its structural and functional units (Fig. 15.11). A lobule consists of many hepatic cells arranged in longitudinal groups that radiate out from a central vein. Hepatic sinusoids separate the groups of cells from each other. Large, fixed phagocytic macrophages called *Kupffer cells* are attached to the lining of the hepatic sinusoids. They remove pathogens and debris that may have entered the hepatic portal vein at the small intestine.

Portal triads consisting of the following three structures are located between the lobules: a bile duct that takes bile away from the liver; a branch of the hepatic artery that brings O_2-rich blood to the liver; and a branch of the hepatic portal vein that transports nutrients from the intestines. The bile ducts merge to form the common hepatic duct.

The central veins from each of the lobules enter a hepatic vein. With the help of Figure 12.18, trace the path of blood from the intestines to the liver via the hepatic portal vein and from the liver to the inferior vena cava via the hepatic veins.

Liver Functions

As the blood from the hepatic portal vein passes through the liver, hepatic cells remove poisonous substances and detoxify them. The liver also removes and stores nutrients and works to keep the contents of the blood constant. It removes and stores iron and the fat-soluble vitamins A, D, E, and K; makes the plasma proteins from amino acids; and helps regulate the quantity of cholesterol in the blood.

The liver maintains the blood glucose level at about 100 mg/100 ml (0.1%), even though a person eats intermittently. When insulin is present, any excess glucose in the blood is removed and stored by the liver as glycogen. Between meals,

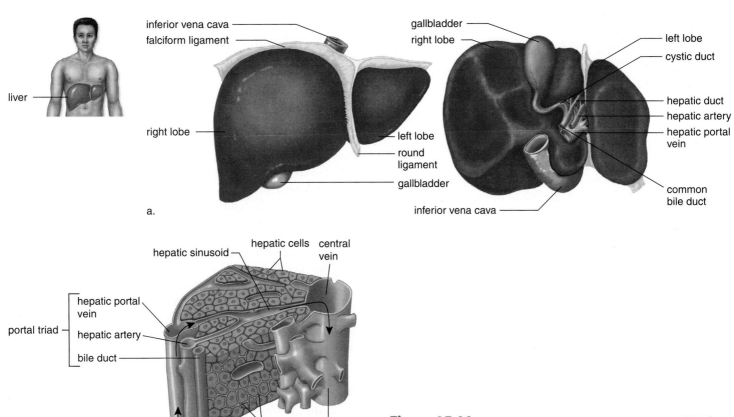

Figure 15.11 Macroscopic and microscopic anatomy of the liver. **a.** The liver has two lobes viewed anteriorly (*left*) and posteriorly from the visceral surface (*right*). **b.** Cross section of a hepatic lobule, illustrating microscopic structure.

glycogen is broken down to glucose, which enters the hepatic veins. In this way, the blood glucose level remains constant.

If the supply of glycogen is depleted, the liver converts glycerol (from fats) and amino acids to glucose molecules. The conversion of amino acids to glucose necessitates *deamination*, the removal of amino groups and the production of ammonia. By a complex metabolic pathway, the liver then combines ammonia with carbon dioxide to form urea:

$$2\ NH_3\ +\ CO_2 \longrightarrow H_2N - \overset{\overset{\textstyle O}{\|}}{C} - NH_2$$
$$\text{ammonia} \quad \text{carbon dioxide} \qquad\qquad \text{urea}$$

Urea is the usual nitrogenous waste product from amino acid breakdown in humans. After its formation in the liver, urea is excreted by the kidneys.

The liver produces bile, which is stored in the gallbladder. Bile has a yellowish-green color because it contains the bile pigment **bilirubin**, which is derived from the breakdown of hemoglobin, the red pigment of red blood cells. Bile also contains **bile salts**. Bile salts are derived from cholesterol. They emulsify fat in the small intestine. When fat is emulsified, it disperses into droplets. Emulsification of fats provides a much larger surface area, which can be acted upon by digestive enzymes from the pancreas.

Altogether, the following are significant ways in which the liver helps maintain homeostasis:

1. Detoxifies blood by removing and metabolizing poisonous substances.
2. Stores iron (Fe^{2+}) and the fat-soluble vitamins A, D, E, and K.
3. Makes plasma proteins, such as albumins and fibrinogen, from amino acids.
4. Stores glucose as glycogen after a meal. Between meals, it breaks down glycogen to glucose to maintain the glucose concentration of blood.
5. Produces urea after breaking down amino acids.
6. Forms and secretes bile. Bile can be secreted directly into the duodenum or into the gallbladder for storage. Bile salts, formed from cholesterol, emulsify fats. Bilirubin, the bile pigment, is a product of the breakdown of hemoglobin from red blood cells.
7. Helps regulate the blood cholesterol level, converting some to bile salts.

Liver Disorders

When a person has a liver ailment, jaundice may occur. **Jaundice** is a yellowish tint to the whites of the eyes and also to the skin of light-pigmented persons. Bilirubin is deposited in the skin due to an abnormally large amount in the blood. In hemolytic jaundice, red blood cells have been broken down in abnormally large amounts; in obstructive jaundice, bile ducts are blocked, or liver cells are damaged.

Jaundice can also result from **hepatitis**, inflammation of the liver. Viral hepatitis occurs in several forms. Hepatitis A is usually acquired from sewage-contaminated drinking water. Hepatitis B, which is usually spread by sexual contact, can also be spread by blood transfusions or contaminated needles. The hepatitis B virus is more contagious than the AIDS virus, which is spread in the same way. Thankfully, however, a vaccine is now available for hepatitis B. Hepatitis C, which is usually acquired by contact with infected blood and for which there is no vaccine, can lead to chronic hepatitis, liver cancer, and death.

Cirrhosis is another chronic disease of the liver. First the organ becomes fatty, and then liver tissue is replaced by inactive fibrous scar tissue. Cirrhosis of the liver is often seen in alcoholics due to malnutrition and the excessive amounts of alcohol (a toxin) the liver is forced to break down.

Hepatitis and cirrhosis affect the entire liver and hinder its ability to repair itself. Therefore, they are life-threatening diseases. The liver has amazing regenerative powers and can recover if the rate of regeneration exceeds the rate of damage. During liver failure, however, there may not be enough time to let the liver heal itself. Liver transplantation is usually the preferred treatment for liver failure, but artificial livers have been developed and tried in a few cases. One type is a cartridge that contains liver cells. The patient's blood passes through the cellulose acetate tubing of the cartridge and is serviced in the same manner as with a normal liver. In the meantime, the patient's liver has a chance to recover.

The Gallbladder

The **gallbladder** is a pear-shaped, muscular sac located in a depression on the inferior surface of the liver (see Fig. 15.11*a*). About 1,000 ml of bile are produced by the liver each day, and any excess is stored in the gallbladder. Water is reabsorbed by the gallbladder so that bile becomes a thick, mucous-like material. When needed, bile leaves the gallbladder by way of the cystic duct. The cystic duct and the common hepatic duct join to form the common bile duct, which enters the duodenum.

The cholesterol content of bile can come out of solution and form crystals. If the crystals grow in size, they form gallstones. The passage of the stones from the gallbladder may block the common bile duct and cause obstructive jaundice. Then the gallbladder may have to be removed.

Function of Bile Salts

Bile salts carry out emulsification; they break up masses of fat into droplets that can be acted on by enzymes that digest fat. Through their ability to make fats interact with water, they also enhance absorption of fatty acids, cholesterol, and the fat-soluble vitamins A, D, E, and K.

4. The protein-digesting enzyme produced by the pancreas is:

   a. pepsin.   c. amylase.

   b. gastrin.   d. trypsin.

5. Which of the following is not a function of the liver?

   a. production of bile

   b. storage of excess glucose as glycogen

   c. deamination of amino acids

   d. secretion of lipase to digest fats

6. From the following list of accessory organs of digestion and their function, choose the correct pair.

   a. pancreas → production of insulin and glucagon

   b. gallbladder → production of bile

   c. salivary glands → production of protein-digesting enzymes

   d. liver → production of nucleases

15.3 Chemical Digestion

The digestive enzymes are **hydrolytic enzymes,** which break down substances by the introduction of water at specific bonds. Digestive enzymes have an optimum pH at which they work best.

In the mouth, saliva from the salivary glands has a neutral pH and contains salivary amylase, the first enzyme to act on carbohydrate:

$$\text{starch} + \text{H}_2\text{O} \xrightarrow{\text{salivary amylase}} \text{maltose}$$

Notice that the name of the enzyme is written above the arrow to indicate that it is not used up. In this case, the enzyme speeds the breakdown of starch to maltose, a disaccharide. Maltose molecules are too large to be absorbed by the alimentary canal; therefore, more digestion is required.

In the stomach, gastric juice secreted by gastric glands has a very low pH—about 2—because it contains hydrochloric acid (HCl). The precursor, pepsinogen, is converted to the enzyme pepsin when exposed to HCl. Pepsin acts on protein to produce peptides:

$$\text{protein} + \text{H}_2\text{O} \xrightarrow{\text{pepsin}} \text{peptides}$$

Peptides vary in length, but they are usually too large to be absorbed and must be broken down further.

In the small intestine, starch, proteins, nucleic acids, and fats are all enzymatically broken down. Pancreatic juice, which enters the duodenum, has a basic pH because it contains sodium bicarbonate (NaHCO$_3$). One pancreatic enzyme, pancreatic amylase, digests starch:

$$\text{starch} + \text{H}_2\text{O} \xrightarrow{\text{pancreatic amylase}} \text{maltose}$$

Recall that there are three pancreatic proteases, or enzymes that digest protein: trypsin, chymotrypsin, and carboxypeptidase. The following chemical equation summarizes their action:

$$\text{protein} + \text{H}_2\text{O} \xrightarrow{\text{pancreatic proteases}} \text{peptides and free amino acids}$$

Free amino acids are ready to be absorbed by the blood. Peptides are too large and must undergo further digestion in the small intestine. It is important to note that all three pancreatic proteases are secreted in an inactive form and cannot digest protein until they are inside the small intestine. This prevents them from accidentally digesting the pancreas itself.

Lipase, a third pancreatic enzyme, digests fat molecules in the fat droplets after they have been emulsified by bile salts:

$$\text{fat} \xrightarrow{\text{bile salts}} \text{fat droplets}$$

$$\text{fat droplets} + \text{H}_2\text{O} \xrightarrow{\text{lipase}} \text{glycerol} + \text{fatty acids}$$

As mentioned previously, glycerol and fatty acids enter the cells of the villi, and within these cells, they are rejoined and packaged as lipoprotein droplets before entering the lacteals.

Peptidases and **disaccharases (maltase, sucrase,** and **lactase)** are brush border enzymes produced by cells lining the small intestine. Intestinal peptidases complete the digestion of protein to amino acids. Intestinal disaccharases digest compound sugar molecules, called *disaccharides,* into single sugar molecules, or *monosaccharides.* Amino acids and monosaccharides are small molecules that are able to cross into the cells of the intestinal villi. Peptides, which result from the first step in protein digestion, are digested to amino acids by peptidases:

$$\text{peptides} + \text{H}_2\text{O} \xrightarrow{\text{peptidases}} \text{amino acids}$$

Maltose, a disaccharide that results from the first step in starch digestion, is digested to glucose by maltase:

$$\text{maltose} + \text{H}_2\text{O} \xrightarrow{\text{maltase}} \text{glucose} + \text{glucose}$$

Similarly, the enzyme sucrase digests table sugar (the disaccharide sucrose) and lactase digests milk sugar (the disaccharide lactose). The monosaccharides that result are absorbed by intestinal cells.

Digestion of the nucleic acids DNA and RNA into free nucleotides is accomplished by a pancreatic nuclease enzyme. The free nucleotides can be further broken down into their components—sugar, phosphate, and organic base—by a nucleotidase enzyme built into the brush border cells.

Table 15.2 lists some of the major digestive enzymes produced by the alimentary canal, salivary glands, or the pancreas. Each type of food is broken down by specific enzymes.

TABLE 15.2 Major Digestive Enzymes

Enzyme	Produced By	Site of Action	Optimum pH	Digestion Process
CARBOHYDRATE-DIGESTING ENZYMES				
Salivary amylase	Salivary glands	Mouth	Neutral	Starch + H_2O → maltose
Pancreatic amylase	Pancreas	Small intestine	Basic	Starch + H_2O → maltose
Maltase	Small intestine	Small intestine	Basic	Maltose → glucose + glucose
Sucrase	Small intestine	Small intestine	Basic	Sucrose → glucose + fructose
Lactase	Small intestine	Small intestine	Basic	Lactose → glucose + galactose
PROTEIN-DIGESTING ENZYMES				
Pepsin	Stomach	Stomach	Acidic	Protein + H_2O → peptides
Trypsin	Pancreas	Small intestine	Basic	Peptides + H_2O → smaller peptides + amino acids
Chymotrypsin	Pancreas	Small intestine	Basic	Peptides + H_2O → smaller peptides + amino acids
Carboxypeptidase	Pancreas	Small intestine	Basic	Peptides + H_2O → smaller peptides + amino acids
Peptidases	Small intestine	Small intestine	Basic	Small peptides → amino acids
LIPID-DIGESTING ENZYMES				
Pancreatic lipase	Pancreas	Small intestine	Basic	Fat droplet + H_2O → glycerol + fatty acids
NUCLEIC ACID-DIGESTING ENZYMES				
Nuclease	Pancreas	Small intestine	Basic	RNA or DNA + H_2O → nucleotides
Nucleotidase	Small intestine	Small intestine	Basic	Nucleotide + H_2O → base + sugar + phosphate

Content CHECK-UP!

7. Choose the correct enzyme to cause this reaction:

starch + water → maltose

a. trypsin   c. lipase

b. pancreatic amylase   d. pepsin

8. Which of the following enzymes is not produced by the pancreas?

a. trypsin   c. nuclease

b. lipase   d. sucrase

9. Which of the following is not a brush border enzyme produced by the small intestine?

a. nuclease   c. sucrase

b. maltase   d. peptidase

15.4 Effects of Aging

The incidence of gastrointestinal disorders increases with age. Periodontitis, which is common in elderly people, leads to the loss of teeth and the need for false teeth.

The esophagus, which rarely causes any difficulties in younger people, is more prone to disorders in the elderly. The portion of the esophagus normally found inferior to the diaphragm can protrude into the thoracic cavity, causing an esophageal hiatal hernia. In some cases, the lower esophageal sphincter opens inappropriately and allows chyme to regurgitate into the esophagus, causing heartburn. Or in some older persons, chest pain may occur when this sphincter fails to open and a bolus cannot enter the stomach. Eventually, the esophagus may develop a diverticulum that allows food to collect abnormally.

Peristalsis generally slows within the alimentary canal as the muscular wall loses tone. Peptic ulcers increase in frequency with age. The failure of older people to consume sufficient dietary fiber can result in diverticulosis and constipation. Constipation and hemorrhoids are frequent complaints among the elderly, as is fecal incontinence.

The liver shrinks with age and receives a smaller blood supply than in younger years. Notably, it needs more time to metabolize drugs and alcohol. With age, gallbladder difficulties occur; there is an increased incidence of gallstones and cancer of the gallbladder. In fact, cancer of the various organs of the gastrointestinal tract is seen more often among the elderly. For example, most cases of pancreatic cancer occur in people over the age of 60.

15.5 Homeostasis

Human Systems Work Together on page 341 tells how the digestive system works with other systems in the body to maintain homeostasis.

Within the alimentary canal, the food we eat is broken down to nutrients small enough to be absorbed by the villi of

Human Systems Work Together

Integumentary System

Digestive tract provides nutrients needed by skin.

Skin helps to protect digestive organs; helps to provide vitamin D for Ca^{2+} absorption.

Skeletal System

Digestive tract provides Ca^{2+} and other nutrients for bone growth and repair.

Bones provide support and protection; hyoid bone assists swallowing.

Muscular System

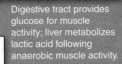

Digestive tract provides glucose for muscle activity; liver metabolizes lactic acid following anaerobic muscle activity.

Smooth muscle contraction accounts for peristalsis; skeletal muscles support and help protect abdominal organs.

Nervous System

Digestive tract provides nutrients for growth, maintenance, and repair of neurons and neuroglia.

Brain controls nerves, which innervate smooth muscle and permit tract movements.

Endocrine System

Stomach and small intestine produce hormones.

Hormones help control secretion of digestive glands and accessory organs; insulin and glucagon regulate glucose storage in liver.

How the Digestive System works with other body systems.

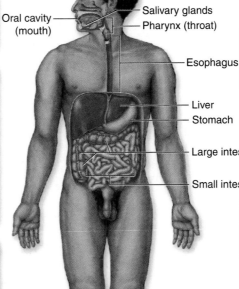

Oral cavity (mouth) —
Salivary glands
Pharynx (throat)
Esophagus
Liver
Stomach
Large intestine
Small intestine

Cardiovascular System

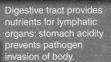

Digestive tract provides nutrients for plasma protein formation and blood cell formation; liver detoxifies blood, makes plasma proteins, destroys old red blood cells.

Blood vessels transport nutrients from digestive tract to body; blood services digestive organs.

Lymphatic System/Immunity

Digestive tract provides nutrients for lymphatic organs; stomach acidity prevents pathogen invasion of body.

Lacteals absorb fats; Peyer patches prevent invasion by pathogens; appendix contains lymphatic tissue.

Respiratory System

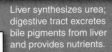

Breathing is possible through the mouth because digestive tract and respiratory tract share the pharynx.

Gas exchange in lungs provides oxygen to digestive tract and excretes carbon dioxide from digestive tract.

Urinary System

Liver synthesizes urea; digestive tract excretes bile pigments from liver and provides nutrients.

Kidneys convert vitamin D to active form needed for Ca^{2+} absorption; compensate for any water loss by digestive tract.

Reproductive System

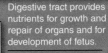

Digestive tract provides nutrients for growth and repair of organs and for development of fetus.

Pregnancy crowds digestive organs and promotes heartburn and constipation.

the small intestine. Digestive enzymes are produced by the salivary glands, gastric glands, and intestinal glands. Three accessory organs of digestion (the pancreas, the liver, and the gallbladder) also contribute secretions that help break down food. The liver produces bile (stored by the gallbladder), which emulsifies fat. The pancreas produces enzymes for the digestion of carbohydrates, proteins, and fat. Secretions from these glands, which are sent by ducts into the small intestine, are regulated by hormones such as secretin produced by the alimentary canal. Therefore, the alimentary canal is also a part of the endocrine system. Other accessory organs, such as the salivary glands and teeth, are also essential to digestion. The skeletal system assists the digestive system in that the teeth sockets are in the mandible and maxillae.

The liver is the most important of the metabolic organs. The liver has a wide variety of functions and is chemically extremely active, which gives it an influence over all other organs. Some actions involve the breakdown of complex chemicals; other important ones involve synthesis, particularly the manufacture of protein molecules. The liver assists the urinary system, producing urea, the main nitrogenous end product of human beings. The liver acts as a cleansing station, inactivating hormones and drugs. The Kupffer cells that line the liver's sinusoids mop up unwanted substances and infectious pathogens reaching it from the small intestine. Because the liver is such an important organ, diseases affecting the liver, such as hepatitis and cirrhosis, are extremely dangerous.

The liver and the cardiovascular system work together. A large amount of the body's blood reaches the liver constantly. Between meals, more than three-quarters of this supply comes to the liver by way of the hepatic portal vein, which drains the intestine. The remainder is from the body's main arterial system via the hepatic artery. When food is eaten, more blood is diverted to the intestine to cope with the tasks of digestion and absorption, and blood flow in the hepatic portal vein increases. The liver assists the cardiovascular system by aiding in the breakdown of red blood cells. It assists the urinary system by excreting bilirubin, a hemoglobin breakdown product.

The nutrients absorbed by the alimentary canal are converted by the body into energy and used for physical activities and for the growth and repair of body tissue. As we shall see in section 15.6, carbohydrates and fats are used to fuel all the body's processes and functions, while protein is mainly used as a building material. Besides these three basic components, the body must also have vitamins and minerals. Vitamins are essential for normal growth and development, and because they cannot be manufactured in the body, they must be supplied ready-made in the diet or as supplements. Minerals assist in many body processes, such as normal nerve and muscle function, but are needed only in small quantities.

The muscular system and digestive system work together. The muscular system benefits from the nutrients absorbed by the alimentary canal, but mechanical digestion is in part dependent upon the muscular walls of the alimentary canal. Also, peristalsis pushes food along from organ to organ.

Peyer patches and Kupffer cells are examples that the digestive system and the lymphatic system work together. Peyer patches in the wall of the small intestine are lymphatic tissue. They are an important way for the small intestine to protect itself from invasion by bacteria. The patches contain large numbers of antibody-secreting lymphocytes.

The endocrine system and the digestive system also work together. The secretion of digestive juices is dependent on hormones produced not by the endocrine glands but by the digestive organs themselves. Thus, these organs become a part of the endocrine system. Certainly the pancreas is counted as an endocrine gland when it produces insulin, which causes cells, including the hepatic cells, to take up glucose. Thereafter, glucose is stored in the liver and muscles for future use.

15.6 Nutrition

Nutrition involves an interaction between food and the living organism. A nutrient is a substance that the body uses to maintain health. A balanced diet contains all the essential nutrients and includes a variety of foods, proportioned as shown in Figure 15.12.

Following digestion, nutrients enter the blood in the cardiovascular system, which distributes them to the tissues, where they are utilized by the body's cells. Mitochondria use glucose to produce a constant supply of ATP for the cell. In other words, glucose is the body's immediate energy source. Because the brain's preferred source of energy is glucose, it needs a constant supply.

The liver is able to chemically alter ingested fats to suit the body's needs, with the exception of linoleic acid and linolenic acid. Both fatty acids are required, for construction of plasma membranes and for synthesis of messenger chemicals. Thus,

Figure 15.12 Newest dietary guidelines, published by the U.S. Department of Agriculture, and available at www.mypyramid.gov. The website can be customized for an individual's age, gender, and activity level.

Antioxidants

Over the past 20 years, numerous statistical studies have been done to determine whether a diet rich in fruits and vegetables protects against cancer. The vitamins C and E and beta-carotene, which is converted to vitamin A in the body, are especially abundant in fruits and vegetables and seem to have a special function in cells.

Cellular metabolism generates free radicals, unstable molecules that can attack and damage other molecules, such as DNA, proteins (e.g., an enzyme), carbohydrates, and lipids, that are found in plasma membranes. The damage to these cellular molecules may lead to disorders, perhaps even cancer. In addition, plaque formation in arteries may begin when arterial linings are injured by damaged cholesterol molecules.

The most common free radical in cells is oxygen in the unstable form O_{32}. Vitamins C and E (and possibly beta-carotene) are believed to defend the body against free radicals, and therefore are termed antioxidants. To receive adequate amounts of these vitamins, you should eat at least five servings of fruits and vegetables daily. Any one of the following is considered "one serving":

- 1 cup of raw leafy greens, such as lettuce or spinach
- ½ cup of raw or cooked vegetables, such as broccoli, cauliflower, peas, green beans, and so on
- one average carrot or one medium potato
- one medium apple, orange, banana, or similar-sized fruit
- ½ cup of grapes or cut fruit, such as diced pineapple
- ¼ cup of dried fruit, such as raisins
- ¾ cup of pure fruit or vegetable juice

Certain minerals form the structure of metalloenzymes, which serve as antioxidants. Glutathione peroxidase, a major intracellular antioxidant, contains the mineral selenium in its structure. Superoxide dismutase, another antioxidant enzyme, contains either magnesium or copper and iron.

Dietary supplements may provide a potential safeguard against cancer and cardiovascular disease, but taking supplements instead of improving your intake of fruits and vegetables is not the solution. Fruits and vegetables provide hundreds of beneficial compounds that cannot be obtained from a vitamin pill. These beneficial compounds include flavonoids and plant phenolics such as those found in red wine. These substances enhance each other's absorption or action and perform independent biological functions.

---

they are considered to be **essential fatty acids.** Essential molecules must be present in food because the body is unable to manufacture them. Other fats, especially saturated fats, should be restricted, as discussed in the Medical Focus on pages 270–271.

If glucose is not available, fats can be metabolized into smaller molecules, which are then used as an alternate energy source. Therefore, fats are said to be a long-term energy source. When adipose tissue cells store fats, the body increases in weight. Cells have the capability of converting excess sugar molecules into fats for storage, which accounts for the fact that carbohydrates can also contribute to weight gain.

Amino acids from protein digestion are used by the cells to construct their own proteins, including the enzymes that carry out metabolism. Protein formation requires 20 different types of amino acids. Of these, nine are required in the diet because the body is unable to produce them. These are termed the **essential amino acids.** The body produces the other 11 amino acids by simply transforming one type into another type. Some protein sources, such as meat, are complete in the sense that they provide all the different types of amino acids. Vegetables and grains supply the body with amino acids, but they are typically incomplete sources because at least one of the essential amino acids is absent. A combination of certain vegetables, however, can provide all of the essential amino acids. In addition, soy protein is a complete protein.

Vitamins

Vitamins are *vital* to life because they play essential roles in cellular metabolism. Because the body is unable to produce them, most vitamins must be present in the diet. Vitamins are organic molecules, but they differ radically from carbohydrates, fats, and proteins. They are much smaller in size and are not broken down to be used as building blocks or as a source of energy. Instead, the body protects them and provides many of them with protein carriers that transport them in the blood to the cells. In the cells, vitamins become helpers in metabolic processes that break down or synthesize other organic molecules. Because vitamins can be used over and over again, they are required in very small amounts.

Vitamins fall into two groups: fat-soluble vitamins (vitamins A, D, E, and K) and water-soluble vitamins (the B-complex vitamins and vitamin C) (Table 15.3). Most of the water-soluble vitamins are coenzymes, or enzyme helpers, that help speed up specific reactions. The functions of the fat-soluble vitamins, some of which have been previously discussed, are more specialized. Vitamin A, as noted in Chapter 9, is used to synthesize the visual pigments. Vitamin D is needed to produce a hormone that regulates calcium and phosphorus metabolism (see Chapter 5). Vitamin E, as discussed in the Medical Focus on this page, is an antioxidant.

TABLE 15.3 Vitamins: Their Role in the Body and Food Sources

Vitamins	Role in Body	Good Food Sources
FAT-SOLUBLE VITAMINS		
Vitamin A	Assists in the formation and maintenance of healthy skin, hair, and mucous membranes; aids in the ability to see in dim light (night vision); is essential for proper bone growth, tooth development, and reproduction	Deep yellow/orange and dark green vegetables and fruits (carrots, broccoli, spinach, cantaloupe, sweet potatoes); cheese, milk, and fortified margarine
Vitamin D	Aids in the formation and maintenance of bones and teeth; assists in the absorption and use of calcium and phosphorus	Milk fortified with vitamin D; tuna, salmon, or cod liver oil; also made in the skin when exposed to sunlight
Vitamin E	Protects vitamin A and essential fatty acids from oxidation; prevents plasma membrane damage	Vegetable oils and margarine; nuts; wheat germ and whole-grain breads and cereals; green, leafy vegetables
Vitamin K	Aids in synthesis of substances needed for clotting of blood; helps maintain normal bone metabolism	Green, leafy vegetables, cabbage, and cauliflower; also made by bacteria in intestines of humans, except for newborns
WATER-SOLUBLE VITAMINS		
Vitamin C	Is important in forming collagen, a protein that gives structure to bones, cartilage, muscle, and vascular tissue; helps maintain capillaries, bones, and teeth; aids in absorption of iron; helps protect other vitamins from oxidation	Citrus fruits, berries, melons, dark green vegetables, tomatoes, green peppers, cabbage, potatoes
B-COMPLEX VITAMINS		
Thiamin	Helps in release of energy from carbohydrates; promotes normal functioning of nervous system	Whole-grain products, dried beans and peas, sunflower seeds, nuts
Riboflavin	Helps body transform carbohydrates, proteins, and fats into energy	Nuts, yogurt, milk, whole-grain products, cheese, poultry, leafy green vegetables
Niacin	Helps body transform carbohydrates, proteins, and fats into energy	Nuts, poultry, fish, whole-grain products, dried fruit, leafy greens, beans; can be formed in the body from tryptophan, an essential amino acid found in protein
Vitamin B_6	Aids in the use of fats and amino acids; aids in the formation of protein	Sunflower seeds, beans, poultry, nuts, bananas, dried fruit, leafy green vegetables
Folic acid	Aids in the formation of hemoglobin in red blood cells; aids in the formation of genetic material	Nuts, beans, whole-grain products, fruit juices, dark green leafy vegetables
Pantothenic acid	Aids in the formation of hormones and certain nerve-regulating substances; helps in the metabolism of carbohydrates, proteins, and fats	Nuts, beans, seeds, poultry, dried fruit, milk, dark green leafy vegetables
Biotin	Aids in the formation of fatty acids; helps in the release of energy from carbohydrates	Occurs widely in foods, especially eggs; made by bacteria in the human intestine
Vitamin B_{12}	Aids in the formation of red blood cells and genetic material; helps in the functioning of the nervous system; requires intrinsic factor from the stomach to be absorbed	Milk, yogurt, cheese, fish, poultry, eggs; not found in plant foods unless fortified (as in some breakfast cereals)

Source: From David C. Nieman, et al., *Nutrition*, Revised 1st ed. Copyright © 1992 Wm. C. Brown Communications, Inc., Dubuque, Iowa. Reprinted by permission of C. V. Mosby, St. Louis, MO.

Vitamin K is required to form *prothrombin*, a substance necessary for normal blood clotting (see Chapter 11).

Minerals

In contrast to vitamins, **minerals** are inorganic elements (Table 15.4). An element, you will recall, is one of the basic substances of matter that cannot be broken down further into simpler substances. Minerals sometimes occur as a single atom, in contrast to vitamins, which contain many atoms. Minerals cannot lose their identity, no matter how they are handled. Because they are indestructible, no special precautions are needed to preserve them when cooking.

Minerals are divided into macronutrients, which are needed in gram amounts per day, and micronutrients (trace

TABLE 15.4 Minerals: Their Role in the Body and Food Sources

Minerals	Role in Body	Good Food Sources
MACRONUTRIENTS		
Calcium	Is used for building bones and teeth and for maintaining bone strength; also involved in muscle contraction, blood clotting, and maintenance of plasma membranes	All dairy products; dark green, leafy vegetables; beans, nuts, sunflower seeds, dried fruit, molasses, canned fish
Phosphorus	Is used to build bones and teeth; to release energy from carbohydrates, proteins, and fats; and to form genetic material, plasma membranes, and many enzymes	Beans, sunflower seeds, milk, cheese, nuts, poultry, fish, lean meats
Magnesium	Is used to build bones, to produce proteins, to release energy from muscle carbohydrate stores (glycogen), and to regulate body temperature	Sunflower and pumpkin seeds, nuts, whole-grain products, beans, dark green vegetables, dried fruit, lean meats
Sodium	Regulates body-fluid volume and blood acidity; aids in transmission of nerve impulses	Most of the sodium in the U.S. diet is added to food as salt (sodium chloride) in cooking, at the table, or in commercial processing; animal products contain some natural sodium
Chloride	Is a component of gastric juice and aids in acid-base balance	Table salt, seafood, milk, eggs, meats
Potassium	Assists in muscle contraction, the maintenance of fluid and electrolyte balance in the cells, and the transmission of nerve impulses; also aids in the release of energy from carbohydrates, proteins, and fats	Widely distributed in foods, especially fruits and vegetables, beans, nuts, seeds, and lean meats
MICRONUTRIENTS (TRACE ELEMENTS)		
Iron	Is involved in the formation of hemoglobin in the red blood cells of the blood and myoglobin in muscles; also is a part of several enzymes and proteins	Molasses, seeds, whole-grain products, fortified breakfast cereals, nuts, dried fruits, beans, poultry, fish, lean meats
Zinc	Is involved in the formation of protein (growth of all tissues), in wound healing, and in prevention of anemia; is a component of many enzymes	Whole-grain products, seeds, nuts, poultry, fish, beans, lean meats
Iodine	Is an integral component of thyroid hormones	Table salt (fortified), dairy products, shellfish, and fish
Fluoride	Is involved in maintenance of bone and tooth structure	Fluoridated drinking water is the best source; also found in tea, fish, wheat germ, kale, cottage cheese, soybeans, almonds, onions, milk
Copper	Is vital to enzyme systems and in manufacturing red blood cells; is needed for utilization of iron	Nuts, oysters, seeds, crab, wheat germ, dried fruit, whole grains, legumes
Selenium	Functions in association with vitamin E; may assist in protecting tissues and plasma membranes from oxidative damage; may also aid in preventing cancer	Nuts, whole grains, lean pork, cottage cheese, milk, molasses, squash
Chromium	Is required for maintaining normal glucose metabolism; may assist insulin function	Nuts, prunes, vegetable oils, green peas, corn, whole grains, orange juice, dark green vegetables, legumes
Manganese	Is needed for normal bone structure, reproduction, and the normal functioning of the central nervous system; is a component of many enzyme systems	Whole grains, nuts, seeds, pineapple, berries, legumes, dark green vegetables, tea
Molybdenum	Is a component of enzymes; may help prevent dental caries	Tomatoes, wheat germ, lean pork, legumes, whole grains, strawberries, winter squash, milk, dark green vegetables, carrots

Source: From David C. Nieman, et al., *Nutrition,* Revised 1st ed. Copyright © 1992 Wm. C. Brown Communications, Inc., Dubuque, Iowa. Reprinted by permission of C. V. Mosby, St. Louis, MO.

elements), which are needed in only microgram amounts per day. The macronutrients sodium, magnesium, phosphorus, chlorine, potassium, and calcium serve as constituents of cells and body fluids, and as structural components of tissues. The micronutrients have very specific functions, as noted in Table 15.4. As research continues, more elements will be added to the list of those considered essential.

Eating Disorders

Authorities recognize three primary eating disorders: obesity, bulimia nervosa, and anorexia nervosa. Although they exist in a continuum as far as body weight is concerned, all represent an inability to maintain normal body weight because of eating habits.

Persons with bulimia nervosa have

- recurrent episodes of binge eating characterized by consuming an amount of food much higher than normal for one sitting and a sense of lack of control over eating during the episode.
- an obsession about their body shape and weight.
- increase in fine body hair, halitosis, and gingivitis.

Body weight is regulated by

- a restrictive diet, excessive exercise.
- purging (self-induced vomiting or misuse of laxatives).

Persons with anorexia nervosa have

- a morbid fear of gaining weight; body weight no more than 85% normal.
- a distorted body image so that person feels fat even when emaciated.
- in females, an absence of a menstrual cycle for at least three months.

Body weight is kept too low by either/or

- a restrictive diet, often with excessive exercise.
- binge eating/purging (person engages in binge eating and then self-induces vomiting or misuses laxatives).

Figure 15.13 Recognizing anorexia nervosa and bulimia nervosa.

Obesity

Obesity is most often defined as a body weight 20% or more above the ideal weight for a person's height. By this standard, 28% of women and 10% of men in the United States are obese. Moderate obesity is 41–100% above ideal weight, and severe obesity is 100% or more above ideal weight.

Obesity is most likely caused by a combination of hormonal, metabolic, and social factors. It is known that obese individuals have more fat cells than normal. When an obese person loses weight, the fat cells simply get smaller; they don't disappear. The social factors that cause obesity include the eating habits of other family members. Consistently eating fatty foods, for example, may cause you to gain weight. Sedentary activities, such as watching television instead of exercising, also determine how much body fat you have. The risk of heart disease is higher in obese individuals, and this alone tells us that excess body fat is not consistent with optimal health.

Treatment depends on the degree of obesity. Surgery to restrict stomach volume may be required for those who are moderately or greatly overweight (see the Medical Focus on page 347). But for most people, a knowledge of good eating habits along with behavior modification may suffice, particularly if a balanced diet is accompanied by a sensible exercise program. A lifelong commitment to a properly planned program is the best way to prevent a cycle of weight gain followed by weight loss. Such a cycle is not conducive to good health.

Bulimia Nervosa

Bulimia nervosa can coexist with either obesity or anorexia nervosa, which is discussed next. People with this condition have the habit of eating to excess (called binge eating) and then purging themselves by some artificial means, such as self-induced vomiting or the use of a laxative. Bulimic individuals are overly concerned about their body shape and weight, and therefore they may be on a very restrictive diet. A restrictive diet may bring on the desire to binge, and typically the person chooses to consume sweets, such as cakes, cookies, and ice cream (Fig. 15.13). The amount of food

Bariatric Surgery for Obesity

It's a final measure sought by increasing numbers of people: **bariatric surgery,** or surgical intervention with the specific goal of causing drastic weight loss. For many overweight people, years of dieting haven't worked. Name a diet plan, and they have tried it—sure, they lose weight only to regain it. Obesity is approaching epidemic levels in the United States, with more than 6 million adults classified as obese. In addition, pediatricians are especially concerned with the dramatic rise in childhood and teenage obesity.

Obesity is defined as a body mass index (BMI) greater than 35,[1] or a weight that is 41% or higher than the ideal weight for one's height. It's a physical and emotional challenge for the overweight patient. Obesity is a primary risk factor for hypertension, type II diabetes mellitus, atherosclerosis, stroke, coronary artery disease, and early heart attack. It has been linked to increased risk of breast, ovarian, uterine, and prostate cancer. Obese individuals suffer disrespect and ridicule from society and discrimination on the job. After years of struggling with the problem, many are willing to undergo surgery as a last-chance option.

However, reputable programs offering bariatric surgery have strict requirements for patients. To be admitted as a surgical candidate, the patient must be morbidly obese (BMI greater than 40; or greater than 100 pounds over ideal weight). Patients cannot have respiratory or cardiac problems that might complicate surgery. Most important, the patient must understand the risks of surgery as well as its benefits. Patients must also understand that lifestyle changes will be necessary even after successful surgery and recovery. Psychological and nutritional counseling is usually required before surgery to prepare the patient for new eating habits and ways of thinking about food. Follow-up counseling tracks the patient's progress in adapting.

The two most commonly used interventions are laparoscopic banding and the Roux-en-Y gastric bypass (Fig. 15D). A more conservative approach, laparoscopic banding, requires making a series of small incisions around the stomach and using an instrument called a laparoscope to illuminate structures in the abdomen. A band is placed around the stomach. Once tightened like a belt, the stomach is divided into a smaller upper chamber, which receives food, and a lower chamber that remains connected to the duodenum. The belt can later be tightened further, or removed if necessary. In Roux-en-Y gastric bypass, the top section of the stomach is cut free and stapled shut to make a pouch about the size of an egg. The small intestine is cut free between the duodenum and jejunum, and the jejunum is sewn to the end of the stomach pouch. Finally, the duodenal segment is sewn back to the jejunum (forming the Y-shaped branch for which the procedure is named). The lower stomach and duodenum remain healthy and continue to secrete digestive enzymes, but never receive food. Regardless of the approach used, the person will only be able to eat small amounts of food, but should feel full due to the small size of the stomach after surgery.

It's important to note that bariatric surgery comes with extremely serious potential complications. Postoperative bleeding and infection are risks of any surgical procedure. Blood clots in the legs can form during hospitalization, causing pulmonary embolism or stroke. After gastric bypass, staple lines in the stomach can leak. In rare cases, the connection between the stomach pouch and jejunum narrows, requiring additional corrective surgery. Worst of all, a small percentage of patients die during the surgery itself. After surgery, vitamin and mineral deficiencies are possible.

Further, bariatric surgery offers no guarantees of permanent weight loss—patients can in fact regain any weight that is lost, even with a drastically smaller stomach. However, with proper nutrition and behavioral changes, bariatric surgery can result in dramatic weight loss and improvement to health.

[1]To calculate BMI, use the following formula: $\dfrac{\text{weight in pounds}}{(\text{height in inches}) \times (\text{height in inches})} \times 703$. BMI is not always accurate in determining obesity. Other factors such as percent body fat may also need to be used.

After laparoscopic banding   **Before surgery**   **After Roux-en-Y gastric bypass**

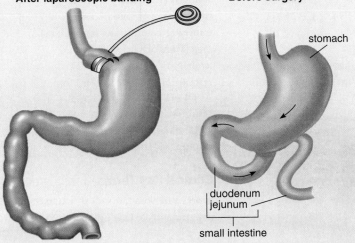

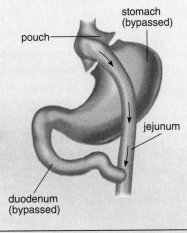

stomach

duodenum
jejunum

small intestine

pouch

stomach
(bypassed)

jejunum

duodenum
(bypassed)

Figure 15D (*far left figure*) A laparascopic band is placed around the stomach and then tightened to decrease stomach size. (*center*) The normal movement of food prior to surgery. (*right*) The flow of food after Roux-en Y gastric bypass surgery.

consumed is far beyond the normal number of calories for one meal, and the person keeps on eating until every bit is gone. Then, a feeling of guilt most likely brings on the next phase, which is a purging of all the calories that have been taken in.

Bulimia is extremely dangerous to your health. Blood composition is altered, leading to an abnormal heart rhythm, and damage to the kidneys can even result in death. At the very least, vomiting can lead to inflammation of the pharynx and esophagus, and stomach acids can cause the teeth to erode. The esophagus and stomach may even rupture and tear due to strong contractions during vomiting.

The most important aspect of treatment is to get the patient on a sensible and consistent diet. Again, behavioral modification is helpful, and so perhaps is psychotherapy to help the patient understand the emotional causes of the behavior. Medications, including antidepressants, have sometimes helped to reduce the bulimic cycle and restore normal appetite.

Anorexia Nervosa

In **anorexia nervosa,** a morbid fear of gaining weight causes the person to be on a very restrictive diet. Athletes such as distance runners, wrestlers, and dancers are at risk of anorexia nervosa because they believe that being thin gives them a competitive edge. In addition to eating only low-calorie foods, the person may induce vomiting and use laxatives to bring about further weight loss. No matter how thin they have become, people with anorexia nervosa think they are overweight. Such a distorted self-image may prevent recognition of the need for medical help.

Actually, the person is starving and has all the symptoms of starvation, including low blood pressure, irregular heartbeat, constipation, and constant chilliness. Bone density decreases, and stress fractures occur. The body begins to shut down; menstruation ceases in females; the internal organs, including the brain, don't function well; and the skin dries up. Impairment of the pancreas and alimentary canal means that any food consumed does not provide nourishment. Death may be imminent. If so, the only recourse may be hospitalization and force-feeding. Eventually, it is necessary to use behavior therapy and psychotherapy to enlist the cooperation of the person to eat properly. Family therapy may be necessary, because anorexia nervosa in children and teens is believed to be a way for them to gain some control over their lives.

SELECTED NEW TERMS

Basic Key Terms

alimentary canal (al-i-men′ter-ē kuh-nal), p. 324

anus (ā-nus), p. 333

ascending colon (uh-send′ing ko′lon), p. 333

bicuspid (bī-cus′-pīd), p. 325

bile (bīl), p. 330

bile salts (bīl sôlt), p. 338

bilirubin (bīl′y-roō′bǐn), p. 338

bolus (bo′lus), p. 326

cecum (se′kum), p. 333

cholecystokinin (kō′lē-sǐs-tō-kī′-nin), p. 331

chyme (kīm), p. 329

colon (kō′lən), p. 333

cuspid (cǔs′-pid), p. 325

descending colon (de-send′ing ko′lon), p. 333

disaccharases (dī-sǎk′oh-rə-ses), p. 339

duodenum (du″o-de′num), p. 330

esophagus (ě-sof′uh-gus), p. 327

essential amino acid (ě-sen′shul uh-me′no as′id), p. 343

essential fatty acid (ě-sen′shul fat′e as′id), p. 343

gallbladder (gawl′blad-er), p. 338

gastric gland (gas′trik gland), p. 329

gastric juice (gǎs′trǐk joōs), p. 329

gastrin (gǎs′trǐn), p. 329

hard palate (hard pal′ut), p. 324

ileum (il′e-um), p. 330

incisor (ĭn-sī-zōr), p. 325

intrinsic factor (in-trǐn′zik fǎk′tər), p. 329

jejunum (jě-ju′num), p. 330

lacteal (lǎk′tē-əl), p. 330

large intestine (larj in-tes′tin), p. 333

lipase (lǐp′as′), p. 336

liver (liv′er), p. 337

lumen (loō′mən), p. 327

mineral (min′er-al), p. 344

mouth (mowth), p. 324

molar (mō′lǎr), p. 325

mucosa (mū-kō′sa), p. 327

muscularis (mǔs-kū-lā′-ris), p. 327

pancreas (pǎng′ krē-əs), p. 336

pancreatic amylase (pǎng′krē-āt′ǐk am′ə-lǎs′), p. 336

pepsin (pěp′-sǐn), p. 329

peptidases (pěp′tǐ-dās′), p. 339

peristalsis (per″ĭ-stal′sis), p. 327

pharynx (fǎr′ǐngks), p. 326

portal triads (pôr′tl trī′ǎd), p. 337

rectum (rek′tum), p. 333

rugae (roo′je), p. 328

salivary amylase (sǎl′ə-věr′ē am′ə-lǎs), p. 324

salivary gland (sal′ǐ-věr-e gland), p. 324

secretin (sǐ-krēt′n), p. 331

serosa (se-rō′sǎ), p. 327

sigmoid colon (sig′moid ko′lon), p. 333

small intestine (smawl in-tes′tin), p. 329

soft palate (sawft pal′ut), p. 324

sphincter (sfingk′ter), p. 327

stomach (stum′ak), p. 327

submucosa (sǔb-moo-kō′sǎ), p. 327

transverse colon (trans-vers′ ko′lon), p. 333

urea (yū-re′uh), p. 338

uvula (yoo′-vyoo-luh), p. 324

vermiform appendix (ver′mǐ-form uh-pen′diks), p. 333

vestibule (ves′ti-byool), p. 324

villi (vil′i), p. 330

vitamin (vi′tuh-min), p. 343

Clinical Key Terms

anorexia nervosa (ǎn′ə-rēk′sē-ə nûr-vō′sə), p. 348

bariatric surgery (bǎr′ē-ǎt′rǐks sûr′jə-rē), p. 347

SUMMARY

15.1 Anatomy of the Digestive System

The alimentary canal consists of the mouth, pharynx, esophagus, stomach, small intestine, and large intestine. Only these structures actually contain food, while the salivary glands, liver, and pancreas supply substances that aid in the digestion of food.

A. The salivary glands send saliva into the mouth, where the teeth chew the food and the tongue forms a bolus for swallowing. Saliva contains salivary amylase, an enzyme that begins the digestion of starch.

B. The air passage and the food passage cross in the pharynx. When a person swallows, the air passage is normally blocked off, and food must enter the esophagus, where peristalsis begins.

C. The stomach expands and stores food. While food is in the stomach, the stomach churns, mixing food with the acidic gastric juices. Gastric juices contain pepsin, an enzyme that digests protein.

D. The duodenum of the small intestine receives bile from the liver and pancreatic juice from the pancreas. Bile, which is produced in the liver and stored in the gallbladder, emulsifies fat and readies it for digestion by lipase, an enzyme produced by the pancreas. The pancreas also produces enzymes that digest starch (pancreatic amylase) and protein (trypsin, chymotrypsin, and carboxypeptidase). The intestinal enzymes finish the process of chemical digestion.

E. The walls of the small intestine have fingerlike projections called villi. The villi have microvilli where small nutrient molecules are absorbed. Amino acids and glucose enter the blood vessels of a villus. Glycerol and fatty acids are joined and packaged as lipoproteins before entering lymphatic vessels called lacteals in a villus.

F. The large intestine consists of the cecum, the colon (including the ascending, transverse, descending, and sigmoid colon), and the rectum, which ends at the anus. The large intestine does not produce digestive enzymes; it does absorb water, salts, and some vitamins. Reduced water absorption results in diarrhea. The intake of water and fiber helps prevent constipation.

15.2 Accessory Organs of Digestion

A. Three accessory organs of digestion—the pancreas, liver, and gallbladder—send secretions to the duodenum via ducts. The pancreas produces pancreatic juice, which contains digestive enzymes for carbohydrate, protein, and fat.

B. The liver produces bile, which is stored in the gallbladder. The liver receives blood from the small intestine by way of the hepatic portal vein. It has numerous important functions, and any malfunction of the liver is a matter of considerable concern.

15.3 Chemical Digestion

A. Digestive enzymes are present in digestive juices and break down food into the nutrient molecules glucose, amino acids, fatty acids, and glycerol (see Table 15.2). Salivary amylase and pancreatic amylase begin the digestion of starch. Pepsin and pancreatic peptidases digest protein to peptides. Lipase digests fat to glycerol and fatty acids. Intestinal enzymes finish the digestion of starch and protein.

B. Digestive enzymes have the usual enzymatic properties. They are specific to their substrate and speed up specific reactions at optimum body temperature and pH.

15.4 Effects of Aging

The structure and function of the digestive system generally decline with age. The various illnesses associated with the digestive system are more likely to be seen among the elderly.

15.5 Homeostasis

The digestive system works with the other systems of the body in the ways described in Human Systems Work Together on page 341.

15.6 Nutrition

A. The nutrients released by the digestive process should provide us with an adequate amount of energy, essential amino acids and fatty acids, and all necessary vitamins and minerals.

B. The diet should be balanced and low in saturated fatty acids and cholesterol molecules, whose intake is linked to cardiovascular disease. Aside from carbohydrates, proteins, and fats, the body requires vitamins and minerals. The vitamins C, E, and A are antioxidants that protect cell contents from damage due to free radicals. The mineral calcium is needed for strong bones.

C. The reasons for eating disorders, including obesity, bulimia nervosa, and anorexia, are being explored in order to help people maintain a normal weight for their height.

STUDY QUESTIONS

1. List the organs of the alimentary canal, and state the contribution of each to the digestive process. (pp. 324–338)
2. Discuss the absorption of the products of digestion into the lymphatic and cardiovascular systems. (p. 330)
3. Name and state the functions of the hormones that assist the nervous system in regulating digestive secretions. (p. 331)
4. Name the accessory digestive organs, and describe the part they play in the digestion of food. (p. 336)
5. Name and discuss any three functions and two serious illnesses of the liver. (p. 338)
6. Discuss the digestion of starch, protein, and fat, listing all the steps that occur with each. (p. 339)
7. How does the digestive system help maintain homeostasis? (pp. 340, 342)
8. How does the cardiovascular system assist the digestive system in maintaining homeostasis? (p. 341)
9. What is the chief contribution of each of these constituents of the diet: (a) carbohydrates; (b) proteins; (c) fats; (d) fruits and vegetables? (pp. 342–343)
10. What role do water-soluble vitamins usually play in the body? (p. 343)
11. Name and discuss three eating disorders. (pp. 345–348)

LEARNING OBJECTIVE QUESTIONS

Fill in the blanks.

1. In the mouth, salivary _____ digests starch.
2. When we swallow, the _____ covers the opening to the larynx.
3. The _____ takes food to the stomach, where _____ is primarily digested.
4. The gastric juices are _____, and therefore, they usually destroy any bacteria in the food.
5. The large intestine has a colon with four _____ and a(n) _____, which leads to the _____.
6. The pancreas transports digestive juices to the _____, the first part of the small intestine.
7. After a meal, the liver stores glucose as _____.
8. The gallbladder stores _____, a substance that _____ fat.
9. Pancreatic juice contains _____ and _____ for digesting protein, _____ for digesting fat, and _____ for digesting starch.
10. The products of digestion are absorbed into the cells of the _____, fingerlike projections of the intestinal wall.

MEDICAL TERMINOLOGY EXERCISE

Consult Appendix B for help in pronouncing and analyzing the meaning of the terms that follow.

1. stomatoglossitis (sto"muh-to-glos-si'tis)
2. glossopharyngeal (glos"o-fah-rin'je-al)
3. esophagectasia (ĕ-sof"ah-jek-ta'se-uh)
4. gastroenteritis (gas"tro-en-ter-i' tis)
5. sublingual (sub-ling'gwal)
6. gingivoperiodontitis (jin"jĭ-vo-pĕr"e-o-don-ti'tis)
7. dentalgia (den-tal'je-uh)
8. pyloromyotomy (pi-lo"ro-mi-ot'o-me)
9. cholangiogram (ko-lan'je-o-gram)
10. cholecystolithotripsy (ko"le-sis"to-lith'o-trip"se)
11. proctosigmoidoscopy (prok"to-sig"moy-dos'kuh-pe)
12. colocentesis (ko-lo'sen-te'sis)
13. trichophagia (tri-ko-fāj'e-uh)
14. duodenorrhaphy (du-o-dĕ-nor'uh-fe)
15. ileocecal (il'e-o-se'kul)

WEBSITE LINK

Visit this text's website at <u>http://www.mhhe.com/maderap6</u> for additional quizzes, interactive learning exercises, and other study tools.

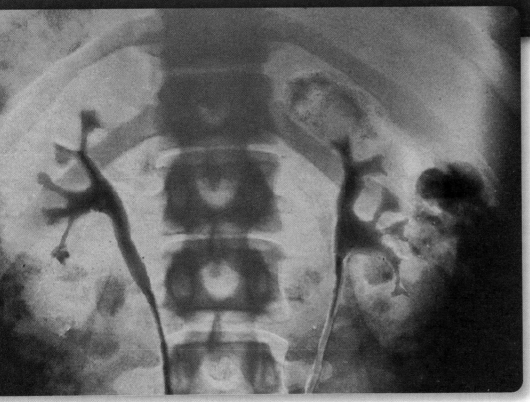

C H A P T E R

16

The Urinary System and Excretion

This artificially colored radiograph of the urinary system shows the kidneys (green). Urine passes from the renal pelvis (red) of each kidney into a ureter (also red).

Chapter Outline & Learning Objectives   *After you have studied this chapter, you should be able to:*

16.1 Urinary System (p. 352)
- List and discuss the functions of the urinary system.
- Name and describe the structure and function of each organ in the urinary system.
- Describe how urination is controlled.

16.2 Anatomy of the Kidney and Excretion (p. 354)
- Describe the macroscopic and microscopic anatomy of the kidney.
- State the parts of a kidney nephron, and relate them to the gross anatomy of the kidney.
- Describe the three steps in urine formation, and relate them to the parts of a nephron.

16.3 Regulatory Functions of the Kidneys (p. 358)
- Describe how the kidneys help maintain the fluid and electrolyte balance of blood.

- Name and explain how three hormones—aldosterone, antidiuretic hormone, and atrial natriuretic hormone—work together to maintain blood volume and pressure.
- Describe three mechanisms, including how the kidneys function, to maintain the acid-base balance of blood.

16.4 Problems with Kidney Function (p. 362)
- State, in general, the normal composition of urine and the benefits of doing a urinalysis.
- Discuss the need for hemodialysis and how hemodialysis functions to bring about the normal composition of urine.

16.5 Effects of Aging (p. 364)
- Describe the anatomical and physiological changes that occur in the urinary system as we age.

16.6 Homeostasis (p. 364)
- Describe how the urinary system works with other systems of the body to maintain homeostasis.

Visual Focus
Steps in Urine Formation (p. 357)

Focus on Forensics
Urinalysis (p. 363)

Medical Focus
Prostate Enlargement and Cancer (p. 365)

Human Systems Work Together
Urinary System (p. 366)

16.1 Urinary System

The kidneys are the primary organs of excretion. **Excretion** is the removal of metabolic wastes from the body. People sometimes confuse the terms excretion and defecation, but they do not refer to the same process. Defecation, the elimination of feces from the body, is a function of the digestive system. Excretion, on the other hand, is the elimination of metabolic wastes, which are the products of metabolism. For example, the undigested food and bacteria that make up feces have never been a part of the functioning of the body, while the substances excreted in urine were once metabolites in the body.

Functions of the Urinary System

The urinary system produces urine and transports it out of the body. As the kidneys produce urine, they carry out four functions that are essential to homeostasis: excretion of metabolic wastes, preservation of water-salt balance in blood and body fluids, maintenance of blood pressure, and maintenance of acid-base balance. As we shall see, the kidneys assist the endocrine system; they also produce erythropoietin (EPO), their own endocrine hormone.

Excretion of Metabolic Wastes

The kidneys excrete metabolic wastes, notably nitrogenous wastes. Urea is the primary nitrogenous end product of metabolism in human beings, but humans also excrete some ammonium, creatinine, and uric acid.

Urea is a by-product of amino acid metabolism. The breakdown of amino acids in the liver releases ammonia, which the liver combines with carbon dioxide to produce urea. Ammonia is very toxic to cells, but urea is much less toxic. Because urea is less toxic than ammonia, it can be safely excreted in the relatively small amounts of water passed daily as urine. (If our bodies had to excrete ammonia instead of urea, the ammonia would have to be diluted in extremely large volumes of urine to avoid being toxic to body cells.)

Creatine phosphate is a high-energy phosphate reserve molecule in muscles. The metabolic breakdown of creatine phosphate results in **creatinine.**

The breakdown of nucleotides, such as those containing adenine and thymine, produces **uric acid.** Uric acid is rather insoluble. If too much uric acid is present in blood, crystals form and precipitate out. Crystals of uric acid sometimes collect in the joints producing a painful ailment called **gout.**

Preservation of Water-Salt Balance

A principal function of the kidneys is to maintain the appropriate water-salt balance of the blood. As we shall see, blood volume is intimately associated with the salt balance of the body. As you know, salts, such as NaCl, have the ability to cause osmosis, the diffusion of water—in this case, into the blood. The more salts there are in the blood, the greater the blood volume. The kidneys also maintain the appropriate level of other ions (electrolytes), such as potassium ions (K^+), bicarbonate ions (HCO_3^-), and calcium ions (Ca^{2+}), in the blood.

Maintenance of Blood Pressure

Recall that blood pressure is the product of cardiac output and peripheral resistance, and the kidneys influence both of these variables. By preserving salt and water balance, the kidneys maintain a normal blood volume. In turn, blood volume determines the heart's stroke volume and, thus, cardiac output. The kidneys also influence peripheral resistance by producing the enzyme renin, which activates the plasma protein angiotensin. Angiotensin constricts blood vessel smooth muscle, increasing resistance.

Maintenance of Acid-Base Balance

The kidneys regulate the acid-base balance of the blood. In order for a person to remain healthy, the blood pH should be just about 7.4. The kidneys monitor and control blood pH, mainly by excreting hydrogen ions (H^+) from the blood into the urine. At the same time, the kidneys reabsorb bicarbonate ions (HCO_3^-) and return them to the blood as needed to keep blood pH at about 7.4. Urine usually has a pH of 6 or lower because our diet often contains acidic foods.

Secretion of Hormones

The kidneys assist the endocrine system in hormone secretion. By releasing the enzyme renin whenever their own blood supply decreases, the kidneys activate angiotensin, which leads to the secretion of the hormone aldosterone. Aldosterone is produced by the adrenal cortex, the outer portion of the adrenal glands, which lie atop the kidneys. As described in section 16.3, aldosterone promotes the reabsorption of sodium ions (Na^+) and water by the kidneys.

Whenever the oxygen-carrying capacity of the blood is reduced, the kidneys secrete the hormone **erythropoietin (EPO)**, which stimulates red blood cell production.

The kidneys also help activate vitamin D from the skin. Vitamin D is the precursor of the hormone calcitriol, which promotes calcium (Ca^{2+}) absorption from the digestive tract.

Organs of the Urinary System

The urinary system consists of the kidneys, ureters, urinary bladder, and urethra. Figure 16.1 shows these organs and also traces the path of urine.

Kidneys

The **kidneys** are paired organs located near the small of the back in the lumbar region on either side of the vertebral column. The kidneys are *retroperitoneal*, which means that they are covered by the parietal peritoneum (the serous membrane that lines the abdominopelvic cavity). The kidneys lie in depressions against the deep muscles of the back, where they receive some protection from the lower rib cage. Each kidney is usually held in place by connective tissue, called renal fascia. Masses of adipose tissue adhere to each kidney. A sharp blow to the back can dislodge a kidney, which is then called a **floating kidney.**

The kidneys are bean-shaped and reddish-brown in color. The fist-sized organs are covered by a tough capsule of fibrous

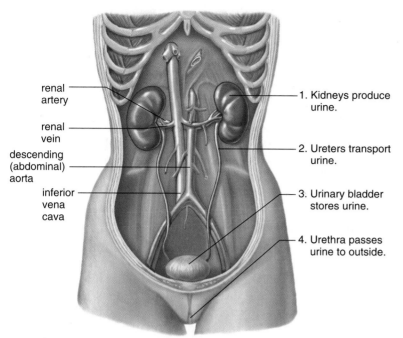

Figure 16.1 The urinary system. Urine is found only within the kidneys, the ureters, the urinary bladder, and the urethra.

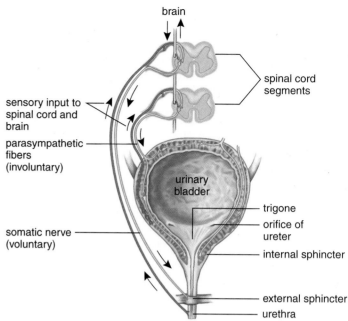

Figure 16.2 Urination. As the bladder fills with urine, sensory impulses go to the spinal cord and then to the brain. The brain can override the urge to urinate. When urination occurs, motor nerve impulses cause the bladder to contract and an internal sphincter to open. Nerve impulses also cause an external sphincter to open.

connective tissue, called a renal capsule. The concave side of a kidney has a depression called the hilum where a **renal artery** enters and a **renal vein** and a ureter exit the kidney.

Ureters

The **ureters,** which extend from the kidneys to the bladder, are small, muscular tubes about 25 cm long and 5 mm in diameter. Each descends behind the parietal peritoneum, from the hilum of a kidney, to enter the bladder posteriorly at its inferior surface.

The wall of a ureter has three layers. The inner layer is a mucosa (mucous membrane), the middle layer consists of smooth muscle, and the outer layer is a fibrous coat of connective tissue. Peristaltic contractions of the ureters cause urine to enter the bladder even if a person is lying down. Urine enters the bladder in spurts that occur at the rate of one to five per minute.

Urinary Bladder

The **urinary bladder** is located in the pelvic cavity, below the parietal peritoneum and just posterior to the pubic symphysis. In males, it is directly anterior to the rectum; in females, it is anterior to the vagina and inferior to the uterus. Its function is to store urine until it is expelled from the body. The bladder has three openings—two for the ureters and one for the urethra, which drains the bladder. The *trigone* is a smooth triangular area at the base of the bladder outlined by these three openings (Fig. 16.2).

Collectively, the muscle layers of the bladder wall are called the *detrusor muscle*. The wall contains a middle layer of circular fiber and two layers of longitudinal muscle, and it can expand. The transitional epithelium of the mucosa becomes thinner, and folds in the mucosa called *rugae* disappear as the bladder fills and enlarges.

The bladder has other features that allow it to retain urine. After urine enters the bladder from a ureter, small folds of bladder mucosa act like a valve to prevent backward flow.

Two sphincters in close proximity are found where the urethra exits the bladder. The internal sphincter occurs around the opening to the urethra. Inferior to the internal sphincter, the external sphincter is composed of skeletal muscle that can be voluntarily controlled.

Urethra

The **urethra** is a small tube that extends from the urinary bladder to an external opening. The urethra is a different length in females than in males. In females, the urethra is only about 4 cm long. The short length of the female urethra makes bacterial invasion easier and helps explain why females are more prone to urinary tract infections than males. In males, the urethra averages 20 cm when the penis is flaccid (limp, nonerect). As the urethra leaves the male urinary bladder, it is encircled by the **prostate gland.** In older men, enlargement of the prostate gland can restrict urination. A surgical procedure can usually correct the condition and restore a normal flow of urine.

In females, the reproductive and urinary systems are not connected. In males, the urethra carries urine during urination and sperm during ejaculation. This double function of the urethra in males does not alter the path of urine.

Urination

When the urinary bladder fills to about 250 ml with urine, stretch receptors send sensory nerve impulses to the spinal cord. Subsequently, motor nerve impulses from the spinal cord cause the urinary bladder to contract and the sphincters to relax so that urination, also called **micturition,** is possible (Fig. 16.2). In older children and adults, the brain controls this reflex, delaying urination until a suitable time.

16.2 Anatomy of the Kidney and Excretion

A sagittal section of a kidney shows that many branches of the renal artery and renal vein reach inside a kidney (Fig. 16.3*a*). Removing the blood vessels shows that a kidney has three regions (Fig. 16.3*b*). The **renal cortex** is an outer, granulated layer that dips down in between a radially striated inner layer called the renal medulla. The **renal medulla** consists of cone-shaped tissue masses called renal pyramids. The **renal pelvis** is a central space, or cavity, that is continuous with the ureter.

Anatomy of a Nephron

Each kidney is composed of over one million **nephrons,** sometimes called renal or kidney tubules, which are microscopic structures (Fig. 16.3*c*). Each nephron has its own blood supply, including two distinct capillary regions (Fig. 16.4). The nephron receives oxygenated arterial blood from an afferent arteriole, which originates from a smaller branch of the renal artery. The afferent arteriole leads to the **glomerulus,** a knot of capillaries inside the glomerular capsule. Blood leaves the glomerulus through the efferent arteriole. The efferent arteriole takes blood to the **peritubular capillary network,**

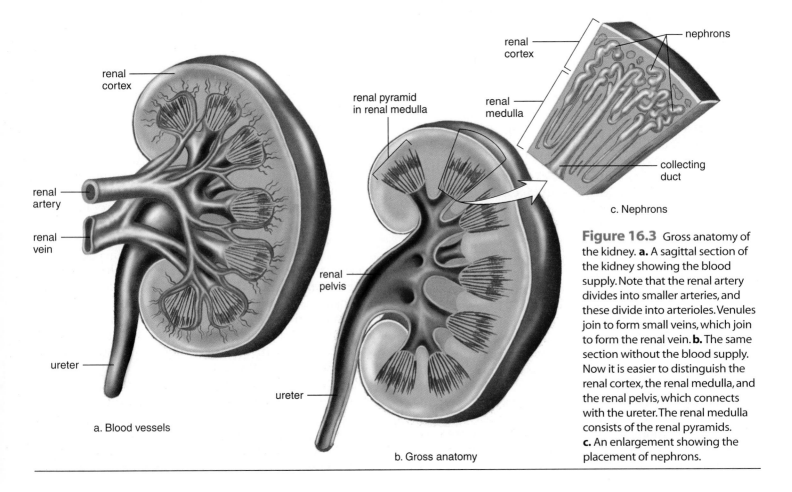

a. Blood vessels

b. Gross anatomy

c. Nephrons

Figure 16.3 Gross anatomy of the kidney. **a.** A sagittal section of the kidney showing the blood supply. Note that the renal artery divides into smaller arteries, and these divide into arterioles. Venules join to form small veins, which join to form the renal vein. **b.** The same section without the blood supply. Now it is easier to distinguish the renal cortex, the renal medulla, and the renal pelvis, which connects with the ureter. The renal medulla consists of the renal pyramids. **c.** An enlargement showing the placement of nephrons.

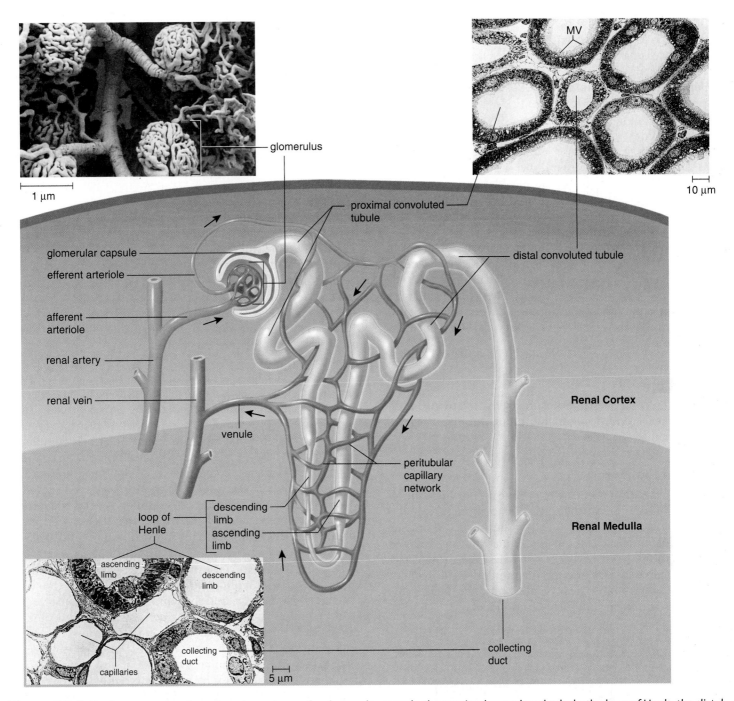

Figure 16.4 Nephron anatomy. A nephron is made up of a glomerular capsule, the proximal convoluted tubule, the loop of Henle, the distal convoluted tubule, and the collecting duct. The photomicrographs show these structures in cross section; MV = microvilli. Trace the path of blood about the nephron by following the arrows.

Source: Top left: © R. G. Kessel and R. H. Kardon, *Tissues and Organs: A Text Atlas of Scanning Electron Microscopy, 1979.*

which surrounds the rest of the nephron. From there, the blood goes into a venule. Venules join larger veins that ultimately empty into the renal vein.

Parts of a Nephron

Each nephron is made up of several parts. First, the closed end of the nephron is pushed in on itself to form a cuplike structure that completely surrounds and attaches to the glomerulus.

This section of the nephron is called the **glomerular capsule** (Bowman's capsule). The inner layer of the glomerular capsule is composed of *podocytes* that have long, cytoplasmic extensions. The podocytes cling to the capillary walls of the glomerulus and leave pores that allow easy passage of small molecules from the glomerulus to the inside of the glomerular capsule.

Next, there is a **proximal convoluted tubule (PCT),** called "proximal" because it is near the glomerular capsule. The cuboidal epithelial cells lining this part of the nephron have numerous microvilli about 1 μm in length. These microvilli are tightly packed and form a brush border, increasing the surface area for reabsorption.

Simple squamous epithelium appears as the tube narrows and makes a U-turn called the **loop of Henle.** Each loop consists of a descending limb and an ascending limb.

The cuboidal epithelial cells of the **distal convoluted tubule (DCT)** have numerous mitochondria, but they lack microvilli. This is consistent with the active role they play in moving molecules from the blood into the tubule, a process called tubular secretion. The distal convoluted tubules of several nephrons enter a single collecting duct. Many **collecting ducts** carry urine to the renal pelvis.

As shown in Figure 16.4, the glomerular capsule and the convoluted tubules always lie within the renal cortex. The loop of Henle dips down into the renal medulla; a few nephrons have a very long loop of Henle, which penetrates deep into the renal medulla. Collecting ducts are also located in the renal medulla, and they give the renal pyramids their lined appearance.

Urine Formation

Figure 16.5 gives an overview of urine formation, which is divided into these steps: glomerular filtration, tubular reabsorption, and tubular secretion.

Glomerular Filtration

Glomerular filtration occurs when whole blood enters the afferent arteriole and the glomerulus. Due to glomerular blood pressure, water and small molecules are pushed from the glomerulus to the inside of the glomerular capsule. This is a filtration process because large molecules and formed elements are unable to pass through the capillary wall. In effect, then, blood in the glomerulus has two portions: the filterable components and the nonfilterable components:

Filterable Blood Components	Nonfilterable Blood Components
Water	Formed elements (blood cells and platelets)
Nitrogenous wastes	Plasma proteins
Nutrients	
Salts (ions)	

The **glomerular filtrate** contains small dissolved molecules in approximately the same concentration as plasma. Small molecules that escape being filtered as well as the nonfilterable components leave the glomerulus by way of the efferent arteriole.

As indicated in Table 16.1, nephrons in the kidneys filter 180 liters of water per day, along with a considerable amount of small molecules (such as glucose) and ions (such as sodium). If the composition of the urine were the same as that of the glomerular filtrate, the body would constantly lose water, salts, and nutrients in the urine. This does not happen in a healthy person, because the composition of the original filtrate is altered significantly as this fluid passes through the remainder of the tubule.

It is important to stress that many small chemicals are freely filtered from the blood. Once in the filtrate, they cannot be reabsorbed and are passed in urine. Thus, urinalysis provides a history of chemicals recently present in the blood, and can be used for detection of illegal drug use, as described in Focus on Forensics (page 363.)

Tubular Reabsorption

Tubular reabsorption occurs as molecules and ions are both passively and actively reabsorbed from the nephron into the blood of the peritubular capillary network. The osmolarity of the blood is maintained by the presence of both plasma proteins and salt. When sodium ions (Na^+) are actively reabsorbed, chloride ions (Cl^-) follow passively. The reabsorption of salt (NaCl) increases the osmolarity of the blood compared to the filtrate, and therefore water moves by osmosis from the tubule into the blood. About 67% of Na^+ is reabsorbed at the proximal convoluted tubule.

Nutrients such as glucose and amino acids also return to the blood at the proximal convoluted tubule. This is a selective process. Only molecules recognized by carrier proteins built into tubular cell membranes will be actively reabsorbed. Glucose is an example of a molecule that ordinarily is completely reabsorbed because there is a plentiful supply of carrier molecules for it. However, every substance has a maximum rate of transport. After all its carriers are in use, any excess in the filtrate will appear in the urine. For example, as reabsorbed levels of glucose approach 1.8–2 mg/ml plasma, the rest appears in the urine. Excess glucose occurs in the blood in untreated diabetes mellitus because of lack of insulin (type I diabetes), or because of failure of insulin receptors to respond normally (type II diabetes). Without insulin function, glucose cannot be transported into cells, and liver and muscle cannot store glucose as glycogen. Excess blood glucose is filtered into the tubular fluid and subsequently passes into the urine because the kidneys cannot reabsorb it all. The presence of glucose in the filtrate increases its osmolarity compared to that of the blood, and therefore less water is reabsorbed into the peritubular capillary network. The frequent urination and increased thirst experienced by untreated diabetics are due to the fact that water is not being reabsorbed.

We have seen that the filtrate that enters the proximal convoluted tubule is divided into two portions: components that are reabsorbed from the tubule into the blood and components that are not reabsorbed and continue to pass through the nephron to be further processed into urine:

Reabsorbed Filtrate Components	Nonreabsorbed Filtrate Components
Most water	Some water
Nutrients	Much nitrogenous waste
Required salts (ions)	Excess salts (ions)

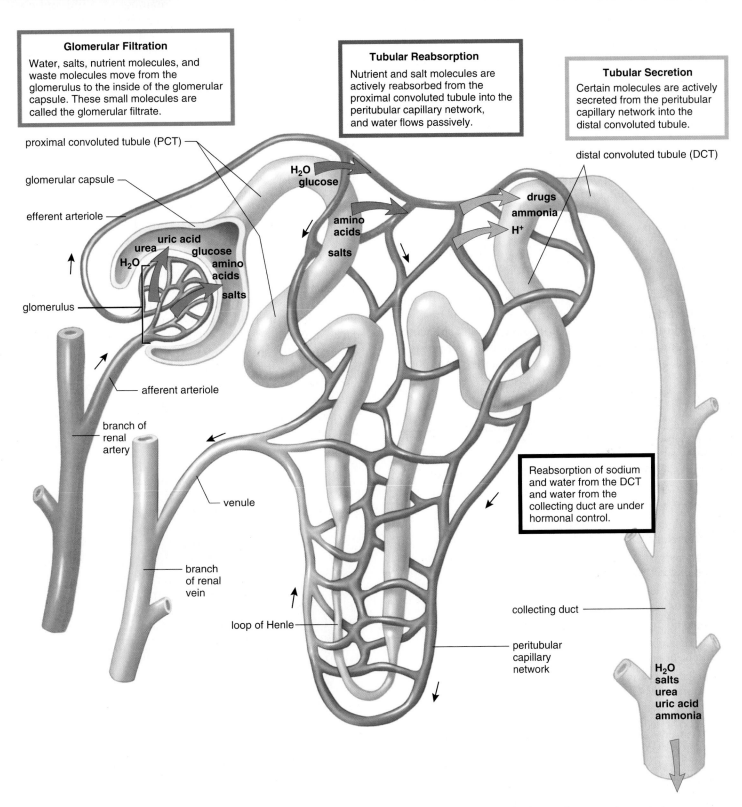

Glomerular Filtration
Water, salts, nutrient molecules, and waste molecules move from the glomerulus to the inside of the glomerular capsule. These small molecules are called the glomerular filtrate.

Tubular Reabsorption
Nutrient and salt molecules are actively reabsorbed from the proximal convoluted tubule into the peritubular capillary network, and water flows passively.

Tubular Secretion
Certain molecules are actively secreted from the peritubular capillary network into the distal convoluted tubule.

proximal convoluted tubule (PCT)

distal convoluted tubule (DCT)

glomerular capsule

efferent arteriole

H_2O
glucose

drugs
ammonia
H^+

uric acid
urea
H_2O
glucose
amino acids
salts

amino acids
salts

glomerulus

afferent arteriole

branch of renal artery

venule

branch of renal vein

Reabsorption of sodium and water from the DCT and water from the collecting duct are under hormonal control.

loop of Henle

collecting duct

peritubular capillary network

H_2O
salts
urea
uric acid
ammonia

Figure 16.5 Steps in urine formation. *Top:* The three main steps in urine formation are described in boxes that are color-coded to arrows showing the movement of molecules into or out of the nephron at specific locations. In the end, urine is composed of the substances within the collecting duct (see gray arrow, lower right).

TABLE 16.1 Reabsorption from Nephrons			
Substance	Amount Filtered (per day)	Amount Excreted (per day)	Reabsorption (%)
Water, (L)	180	1.8	99.0
Sodium, (g)	630	3.2	99.5
Glucose, (g)	180	0.0	100.0
Urea, (g)	54	30.0	44.0
L = Liter; g = grams			

The substances that are not reabsorbed become the tubular fluid, which enters the loop of Henle.

Tubular Secretion

Tubular secretion is a second way by which substances are removed from blood and actively transported into the tubular fluid. Hydrogen ions, potassium ions, creatinine, and drugs such as penicillin are some of the substances that are moved by active transport from the blood into the distal convoluted tubule. In the end, urine contains (1) substances that have undergone glomerular filtration but have not been reabsorbed, and (2) substances that have undergone tubular secretion.

Content CHECK-UP!

4. Select the correct order of the blood supply for a nephron:
   a. afferent arteriole → efferent arteriole → glomerulus → peritubular capillary
   b. efferent arteriole → glomerulus → afferent arteriole → peritubular capillary
   c. afferent arteriole → glomerulus → efferent arteriole → peritubular capillary
   d. glomerulus → afferent arteriole → efferent arteriole → peritubular capillary

5. Select the correct order of the parts of a nephron:
   a. glomerular capsule → proximal convoluted tubule → loop of Henle → distal convoluted tubule
   b. glomerular capsule → proximal convoluted tubule → distal convoluted tubule → loop of Henle
   c. proximal convoluted tubule → glomerular capsule → loop of Henle → distal convoluted tubule
   d. proximal convoluted tubule → loop of Henle → distal convoluted tubule → glomerular capsule

6. The tubular fluid in the glomerular capsule contains all of the following except:
   a. glucose.
   b. water.
   c. sodium chloride.
   d. red blood cells.

16.3 Regulatory Functions of the Kidneys

The kidneys are involved in maintaining the fluid and electrolyte balance, and also the acid-base balance, of the blood. If the kidneys fail to carry out these vital functions, either hemodialysis or a kidney transplant is needed.

Fluid and Electrolyte Balance

The average adult male body is about 60% water by weight. The average adult female body is only about 50% water by weight because females generally have more subcutaneous adipose tissue, which contains less water. About two-thirds of this water is inside the cells (called intracellular fluid), and the rest is largely distributed in the plasma, tissue fluid, and lymph (called extracellular fluid). Water is also present in such fluids as cerebrospinal fluid and synovial fluid; in Figure 16.6, these fluids are referred to as "other" fluids.

For body fluids to be normal, it is necessary for the body to be in fluid balance. The total water intake should equal the total water loss. Table 16.2 shows how water enters the body—namely, in liquids we drink, in foods we eat, and as a by-product of metabolism. We drink water when the osmolarity of the blood rises as determined by the hypothalamus. Table 16.2 also shows how water exits the body—namely, in urine, sweat, exhaled air, and feces. Similar to the gain and loss of water, the body also gains and loses electrolytes. Despite these changes, the kidneys keep the fluid and electrolyte balance of the blood within normal limits. In this way, they also maintain the blood volume and blood pressure.

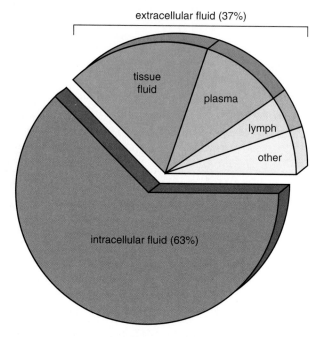

Figure 16.6 Location of fluids in the body. Most of the body's water is inside cells (intracellular fluid), and only about one-third is located outside cells (extracellular fluid).

TABLE 16.2 Fluid Balance

Water Input	Average ml/day and % of Total	Water Output	Average ml/day and % of Total
In liquids	1500; 60%	In urine	1,300; 52%
In moist food	750; 30%	In sweat	650; 26%
From metabolism	250; 10%	In exhaled air	450; 18%
		In feces	100; 4%
	Total 2,500; 100%		Total 2,500; 100%

Reabsorption of Water

Because of the process of osmosis (see Chapter 3, page 50), the reabsorption of salt (NaCl) automatically leads to the reabsorption of water until the osmolarity is the same on both sides of a plasma membrane. Most of the salt, and therefore water, present in the filtrate are reabsorbed across the plasma membranes of the cells lining the proximal convoluted tubule. But the amount of salt and water reabsorbed is not sufficient to result in a hypertonic urine—one in which the osmolarity is higher than that of blood. How is it, then, that humans produce a hypertonic urine? We now know that the excretion of a hypertonic urine is dependent upon the reabsorption of water from the loop of Henle and the collecting duct.

Loop of Henle and Collecting Duct A loop of Henle has a descending limb and an ascending limb. A long loop of Henle penetrates deep into the renal medulla. In the ascending limb, salt (NaCl) passively diffuses out of the lower portion and is actively transported out of the upper portion into the tissue of the outer renal medulla (Fig. 16.7). Less and less salt is available for active transport as fluid moves up the thick portion of the ascending limb. Therefore, the concentration of salt is greater in the inner medulla than in the outer medulla. (It is important to realize that water cannot leave the thick segment of the ascending limb because the ascending limb is impermeable to water.)

The large arrow to the side in Figure 16.7 indicates that the *lowest* portion of the inner medulla has the highest concentration of solutes. You can see that this is due not only to the presence of salt, but also to the presence of urea. Urea is believed to leak from the lower portion of the collecting duct, and it is this molecule that contributes to the high solute concentration of the lowest portion of the inner medulla.

Because of the osmotic gradient within the renal medulla, water leaves the descending limb along its entire length. There is a higher concentration of water at the top of the descending limb, and lesser amount of solute in the medulla is needed to pull water out. The remaining fluid within the descending limb encounters an even greater osmotic concentration of solute as it moves along; therefore, water continues to leave the descending limb from the top to the bottom. Such a mechanism is called a countercurrent mechanism.

At the top of the ascending limb, any remaining water enters the distal

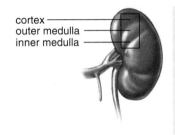
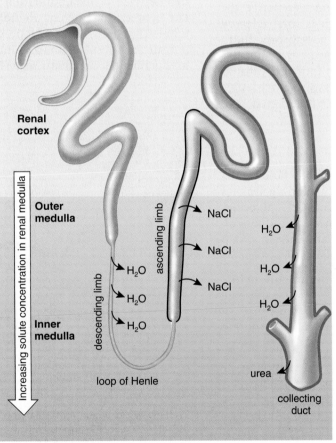

Figure 16.7 Reabsorption of water at the loop of Henle and the collecting duct. A hypertonic environment in the tissues of the medulla of a kidney draws water out of the descending limb and the collecting duct. This water is returned to the cardiovascular system. (The thick black line means the ascending limb is impermeable to water. The thick red line shows portions of the tubule where water permeability is hormone-dependent.)

convoluted tubule. Surprisingly, the fluid inside the nephron is still not hypertonic—the net effect of reabsorption of salt and water so far is the production of a fluid that has the same tonicity as blood plasma. However, the collecting duct also encounters the same osmotic gradient as did the descending limb of the loop of the nephron (Fig. 16.7). Therefore, water diffuses out of the collecting duct into the renal medulla, and the urine within the collecting duct becomes hypertonic to blood plasma.

Antidiuretic Hormone (ADH) ADH produced by the hypothalamus, and released by the posterior lobe of the pituitary, plays a role in water reabsorption at the distal convoluted tubule collecting duct. In order to understand the action of this hormone, consider its name. *Diuresis* means flow of urine, and *antidiuresis* means against a flow of urine. In practical terms, if an individual does not drink much water on a given day, the hypothalamus will detect an increase in the osmotic pressure of the blood; i.e., the solutes in the blood will become more concentrated. (As an analogy, imagine leaving a glass of salt water on a windowsill for a week or more. The salt will become more concentrated over time because water is lost by evaporation. In the same way, blood solutes become more concentrated if the water lost daily isn't replaced by drinking.) In response to increased blood osmolarity, the hypothalamus will produce ADH, which is then released from the posterior pituitary. More water is reabsorbed into the blood, blood volume and blood pressure rise, and less urine is formed. On the other hand, if an individual drinks a large amount of water and does not perspire much, ADH is not released. In that case, more water is excreted, and more urine forms.

Reabsorption of Electrolytes

As previously discussed, the osmolarity of body fluids is dependent upon the concentration of particular electrolytes within the fluids. **Electrolytes** are compounds and molecules that are able to ionize and, thus, carry an electrical current. The kidneys regulate electrolyte excretion and therefore help control blood composition.

The Electrolytes The most common electrolytes in the plasma are sodium (Na^+), potassium (K^+), and bicarbonate ion (HCO_3^-). Na^+ and K^+ are termed *cations* because they are positively charged, and HCO_3^- is termed an *anion* because it is negatively charged.

Sodium The movement of Na^+ across an axon membrane, you will recall, is necessary to the formation of a nerve action potential signal and also to muscle contraction. The concentration of Na^+ in the blood is also the best indicator of the blood's osmolarity.

Potassium The movement of K^+ across an axon membrane is also necessary to the formation of a nerve action potential and muscle contraction. Abnormally low K^+ concentrations in the blood, as might occur if diuretics are abused, can lead to cardiac arrest.

Bicarbonate Ion HCO_3^- is the form in which carbon dioxide is carried in the blood. The bicarbonate ion has the very important function of helping maintain the pH of the blood, as will be discussed later in this section.

Other Ions The plasma contains many other ions. For example, calcium ions (Ca^{2+}) and phosphate ions (HPO_4^{2-}) are important to bone formation and cellular metabolism. Their absorption from the intestine and excretion by the kidneys is regulated by hormones, as discussed in Chapter 10.

The Kidneys More than 99% of sodium (Na^+) filtered at the glomerulus is returned to the blood. Most sodium (67%) is reabsorbed at the proximal convoluted tubule, and a sizable amount (25%) is actively transported from the tubule by the ascending limb of the loop of Henle. The rest is reabsorbed from the distal convoluted tubule and collecting duct.

Aldosterone Hormones control the reabsorption of sodium and water at the distal convoluted tubule. **Aldosterone,** a hormone secreted by the adrenal cortex, promotes the excretion of potassium ions (K^+) and the reabsorption of sodium ions (Na^+). Water is subsequently reabsorbed by osmosis.

The release of aldosterone is set in motion by the kidneys themselves. The **juxtaglomerular apparatus** is a region of contact between the afferent arteriole and the distal convoluted tubule (Fig. 16.8). When blood volume, and therefore blood pressure, is not sufficient to promote glomerular filtration, the juxtaglomerular apparatus secretes renin. **Renin** is an enzyme

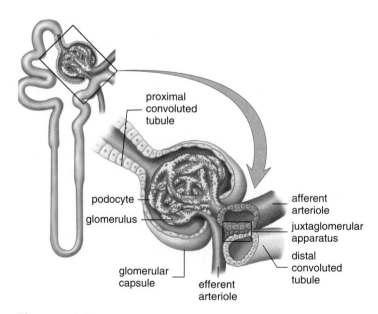

Figure 16.8 Juxtaglomerular apparatus. This drawing shows that the afferent arteriole and the distal convoluted tubule usually lie next to each other. The juxtaglomerular apparatus occurs where they touch. The juxtaglomerular apparatus secretes renin, a substance that leads to the release of aldosterone by the adrenal cortex. Reabsorption of sodium ions followed by water then occurs. Therefore, blood volume and blood pressure increase.

that changes angiotensinogen (a large plasma protein produced by the liver) into angiotensin I. Later, angiotensin I is converted to angiotensin II, a powerful vasoconstrictor that also stimulates the adrenal cortex to release aldosterone. The reabsorption of sodium ions is followed by the reabsorption of water. Therefore, blood volume and blood pressure increase.

Atrial Natriuretic Hormone (ANH) ANH is a hormone secreted by the atria of the heart when cardiac cells are stretched due to increased blood volume. ANH inhibits the secretion of renin by the juxtaglomerular apparatus and the secretion of aldosterone by the adrenal cortex. Its effect, therefore, is to promote the excretion of Na^+, called *natriuresis*. When Na^+ is excreted, so is water, and therefore blood volume and blood pressure decrease.

Diuretics

Diuretics are chemicals that increase the flow of urine. Drinking alcohol causes diuresis because it inhibits the secretion of ADH. The dehydration that follows is believed to contribute to the symptoms of a hangover. Caffeine is a diuretic because it increases the glomerular filtration rate and decreases the tubular reabsorption of Na^+. Diuretic drugs developed to counteract high blood pressure inhibit active transport of Na^+ at the loop of Henle or at the distal convoluted tubule into the blood. A decrease in water reabsorption and a decrease in blood volume follow. Subsequently, blood pressure decreases because cardiac output is diminished.

Acid-Base Balance

The pH scale, as discussed in Chapter 2, can be used to indicate the basicity (alkalinity) or the acidity of body fluids. A basic solution has a lower hydrogen ion concentration $[H^+]$ than the neutral pH of 7.0. An acidic solution has a greater $[H^+]$ than neutral pH. The normal pH for body fluids is about 7.4. This is the pH at which our proteins, such as cellular enzymes, function properly. If the blood pH rises above 7.4, a person is said to have **alkalosis,** and if the blood pH decreases below 7.4, a person is said to have **acidosis.** Alkalosis and acidosis are abnormal conditions that may need medical attention.

The foods we eat add basic or acidic substances to the blood, and so does metabolism. For example, cellular respiration adds carbon dioxide that combines with water to form carbonic acid, and fermentation adds lactic acid. The pH of body fluids stays at just about 7.4 via several mechanisms, primarily acid-base buffer systems, the respiratory center, and the kidneys.

Acid-Base Buffer Systems

The pH of the blood stays near 7.4 because the blood is buffered. A **buffer** is a chemical or a combination of chemicals that can take up excess hydrogen ions (H^+) or excess hydroxide ions (OH^-). One of the most important buffers in the blood is a combination of carbonic acid (H_2CO_3) and

bicarbonate ions (HCO_3^-). Carbonic acid is a weak acid that minimally dissociates and re-forms in the following manner:

$$H_2CO_3 \underset{\text{re-forms}}{\overset{\text{dissociates}}{\rightleftharpoons}} H^+ + HCO_3^-$$
$$\text{carbonic acid} \qquad \text{hydrogen ion} \qquad \text{bicarbonate ion}$$

When hydrogen ions (H^+) are added to blood, the following reaction occurs:

$$H^+ + HCO_3^- \rightarrow H_2CO_3$$

When hydroxide ions (OH^-) are added to blood, this reaction occurs:

$$OH^- + H_2CO_3 \rightarrow HCO_3^- + H_2O$$

These reactions temporarily prevent any significant change in blood pH. A blood buffer, however, can be overwhelmed unless some more permanent adjustment is made. The next adjustment to keep the pH of the blood constant occurs at pulmonary capillaries.

Respiratory Center

As discussed in Chapter 14, the respiratory center in the medulla oblongata increases the breathing rate if the hydrogen ion concentration of the blood rises. Increasing the breathing rate rids the body of hydrogen ions because the following reaction takes place in pulmonary capillaries:

$$H^+ + HCO_3^- \rightleftharpoons H_2CO_3 \rightleftharpoons H_2O + CO_2$$

In other words, when carbon dioxide is exhaled, hydrogen ions recombine with bicarbonate ions to form carbonic acid. Carbonic acid then dissociates to form the carbon dioxide that is eliminated from the body. Now the hydrogen ions are no longer free in solution, but tied up harmlessly with water. Increasing the respiratory rate provides the body with a rapid way to remove excess free hydrogen ions from the blood.

It is important to have the correct proportion of carbonic acid and bicarbonate ions in the blood. Breathing readjusts this proportion so that this particular acid-base buffer system can continue to absorb both H^+ and OH^- as needed.

The Kidneys

As powerful as the acid-base buffer and the respiratory center mechanisms are, only the kidneys can rid the body of a wide range of acidic and basic substances and otherwise adjust the pH. The kidneys are slower acting than the other two mechanisms, but they have a more powerful effect on pH. For the sake of simplicity, we can think of the kidneys as reabsorbing bicarbonate ions and excreting hydrogen ions as needed to maintain the normal pH of the blood (Fig. 16.9). If the blood is acidic, hydrogen ions are excreted and bicarbonate ions are reabsorbed. If the blood is basic, hydrogen ions are not excreted and

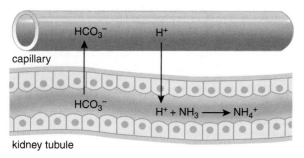

Figure 16.9 Acid-base balance. In the kidneys, bicarbonate ions (HCO_3^-) are reabsorbed and hydrogen ions (H^+) are excreted as needed to maintain the pH of the blood. Excess hydrogen ions are buffered, for example, by ammonia (NH_3), which becomes ammonium (NH_4^+). Ammonia is produced in tubule cells by the deamination of amino acids.

bicarbonate ions are not reabsorbed. Because the urine is usually acidic, it follows that an excess of hydrogen ions is usually excreted. Ammonia (NH_3) provides a means for buffering these hydrogen ions in urine. Ammonia (whose presence is quite obvious in the diaper pail or kitty litter box) is produced in kidney tubule cells by the deamination of amino acids. Ammonia is a basic molecule that diffuses easily through tubule cells into the urine. Once there, it combines with hydrogen ions to form ammonium ions (NH_4^+) by this reaction:

$$NH_3 + H^+ \rightarrow NH_4^+$$

Ammonium ions cannot diffuse back out of the kidney tubules and pass out of the body in the urine. Phosphate provides another means of buffering hydrogen ions in urine.

The importance of the kidneys' ultimate control over the pH of the blood cannot be overemphasized. As mentioned, the enzymes of cells cannot continue to function if the internal environment does not have near-normal pH.

Content CHECK-UP!

7. Which of the following statements about body water is incorrect?

  a. Males have a higher percentage of body water by weight than females.

  b. Most of the body's water is found in blood plasma.

  c. Water can be lost from the body in exhaled air.

  d. A portion of daily water intake comes from water produced during metabolism.

8. Which hormone is produced by the hypothalamus and released from the posterior pituitary?

  a. atrial natriuretic hormone (ANP)   c. antidiuretic hormone (ADH)

  b. aldosterone   d. angiotensin

9. Which hormone causes excretion of sodium and water from the body?

  a. atrial natriuretic hormone (ADH)   c. antidiuretic hormone (ADH)

  b. aldosterone   d. angiotensin

16.4 Problems with Kidney Function

The composition of normal urine is given in Table 16.3. Water accounts for almost all of the volume of urine (95%). The remaining 5% consists of electrolytes and various solutes, including nitrogenous end products and substances derived from drugs. Notice that urine is typically free of proteins and blood cells because they are not filtered at the glomerulus.

Urinalysis is an examination of the physical, chemical, and microscopic properties of the urine. A urinalysis is done to help determine the state of the body. As discussed in the Focus on Forensics on page 363, the composition of the urine changes if disease has altered body metabolism or if kidney function is abnormal. Abnormal substances in urine and abnormal quantities of normal constituents are both matters of concern.

Many types of illnesses, especially diabetes, hypertension, and inherited conditions, cause progressive renal disease and renal failure. Infections of the urinary tract are fairly common, particularly in females because the urethra is considerably shorter than that of the male. If the infection is localized in the urethra, it is called **urethritis.** If the infection invades the urinary bladder, it is called **cystitis.** Finally, if the kidneys are affected, the infection is called **pyelonephritis.**

Glomerular damage sometimes leads to blockage of the glomeruli so that glomerular filtration either does not occur or allows large substances to pass through. This is detected when a urinalysis is done. If the glomeruli are too permeable, albumin, white blood cells, or even red blood cells appear in the urine. A trace amount of protein in the urine is usually not a matter of concern.

When glomerular damage is so extensive that more than two-thirds of the nephrons are inoperative, urea and other waste substances accumulate in the blood. This condition is called **uremia.** Although nitrogenous wastes can cause serious damage, the retention of water and salts is of even greater concern. The latter causes edema, fluid accumulation in the body

TABLE 16.3 Composition of Urine		
Water		95%
Solids		5%
Organic nitrogenous wastes (per 1,500 ml of urine)		
Urea		30 g
Creatinine		1–2 g
Ammonia		1–2 g
Uric acid		1 g
Electrolytes		25 g
Positive	*Negative*	
Sodium (Na^+)	Chlorides (Cl^-)	
Potassium (K^+)	Sulfates (SO_4^{2-})	
Magnesium (Mg^{2+})	Phosphates (PO_4^{3-})	
Calcium (Ca^{2+})		

Urinalysis

Since ancient times, urinalysis has been used to diagnose disease. As early as 600 B.C., Hindu physicians in India noted that the urine of a diabetic was sweet to the taste. In diabetes mellitus, blood glucose is abnormally high, either because insulin-secreting cells have been destroyed or because cell receptors don't respond to the insulin that is present. Thus, the filtrate level of glucose is extremely high, and the proximal convoluted tubule can't reabsorb it all. Tasting urine to diagnose diabetes persisted through the 1800s (thankfully, modern techniques have made it obsolete). Similarly, the great Greek physician Hippocrates studied and wrote about urinalysis, producing perhaps the first written work about renal failure. Hippocrates noted that shaking the urine of renal failure patients produced frothy bubbles at the surface of the sample. Today, we know that the froth is a sign of *proteinuria*, or protein in the urine. Proteinuria indicates that the glomerulus is more permeable than normal, and is an early indicator of chronic renal failure.

Today, the use of urinalysis has expanded beyond medical applications to include forensic diagnosis of drug abuse. Screening for illegal drug use is now mandated by federal and state agencies as a condition of employment, and most private employers now require it as well. Urinalysis can be court ordered if drug abuse is involved in the commission of a crime. The National Collegiate Athletic Association requires all student athletes to undergo testing, as do many high school athletic programs.

Urinalysis is not used to screen for the drugs themselves, but for drug metabolites—the breakdown products of drugs that are consumed or injected. Once in the body, drugs are metabolized by the liver and filtered by the kidney. Thus, metabolites will be present in the urine of a drug abuser. Two types of techniques can be used for metabolite detection. The first, a screening exam, involves a test strip placed into freshly voided urine. The test strip contains monoclonal antibodies (see pages 292 and 294) specific for metabolites of street drugs. Strips can test for 12 or more different drugs at once, but most screen for 5 commonly abused drugs: marijuana, amphetamine, PCP, cocaine, and opiates such as heroin. Urine strip testing will give results within minutes, but certain legal, over-the-counter medications can give false-positive results. If a sample tests positive, it can be immediately sent for a sophisticated chemical analysis such as gas chromatography. Tracking long-term drug use may require using hair samples as well as urine samples because drug metabolites can be incorporated into the hair.

The urine specimen must be properly collected to avoid possible tampering by the individual being tested. Specimens can be altered by simple dilution with sink or toilet water, or by contaminating the sample with any number of additives. Bleach, drain cleaner, soft drinks, etc., can be used, as well as products specifically sold for the express purpose of "helping to beat a drug screen." Drug abusers may also attempt to substitute the urine of a "clean" individual for their own. Tampering may be prevented by simply requiring the presence of a witness at all times while the sample is being collected and stored. In addition, proper documentation must accompany any urine sample. Chain-of-custody forms record each step of the handling of a specimen, from collection to disposal. This provides proof of everything that happens to the specimen, and prevents the possibility of the specimen from being rejected as evidence in court proceedings.

The goal of any forensic urinalysis testing program should go beyond mere *detection* of illegal drug use. The more important aim of screening should be *intervention*. With proper treatment and counseling, drug abusers can overcome addiction and lead healthier and more productive lives.

tissues. Imbalance in the ionic composition of body fluids can lead to loss of consciousness and to heart failure.

Hemodialysis

Patients with renal failure can undergo **hemodialysis**, utilizing either an artificial kidney machine or *continuous ambulatory peritoneal dialysis* (CAPD). *Dialysis* is defined as the diffusion of dissolved molecules through a semipermeable natural or synthetic membrane having pore sizes that allow only small molecules to pass through. In an artificial kidney machine (Fig. 16.10), the patient's blood is passed through a membranous tube, which is in contact with a dialysis solution, or **dialysate.** Substances more concentrated in the blood diffuse into the dialysate, and substances more concentrated in the dialysate diffuse into the blood. The dialysate is continuously replaced to maintain favorable concentration gradients. In this way, the artificial kidney can be utilized either to extract substances from blood, including waste products or toxic chemicals and drugs, or to add substances to blood—for example, bicarbonate ions (HCO_3^-) if the blood is acidic. In the course of a three- to six-hour hemodialysis, from 50 to 250 grams of urea can be removed from a patient, which greatly exceeds the amount excreted by normal kidneys. Therefore, a patient needs to undergo treatment only about twice a week.

CAPD is so named because the peritoneal lining of the peritoneal (abdominal) cavity is the dialysis membrane. A fresh amount of dialysate is introduced directly into the abdominal cavity from a bag that is temporarily attached to a permanently implanted plastic tube. The dialysate flows into the peritoneal cavity by gravity. Waste and salt molecules pass

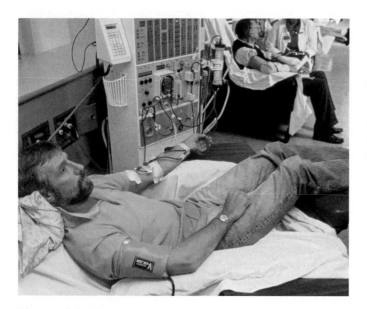
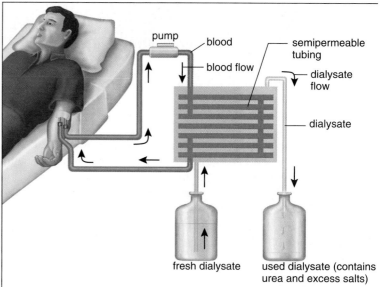

Figure 16.10 An artificial kidney machine. As the patient's blood is pumped through dialysis tubing, it is exposed to a dialysate (dialysis solution). Wastes exit from blood into the solution because of a preestablished concentration gradient. In this way, blood is not only cleansed, but its water-salt and acid-base balances can also be adjusted.

from the blood vessels in the abdominal wall into the dialysate before the fluid is collected four or eight hours later. The solution is drained into a bag from the abdominal cavity by gravity, and then it is discarded. One advantage of CAPD over an artificial kidney machine is that the individual can go about his or her normal activities during CAPD. However, CAPD is not appropriate for all patients with renal failure. Because the abdominal cavity has a permanent opening, the patient must be carefully trained and very diligent with sterile technique to avoid causing abdominal infection (peritonitis).

Replacing a Kidney

Patients with renal failure sometimes undergo a kidney transplant operation during which a functioning kidney from a donor is received. As with all organ transplants, there is the possibility of organ rejection. Receiving a kidney from a close relative has the highest chance of success. The current one-year survival rate is 97% if the kidney is received from a relative and 90% if it is received from a non-relative. Even with a successful surgery, the transplant patient must take antirejection medication for the rest of his or her life.

16.5 Effects of Aging

Urinary disorders are significant causes of illness and death among the elderly. Total renal function in an elderly individual may be only 50% of that of the young adult. With increasing age, the kidneys decrease in size and have significantly fewer nephrons. However, vascular changes may play a more significant role in declining renal efficiency than renal tissue loss. Microscopic examination shows many degenerate

glomeruli through which blood no longer flows and many other glomeruli that are completely destroyed.

Kidney stones occur more frequently with age, possibly as a result of improper diet, inadequate fluid intake, and kidney infections. Infections of the urethra, bladder, ureters, and kidneys increase in frequency among the elderly. Enlargement of the prostate occurs in males and, as is discussed in the Medical Focus on page 365, this can lead to urine retention and kidney disease. Cancer of the prostate and bladder are the most common cancers of the urogenital system.

The involuntary loss of urine, called **incontinence**, increases with age. The bladder of an elderly person has a capacity of less than half that of a young adult and often contains residual urine. Therefore, urination is more urgent and frequent.

16.6 Homeostasis

The illustration in Human Systems Work Together on page 366 tells how the urinary system works with the other systems of the body to help maintain homeostasis.

Recall that excretion means to rid the body of a metabolic waste. Using this definition, it is possible to classify three other organs in addition to the kidneys as excretory organs:

1. The sweat glands in the skin excrete perspiration, which is a solution of water, salt, and some urea. Although perspiration is a form of excretion, we perspire not so much to rid the body of wastes as to cool it. In times of kidney failure, urea is excreted by the sweat glands and forms a so-called uremic frost on the skin.
2. The liver breaks down hemoglobin and excretes bile pigments, which are derived from heme. Bile pigments are

Prostate Enlargement and Cancer

The prostate gland, which is part of the male reproductive system, surrounds the urethra at the point where the urethra leaves the urinary bladder (Fig. 16A). The prostate gland produces and adds a fluid to semen as semen passes through the urethra within the penis. At about age 50, the prostate gland often begins to enlarge, growing from the size of a walnut to that of a lime or even a lemon. This condition is called **benign prostatic hyperplasia (BPH)**. As it enlarges, the prostate squeezes the urethra, causing urine to back up—first into the bladder, then into the ureters, and finally, perhaps, into the kidneys.

The treatment for BPH can involve (1) taking a drug that is expected to shrink the prostate and/or improve urine flow, or (2) a more invasive procedure to reduce the size of the prostate. Prostate tissue can be destroyed by applying microwaves to a specific portion of the prostate. In many cases, however, a physician may decide that prostate tissue should be removed surgically. In some cases, rather than performing abdominal surgery, which requires an incision in the abdomen, the physician gains access to the prostate via the urethra. This operation, called transurethral resection of the prostate (TURP), requires careful consideration because one study found that the death rate during the 5 years following TURP is much higher than that following abdominal surgery.

Prostate enlargement is due to a prostate enzyme (5a-reductase) that acts on the male sex hormone testosterone, converting it into a substance that promotes prostate growth. That growth is fine during puberty, but continued growth in an adult is undesirable. Two substances, one a nutrient supplement and the other a prescription drug, interfere with the action of this enzyme. Saw palmetto, which is sold in tablet form as an over-the-counter nutrient supplement, is derived from a plant of the same name. This drug should not be taken unless the need for it is confirmed by a physician, but it is particularly effective during the early stages of prostate enlargement. Finasteride, a prescription drug, is a more powerful inhibitor of the enzyme, but patients complain of erectile dysfunction and loss of libido while on the drug.

Two other medications have a different mode of action. Nafarelin prevents the release of luteinizing hormone, which leads to testosterone production. When it is administered, approximately half of the patients report relief of urinary symptoms even after drug treatment is halted. However, again, the patients experience erectile dysfunction and other side effects, such as hot flashes. The drug terazosin, which is on the market for hypertension because it relaxes arterial walls, also relaxes muscle tissue in the prostate. Improved urine flow was experienced by 70% of the patients taking this drug. However, the drug has no effect on the prostate's overall size.

Many men are concerned that BPH may be associated with prostate cancer, but the two conditions are not necessarily related. BPH occurs in the inner zone of the prostate, while cancer tends to develop in the outer area. If prostate cancer is suspected, blood tests and a biopsy, in which a tiny sample of prostate tissue is surgically removed, will confirm the diagnosis.

Although prostate cancer is the second most common cancer in men, it is not a major killer. Typically, prostate cancer is so slow growing that the survival rate is about 98% if the condition is detected early.

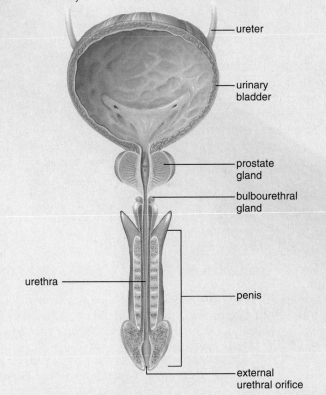

Figure 16A Longitudinal section of a male urethra leaving the bladder. Note the position of the prostate gland, which can enlarge to obstruct urine flow.

incorporated into bile, a substance stored in the gallbladder before it passes into the small intestine by way of ducts.

The yellow pigment found in urine, called urochrome, also is derived from the breakdown of heme, but this pigment is deposited in blood and is subsequently excreted by the kidneys. As you know, the liver also produces urea—our main nitrogenous end product—which is excreted by the kidneys.

3. The lungs excrete carbon dioxide. The process of exhalation not only removes carbon dioxide, but also

Human Systems Work Together

URINARY SYSTEM

Integumentary System

Kidneys compensate for water loss due to sweating; activate vitamin D precursor made by skin.

Skin helps regulate water loss; sweat glands carry on some excretion.

Skeletal System

Kidneys provide active vitamin D for Ca^{2+} absorption and help maintain blood level of Ca^{2+}, needed for bone growth and repair.

Bones provide support and protection.

Muscular System

Kidneys maintain blood levels of Na^+, K^+, and Ca^{2+}, which are needed for muscle innervation, and eliminate creatinine, a muscle waste.

Smooth muscular contraction assists voiding of urine; skeletal muscles support and help protect urinary organs.

Nervous System

Kidneys maintain blood levels of Na^+, K^+, and Ca^{2+}, which are needed for nerve conduction.

Brain controls nerves, which innervate muscles that permit urination.

Cardiovascular System

Kidneys keep blood values within normal limits so that transport of hormones continues.

ADH and aldosterone, and atrial natriuretic hormone regulate reabsorption of Na^+ by kidneys.

How the Urinary System works with other body systems.

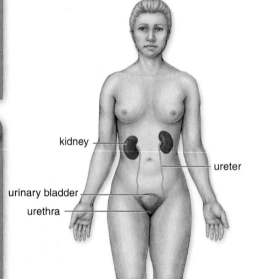

kidney

ureter

urinary bladder

urethra

Reproductive System

Semen is discharged through the urethra in males; kidneys excrete wastes and maintain electrolyte levels for mother and child.

Penis in males contains the urethra and performs urination; prostate enlargement hinders urination.

Cardiovascular System

Kidneys filter blood and excrete wastes; maintain blood volume, pressure, and pH; produce renin and erythropoietin.

Blood vessels deliver waste to be excreted; blood pressure aids kidney function; heart produces atrial natriuretic hormone.

Lymphatic System/Immunity

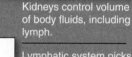

Kidneys control volume of body fluids, including lymph.

Lymphatic system picks up excess tissue fluid, helping to maintain blood pressure for kidneys to function; immune system protects against infections.

Respiratory System

Kidneys compensate for water lost through respiratory tract; work with lungs to maintain blood pH.

Lungs excrete carbon dioxide, provide oxygen, and convert angiotensin I to angiotensin II, leading to kidney regulation.

Digestive System

Kidneys convert vitamin D to active form needed for Ca^{2+} absorption; compensate for any water loss by digestive tract.

Liver synthesizes urea; digestive tract excretes bile pigments from liver and provides nutrients.

results in the loss of water. The air we exhale contains moisture, as demonstrated by breathing onto a cool mirror.

The kidneys are the primary organs of excretion. They excrete almost all of our nitrogenous wastes, namely urea, creatinine, and uric acid. The liver makes urea, and muscles make creatinine. Urea is the end product of protein metabolism, and creatinine is a breakdown product of creatine phosphate, a molecule that stores energy in muscles. Uric acid is produced by cells when they break down nucleotides. Nitrogenous wastes are carried by the cardiovascular system to the kidneys. In this way, the cardiovascular system and the kidneys work together to clear the blood of nitrogenous end products. The excretion of nitrogenous wastes may not be as critical as maintaining the water-salt and the acid-base balances, but it is still necessary because urea is a toxic substance.

The kidneys are primary organs of homeostasis because they maintain the water-salt (electrolyte) and the acid-base balance of the blood. If blood does not have the usual water-salt balance, blood volume and blood pressure are affected. Without adequate blood pressure, exchange across capillary walls cannot take place, nor is glomerular filtration possible in the kidneys themselves. The kidneys and the endocrine system work together to help maintain blood pressure. The production of renin by the kidneys and subsequently the renin-angiotensin-aldosterone sequence help ensure that the sodium (Na^+)

concentration of the blood, and therefore osmolarity and blood pressure, stay normal. The lymphatic system assists the urinary system because it makes a significant contribution to blood pressure by picking up excess tissue fluid and returning it to the cardiovascular veins.

Aside from producing renin, the kidneys assist the endocrine system and also the cardiovascular system by producing erythropoietin. Erythropoietin stimulates red bone marrow to produce red blood cells. The kidneys assist the skeletal, nervous, and muscular systems by helping to regulate the amount of calcium ions (Ca^{2+}) in the blood. The kidneys convert vitamin D to its active form needed for Ca^{2+} absorption by the digestive tract, and they regulate the excretion of electrolytes, including Ca^{2+}. The kidneys also regulate the sodium (Na^+) and potassium (K^+) content of the blood. These ions are necessary to the contraction of the heart and other muscles in the body, and are also needed for nerve conduction.

We have already described how the kidneys work with the cardiovascular system and the respiratory system to maintain the acid-base balance (page 361). This is a critical function to prevent the occurrence of alkalosis, or acidosis, which are life-threatening conditions. This function must be performed by a machine when people undergo hemodialysis. So, while we tend to remember that the kidneys excrete urea, we must also keep in mind all the other functions of the kidneys that are absolutely essential to homeostasis.

SELECTED NEW TERMS

Basic Key Terms

acidosis (ăs′ĭ-dō′sĭs) p. 361

aldosterone (ăl-dŏs′tə-rōn) p. 360

alkalosis (ăl′kə-lō′sĭs), p. 361

antidiuretic hormone (ADH), (an′tĭ-dī′u-ret′ik hôr′mōn′), p. 360

atrial natriuretic peptide hormone (ANH), (a′tre-al na′tre-u-ret′ik pĕp′tid′ hôr′mon′), p. 361

buffer (bŭf′ər), p. 361

collecting duct (kuh-lek′ting dukt), p. 356

creatinine (kre-at′ĭ-nēn), p. 352

distal convoluted tubule (dis′tal kon′vo-lūt-ed tu′byūl), p. 356

electrolytes (ĭ-lĕk′trə-lit′), p. 360

erythropoietin (ĭ-rĭth′rō-poi-ē′tĭn), p. 352

excretion (ek-skre′shun), p. 352

glomerular capsule (glo-mĕr′yū-ler kap′sul), p. 355

glomerular filtrate (glo-mĕr′yū-ler fil′trāt), p. 356

glomerular filtration (glo-mĕr′yū-ler fil-tra′shun), p. 356

glomerulus (glo-mĕr′yū-lus), p. 354

gout (gout), p. 352

juxtaglomerular apparatus (juks″tuh-glo-mĕr′yū-ler ap″uh-ra′tus), p. 360

kidney (kid′ne), p. 352

loop of Henle (lōōp of Hĕn′lē), p. 356

micturition (mik″tu-rish′un), p. 354

nephron (nef′ron), p. 354

peritubular capillary network (per″ĭ-tu′byū-ler kap′ĭ-lār″e net′werk), p. 354

proximal convoluted tubule (prok′sī-mal kon′vo-lū-ted tu′byūl), p. 355

prostate gland (pros′tāt glănd), p. 353

renal cortex (re′nul kor′teks), p. 354

renal medulla (re′nul mĕ-du′luh), p. 354

renal pelvis (re′nul pel′vis), p. 354

renin (rĕ′nĭn), p. 360

tubular reabsorption (tu′byū-ler re″ab-sorp′shun), p. 356

tubular secretion (tu′byū-ler sĕ-kre′shun), p. 358

urea (yoō-rē′ə), p. 352

ureters (yū-re′terz), p. 353

urethra (yū-re′thruh), p. 353

uric acid (yū-rik as′id), p. 352

urinary bladder (yūr′in-ār-e blad′er), p. 353

Clinical Key Terms

benign prostatic hyperplasia (bĭ-nīn′ prahs-tat′ik hi″per-pla′ze-uh), p. 365

cystitis (sis′ti′tis), p. 362

dialysate (dī-al′ĭ-sĭs), p. 363

diuretic (di-yū-ret′ik), p. 361

floating kidney (flōt′ing kid′ne), p. 352

hemodialysis (he-mo-di-al′ĭ-sis), p. 363

incontinence (in-con′tin-ents), p. 364

pyelonephritis (pi-lo-nef-ri′tis), p. 362

uremia (yū-re′me-uh), p. 362

urethritis (yū-re-thri′tis), p. 362

urinalysis (yū-rĭ-nal′ĭ-sis), p. 362

SUMMARY

16.1 Urinary System

A. The kidneys excrete nitrogenous wastes, including urea, uric acid, and creatinine. They maintain the normal water-salt balance, blood pressure, and acid-base balance of the blood, as well as influencing the secretion of certain hormones.

B. The kidneys produce urine, which is conducted by the ureters to the bladder, where it is stored before being released by way of the urethra.

16.2 Anatomy of the Kidney and Excretion

A. Macroscopically, the kidneys are divided into the renal cortex, renal medulla, and renal pelvis. Microscopically, they contain the nephrons.

B. Each nephron has its own blood supply; the afferent arteriole approaches the glomerular capsule and divides to become the glomerulus, a knot of capillaries. The efferent arteriole leaves the capsule and immediately branches into the peritubular capillary network.

C. Each region of the nephron is anatomically suited to its task in urine formation. The spaces between the podocytes of the glomerular capsule allow small molecules to enter the capsule from the glomerulus. The cuboidal epithelial cells of the proximal convoluted tubule have many mitochondria and microvilli to carry out active transport (following passive transport) from the tubule to the blood. In contrast, the cuboidal epithelial cells of the distal convoluted tubule have numerous mitochondria but lack microvilli. They carry out active transport from the blood to the tubule.

D. The steps in urine formation are glomerular filtration, tubular reabsorption, and tubular secretion.

16.3 Regulatory Functions of the Kidneys

A. The kidneys regulate the fluid and electrolyte balance of the body. Water is reabsorbed from certain parts of the tubule, and the loop of Henle establishes an osmotic gradient that draws water from the descending loop of the nephron and also from the collecting duct. The permeability of the collecting duct is under the control of the hormone ADH. The reabsorption of salt increases blood volume and pressure because more water is also reabsorbed. Two other hormones, aldosterone and ANH, control the kidneys' reabsorption of sodium (Na^+) and water.

B. The kidneys regulate the acid-base balance of the blood. Before the work of the kidneys begins, the acid-base buffer systems of the blood have functioned to keep the pH temporarily under control; also, the respiratory center has regulated the breathing rate to control the excretion of carbon dioxide at pulmonary capillaries. The kidneys largely function by excreting hydrogen ions and reabsorbing bicarbonate ions as needed.

16.4 Problems with Kidney Function

Various types of problems, including repeated urinary infections, can lead to renal failure, which necessitates receiving a kidney from a donor or undergoing hemodialysis by utilizing a kidney machine or CAPD.

16.5 Effects of Aging

Kidney function declines with age. Also, kidney stones, infections, and urination problems are more common.

16.6 Homeostasis

The urinary system works with the other systems of the body to maintain homeostasis in the ways described in the Human Systems Work Together on page 366.

STUDY QUESTIONS

1. List and explain four functions of the urinary system. (p. 352)
2. Trace the path of urine, and describe the function of each organ mentioned. (pp. 352–353)
3. Explain how urination is controlled. (pp. 353–354)
4. Describe the detailed anatomy of a kidney. (p. 354)
5. Trace the path of blood about a nephron. (pp. 354–355)
6. Name the parts of a nephron, and tell how the structure of the convoluted tubules suits their respective functions. (pp. 355–356)
7. State and describe the three steps of urine formation. (pp. 356–358)
8. Where in particular are salt and water reabsorbed along the length of the nephron? Describe the contribution of the loop of the nephron. (pp. 358–360)
9. Name and describe the action of antidiuretic hormone (ADH), the reninaldosterone connection, and atrial natriuretic hormone (ANH). (pp. 360–361)
10. How do the kidneys maintain the pH of the blood within normal limits? (pp. 361–362)
11. Explain how the artificial kidney machine works. (p. 363)

LEARNING OBJECTIVE QUESTIONS

Fill in the blanks.

1. The lungs are organs of excretion because they rid the body of _____.

2. The capillary tuft inside the glomerular capsule is called the _____.

3. Urine leaves the bladder in the _____.

4. _____ is a substance that is found in the filtrate, is reabsorbed, and is present in urine.

5. Tubular secretion takes place at the _____, a portion of the nephron.

6. The primary nitrogenous end product of humans is _____.

7. _____ is a substance that is found in the filtrate, is not reabsorbed, and is concentrated in urine.

8. In addition to excreting nitrogenous wastes, the kidneys adjust the _____, _____, and _____ balance of the blood.

9. Reabsorption of water from the collecting duct is regulated by the hormone _____.

10. A _____ is a chemical that can combine with either hydrogen ions or hydroxide ions, depending on the pH of the solution.

11. The accumulation of uric acid crystals in a joint cavity produces a condition called _____.

12. Urine is carried from the kidneys to the urinary bladder by a pair of organs called _____.

13. The outer granulated layer of the kidney is the renal _____, whereas the inner striated layer is the renal _____.

MEDICAL TERMINOLOGY EXERCISE

Consult Appendix B for help in pronouncing and analyzing the meaning of the terms that follow.

1. hematuria (hem″uh-tu-re′uh)
2. oliguria (ol″ĭ-gu′re-uh)
3. polyuria (pol″e-yū′re-uh)
4. extracorporeal shock wave lithotripsy (ESWL) (eks″truh-kor-po′re-al lith″o-trip′se)
5. antidiuretic (an″tĭ-di″yū-ret′ik)
6. urethratresia (yū-re″thruh-tre′ze-uh)
7. cystopyelonephritis (sis″to-pi″e-lo-nĕ-fri′tis)
8. nocturia (nok-tu′re-uh)
9. glomerulonephritis (glo-mĕr″yū-lo-nĕ-fri′tis)
10. ureterovesicostomy (yū-re″ter-o-ves″ĭ-kos′to-me)

WEBSITE LINK

Visit this text's website at http://www.mhhe.com/maderap6 for additional quizzes, interactive learning exercises, and other study tools.

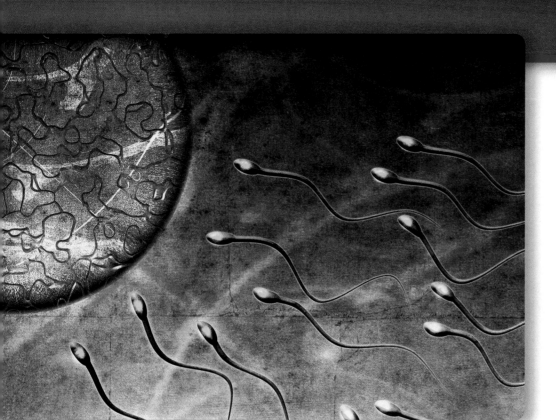

PART V

CHAPTER

17

The Reproductive System

Though millions of sperm cells are released by the male during intercourse, only a single sperm can fertilize the female ovum if development is to continue.

Chapter Outline & Learning Objectives

After you have studied this chapter, you should be able to:

17.1 Human Life Cycle

Unlike the other systems of the body, which are the same in males and females, the reproductive system is quite different in males and females. Development of a child into a sexually mature young adult is a sequence of events called **puberty.** The reproductive system does not begin to fully function until puberty is complete. Sexual maturity occurs between the ages of 11 and 13 in girls, and 14 and 16 in boys. At the completion of puberty, the individual is capable of producing children.

The reproductive organs (**genitals**) have the following functions:

1. Males produce sperm within testes, and females produce ova (eggs) within ovaries.
2. In males, sperm mature and are then transported in a duct system until they exit through the penis. In females, ova are transported in the uterine tubes to the uterus.
3. The male penis functions to deliver sperm to the female vagina, which functions to receive the sperm. The vagina also transports menstrual fluid to the exterior and is the birth canal.

4. The uterus of the female allows the fertilized egg to develop within her body. After birth, the female breast provides nourishment in the form of milk.
5. The testes and ovaries produce the sex hormones that maintain the testes and ovaries and have a profound effect on the body because they bring about masculinization and feminization of various features. In females, ovarian hormones allow a pregnancy to continue.

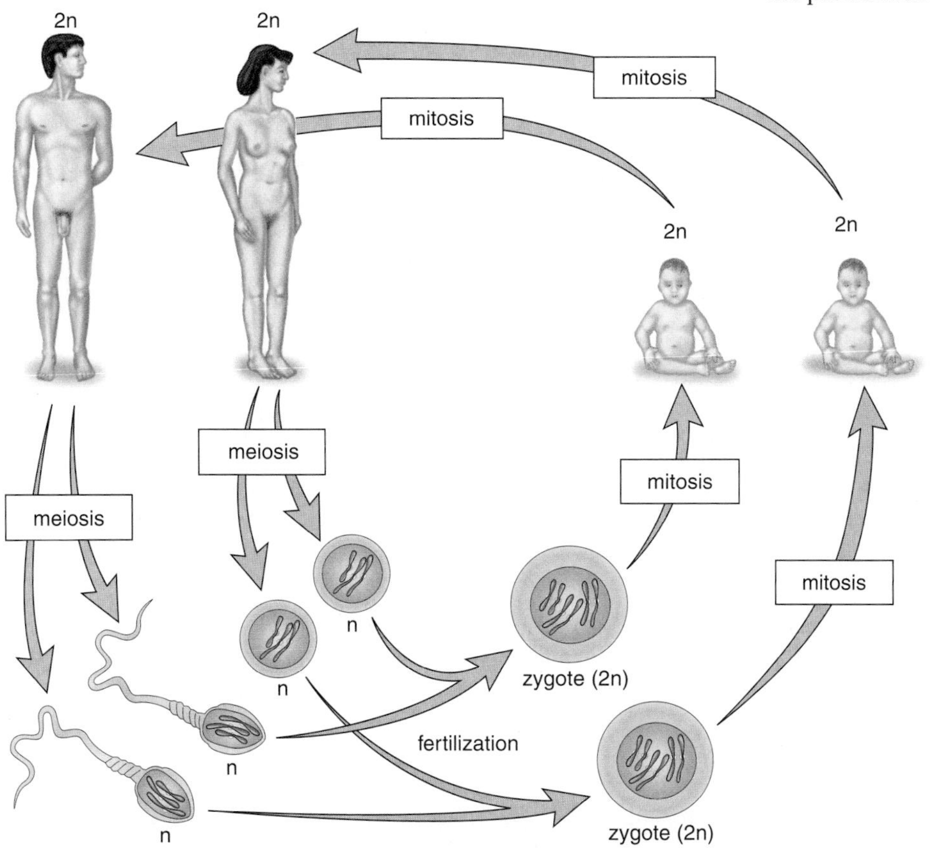

Meiosis

We studied the type of cell division called mitosis in Chapter 3 and learned that mitosis occurs during growth and repair of the body's tissues. As a result of mitosis, the chromosome number stays constant, and every cell in your body has 46 identical chromosomes. Mitosis is *duplication division*. (As an analogy, imagine the cell producing exact copies of itself during mitosis, much like a duplicating machine does with a page of notes.)

In addition to mitosis, the human life cycle (Fig. 17.1) includes a type of cell division called **meiosis,** which is *reduction division*. Meiosis only takes place in the testes of males during the production of sperm and in the ovaries of females during the production of ova. During meiosis, the chromosome number is reduced from the normal 46 chromosomes, called the **diploid** or **2n** number, down to 23 chromosomes, called the **haploid** or **n** number of chromosomes. This occurs in two successive divisions, called meiosis I and meiosis II (Fig. 17.2a). The stages of division in meiosis and mitosis are the same—prophase, metaphase, anaphase, and telophase.

To begin meiosis, the cell (called a primary cell) duplicates each chromatid (individual DNA strand), so there are 92 individual chromatids, which remain joined together (Fig. 17.2b, step 1). (This is the same process that occurs in body cells during the S phase of the cell cycle, which precedes mitosis; see page 52). Next, chromosomes of the same shape and size, called homologous pairs of chromosomes, line up side-by-side in prophase 1 (Fig. 17.2b, steps 2, 3). Each set of four chromatids is called a tetrad. As they line up, each of the chromatids in a tetrad can swap bits and pieces of DNA with the others (Fig 17.2b, steps 3, 4). The effect of this "gene-swap" (the technical term is *crossing over*) is that one's **gametes** (ova or sperm) are *not* exact copies of body cells. Instead, new and different gene combinations are created and passed down to one's children. As meiosis I is completed, two secondary cells are created, each with 46 chromatids, which are still joined together (Fig 17.2b,

Figure 17.1 The human life cycle has two types of cell divisions: mitosis, in which the chromosome number stays constant, and meiosis, in which the chromosome number is reduced. Meiosis only occurs in the testes of males during the production of sperm and in the ovaries of females during the production of ova (eggs). In human beings, the ova (egg) and sperm have 23 chromosomes each, called the n number. Following fertilization, the new individual has 46 chromosomes, called the 2n number.

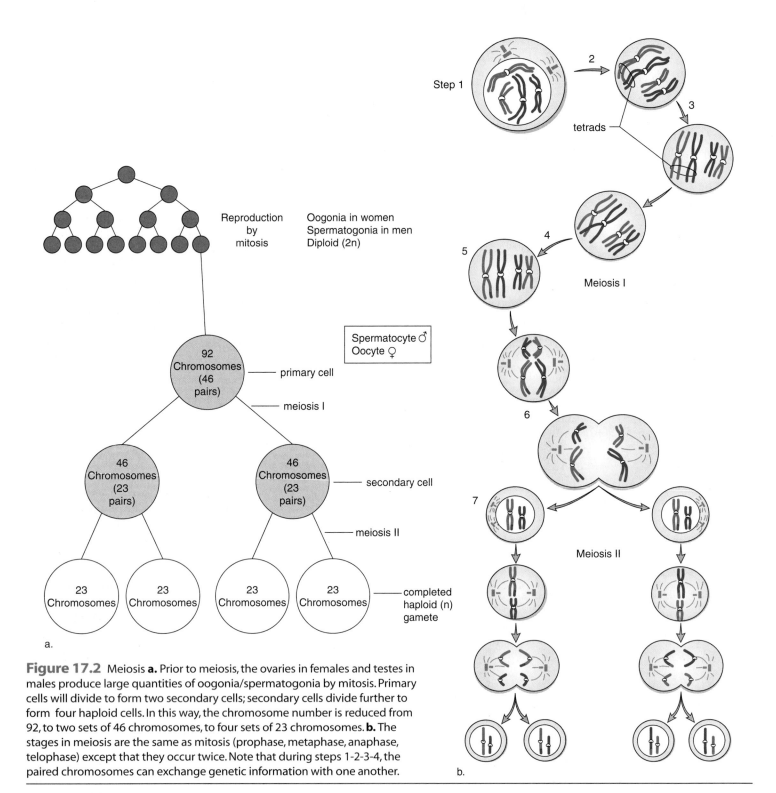

Figure 17.2 Meiosis **a.** Prior to meiosis, the ovaries in females and testes in males produce large quantities of oogonia/spermatogonia by mitosis. Primary cells will divide to form two secondary cells; secondary cells divide further to form four haploid cells. In this way, the chromosome number is reduced from 92, to two sets of 46 chromosomes, to four sets of 23 chromosomes. **b.** The stages in meiosis are the same as mitosis (prophase, metaphase, anaphase, telophase) except that they occur twice. Note that during steps 1-2-3-4, the paired chromosomes can exchange genetic information with one another.

steps 4, 5, 6). In meiosis II, only one member of each pair goes into the daughter cells (Fig. 17.2*b*, step 7). To summarize, each primary cell starts with 92 chromatids, which are divided during meiosis I into two cells with 46 chromatids, and then divided once more during meiosis II into four cells with 23 chromatids.

A **zygote,** the first cell of a new human being, forms following *fertilization,* when a sperm joins with an egg. Because the sperm has 23 chromosomes and the egg has 23 chromosomes, the zygote has 23 pairs of homologous chromosomes, or 46 chromosomes altogether. Without meiosis, the chromosome number in each generation of human beings would double, and cells would no longer be able to function. The zygote undergoes mitosis during development to produce the many cells of a newborn, and mitosis also occurs as a child becomes an adult.

17.2 Male Reproductive System

The male reproductive system includes the organs depicted in Figure 17.3. The *primary sex organs* of a male are the paired testes (sing, **testis**), which are suspended within the sacs of the **scrotum**. The testes are the primary sex organs because they produce sperm and the male sex hormones (**androgens**).

The other organs depicted in Figure 17.3 are the *accessory* (or secondary) *sex organs* of a male. Sperm produced by the testes are stored within the **epididymis** (pl., epididymides). Then they enter a **vas deferens** (pl., vasa deferentia), which transports them to an **ejaculatory duct**. The ejaculatory ducts enter the **urethra**. (The urethra in males is a part of both the urinary system and the reproductive system.) The urethra passes through the penis and transports sperm to outside the body.

At the time of **ejaculation**, sperm leave the penis in a fluid called **semen** (seminal fluid). The seminal vesicles, the prostate gland, and the bulbourethral glands (Cowper glands) add secretions to seminal fluid. The **seminal vesicles** lie lateral to the vas deferens, and their ducts join to form an ejaculatory duct. The **prostate gland** is a single, walnut-sized gland that surrounds the upper portion of the urethra just inferior to the

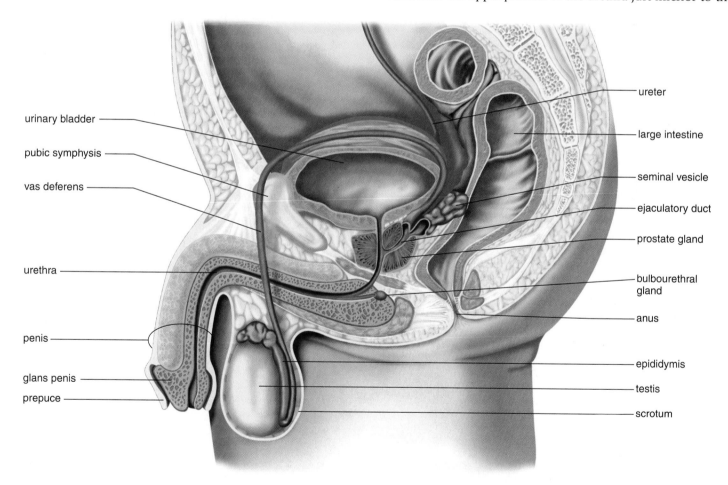

Figure 17.3 The male reproductive system. The testes produce sperm. The seminal vesicles, the prostate gland, and the bulbourethral glands provide a fluid medium for the sperm, which move from a testis to an epididymis to a vas deferens and through the ejaculatory duct to the urethra in the penis. The foreskin (prepuce) is removed when a penis is circumcised.

bladder. **Bulbourethral glands** are pea-sized organs that lie inferior to the prostate on either side of the urethra.

Each component of seminal fluid seems to have a particular function. Sperm are more viable in a basic solution, and seminal fluid, which is milky in appearance, has a slightly basic pH (about 7.5). Swimming sperm require energy and seminal fluid contains the sugar, fructose, which presumably serves as an energy source. Semen also contains **prostaglandins**, chemicals that cause the uterus in females to contract. Uterine contractions help propel the sperm toward the egg.

The Testes

The testes, which produce sperm and also the male sex hormones, lie outside the abdominal cavity of the male within the scrotum. The testes begin their development inside the abdominal cavity but descend into the scrotal sacs during the last two months of fetal development. If, by chance, the testes do not descend (a condition called **cryptorchidism**), and the male is not operated on to place the testes in the scrotum, sterility—the inability to produce offspring—usually follows. This is because the internal temperature of the body is too high to produce viable sperm. A subcutaneous muscle along with an adjoining muscle raise the scrotum during sexual excitement and when a higher temperature is need to warm the testes.

Anatomy of a Testis

A sagittal section of a testis shows that it is enclosed by a tough, fibrous capsule. The connective tissue of the capsule extends into the testis, forming septa that divide the testis into compartments called lobules. Each lobule contains one to three tightly coiled **seminiferous tubules** (Fig. 17.4a). Altogether,

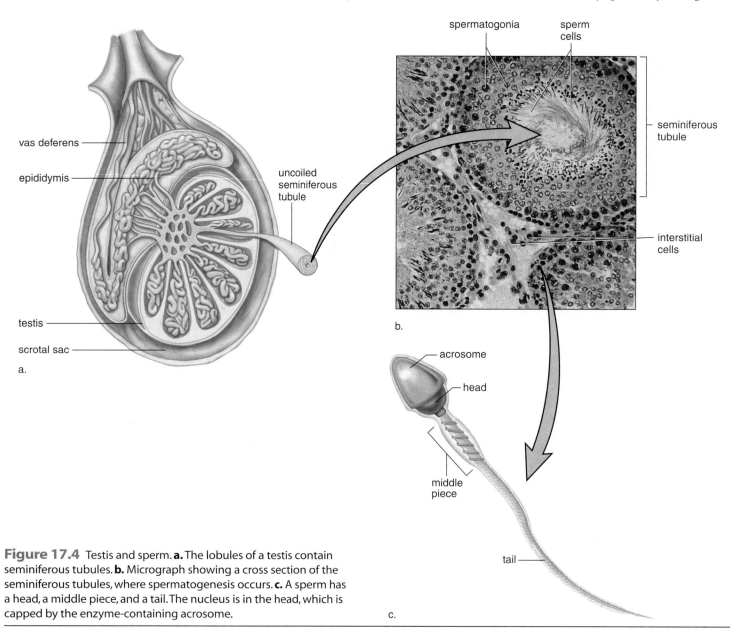

Figure 17.4 Testis and sperm. **a.** The lobules of a testis contain seminiferous tubules. **b.** Micrograph showing a cross section of the seminiferous tubules, where spermatogenesis occurs. **c.** A sperm has a head, a middle piece, and a tail. The nucleus is in the head, which is capped by the enzyme-containing acrosome.

these tubules have a combined length of approximately 250 m. A microscopic cross section of a seminiferous tubule reveals that it is packed with cells undergoing spermatogenesis (Fig. 17.4b), the production of sperm.

Delicate connective tissue surrounds the seminiferous tubules. Cells that secrete the male sex hormones, the androgens, are located here between the seminiferous tubules. Therefore, these endocrine cells are called **interstitial cells.** The most important of the androgens is **testosterone,** whose functions are discussed later in this section.

Testicular cancer, or cancer of the testes, is one type of cancer that can be detected by self-examination, as explained in the Medical Focus on page 391.

Spermatogenesis

Spermatogenesis, the production of sperm, includes the process of meiosis as the sperm form. Before puberty, the testes, including the seminiferous tubules, are small and nonfunctioning. At the time of puberty, the interstitial cells become larger and start producing androgens. Then, the seminiferous tubules also enlarge, and they start producing sperm.

The seminiferous tubules contain two types of cells: **germ cells,** which are involved in spermatogenesis, and **sustentacular (Sertoli) cells.** Sustentacular cells are large; they extend from the capsule to the lumen of the seminiferous tubule. The sustentacular cells support, nourish, and regulate the development of cells undergoing spermatogenesis.

The germ cells near the capsule are called **spermatogonia.** The spermatogonia divide, producing more cells by mitosis. Some of these cells remain as spermatogonia, which continue to divide. These provide a constant source of sperm cells throughout a normal man's lifetime. Other spermatogonia replicate their DNA and become **primary spermatocytes,** containing 92 chromatids (strands of DNA) held together in 46 pairs (Fig. 17.5). Primary spermatocytes are termed *diploid,* or 2n cells. The primary spermatocytes start the process of meiosis, which requires two divisions. Following meiosis I,

two **secondary spermatocytes** are formed, each containing 46 chromatids in 23 pairs. When meiosis II has been completed, there are four spermatids, cells that are termed *haploid* because they possess *half* the normal chromosome number. Spermatids then mature into sperm.

Mature **sperm,** or spermatozoa, have three distinct parts: a head, a middle piece, and a tail (see Fig. 17.4c). Mitochondria in the middle piece provide energy for the movement of the tail, which has the structure of a flagellum. The head contains a nucleus covered by a cap called the **acrosome,** which stores enzymes needed to penetrate the egg. Notice in Figure 17.4b, that the sperm are situated so that their tails project into the lumen of the seminiferous tubules.

When formed, the sperm are transported to the epididymis because the seminiferous tubules unite to form a complex network of channels that join, forming ducts. When the ducts join, an epididymis is formed.

The ejaculated semen of a normal human male contains several hundred million sperm, but only one sperm normally enters an egg. Sperm typically survive in the female reproductive tract for approximately 48 hours after sexual intercourse. However, under optimal conditions sperm can survive for 4–6 days after intercourse.

Male Internal Accessory Organs

Table 17.1 lists and Figure 17.6 depicts the internal accessory organs, as well as the other reproductive organs, of the male. Sperm are transported to the urethra by a series of ducts. Along the way, various glands add secretions to seminal fluid.

Epididymides

Each epididymis is a tightly coiled, threadlike tube that would stretch about 6 meters if uncoiled. A epididymis runs posteriorly down along a testis and becomes a vas deferens that ascends a testis medially.

The lining of an epididymis consists of pseudostratified columnar epithelium with long cilia. Sperm are stored in the

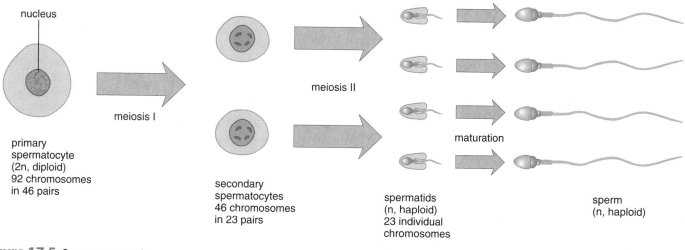

nucleus

primary
spermatocyte
(2n, diploid)
92 chromosomes
in 46 pairs

meiosis I

secondary
spermatocytes
46 chromosomes
in 23 pairs

meiosis II

spermatids
(n, haploid)
23 individual
chromosomes

maturation

sperm
(n, haploid)

Figure 17.5 Spermatogenesis.

TABLE 17.1 Male Internal Accessory Organs

Organ	Function
Epididymides	Ducts where sperm mature and some sperm are stored
Vas deferens	Transport and store sperm
Seminal vesicles	Contribute nutrients and fluid to semen
Ejaculatory ducts	Transport sperm
Prostate gland	Contributes basic fluid to semen
Urethra	Transports sperm
Bulbourethral glands	Contribute mucoid fluid to semen

structure called the **spermatic cord.** The spermatic cord consists of connective tissue and muscle fibers, and contains the testicular artery, vein, and nervous supply in addition to the vas deferens. As each spermatic cord ascends into the abdomen, it passes through the inguinal canal. This is the passageway by which the testis descended from the abdomen into the scrotum during fetal life. The inguinal canal remains a weak point in the abdominal wall. As such, it is frequently a site of **inguinal hernia.** (A **hernia** is an opening or separation of some part of the abdominal wall through which a portion of an internal organ, usually the intestine, protrudes. Hernias can also occur in the umbilical region, diaphragm or elsewhere in the abdomen.)

After each vas deferens enters the abdomen, it passes over the superior surface of the bladder to reach the posterior side of the urinary bladder. Each vas deferens widens at its ampulla, located at the posterior base of the urinary bladder, then joins with the duct of a seminal vesicle to form an ejaculatory duct. The ejaculatory ducts pass through the prostate gland to join the urethra.

epididymides, and the lining secretes a fluid that supports them. The wall of an epididymis contains a thin layer of smooth muscle. Peristaltic contractions move the sperm along as they mature. By the time the sperm leave the epididymides, they are capable of fertilizing an egg even though they do not "swim" until they enter the vagina.

Vas Deferens

Each vas deferens is a continuation of the epididymis. Like the epididymis, the vas deferens is lined by pseudostratified columnar epithelium that is ciliated at the testicular end of the tube. The vas deferens is contained within a protective

Seminal Vesicles

The two seminal vesicles lie lateral to each vas deferens on the posterior side of the bladder. They are coiled, membranous pouches about 5 cm long. The glandular lining of the seminal vesicles secretes an alkaline fluid that contains fructose and prostaglandins into an ejaculatory duct. The pH of the fluid helps modify the pH of seminal fluid; the fructose provides energy for sperm; and the prostaglandins promote muscular contractions of the female genital tract that help move sperm along.

Prostate

The prostate gland encircles the urethra just inferior to the bladder. The walnut-sized gland is about 4 cm across, 2 cm thick, and 3 cm in length. The fibrous connective tissue of its capsule extends inward to divide the gland into lobes, each of which contains about 40 to 50 tubules. The epithelium lining the tubules secretes a fluid that is thin, milky, and alkaline. In addition to adjusting the pH of seminal fluid, prostatic fluid enhances the motility of sperm. The secretion of the prostate gland enters the urethra when the smooth muscle in its capsular wall contracts.

As discussed in the Medical Focus on page 365, the prostate gland is a frequent site for cancer.

Bulbourethral Glands

The bulbourethral glands (Cowper glands) are two small glands about the size of peas. They are located inferior to the prostate gland and enclosed by fibers of the external urethral sphincter. These glands also contain many tubules that secrete a mucuslike fluid.

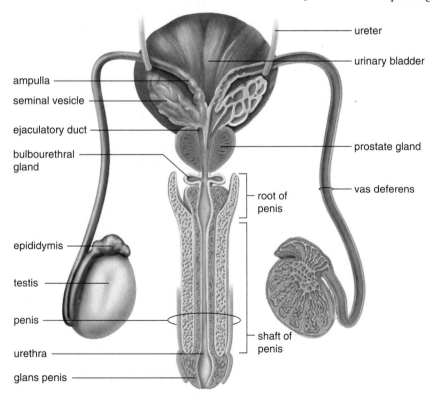

Figure 17.6 Male reproductive system, posterior view. This view shows the duct system that transports sperm from each testis to the urethra, which continues in the penis.

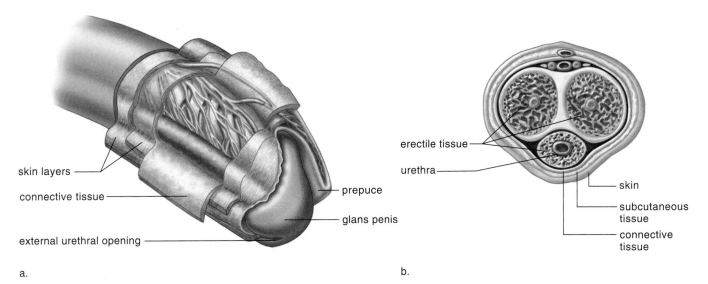

skin layers

connective tissue

external urethral opening

prepuce

glans penis

erectile tissue

urethra

skin

subcutaneous tissue

connective tissue

a.

b.

Figure 17.7 Penis anatomy. **a.** Beneath the skin and the connective tissue lies the urethra surrounded by erectile tissue. This tissue expands to form the glans penis, which in uncircumcised males is partially covered by the foreskin (prepuce). **b.** Two other columns of erectile tissue in the penis are located posteriorly.

This fluid lubricates the end of the penis preparatory to sexual intercourse.

Male Sexual Response

The **external genitals** are the sex organs that can be easily observed because they are located outside the body. The **penis** and the scrotum are the external genitals of the male. The penis is the male organ of sexual intercourse by which sperm are introduced into the female reproductive tract.

The penis has an internal root and an external shaft (see Fig. 17.6). At the glans penis, the skin folds back on itself to form the *prepuce,* or foreskin (Fig. 17.7a). This is the structure that is removed in the surgical procedure called **circumcision.** Internally, the penis contains three cylindrical bodies of erectile tissue; the urethra passes through one of them. These three columns are supported by fibrous connective tissue, and the whole is covered with a thin, loose skin (Fig. 17.7b).

The erectile tissues contain distensible blood spaces. During sexual arousal, autonomic nerves release nitric oxide, NO. This stimulus leads to the production of cGMP (cyclic guanosine monophosphate), causing the smooth muscle walls of incoming arteries to relax and the erectile tissue to fill with blood. The veins that take blood away from the penis are compressed, and the penis becomes erect. **Erectile dysfunction** (formerly called impotency) exists when the erectile tissue doesn't expand enough to compress the veins. Medications for treatment of erectile dysfunction inhibit an enzyme that breaks down cGMP, ensuring that a full erection will take place. However, individuals who take these medications may experience vision problems because the same enzyme occurs in the retina.

Orgasm (climax) in males is marked by ejaculation, which has two phases: emission and expulsion. During emission, sperm enter the urethra from each ejaculatory duct, and the prostate, seminal vesicles, and bulbourethral glands contribute secretions to the seminal fluid. Once seminal fluid is in the urethra, rhythmic muscle contractions cause seminal fluid to be expelled from the penis in spurts. During ejaculation, a sphincter closes off the bladder so that no urine enters the urethra. (Notice that the urethra carries either urine or semen at different times.)

Male orgasm includes expulsion of seminal fluid from the penis but also the physiological and psychological sensations that occur at the climax of sexual stimulation. The psychological sensation of pleasure is centered in the brain, but the physiological reactions involve the genital organs and associated muscles, as well as the entire body. Marked muscular tension is followed by contraction and relaxation.

Following ejaculation and/or loss of sexual arousal, the penis returns to its normal flaccid state. After ejaculation, a male typically experiences a period of time, called the refractory period, during which stimulation does not bring about an erection. The length of the refractory period increases with age.

There may be in excess of 400 million sperm in the 3.5 ml of semen expelled during ejaculation. The sperm count can be much lower than this, however, and fertilization of the egg by a sperm can still take place.

Regulation of Male Hormone Levels

At the time of puberty, the sex organs mature, and then changes occur in the physique of males. The cause of puberty is related to the level of sex hormones in the body, as regulated by the negative feedback system described in Figure 17.8. We now know that this feedback system functions long before puberty, but the level of hormones is low because the hypothalamus is supersensitive to feedback control. At the start of puberty, the hypothalamus becomes less sensitive to feedback control and begins to increase its production of gonadotropin-releasing hormone (GnRH), which stimulates the anterior pituitary to produce the gonadotropic hormones.

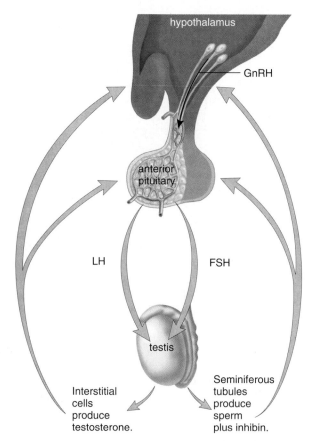

Figure 17.8 Negative feedback. Regulation of testosterone secretion involves negative feedback (reverse arrows) by testosterone on GnRH and LH. Regulation of sperm production involves negative feedback by inhibin on GnRH and FSH.

Two gonadotropic hormones, **FSH (follicle-stimulating hormone)** and **LH (luteinizing hormone),** are named for their function in females but exist in both sexes, stimulating the appropriate organs in each. FSH promotes spermatogenesis in the seminiferous tubules, and LH promotes androgen (e.g., testosterone) production in the interstitial cells. LH in males is also called interstitial cell-stimulating hormone (ICSH).

Negative Feedback Mechanisms

As mentioned, the hypothalamus, anterior pituitary, and testes are involved in a negative feedback system. The system maintains testosterone production at a fairly constant level. When the amount of testosterone in the blood rises to a certain level, it causes the hypothalamus and anterior pituitary to decrease their respective secretion of GnRH and LH. As the level of testosterone begins to fall, the hypothalamus increases its secretion of GnRH, and the anterior pituitary increases its secretion of LH. Thus, stimulation of the interstitial cells occurs. Only minor fluctuations of the testosterone level occur in the male, and the feedback mechanism in this case acts to maintain testosterone at a normal level.

A similar feedback mechanism maintains the continuous production of sperm. The sustentacular cells in the wall of the seminiferous tubules produce a hormone called **inhibin,** which blocks GnRH and FSH secretion when appropriate.

Testosterone

The male sex hormone, testosterone, has many functions. It is essential for normal development and function of the sex organs in males. For example, greatly increased testosterone secretion at the time of puberty stimulates maturation of the penis and the testes.

Secondary Sex Characteristics Testosterone also brings about and maintains the male **secondary sex characteristics,** which develop at the time of puberty and visibly distinguish males from females. These characteristics include a pattern of male hair growth, activity of cutaneous glands, deeper pitch to the voice, and muscle strength.

At puberty, males experience growth of a beard, axillary (underarm) hair, and pubic hair. In males, pubic hair tapers toward the navel. A side effect of testosterone activity is baldness. Genes for baldness are probably inherited by both sexes, but baldness is seen more often in males because of the presence of testosterone. This makes baldness a sex-influenced trait.

Testosterone also causes oil and sweat glands in the skin to secrete, thereby contributing to acne and body odor. The larynx and vocal cords enlarge, causing the voice to change. The "Adam's apple" is a part of the larynx, and it is usually more prominent in males than in females.

Testosterone is responsible for the greater muscular strength of males, which is why some athletes take a supplemental anabolic steroid, which is either testosterone or a related chemical. The disadvantages of anabolic steroid use are discussed in a Medical Focus in Chapter 10, page 220.

Content **CHECK-UP!**

4. In males, which structure lies alongside the testis and stores sperm?
   a. ejaculatory duct   c. epididymis
   b. scrotum   d. seminal vesicle

5. Which of the following structures produces a basic solution containing fructose that is added to the semen?
   a. seminal vesicle   c. bulbourethral gland
   b. prostate gland   d. glans penis

6. Identify, in order, the structures a sperm cell must pass through as it leaves the testis:
   a. epididymis → vas deferens → ejaculatory duct → urethra → penis
   b. vas deferens → epididymis → ejaculatory duct → urethra → penis
   c. ejaculatory duct → vas deferens → epididymis → urethra → penis
   d. epididymis → vas deferens → urethra → ejaculatory duct → penis

17.3 Female Reproductive System

The female reproductive system includes the organs depicted in Figure 17.9. The primary sex organs of a female are the paired **ovaries** that lie in shallow depressions, one on each side of the upper pelvic cavity. The ovaries are the primary sex organs because they produce **ova** and the female sex hormones, estrogen and progesterone.

The other organs depicted in Figure 17.9 are the accessory (or secondary) sex organs of a female. When an **ovum** leaves an ovary, it is usually swept into a uterine (fallopian) tube by the combined action of the fimbriae (fingerlike projections of a uterine tube) and the beating of cilia that line the uterine tube.

Once in a uterine tube, the ovum is transported toward the uterus. Fertilization, and therefore zygote formation, usually takes place in the uterine tube. The developing embryo normally arrives at the **uterus** several days later, and then **implantation** occurs as the embryo embeds in the uterine lining, which has been prepared to receive it.

Development of the embryo and fetus normally takes place in the uterus. The lining of the uterus, called the **endometrium,** participates in the formation of the placenta (see Chapter 18, page 408), which supplies nutrients needed for embryonic and fetal development.

The uterine tubes join the uterus at its upper end, while at its lower end, the **cervix** enters the vagina nearly at a right angle. A small opening in the cervix leads from the uterus to the vagina.

The **vagina** is the birth canal and organ of sexual intercourse in females. The vagina also acts as an exit for menstrual flow. If fertilization and implantation do not occur, the endometrium is sloughed off during menstruation.

The external genital organs of the female are known collectively as the **vulva.** The vulva consists of the mons pubis, clitoris, labia majora, and labia minora. The area between the labia minora, which is termed the vestibule, contains the openings of the urethra and the vagina.

Notice that the urinary and reproductive systems in the female are entirely separate. For example, the urethra carries only urine, and the vagina has functions that pertain only to reproduction.

The Ovary

The ovaries are paired, oval bodies about 3 cm in length by 1 cm in width and less than 1 cm thick. They lie to either side of the uterus on the lateral walls of the pelvic cavity.

Several ligaments hold the ovaries in place (see Fig. 17.12). The largest of these, the broad ligament, is also attached to the

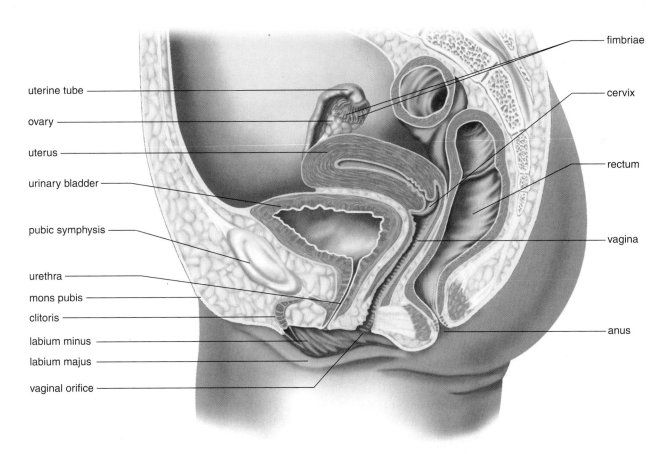

Figure 17.9 The female reproductive system. The ovaries release one egg per month. Fertilization occurs in the uterine tube and development occurs in the uterus. The vagina is the birth canal as well as the organ of sexual intercourse.

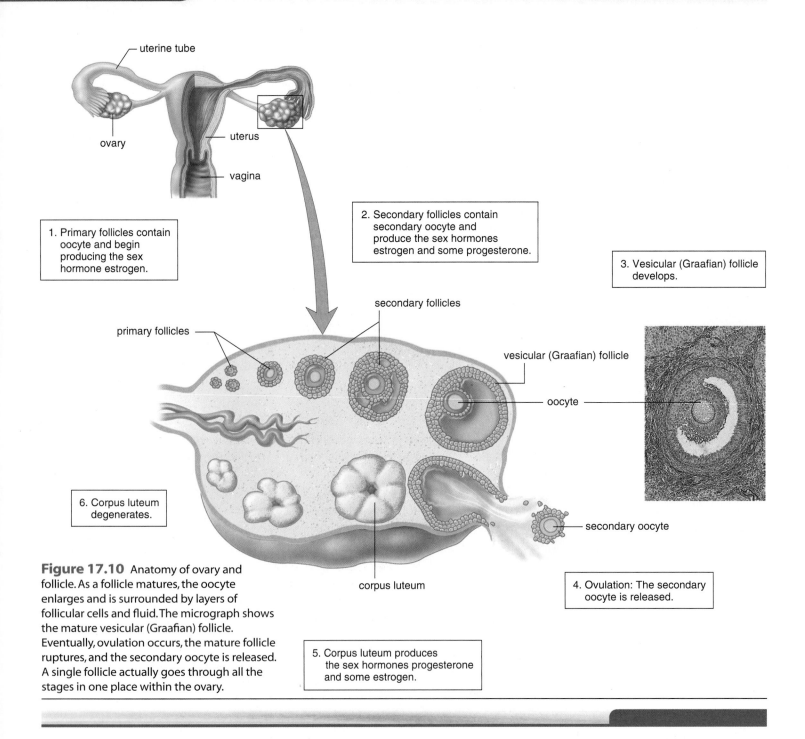

1. Primary follicles contain oocyte and begin producing the sex hormone estrogen.

2. Secondary follicles contain secondary oocyte and produce the sex hormones estrogen and some progesterone.

3. Vesicular (Graafian) follicle develops.

4. Ovulation: The secondary oocyte is released.

5. Corpus luteum produces the sex hormones progesterone and some estrogen.

6. Corpus luteum degenerates.

uterine tube

ovary

uterus

vagina

primary follicles

secondary follicles

vesicular (Graafian) follicle

oocyte

secondary oocyte

corpus luteum

Figure 17.10 Anatomy of ovary and follicle. As a follicle matures, the oocyte enlarges and is surrounded by layers of follicular cells and fluid. The micrograph shows the mature vesicular (Graafian) follicle. Eventually, ovulation occurs, the mature follicle ruptures, and the secondary oocyte is released. A single follicle actually goes through all the stages in one place within the ovary.

uterine tubes and the uterus. The suspensory ligament holds the upper end of the ovary to the pelvic wall, and the ovarian ligament attaches the lower end of the ovary to the uterus.

A sagittal section through an ovary shows that it is made up of an outer cortex and an inner medulla. In the cortex are many **follicles,** each one containing an immature ovum, called an **oocyte** (Fig. 17.10). A female is born with as many as 2 million follicles, but the number is reduced to 300,000–400,000 by the time of puberty. Only a small number of follicles (about 400) ever mature because a female usually produces only one ovum per month during her reproductive years. Because oocytes are present at birth, they age as the woman ages. This may be one

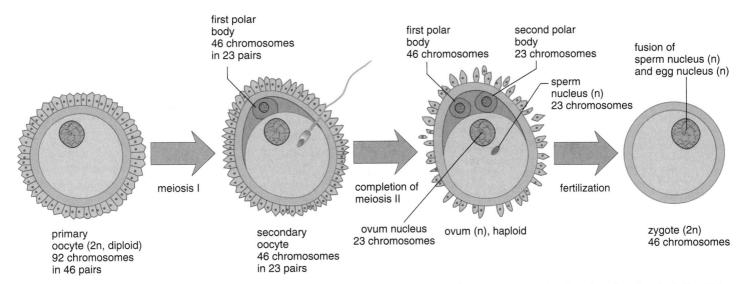

first polar
body
46 chromosomes
in 23 pairs

first polar
body
46 chromosomes

second polar
body
23 chromosomes

fusion of
sperm nucleus (n)
and egg nucleus (n)

sperm
nucleus (n)
23 chromosomes

meiosis I

completion of
meiosis II

fertilization

ovum nucleus
23 chromosomes

ovum (n), haploid

primary
oocyte (2n, diploid)
92 chromosomes
in 46 pairs

secondary
oocyte
46 chromosomes
in 23 pairs

zygote (2n)
46 chromosomes

Figure 17.11 Oogenesis in an ovary. Oogenesis involves meiosis I, during which the chromosome number is reduced, and meiosis II, which results in a single egg. Meiosis II takes place after a sperm enters the secondary oocyte. At the end of oogenesis, there are also at least two polar bodies, nonfunctional cells that later disintegrate.

reason older women are more likely to produce children with chromosome defects.

Cancer of an ovary, or ovarian cancer, which is discussed in the Medical Focus on page 383, causes more deaths than cervical and uterine cancer.

Oogenesis

Oogenesis, the production of an ovum, includes the process of meiosis. Similar to spermatogenesis, oogenesis begins with a primary oocyte that undergoes meiosis I to become a secondary oocyte having 23 paired chromosomes. The secondary oocyte undergoes meiosis II, but only if it is first fertilized by a sperm cell. If the secondary oocyte remains unfertilized, it never completes meiosis and will die shortly after being released from the ovary.

Oogenesis begins within a follicle. As the follicle matures, it develops from a primary follicle to a secondary follicle to a **vesicular (Graafian) follicle** (see Fig. 17.10). The epithelium of a primary follicle surrounds a primary oocyte. Pools of follicular fluid surround the oocyte in a secondary follicle. In a vesicular follicle, a fluid-filled cavity increases to the point that the follicle wall balloons out on the surface of the ovary.

Figure 17.11 traces the steps of oogenesis. As a follicle matures, the primary oocyte divides, producing two cells. One cell is a secondary oocyte, and the other is a polar body. A **polar body** is formed only during oogenesis, and its function is simply to hold the discarded chromosomes, as a sort of cellular "trash can." Because it is almost completely lacking in cytoplasm, the polar body can never be fertilized or develop further. The vesicular follicle bursts, releasing the secondary oocyte surrounded by a clear membrane and attached follicular cells. This process is referred to as **ovulation.**

The secondary oocyte, often called an egg for convenience, enters a uterine tube. If fertilization occurs, a sperm enters the secondary oocyte, which then completes meiosis II. A haploid ovum with 23 chromosomes and a second polar body result. Thus, formation of sperm cells (spermatogenesis) and formation of the ovum (oogenesis) both result in cells that are haploid, containing only 23 chromosomes. However, spermatogenesis produces four sperm cells for every primary cell. In oogenesis, only one haploid ovum is formed for each primary oocyte (the extra chromosomes are contained in polar bodies). When the sperm nucleus unites with the egg nucleus during conception, a diploid zygote with 46 chromosomes is produced.

A follicle that has lost its egg develops into a **corpus luteum,** a glandlike structure. If implantation does not occur, the corpus luteum begins to degenerate after about 10 days. The remains of a corpus luteum is a white scar called the **corpus albicans.** If implantation does occur, the corpus luteum continues to secrete for about six months. Hormones from the corpus luteum help keep the uterine lining intact and allow a pregnancy to continue.

Although a number of follicles grow during each month, only one reaches full maturity and ruptures to release a secondary oocyte. Presumably the ovaries alternate in producing functional ova. The number of secondary oocytes produced by a female during her lifetime is minuscule compared to the number of sperm produced by a male.

Female Internal Accessory Organs

Table 17.2 lists and Figure 17.12 depicts the internal accessory organs, as well as the other reproductive organs, of a female.

Uterine Tubes

The **uterine tubes,** also called fallopian tubes or oviducts, extend from the uterus to the ovaries. Usually the secondary

TABLE 17.2 Female Internal Accessory Organs	
Organ	**Function**
Uterine tubes (fallopian tubes, oviducts)	Transport ovum; location of fertilization
Uterus (womb)	Houses developing fetus
Cervix	Contains opening to uterus
Vagina	Receives penis during sexual intercourse; serves as birth canal and as an exit for menstrual flow

Uterus

The *uterus* is a thick-walled, muscular organ about the size and shape of an inverted pear. Normally, it lies above and is tipped over the urinary bladder. The uterus has three sections. The *fundus* is the region superior to the entrance of the uterine tubes. The *body* of the uterus is the major region. The *cervix* is the narrow end of the uterus that projects into the vagina. A cervical orifice leads to the lumen of the vagina.

Development of the embryo normally takes place in the uterus. This organ, sometimes called the womb, is approximately 5 cm wide in its usual state but is capable of stretching to over 30 cm to accommodate the growing baby. The lining of the uterus, called the *endometrium*, participates in the formation of the placenta (see Chapter 18), which supplies nutrients needed for embryonic and fetal development. In the nonpregnant female, the endometrium varies in thickness during a monthly menstrual cycle, discussed later in this chapter.

Cancer of the cervix is a common form of cancer in women. Early detection is possible by means of a **Pap smear,** which entails the removal of a few cells from the region of the cervix for microscopic examination. If the cells are cancerous, a hysterectomy (the removal of the uterus) may be recommended. Removal of the ovaries in addition to the uterus is termed an **ovariohysterectomy.** Because the vagina remains intact, the woman can still engage in sexual intercourse.

oocyte enters a uterine tube because the **fimbriae** sweep over the ovary at the time of ovulation, and the beating of the cilia that line uterine tubes creates a suction effect. Once in the uterine tube, the egg is propelled slowly toward the uterus by action of the cilia and by muscular contractions in the wall of the uterine tubes.

Fertilization, the completion of meiosis, and zygote formation normally occur in the upper one-third of a uterine tube. The developing embryo usually does not arrive at the uterus for several days. Once in the uterus, the embryo embeds itself in the uterine lining, which has been prepared to receive it.

Occasionally, the embryo becomes embedded in the wall of a uterine tube, where it begins to develop. Tubular pregnancies cannot succeed and must be surgically removed because the tubes are not anatomically capable of allowing full development to occur. Any pregnancy that occurs outside the uterus is called an **ectopic pregnancy.**

Vagina

The vagina is a tube that makes a 45° angle with the small of the back. The mucosal lining of the vagina lies in folds that extend when the fibromuscular wall stretches. This capacity to

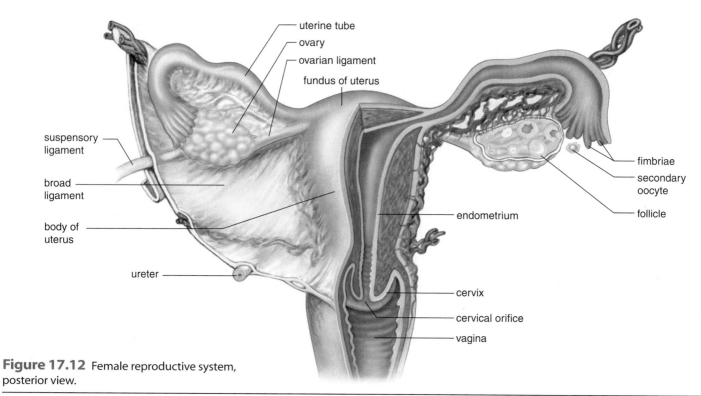

Figure 17.12 Female reproductive system, posterior view.

Ovarian cancer is often "silent," showing no obvious signs or symptoms until late in its development. The most common sign is enlargement of the abdomen, which is caused by the accumulation of fluid. Rarely is there abnormal vaginal bleeding. In women over 40, vague digestive disturbances (stomach discomfort, gas, distention) that persist and cannot be explained by any other cause may indicate the need for a thorough evaluation for ovarian cancer.

The risk for ovarian cancer increases with age. The highest rates are for women over age 60. Women who have never had children are twice as likely to develop ovarian cancer as those who have. Early age at first pregnancy, early menopause, and the use of oral contraceptives, which reduces ovulation frequency, appear to be protective against ovarian cancer. If a woman has had breast cancer, her chances of developing ovarian cancer double. Certain rare genetic disorders are associated with increased risk. The highest incidence rates are reported in the more industrialized countries, with the exception of Japan.

Early detection requires periodic, thorough pelvic examinations. The Pap smear, useful in detecting cervical cancer, does not reveal ovarian cancer. Women over age 40 should have a cancer-related checkup every year. Researchers are currently trying to develop a blood test that will allow for consistent diagnosis of early ovarian cancer. Testing for high levels of tumor marker CA-125, a protein antigen, is sometimes used to screen for ovarian cancer. However, CA-125 is unreliable because it gives false positive results for many other disorders, including benign ovarian cysts and pregnancy. A combination of blood testing and ultrasound examination of the ovaries is currently the most accurate way to diagnose ovarian cancer.

Surgery, radiation therapy, and drug therapy are treatment options. Surgery usually includes the removal of one or both ovaries (**oophorectomy**), the uterus (**hysterectomy**), and the uterine tubes (**salpingectomy**). In some very early tumors, only the involved ovary is removed, especially in young women. In advanced disease, an attempt is made to remove all intra-abdominal cancerous tissue to enhance the effect of chemotherapy.

---

extend is especially important when the vagina serves as the birth canal, and it can also facilitate intercourse, when the vagina receives the penis.

External Genitals

The female external genitals (Fig. 17.13) are known collectively as the vulva. The vulva includes two large, hair-covered folds of skin called the **labia majora** (sing., labium majus). They extend posteriorly from the **mons pubis,** a fatty prominence underlying the pubic hair. The **labia minora** (sing., labium minus) are two small folds of skin lying just inside the labia majora. They extend forward from the vaginal opening to encircle and form a foreskin for the **clitoris,** an organ that is homologous to the penis. Although quite small, the clitoris has a shaft of erectile tissue and is capped by a pea-shaped glans. The clitoris also has sensory receptors that allow it to function as a sexually sensitive organ.

The **vestibule,** a cleft between the labia minora, contains the orifices of the urethra and the vagina. The vagina can be partially closed by a ring of tissue called the hymen. The hymen ordinarily is ruptured by initial sexual intercourse; however, it can also be disrupted by other types of physical activities. If the hymen persists after sexual intercourse, it can be surgically ruptured.

The urinary and reproductive systems in the female are entirely separate: The urethra carries only urine, and the vagina serves only as the birth canal and as the organ for sexual intercourse.

Female Sexual Response

Upon sexual stimulation, the labia minora, the vaginal wall, and the clitoris become engorged with blood. The breasts also swell, and the nipples become erect.

The vagina expands and elongates. Blood vessels in the vaginal wall release small droplets of fluid that seep into the

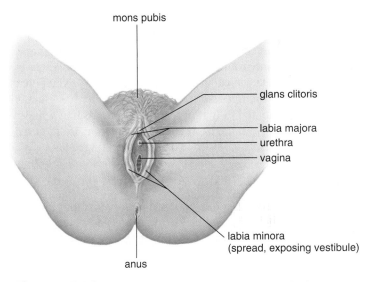

Figure 17.13 External genitals of the female. At birth, the opening of the vagina is partially blocked by a membrane called the hymen. Physical activities and sexual intercourse disrupt the hymen.

vagina and lubricate it. Mucus-secreting glands beneath the labia minora on either side of the vagina also provide lubrication for entry of the penis into the vagina. Although the vagina is the organ of sexual intercourse in females, the clitoris plays a significant role in the female sexual response. The extremely sensitive clitoris can swell to two or three times its usual size. The thrusting of the penis and the pressure of the pubic symphyses of the partners act to stimulate the clitoris.

Orgasm occurs at the height of the sexual response. Blood pressure and pulse rate rise, breathing quickens, and the walls of the uterus and uterine tubes contract rhythmically. A sensation of intense pleasure is followed by relaxation when organs return to their normal size. Females have no refractory period, and multiple orgasms can occur during a single sexual experience.

Regulation of Female Hormone Levels

At the time of puberty in females, the hypothalamus increases its secretion of GnRH, and the anterior pituitary releases larger amounts of the gonadotropins, FSH and LH. These hormones stimulate the ovaries to produce and release ova. FSH and LH also cause elevated estrogen and progesterone production by the ovaries.

Estrogen and Progesterone

In particular, estrogen stimulates the growth of the uterus and the vagina. **Estrogen** is also necessary for maturation of the ovum and for the onset of the menstrual cycle. Development of the secondary sex characteristics in the female is caused by increased estrogen production during puberty. Along with sexual development, girls going through puberty experience an early increase in long bone growth that leads to increased height. However, estrogen also leads to early fusion of the growth plates in long bones, so that most young women have achieved their full adult height by about age 16.

Secondary Sex Characteristics These characteristics include the female pattern of body hair and fat distribution. In general, females have a more rounded appearance than males because of a greater accumulation of fat beneath the skin. Also, the pelvic girdle enlarges in females so that the pelvic cavity has a larger relative size compared to that of males; this means that females have wider hips. Both estrogen and **progesterone** are required for breast development, which is discussed on page 386.

Menstrual Cycle

The **menstrual cycle** is a monthly series of events that involve the ovaries and uterus plus the hormones already mentioned. The cycle is about 28 days long, but it can be as short as 18 days or as long as 40 days (Fig. 17.14).

Pre-Ovulation Events

Under the influence of follicle-stimulating hormone (FSH) from the anterior pituitary, several follicles begin developing in the ovary. Therefore, this period of time (days 1–14) is called the *follicular phase* of the ovary (Fig. 17.14). Although several follicles begin growing, only one follicle continues developing, and it secretes increasing amounts of estrogen. This particular follicle becomes more and more sensitive to FSH and then LH. Eventually, the very high level of estrogen exerts *positive feedback control* over the hypothalamus so that it secretes ever greater amounts of GnRH. GnRH induces a surge in FSH and LH secretion by the pituitary. The LH level rises to a greater extent than does the FSH level. Ovulation is triggered by the very high level of hormonal stimulation, and in particular the spike in LH, called the LH surge. (It is interesting to note that over-the-counter ovulation prediction test kits work by detecting high levels of LH metabolites in the urine, combining the metabolites with monoclonal antibodies to produce a color change.)

While the ovary is experiencing its follicular phase, first menstruation and then the proliferative phase occur in the uterus. During menstruation (days 1–5), a low level of female sex hormones in the body causes the endometrial tissue to disintegrate and its blood vessels to rupture. A flow of blood and tissues, known as the **menses,** passes out of the vagina during menstruation, also called the menstrual period.

Under the influence of estrogen released by the new follicle, the endometrium thickens and becomes vascular and glandular. This is the *proliferative phase* of the uterus, which ends when ovulation occurs.

Post-Ovulation Events

Under the influence of LH from the anterior pituitary, the emptied ovarian follicle tissue becomes a hormone-secreting tissue called the corpus luteum. Therefore, this period of time (days 15–28) is known as the *luteal phase* of the ovary (Fig. 17.14). The corpus luteum secretes progesterone and some estrogen. As the blood level of progesterone rises, it exerts *negative feedback control* over the anterior pituitary's secretion of LH so that the corpus luteum in the ovary begins to degenerate. If fertilization of the egg does occur, the corpus luteum persists for reasons that will be discussed shortly.

Under the influence of progesterone secreted by the corpus luteum, a secretory phase (days 15–28) begins in the uterus. During the *secretory phase* of the uterus, the endometrium of the uterus doubles or even triples in thickness (from 1 mm to 2–3 mm), and the uterine glands mature, producing a thick, mucoid secretion. The endometrium is now prepared to receive the embryo. If implantation of an embryo does not take place, the corpus luteum disintegrates, and menstruation occurs. The disintegrated corpus luteum in the ovary becomes a patch of white scar tissue called the corpus albicans. With age, more of a woman's ovarian tissue becomes nonfunctional corpus albicans. This is why ovulation and menstruation eventually stop during menopause.

If fertilization occurs and is followed by implantation, the developing placenta produces **human chorionic gonadotropin (HCG),** which maintains the corpus luteum in the ovary until

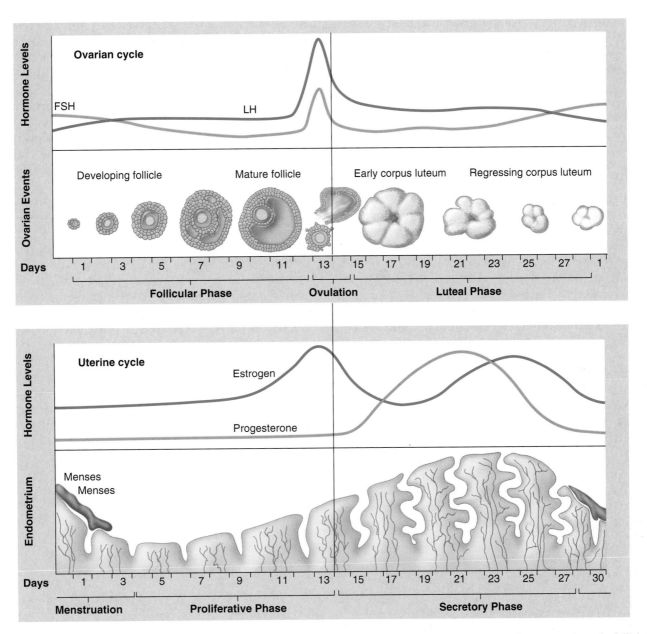

Figure 17.14 During the menstrual cycle, FSH and LH are released by the anterior pituitary. FSH promotes the maturation of a follicle in the ovary. The follicle produces increasing levels of estrogen, which cause the endometrium to thicken during the proliferative phase in the uterus. An LH surge causes ovulation. After ovulation, LH promotes the development of the corpus luteum. This structure produces increasing levels of progesterone, which causes the endometrial lining to become secretory. Menses due to the breakdown of the endometrium begins when progesterone production declines to a low level due to corpus luteum disintegration.

the placenta begins to produce progesterone and estrogen. The placental hormones shut down the anterior pituitary so that no new follicle in the ovaries matures. Placental hormones also maintain the endometrium so that the corpus luteum in the ovary is no longer needed. Usually, no menstruation occurs during pregnancy. After human chorionic gonadotropin is metabolized and excreted in the urine, it can be detected using prepared monoclonal antibodies. Over-the-counter home pregnancy test kits function in this fashion, as discussed in Chapter 18.

Menopause

Menopause, the period in a woman's life during which the menstrual cycle ceases, is likely to occur between ages 45 and 55. The ovaries are no longer responsive to the gonadotropic hormones produced by the anterior pituitary, and the ovaries secrete very low amounts of estrogen and progesterone. At the onset of menopause, the uterine cycle becomes irregular, but as long as menstruation occurs, it is still possible for a woman to conceive. Therefore, a woman is usually not considered to have completed menopause until menstruation has been absent for a year.

The hormonal changes during menopause often produce physical symptoms, such as "hot flashes" (caused by circulatory irregularities), dizziness, headaches, insomnia, sleepiness, and depression. These symptoms may be mild or even absent. If they are severe, medical attention should be sought. Women sometimes report an increased sex drive following menopause. It has been suggested that this may be due to androgen production by the adrenal cortex.

Female Breast and Lactation

Early growth of the female breasts during puberty is referred to as *budding* of the breasts. Budding is followed by development of lobes, the functional portions of the breast, and deposition of adipose tissue, which gives breasts their adult shape.

A breast contains 15 to 25 lobules, each with a milk duct that begins at the nipple. The nipple is surrounded by a pigmented area called the **areola**. Hair and glands are absent from the nipples and areola, but glands are present that secrete a saliva-resisting lubricant to protect the nipples, particularly during nursing. Smooth muscle fibers in the region of the areola may cause the nipple to become erect in response to sexual stimulation or cold.

Within each lobe, the mammary duct divides into numerous alveolar ducts that end in blind sacs called alveoli (Fig. 17.15). The alveoli are made up of the cells that can produce milk. Estrogen and progesterone are required for lobe development. It is believed that estrogen causes proliferation of ducts and that both estrogen and progesterone bring about alveolar development. The abundance of these hormones during pregnancy means that the alveoli proliferate at this time. A nonlactating breast has ducts but few alveoli, while a lactating breast has many ducts and alveoli.

During pregnancy, the breasts enlarge as the ducts and alveoli increase in number and size. The same hormones that affect the mother's breasts can also affect the child's. Some newborns, including males, even secrete a small amount of milk for a few days.

Usually, no milk is produced during pregnancy. The hormone prolactin is needed for **lactation** to begin, and the production of this hormone is suppressed during pregnancy. Negative feedback control by high levels of estrogen and progesterone produced during pregnancy shuts down prolactin secretion by the anterior pituitary. Once the baby is delivered, however, the pituitary begins secreting prolactin. It takes a couple of days for milk production to begin, and in the meantime, the breasts produce **colostrum**, a thin, yellow, milky fluid rich in protein, including antibodies.

The continued production of milk requires a suckling child. When a breast is suckled, the nerve endings in the areola are stimulated, and a nerve impulse travels along neural pathways from the nipples to the hypothalamus, which directs the posterior pituitary gland to release the hormone oxytocin. When this hormone arrives at the breast, it causes the

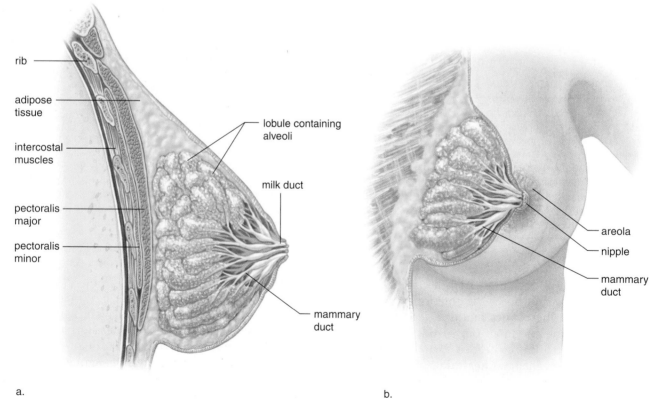

Figure 17.15 Structure of the female breast and mammary glands. **a.** Sagittal section. **b.** Anterior view.

lobules to contract so that milk flows into the ducts (called milk letdown), where it may be drawn out of the nipple by the suckling child.

Breast cancer is one of the few types of cancer that can be detected by the female herself. The Medical Focus on page 391 tells how to do a shower check for breast cancer.

Content CHECK-UP!

7. Which structure in the ovary eventually releases the secondary oocyte or ovum?

    a. vesicular follicle   c. corpus albicans

    b. corpus luteum   d. polar body

8. The second stage of meiosis in an ovum occurs:

    a. in the ovary, before the ovum is released at ovulation.

    b. after the ovum is released and travels through the uterine tube.

    c. only if the ovum is fertilized by a sperm cell.

    d. just after the ovum is released from the ovary, but before it enters the uterine tube.

9. During which stage of the monthly cycle of the uterus does the endometrium thicken and become glandular and vascular?

    a. menstrual phase   c. proliferative phase

    b. secretory phase   d. ovulation phase

17.4 Control of Reproduction and Sexually Transmitted Diseases

Contraception

Today, there are more options than ever before for **contraception,** or the prevention of pregnancy. Although many contraceptives are available without prescription, women should consult a health-care provider before choosing a birth control method. Health-care providers can give patients advice on the best method to use based on lifestyle, and can also ensure that the patient uses the product correctly. Table 17.3 lists the means of birth control used in the United States and rates their effectiveness. For example, consider any of the prescription combined hormone methods, which are among the most effective methods listed in the table. Based on statistics, we expect that within a year, 98 or 99 out of 100, or 98–99%, of sexually active women will *not* get pregnant using one of these methods, while one or two women will get pregnant. (It is important to note that the failure rate for any contraception method might be a bit high—it also includes those users who don't follow directions and therefore use the product incorrectly.)

Figure 17.16 features some of the most effective and commonly used means of birth control. Combination hormone methods can be in the form of pills taken orally, and a new regimen of combined pill allows a woman to have only 3 or 4 menstrual cycles a year (Fig. 17.6a). In addition, combination hormones can be introduced into the blood via patches that allow hormone absorption through the skin (Fig. 17.16e), or by a plastic ring placed in the vagina. The ring slowly releases hormones into the bloodstream. Hormone injections can also be administered monthly. Regardless of delivery method, the estrogen and progesterone in the combined hormone form effectively shut down the pituitary's production of both FSH and LH so that no follicle begins to develop in the ovary. Because ovulation does not occur, pregnancy cannot take place. Minor side effects of combined hormone methods include dizziness, nausea, weight gain, and changes in menstruation and mood. Rarely, serious complications can occur in women who use combined hormonal methods, including high blood pressure, cardiovascular disease, blood clots, heart attack, and strokes. Serious complications are more likely in women who smoke; therefore, smokers should use other birth control methods. Because of their possible side effects, these methods are offered by prescription only. Women using combined hormonal methods should see a physician or other health-care provider on a regular basis to monitor their use.

A second hormonal method of contraception involves use of only a synthetic progesterone, without estrogen. The oral form is commonly referred to as the "mini-pill" (because it contains only one hormone) and is taken daily. An injectable form can be taken once every three months (Fig. 17.16f). Both work by inhibiting ovulation and making the mucus of the uterine cervix thick and sticky. Sperm cells cannot swim through the thick mucus, thus preventing fertilization of the ovum. If fertilization does occur, the uterine lining is changed by this medication so that an embryo will not be able to implant. As with other hormonal forms of contraception, progesterone-only forms are available by prescription only and do have some side effects and health risks (see Table 17.3).

An intrauterine device (IUD) is a small piece of molded plastic that is inserted into the uterus by a physician (Fig. 17.16b). IUDs are believed to alter the environment of the uterus and uterine tubes so that fertilization probably will not occur—but if fertilization does occur, implantation cannot take place. The type of IUD shown in Figure 17.16b has copper wire wrapped around the plastic. Other forms will slowly release progesterone over time into the bloodstream. IUDs are very effective. However, because of their possible side effects—cramps, bleeding, pelvic inflammatory disease, infertility, and a slight risk of perforation of uterus—intrauterine devices currently have a limited availability.

The diaphragm is a soft latex cup with a flexible rim (Fig. 17.16c) that lodges behind the pubic bone and fits over the cervix. The diaphragm can be inserted into the vagina no more than two hours before sexual relations. Also, it must be used with spermicidal jelly or cream and should be left in place at least six hours after sexual relations. A cervical shield is smaller than a diaphragm, and the cervical cap is even smaller still. All three devices must be fitted by a physician or other health-care provider. Each carries a slight risk of urinary tract infection in the user. Rarely, incidents of a very

TABLE 17.3 Common Methods of Contraception

Name	Procedure	How Does It Work?	How Effective Is It?	Side Effects and Other Health Risks
NATURAL METHODS				
Abstinence	Refrain from sexual intercourse	No sperm in vagina	100%	None; also protects against sexually transmitted diseases
Natural family planning	Determine day of ovulation by keeping records; various testing methods	Intercourse is avoided only during the time period while ovum is viable	80%	None
Withdrawal method	Penis withdrawn from vagina just before ejaculation	Ejaculation outside the woman's body; no sperm in vagina	75%	None
Douching	Vagina cleansed after intercourse	Washes sperm out of woman's body	Less than 70%	May cause infection or inflammation
NONPRESCRIPTION METHODS				
Male condom	Sheath of latex, polyurethane, or natural material fitted over erect penis	Traps sperm and prevents entry into vagina; latex and polyurethane forms protect against STDs.	89%	With latex: latex allergy; natural material condoms give no protection against STDs.
Female condom	Polyurethane liner fitted inside vagina	Traps sperm and prevents entry into vagina; some protection against STDs	79%	Possible allergy or irritation; urinary tract infection
Spermicide: jellies, foams, films, creams, tablets	Spermicidal products inserted into the vagina before intercourse	Spermicidal chemical nonoxynol-9 kills large numbers of sperm cells	50–80%	Irritation, inflammation; allergic reaction; urinary tract infection
Contraceptive sponges	Sponge containing spermide inserted into vagina and placed against cervix	Spermicidal chemical nonoxynol-9 kills large numbers of sperm cells	72–86%	Irritation, inflammation; allergic reaction; urinary tract infection; toxic shock syndrome; limited availability
PRESCRIPTION HORMONAL METHODS				
Combined hormone, vaginal ring	Flexible plastic ring inserted into vagina; releases hormones that are absorbed into the bloodstream	Suppresses ovulation by the combined actions of the hormones estrogen and progestin	98%	Vaginal irritation, inflammation; dizziness, nausea; changes in menstruation, mood, and weight; rarely: cardiovascular disease, including high blood pressure, blood clots, heart attack, and strokes
Combined hormone, injection (Lunelle)	Injection of long-acting hormone given once a month	Suppresses ovulation by the combined actions of the hormones estrogen and progestin	99%	Dizziness; nausea; changes in menstruation, mood, and weight; rarely: cardiovascular disease, including high blood pressure, blood clots, heart attack, and strokes
Progesterone-only injection (Depo-Provera)	Injection of progestin once every three months	Inhibits ovulation; prevents sperm from reaching the egg and prevents the fertilized egg from implanting in the uterus	99%	Irregular bleeding; weight gain, breast tenderness; headaches; osteoporosis if used for extended period
Emergency contraception	Must be taken within 72 hours after unprotected intercourse	Suppresses ovulation by the combined actions of the hormones estrogen and progestin; prevents implantation by embryo	80%	Nausea; vomiting; abdominal pain; fatigue; headache

PRESCRIPTION HORMONAL METHODS

Method	Description	Mechanism	Effectiveness	Risks/Side effects
Combined hormone pill	Pills are swallowed daily; chewable form also available	Suppresses ovulation by the combined actions of the hormones estrogen and progestin	98%	Dizziness, nausea; changes in menstruation, mood, and weight; rarely: cardiovascular disease, including high blood pressure, blood clots, heart attack, and strokes
Progestin-only mini-pill	Pills are swallowed daily	Thickens cervical mucus, preventing sperm from contacting ovum	98%	Irregular bleeding; weight gain; breast tenderness
Combined hormone 91, daily regimen	Pills are swallowed daily; user has 3–4 menstrual periods a year	Suppresses ovulation by the combined actions of the hormones estrogen and progestin	98%	Dizziness, nausea; changes in menstruation, mood, and weight; rarely: cardiovascular disease, including high blood pressure, blood clots, heart attack, and strokes
Combined hormone patch	Patch is applied to skin and left in place for 1 week; new patch applied	Suppresses ovulation by the combined actions of the hormones estrogen and progestin	98%	Skin irritation, allergy: dizziness, nausea; changes in menstruation, mood, and weight; rarely: cardiovascular disease, including high blood pressure, blood clots, heart attack, and strokes

PRESCRIPTION BARRIER METHODS

Method	Description	Mechanism	Effectiveness	Risks/Side effects
Diaphragm	Latex cup, placed into vagina to cover cervix before intercourse	Blocks entrance of sperm into uterus; spermicide kills sperm	90% with spermicide	Irritation, inflammation; allergic reaction; urinary tract infection; toxic shock syndrome
Cervical cap	Latex cap held over cervix	Blocks entrance of sperm into uterus; spermicide kills sperm	90% with spermicide	Irritation, inflammation; allergic reaction; toxic shock syndrome; abnormal Pap smear
Cervical shield	Latex cap placed in upper vagina, held in place by suction	Blocks entrance of sperm into uterus; spermicide kills sperm	90% with spermicide	Irritation, inflammation; allergic reaction; toxic shock syndrome; urinary tract infection
Intrauterine device, copper T	Placed in uterus	Causes cervical mucus to thicken; fertilized embryo cannot implant	99%	Cramps; bleeding; pelvic inflammatory disease; infertility; perforation of uterus; limited availability
Intrauterine device, progesterone-releasing type	Placed in uterus	Prevents ovulation; causes cervical mucus to thicken; fertilized embryo cannot implant	99%	Cramps; bleeding; pelvic inflammatory disease; infertility; perforation of uterus; limited availability

SURGICAL STERILIZATION

Method	Description	Mechanism	Effectiveness	Risks/Side effects
Vasectomy	Vasa deferentia are cut and tied	No sperm in seminal fluid	Almost 100%	May not be reversible; minor risk of surgical complications such as infection or reaction to anesthetic
Tubal ligation: trans-abdominal surgery	Oviducts are cut and tied, or a clip is placed on the oviduct	Sperm cannot enter oviduct; ova cannot pass through oviduct	Almost 100%	May not be reversible; pain, bleeding, infection; other post-surgical complications; ectopic (tubal) pregnancy
Sterilization implant (Essure system)	Small metallic implant is placed into uterine tubes; inserted through the vagina using a catheter	Causes scar tissue to form, blocking uterine tubes and preventing conception	Almost 100%	May not be reversible; pain after placement; ectopic pregnancy

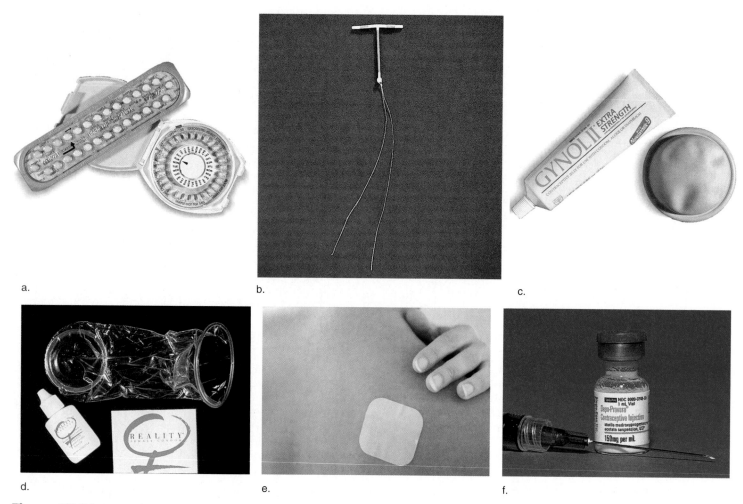

a.   b.   c.

d.   e.   f.

Figure 17.16 Various birth control devices. **a.** Oral contraception (birth control pills). **b.** Intrauterine device. **c.** Spermicidal jelly and diaphragm. **d.** Female condom. **e.** Contraceptive patch. **f.** Contraceptive injection.

serious illness—toxic shock syndrome—have been reported in users of these methods. Use of proper hygiene methods, including handwashing before inserting the device, and prompt removal of the device, can minimize the risk of serious illness.

There has been a revival of interest in barrier methods of birth control because these methods also offer some protection against sexually transmitted diseases. A female condom, now available, consists of a large polyurethane tube with a flexible ring that fits onto the cervix (Fig. 17.16d). The open end of the tube has a ring that covers the external genitals. A male condom is most often a latex or polyurethane sheath that fits over the erect penis. The ejaculate is trapped inside the sheath and, thus, does not enter the vagina. When used in conjunction with a spermicide, protection is better than with the condom alone. So-called "natural" condoms made of lambskin or other material do not provide any protection against sexually transmitted disease.

Research continues into other forms of contraception, including contraceptive vaccines. One form, which immunizes against human chorionic gonadotropin, the hormone

of pregnancy, was successful in one limited clinical trial. However, such vaccines have limited practical application currently because they have not been proven to be completely effective in some users. There is also question as to whether their effects are reversible. Current hormonal contraceptives designed for use in males cause undesirable side effects such as erectile dysfunction. More study needs to be done into alternative methods that will be effective, safe, and inexpensive.

Emergency Contraception

Emergency contraception, sometimes referred to as "morning-after pills," refers to a medication that will prevent pregnancy after unprotected intercourse. The name is a misnomer because the medication does not have to be taken the next morning—the woman can begin the medication one to several days after having unprotected intercourse.

One type of emergency contraception contains four synthetic estrogen-progesterone pills; a second type contains progesterone-only pills. Two pills are taken up to 72 hours after unprotected intercourse, and two more are taken 12 hours later.

Shower Check for Cancer

The American Cancer Society urges women to do a breast self-exam and men to do a testicle self-exam every month. Breast cancer and testicular cancer are far more curable if found early, and we must all take the responsibility of checking for one or the other.

Breast Self-Exam for Women

1. Check your breasts for any lumps, knots, or changes about one week after your period.
2. Place your right hand behind your head. Move your *left* hand over your *right* breast in a circle. Press firmly with the pads of your fingers (Fig. 17A). Also check the armpit.
3. Now place your left hand behind your head and check your *left* breast with your *right* hand in the same manner as before. Also check the armpit.
4. Check your breasts while standing in front of a mirror right after you do your shower check. First, put your hands on your hips and then raise your arms above your head (Fig. 17B). Look for any changes in the way your breasts look: dimpling of the skin, changes in the nipple, or redness or swelling.
5. If you find any changes during your shower or mirror check, see your doctor right away.

You should know that the best check for breast cancer is a mammogram. When your doctor checks your breasts, ask about getting a mammogram. Yearly mammograms are recommended for women over 40 years of age. Women with a personal or family history of breast cancer (mother, sister, or aunt who have or had the disease) may be advised to start such screening exams at an earlier age, and to have the exams more frequently. Additional exams such as ultrasound or MRI of the breast may also be advised.

Testicle Self-Exam for Men

1. Check your testicles once a month.
2. Roll each testicle between your thumb and finger as shown in Figure 17C. Feel for hard lumps or bumps.
3. If you notice a change or have aches or lumps, tell your doctor right away so he or she can recommend proper treatment.

Cancer of the testicles can be cured if you find it early. You should also know that prostate cancer is the most common cancer in men. Men over age 50 should have an annual health checkup that includes a prostate examination.

Information provided by the American Cancer Society. Used by permission.

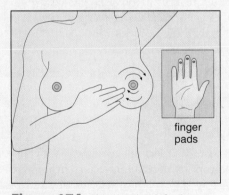

Figure 17A Shower check for breast cancer.

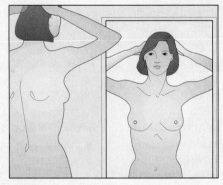

Figure 17B Mirror check for breast cancer.

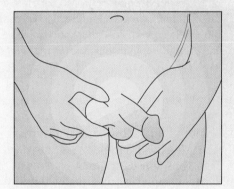

Figure 17C Shower check for testicular cancer.

The medication upsets the normal uterine cycle, making it difficult for an embryo to implant in the endometrium. A recent study estimated that the medication was 80% effective in preventing unintended pregnancies. The medication should not be used for regular contraception. Side effects may include nausea, vomiting, abdominal pain, fatigue, and headache. Other, more serious complications include hypertension, cardiovascular disease, and stroke.

Mifepristone, better known as RU-486, does not prevent conception. Rather, this medication is presently used to cause the loss of an implanted embryo by blocking the progesterone receptor proteins of endometrial cells. Without functioning receptors for progesterone, the endometrium sloughs off, carrying the embryo with it. When taken in conjunction with a prostaglandin to induce uterine contractions, RU-486 is 95% effective in terminating an early pregnancy. However, cases of

MEDICAL FOCUS

Endocrine-Disrupting Contaminants

Rachel Carson's book *Silent Spring*, published in 1962, predicted that pesticides would have a deleterious effect on animal life. Soon thereafter, it was found that pesticides caused the eggshells of bald eagles to become so thin that their eggs broke and the chicks died. Additionally, populations of terns, gulls, cormorants, and lake trout declined after they ate fish contaminated by high levels of environmental toxins. The concern was so great that the United States Environmental Protection Agency (EPA) came into existence. The efforts of this agency and civilian environmental groups have brought about a reduction in pollutant release and a cleaning up of emissions. Even so, we are now aware of more subtle effects that pollutants can have.

Hormones influence nearly all aspects of physiology and behavior in animals, including tissue differentiation, growth, and reproduction. Therefore, when wildlife in contaminated areas began to exhibit certain types of abnormalities, researchers began to think that certain pollutants can affect the endocrine system. In England, male fish exposed to sewage developed ovarian tissue and produced a metabolite normally found only in females during egg formation. In California, western gulls displayed abnormalities in gonad structure and nesting behaviors. Hatchling alligators in Florida possessed abnormal gonads and hormone concentrations linked to nesting.

At first, such effects seemed to indicate only the involvement of the female hormone estrogen, and researchers therefore called the contaminants ecoestrogens. However, further study brought more information to light. Many of the contaminants interact with hormone receptors, and in that way cause developmental effects. Others bind directly with sex hormones such as testosterone and estradiol. Still others alter the physiology of the growth hormones and neurotransmitters responsible for brain development and behavior. Therefore, the preferred term today for these pollutants is endocrine-disrupting contaminants (EDCs).

Many EDCs are chemicals used as pesticides and herbicides in agriculture, and some are associated with the manufacture of various other synthetic organic compounds such as PCBs (polychlorinated biphenyls). Some chemicals shown to influence hormones are found in plastics, food additives, and personal hygiene products. Metals such as mercury and lead are also endocrine disrupters. In mice, phthalate esters, which are plastic components, affect neonatal development when present in the part-per-trillion range. It is, therefore, of great concern that EDCs have been found at levels one thousand times greater than this—even in amounts comparable to functional hormone levels in the human body. Furthermore, it is not surprising that EDCs are affecting the endocrine systems of a wide range of organisms (Fig. 17D).

Scientists and those representing industrial manufacturers continue to debate whether EDCs pose a health risk to humans. Some suspect that EDCs lower sperm counts reduce male and female fertility, and increase the rates of certain cancers (breast, ovarian, testicular, and prostate). Additionally, some studies suggest that EDCs contribute to learning deficits and behavioral problems in children. Laboratory and field research continues to identify chemicals that have the ability to influence the endocrine system. Millions of tons of potential EDCs are produced

---

blood-borne infection and hemorrhage have occurred in women using the drug, and it has caused fatalities in Europe and the U.S. Thus, it is critical that patients using RU-486 be under the constant supervision of qualified health care professionals.

Although it may seem an impossibly old-fashioned, outmoded ideal, it is still ultimately true—and therefore worth mentioning in any textbook dealing with human anatomy and physiology. The most reliable method of birth control is *still* abstinence—that is, not engaging in sexual intercourse. This form of birth control has the added advantage of preventing transmission of a sexually transmitted disease.

Infertility

Infertility is the failure of a couple to achieve pregnancy after one year of regular, unprotected intercourse. The American Medical Association estimates that 15% of all couples are infertile. The cause of infertility can be attributed to the male (40%), the female (40%), or both (20%).

Causes of Infertility

The most frequent cause of infertility in males is low sperm count and/or a large proportion of abnormal sperm. There are numerous causes for low sperm count. Herbicides and pesticides may cause feminizing effects in the man's body, as discussed in the Medical Focus on this page. Chemicals used on the job, such as paints, varnishes, and degreasers, may cause infertility as well. Smoking, alcohol and/or drug abuse decrease the numbers and quality of a man's sperm cells. In particular, abuse of anabolic (body-building) steroids can shrink the testes and decrease sperm count. Exposing the testicles to overheating must be avoided, as it impairs sperm production and lowers sperm count. Common causes for overheating include obesity and/or remaining seated for very long periods (as in long-distance driving, for example). Frequent use of saunas or hot tubs, wearing extremely tight clothing such as biking shorts, and long periods of using a very warm laptop computer can also cause overheating and a subsequent low sperm count.

annually in the United States, and the EPA is under pressure to certify these compounds as safe. Currently, the EPA regulates the concentration of 23 inorganic chemicals including mercury, lead, and arsenic. Maximum acceptable concentrations have been established, and penalties exist for industries guilty of pollution. More that 50 organic chemicals are also regulated. Doubtless other inorganic and organic chemicals will be added to the list as research continues. The European Economic Community has already restricted the use of certain EDCs, and has banned the production of specific plastic components that are found in items intended for use by children, specifically toys. Only through continued scientific research and the cooperation of industry can we identify the risks that EDCs pose to the environment, wildlife, and humans.

Figure 17D Exposure to endocrine-disrupting contaminants. Various types of wildlife, as well as humans, are exposed to endocrine-disrupting contaminants that can seriously affect their health and reproductive abilities.

Body weight appears to be the most significant factor in causing female infertility. In women of normal weight, fat cells produce a hormone called leptin that stimulates the hypothalamus to release GnRH. In overweight women, the ovaries often contain many small follicles, and the woman fails to ovulate. Other causes of infertility in females are blocked uterine tubes due to pelvic inflammatory disease (see page 396) and endometriosis. **Endometriosis** is the presence of uterine tissue outside the uterus, particularly in the uterine tubes and on the abdominal organs. Backward flow of menstrual fluid allows living uterine cells to establish themselves in the abdominal cavity, where they go through the usual uterine cycle, causing pain and structural abnormalities that make it more difficult for a woman to conceive.

Sometimes the causes of infertility can be corrected by medical intervention so that couples can have children. If no obstruction is apparent and body weight is normal, it is possible to give females fertility drugs, which are gonadotropic hormones that stimulate the ovaries and bring about ovulation.

Such hormone treatments may cause multiple ovulations and multiple births.

When reproduction does not occur in the usual manner, many couples adopt a child. Others sometimes try one of the assisted reproductive technologies discussed in the following paragraphs.

Assisted Reproductive Technologies

Assisted reproductive technologies (ART) consist of techniques used to increase the chances of pregnancy. Often, sperm and/or eggs are retrieved from the testes and ovaries, and fertilization takes place in a clinical or laboratory setting.

Artificial Insemination During **artificial insemination,** sperm are placed in the vagina by a physician. Sometimes a woman is artificially inseminated by her partner's sperm. This is especially helpful if the partner has a low sperm count, because the sperm can be collected over a period of time and concentrated so that the sperm count is sufficient to result in fertilization. Often, however, a woman is inseminated by sperm acquired from a donor who is a complete stranger to her. At times, a combination of partner and donor sperm is used.

During *intrauterine insemination (IUI)*, fertility drugs are given to stimulate the ovaries, and then the donor's sperm is placed in the uterus rather than in the vagina.

If the prospective parents wish, sperm can be sorted into those believed to be X-bearing or Y-bearing to increase the chances of having a child of the desired sex.

In Vitro Fertilization During **in vitro fertilization (IVF)**, conception occurs in the laboratory. Ultrasound machines can now spot follicles in the ovaries that hold immature eggs; therefore, the latest method is to forgo the administration of fertility drugs and retrieve immature eggs by using a needle. The immature eggs are then brought to maturity in the laboratory before concentrated sperm are added. After about two to four days, the embryos are ready to be transferred to the uterus of the woman, who is now in the secretory phase of her uterine cycle. If desired, preimplantation genetic analysis can be done and only embryos found to be free of genetic disorders are used. If implantation is successful, development is normal and continues to term.

Intracytoplasmic Sperm Injection In **intracytoplasmic sperm injection (ICSI)**, a highly sophisticated procedure, a single sperm is injected into an egg. This method is used effectively when a man has severe infertility problems.

Sexually Transmitted Diseases

Sexually transmitted diseases (STDs) are caused by organisms ranging from viruses to arthropods; however, we will discuss only certain STDs caused by viruses and bacteria. Unfortunately, for unknown reasons, humans cannot develop effective immunity to any STDs. Therefore, any person exposed to an STD should seek prompt medical treatment. To prevent the spread of STDs, a latex or polyurethane condom can be used; the concomitant use of a spermicide containing nonoxynol-9 gives added protection.

Among those STDs caused by viruses, treatment is available for AIDS and genital herpes. However, it is important to note that treatment for HIV/AIDS and genital herpes cannot presently eliminate the virus from the person's body. Drugs used for treatment can merely slow replication of the viruses. Thus, neither viral disease is presently curable. Further, antiviral drugs have serious, debilitating side effects on the body.

Only STDs caused by bacteria (e.g., chlamydia, gonorrhea, and syphilis) are curable with antibiotics. Bacteria that acquire antibiotic resistance may necessitate treatment with extremely strong drugs for an extended period to achieve a cure.

Genital Warts

Genital warts are caused by the human papillomaviruses (HPVs). Many times, carriers either do not have any sign of warts or merely have flat lesions. When present, the warts commonly are seen on the penis and foreskin of men and near the vaginal opening in women. A newborn can become infected while passing through the birth canal.

Individuals who are currently infected with visible growths may have those growths removed by surgery, freezing, or burning with lasers or acids. However, visible warts that are removed may recur. A new vaccine has been released for the human papillomaviruses that most commonly cause genital warts. This development is an extremely important step in the prevention of cancer, as well as the warts themselves. Genital warts are associated with cancer of the cervix, as well as tumors of the vulva, vagina, anus, and penis. Researchers believe that these viruses may be involved in up to 95% of all cases of cancer of the cervix. Vaccination might make such cancers a thing of the past.

Genital Herpes

Genital herpes is caused by herpes simplex virus. Type 1 usually causes cold sores and fever blisters, while type 2 more often causes genital herpes.

Persons usually get infected with herpes simplex virus type 2 when they are adults. Some people exhibit no symptoms; others may experience a tingling or itching sensation before blisters appear on the genitals. Once the blisters rupture, they leave painful ulcers that may take as long as three weeks or as little as five days to heal. The blisters may be accompanied by fever, pain on urination, swollen lymph nodes in the groin, and in women, a copious discharge. At this time, the individual has an increased risk of acquiring an HIV infection.

After the ulcers heal, the disease is only latent, and blisters can recur, although usually at less frequent intervals and with milder symptoms. Fever, stress, sunlight, and menstruation are associated with recurrence of symptoms. Exposure to herpes in the birth canal can cause an infection in the newborn, which leads to neurological disorders and even death. Birth by cesarean section prevents this possibility.

Hepatitis

Hepatitis infects the liver and can lead to liver failure, liver cancer, and death. The type of hepatitis and the virus that causes it are designated by the same letter. There are six known viruses that cause hepatitis, designated A-B-C-D-E-G. Hepatitis A is usually acquired from sewage-contaminated drinking water, but this infection can also be transmitted sexually through oral/anal contact. Hepatitis B is spread through sexual contact and by blood-borne transmission (accidental needle stick on the job, receiving a contaminated blood transfusion, a drug abuser sharing infected needles while injecting drugs, from mother to fetus, etc.). Simultaneous infection with hepatitis B and HIV is common because both share the same routes of transmission. Fortunately, a combined vaccine is available for hepatitis A and B; it is currently recommended that all children receive the vaccine to prevent infection (see pages 290–291). Hepatitis C (also called non-A, non-B hepatitis) causes most cases of post-transfusion hepatitis. Hepatitis D and hepatitis G are transmitted sexually, and hepatitis E is acquired from contaminated water. Screening of blood and blood products can prevent transmission of hepatitis viruses during a transfusion. Proper water treatment techniques can prevent contamination of drinking water.

Preventing Transmission of STDs

It is wise to protect yourself from getting a sexually transmitted disease (STD). Some of the STDs, such as gonorrhea, syphilis, and chlamydia, can be cured by taking an antibiotic, but medication for the ones transmitted by viruses is much more problematic. In any case, it is best to prevent the passage of STDs from person to person so that treatment becomes unnecessary.

Sexual Activities Transmit STDs (Fig. 17E)

Abstain from sexual intercourse or develop a long-term monogamous (always the same partner) *sexual relationship* with a partner who is free of STDs.

Refrain from multiple sex partners or having relations with someone who has multiple sex partners. If you have sex with two other people and each of these has sex with two people and so forth, the number of people who are relating is quite large.

Remember that, although the prevalence of AIDS is presently higher among homosexuals and bisexuals, the highest rate of increase is now occurring among heterosexuals. The lining of the uterus is only one cell thick, and it does allow infected cells from a sexual partner to enter.

Be aware that having relations with an intravenous drug user is risky because the behavior of this group risks hepatitis and an HIV infection. Be aware that anyone who already has another sexually transmitted disease is more susceptible to an HIV infection.

Uncircumcised males are more likely to become infected with an STD than circumcised males because vaginal secretions can remain under the foreskin for a long period of time.

Avoid anal-rectal intercourse (in which the penis is inserted into the rectum) because the lining of the rectum is thin and cells infected with HIV can easily enter the body there.

Unsafe Sexual Practices Transmit STDs

Always use a latex condom during sexual intercourse if you do not know for certain that your partner has been free of STDs for some time. Be sure to follow the directions supplied by the manufacturer. Use of a water-based spermicide containing nonoxynol-9 in addition to the condom can offer further protection because nonoxynol-9 immobilizes viruses and virus-infected cells.

Avoid fellatio (kissing and insertion of the penis into a partner's mouth) *and cunnilingus* (kissing and insertion of the tongue into the vagina) because they may be a means of transmission. The mouth and gums often have cuts and sores that facilitate the entrance of infected cells.

Be cautious about using alcohol or any drug that may prevent you from being able to control your behavior.

Drug Use Transmits Hepatitis and HIV (Fig. 17F)

Stop, if necessary, or do not start the habit of injecting drugs into your veins. Be aware that hepatitis and HIV can be spread by blood-to-blood contact.

Always use a new, sterile needle for injection or one that has been cleaned in bleach if you are a drug user and cannot stop your behavior.

Figure 17E Sexual activities transmit STDs.

Figure 17F Sharing needles transmits STDs.

Chlamydia

Chlamydia is named for the tiny bacterium that causes it (*Chlamydia trachomatis*). The incidence of new chlamydia infections has steadily increased since 1984.

Chlamydia infections of the lower reproductive tract are usually mild or asymptomatic, especially in women. About 18 to 21 days after infection, men may experience a mild burning sensation on urination and a mucoid discharge. Women may have a vaginal discharge along with the symptoms of a urinary tract infection. Chlamydia also causes cervical ulcerations, which increase the risk of acquiring HIV.

If the infection is misdiagnosed or if a woman does not seek medical help, there is a particular risk of the infection spreading from the cervix to the uterine tubes so that **pelvic inflammatory disease (PID)** results. This very painful condition can result in blockage of the uterine tubes with the possibility of sterility and infertility. If a baby comes in contact with chlamydia during birth, inflammation of the eyes or pneumonia can result.

Gonorrhea

Gonorrhea is caused by the bacterium *Neisseria gonorrhoeae*. Diagnosis in the male is not difficult, since typical symptoms are pain upon urination and a thick, greenish-yellow urethral discharge. In males and females, a latent infection leads to pelvic inflammatory disease (PID), which can also cause sterility in males. If a baby is exposed during birth, an eye infection leading to blindness can result. All newborns are given eyedrops to prevent this possibility.

Gonorrhea proctitis, an infection of the anus characterized by anal pain and blood or pus in the feces, also occurs in patients. Oral/genital contact can cause infection of the mouth, throat, and tonsils. Gonorrhea can spread to internal parts of the body, causing heart damage or arthritis. If, by chance, the person touches infected genitals and then touches his or her eyes, a severe eye infection can result. Up to now, gonorrhea was curable by antibiotic therapy, but resistance to antibiotics is becoming more and more common, and 40% of all strains are now known to be resistant to therapy.

Syphilis

Syphilis is caused by a bacterium called *Treponema pallidum*. As with many other bacterial diseases, penicillin is an effective antibiotic. Syphilis has three stages, often separated by latent periods, during which the bacteria are resting before multiplying again. During the *primary stage*, a hard **chancre** (ulcerated sore with hard edges) indicates the site of infection. The chancre usually heals spontaneously, leaving little scarring. During the *secondary stage*, the victim breaks out in a rash that does not itch and is seen even on the palms of the hands and the soles of the feet. Hair loss and infectious gray patches on the mucous membranes may also occur. These symptoms disappear of their own accord.

During the *tertiary stage*, which lasts until the patient dies, syphilis may affect the cardiovascular system by causing aneurysms, particularly in the aorta. In other instances, the dis-ease may affect the nervous system, resulting in psychological disturbances. Also **gummas,** large destructive ulcers, may develop on the skin or within the internal organs.

Congenital syphilis is caused by syphilitic bacteria crossing the placenta. The child is born blind and/or with numerous anatomical malformations. Control of syphilis depends on prompt and adequate treatment of all new cases; therefore, it is crucial for all sexual contacts to be traced so they can be treated. Diagnosis of syphilis can be made by blood tests or by microscopic examination of fluids from lesions.

Content CHECK-UP!

10. A combined hormonal method of contraception is available through the following delivery method(s).

    a. pills taken orally   c. injections received monthly

    b. patches applied to the skin   d. all of the above

11. Which of the following is not a bacterial STD?

    a. syphilis   c. gonorrhea

    b. chlamydia   d. genital warts

12. Which of the following is a characteristic of tertiary syphilis?

    a. large, destructive ulcers called gummas

    b. formation of a hard chancre

    c. a skin rash that does not itch

    d. hair loss

17.5 Effects of Aging

Sex hormone levels decline with age in both men and women. Menopause, the period in a woman's life during which menstruation ceases, is likely to occur between the ages of 45 and 55. The ovaries are no longer responsive to pituitary gonadotropic hormones because after years of ovulating, much of functional ovarian tissue has been replaced by corpus albicans. After menopause, the ovaries stop producing ova and produce only minimal amounts of estrogen and progesterone. At the onset of menopause, the menstrual cycle becomes irregular and then eventually ceases. Hormonal imbalance often produces physical symptoms, such as dizziness, headaches, insomnia, sleepiness, depression, and "hot flashes" that are caused by circulatory irregularities. Menopausal symptoms vary greatly among women, and some symptoms may be absent altogether.

Following menopause, atrophy of the uterus, vagina, breasts, and external genitals is likely. The lack of estrogen also promotes changes in the skin (e.g., wrinkling; see Chapter 5) and in the skeleton (e.g., osteoporosis; see Chapter 6).

In men, testosterone production diminishes steadily after age 50, which may be responsible for the enlargement of the prostate gland. Sperm production declines with age, yet men can remain fertile well into old age. With age, however, the chance of erectile dysfunction due to degenerative vascular changes in the penis increases.

FOCUS on FORENSICS

Rape

Rape is a crime of violence, and not a crime of sexuality. It happens to all manner of victims: women, men, and children; heterosexual and homosexual; even babies and the elderly. It is a crime that reportedly occurs once every 2 minutes—but that statistic may be much lower than the number of actual incidents. An estimated 25% of reported cases are committed by a complete stranger. Family members carry out approximately 12% of rapes, but the highest numbers of such crimes—63%—are committed by an acquaintance of the victim. Many rapes go unreported, their victims too afraid or too ashamed to seek medical and legal assistance. When rape victims do seek emergency medical care, it is imperative that their needs for physical, psychological, and legal assistance are met promptly and compassionately.

Many communities have formed Sexual Assault Response Teams (SARTs). A key member of such a team is a trained sexual assault nurse examiner (SANE). Forensic nursing is a subspecialty of nursing recognized by the American Nursing Association, which provides training for nurses desiring SANE certification. The team also includes EMS personnel, physicians, psychologists, social workers, and law enforcement officials.

The first step in care for the rape victim is taking a thorough medical history. The patient should be made as comfortable as possible, and may prefer to remain clothed during the initial assessment. A physical exam will follow during which a standardized rape kit may be used to collect evidence. The patient should disrobe while standing on a sheet of exam-table paper in order to catch any hairs from the suspect that might remain on the patient's body. All clothing should be placed in paper evidence bags, and any fluid on the clothing (blood, urine, semen, etc.) should be allowed to air-dry before the paper bags are sealed. This will prevent mold growth, which might destroy evidence. Bruises and abrasions on the patient's body must be documented and photographed. Vaginal exams of females and oral-rectal exams for either gender may also give evidence of injury. The pubic hair is combed through with a fine-tooth comb to capture any pubic hair from the perpetrator.

Fluids found on the patient's body should be collected on swabs for DNA and other chemical testing, and their presence recorded. The victim's blood, hair, and urine will also be sampled for comparison. A high-frequency ultraviolet lamp called a Wood's lamp can detect the presence of semen on the patient's body. Seminal fluid will fluoresce at specific wavelengths of ultraviolet light. Other chemical testing for the components of semen can be done as well. However, it is important to note that absence of semen does not mean that rape did not occur. The perpetrator may have had a vasectomy. Further, a study of assailants documented that 50% reported erectile dysfunction and inability to ejaculate while attempting a rape.

Medical treatment for the victim will include testing for sexually transmitted disease, including gonorrhea, syphilis, and HIV. Antibiotics and anti-retroviral drugs can be prescribed. If the patient is not immunized for hepatitis B, he or she should receive the vaccine. Testing for HIV and hepatitis should be repeated approximately 6 months after the incident. Female victims should also be tested for a pre-existing pregnancy. Nonpregnant women may elect to receive emergency contraception to prevent pregnancy. Such treatment can damage an existing pregnancy and should not be given to a pregnant woman. The patient should also receive urinalysis, especially if he or she does not recall details of the rape. Testing can show evidence of so-called "date-rape" drugs, which induce short-term memory loss.

Psychological counseling is essential if a rape victim is to recover. A rape victim may initially feel anger, fear, or anxiety—although he or she may express no emotion at all. Many victims will internalize blame for the incident, believing that some past incident may have invited the assault. Flashbacks—painful, detailed memories of the incident—are common after rape. Anxiety disorders and post-traumatic stress disorder may occur years after the incident, and also require psychological counseling.

By properly documenting history and evidence, EMS, nursing, and medical professionals can enable successful prosecution of the rapist. Detailed evidence is particularly important in cases of nonstranger sexual assault, such as child, spousal, or elder abuse. Evidence from health-care professionals in these cases can support a victim's claims of lack of consent to sexual activity. Just as important, careful medical and psychological care can enable the victim to recover from his or her ordeal.

Sexual desire and activity need not decline with age, and many older men and women enjoy sexual relationships. Men are likely to experience reduced erection until close to ejaculation, and women may experience a drier vagina.

17.6 Homeostasis

Regulation of sex hormone blood level is an example of homeostatic control. Figure 17.8 shows how the blood level of testosterone is maintained within normal limits. Negative feedback results in a self-regulatory mechanism that maintains the appropriate level of these hormones in the blood.

The illustration in Human Systems Work Together on page 398 shows how the reproductive system works with the other systems of the body to maintain homeostasis. Usually we stress that the function of sex hormones produced not only by the gonads but also by the adrenal glands is to foster

Human Systems Work Together

Integumentary System

Androgens activate oil gland; sex hormones stimulate fat deposition, affect hair distribution in males and females.

Skin receptors respond to touch; modified sweat glands produce milk; skin stretches to accommodate growing fetus.

Skeletal System

Sex hormones influence bone growth and density in males and females.

Bones provide support and protection of reproductive organs.

Muscular System

Androgens promote growth of skeletal muscle.

Muscle contraction occurs during orgasm and moves gametes; abdominal and uterine muscle contractions occur during childbirth.

Nervous System

Sex hormones masculinize or feminize the brain, exert feedback control over the hypothalamus, and influence sexual behavior.

Brain controls onset of puberty; nerves are involved in erection of penis and clitoris, movement of gametes along ducts, and contraction of uterus.

Endocrine System

Gonads produce the sex hormones.

Hypothalamic, pituitary, and sex hormones control sex characteristics and regulate reproductive processes.

How the Reproductive System works with other body systems.

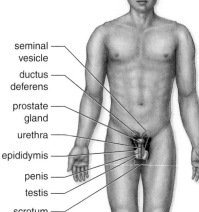
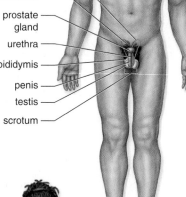

seminal vesicle
ductus deferens
prostate gland
urethra
epididymis
penis
testis
scrotum

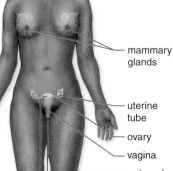

mammary glands

uterine tube
ovary
vagina
external genitalia (clitoris, labia)
uterus

Cardiovascular System

Sex hormones influence cardiovascular health; sexual activities stimulate cardiovascular system.

Blood vessels transport sex hormones; vasodilation causes genitals to become erect; blood services the reproductive organs.

Lymphatic System/Immunity

Sex hormones influence immune functioning; acidity of vagina helps prevent pathogen invasion of body; milk passes antibodies to newborn.

Immune system does not typically attack sperm or fetus, even though they are foreign to the body.

Respiratory System

Sexual activity increases breathing; pregnancy causes breathing rate and vital capacity to increase.

Gas exchange increases during sexual activity.

Digestive System

Pregnancy crowds digestive organs and promotes heartburn and constipation.

Digestive tract provides nutrients for growth and repair of organs and for development of fetus.

Urinary System

Penis in males contains the urethra and performs urination; prostate enlargement hinders urination.

Semen is discharged through the urethra in males; kidneys excrete wastes and maintain electrolyte levels for mother and child.

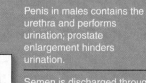

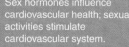
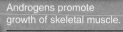

the maturation of the reproductive organs and to maintain the secondary sex characteristics. However, these functions of sex hormones have nothing to do with homeostasis. Why? Because homeostasis pertains to the constancy of the internal environment of cells. Other activities of the sex hormones do affect the internal environment. For example, estrogen promotes fat deposition, which serves as a source of energy for cells and also helps the body maintain a normal temperature because of its insulating effect.

In recent years, it's been discovered that the sex hormones perform other activities that affect homeostasis even more directly. We are just now beginning to discover the role that estrogen and androgens play in the metabolism of cells and therefore their role in homeostasis in general. Estrogen induces the liver to produce many types of proteins that transport substances in the blood. These include proteins that bind iron and copper and lipoproteins that transport cholesterol. Iron and copper are enzyme cofactors necessary to cellular metabolism. Although we associate cholesterol with cardiovascular diseases, in fact, it is also a substance that contributes to the functioning of the plasma membrane.

Estrogen also induces synthesis of bone matrix proteins and counteracts the loss of bone mass. At menopause, when the rate of estrogen secretion is drastically reduced, osteoporosis (decrease in bone density) may develop. Nevertheless, long-term estrogen therapy is no longer recommended. The National Institutes of Health conducted a study of 16,608 healthy women who were taking both estrogen and progesterone—that is, hormone replacement therapy (HRT)—or a placebo. The study was halted after 5.2 years because physicians concluded that the risks for the group on HRT outweighed the benefits. The women on HRT had a small but significant increased risk for breast cancer, coronary heart disease, stroke, and blood clots. The benefits of HRT included lower risk of hip fractures and colon cancer.

Similarly, besides the action of androgens (e.g., testosterone) on the sexual organs and functions of males, androgens play a metabolic role in cells. They stimulate synthesis of structural proteins in skeletal muscles and bone, and they also affect the activity of various enzymes in the liver and kidneys. For example, in the kidney, androgens stimulate synthesis of erythropoietin, the protein that signals the bone marrow to increase the production of red blood cells.

SELECTED NEW TERMS

Basic Key Terms

acrosome (ak′ro-sōm), p. 375

androgens (ăn′drə-jən)

areola (ă-rē′ō-lă), p. 386

assisted reproductive technologies (uh-sis′ted re″pro-duk′tiv tek-nah′lo-jēz), p. 393

bulbourethral gland (bul″bo-yū-re′thral gland), p. 374

cervix (ser′viks), p. 379

clitoris (klĭ′to-ris), p. 383

colostrum (ko-los′trum), p. 386

corpus albicans (kor′pus al′bĭ-kanz), p. 381

corpus luteum (kor′pus lu′te-um), p. 381

diploid (dĭp′loid′), p. 371

ejaculation (e-jak″yū-la′shun), p. 373

ejaculatory duct (e-jak′yū-luh-to″re dukt), p. 373

endometrium (en-do-me′tre-um), p. 379

epididymis (ep″ĭ-did′ĭ-mis), p. 373

estrogen (ĕs′trə-jən), p. 384

external genitals (eks-ter′nal jen′ĭ-talz), p. 377

fimbriae (fim′bre-e), p. 382

follicle (fol′ĭ-kl), p. 380

FSH (follicle-stimulating hormone) (fol′ĭ-kl stim′yū-la-ting hor′mōn), p. 378

gametes (găm′ēt′), p. 371

genitals (jen′i-tălz), p. 371

germ cells (jûrm sēl), p. 375

haploid (hăp′loid′), p. 371

human chorionic gonadotropin (hyū′man ko″re-on′ik go′nad-o-tro′pin), p. 384

implantation (im″plan-ta′shun), p. 379

inhibin (in-hib′in), p. 378

interstitial cell (in″ter-stish′ul sel), p. 375

labia majora (la′be-uh muh-jor′uh), p. 383

labia minora (la′be-uh mi-nor′uh), p. 383

lactation (lak-tā′-shun), p. 386

LH (luteinizing hormone) (lu′te-ĭ-nīz-ing hor′mōn), p. 378

meiosis (mi-o′sis), p. 371

menopause (men′o-pawz), p. 385

menses (mĕn′sēz′), p. 384

menstrual cycle (men″stru-ăl), p. 384

mons pubis (mŏnz pyoo′bis), p. 383

oocyte (o′o-sīt), p. 380

oogenesis (o″o-jen′ĕ-sis), p. 381

orgasm (ŏr-gazm′), p. 377

ova (ō-vŭh), p. 379

ovaries (ō′və-rēs), p. 379

ovulation (ov′yū-la′shun), p. 381

ovum (o-vŭm), p. 379

pap smear (păp smîr), p. 382

penis (pe′nis), p. 377

polar body (pō′lĕr bŏd-ē), p. 381

progesterone (prŏ-jēs′tə-rōn′), p. 384

prostaglandins (prŏs′tə-glăn′dĭn), p. 374

prostrate gland (prŏs′trăt′glănd), p. 373

puberty (pyu′ber-te), p. 371

scrotum (skro′tum), p. 373

semen (se′men), p. 373

seminal vesicle (sem′i-năl vĕs-i- kul), p. 373

seminiferous tubule (se″mĭ-nif′er-us tu′byūl), p. 374

sperm (sperm), p. 375

spermatic cord (sperm-ă′-tik kŏrd), p. 376

spermatogenesis (sper″muh-to-jen′ĕ-sis), p. 375

spermatogonia (spĕr′-mă-tō-gō-nē-ŭm), p. 375

sustentacular (Sertoli) cells (sus′tĕn-tăk-yoo-lĕr sells), p. 375

testis (tĕs′tĭs)

testosterone (tĕs-tŏs′tə-rōn′), p. 375

urethra (yōō-rē′thrə)

uterine tube (yū′ter-in tūb), p. 381

uterus (yoo′-tĕr-ŭs), p. 382

vagina (vuh-ji′nuh), p. 379

vas deferens (vas def′er-ens), p. 373

vesicular (Graafian) follicle (ves-ik′yū-ler [graf′e-un] fol′ĭ-kl), p. 381

vestibule (vĕs′tə-byōō-l′), p. 383

vulva (vul′vuh), p. 379

zygote (zi′gōt), p. 372

Clinical Key Terms

artificial insemination (ar″tĭ-fĭ′shul in-sem-ĭ-na′shun), p. 393

chancre (shan′ker), p. 396

chlamydia (kluh-mĭ′de-uh), p. 396

circumcision (ser″kum-sizh′un), p. 377

contraception (kon-tră-sep′shun), p. 387

cryptorchidism (krĭp-tōr′ki-dizm), p. 374

ectopic pregnancy (ek-top′ik preg′nun-se), p. 382

endometriosis (en-do-me″tre-o′sis), p. 393

erectile dysfunction (e-rek′til dis-funk′shun), p. 377

SUMMARY

17.1 Human Life Cycle

The life cycle of humans requires two types of nuclear division: mitosis and meiosis. Meiosis is involved in sperm production in males and ova production in females.

17.2 Male Reproductive System

A. In males, spermatogenesis, occurring in seminiferous tubules of the testes, produces sperm that mature and are stored in the epididymides. Sperm may also be stored in the vasa deferentia before entering the urethra, along with secretions produced by the seminal vesicles, prostate gland, and bulbourethral glands. Sperm and these secretions are called semen, or seminal fluid.

B. The external genitals of males are the penis, the organ of sexual intercourse, and the scrotum, which contains the testes. Orgasm in males is a physical and emotional climax during sexual intercourse that results in ejaculation of semen from the penis.

C. Hormonal regulation, involving secretions from the hypothalamus, the anterior pituitary, and the testes, maintains testosterone, produced by the interstitial cells of the testes, at a fairly constant level. FSH from the anterior pituitary promotes spermatogenesis in the seminiferous tubules, and LH promotes testosterone production by the interstitial cells.

17.3 Female Reproductive System

A. In females, oogenesis occurring within the ovaries typically produces one mature follicle each month. This follicle balloons out of the ovary and bursts, releasing a secondary oocyte, which enters a uterine tube. The uterine tubes lead to the uterus, where implantation and development occur. The external genital area includes the vaginal opening, mons pubis, the clitoris, the labia minora, and the labia majora.

B. The vagina is the organ of sexual intercourse and the birth canal in females. The vagina and the external genitals, especially the clitoris, play an active role in orgasm, which culminates in uterine and uterine tube contractions.

C. Hormonal regulation in females involves the production of FSH and LH by the anterior pituitary and also production of estrogen and progesterone by the ovaries.

D. The menstrual cycle is a monthly series of events that can be divided into the pre-ovulation events and the post-ovulation events. Before ovulation, FSH from the anterior pituitary causes an estrogen-producing follicle to begin developing in the ovary. Meanwhile in the uterus, menstruation occurs before the proliferation phase occurs. During the proliferation phase, estrogen causes the uterine lining to thicken.

Ovulation is caused by a positive feedback cycle in which an abundance of estrogen brings about an FSH and LH surge. After ovulation, the corpus luteum in the ovary secretes primarily progesterone that causes the uterine lining to become secretory. If the egg is fertilized, it implants itself in the uterine lining and the corpus luteum does not disintegrate.

E. If fertilization takes place, the embryo implants itself in the thickened endometrium. If fertilization and implantation occur, the corpus luteum in the ovary is maintained because of HCG production by the placenta, and therefore progesterone production does not cease. Menstruation usually does not occur during pregnancy.

17.4 Control of Reproduction and Sexually Transmitted Diseases

A. Numerous contraceptive methods and devices, such as the birth control pill, diaphragm, and condom, are available for those who wish to prevent pregnancy.

Emergency contraception can be taken before there is any indication of pregnancy. Mifepristone blocks progesterone receptors, causing an implanted embryo to be shed with the uterine lining.

B. Some couples are infertile, and if so, they may use assisted reproductive technologies in order to have a child. Artificial insemination and in vitro fertilization have been followed by more sophisticated techniques such as intracytoplasmic sperm injection.

17.5 Effects of Aging

Menopause occurs between the ages of 45 and 55 in women. Following menopause, atrophy of the genitals is likely. In men, testosterone production decreases after age 50, and the incidence of erectile dysfunction increases.

17.6 Homeostasis

The reproductive system works with the other systems of the body in the ways described in Human Systems Work Together on page 398.

STUDY QUESTIONS

1. Outline the path of sperm. What glands contribute fluids to semen? (p. 371)
2. Discuss the anatomy and physiology of the testes. Describe the structure of sperm. (pp. 374–375)
3. Name the endocrine glands involved in maintaining the sex characteristics of males and the hormones produced by each. (p. 376)
4. Describe the organs of the female genital tract. Where do fertilization and implantation occur? Name two functions of the vagina. (pp. 379–380)
5. Name and describe the external genitals in females. (p. 383)
6. Discuss the anatomy and the physiology of the ovaries. (pp. 379–380)
7. Describe the pre- and post-ovulation events of the menstrual cycle. In what way is menstruation prevented if pregnancy occurs? (p. 384)
8. Name three functions of the female sex hormones. (p. 384)
9. Discuss the various means of birth control and their relative effectiveness in preventing pregnancy. (pp. 387–390)
10. Describe how in vitro fertilization is carried out. (p. 394)

LEARNING OBJECTIVE QUESTIONS

Fill in the blanks.

1. In tracing the path of sperm, the structure that follows the epididymis is the _____.
2. The prostate gland, the bulbourethral glands, and the _____ all contribute to seminal fluid.
3. An erection is caused by the entrance of _____ into erectile tissue of the penis.
4. The primary male sex hormone is _____.
5. In the female reproductive system, the uterus lies between the uterine tubes and the _____.
6. The female sex hormones are _____ and _____.
7. In the menstrual cycle, once each month a _____ produces an egg, and the _____ is prepared to receive the pre-embryo.
8. The most frequent causes of male infertility are _____ and _____.
9. Spermatogenesis occurs within the _____ of the testes.
10. Androgens are secreted by the _____ cells that lie between seminiferous tubules.
11. Once a vesicular follicle has released an egg, it develops into a glandlike structure called a _____.
12. The release of an egg from a vesicular follicle is called _____.
13. Whereas AIDS and genital herpes are caused by _____, gonorrhea and chlamydia are caused by _____.

MEDICAL TERMINOLOGY EXERCISE

Consult Appendix B for help in pronouncing and analyzing the meaning of the terms that follow.

1. orchidopexy (or″kĭ-do-pek′se)
2. transurethral resection of prostate (TURP) (trans″yū-re′thral re-sek′shun ov pros′tāt)
3. gonadotropic (go″nad-o-trōp′ik)
4. contraceptive (kon″truh-sep′tiv)
5. gynecomastia (jin″ĕ-ko-mas′te-uh)
6. hysterosalpingo-oophorectomy (his″ter-o-sal-ping′go-o″ahf-or-ek′to-me)
7. colporrhaphy (kol-por′uh-fe)
8. menometrorrhagia (men′o-me-tro-ra-je-uh)
9. multipara (mul-tip′uh-ruh)
10. balanitis (bal″uh-ni′tis)
11. seminoma (sem′ĭ-no′muh)
12. genitourinary (jen′ĭ-to-yū′rĭ-nār′e)
13. prostatic hypertrophy (pros-tat′ik hy′per-tro′fe)
14. azoospermia (ā-zo′o-sper′me-uh)

WEBSITE LINK

Visit this text's website at http://www.mhhe.com/maderap6 for additional quizzes, interactive learning exercises, and other study tools.

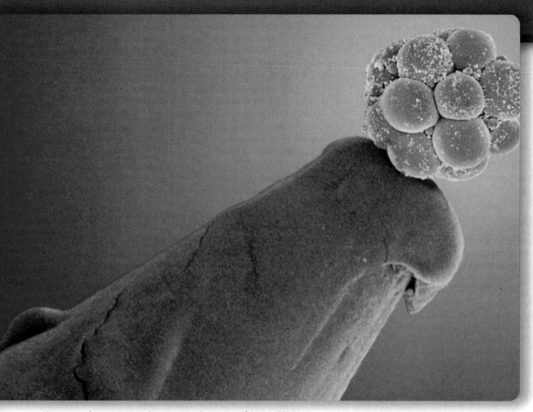

CHAPTER

18

Human Development and Birth

Human embryo at an early stage on the point of a pin (SEM).

Chapter Outline & Learning Objectives

After you have studied this chapter, you should be able to:

18.1 Fertilization (p. 403)
- Explain the events of fertilization and the conversion of the egg into a zygote.

18.2 Development (p. 404)
- Discuss the processes of development.
- Name the four extraembryonic membranes, and give a function for each.
- Describe the events that occur during pre-embryonic and embryonic development.

- Describe the structures and functions of the placenta and the umbilical cord.
- Describe the events that occur during fetal development.

18.3 Birth (p. 415)
- Describe the three stages of birth.
- In general, describe the physical and physiological changes in the mother during pregnancy.

What's New
Therapeutic Cloning (p. 407)

Medical Focus
Premature Babies (p. 413)
Preventing Birth Defects (pp. 416–418)

18.1 Fertilization

Fertilization of an ovum at conception requires that a single sperm penetrate the ovum. When the genetic materials from ovum and sperm are combined, the **zygote** is formed. This single cell is the foundation for a new, unique individual. Figure 18.1 shows the manner in which an ovum is fertilized by a sperm in humans.

Sperm and Ovum Anatomy

A sperm has three distinct parts: a head, a middle piece, and a tail. The tail is a flagellum, which allows the sperm to swim through the female reproductive tract toward the ovum. The middle piece contains energy-producing mitochondria. The head contains the sperm nucleus, which encloses the 23 chromosomes from the male. Capping the sperm head is a membrane-bound **acrosome,** which contains digestive enzymes. Very little cytoplasm is present in a sperm cell. Notice that only the nucleus from the sperm head fuses with the ovum nucleus. Therefore, the zygote receives cytoplasm and organelles only from the mother.

The plasma membrane of the ovum is surrounded by an extracellular matrix termed the **zona pellucida.** In turn, the zona pellucida is surrounded by a few layers of adhering follicular cells, collectively called the **corona radiata.** These cells nourished the ovum when it was in a follicle of the ovary.

Steps of Fertilization

During fertilization, (1) several sperm penetrate the corona radiata, (2) several sperm attempt to penetrate the zona pellucida, and (3) one sperm enters the ovum. The acrosome plays a role in allowing sperm to penetrate the zona pellucida. After a sperm head binds tightly to the zona pellucida, the acrosome releases digestive enzymes that forge a pathway for the sperm through the zona pellucida. When a sperm binds to the ovum, their plasma membranes fuse, and this sperm (the head, the middle piece, and usually the tail) enters the ovum. Fusion of the sperm nucleus and the ovum nucleus follows.

To ensure proper development, only one sperm should enter an ovum. Accidental entry of more than one sperm cell will halt any further development by the zygote. Cell division cannot take place because additional chromosomes in the zygote will prevent the pairing of chromosomes essential for mitosis. Prevention of polyspermy (entrance of more than one sperm) depends on changes in the ovum's plasma membrane and in the zona pellucida. As soon as a sperm touches

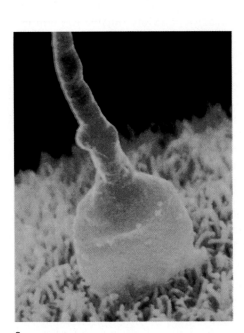

a.

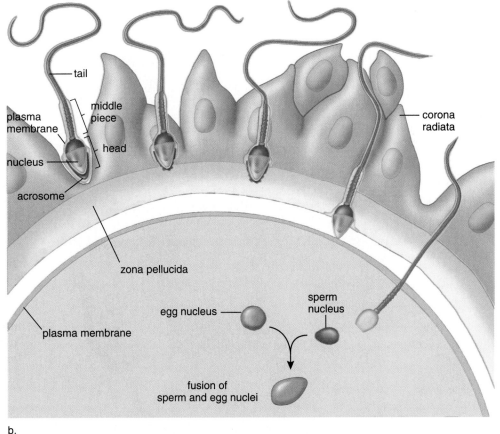

b.

Figure 18.1 Fertilization. **a.** During fertilization, a single sperm enters the egg. **b.** The head of a sperm has a membrane-bound acrosome filled with enzymes. When released, these enzymes digest a pathway for the sperm through the zona pellucida. After it binds to the plasma membrane of the egg, a sperm enters the egg. When the sperm nucleus fuses with the egg nucleus, fertilization is complete.

an ovum, the ovum's plasma membrane depolarizes (from –65 mV to 10 mV), and this prevents the binding of any other sperm. Then the ovum releases substances that lead to a lifting of the zona pellucida away from the surface of the ovum. Now sperm cannot bind to the zona pellucida either.

Content CHECK-UP!

1. The nonliving extracellular matrix material surrounding the ovum is called the:
   a. corona radiata.   c. corpus albicans.
   b. zona pellucida.   d. acrosome.

2. Put the following events of fertilization in the correct order as they occur:
   1. Sperm nucleus and ovum nucleus fuse.
   2. Several sperm penetrate the corona radiata.
   3. One sperm penetrates the ovum.
   4. Barriers are set up to prevent polyspermy.

   a. 2-3-4-1   c. 3-2-4-1
   b. 2-4-3-1   d. 3-4-2-1

3. Which of the following is the first layer that a sperm must pass through to fertilize the ovum?
   a. corona radiata   c. cell membrane
   b. zona pellucida

18.2 Development

Before we discuss the stages of development, you will want to become familiar with the processes of development and the names and functions of the extraembryonic membranes.

Processes of Development

As a human being develops, these processes occur:

Cleavage. Immediately after fertilization, the zygote begins to divide so that there are first 2, then 4, 8, 16, 32 cells, and so forth. During these first divisions, the ball of cells that is formed does not increase in size; it remains the same size as the zygote (see Fig. 18.3). Cell division during cleavage is mitotic, and each cell receives a full complement of chromosomes and genes.

Growth. During embryonic development, cell division is accompanied by an increase in size of the daughter cells.

Morphogenesis. Morphogenesis refers to the shaping of the embryo and is first evident when certain cells are seen to move, or migrate, in relation to other cells. By these movements, the embryo begins to take on a defined shape.

Differentiation. When cells take on a specific structure and function, differentiation occurs. The first system to become visibly differentiated is the nervous system.

Extraembryonic Membranes

The **extraembryonic membranes** are not part of the embryo and fetus; instead, as implied by their name, they are outside the embryo (Fig. 18.2). The names of the extraembryonic membranes in humans are strange to us because they are named for their function in shelled animals! In shelled animals, the chorion lies next to the shell and carries on gas exchange. The amnion contains the protective amniotic fluid, which bathes the developing embryo. The allantois collects nitrogenous wastes, and the yolk sac surrounds the yolk, which provides nourishment.

The functions of the extraembryonic membranes are different in humans because humans develop inside the uterus. The extraembryonic membranes have these functions in humans:

1. **Chorion.** The chorion develops into the fetal half of the **placenta,** the organ that provides the embryo/fetus with nourishment and oxygen and takes away its waste.

2. **Yolk Sac.** The yolk sac has little yolk and is the first site of blood cell formation.

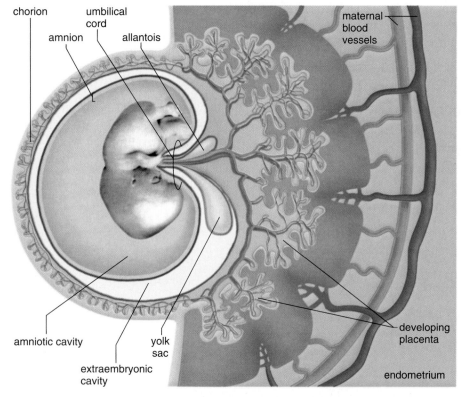

Figure 18.2 The extraembryonic membranes. The chorion and amnion surround the embryo. The two other extraembryonic membranes, the yolk sac and allantois, contribute to the umbilical cord.

TABLE 18.1 Human Development

Time	Events for Mother	Events for Baby
PRE-EMBRYONIC DEVELOPMENT		
First week	Ovulation occurs.	Fertilization occurs. Cell division begins and continues. Chorion appears.
EMBRYONIC DEVELOPMENT		
Second week	Symptoms of early pregnancy (nausea, breast swelling and tenderness, fatigue) are present. Blood pregnancy test is positive.	Implantation occurs. Amnion and yolk sac appear. Embryo has tissues. Placenta begins to form.
Third week	First menstruation is missed. Urine pregnancy test is positive. Symptoms of early pregnancy continue.	Nervous system begins to develop. Allantois and blood vessels are present. Placenta is well formed.
Fourth week		Limb buds form. Heart is noticeable and beating. Nervous system is prominent. Embryo has tail. Other systems form.
Fifth week	Uterus is the size of a hen's egg. Mother feels frequent need to urinate due to pressure of growing uterus on bladder.	Embryo is curved. Head is large. Limb buds show divisions. Nose, eyes, and ears are noticeable.
Sixth week	Uterus is the size of an orange.	Fingers and toes are present. Skeleton is cartilaginous.
Two months	Uterus can be felt above the pubic bone.	All systems are developing. Bone is replacing cartilage. Facial features are becoming refined. Embryo is about 38 mm (1½ in.) long.
FETAL DEVELOPMENT		
Third month	Uterus is the size of a grapefruit.	Gender can be distinguished by ultrasound. Fingernails appear.
Fourth month	Fetal movement is felt by a mother who has previously been pregnant.	Skeleton is visible. Hair begins to appear. Fetus is about 150 mm (6 in.) long and weighs about 170 g (6 oz).
Fifth month	Fetal movement is felt by a mother who has not previously been pregnant. Uterus reaches up to level of umbilicus, and pregnancy is obvious.	Protective cheesy coating, called vernix caseosa, begins to be deposited. Heartbeat can be heard.
Sixth month	Doctor can tell where baby's head, back, and limbs are. Breasts have enlarged, nipples and areolae are darkly pigmented, and colostrum is produced.	Body is covered with fine hair called lanugo. Skin is wrinkled and reddish.
Seventh month	Uterus reaches halfway between umbilicus and rib cage.	Testes descend into scrotum. Eyes are open. Fetus is about 300 mm (12 in.) long and weighs about 1,350 g (3 lb).
Eighth month	Weight gain is averaging about a pound a week. Standing and walking are difficult because center of gravity is thrown forward.	Body hair begins to disappear. Subcutaneous fat begins to be deposited.
Ninth month	Uterus is up to rib cage, causing shortness of breath and heartburn. Sleeping becomes difficult.	Fetus is ready for birth. It is about 530 mm (20½ in.) long and weighs about 3,400 g (7½ lb).

3. **Allantois.** The allantois blood vessels become the umbilical blood vessels.
4. **Amnion.** The amnion contains fluid to cushion and protect the embryo, which develops into a fetus.

Stages of Development

Development encompasses the events that occur from fertilization to birth. In humans, this **gestation** period is usually calculated by adding 280 days to the start of the last menstruation, a date that is usually known. However, only about 5% of babies actually arrive on the predicted date.

Pre-Embryonic Development

Table 18.1 shows that we can subdivide development into pre-embryonic, embryonic, and fetal development. **Pre-embryonic development** encompasses the events of the first week, as shown in Figure 18.3.

Immediately after fertilization, the zygote divides repeatedly as it passes down the uterine tube to the uterus. A **morula** is a compact ball of embryonic cells that becomes a **blastocyst.** The many cells of the blastocyst arrange themselves so that there is an **inner cell mass** surrounded by a layer of cells, the **trophoblast.** The trophoblast will become the *chorion.* The early appearance of the chorion emphasizes the complete

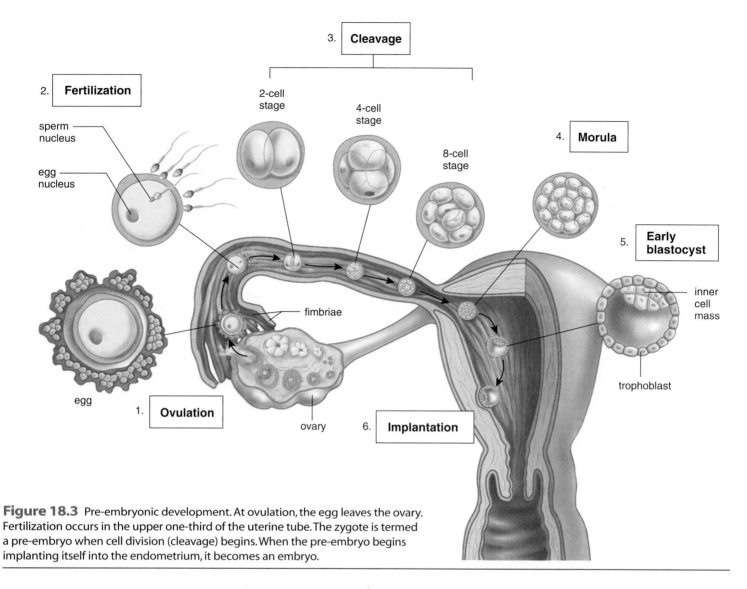

3. **Cleavage**

2. **Fertilization**

sperm
nucleus

egg
nucleus

2-cell
stage

4-cell
stage

8-cell
stage

4. **Morula**

5. **Early blastocyst**

inner
cell
mass

fimbriae

trophoblast

egg

1. **Ovulation**

ovary

6. **Implantation**

Figure 18.3 Pre-embryonic development. At ovulation, the egg leaves the ovary. Fertilization occurs in the upper one-third of the uterine tube. The zygote is termed a pre-embryo when cell division (cleavage) begins. When the pre-embryo begins implanting itself into the endometrium, it becomes an embryo.

dependence of the developing embryo on this extraembryonic membrane. The inner cell mass will become the embryo.

Each cell within the morula and blastocyst has the genetic capability of becoming any tissue. This recognition has recently led to a new procedure called therapeutic cloning, as discussed in the What's New reading on page 407. Sometimes during development, the cells of the morula separate, or the inner cell mass splits, and two pre-embryos are present rather than one. If each of the two pre-embryos is able to complete development, the two babies formed will be *identical twins* because they have inherited exactly the same chromosomes. Should separation of the cell mass be incomplete, the identical twins formed are called *conjoined twins* (formerly referred to as Siamese twins). Conjoined twins may be joined at any part of the body. *Fraternal twins,* who arise when two different eggs are fertilized by two different sperm, do not have identical chromosomes.

Embryonic Development

Embryonic development begins with the second week and lasts until the end of the second month of development.

Second Week

At the end of the first week, the **embryo** usually begins the process of implanting itself in the wall of the uterus. If **implantation** is successful, the woman is clinically pregnant. On occasion, it happens that the embryo implants itself in a location other than the uterus—most likely, the uterine tube. Implantation of the embryo in any location outside the uterus is called extrauterine or **ectopic pregnancy.** An ectopic pregnancy in the uterine tube must be surgically removed, because the tube is too narrow to allow the continuing growth of the embryo. Should the pregnancy continue, the uterine tube will eventually rupture, causing a fatal hemorrhage to both mother and embryo. In rare circumstances, ectopic pregnancies elsewhere in the abdominal cavity (for example, on the outside of the uterine wall or on the wall of the urinary bladder) can continue to term, producing a live infant. Of course, birth is by Caesarian section in this circumstance.

During implantation, the trophoblast secretes enzymes to digest away some of the tissue and blood vessels of the endometrium of the uterus. The trophoblast also begins to secrete **human chorionic gonadotropin (HCG),** the hormone

Therapeutic Cloning

The term *cloning* means making exact multiple copies of genes, a cell, or an organism. For example, identical twins are clones of a single zygote. The cloning of human beings may some day be possible. The procedure would begin as described in Figure 18A. In other words, the person being cloned need not contribute sperm or an egg to the process. Instead, a 2n (diploid) nucleus from, say, a fibroblast can be placed in an enucleated egg, and that egg begins developing. The pre-embryo (blastocyst) would be implanted in the uterus of a surrogate mother where it would develop until birth. Although subject to environmental influences, the clone would be expected to closely resemble its "parent."

Therapeutic cloning is not the same as cloning a human being because the cells of the pre-embryo are separated and treated to become particular tissues, which can be used to treat the patient. The separated cells of a pre-embryo are called stem cells because they divide repeatedly and can become various tissues, as shown in Figure 18A. The tissues resulting from this procedure will not be subject to rejection by the patient because they bear the same surface molecules as the patient's cells. However, there is another way to carry out therapeutic cloning. Fertility clinics store extra pre-embryos prepared by in vitro fertilization but not used. These pre-embryos could be a source of stem cells to make tissues that could be stored and used when needed by any patient, if they were stripped of rejection-causing surface molecules.

So far, therapeutic cloning is experimental and has not been perfected. (Claims of success by a Korean researcher in autumn 2005 were later proven to be a hoax.) However, one day it may provide insulin-secreting cells for diabetics, nerve cells for stroke patients or those with Parkinson's disease, cardiac cells for those with heart disease, and so forth. Yet, it is important to recognize that ethical concerns about therapeutic cloning remain—after all, any pre-embryo is potentially a living, breathing human being.

Anticipating intense interest in therapeutic cloning, the U.S. National Academy of Sciences proposed strict new guidelines for federally funded research in 2005. Among the committee's recommendations was the establishment of Embryonic Stem Cell Research Oversight, or ESCRO, committees, to approve the research before it is begun. An ESCRO committee review would be in addition to the current review required by an Institutional Review Board. ESCRO committees would include bioethicists and legal experts, as well as members of the general public.

Further, the Academy recommended that prior to beginning any research, informed consent from the donors of ova or sperm must be obtained. Stem cells created for therapeutic cloning would never be used for reproductive purposes under the proposed guidelines. The guidelines also would require that the embryos created could not be grown in culture for longer than 14 days. At that point, the primitive streak of the developing nervous system begins to form. Finally, the Academy called for the formation of a national agency charged with regularly reviewing all guidelines on stem cell research.

Because of the controversy over the cloning of stem cells derived from pre-embryos, some researchers have begun searching for other sources. It turns out that the adult body has not only blood stem cells, but also neural stem cells in the brain. It has even been possible to coax blood stem cells and neural stem cells to become other types of human tissues in the body. Another potential source of blood stem cells is a baby's umbilical cord and umbilical blood can now be stored for future use. Once researchers have the know-how, they may be able to use any type of stem cell to cure many of the disorders afflicting human beings.

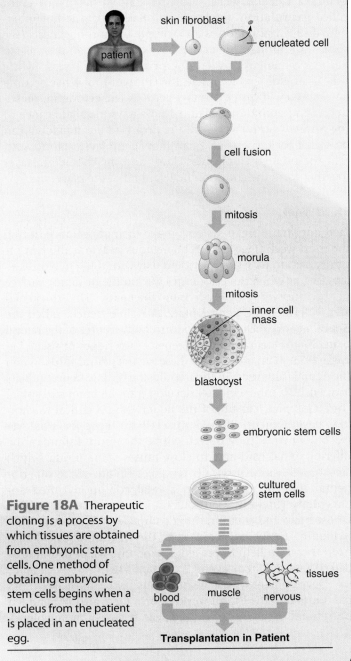

Figure 18A Therapeutic cloning is a process by which tissues are obtained from embryonic stem cells. One method of obtaining embryonic stem cells begins when a nucleus from the patient is placed in an enucleated egg.

that is the basis for the pregnancy test. HCG acts like luteinizing hormone (LH) in that it serves to maintain the corpus luteum past the time it normally disintegrates. Because it is being stimulated, the corpus luteum secretes progesterone, the endometrium is maintained, and the expected menstruation does not occur.

The embryo is now about the size of the period at the end of this sentence. As the week progresses, further growth of cells below the inner cell mass causes it to separate from the trophoblast. The inner cell mass now becomes the **embryonic disk,** and two more extraembryonic membranes form. The yolk sac is the first site of blood cell formation. The amniotic cavity surrounds the embryo (and then the fetus) as it develops. In humans, amniotic fluid acts as an insulator against cold and heat and also absorbs shock, such as that caused by the mother exercising.

Primary Germ Layers With the start of the major event called **gastrulation,** the inner cell mass becomes the embryonic disk. Gastrulation is an example of morphogenesis (see page 404) during which cells move or migrate, in this case to become tissue layers called the **primary germ layers.** By the time gastrulation is complete, the embryonic disk has become an embryo with three primary germ layers: **ectoderm, mesoderm,** and **endoderm.** Figure 18.4 shows the significance of the primary germ layers—all the organs of an individual can be traced back to one of the primary germ layers. Notice also that the original trophoblast layer and the mesoderm layer together form the chorion, the fetal half of the placenta.

Third Week

Two important organ systems make their appearance during the third week (Fig. 18.5). The nervous system is the first organ system to be visually evident. Growth of the nervous system begins when the ectoderm layer on the posterior surface of the embryonic disk becomes thickened. This thickened area is termed the **neural plate.** A visible trench, called the **neural groove,** next begins to form in the center of the neural plate. As the neural groove deepens, two ridges of tissue become evident on either side; these are the **neural folds.** When the neural folds meet in the middle, the hollow **neural tube** is formed. This tube will later develop into the brain and spinal cord. Complete closure of the neural tube is critical for normal development. Should this fail to occur for whatever reason, a **neural tube defect** will be the result. Failure of the inferior spinal cord area to close causes **spina bifida,** which can sometimes be surgically repaired. Failure of the superior neural tube to close causes a fatal condition called **anencephaly** (literally, "without brain"), resulting in a fetus whose brain and cranium never complete development. Multivitamin supplements containing the vitamin folic acid have been shown to help to prevent the occurrence of neural tube defects (see the Medical Focus on page 416).

Development of the heart begins in the third week and continues into the fourth week. At first, there are right and left heart tubes; when these fuse, the heart begins pumping blood, even though the chambers of the heart are not fully formed. The veins enter posteriorly, and the arteries exit anteriorly from this largely tubular heart, but later the heart twists so that all major blood vessels are located anteriorly.

Fourth and Fifth Weeks

At four weeks, the embryo is barely larger than the height of this print. A body stalk connects the caudal (tail) end of the embryo with the chorion. Branching projections of the chorion called **chorionic villi** extend into the endometrial wall of the uterus (see Fig. 18.5). Within each chorionic villus is a capillary network. Exchange of gases, nutrients, and wastes from mother to fetus can take place across this capillary network. The fourth extraembryonic membrane, the allantois, lies within the body stalk, and its blood vessels become the umbilical blood vessels. The head and the tail then lift up, and the body stalk moves anteriorly by constriction. Once this process is complete, the **umbilical cord,** which connects the developing embryo to the placenta, is fully formed (see Fig. 18.5d).

Little flippers called limb buds appear (Fig. 18.6); later, the arms and the legs develop from the limb buds, and even the hands and the feet become apparent. At the same time—during the fifth week—the head enlarges and the sense organs become more prominent. It is possible to make out the developing eyes and ears, and even the nose.

Sixth Through Eighth Weeks

During the sixth through eighth weeks of development, the embryo changes to a form that is easily recognized as a human being. Concurrent with brain development, the head achieves its normal relationship with the body as a neck region develops. The nervous system is developed well enough to permit reflex actions, such as a startle response to touch. At the end of this period, the embryo is about 38 mm (1.5 in.) long and weighs no more than an aspirin tablet, even though all organ systems have been established.

Placenta

The placenta is shaped like a pancake, measuring 15 to 20 cm in diameter and 2.5 cm thick. The placenta is normally fully formed and functional by the end of the embryonic period and before the fetal period begins. The placenta is expelled as the **afterbirth** following the birth of an infant.

The placenta has two portions, a fetal portion composed of chorionic tissue and a maternal portion composed of uterine tissue. Chorionic villi cover the entire surface of the chorion until about the eighth week when they begin to disappear, except in one area. These villi are surrounded by maternal blood, and it is here that exchanges of materials take place across the placental membrane. The **placental membrane** consists of the epithelial wall of an embryonic capillary and the epithelial wall of a chorionic villus. Maternal blood rarely mingles with fetal blood. Instead, oxygen and nutrient molecules, such as glucose and amino acids, diffuse from maternal blood across the placental membrane into fetal blood, and

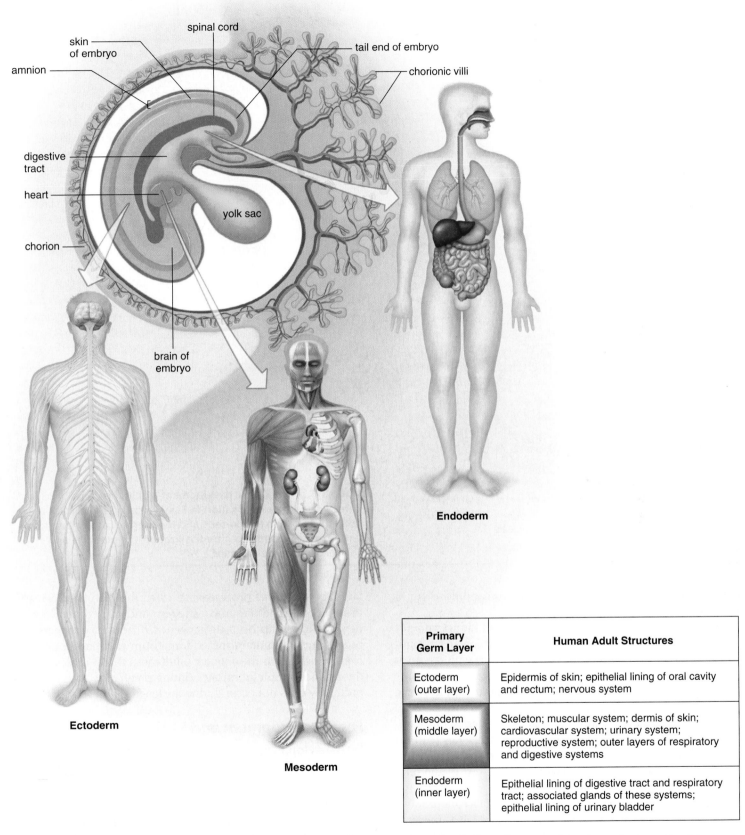

Primary Germ Layer	Human Adult Structures
Ectoderm (outer layer)	Epidermis of skin; epithelial lining of oral cavity and rectum; nervous system
Mesoderm (middle layer)	Skeleton; muscular system; dermis of skin; cardiovascular system; urinary system; reproductive system; outer layers of respiratory and digestive systems
Endoderm (inner layer)	Epithelial lining of digestive tract and respiratory tract; associated glands of these systems; epithelial lining of urinary bladder

Figure 18.4 An embryo has three primary germ layers: ectoderm, mesoderm, and endoderm. Organs and tissues can be traced back to a particular germ layer as indicated in this illustration.

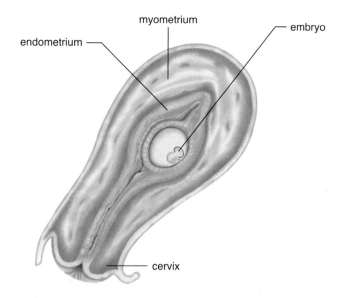

a. 3 weeks

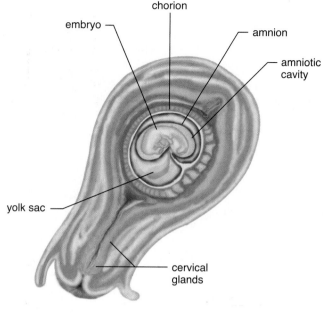

b. 5 weeks

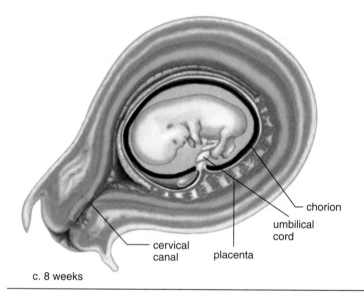

c. 8 weeks

Figure 18.5 Embryonic development within the uterus. **a.** Three weeks after fertilization. **b.** Five weeks after fertilization, amnion and chorion are present, and the uterus is about the size of a hen's egg. **c.** Two months after fertilization, the placenta and umbilical cord are well formed.

carbon dioxide and other wastes, such as urea, diffuse out of fetal blood into maternal blood.

Note that the digestive system, lungs, and kidneys do not function in the fetus. The functions of these organs are not needed because the placenta supplies the fetus with its nutritional and excretory needs.

The umbilical cord transports fetal blood to and from the placenta (Fig. 18.7; see Fig. 12.19). The umbilical cord is the fetal lifeline because it contains the umbilical arteries and a single umbilical vein, which transport waste molecules (carbon dioxide and urea) to the placenta for disposal and oxygen and nutrient molecules from the placenta to the rest of the fetal circulatory system.

As mentioned, the chorion and then the placenta produce HCG, the hormone detected by a pregnancy test. HCG prevents the normal degeneration of the corpus luteum of the ovary. Instead, HCG stimulates the corpus luteum to secrete even

larger quantities of progesterone. Later, the placenta begins to produce progesterone and estrogen, and the corpus luteum degenerates—it is no longer needed. Placental estrogen and progesterone maintain the endometrium and have a negative feedback effect on the anterior pituitary so that it ceases to produce gonadotropic hormones during pregnancy. Menstruation ordinarily does not occur during the length of pregnancy.

Fetal Development and Birth

Fetal development includes the third through the ninth months of development. At this time, the fetus looks human (Fig. 18.8).

Third and Fourth Months

At the beginning of the third month, the fetal head is still very large, the nose is flat, the eyes are far apart, and the ears are well formed. Head growth now begins to slow down as the

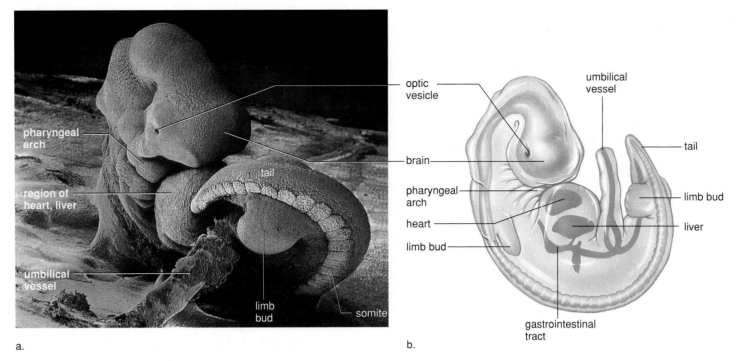

Figure 18.6 Human embryo at beginning of fifth week. **a.** Scanning electron micrograph. **b.** The embryo is curled so that the head touches the region of the heart and liver. The organs of the gastrointestinal tract are forming, and the arms and the legs develop from the bulges called limb buds. The tail is an evolutionary remnant; its bones regress and become those of the coccyx (tailbone).

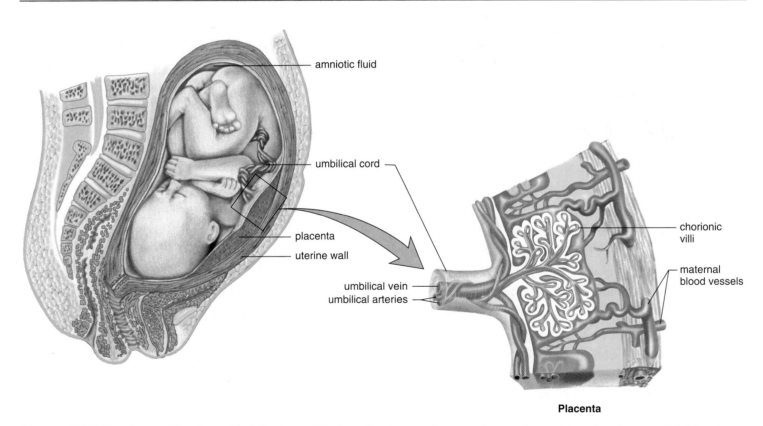

Figure 18.7 The placenta. Blood vessels within the umbilical cord lead to the placenta where exchange takes place between fetal blood and maternal blood. Note that the umbilical vein is colored red here because it is bringing oxygenated blood from the placenta to the fetus. Likewise, the umbilical arteries are colored blue to reflect the deoxygenated blood flowing from the fetus to the placenta for gas exchange with the maternal blood vessels.

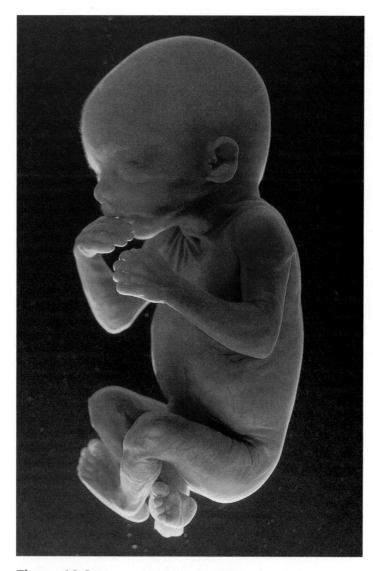

Figure 18.8 Five- to seven-month-old fetus.

rest of the body increases in length. Epidermal refinements, such as fingernails, nipples, eyelashes, eyebrows, and hair on the head, appear.

Cartilage begins to be replaced by bone as ossification centers appear in most of the bones. Cartilage remains at the ends of the long bones, and ossification is not complete until age 18 or 20 years. The skull has six large membranous areas called **fontanels**, which permit a certain amount of flexibility as the head passes through the birth canal and allow rapid growth of the brain during infancy. Progressive fusion of the skull bones causes the fontanels to close, usually by 2 years of age.

Sometime during the third month, it is possible to distinguish males from females. Researchers have discovered a series of genes on the X and Y chromosomes that cause the differentiation of gonads into testes and ovaries. Once these have differentiated, they produce the sex hormones that influence the differentiation of the genital tract.

At this time, either testes or ovaries are located within the abdominal cavity, but later, in the last trimester of fetal development, the testes descend into the scrotal sacs (scrotum). Sometimes the testes fail to descend, and in that case, an operation may be done later to place them in their proper location.

During the fourth month, the fetal heartbeat is loud enough to be heard when a physician applies a stethoscope to the mother's abdomen. By the end of this month, the fetus is about 152 mm (6 in.) in length and weighs about 171 g (6 oz).

Fifth Through Seventh Months

During the fifth through seventh months (Fig. 18.8), the mother begins to feel movement. At first, there is only a fluttering sensation, but as the fetal legs grow and develop, kicks and jabs are felt. The fetus remains tightly curled in the uterus, with the head bent down and in contact with the flexed knees. This posture is aptly termed the fetal position.

The wrinkled, translucent, pink-colored skin is covered by a fine down called **lanugo.** This in turn is coated with a white, greasy, cheeselike substance called **vernix caseosa,** which probably protects the delicate skin from the amniotic fluid. The eyelids are now fully open.

At the end of this period, the fetus's length has increased to about 300 mm (12 in.), and it weighs about 1,380 g (3 lb). It is possible that, if born now, the baby will survive.

Eighth Through Ninth Months

As the end of development approaches, the fetus usually rotates so that the head is pointed toward the cervix. However, if the fetus does not turn, a **breech birth** (rump first) is likely. It is very difficult for the cervix to expand enough to accommodate this form of birth, and asphyxiation of the baby is more likely to occur. Thus, a **cesarean section** may be prescribed for delivery of the fetus (incision through the abdominal and uterine walls).

At the end of nine months, the fetus is about 530 mm (20½ in.) long and weighs about 3,400 g (7½ lb). Weight gain is due largely to an accumulation of fat beneath the skin. Full-term babies have the best chance of survival; as discussed in the Medical Focus on page 413, premature babies are subject to various challenges.

Development of Male and Female Sex Organs

The gender of an individual is determined at the moment of fertilization. Both males and females have 23 pairs of chromosomes; in males, one of these pairs is an X and Y, while females have two X chromosomes. During the first several weeks of development, it is impossible to tell by external inspection whether the unborn child is a boy or a girl. Gonads don't start developing until the seventh week of development. The tissue that gives rise to the gonads is called *indifferent* because it can become testes or ovaries depending on the action and concentration of hormones. Genes on the Y chromosome cause the

Premature Babies

In Memoriam: Brandon Nelson, 1982–1983

Each and every day in the United States, approximately 1,300 babies will be born too soon. **Premature birth** is defined as birth prior to 37 weeks gestation (normal gestation is approximately 40 weeks). Many other babies, although full-term, have a low birth weight (less than 5.5 pounds at birth). These babies face weeks to months in the neonatal intensive care unit, or NICU. Fortunately, advances in the care of these tiniest patients allow most of them to continue to grow and develop while slowly gaining weight. Most eventually go home. However, "premies" and term babies with low birth weight face many serious challenges to their survival:

Respiratory distress syndrome (hyaline membrane disease) The lungs do not produce enough of a chemical surfactant that helps the alveoli stay open. Therefore, the lungs tend to collapse, instead of expanding to be filled with air.

Retinopathy of prematurity The high level of oxygen needed to ensure adequate gas exchange by the immature lungs can lead to proliferation of blood vessels within the eyes, with ensuing blindness.

Intracranial hemorrhage The delicate blood vessels in the brain are apt to break, causing swelling and inflammation of the brain. If not fatal, this can lead to brain damage.

Jaundice The immature liver fails to excrete the waste product bilirubin, which instead builds up in the blood, possibly causing brain damage.

Infections The level of antibodies in the body is low, and the various medical procedures performed could possibly introduce germs. Also, bowel infection is common, along with perforation, bleeding, and shock.

Circulatory disorders Fetal circulation, discussed in Chapter 12, has two features: (1) the oval opening between the atria, and (2) the arterial duct that allows blood to bypass the lungs. If these features persist in the newborn, oxygen-rich blood will mix with oxygen-poor blood. Blood circulation will be impaired, perhaps leading to the delivery of a "blue baby"—that is, a baby with cyanosis, a bluish cast to the skin. Heart failure can also result from these conditions.

Risk for permanent disability Many premies grow up to be normal and healthy, but many others face permanent disability.

Mental retardation, cerebral palsy, learning disorders, chronic lung disease, blindness, and deafness are all potential consequences of premature birth. Further, pediatricians estimate that fully half of all neurological disabilities in children are related to premature birth.

Why do women deliver prematurely? Obstetricians (physicians who specialize in the treatment of women during and after pregnancy) have come up with four general causes for pre-term labor and delivery. The placenta may be disrupted by a bacterial infection; this triggers uterine contractions. Bleeding from the placenta or from the uterus itself may trigger labor. The uterus may be stretched excessively, as often occurs in multiple births. Finally, maternal or fetal stress may trigger a hormone cascade that leads to uterine contraction and pre-term delivery. Pregnant women who have already had a previous pre-term delivery are known to be at risk for subsequent pre-term delivery, as are women pregnant with twins, triplets, or more. Structural abnormalities of the uterus or the cervix may also lead to delivery of a premature baby.

How can a pregnant woman maximize her chances for a full-term pregnancy? First, she should take those precautions needed for all healthy pregnancies: continuing prenatal care throughout the pregnancy; abstaining from alcohol, smoking, and recreational drugs; and early antibiotic therapy for bacterial infections, especially sexually transmitted diseases (see the Medical Focus on page 395). If she has had a previous pre-term delivery, or is carrying more than one fetus, evaluation by an obstetrician who specializes in high-risk pregnancy is an appropriate precaution. Stress in her life also needs to be kept to a minimum, if possible. Being in an abusive relationship, lacking a social network, and working long hours with long periods of standing have all been linked to an increased risk for pre-term labor and delivery.

Unfortunately, despite increased research and improvements in medical care, the rate of premature birth has increased by 31% in the past 20 years. Obviously, more needs to be done to prevent premature birth. The March of Dimes supports educational efforts and research into the causes of birth defects and prematurity. More information can be obtained at www.marchofdimes.com.

---

production of testosterone, and then the indifferent tissue becomes testes.

In Figure 18.9*a*, notice that at six weeks both males and females have the same types of ducts. During this indifferent stage, an embryo has the potential to develop into a male or a female. If a Y chromosome is present, testosterone stimulates

the Wolffian ducts to become male genital ducts. The Wolffian ducts enter the urethra, which belongs to both the urinary and reproductive systems in males. The testes secrete an anti-müllerian hormone that causes the Müllerian ducts to regress.

In the absence of a Y chromosome, ovaries develop instead of testes from the same indifferent tissue. Now the Wolffian

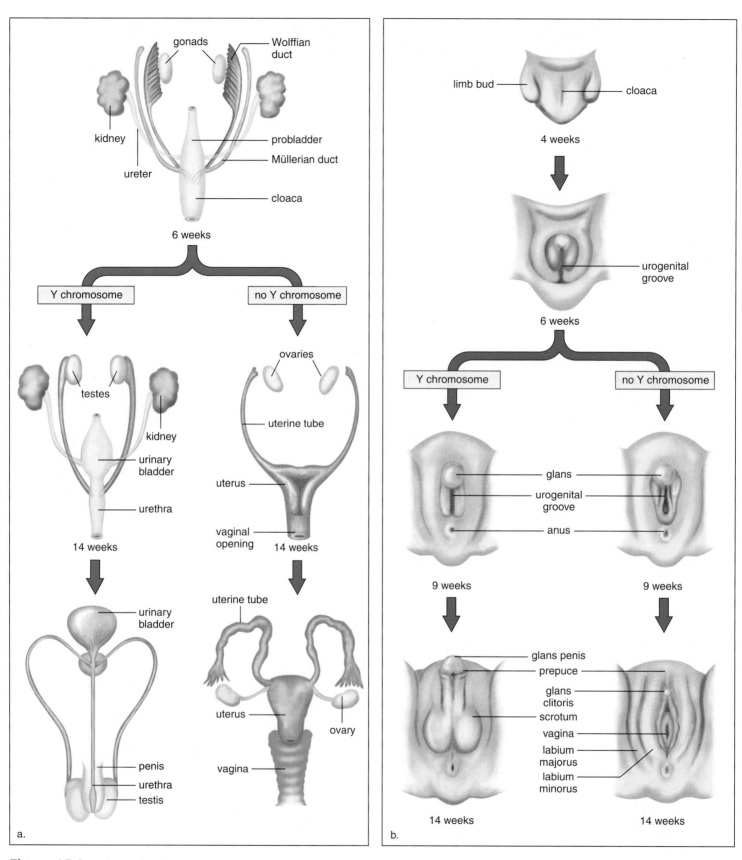

Figure 18.9 Male and female organs. **a.** Development of gonads and ducts. **b.** Development of external genitals.

ducts regress, and the Müllerian ducts develop into the uterus and uterine tubes. A developing vagina also extends from the uterus. There is no connection between the urinary and genital systems in females.

At fourteen weeks, both the primitive testes and ovaries are located deep inside the abdominal cavity. An inspection of the interior of the testes would show that sperm are even now starting to develop, and similarly, the ovaries already contain large numbers of tiny follicles, each having an ovum. Toward the end of development, the testes descend into the scrotal sac; the ovaries remain in the abdominal cavity.

Figure 18.9*b* shows the development of the external genitals. These tissues are also indifferent at first—they can develop into either male or female genitals. At six weeks, a small bud appears between the legs; this can develop into the male penis or the female clitoris, depending on the presence or absence of the Y chromosome and testosterone. At nine weeks, a urogenital groove bordered by two swellings appears. By fourteen weeks, this groove has disappeared in males, and the scrotum has formed from the original swellings. In females, the groove persists and becomes the vaginal opening. Labia majora and labia minora are present instead of a scrotum.

Content CHECK-UP!

4. The site of red blood cell production in an embryo is the:
   a. bone marrow.   c. yolk sac.
   b. chorion.   d. allantois.

5. Which of the following statements regarding the umbilical cord is false?
   a. Blood vessels of the allantois become the umbilical blood vessels.
   b. A body stalk that connects the embryo to the chorion becomes the umbilical cord.
   c. There are two umbilical veins and two umbilical arteries in the cord.
   d. Umbilical arteries carry blood to the placenta, where it is oxygenated.

6. If an embryo has a Y chromosome, which of the following will happen?
   a. The embryo will develop into a male.
   b. The Wolffian duct will develop fully, whereas the Müllerian duct regresses.
   c. The Müllerian duct will develop fully, whereas the Wolffian duct regresses.
   d. a and b   e. a and c

18.3 Birth

The uterus has contractions throughout pregnancy. At first, these are light, lasting about 20–30 seconds and occurring every 15–20 minutes. Near the end of pregnancy, the contractions may become stronger and more frequent so that a woman may think she is in labor. "False-labor" contractions are called **Braxton Hicks** contractions. However, the onset of true labor

is marked by uterine contractions that occur regularly every 15–20 minutes (and more frequently in labor's final stages) and last for 40 seconds or longer.

A positive feedback mechanism can explain the onset and continuation of labor. Uterine contractions are induced by a stretching of the cervix, which also brings about the release of oxytocin, which is produced by the hypothalamus and released from the posterior pituitary. Oxytocin stimulates the uterine muscles, both directly and through the action of prostaglandins. Uterine contractions push the fetus downward, and the cervix stretches even more. This cycle keeps repeating itself until birth occurs.

Prior to or at the first stage of **parturition,** which is the process of giving birth to an offspring, there can be a "bloody show." This is caused by expulsion of the mucous plug from the cervix. During pregnancy, this thick, sticky mucus effectively sealed the uterine cervix, preventing bacteria and sperm from entering the uterus. Loss of the plug is often a first sign that the baby's birth is imminent.

Stage 1

During the first stage of labor, the uterine contractions of labor occur in such a way that the cervical canal slowly disappears as the lower part of the uterus is pulled upward toward the baby's head. This process is called effacement, or "taking up the cervix." With further contractions, the baby's head acts as a wedge to assist cervical dilation (Fig. 18.10*b*). If the amniotic membrane has not already ruptured, it is apt to do so during this stage, releasing the amniotic fluid, which leaks out the vagina (an event sometimes referred to as "breaking water"). The first stage of parturition ends once the cervix is dilated completely.

Stage 2

During the second stage of parturition, the uterine contractions occur every 1–2 minutes and last about one minute each. They are accompanied by a desire to push, or bear down. As the baby's head gradually descends into the vagina, the desire to push becomes greater. When the baby's head reaches the exterior, it turns so that the back of the head is uppermost (Fig. 18.10*c*). Since the vaginal orifice may not expand enough to allow passage of the head, an **episiotomy** is often performed to prevent the mother's tissues from tearing. This incision, which enlarges the opening, is sewn together later. As soon as the head is delivered, the baby's shoulders rotate so that the baby faces either to the right or the left. At this time, the physician may hold the head and guide it downward, while one shoulder and then the other emerges. The rest of the baby follows easily.

Once the baby is breathing normally, the umbilical cord is cut and tied, severing the child from the placenta. The stump of the cord shrivels and falls off to create a scar, which is the umbilicus.

Stage 3

The placenta, or afterbirth, is delivered during the third stage of parturition (Fig. 18.10*d*). About 15 minutes after delivery

MEDICAL FOCUS

Preventing Birth Defects

Before the nursery room gets painted pink or blue, before the baby registry is completed at the department store, before deciding whether the baby gets named after Great-Aunt Adelaide or Grandma Gertrude, a couple that has decided to start (or add to) a family must take steps to ensure a healthy pregnancy for both baby and mother. One way to maximize the odds of having a healthy baby is to take precautions to prevent **congenital,** or **birth, defects.** Some birth defects are unavoidable—those caused by chromosomal or genetic abnormalities, for example. Tragically, however, many other birth defects are completely preventable.

For decades, the placenta was believed to be an effective barrier between mother and fetus. Continuing research has shown instead that the placenta is not a barrier at all; rather, it functions as a *filter* through which many substances can enter the baby's bloodstream from the mother's circulation. Thus, almost anything that mom eats, drinks, smokes, or injects—good and bad—is potentially capable of winding up in the baby's body. Chemicals, bacteria, viruses, parasites, antibodies from the mother's immune system (both harmful and beneficial) can all act as **teratogens**—agents capable of causing birth defects.

Ideally, preventing birth defects, as well as ensuring a healthy pregnancy for mom-to-be, is a process that should begin before a woman ever becomes pregnant. A complete physical exam is an important first step. At that time, she can receive any necessary immunization boosters, or "catch up" with immunizations she may have missed as a child (see page 291). It is especially important to immunize against **rubella (German measles).** If a pregnant woman contracts rubella, it can cause blindness, deafness, mental retardation, heart malformations, and other serious problems in an unborn child. A woman who has neither contracted chickenpox nor been immunized against it should be vaccinated at this time as well. Screening for sexually transmitted infections should also take place during this initial physical. Antibiotic treatment can cure bacterial infections (syphilis, gonorrhea, chlamydia, etc.) and prevent serious harm to the embryo. Although there are no cures for viral STDs, it is still important to know if a woman is infected. HIV (the causative agent for AIDS) can be transmitted across the placenta to infect the fetus. Anti-retroviral therapy can prevent this type of infection in many cases (see pages 289 and 290). During a pre-pregnancy physical, health-care providers will also recommend prenatal vitamin/mineral supplements. Studies have shown that taking vitamins with high levels of vitamin B_{12} and folate, before and after conception, minimize the risk of neural tube defects.

Couples trying to conceive should also adapt their lifestyle, if necessary. If either or both partners smoke, there will never be a more perfect time to quit. It has been shown that the best way to quit smoking is by having supportive family and friends around (see pages 316 and 317). Males who smoke are more likely to have erectile dysfunction than nonsmoking males, making a pregnancy more difficult to achieve. Further, it will later be essential for both the pregnant mother and the fetus to be in a smoke-free environment at all times. Lifestyle changes must also include giving up any use of recreational drugs because *all* will decrease a man's sperm count. Use of recreational drugs by the pregnant woman can harm the fetus.

Once pregnancy occurs, continuing prenatal medical care and good health habits are a must. Basic nutrients are required in adequate amounts to meet the demands of both fetus and mother. This includes increasing calories and protein intake. The pregnant woman of normal weight should expect to gain 25–35 pounds during her pregnancy. Without adequate weight gain during pregnancy, the woman risks giving birth to an infant with low birth weight. Low birth weight increases the risk of complications, such as respiratory problems, immediately after birth. Furthermore, growing number of studies confirm that newborns of low birth weight (less than 5 pounds, 8 ounces at birth) are more likely to develop certain chronic diseases, such as diabetes and high blood pressure, when they become adults.

Certain foods must be avoided during pregnancy, however. Game fish species, such as shark, mackerel, swordfish, and salmon, should be passed up because of potentially high levels of mercury that might be in these fish. Mom should also turn down unpasteurized milk or cheese/yogurt products made from unpasterized milk. Raw or undercooked shellfish, poultry, pork, beef, or eggs are also off the menu for 9 months (and longer if mom decides to breast-feed her baby). These foods may contain bacteria or parasites that could cause illness and, in some cases, also infect the unborn baby.

It is important that mom continues to take prenatal vitamin/mineral supplements throughout her pregnancy. Calcium is needed for bone growth, iron for red blood cell formation, and vitamin B_6 for proper metabolism. Perhaps most important of all, pregnant women need more folate (folic acid) each day to meet the increased rate of cell division and DNA synthesis in their own bodies and that of the developing child. As mentioned previously, maternal deficiency of folate has been linked to development of neural tube defects in the fetus. These defects include spina bifida (spinal cord or spinal fluid bulge through the back) and anencephaly (absence of the brain). Perhaps as many as 75% of these defects could be avoided by adequate folate intake even before a pregnancy occurs. Consuming fortified breakfast cereals is a good way to meet folate needs because they contain a more absorbable form of folate.

Because the placenta functions as a filter, if mom smokes or exposes herself to secondhand smoke, her baby "smokes" too. Cleft lip/cleft palate may occur because of exposure to toxins in cigarette smoke. Children of smoking mothers are more likely to be stillborn, or die shortly after birth. Death from sudden infant

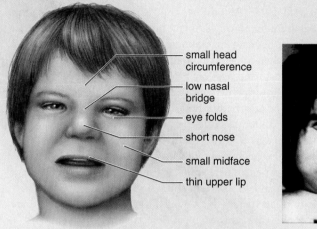

small head
circumference

low nasal
bridge

eye folds

short nose

small midface

thin upper lip

a.

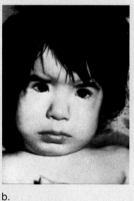

b.

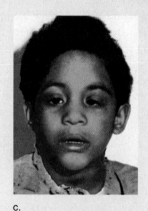

c.

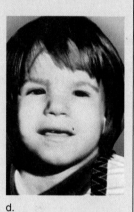

d.

Figure 18B Fetal alcohol syndrome is characterized by certain features. **a.** These features include malformations of the head and face. **b–d.** Photos of children who have fetal alcohol syndrome.

death syndrome (SIDS—unexplained death of an otherwise healthy baby) is more common in babies of smoking mothers. The occurrence of preterm babies with low birth weight is doubled for women who smoke.

Likewise, alcohol easily crosses the placenta. There is no safe dosage during pregnancy. Even one drink a day appears to increase the chances of a spontaneous abortion. The more alcohol consumed, the greater are the chances of physical abnormalities if the pregnancy continues. **Fetal alcohol syndrome (FAS)** is the term given to the collection of physical, mental, and behavioral abnormalities seen in babies born to heavy drinkers. Babies with FAS have decreased weight, height, and head size, with malformation of the head and face (Fig. 18B). Later, mental retardation is common, as are numerous other physical malformations. Babies born to heavy drinkers are apt to undergo an extremely painful withdrawal, called **delirium tremens**, after birth—shaking, vomiting, and extreme irritability.

Certainly, illegal drugs such as marijuana, cocaine, and heroin are teratogens that must be completely avoided during pregnancy. Cocaine babies now make up 60% of drug-affected babies. Cocaine use causes severe fluctuations in a mother's blood pressure that temporarily deprive the developing fetus's brain of oxygen. Cocaine babies have visual problems, lack coordination, and are often mentally retarded.

Mom has to be careful at home and on the job as well. Toxic chemicals, such as pesticides, and many organic industrial chemicals, such as vinyl chloride, formaldehyde, asbestos, and benzenes, are teratogens and can cross the placenta, resulting in fetal abnormalities. Cleaning the kitty litter box should become someone else's job during the mom's entire pregnancy—cat feces can cause her to contract **toxoplasmosis,** a dangerous parasitic infection. Lead (from old-fashioned, lead-based paint or pipes) circulating in a pregnant woman's blood can cause her child to be mentally retarded. If mom works in a health-care setting (hospital, doctor or dentist office, etc.), she must avoid accidental X-ray exposure. It takes a lower amount of exposure to X rays to cause mutations in a developing embryo or fetus than in an adult. X rays and other ionizing radiation increase the baby's risk of developing leukemia after birth. If a woman must have diagnostic X rays during pregnancy (if she has broken a bone, for example), she should inform her physician that she is pregnant. Shielding can minimize or eliminate danger to the fetus.

The medicine cabinet can be a pretty dangerous spot if you're pregnant. Many medications, including herbal supplements, nonprescription medications, and prescription drugs, can be teratogenic. For example, the isotretinoins, a group of drugs chemically similar to vitamin A and prescribed for severe acne, are known to cause severe birth defects. A woman must have a negative pregnancy test and sign a contract agreeing to use two forms of birth control simultaneously before she can receive these drugs. In the 1950s and 1960s, DES (diethylstilbestrol), a synthetic hormone related to the natural female hormone estrogen, was given to pregnant women to prevent cramps, bleeding, and threatened miscarriage. But in the 1970s and 1980s, some adolescent girls and young women whose mothers had been treated with DES showed various abnormalities of the reproductive organs and increased tendency

—*Continued*

Continued—

toward cervical cancer. Other sex hormones, including birth control pills, can possibly cause abnormal fetal development, including abnormalities of the sex organs. The drug thalidomide was a popular tranquilizer during the 1950s and 1960s in many European countries and, to a degree, in the United States. The drug, which was taken to prevent nausea in pregnant women, arrested the development of arms and legs in some children and also damaged heart and blood vessels, ears, and the digestive tract. Some mothers of affected children report that they took the drug for only a few days. Because of these situations and others, physicians are generally very cautious about prescribing drugs during pregnancy. No pregnant woman should take any drug—including ordinary cold remedies and herbal supplements—without checking first with her physician.

As her big day draws nearer, mom must continue to be careful to prevent birth defects and illness in her newborn. When a mother has HIV, herpes, gonorrhea, or chlamydia, newborns can become infected as they pass through the birth canal. Blindness and other physical and mental defects may develop. Birth by cesarean section can prevent these occurrences. An Rh-negative woman who has given birth to an Rh-positive child should receive an Rh immunoglobulin injection within 72 hours after birth to prevent her body from producing Rh antibodies. She will start producing these antibodies when some of the child's Rh-positive red blood cells enter her bloodstream, possibly before, but particularly at, birth. Rh antibodies can cause nervous system and heart defects in a fetus. The first Rh-positive baby is not usually affected. But in subsequent pregnancies, antibodies created at the time of the first birth cross the placenta and begin to destroy the blood cells of the fetus, thereby causing anemia and other complications in the baby.

With proper prenatal care, a woman's chances of having a normal pregnancy and healthy baby are better now than ever before in human history. Now that physicians and laypeople are aware of the various ways birth defects can be prevented, it is hoped that the incidence of birth defects will decrease in the future. For more information on all aspects of pregnancy, consult www.marchofdimes.com.

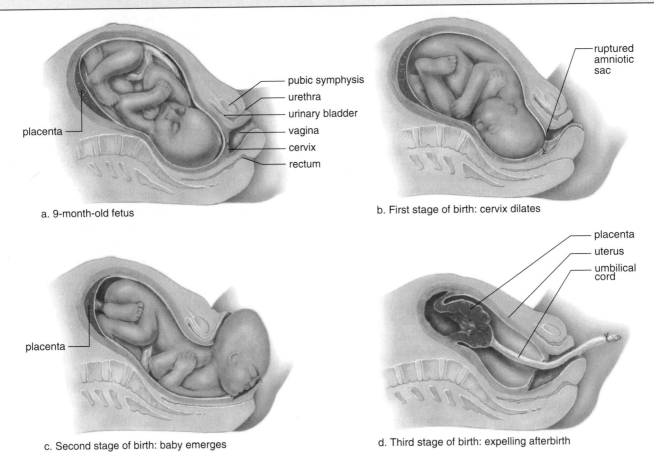

a. 9-month-old fetus

b. First stage of birth: cervix dilates

c. Second stage of birth: baby emerges

d. Third stage of birth: expelling afterbirth

Figure 18.10 Three stages of parturition (birth). **a.** Position of fetus just before birth begins. **b.** Dilation of cervix. **c.** Birth of baby. **d.** Expulsion of afterbirth.

of the baby, uterine muscular contractions shrink the uterus and dislodge the placenta. The placenta then is expelled into the vagina. As soon as the placenta and its membranes are delivered, the third stage of parturition is complete.

Effects of Pregnancy on the Mother

Major changes take place in the mother's body during pregnancy. When first pregnant, the mother may experience nausea and vomiting, loss of appetite, and fatigue. These symptoms subside, and some mothers report increased energy levels and a general sense of well-being despite an increase in weight. During pregnancy, the mother gains weight due to breast and uterine enlargement, weight of the fetus, amount of amniotic fluid, size of the placenta, her own increase in total body fluid; and an increase in storage of proteins, fats, and minerals. The increased weight can lead to lordosis (swayback) and lower back pain.

Aside from an increase in weight, many of the physiological changes in the mother are due to the presence of the placental hormones that support fetal development (Table 18.2). Progesterone decreases uterine motility by relaxing smooth muscle, including the smooth muscle in the walls of arteries. The arteries expand, and this leads to a low blood pressure. Hypotension causes the kidneys to release their enzyme, renin. The renin-angiotensin-aldosterone mechanism is thus activated, and is promoted by estrogen. Aldosterone activity promotes sodium and water retention, and blood volume increases until it reaches its peak sometime during weeks 28–32 of pregnancy. Altogether, blood volume increases from 5 liters to 7 liters—a 40% rise. An increase in the number of red blood cells follows. With the rise in blood volume, cardiac output increases by 20–30%. Blood flow to the kidneys, placenta, skin, and breasts rises significantly. Smooth muscle relaxation also explains the common gastrointestinal effects of pregnancy. The heartburn experienced by many is due to relaxation of the esophageal sphincter and reflux of stomach contents into the esophagus. Constipation is caused by a decrease in intestinal tract motility.

Of interest is the increase in pulmonary values in a pregnant woman. The bronchial tubes relax, but this alone cannot explain the typical 40% increase in vital capacity and tidal volume. The increasing size of the uterus from a nonpregnant weight of 60–80 g to 900–1,200 g contributes to an improvement in respiratory functions. The uterus comes to occupy most of the abdominal cavity, reaching nearly to the xiphoid process of the sternum. This increase in size not only pushes the intestines, liver, stomach, and diaphragm superiorly, but it also widens the thoracic cavity. Compared to nonpregnant values, the maternal oxygen level changes little, but blood carbon dioxide levels fall by 20%, creating a concentration gradient favorable to the flow of carbon dioxide from fetal blood to maternal blood at the placenta.

The enlargement of the uterus does result in some problems. In the pelvic cavity, compression of the ureters and urinary bladder can result in stress incontinence. Compression of the inferior vena cava, especially when lying down, decreases venous return, and the result is edema and varicose veins.

Aside from the steroid hormones progesterone and estrogen, the placenta also produces some peptide hormones. One of these makes cells resistant to insulin, and the result can be pregnancy-induced diabetes. Some of the integumentary changes observed during pregnancy are also due to placental hormones. **Striae gravidarum,** commonly called "stretch marks," typically form over the abdomen and lower breasts in response to increased steroid hormone levels rather than stretching of the skin. Melanocyte activity also increases during pregnancy. Darkening of the areolae, skin in the line from the navel to the pubis, areas of the face and neck, and vulva is common.

Changes in breast anatomy and the occurrence of lactation are discussed in Chapter 17, page 386.

Content **CHECK-UP!**

7. Which of the following steps describing the positive feedback loop of labor and delivery is false?
   a. Oxytocin causes uterine smooth muscle contraction.
   b. Oxytocin is manufactured by the posterior pituitary.
   c. Stretching of the uterine cervix stimulates oxytocin production and release.
   d. The baby's head pushes against the cervix, stimulating uterine contractions.

8. Effacement is the technical term for:
   a. leakage of amniotic fluid.
   b. false labor contractions.
   c. the process of gradually stretching the cervix until it slowly disappears.
   d. loss of the mucous plug that blocked the cervical canal.

9. Which of the following is an effect of placental estrogen?
   a. reduced uterine motility
   b. relaxation of smooth muscle
   c. increased insulin resistance
   d. increased uterine blood flow

TABLE 18.2 Effect of Placental Hormones on Mother	
Hormone	**Chief Effects**
Progesterone	Relaxation of smooth muscle; reduced uterine motility; reduced maternal immune response to fetus
Estrogen	Increased uterine blood flow; increased renin-angiotensin-aldosterone activity; increased protein biosynthesis by the liver
Peptide hormones	Increased insulin resistance

Source: Moore, Thomas R., *Gestation Encyclopedia of Human Biology*, Vol. 7, 7th edition. Copyright © 1997 Academic Press.

SELECTED NEW TERMS

Basic Key Terms

acrosome (ak′ro-sōm), p. 403
afterbirth (af′ter-berth), p. 408
allantois (uh-lan′to-is), p. 405
amnion (am′ne-on), p. 405
blastocyst (blas′to-sist), p. 405
chorion (ko′re-on), p. 404
chorionic villi (ko″re-on′ik vil′l), p. 408
cleavage (klěv′ij), p. 404
corona radiata (kō-rō′nă rā dě ă-toh), p. 403
differentiation (dif′er-en″she-a′shun), p. 404
ectoderm (ek′tō-derm), p. 408
embryo (em′bre-o), p. 406
embryonic development (em″bre-on′ik
 de-vel′op-ment), p. 406
embryonic disk (em″bre-on′ik disk), p. 408
endoderm (en′dō-derm), p. 408
extraembryonic membrane (eks″truh-em″bre-
 on′ik mem′brān), p. 404
fertilization (fer′tĭ-lĭ-za′shun), p. 403
fetal development (fe′tal de-vel′op-ment), p. 410

fontanels (fŏn′tn-ěl′), p. 412
gastrulation (gas′tru-la′shun), p. 408
gestation (jes-ta′shun), p. 405
growth (grōth), p. 404
human chorionic gonadotropin (HCG)
 (hyōō-mān′ kôr′ē-ŏn′ik gō-nād′ə-trō′pĭn),
 p. 406
inner cell mass (in′er sel mas), p. 405
implantation (im′plăn-tăshən), p. 406
lanugo (luh-nu′go), p. 412
mesoderm (mez′o-derm), p. 408
morphogenesis (morf-o-jen′ě-sis), p. 404
morula (mor′u-luh), p. 405
parturition (par″tu-rish′un), p. 415
placenta (pluh-sen′tuh), p. 404
primary germ layers (prī′-mă-ree gěrm
 lā′-yerz), p. 408
trophoblast (trof′o-blast), p. 405
umbilical cord (um-bil′ĭ-kl kord), p. 408
vernix caseosa (ver′niks ka″se-o′suh), p. 412
yolk sac (yōk sak), p. 404

zona pellucida (zō′-nă pěl-loo-sĭ-duh), p. 403
zygote (zĭ′gōt′), p. 403

Clinical Key Terms

anencephaly (an′en-sef′uh-le), p. 408
Braxton Hicks contraction (braks′ton hiks
 con-trak′shun), p. 415
breech birth (brěch berth), p. 412
cesarean section (sĭ-zăr′e-un sek′shun), p. 412
congenital defect (kon-jen′ĭ-tul de′fekt), p. 416
delirium tremens (de-lēr′e-um tre′mens), p. 417
ectopic pregnancy (ěk-tŏp′ĭk prěg′nən-sē), p. 406
episiotomy (e-piz″e-ot′o-me), p. 415
fetal alcohol syndrome (fe′tal al′cuh-hol
 sin′drōm), p. 417
premature birth (pre-mah-tyūr′), p. 413
rubella (German measles) (ru-bel′uh), p. 416
spina bifida (spi′nuh bif′ĭ-duh), p. 408
striae gravidarum (strī′e grăv-ĭ-dăr′-um), p. 419
teratogen (ter-ah′to-jen), p. 416
toxoplasmosis (tŏk′sō-plăz-mŏ′sĭs), p. 417

SUMMARY

18.1 Fertilization
During fertilization, a sperm nucleus fuses with the egg nucleus. The resulting zygote begins to develop into a mass of cells, which travels down the uterine tube and embeds itself in the endometrium. Cells surrounding the embryo produce HCG, the hormone whose presence indicates that the female is pregnant.

18.2 Development
A. The extraembryonic membranes, placenta, and umbilical cord allow humans to develop internally within the uterus. These structures protect the embryo and allow it to exchange waste for nutrients and oxygen with the mother's blood.
B. At the end of the embryonic period, all organ systems are

established. There is a mature and functioning placenta. The embryo is only about 38 mm (1½ in.) long.
C. Fetal development extends from the third through the ninth months. During the third and fourth months, the skeleton is becoming ossified. The sex of the fetus becomes distinguishable.
D. During the fifth through the ninth months, the fetus continues to grow and to gain weight. Babies born after six or seven months may survive, but full-term babies have a better chance of survival.
E. During pregnancy, the mother's uterus enlarges greatly, resulting

in weight gain, standing and walking difficulties, and general discomfort.

18.3 Birth
A. During stage 1 of parturition, the cervix dilates. During stage 2, the child is born. During stage 3, the afterbirth is expelled.
B. During pregnancy, the mother gains weight as the uterus comes to occupy most of the abdominal cavity with resultant annoyances such as incontinence. Many of the complaints of pregnancy, such as constipation, heartburn, darkening of certain skin areas, and diabetes of pregnancy, are due to the presence of the placental hormones.

STUDY QUESTIONS

1. Describe the process of fertilization and the events immediately following it. (p. 403)
2. Name the four extraembryonic membranes, and give a function for each. (pp. 404–405)
3. Label the diagram on page 421 to review the events that occur during pre-embryonic development. (p. 406)
4. What is the basis of the pregnancy test? (p. 406)
5. Specifically, what events normally occur during embryonic development? What events normally occur during fetal development? (pp. 406–408)

6. Describe the structure and function of the umbilical cord. (p. 410)
7. Describe the structure and function of the placenta. (p. 408)

8. What are the three stages of birth? Describe the events of each stage. (p. 415)

9. In general, describe the physical changes in the mother during pregnancy. (p. 419)

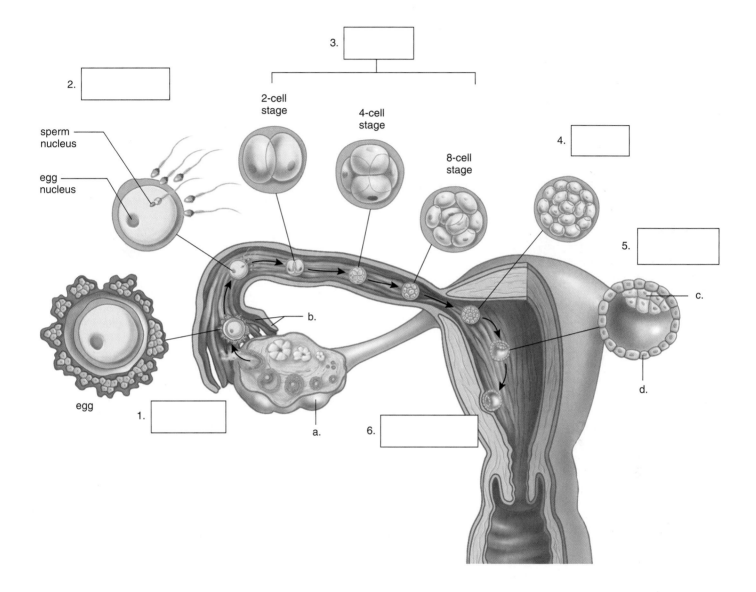

sperm nucleus

egg nucleus

2-cell stage

4-cell stage

8-cell stage

2.

3.

4.

5.

b.

egg

1.

a.

6.

c.

d.

LEARNING OBJECTIVE QUESTIONS

Fill in the blanks.

1. Fertilization occurs when the _____ nucleus fuses with the _____ nucleus.
2. The _____ membranes include the chorion, the _____, the yolk sac, and the allantois.
3. During development, the nutrient needs of the developing embryo (fetus) are served by the _____ .

4. The zygote divides as it passes down a uterine tube. This process is called _____.
5. When cells take on a specific structure and function, _____ occurs.
6. Once the blastocyst arrives at the uterus, it begins to _____ itself in the endometrium.
7. During embryonic development, all major _____ form.

8. Fetal development begins at the end of the _____ month.
9. In most deliveries, the _____ appear(s) before the rest of the body.
10. During pregnancy, constipation, darkening of certain areas of the skin, and diabetes are all due to _____ produced by the placenta.

MEDICAL TERMINOLOGY EXERCISE

Consult Appendix B for help in pronouncing and analyzing the meaning of the terms that follow.

1. morphogenesis (mor″fo-jen′ĕ-sis)
2. neonatologist (ne″o-na-tol′o-jist)
3. prenatal (pre-na′tal)
4. hyperemesis gravidarum (hi″per-em′ĕ-sis grav-id-ar′um)

5. dysmenorrhea (dis″men-o-re′uh)
6. pseudocyesis (su″do-si′ĕ-sis)
7. primigravida (pri″mĭ-grav′ ĭ-duh)
8. cryptorchidism (krip-tor′kĭ-dizm)
9. oligospermia (ol″i-go-sper′me-uh)
10. perineorrhaphy (per″i-ne-or′uh-fe)
11. abruptio placentae (ab-rup′she-o pluh-sen′te)

12. dystocia (dis-to′se-uh)
13. galactostasis (gal″ak-tos′tuh-sis)
14. polyhydramnios (pol″e-hi-dram′ ne-os)
15. amniorrhea (am′ne′-o-re′uh)
16. placenta previa (pla-se′n′tuh pre′ve-uh)

WEBSITE LINK

Visit this text's website at http://www.mhhe.com/maderap6 for additional quizzes, interactive learning exercises, and other study tools.

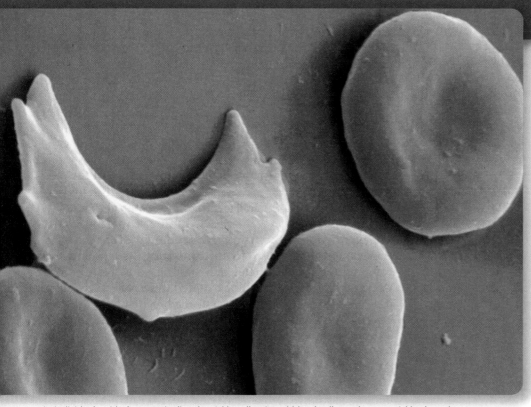

In individuals with the genetic disorder sickle-cell trait, red blood cells can become sickle-shaped.

CHAPTER

19

Human Genetics

Chapter Outline & Learning Objectives

After you have studied this chapter, you should be able to:

19.1 Chromosomal Inheritance (p. 424)
- Explain the normal chromosomal inheritance of humans.
- Describe how a karyotype is prepared and two ways to obtain fetal chromosomes.
- Explain how nondisjunction results in inheritance of an abnormal chromosome number.
- Describe Down syndrome and various syndromes that result from the inheritance of an abnormal sex chromosome number.

19.2 Genetic Inheritance (p. 429)
- Explain autosomal dominant and recessive allele inheritance.

- Explain X-linked inheritance and why males have more X-linked disorders than females.
- Relate the inheritance of an allele to protein synthesis.
- Tell how a genetic counselor could help a couple who are carriers for cystic fibrosis.

19.3 DNA Technology (p. 434)
- Explain how gene therapy is being used to treat genetic disorders.
- Discuss genomics, including how genomics might lead to better treatments for illnesses.

Medical Focus
Living with Klinefelter Syndrome (p. 428)
Preimplantation Genetic Studies (p. 433)

What's New
Chimeras (p. 435)

19.1 Chromosomal Inheritance

Normally, both males and females have 23 pairs of chromosomes, for a total of 46 chromosomes. The first 22 pairs, or chromosomes 1–44, are called **autosomes** (literally, "self" chromosomes). The final pair, chromosomes 45 and 46, are the **sex chromosomes.** These chromosomes determine the gender of the individual. In humans, males have a **Y chromosome** and an X chromosome. Females have two **X chromosomes.**

Various human disorders result from the inheritance of an abnormal chromosome number, or from inheriting chromosomes that are abnormal in size or shape. Such a disorder may be a **syndrome,** which is a group of symptoms that always occur together. Table 19.1 lists several syndromes that are due to an abnormal chromosome number. It is possible to view an individual's chromosomes by constructing a **karyotype,** a display of the chromosomes arranged by size, shape, and banding pattern. Karyotypes reveal whether an individual has inherited an abnormal number of chromosomes (too many or too few). Karyotyping can also often reveal if a chromosome has an abnormal length (too long or too short) or is otherwise abnormal in shape.

Karyotyping

Any cell in the body except red blood cells, which lack a nucleus, can be a source of chromosomes for karyotyping. In adults, it is easiest to use white blood cells separated from a blood sample for this purpose. The chromosomes of embryos or fetuses can be obtained by either amniocentesis or chorionic villi sampling (CVS). Because each carries a risk of spontaneous abortion (miscarriage), these procedures are not routinely performed for every pregnancy, but only when maternal history and/or prenatal testing indicate a need. For example, a pregnant woman who is over the age of 40 may elect to have CVS or amniocentesis because older women are at increased risk for chromosomal abnormalities in the embryo. Blood testing for pregnant women of any age may indicate the need for follow-up chromosomal studies. A blood test routinely offered to pregnant women measures the levels of four chemicals present in the blood: alpha-fetoprotein, produced by the fetal liver; estriol, an estrogen-like hormone produced by the placenta; human chorionic gonadotropin, the pregnancy hormone, and inhibin-A, a hormone produced by the fetus and placenta. Abnormally high or low blood levels of these chemicals in the blood are linked to certain chromosomal and structural abnormalities in the embryo and fetus. Amniocentesis or CVS may be recommended if abnormal levels are detected. It is important to note that this blood test does not *confirm* any abnormality; it just *suggests* that possibility.

During **amniocentesis,** a sample of amniotic fluid is withdrawn from the uterus of a pregnant woman. Blood tests and the age of the mother are used to determine whether the procedure should be done. Amniocentesis is not usually performed until about the fourteenth to the seventeenth week of pregnancy. A long needle is passed through the abdominal and uterine walls to obtain a small amount of fluid, which also contains fetal cells (Fig. 19.1a). Testing the cells and karyotyping the chromosomes may be delayed as long as four weeks so that the cells can be cultured to increase their number. As many as 400 chromosomal and biochemical problems can be detected by testing the cells and the amniotic fluid.

The risk of spontaneous abortion increases by about 0.3% due to amniocentesis, and doctors only use the procedure if it is medically warranted.

Chorionic villi sampling (CVS) is a procedure for obtaining chorionic cells in the region where the placenta will develop. This procedure can be done as early as the fifth week of pregnancy. A long, thin catheter tube is inserted through the vagina into the uterus (Fig. 19.1b). Ultrasound, which gives a picture of the uterine contents, is used to place the tube between the uterine lining and the chorionic villi. Remember that the chorion is the most superficial membrane surrounding the embryo, and it is formed from trophoblast cells of the embryo (see Chapter 18, page 404). Thus, the chromosomes in the cells of chorionic villi are the same as those of the embryo, although chorionic villi cells do not contribute to forming the embryo itself. Once the catheter is in place, a tiny sample of chorionic cells is withdrawn by suction. The cells do not have to be cultured, and karyotyping can be done immediately. However, this sampling procedure does not gather any amniotic fluid, so the biochemical tests done on the amniotic fluid following amniocentesis are not

TABLE 19.1 Syndromes from Abnormal Chromosome Numbers					
Syndrome	Sex	Disorder	Chromosome Number	Frequency	
				SPONTANEOUS ABORTIONS	LIVE BIRTHS
Down	M or F	Trisomy 21	47	1/40	1/800
Poly-X	F	XXX (or XXXX)	47 or 48	0	1/1,500
Klinefelter	M	XXY (or XXXY)	47 or 48	1/300	1/800
Jacobs	M	XYY	47	?	1/1,000
Turner	F	XO	45	1/18	1/2,500

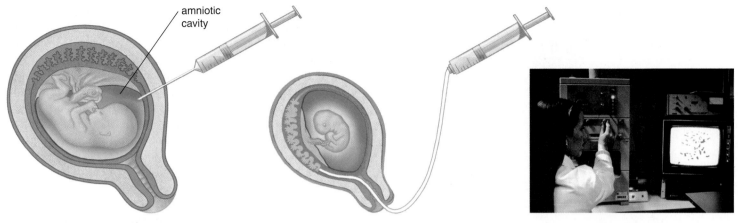

a. During amniocentesis, a long needle is used to withdraw amniotic fluid containing fetal cells.

b. During chorionic villi sampling, a tube is used to remove cells from the chorion, where the placenta will develop.

c. Cells are microscopically examined and photographed.

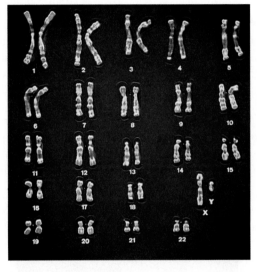

d. Normal male karyotype with 46 chromosomes

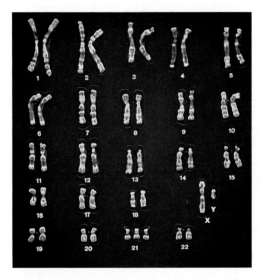

e. Down syndrome karyotype with an extra chromosome 21

Figure 19.1 Human karyotype preparation. A karyotype is an arrangement of an individual's chromosomes into numbered pairs according to their size, shape, and banding pattern. **a.** Amniocentesis and **(b)** chorionic villi sampling provide cells for karyotyping to determine if the unborn child has a chromosomal abnormality. **c.** After cells are treated as described in the text, the karyotype can be constructed. **d.** Karyotype of a normal male. **e.** Karyotype of a male with Down syndrome. A Down syndrome karyotype has three number 21 chromosomes.

possible. Also, CVS carries a greater risk of spontaneous abortion than amniocentesis—0.8% compared to 0.3%. The advantage of CVS is that the results of karyotyping are available at an earlier date.

Preparing the Karyotype

After a sample of cells has been obtained, the cells are stimulated to divide in a culture medium. A chemical is used to stop mitosis during metaphase when chromosomes are the most highly compacted and condensed. The cells are then spread on a microscope slide and dried. Stains are applied to the slides, and the cells are photographed. Staining produces dark and light cross-bands of varying widths, and these can be used in addition to size and shape to help pair up the chromosomes. Today, a computer may be used to arrange the chromosomes in pairs. It is possible to photograph the nucleus of a cell that is about to divide (the chromosomes are more visible then), so that a picture of the chromosomes is obtained. The picture may be entered into a computer, and the chromosomes electronically arranged by pairs (Fig. 19.1c). The resulting display of chromosomes is the karyotype. Figure 19.1d,e compares a

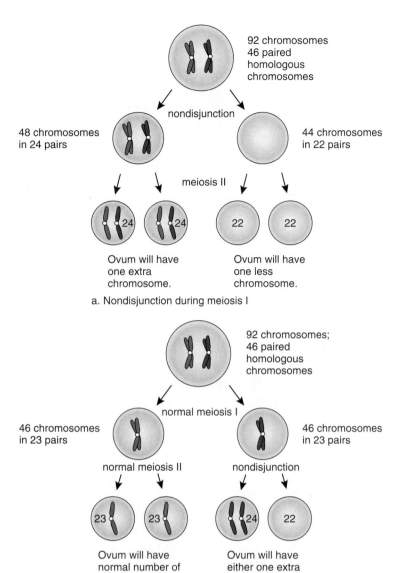

92 chromosomes
46 paired
homologous
chromosomes

nondisjunction

48 chromosomes
in 24 pairs

44 chromosomes
in 22 pairs

meiosis II

24   24   22   22

Ovum will have
one extra
chromosome.

Ovum will have
one less
chromosome.

a. Nondisjunction during meiosis I

92 chromosomes;
46 paired
homologous
chromosomes

normal meiosis I

46 chromosomes
in 23 pairs

46 chromosomes
in 23 pairs

normal meiosis II   nondisjunction

23   23   24   22

Ovum will have
normal number of
chromosomes.

Ovum will have
either one extra
or one less
chromosome.

b. Nondisjunction during meiosis II

Figure 19.2 Nondisjunction during oogenesis. Nondisjunction can occur **(a)** during meiosis I if homologous chromosomes fail to separate or **(b)** during meiosis II if the sister chromatids fail to separate completely. In either case, abnormal gametes have an extra chromosome or lack a chromosome.

normal karyotype with that of a person who has Down syndrome, the most common autosomal abnormality.

Nondisjunction

An abnormal chromosomal makeup in an individual can be due to nondisjunction. **Nondisjunction** occurs during meiosis I when both members of a homologous pair go into the same daughter cell or during meiosis II when the sister chromatids fail to separate and both daughter chromosomes go into the same gamete (Fig. 19.2). If an ovum with 24 chromosomes is

fertilized by a normal sperm cell with 23 chromosomes, the zygote formed has a **trisomy:** One type of chromosome is present in three copies instead of the normal two, and the total chromosome count is 47 instead of the normal 46. Conversely, if an ovum with 22 chromosomes is fertilized with a normal sperm, the zygote formed carries a **monosomy:** One type of chromosome is present in a single copy, and the zygote has only 45 chromosomes. It is important to note that nondisjunction can occur during the formation of gametes in either gender. Thus, although Figure 19.2 shows nondisjunction during the formation of ova, it could also happen during sperm formation as well.

Researchers believe that when the zygote carries either a monosomy or a trisomy, most nondisjunctions produce embryos that cannot survive and develop normally. The result is an early spontaneous abortion (miscarriage), which may occur before a woman even realizes that she is pregnant. Note on Table 19.1 the high rate of miscarriages occurring from monosomy or trisomy.

Down Syndrome

Down syndrome is also called trisomy 21 because the individual usually has three copies of chromosome 21. In most instances, the egg had two copies instead of one of this chromosome. (In 23% of the cases studied, however, the sperm had the extra chromosome 21.)

Individuals who have Down syndrome share characteristic physical features (Fig. 19.3). Their heads are rounded; they have a short, webbed neck, and eyes that slant upward and have an inner eyelid fold. Facial features are typically small: a

Figure 19.3 Down syndrome. Down syndrome occurs when the egg or the sperm has an extra chromosome 21 due to nondisjunction in either meiosis I or meiosis II. Characteristics include a wide, rounded face and narrow, slanting eyelids. Mental retardation to varying degrees is usually present.

small, flattened nose; small ears that may fold over at the top; and a small mouth that often makes the tongue seem large and protruding. The Down syndrome individual is short statured, with small, short hands that have a large, single palmar crease (the so-called simian line). Mild to moderate mental retardation is typical of the syndrome. In rare cases, mental retardation can be severe.

The chance of a woman having a Down syndrome child increases rapidly with age, starting at about age 40. The frequency of Down syndrome is 1 in 800 births for mothers under 40 years of age and 1 in 80 for mothers over 40 years of age. Most Down syndrome babies are born to women younger than age 40, however, because this is the age group having the most babies. Maternal blood testing can screen for abnormal levels of chemicals associated with Down syndrome, but this type of test is not a positive confirmation. Amniocentesis followed by karyotyping is needed to confirm the diagnosis of Down syndrome in the fetus.

It is known that the genes that cause Down syndrome are located on the bottom third of chromosome 21, and extensive investigative work has been directed toward discovering the specific genes responsible for the characteristics of the syndrome. One day it might be possible to control the expression of these genes even before birth, so that the symptoms of Down syndrome do not appear.

Sex Chromosome Inheritance

As stated, the sex chromosomes in humans are called X and Y. Because women are XX, an ovum always bears an X, but since males are XY, a sperm can bear an X or a Y. Therefore, the gender of the newborn child is determined by the father. If a Y-bearing sperm fertilizes the egg, then the XY combination results in a male. On the other hand, if an X-bearing sperm fertilizes the egg, the XX combination results in a female. All factors being equal, there is a 50% chance of having a girl or a boy.

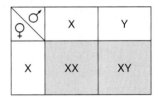

Nondisjunction also occurs with regard to the sex chromosomes. Ova or sperm with too many or too few sex chromosomes can occur. Therefore, nondisjunction accounts for the birth of individuals with too few or too many sex chromosomes.

Too Many/Too Few Sex Chromosomes

From birth, an XO individual with **Turner syndrome** has only one sex chromosome, an X; the O signifies the absence of a second sex chromosome. Turner females are short, with a broad chest and a webbed neck. The ovaries, uterine tubes, and uterus are very small and nonfunctional. Turner females do not undergo puberty or menstruate, and their breasts do not develop (Fig. 19.4a). They are usually of normal intelligence and can lead fairly normal

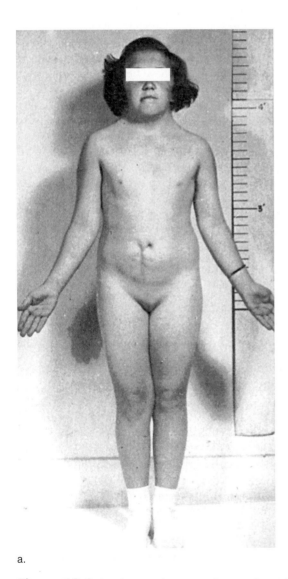

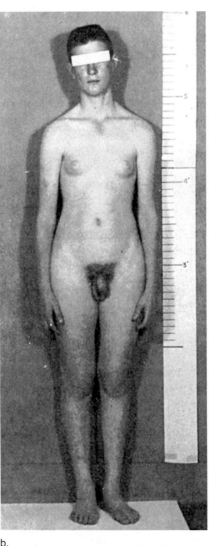

a.   b.

Figure 19.4 Syndromes due to an abnormal sex chromosome number. **a.** A female with Turner (XO) syndrome has a short thick neck, short stature, and lack of breast development. **b.** A male with Klinefelter (XXY) syndrome has immature sex organs and some development of the breasts.

Living with Klinefelter Syndrome

In 1996, at the age of 25, I was diagnosed with Klinefelter syndrome (KS). Being diagnosed has changed my life for the better.

I was a happy baby, but when I was still very young, my parents began to believe that there was something wrong with me. I knew something was different about me, too, as early on as five years old. I was very shy and had trouble making friends. One minute I'd be well behaved, and the next I'd be picking fights and flying into a rage. Many psychologists, therapists, and doctors tested me because of school and social problems and severe mood changes. Their only diagnosis was "learning disabilities" in such areas as reading comprehension, abstract thinking, word retrieval, and auditory processing. In the seventh grade, a psychologist told me that I was stupid and lazy, I would probably live at home for the rest of my life, and I would never amount to anything. For the next five years, he was basically right, and I barely graduated from high school.

I believe, though, that I have succeeded because I was told that I would fail. When I enrolled at a community college, I decided I could figure things out on my own and did not need tutoring. I received an associate degree there, then transferred to a small liberal arts college. However, I never had a semester below a 3.0, and I graduated with two B.S. degrees. I was accepted into a graduate program but decided instead to accept a job as a software engineer even though I did not have an educational background in this field. As I later learned, many KS'ers excel in computer skills. I had been using a computer for many years and had learned everything I needed to know on my own, through trial and error.

Around the time I started the computer job, I went to my physician for a physical. He sent me for blood tests because he noticed that my testes were smaller than usual. The results were conclusive: Klinefelter syndrome with sex chromosomes XXY. I initially felt denial, depression, and anger, even though I now had an explanation for many of the problems I had experienced all my life. But then I decided to learn as much as I could about the condition and the treatments available. I now give myself a testosterone injection once every two weeks, and it has made me a different person, with improved learning abilities and stronger thought processes in addition to a more outgoing personality.

I found, though, that the best possible path I could take was to help others live with the condition. I attended my first support group meeting four months after I was diagnosed. I had decided to work diligently to help people with KS forever. I have been very involved in KS conferences and have helped start support groups in the U.S., Spain, and Australia.

Since my diagnosis, it has been my dream to have a son with KS, although when I was diagnosed, I found out it was unlikely that I could have biological children. Through my work with KS, I had the opportunity to meet my fiancee, Chris. She has two wonderful children: a daughter, and a son who has the same condition that I do. There are a lot of similarities between my stepson and me, and I am happy I will be able to help him get the head start in coping with KS that I never had. I also look forward to many more years of helping other people seek diagnosis and live a good life with Klinefelter syndrome.

Stefan Schwarz
stefan13@mail.ptd.net

---

lives. Some Turner females have even given birth following in vitro fertilization using donor ova.

A male with **Klinefelter syndrome** has two or more X chromosomes in addition to a Y chromosome, and is sterile. The testes and prostate gland are underdeveloped, and the individual has no facial hair. Also, some breast development may occur (Fig. 19.4b). Affected individuals have large hands and feet and very long arms and legs. They are usually slow to learn but not mentally retarded unless they inherit more than two X chromosomes. The Medical Focus on this page suggests that much can be done to help a person with Klinefelter syndrome lead a normal life.

A **poly-X female** has more than two X chromosomes. Females with three X chromosomes have no distinctive physical appearance, aside from a tendency to be tall and thin. Although some exhibit delayed motor and language development, most poly-X females are not mentally retarded. Some may have menstrual difficulties, but many menstruate regularly and are fertile. Their children usually have a normal karyotype.

Females with more than three X chromosomes occur rarely. Unlike XXX females, XXXX females are usually tall and severely retarded. Various physical abnormalities are seen, but they may menstruate normally.

XYY males with **Jacobs syndrome** can only result from nondisjunction during spermatogenesis. This is because only males have a Y chromosome. Affected males are usually taller than average, suffer from persistent acne, and tend to have speech and reading problems. At one time, it was suggested that these men were likely to be criminally aggressive, but it has since been shown that the incidence of such behavior among them may be no greater than among XY males.

Notice that there are no YO males. This shows that at least one X chromosome is needed for survival. However, XXY individuals are males, not females.

1. Select one false statement about amniocentesis:
   a. A long, thin needle is used to penetrate the abdomen and uterus of the pregnant woman.
   b. A sample of amniotic fluid is taken from the amniotic sac surrounding the fetus.
   c. Four hundred or more biochemical and chromosomal tests can be carried out.
   d. The sample can be tested immediately and results are available within days

2. If a sperm cell that has 24 chromosomes fertilizes a normal ovum, the resulting zygote will:
   a. have 47 chromosomes and a trisomy.
   b. have 45 chromosomes and a monosomy.
   c. have 46 chromosomes and be normal.

3. This is the chromosome designation for Klinefelter's syndrome.
   a. 47, XXX
   c. 47, XYY
   b. 45, X0
   d. 47, XXY

19.2 Genetic Inheritance

Genes, which are sections of chromosomes, control body traits, such as height, eye color, or type of earlobe. Genes can also control the production of specific proteins, such as structural proteins, cell membrane proteins, and enzymes. Recall that as described in Chapter 17, page 371, each cell contains homologous pairs of chromosomes. Alternate forms of a gene having the same position (called the **locus**), on a pair of homologous chromosomes and affecting the same trait are called **alleles.** It is customary to designate an allele by a letter, which represents the specific characteristic it controls. A **dominant allele** is assigned an uppercase (capital) letter, while a **recessive allele** is given the same letter but in lowercase. In humans, for example, unattached (free) earlobes are dominant over attached earlobes, so a suitable key would be *E* for unattached earlobes and *e* for attached earlobes.

Inheritance of Genes on Autosomal Chromosomes

An individual normally has two alleles for an autosomal trait. Just as one member of each pair of chromosomes is inherited from each parent, so too is one of each pair of alleles inherited from each parent.

The term **genotype** refers to the genes of the individual. Figure 19.5 shows three possible fertilizations and the resulting genotype of the individual for earlobe attachment. In the first instance, the chromosomes of both the sperm and the egg carry an *E.* Consequently, the zygote and subsequent individual have the alleles *EE,* which may be called a **homozygous dominant** genotype. A person with genotype *EE*

obviously has unattached earlobes. The physical appearance of the individual—in this case, unattached earlobes—is called the **phenotype.**

In the second fertilization, the zygote has received two recessive alleles (*ee*), and the genotype is called **homozygous recessive.** An individual with this genotype has the recessive phenotype, which is attached earlobes. In the third fertilization, the resulting individual has the alleles *Ee,* which is called a **heterozygous** genotype. A heterozygote shows the dominant characteristic; therefore, the phenotype of this individual is unattached earlobes.

How many dominant alleles does an individual need to inherit to have a dominant phenotype? These examples show that a dominant allele contributed from only one parent can bring about a particular dominant phenotype. How many recessive alleles does an individual need to inherit to have the recessive phenotype? A recessive allele must be received from both parents to bring about the recessive phenotype.

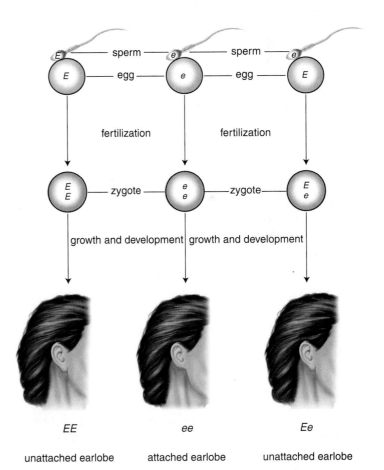

EE — unattached earlobe
ee — attached earlobe
Ee — unattached earlobe

Figure 19.5 Genetic inheritance. Individuals inherit a minimum of two alleles for every characteristic of their anatomy and physiology. The inheritance of a single dominant allele (*E*) causes an individual to have unattached earlobes; two recessive alleles (*ee*) cause an individual to have attached earlobes. Notice that each individual receives one allele from the father (by way of a sperm) and one allele from the mother (by way of an egg).

Sex-Linked Inheritance

The sex chromosomes contain genes just as the autosomal chromosomes do. Some of these genes determine whether the individual is a male or a female. Investigators have now discovered a series of genes on the Y chromosome that determine the development of male genitals, and at least one on the X chromosome that seems to be necessary for the development of female genitals.

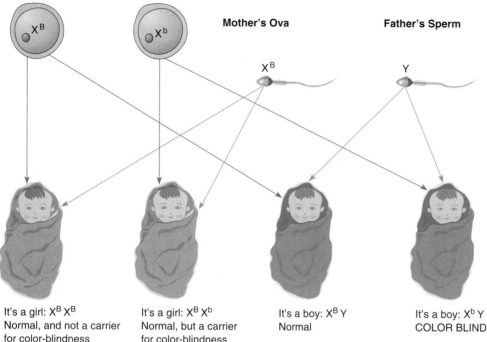

It's a girl: $X^B X^B$
Normal, and not a carrier for color-blindness

It's a girl: $X^B X^b$
Normal, but a carrier for color-blindness

It's a boy: $X^B Y$
Normal

It's a boy: $X^b Y$
COLOR BLIND

Some traits controlled by alleles on the sex chromosomes have nothing to do with the gender of the individual. These traits are said to be **sex-linked.** An allele that is only on the X chromosome is **X-linked** and an allele that is only on the Y chromosome is **Y-linked.** Most sex-linked traits are controlled by alleles on the larger X chromosome. The very small Y chromosome does not have the corresponding allele, simply because it is so short. Thus, the sex chromosomes in males—X and Y—are not truly homologous chromosomes.

As you know, a female has two X chromosomes, and thus her ova will contain either of her two X chromosomes. A male always receives a sex-linked recessive condition from his mother, from whom he inherited his single X chromosome. The Y chromosome from his father does not carry a corresponding allele for the trait. Therefore, males need receive only one recessive allele to have the X-linked disorder. Consider red-green color blindness, for example (Fig. 19.6a). This genetic disorder results in an individual who cannot see red or green because the cone cells of the eye are abnormal. It is caused by a recessive gene on the X chromosome. Note that although the mother has one recessive gene on an X chromosome, she has normal vision due to the dominant gene on her other X chromosome. She is referred to as a **carrier** for the trait. A carrier is an individual who is heterozygous for a recessive genetic disorder and therefore has no symptoms. If the carrier's ova are fertilized by sperm from a male with normal vision, any male children she bears will have a 50% chance of inheriting the disorder. Similarly, any female children will have a 50% chance of being a carrier for the disorder.

Would it ever be possible for a female to have red-green color blindness?

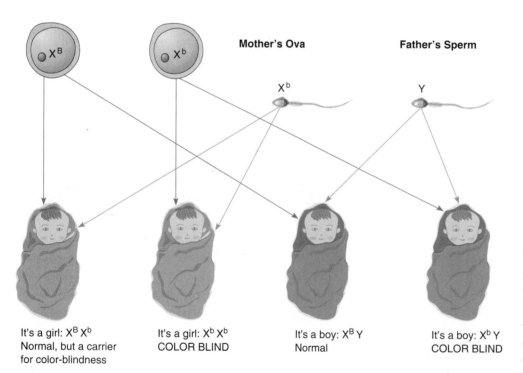

It's a girl: $X^B X^b$
Normal, but a carrier for color-blindness

It's a girl: $X^b X^b$
COLOR BLIND

It's a boy: $X^B Y$
Normal

It's a boy: $X^b Y$
COLOR BLIND

Figure 19.6 Sex-linked inheritance. **a.** A typical pattern of sex-linked inheritance for the color blindness trait. The mother is a carrier for the trait, and her sons have a 50% chance of inheriting the trait. **b.** A sex-linked trait can be inherited by a daughter only if her father is affected and her mother is a carrier or color blind also.

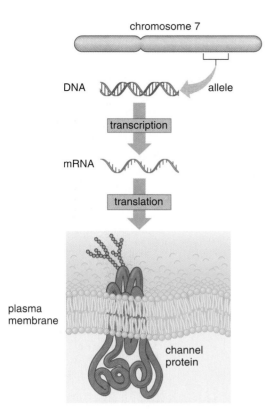

chromosome 7

DNA   allele

transcription

mRNA

translation

plasma
membrane

channel
protein

Figure 19.7 A person with cystic fibrosis has an abnormal allele, and the result is an abnormal channel protein in the plasma membrane. Chloride ions cannot exit the cell, resulting in a very thick mucus, particularly in the bronchial tubes and pancreatic duct.

It is possible, although unlikely. A female must receive two recessive X-linked alleles, one from each parent, before she has an X-linked recessive condition. The inheritance of a dominant allele from either her mother or her father can offset the inheritance of a recessive X-linked allele. A female with red-green color blindness will have inherited a recessive X chromosome from both her carrier mother and her color-blind father (Fig. 19.6b).

Genetic Counseling

Couples who believe that they may be at risk for having a child with a chromosomal or genetic disorder may choose to seek the advice of a genetic counselor. Genetic counselors are professionals with special training and certification. They help families to understand chromosomal and genetic inheritance and to make informed decisions about pregnancy and childbearing. In order to understand genetic counseling, we will consider the inheritance of cystic fibrosis. **Cystic fibrosis (CF)** is the most common lethal genetic disorder among Caucasians in the United States. About one in 20 Caucasians is a carrier for CF, and about one in 3,000 newborns has the disorder. In children with CF, the mucus

in the bronchial tubes and pancreatic ducts is particularly thick and viscous, interfering with the function of the lungs and pancreas. In the past few years, new treatments have raised the life expectancy for CF patients to as much as 35 years of age.

Normal individuals have at least one dominant allele for a plasma membrane channel protein. Figure 19.7 shows the relationship between inheritance of a normal CF allele on a chromosome (chromosome 7) and development of the channel protein. The illustration emphasizes that alleles are actually a segment of DNA and that alleles cause the production of certain proteins in a cell.

As shown in Figure 19.8a, if a genetic counselor knows the genotype of the potential parents, the counselor can predict the chances that a couple will have a child having a recessive autosomal disorder such as CF. If the parents are both carriers, each offspring has a 25% chance of receiving two recessive alleles and having CF.

If the counselor does not know the inheritance pattern of a disorder, it is sometimes possible to deduce it by studying a

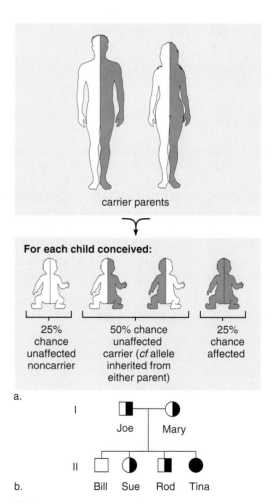

carrier parents

For each child conceived:

| 25% chance unaffected noncarrier | 50% chance unaffected carrier (*cf* allele inherited from either parent) | 25% chance affected |

a.

b.   Joe   Mary   Bill   Sue   Rod   Tina

Figure 19.8 Inheritance pattern for CF, an autosomal recessive disorder. **a.** The figures below the parents show four possible combinations of inherited alleles. Therefore, each offspring has a 25% chance of inheriting two recessive alleles and having CF.

TABLE 19.2 Inheritance of Some Genetic Disorders

Dominant	Recessive	X-Linked	Multiple Genes
Examples of dominantly inherited disorders include: • Neurofibromatosis—benign tumors in skin or deeper • Achondroplasia—a form of dwarfism • Chronic simple glaucoma (some forms)—a major cause of blindness if untreated • Huntington disease—progressive nervous system degeneration • Familial hypercholesterolemia—high blood cholesterol levels, propensity to heart disease • Polydactyly—extra fingers or toes	Examples of recessive inherited disorders include: • Cystic fibrosis—disorder affecting function of mucous and sweat glands • Galactosemia—inability to metabolize milk sugar • Phenylketonuria—essential liver enzyme deficiency • Sickle-cell disease—blood disorder primarily affecting blacks • Thalassemia—blood disorder primarily affecting persons of Mediterranean ancestry • Tay-Sachs disease—lysosomal storage disease leading to nervous system destruction	Examples of X-linked disorders include: • Agammaglobulinemia—lack of immunity to infections • Color blindness—inability to distinguish certain colors • Hemophilia (some forms)—defect in blood-clotting mechanisms • Muscular dystrophy (some forms)—progressive wasting of muscles • Spinal ataxia (some forms)—spinal cord degeneration	Examples of multifactorial inheritance include: • Cleft lip and/or palate • Clubfoot • Congenital dislocation of the hip • Spina bifida—open spine • Hydrocephalus (with spina bifida)—water on the brain • Pyloric stenosis—narrowed or obstructed opening from stomach into small intestine

Source: Data from the National Foundation/March of Dimes.

pedigree. A **pedigree** is a diagram that depicts the inheritance of a particular trait: Circles are females, and squares are males; shaded-in symbols represent those who have a trait; half-shaded symbols represent carriers; and roman numerals indicate generations. Notice that the pedigree in Figure 19.8*b* has to be for a recessive disorder because unaffected parents have a child with the disorder. This can only happen when a condition is recessive. If the condition were dominant, one of the parents would be affected!

Prenatal Testing for Genetic Disorders

Commonly inherited genetic disorders are listed in Table 19.2. Several of these disorders do not appear unless multiple abnormal genes are inherited. The inheritance of these conditions, listed in the "Multiple Genes" column in Table 19.2, is complex.

Sometimes parents want to improve their chances of having a child free of a particular genetic disorder that runs in their families. Until recently, the best that could be done was to test for a genetic disorder following amniocentesis or chorionic villi sampling. If the embryo had the disorder, the parents could consider an abortion. Now it is possible to retrieve ova from the ovary to make sure the ovum does not carry an abnormal allele prior to in vitro fertilization or to screen morulas following in vitro fertilization. The Medical Focus reading on page 433 discusses the latter technique. In the future, it might be possible to use gene therapy to cure any genetic defects found in an ovum or embryo.

Content CHECK-UP!

4. The outward, or physical expression of a person's genes for a trait is the:

a. allele.

b. phenotype.

c. genotype.

d. recessive gene.

5. A couple has just had a child diagnosed with sickle-cell anemia, which is an autosomal recessive trait. Neither parent has the disease. What are the parents' genotypes? (Use the letter *A* to represent the dominant trait, and the letter *a* to represent the recessive trait.)

a. father *Aa*, mother *aa*

b. father *aa*, mother *aa*

c. father *aa*, mother *Aa*

d. father *Aa*, mother *Aa*

6. Which of the following statements about sex-linked recessive disorders is false?

a. The recessive trait is carried on the X chromosome.

b. The mother is typically the carrier for the disorder.

c. If the mother is a carrier and the father is normal, female children have a 50% chance of inheriting the disorder.

d. If the mother is a carrier and the father is normal, male children have a 50% chance of inheriting the disorder.

We have already seen that in vitro fertilization can lead to thera-peutic cloning (see page 407) for the production of various tis-sues. Now, we will consider the medical implications of the fact that each cell of the pre-embryo is totipotent, meaning that each cell can become a complete embryo because the cells have not become specialized yet. Not surprisingly, then, if a single cell is removed from an eight-celled cleavage pre-embryo, the seven-celled pre-embryo will still go on and develop into a normal newborn (Figure 19A). Researchers are using this knowledge to allow them to carry out preimplantation genetic analysis.

Consider that a woman might have one of the autosomal dominant genetic disorders listed in Table 19.2 or be a carrier for one of the recessive disorders. Her partner might have one of the autosomal dominant genetic disorders, be a carrier for one of the recessive autosomal disorders, or have an X-linked recessive dis-order. As potential parents, wouldn't they want the assurance that the pre-embryo is free of genetic disorders and will develop into a normal adult? In some instances, it is possible to perform in vitro fertilization and then **preimplantation genetic diagnosis (PGD)** in order to determine the genotype of the pre-embryo with regard to particular genetic disorders. Then, only a normal pre-embryo is implanted in the female. Currently, PGD can be used to screen for

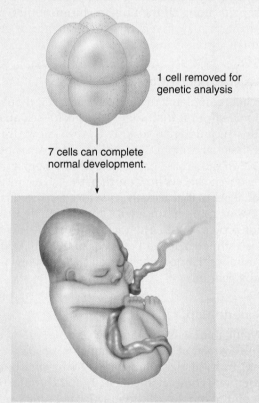

1 cell removed for genetic analysis

7 cells can complete normal development.

Figure 19A One cell from an eight-celled pre-embryo can be tested for abnormal alleles; if all tested alleles are normal, the seven-celled pre-embryo can complete normal development.

over 30 chromosomal and genetic abnormalities, including Down syndrome, cystic fibrosis, hemophilia, and sickle-cell ane-mia. More than 1,000 children who were screened as embryos by PGD have been born worldwide. All are free of the genetic disor-ders that run in their families. Proponents of the procedure cite its advantages: By selecting only healthy pre-embryos for implanta-tion, a couple is less likely to suffer a miscarriage that results from chromosomal abnormality. Multiple embryos will not have to be implanted, and multiple pregnancy, with its increased risk of pre-mature birth, can be avoided. Furthermore, the couple need not go through chorionic villus sampling or amniocentesis if preg-nancy is achieved, and will not have to face the painful decision to terminate an existing pregnancy because of a genetic or chromo-somal disorder. Although the PGD procedure is expensive, propo-nents argue that lifetime medical care for a child with a genetic or chromosomal disorder is far more expensive. In the future, it is possible that PGD might be coupled with gene therapy so that any pre-embryo will be suitable for implantation.

PGD has its opponents as well, and experts in medical ethics continue to question its use. Pointing to the fact that each cell in an eight-cell pre-embryo is totipotent, they argue that the single cell used for testing is a potential human being that is destroyed by the testing process. Likewise, defective embryos are potential human beings that will ultimately be destroyed. Further, it is known that many genetic disorders do not appear in the affected individual be-fore the third or fourth decade of life. It is feasible that a cure for the disorder could be found during that time period. Finally, medical ethicists stress that although children with genetic or chromosomal abnormalities may have medical issues that must be overcome, their lives can have tremendous meaning and value to their families and to society. Moral philosophers contend that ending a life shortly after conception is an act whose morality must be questioned.

In several instances worldwide, families with a child suffer-ing from a serious blood disorder (leukemia or genetic anemias) have already used PGD to create a "savior baby"—a second child free of the genetic disorder—whose umbilical cord blood was then transfused into the ill sibling. This practice creates further concern that PGD could be used to create babies for their "spare parts." Perhaps most disturbing is the notion that PGD could eventually be used to select only for pre-embryos with a couple's desired characteristics: gender, intelligence, physical appearance, or athletic ability. For example, it is possible that couples who al-ready have a daughter could use PGD to select only male em-bryos for implantation—and any female embryos would be sub-sequently destroyed. In societies worldwide where male children are valued more than females, unethical practitioners of PGD could find a lucrative market for their skills.

The debate will undoubtedly continue over the uses of this type of technology, and others like it that will develop in the future. It is important that all citizens are well informed and understand the scientific issues so that they can make informed decisions.

19.3 DNA Technology

Previously, you studied the structure of DNA and how it replicates and carries out protein synthesis. **DNA technology** includes our ability to work directly with DNA to determine the relatedness of individuals, to assist forensics in determining whether a person has committed a crime, and to develop new treatments for human illnesses, called gene therapy.

Gene Therapy

Gene therapy is the insertion of genetic material into human cells for the treatment of a disorder. It includes procedures that give a patient healthy genes to make up for faulty genes and also includes the use of genes to treat various other human illnesses, such as cancer and cardiovascular disease. Currently, approximately 1,700 patients are enrolled in over 300 approved gene therapy trials in the United States.

How can a molecule as large as a gene be introduced, intact, into a cell? The most common approach used by scientists is to employ a **vector**—a carrier whose job is to transport the DNA directly into the cell. Viruses are nature's own vectors because they replicate themselves by inserting their own genetic material into a host cell. Scientists studying gene therapy techniques have adapted viruses to function as vectors. The viruses are first modified: Viral genetic material is removed or altered so that it does not cause disease, and human genes are inserted into the virus instead. Next, the virus is allowed to infect human tissue, where it inserts the healthy human genes into tissue cells. Four classes of viruses have been employed in gene therapy research: (1) retroviruses, like the virus that causes AIDS; (2) adenoviruses, which are a form of cold virus; (3) herpes viruses, which cause cold sores on the mouth; and (4) the most commonly used virus vector, the adeno-associated virus, a very small virus that can have all of its own genetic material removed and still be capable of infecting cells.

Other mechanisms to insert genes into cells are also being investigated. One way is to introduce DNA directly into cells. A second mechanism involves attaching the gene to a liposome. Liposomes are microscopic vesicles that form spontaneously when lipoproteins are put into a solution. Still other scientists attach the DNA to molecules that have specific receptors on the cell membrane, then allowing the cell to incorporate the DNA by phagocytosis.

Among the many gene therapy trials, one is for the treatment of familial hypercholesterolemia, a condition that develops when liver cells lack a receptor for removing cholesterol from the blood. The high levels of blood cholesterol make the patient subject to fatal heart attacks at a young age. In a newly developed procedure, a small portion of the liver is surgically excised and infected with a virus containing a normal gene for the receptor. Several patients have experienced lowered serum cholesterol levels following this procedure.

Cystic fibrosis patients have an abnormal gene for a transmembrane carrier of the chloride ion. Patients often die due to numerous infections of the respiratory tract. In a newly developed procedure, liposomes have been coated with the gene needed to cure cystic fibrosis. The liposomes have then been delivered to the lower respiratory tract.

Genes are also being used to treat medical conditions other than the known genetic disorders. VEGF (vascular endothelial growth factor) can cause the growth of blood vessels. The gene that codes for this growth factor can be injected alone or within a virus into the heart to stimulate branching of coronary blood vessels. Coronary patients report that they have less chest pain and can run longer on a treadmill.

Gene therapy is increasingly used as part of cancer therapy. Genes are being used to make healthy cells more tolerant of, and tumors more vulnerable to, chemotherapy. The gene *p53* brings about the death of cells, and there is much interest in introducing it into cancer cells, and in that way killing them off.

Genomics

Genomics is the molecular analysis of a **genome,** which is all the genetic information in all the chromosomes of an individual. Recall from Chapter 2, Figure 2.15 that the two DNA strands of a double helix are held together by bonding between their bases, designated as A, T, G, and C. Base A is always paired with base T, and base G is always paired with base C. In a worldwide effort known as the Human Genome Project, researchers set out to map the human chromosomes in two ways: (1) by constructing a map that shows the sequence of base pairs along all the human chromosomes, and (2) by constructing a map that shows the sequence of genes along the human chromosomes.

The Base Sequence Map

Researchers completed the first goal in 2003, thanks to an international effort. It took some 15 years for scientists to complete this monumental task. It is now known that human genetic material contains 3.16 billion base pairs, and the order of 99.9% of those base pairs is exactly the same in all people. To date, no one knows how much of the DNA consists of truly functional genes that actually code for proteins. It is estimated that 50% or more is so-called "junk DNA," which does not code for protein at all. The total number of truly functional genes is estimated to be between 30,000–50,000—a much lower number than originally projected, and less than 2% of the entire genome.

The Genetic Map

The genetic map tells the location of genes along each chromosome. Sixteen of the 46 human chromosomes have been mapped, including chromosome 22 (the first human chromosome to be completely mapped) and chromosome 4 (the most recently completed). Completing the genetic map should accelerate now that the base sequence map is done. Researchers need only know a short sequence of bases in a gene

What's New

Chimeras

In Greek mythology, it was the unlikely combination of a goat's body, a lion's head, and a serpent's tail. In biotechnology, a **chimera** (pronounced ky-mare-uh) is the term used for an animal that bears the cells of two or more species in its bodies. The newest genetically engineered chimeras are animals that bear human genes and are able to produce human proteins.

Scientists have been employing bacterial chimeras for several decades as "biological factories" to synthesize human proteins. Recombinant DNA technology introduces the desired human gene into the single, ring-shaped bacterial chromosome. Bacterial enzymes do the molecular "cut-and-paste" to incorporate the human DNA into bacterial DNA. By introducing human genes into harmless strains of *Escherichia coli*, a common microbe, scientists have created bacterial colonies that can produce numerous human proteins: insulin, growth hormone, parathyroid hormone (the hormone that increases plasma calcium), and erythropoietin (the hormone that promotes red blood cell growth and development), to name only a few. So far, however,

this technique has been limited to the synthesis of small peptides and proteins.

Now, thanks to the technique that allowed the cloning of Dolly the sheep, scientists are creating animals that can produce large quantities of complex proteins and even human-compatible organs (see Therapeutic Cloning, page 407). First, the desired human gene or genes are inserted into chromosomes within the nucleus of a single cell from the animal. Next, the nucleus is removed from one of the animal's ova, and replaced with the altered nucleus. Electrical stimulation will trigger the growth of an embryo, which can then be implanted into a female to complete its development.

Researchers have cloned chimeric goats that express a human protein called antithrombin in their milk. This powerful anticoagulant is used to prevent potentially fatal thrombi, or blood clots, inside blood vessels. Likewise, cattle and rabbits have been bioengineered to make human antibodies in their blood. Pigs engineered with human genes incorporated into their kidneys and hearts are being developed to provide a source for transplantable organs (Fig. 19B). Chickens, mice, and sheep have also been altered for biomedical applications.

As with all developing biotechnology, the creation of chimeras raises perplexing questions about the ethics of the procedures involved. Medical ethicists worry that scientists will be tempted to create chimeric animals with human nervous capability—and human conciousness and reasoning ability as a result. Animal-rights activists argue that altering animals to produce chimeras is inherently abusive. Finally, many researchers fear that introducing chimeric products, whether proteins or entire organs, could also introduce animal viruses or other pathogens into humans. A new animal virus, to which humans have never been exposed, has the potential for causing catastrophic epidemics (see What's New on page 293).

Regulatory approval for chimera-produced antithrombin is being sought as of this writing. Doubtless, it will be only the first of many such applications. It is important that all of us, as potential consumers of these products, remain informed about their benefits and hazards.

Figure 19B Products produced by chimeric animals could be used to save human lives.

of interest in order for computers to search the genome for a match. Then, computers can help to pinpoint where this gene is located. However, decades—if not centuries—of research remains to be completed about the genome. Many questions remain unanswered: What does each gene do? How is the gene turned on and off? What is all the junk DNA for? How do genes interact to make each unique individual? What gene, or combination of genes, causes disease? For each question answered, many others will continue to arise.

Many ethical questions also arise regarding how we should use our knowledge of the human genome. Who owns genetic information? How should it be used and accessed? Should humans attempt to control and/or manipulate human genes and, if so, how? Manipulation of human genes is already underway, through the formation of human-animal hybrids, as described in the What's New on this page. It is imperative that you be knowledgeable about the human genome so that you can help decide these issues.

SELECTED NEW TERMS

Basic Key Terms

allele (uh-lēl'), p. 429

amniocentesis (am″ne-o-sen-te′sis), p. 424

autosome (aw′to-sōm), p. 424

carrier (kār′e-er), p. 430

chorionic villi sampling (ko-re-on′ik vil′ī sam′pling), p. 424

cystic fibrosis (sĭs′tĭh′ fī′bras), p. 431

dominant allele (dom′ĭ-nant uh-lēl′), p. 429

genes (jēn), p. 429

genome (je′nōm), p. 434

genotype (je′-no-tīp), p. 429

heterozygous (het-er-o-zi′gus), p. 429

homozygous dominant (ho-mo-zi′gus dom′ĭ-nant), p. 429

homozygous recessive (ho-mo-zi′gus re-ses′iv), p. 429

karyotype (kār′e-o-tīp), p. 424

locus (lōh-kŭs), p. 429

monosomy (mŏ-nōh-sōh-mē), p. 426

nondisjunction (non″dis-junk′shun), p. 426

pedigree (ped-ĭ-gre), p. 432

phenotype (fe′no-tīp), p. 429

recessive allele (re-ses′iv uh-lēl′), p. 429

sex chromosome (seks kro′mo-sōm), p. 424

sex-linked (seks-linkt), p. 430

syndrome (sin′drōm), p. 424

trisomy (trī′sōh-mē), p. 426

X chromosome (x kro′mo-sōm), p. 424

X-linked (x-linkt), p. 430

Y chromosome (y kro′mo-sōm), p. 424

Clinical Key Terms

Down syndrome (down sin′drōm), p. 426

gene therapy (jen thēr′uh-pe), p. 434

Jacobs syndrome (ja′kubs sin′drōm), p. 428

Klinefelter syndrome (klīn′fel-ter sin′drōm), p. 428

poly-X female (pah′le-x fe′māl), p. 428

Turner syndrome (tur′ner sin′drōm), p. 427

XYY male (xyy māl), p. 428

SUMMARY

19.1 Chromosomal Inheritance

A. Normally, an individual inherits 22 pairs of autosomal chromosomes and one pair of sex chromosomes. Females are XX and males are XY.

B. Amniocentesis and chorionic villi sampling are used to provide cell samples for karyotyping fetal chromosomes.

C. Nondisjunction during oogenesis or spermatogenesis explains the inheritance of an abnormal number of chromosomes.

D. The most common autosomal abnormality is Down syndrome, which is due to the inheritance of an extra chromosome 21.

E. Abnormal combinations of sex chromosomes include XO (Turner syndrome), XXX (poly-X), XXY (Klinefelter syndrome), and XYY (Jacobs syndrome).

19.2 Genetic Inheritance

A. Genes control human traits. Uppercase letters designate dominant alleles; lowercase letters designate recessive alleles.

B. The genotype represents the genes of an individual, and the phenotype refers to outward expression of the gene. A homozygous dominant individual inherited two dominant alleles and has the dominant phenotype; a heterozygous individual inherited one dominant and one recessive allele and has the dominant phenotype; a homozygous recessive individual inherited two recessive alleles and has the recessive phenotype.

C. The inheritance of X-linked alleles differs in males and females. Males require only one recessive allele to have an X-linked trait; females require two recessive alleles. This

means that males are more likely to inherit an X-linked disorder

D. Genetic counselors help couples determine the chances of having children with a genetic disorder, such as cystic fibrosis. They can also determine the pattern of inheritance from studying a family's pedigree.

19.3 DNA Technology

A. Gene therapy, which involves replacing defective genes with healthy genes, is now a reality. Researchers are envisioning various applications aimed at curing human genetic disorders, as well as many other types of illnesses.

B. Genomics is an actively growing field. Because the sequence of base pairs along the human chromosomes has now been determined, new treatments for genetic disorders are expected.

STUDY QUESTIONS

1. What is the normal chromosome inheritance of humans? (p. 424)

2. How is a karyotype prepared? What are the possible sources for cell samples in an adult? In the fetus? (p. 425)

3. What is nondisjunction, and when does nondisjunction occur during meiosis? (p. 426)

4. What are the characteristics of a person with Down syndrome? (p. 426)

5. What are the characteristics of the most common human conditions resulting from inheritance of abnormal numbers of sex chromosomes? (pp. 427-428)

6. Explain autosomal dominant and recessive genetic inheritance. (p. 429)

7. Explain X-linked allele inheritance in humans. (p. 430)

8. What type of information does a genetic counselor give parents who might pass on a genetic disorder? (pp. 431-432)

9. Give examples of dominant, recessive, and X-linked genetic disorders in humans. (p. 430)

10. Describe the function of an allele using the CF allele as an example. (p. 431)

11. What is gene therapy, and what types of genetic disorders have been treated thus far? (p. 434)

12. What is a genome? What is genomics, and what might be the benefits of genomics in the future? (pp. 434-435)

LEARNING OBJECTIVE QUESTIONS

Fill in the blanks.

1. The genes are on the _____.
2. A karyotype shows the individual's _____.
3. The sex chromosomes of a male are labeled _____.
4. A person with Down syndrome has inherited _____ copies of chromosome 21.
5. A person with Klinefelter syndrome has the chromosomes _____.
6. A dominant autosomal genetic disorder only requires the inheritance of _____ (one or two) abnormal gene(s).
7. A male with an X-linked recessive disorder inherited a _____ allele for the disorder.
8. If a person inherits an autosomal genetic disorder and both parents are unaffected by the disorder, the disorder is _____.
9. Replacing defective genes with healthy genes is the goal of _____.
10. The molecular analysis of a genome is called _____.

MEDICAL TERMINOLOGY EXERCISE

Consult Appendix B for help in pronouncing and analyzing the meaning of the terms that follow.

1. neogenesis (ne"o-jen'ĕ-sis)
2. regeneration (re-jen"er-a'shun)
3. fetoscope (fe'to-skōp)
4. polydysplasia (pol"e-dis-pla'ze-uh)
5. congenital (kon-jen'ĭ-tal)

WEBSITE LINK

Visit this text's website at <u>http://www.mhhe.com/maderap6</u> for additional quizzes, interactive learning exercises, and other study tools.

APPENDIX A

Reference Figures | THE HUMAN ORGANISM

The following series of reference figures show the major organs of the human torso. The first plate illustrates the anterior surface and reveals the superficial muscles on one side. Each subsequent plate exposes deeper organs, including those in the thoracic, abdominal, and pelvic cavities.

Chapters 5–17 of this textbook describe the organ systems of the human organism in detail. As you read them, you may want to refer to these plates to help visualize the locations of organs and the three-dimensional relationships among them

Plate 1 Anterior view of the human torso with the superficial muscles exposed. (m. = muscles; v. = vein.)

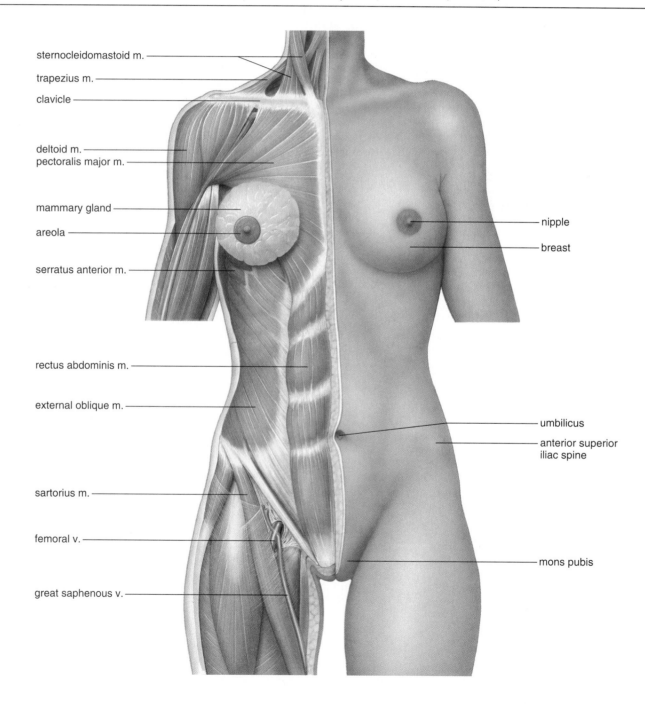

sternocleidomastoid m.

trapezius m.

clavicle

deltoid m.

pectoralis major m.

mammary gland

areola

serratus anterior m.

rectus abdominis m.

external oblique m.

sartorius m.

femoral v.

great saphenous v.

nipple

breast

umbilicus

anterior superior iliac spine

mons pubis

Plate 2 The torso, with the deep muscles exposed. (m. = muscle; n. = nerve; a. = artery; v. = vein.)

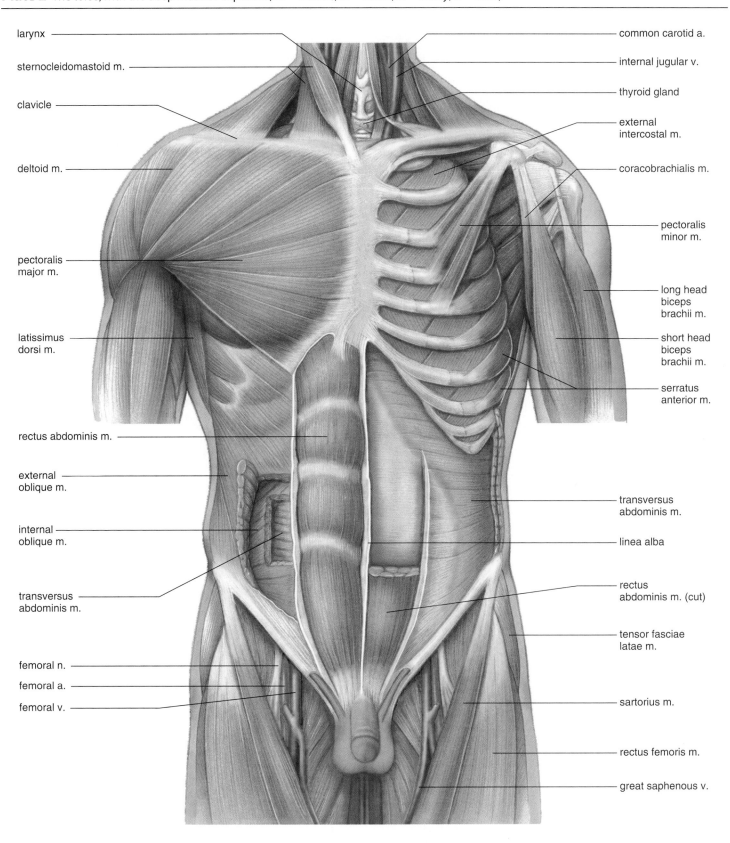

larynx

sternocleidomastoid m.

clavicle

deltoid m.

pectoralis major m.

latissimus dorsi m.

rectus abdominis m.

external oblique m.

internal oblique m.

transversus abdominis m.

femoral n.

femoral a.

femoral v.

common carotid a.

internal jugular v.

thyroid gland

external intercostal m.

coracobrachialis m.

pectoralis minor m.

long head biceps brachii m.

short head biceps brachii m.

serratus anterior m.

transversus abdominis m.

linea alba

rectus abdominis m. (cut)

tensor fasciae latae m.

sartorius m.

rectus femoris m.

great saphenous v.

Plate 3 The torso, with the anterior abdominal wall removed to expose the abdominal viscera. (a. = artery; v. = vein; m. = muscle; n. = nerve.)

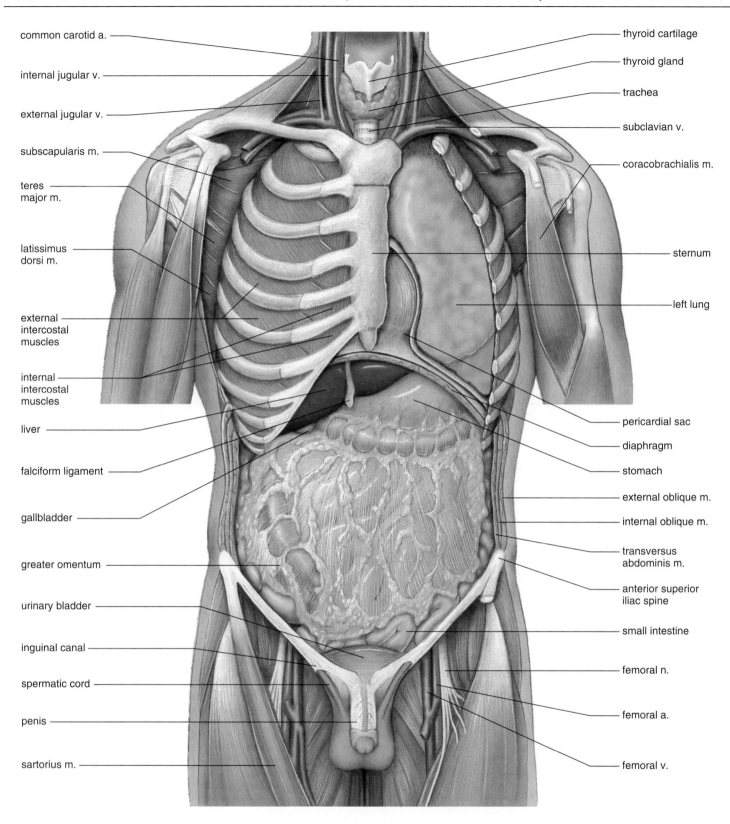

common carotid a.

internal jugular v.

external jugular v.

subscapularis m.

teres major m.

latissimus dorsi m.

external intercostal muscles

internal intercostal muscles

liver

falciform ligament

gallbladder

greater omentum

urinary bladder

inguinal canal

spermatic cord

penis

sartorius m.

thyroid cartilage

thyroid gland

trachea

subclavian v.

coracobrachialis m.

sternum

left lung

pericardial sac

diaphragm

stomach

external oblique m.

internal oblique m.

transversus abdominis m.

anterior superior iliac spine

small intestine

femoral n.

femoral a.

femoral v.

Plate 4 The torso, with the anterior thoracic wall removed to expose the thoracic viscera. (a. = artery; m. = muscle; n. = nerve; v. = vein.)

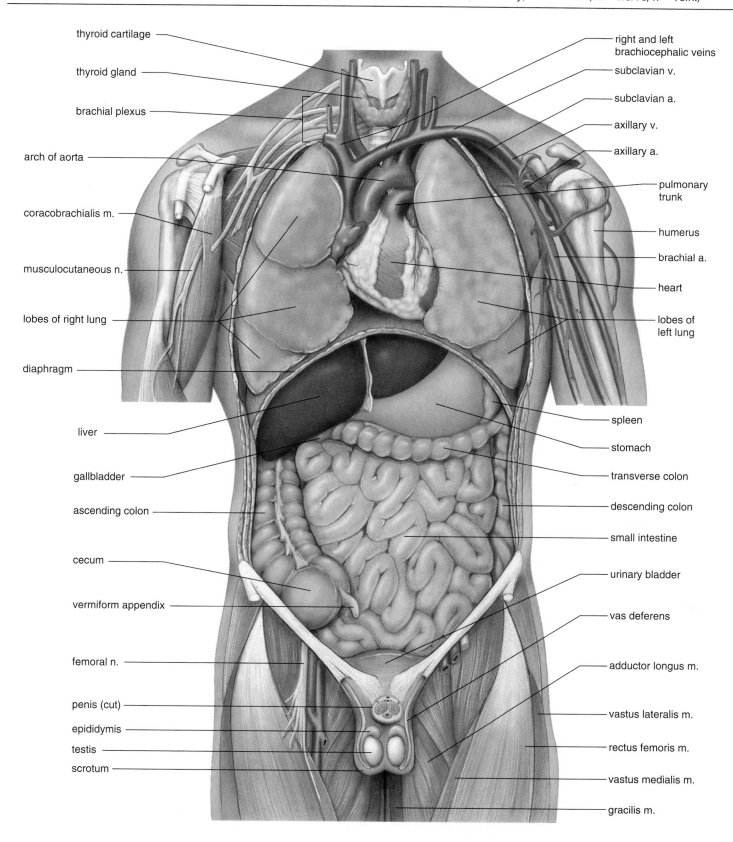

thyroid cartilage

thyroid gland

brachial plexus

arch of aorta

coracobrachialis m.

musculocutaneous n.

lobes of right lung

diaphragm

liver

gallbladder

ascending colon

cecum

vermiform appendix

femoral n.

penis (cut)

epididymis

testis

scrotum

right and left brachiocephalic veins

subclavian v.

subclavian a.

axillary v.

axillary a.

pulmonary trunk

humerus

brachial a.

heart

lobes of left lung

spleen

stomach

transverse colon

descending colon

small intestine

urinary bladder

vas deferens

adductor longus m.

vastus lateralis m.

rectus femoris m.

vastus medialis m.

gracilis m.

Plate 5 The torso as viewed with the thoracic viscera sectioned in a coronal plane; the abdominal viscera as viewed with most of the small intestine removed. (a. = artery; m. = muscle; v. = vein.)

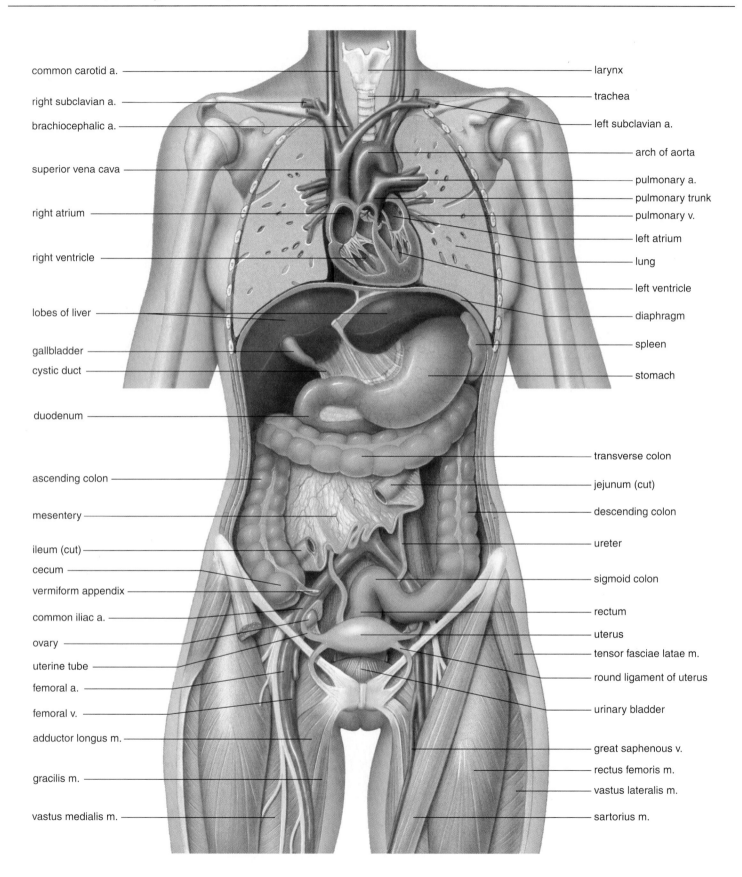

common carotid a.

right subclavian a.

brachiocephalic a.

superior vena cava

right atrium

right ventricle

lobes of liver

gallbladder

cystic duct

duodenum

ascending colon

mesentery

ileum (cut)

cecum

vermiform appendix

common iliac a.

ovary

uterine tube

femoral a.

femoral v.

adductor longus m.

gracilis m.

vastus medialis m.

larynx

trachea

left subclavian a.

arch of aorta

pulmonary a.

pulmonary trunk

pulmonary v.

left atrium

lung

left ventricle

diaphragm

spleen

stomach

transverse colon

jejunum (cut)

descending colon

ureter

sigmoid colon

rectum

uterus

tensor fasciae latae m.

round ligament of uterus

urinary bladder

great saphenous v.

rectus femoris m.

vastus lateralis m.

sartorius m.

Plate 6 The torso as viewed with the heart, liver, stomach, and portions of the small and large intestines removed. (a. = artery; m. = muscle; v. = vein.)

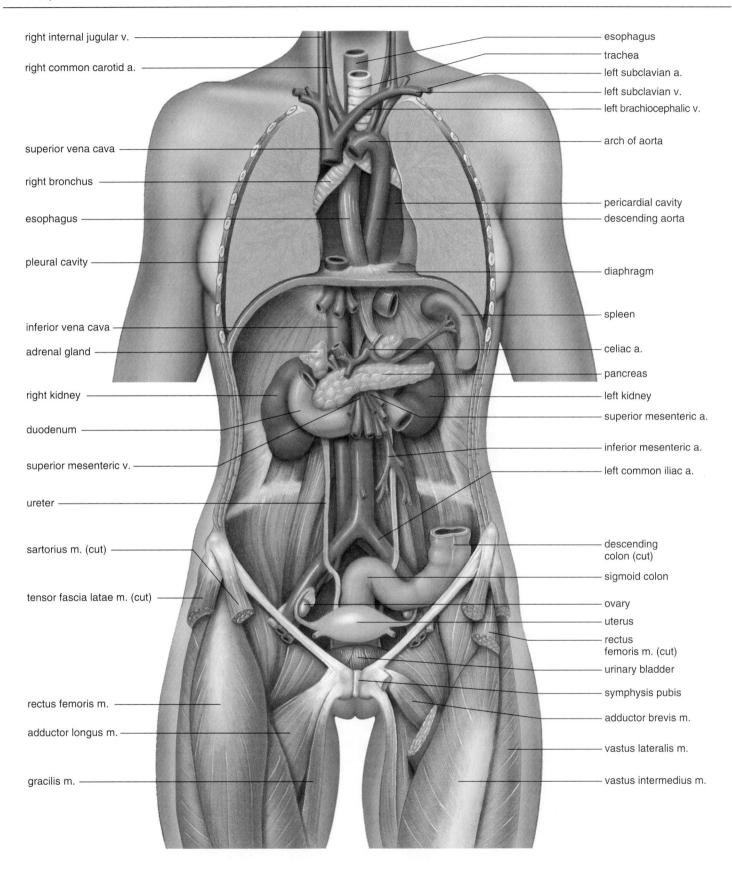

right internal jugular v.

right common carotid a.

superior vena cava

right bronchus

esophagus

pleural cavity

inferior vena cava

adrenal gland

right kidney

duodenum

superior mesenteric v.

ureter

sartorius m. (cut)

tensor fascia latae m. (cut)

rectus femoris m.

adductor longus m.

gracilis m.

esophagus

trachea

left subclavian a.

left subclavian v.

left brachiocephalic v.

arch of aorta

pericardial cavity

descending aorta

diaphragm

spleen

celiac a.

pancreas

left kidney

superior mesenteric a.

inferior mesenteric a.

left common iliac a.

descending colon (cut)

sigmoid colon

ovary

uterus

rectus femoris m. (cut)

urinary bladder

symphysis pubis

adductor brevis m.

vastus lateralis m.

vastus intermedius m.

Plate 7 The torso, with the anterior thoracic and abdominal walls removed, along with the viscera, to expose the posterior walls and body cavities. (a. = artery; m. = muscle.)

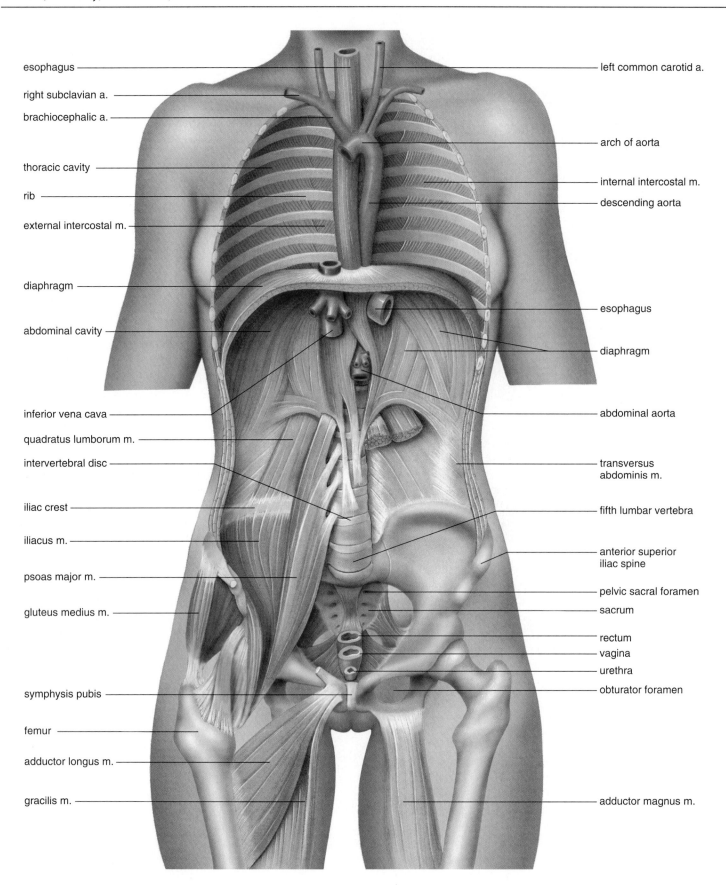

esophagus

right subclavian a.

brachiocephalic a.

thoracic cavity

rib

external intercostal m.

diaphragm

abdominal cavity

inferior vena cava

quadratus lumborum m.

intervertebral disc

iliac crest

iliacus m.

psoas major m.

gluteus medius m.

symphysis pubis

femur

adductor longus m.

gracilis m.

left common carotid a.

arch of aorta

internal intercostal m.

descending aorta

esophagus

diaphragm

abdominal aorta

transversus abdominis m.

fifth lumbar vertebra

anterior superior iliac spine

pelvic sacral foramen

sacrum

rectum

vagina

urethra

obturator foramen

adductor magnus m.

APPENDIX B
Understanding Medical Terminology

Learning Objectives

Upon completion of this section, you should be able to:

1. Discuss the importance of medical terminology and how it can be incorporated into the study of the human body.
2. Differentiate between a root word, a prefix, a suffix, and a compound word.
3. Link word parts to form medical terms.
4. Differentiate between singular and plural endings of medical terms.
5. Pronounce medical words.
6. Separate compound medical words into parts, and analyze their meaning.
7. Recognize root words, prefixes, and suffixes commonly used in medical terminology.

Introduction to Medical Terminology

As students of medical science, we have inherited a wealth of knowledge from scholars of the past. This knowledge was nurtured largely in the atmospheres of European universities, in which Latin and Greek were the languages of lecture and writing. Past scientists sought to define a universal language in which to communicate their findings. Latin and Greek, studied throughout Europe, became the languages of choice for university scholars whose native tongues were English, German, French, Spanish, and so on. Many important early works in medicine were first written in Latin, and their vocabularies remain to this day.

Anatomy and physiology, like many other scientific disciplines, have their own particular language. As the language was developed, names for each new structure to be described were created from existing words. Words were then combined, until they approximated an acceptable description. Medical terminology is simply a catalog of parts that allows us to take apart and reassemble the special language of medicine. The study of medical terminology is easier than it first seems. Relax; you won't need to walk around with a medical dictionary to make sense of medical terminology!

Medical words have three basic parts: a root word, a prefix, and a suffix. The **root word** is the main part of each medical term. Most describe parts of the body, using a Latin or Greek term. Mastering the list of root words in this appendix is an important first step in successfully translating medical terminology. Many of these words will be familiar to you from everyday language: For example, the root word *pyr-* for fire, shows up in the word *pyrotechnics* (fireworks), and the root word for produce, *gen-*, is familiar from the word *generate*.

A **prefix** comes before a root word and alters its meaning. For example, the prefix *hyper-* means over or above. Hyper/kinetic means overactive, hyper/esthesia is overly sensitive, hyper/tension is high blood pressure, and hyper/trophy is overdevelopment.

A **suffix** is attached to the end of a root word and also changes the meaning of the word. For example, the suffix *-itis* means inflammation. Inflammation can occur at almost any part of the body, so *-itis* can be added to root words to make hundreds of words. Dermat/itis is inflammation of the skin, rhin/itis is inflammation of the nose, gastr/itis is inflammation of the stomach, and so on.

Once the root word is known for each part of the anatomy, the prefixes and suffixes can be used to analyze and/or build many medical words. For example, the root word for heart is *cardi*. A few terms in which *cardi* appears are: cardi/algia (pain in the heart), cardio/megaly (enlarged heart), brady/cardia (slow heart), and peri/cardio/centesis (puncture to aspirate fluid from around the heart).

Many medical words have, in addition to a prefix and/or a suffix, more than one root word. These are called **compound words,** and can be analyzed by breaking them into parts. For example, hysterosalpingo-oophorectomy is made up of three root words and a suffix. *Hyster* is the root word for uterus, *salping* is the root word for tube, *oophor* is the root word for ovary, and *-ectomy* is the suffix for cut out. Now we know that hysterosalpingo-oophorectomy means the surgical removal of the uterus, uterine (fallopian) tube, and ovary. (A woman with uterine or ovarian cancer might undergo such a surgery.)

To facilitate pronunciation, word parts need to be linked together. The linkage for word parts is *o*, which may be referred to as a **combining form.** For example, linking the root *cardi* with the suffix *-pathy* would produce a word that would be difficult to pronounce; therefore, an *o* is used to link the root word with the suffix. The complete word is written *cardiopathy* and pronounced kar"de-op'uh-the; the combining form is cardi/o.

When a word is only a root or ends with a root, the word ending depends on whether the word is a noun or an adjective. For example, duodenum (noun) is a part of the small intestine, while duodenal (adjective) is related to the duodenum (for example, duodenal ulcer).

Accurate spelling of each word part is essential:

1. Changing one letter may change the word part. For example, *ileum* is a part of the small intestine, whereas *ilium* is a pelvic bone.
2. Finding a word in the dictionary requires a knowledge of spelling—at least of the beginning of the word. For

example, *pneumonia* and *psychology* have a silent *p*, *rhinitis* (inflammation of the nose) has a silent *h,* and *eupnea* (easy breathing) has an initial silent *e.*

Plural Endings

In many English words, the plurals are formed by adding *s* or *es,* but in Greek and Latin, the plural may be designated by changing the ending:

Singular Ending	Plural Ending	Examples
-a	-ae	aorta—aortae
-ax	-aces	thorax—thoraces
-en	-ina	lumen—lumina
-ex	-ices	cortex—cortices
-ix	-ices	appendix—appendices
-is	-es	testis—testes
-on	-a	phenomenon—phenomena
-um	-a	medium—media
-ur	-ora	femur—femora
-us	-i	bronchus—bronchi
-x	-ces	calyx—calyces
-y	-ies	anomaly—anomalies
-ma	-mata	adenoma—adenomata

If a word ends in *s* and the vowel in the last syllable is short, the word is singular. If the word ends in *s* and the vowel in the last syllable is long, the word is plural. Any word ending in a consonant is singular (for example, *-urn, -us, -at).*

Pronunciation Key

1. Words are made up of syllables.
2. Syllables are made up of letters—consonants and the vowels *a,e,i,o,u.*
3. Diacritical marks are placed over a vowel to help you understand how to pronounce that syllable. The two diacritical marks you should know are the macron (¯) and the breve (˘). The macron designates a long vowel sound, and the breve indicates a short vowel sound. Examples are given below.
4. In this book, vowels standing alone at the end of a syllable with no diacritical mark above are understood to have the "long" sound:

a as in "day"	(matrix = ma'triks)
e as in "be"	(amino = uh-me'no)
i as in "hi"	(ion = i'on)
o as in "no"	(acidosis = as-i-do'sis)
u as in "blue"	(nucleus = nu'kle-us)

5. Vowels embedded within a syllable with no diacritical mark above are understood to have the "short" sound:

a as in "cat"	(lateral = lat'er-al)
e as in "yes"	(pelvic = pel'vik)
i as in "sit"	(distal = dis'tal)
o as in "not"	(abdominal = ab-dom'i-nal)
u as in "cut"	(buffer = buf'er)

6. Vowels in other positions are marked as long (¯) or short (˘) with diacritical marks:

membrane = mem'brān
feedback = fēd'bak
peptide = pep'tīd
disease = di-zēz'
cytokinesis = si'to-kī-ne'sis
vesicle = ves'ĭ-kl
centriole = sen'tre-ōl
ribosome = ri'bo-sōm

7. Other vowel sounds are expressed as follows:

ah the sound of **a** as in "father"
aw the sound of **a** as in "fall"
ar as in "large" (sarcoma = sar-ko'muh)
ār as in bare (larynx = lār'inks; paranasal = pār-uh-na'zal)
uh the sound of **a** as in "about" (anatomy = uh-nat'o-me; negative = neg'uh-tiv)
er as in "her"
ĕr as in "very" (therapy = thĕr'uh-pe; glomerulus = glo-mĕr'yū-lus)
ēr as in peer (delirium = de-lēr'e-um; sarcomere = sar'ko-mēr)
ow as in "cow"
oy as in "boy"
yū the sound of **u** as in "cute" (mucus = myū'kus; tuberculosis = tu-ber-kyū-lo'sis)

8. The primary accent in a word is indicated by a single accent mark; for example, ostomy (os'to-me).
9. The secondary accent is indicated by a double accent; for example, duodenostomy (du"o-dĕ-nos'to-me).
10. The accent on medical terms is generally on the third from the last syllable.

Practice pronouncing the following words:

1. hematemesis (hem"uh-tem'ĕ-sis), vomiting blood
2. hysterosalpingo-oophorectomy (his"ter-o-sal-ping"go-o" of-or-ek'to-me), surgical excision of the uterus, uterine tube, and ovary
3. phrenohepatic (fren"o-he-pat'ik), pertaining to the diaphragm and liver
4. gastropathy (gas"trop'uh-the), disease of the stomach
5. metatarsus (met"uh-tar'sus), part of the foot between the tarsus (ankle) and toes

Commonly Used Prefixes

Prefix	Meaning	Example
a-, an-, in-	without, negative	a/men/orrhea, without a monthly flow
ab-	from, away from	ab/normal, away from normal
ad-, ac-, as-, at-	to, toward	ad/duct, carry toward
allo-	other	allo/graft, grafted tissue donated from one person to another
amphi-	both sides, surrounding	amphi/arthrosis, union between two bones at a joint
aniso-	unequal	an/iso/cyt/osis, abnormal condition of unequal cells
ante-, pre-	before	anterior, front; pre/natal, before birth
anti-, ant-, ob-	against	anti/pyre/tic, agent used against fever
bi-	two	bi/lateral, two sides
bio-	life	bio/logy, study of life
brachy-	short	brachy/dactyl/ism, short fingers and toes
brady-	slow	brady/cardia, slow heart rate
cata-	down	cata/bolism, breakdown of nutrients
cent-	hundred	centi/meter, 1/100 of a meter
chem-	chemical	chemo/therapy, chemical treatment for cancer
chlor-	green	chlor/opsia, green vision, as in certain cases of poisoning
circum-	around	circum/cis/ion, to cut around
co-, com-, con-	with, together	con/genital, born with
contra-	against	contra/indicated, against indication
cyano-	blue	cyano/sis, blue color to skin
de-	away from	de/hydrate, loss of water
dextr-	right	dextr/o/cardia, heart displaced to the right
dia-	through	dia/rrhea, flow through
diplo-	double, twofold	diplo/pia, double vision
dis-	apart	dis/sect, to cut apart
dys-	bad, difficult	dys/pnea, difficult breathing
e-, ex-	out, out from	ex/cise, to cut out
ect-, exo-, extra-	outside	extra/corporeal, outside the body
en-	in, on	en/capsulated, in a capsule
end-	within	endo/scopy, visualization within
epi-	upon	epi/dermis, upon the skin
eu-	good	eu/phonic, good sound
eury-	broad, wide	eury/gnathous, having a wide jaw
gero-, geronto-	old, elderly	geronto/logy, study of diseases of the elderly
hem-, semi-	half	hemi/gastr/ectomy, surgical removal of half of the stomach
hydro-	water	hydro/cephalus, water on the brain
hyper-	over, above	hyper/kinetic, overactive
hypo-	under, below	hypo/glossal, under the tongue
immun-	free, exempt	immun/ity, exempt from the effects of specific disease-causing agents
in-	not	in/soluble, not able to dissolve
infra-	beneath	infra/mammary, beneath the breast
inter-	between	inter/cellular, between the cells
intra-	within	intra/cranial, within the cranium
iso-	same	iso/coria, pupils of equal size
kil-	thousand	kilo/gram, 1,000 grams
leio-	smooth	leio/myoma, smooth muscle tumor
lepto-	high, thin	lepto/cephaly, having a long, thin head and neck

Prefix	Meaning	Example
levo-	left	levo/duction of eye, movement of eye to the left
lys-	dissolution, disintegration	lyso/some, organelle that degrades worn cell parts
macr-	large	macro/cyte, large cell
mal-	bad	mal/nutrition, bad nourishment
mes-	middle	mes/entery, middle of intestine
meta-	after, beyond	meta/carpals, beyond the carpals (wrist)
micr-	small	micro/cephal/ic, having a small head
milli-	one-thousandth	milli/liter, 1/1,000 of a liter
multi-	many	multi/para, one who has many children
neo-	new	neo/plasm, new growth
normo-	normal	normo/cyte, normal, healthy red blood cell
olig-	scanty, few	olig/uria, scanty amount of urine
onc-	tumor	onc/ology, study of tumors
opistho-	behind, backward	opistho/tonos, bending backward
pachy-	thick	pachy/derma, thick skin
per-	through	per/cutaneous, through the skin
peri-	around	peri/tonsillar, around the tonsil
photo-	light	photo/phobia, avoidance of light
pleur-	rib, side	pleur/al membranes, serous membranes that enclose the lungs
poly-	much, many	poly/cystic, many cysts
post-	after	post/mortem, after death
pre-	before	pre/natal, before birth
presby-	old	presby/opia, old vision
primi-	first	primi/gravida, first pregnancy
pro-	before	pro/gnosis, prediction of the probable outcome of a disease
re-	back, again	re/generate, produce, develop again
retr-, retro-	behind	retro/sternal, behind the sternum
rhod-	red	rhod/opsin, visual pigment in the eye
semi-	half	semi/comatose, in a partial comatose state; semiconscious
sub-	under	sub/lingual, under the tongue
super-, supra-	above	supra/spinal, above the vertebral column
syn-, sym-	with, together	syn/ergism, working together
tachy-	fast	tachy/phasia, fast speech
un-	not	un/conscious, not conscious

Commonly Used Suffixes

Suffix	Meaning	Example
-algia	pain	dent/algia, pain in the tooth
-ase	suffix for an enzyme	protein/ase, protein-digesting enzyme
-atresia	without an opening	proct/atresia, rectum without an opening
-blast	immature cell form	erythro/blast, immature red blood cell
-cele	hernia	omphalo/cele, umbilical hernia
-centesis	puncture to aspirate fluid	arthro/centesis, puncture to aspirate fluid from a joint
-cept	take, receive	re/cept/or, something that receives again
-cide	kill	bacteri/cidal, able to kill bacteria
-cis	cut	circum/cis/ion, cutting around
-cyte	cell	erythro/cyte, red cell
-denia	pain	cephalo/denia, pain in the head
-desis	fusion	arthro/desis, fusion of a joint
-ectasia	expansion	cor/ectasis, expanding or dilating pupil
-ectomy	cut out, excise	nephr/ectomy, surgically remove kidney
-edema	swelling	cephal/edema, swelling of head
-emesis	vomiting	hyper/emesis, excessive vomiting
-emia	blood	hyper/glyc/emia, elevated blood sugar
-gen	precursor of	terato/gen, agent that causes a malformed fetus
-gnosis	knowledge	dia/gnosis, knowledge through examination (determining cause of disease)
-gram	record	myelo/gram, X ray of the spinal cord
-graphy	making a record	angio/graphy, making a record of vessels
-iasis	condition	chole/lith/iasis, condition of gallstones
-ism	condition, disease	alcohol/ism, addiction to alcohol
-ist	one who	opto/metr/ist, one who measures vision
-itis	inflammation	aden/itis, inflammation of a gland
-lepsy	seizures	narco/lepsy, seizures of numbness, attacks of deep sleep
-lexy	speech, words	dys/lexia, difficulty in reading words
-logist	one who specializes	ophthalmo/logist, one specializing in eyes
-logy	study of	bio/logy, study of life
-lysis, -lytic, -lyze	break down, dissolve	teno/lysis, destruction of tendons
-lyt	dissolvable	electro/lyte, substance that ionizes in water solution
-malacia	abnormal softening	osteo/malacia, abnormal softening of bone
-mania	madness	pyro/mania, irresistible urge to set fires
-megaly	enlargement	spleno/megaly, enlargement of spleen
-meter	measure	thermo/meter, instrument to measure temperature
-oid	resembling	muc/oid, resembling mucus
-oma	tumor	neur/oma, nerve tumor
-opia	vision	ambly/opia, dim vision
-osis	abnormal condition	nephr/osis, abnormal condition of kidney
-osme, -osmia	smell	an/osmia, inability to smell
-ostomy	create an opening	col/ostomy, create an opening in the colon
-otia	ear	macr/otia, large ear
-pathy	disease	encephalo/pathy, disease of the brain
-penia	deficiency, poor	leuko/cyto/penia, deficiency of white cells
-pepsia	digestion	dys/pepsia, bad digestion
-pexy	surgical fixation	nephro/pexy, surgical fixation of kidney

Suffix	Meaning	Example
-phasia	speak, say	a/phasia, without ability to speak
-philia	love, attraction	chromo/philic, attracted to color
-phobia	abnormal fear	agora/phobia, abnormal fear of crowds
-plasia	formation	hyper/plasia, excessive formation
-plasm	substance	proto/plasm, original substance
-plasty	make, shape	rhino/plasty, to shape the nose
-plegia	paralysis	hemi/plegia, paralysis of one half of body
-pnea	breath	tachy/pnea, fast breathing
-ptosis	prolapse, dropping	hystero/ptosis, prolapse of uterus
-rrhagia	burst forth	metro/rrhagia, hemorrhage from uterus
-rrhaphy	suture, sew	hernio/rrhaphy, suture a hernia
-rrhea	flow, discharge	oto/rrhea, discharge from ear
-rrhexis	rupture	spleno/rrhexis, rupture of the spleen
-scope	instrument for viewing	oto/scope, instrument to look in ears
-scopy	visualization	laryngo/scopy, visualization of larynx
-some, -soma	body	lyso/some, body that lyses or dissolves
-spasm	twitching	blepharo/spasm, twitching of eyelid
-stasis	stop, control	hemo/stasis, control bleeding
-therapy	treatment	hydro/therapy, treatment with water
-tome	instrument to cut	osteo/tome, instrument to cut bone
-tomy	to cut	laparo/tomy, to cut into the abdomen
-tripsy	crushing	nephro/litho/tripsy, crushing stone in kidney
-trophy, -trophic, -trophin	development	hyper/trophy, overdevelopment
-uria	urine	hemat/uria, blood in the urine

Commonly Used Root Words*

Root	Meaning	Example
acro-	extremity, peak	acro/megaly, enlarged extremities; acro/phobia, abnormal fear of heights
aden-	gland	adeno/pathy, disease of a gland
adip-	fat, lipid	adip/ose tissue, fat storage tissue
aer-, aero-	air	aero/phagia, swallowing air
andro-	man, male	andro/gen, agent that causes male development
angi-	vessel	angi/oma, tumor of a vessel
arthr-	joint	arthr/algia, pain in the joint
athero-	soft, pasty material	athero/sclerosis, pasty hardening of an artery
balano-	penis	balano/plasty, surgical repair of penis
blast-	bud, growing thing	neuro/blast, growing nerve cell
blephar-	eyelid	blephar/op/tosis, drooping eyelid
brachi-	arm	brachi/al, pertaining to the arm
bronch-	windpipe	bronch/us, a branch of the trachea (windpipe)
bucco-	cheek	bucco/labial, referring to cheek and lip
calcaneo-	heel	calcaneo/dynia, painful heel
carcin-	cancer	adeno/carcin/oma, cancerous tumor of a gland
cardi-	heart	myo/cardi/tis, inflammation of heart muscle
carp-	wrist	flexor carp/i, muscle to bend wrist
caud-	tail	caud/al, pertaining to tail
celio-	abdomen	celio/tomy, incision of the abdomen
cephal-	head	cephalo/dynia, pain in the head
cervic-	neck, cervix	cervic/itis, inflammation of the neck of uterus
cheil-	lip	cheilo/plasty, shaping the lip
cheir-, chir-	hand	chiro/megaly, large hands
chol-	bile, gall	chole/cyst/ectomy, surgical removal of the gallbladder
chondr-	cartilage	chondro/malacia, softening of cartilage
chrom-	color	poly/chromatic, having many colors
chron-	time	syn/chron/ous, occurring at the same time
cleid-	clavicle	cleido/costal, referring to clavicle and ribs
col-	colon	mega/colon, enlarged colon
colp-	vagina	colp/orrhaphy, suture of vagina
core-	pupil	core/pexy, suturing the iris to change the pupil
cost-	ribs	inter/costal, between the ribs
crani-	skull	crani/otomy, incision into the skull
cry-	cold	cryo/philic, cold loving
crypt-	hidden	crypt/orchid/ism, hidden (undescended) testicle
cutan-, cut-	skin	sub/cutaneous, below the skin
cyan-	blue	acro/cyan/osis, abnormal condition of blueness of extremities
cyst-	bladder	cysto/cele, bladder hernia
cyt-	cell	thrombo/cyte, clotting cell (platelet)
dacry-	tear	dacryo/rrhea, flow of tears
dactyl-	fingers, toes	poly/dactyl/ism, too many fingers and toes
dent-, -odont	teeth	peri/odontal, around the teeth; dent/algia, toothache
derm-, dermat-	skin	intra/dermal, within the skin
desmo-	ligament	desmo/dentium, ligament anchoring tooth
dextr-	right	dextro/cardia, heart displaced to the right
dips-	thirst	poly/dipsia, excessive thirst

Commonly Used Root Words—continued

Root	Meaning	Example
dors-	back	dors/al, pertaining to the back
duct-	carry	ovi/duct, tube to carry ova (eggs)
encephal-	brain	encephalo/cele, hernia of the brain
enter-	intestine	gastro/enter/itis, inflammation of stomach and intestine
erg-	work	en/ergy, working with
erthyr-	red	erythro/cyto/penia, deficiency of red cells
esthe-	sensation	an/esthe/tic, agent to eliminate sensation
esthen-	weakness	my/esthenia, muscle weakness
facio-	facial, the face	facio/plasty, plastic surgery on the face
febr-	fever	a/febrile, without a fever
flex-	bend	dorsi/flex, bend backward
gastr-	stomach	gastro/scopy, visualization of the stomach
gen-	produce	patho/genic, agent that produces disease
gingiv-	gums	gingiv/ectomy, removal of gums
gloss-	tongue	hypo/glossal, under the tongue
glyc-, glu-	sugar	hypo/glyc/emia, low blood sugar
gnath-	jaw	micro/gnath/ism, small jaw
gonado-	sex organs	gonado/trophin, causes development of sex organs
grav-	heavy, pregnancy	secundi/gravida, second pregnancy
gynec-	female	gyneco/logy, study of female conditions
hem-, hemat-	blood	hemat/emesis, vomiting blood
hepat-	liver	hepato/megaly, enlarged liver
heter-	different	hetero/genous, different origins
hidr-	perspiration	hidro/rrhea, flow of perspiration
hist-	tissue	histo/logy, study of tissue
home-, hom-	same	homeo/stasis, stay same, equilibrium
hydro, hydra-	water	de/hydra/tion, process of losing water
hyster-	uterus	hyster/ectomy, removal of the uterus
iatr-	physician	iatro/genic, produced by the physician
irid-	iris	irid/ectomy, surgical removal of iris
is-	equal	iso/tonic, equal in pressure
kary-	nut, nucleus	mega/karyo/cyte, cell with large nucleus
kerat-	cornea	kerato/plasty, repair of cornea
kin-	move	kinesio/logy, study of movement
labio-	lips	labio/gingival, refers to lips and gums
lacrim-	tear	lacrima/tion, crying
lact-, galact-	milk	lacto/genic, milk-producing
lapar-	abdomen	laparo/rrhaphy, suture of the abdomen
laryng-	larynx	laryngo/scopy, visualization of larynx
later-	side	bi/lateral, two sides
leuk-, leuc-	white	leuko/rrhea, white discharge
lingu-	tongue	sub/lingual, under the tongue
lip-	fat	lip/oma, tumor of fat
lith-	stone	litho/tripsy, crushing a stone
lymphadeno-	lymph node	lymphadeno/pathy, disease affecting lymph nodes
lymphangio-	lymph vessel	lymphangio/graphy, X rays of lymph vessels and nodes
mast-, mammo-	breast	mast/itis, inflammation of the breast; mammo/gram, X ray of breast
melan-	black	melan/oma, black tumor

Root	Meaning	Example
men-	monthly, menstrual	dys/meno/rrhea, difficult monthly flow
mening-	membrane	mening/es, membranes that cover the brain and spinal cord
metr-	uterus	endo/metr/ium, lining of uterus
morph-	shape, form	poly/morphic, pertaining to many shapes
my-	muscle	myo/metr/itis, inflammation of muscle of uterus
myc-	fungus	onycho/myc/osis, fungus condition of the nails
myel-	marrow, spinal cord	myelo/gram, X-ray record of spinal cord
myring-	eardrum	myringo/tomy, opening into eardrum
nas-	nose	naso/pharyng/eal, pertaining to nose and throat
nat-	to be born	pre/nat/al, before birth
necr-	dead	necr/opsy, examining dead bodies; autopsy
nephr-, -ren	kidney	hydro/nephr/osis, abnormal condition of water in the kidney
neur-	nerve	neur/algia, nerve pain
noct-, -nyct	night	noct/uria, voiding at night
nucle-	kernel	nucle/us, dense core (kernel) of an atom
null-	none	nulli/gravida, woman who has had no pregnancies
ocul-	eye	mon/ocul/ar, pertaining to one eye
omphal-	umbilicus	omphalo/rrhea, discharge from the navel
onych-	nail	onycho/crypt/osis, condition of hidden nail (ingrown)
oo-	ova, egg	oo/genesis, producing eggs
oophor-	ovary	oophoro/cyst/ectomy, removal of cyst from ovary
ophthalm-	eye	ex/ophthalmos, condition of protruding eyes
or-	mouth	oro/pharyngeal, pertaining to mouth and throat
orchid-	testis	orchid/ectomy, removal of testis
orexis-	appetite	an/orexis, absence of appetite
orth-	straight	orth/odont/ist, one who straightens teeth
oste-, oss-	bone	osteo/chondr/oma, tumor of bone and cartilage
ot-, aur-	ear	ot/itis, inflammation of the ear; post/auricular, behind the ear
para-	to bear	primi/para, to bear first child.
path-	disease	patho/physio/logy, study of effects of disease on body functioning
pect-	chest	pecto/ralis, chest muscle
ped-	child	ped/iatrician, doctor who specializes in children
peps-	digest	dys/pepsia, bad digestion
phag-	swallow, eat	a/phagia, inability to swallow
phallo-	penis	phallo/dynia, pain in the penis
pharmac-	drug	pharmaco/logy, study of drugs
pharyng-	throat	pharyng/itis, inflammation of the throat
phas-	speak, say	tachy/phasia, speaking rapidly
phleb-	vein	phlebo/thromb/osis, abnormal condition of clot in vein
phon-	voice	a/phonic, absence of voice
phren-	diaphragm	phreno/hepatic, pertaining to the diaphragm and liver
pil-, trich-	hair	tricho/glossia, hairy tongue
pleur-	lining of lung, thorax	pleur/isy, inflammation of the pleura
pneum-	air, breath	pneumo/thorax, air in the chest
pneumon-	lung	pneumon/ectomy, surgical removal of the lung
pod-	foot	pod/iatrist, one who specializes in foot problems
proct-	rectum	procto/scopy, visualization of the rectum
pseud-	false	pseudo/cyesis, false pregnancy
psych-	mind	psycho/somatic, pertaining to the mind and the body
pulmon-	lung	cardio/pulmonary, pertaining to heart and lungs

Root	Meaning	Example
py-	pus	pyo/rrhea, flow of pus
pyel-	kidney	pyelo/nephr/itis, inflammation of the kidney
pyl-	door, orifice	pyl/oric sphincter, ring of muscle that controls food entry into duodenum
pyr-	fire, fever	anti/pyretic, agent used against fever
quadri-	four	quadri/plegia, paralysis of all four extremities
rhin-	nose	rhino/plasty, revision of the nose
salping-	tube	salping/itis, inflammation of the uterine tube
sanguin-	blood	ex/sanguina/tion, process of bleeding out (bleed to death)
sarc-, sarco-	flesh, striated muscle	sarco/lemma, cell membrane of a muscle fiber
scler-	hard	arterio/scler/osis, condition of hardening of arteries
scot-	dark	scot/oma, blind spot in the visual field
sect-	cut	dis/section, cutting apart
sept-	contamination	anti/septic, agent used against contamination
sial-	saliva	sialo/rrhea, excessive salivation
spondyl-	vertebra	spondyl/itis, inflammation of the vertebrae
steato-	fat, lipid	steato/rrhea, fat in the fecal material
sten-	narrow, constricted	pyloric sten/osis, narrowing of pylorus
stomat-	mouth	stomat/itis, inflammation of the mouth
strict-	draw tight	vaso/con/strict/or, agent that compresses vessels
tax-	order, arrange	a/taxic, uncoordinated
ten-	tendon	teno/rrhaphy, suture a tendon
terato-	malformed fetus	terato/gen, agent that can harm a fetus
therm-	heat	hyper/thermia, raising body heat
thorac-	chest	thoraco/centesis, puncture to aspirate fluid from chest
thromb-	clot	thrombo/cyte, clotting cell
tox-	poison	tox/emia, poison in the blood
trache-	windpipe	tracheo/malacia, softening of tracheal cartilages
trachel-	neck	trachel/orrhaphy, suture of cervix (neck of uterus)
traumat-	wound	traumat/ology, study of trauma
tri-	three	tri/geminal, having three beginnings
trop-	turn	ec/tropion, turned out
ur-	urine	ur/emia, urine constituents in the blood
vas-	vessel	vaso/constriction, narrowing of a vessel
veni-	vein	veni/puncture, puncture of a vein
vert-	turn	retro/vert/ed, turned backward
vesic-	bladder	vesico/cele, hernia of the bladder
viscer-	internal organs	e/viscera/tion, process of viscera protruding from abdominal wall
vita-	life	vital, necessary for life

*Words that are the same as anatomical terms used in English (for example, pancreas, tonsil, and so on) have been omitted.

Now you are ready to apply your knowledge! Each time you find a new word in the text, try to analyze its meaning. Break each words into its parts: root, prefix, and suffix. You'll be able to translate the words without using a dictionary. At the end of each chapter, in the Medical Terminology Reinforcement Exercise, you are given an opportunity to reinforce your knowledge. Pronounce the words, and dissect them into parts to arrive at a meaning. You can also begin to build medical words and use them in your everyday conversation. You will be amazed at how rapidly your vocabulary will grow, and how your study of the human body will become easier and more enjoyable.

GLOSSARY

A

abdominal cavity Portion of the body between the diaphragm and the pelvis. 7

abdominopelvic cavity Pertaining to the abdominal and pelvic regions. 6

abduction Movement of a body part away from the midline. 106

acetabulum Socket in the lateral surface of the hipbone into which the head of the femur articulates. 101

acetylcholine (ACh) Neurotransmitter secreted at the ends of many neurons; responsible for the transmission of a nerve impulse across a synaptic cleft. 145

acetylcholinesterase (AChE) Enzyme in the membrane of postsynaptic cells that breaks down acetylcholine; this enzymatic reaction inactivates the neurotransmitter. 145

ACh See *acetylcholine.* 145

AChE See *acetylcholinesterase.* 145

acid Solution in which pH is less than 7; substance that contributes or liberates hydrogen ions in a solution; opposite of *base.* 123

acidosis Excessive accumulation of acids in body fluids. 23, 197, 333

acne vulgaris Inflammation of sebaceous glands; the common form of acne. 73

acquired immunodeficiency syndrome See *AIDS.* 264, 362

acromegaly Condition resulting from an increase in growth hormone production after adult height has been achieved. 190

acrosome Covering on the tip of a sperm cell's nucleus that is believed to contain enzymes necessary for fertilization. 345

ACTH See *adrenocorticotropic hormone.* 188

actin One of the two major proteins of muscle; makes up thin myofilaments in myofibrils of muscle cells. See *myosin.* 42, 116

action potential Change in potential propagated along the membrane of a neuron; the nerve impulse. 143

active immunity Resistance to disease due to the immune system's response to a microorganism or a vaccine. 266

active transport Transfer of a substance into or out of a cell against a concentration gradient by a process that requires a plasma membrane carrier protein and an expenditure of energy. 44

acute bronchitis Infection of the primary and secondary bronchi. 286

acute disease Sudden in onset and severe. 12

acute lymphoblastic leukemia (ALL) Cancer of the blood in which immature lymphocytes proliferate in bone marrow, the thymus, and lymph nodes. 213

Addison's disease Condition resulting from a deficiency of adrenal cortex hormones. 195

adduction Movement of a body part toward the midline. 106

adenosine diphosphate (ADP) Molecule produced when the terminal phosphate is lost from a molecule of adenosine triphosphate; ATP. 32

adenosine triphosphate (ATP) Molecule used by cells when energy is needed. 32

ADH See *antidiuretic hormone.* 188

adhesion junction Junction between cells in which the adjacent plasma membranes do not touch but are held together by intercellular filaments attached to buttonlike thickenings. 65

ADP See *adenosine diphosphate.* 32

adrenal cortex Outer portion of the adrenal gland. 193

adrenal gland Endocrine gland located on the superior portion of each kidney. 193

adrenaline See *epinephrine.* 193

adrenal medulla Inner portion of the adrenal gland. 193

adrenocorticotropic hormone (ACTH) Hormone secreted by the anterior lobe of the pituitary gland that stimulates the adrenal cortex to produce cortisol. 188

afterbirth Placenta and the extraembryonic membranes, which are delivered (expelled) during the third stage of parturition. 385

agglutination Clumping of cells, particularly in reference to red blood cells involved in an antigen-antibody reaction. 218

agranular leukocyte White blood cell with poorly visible cytoplasmic granules. 212

AID See *artificial insemination by a donor.* 360

AIDS (acquired immunodeficiency syndrome) Disease caused by a retrovirus and transmitted via body fluids; characterized by failure of the immune system. 264, 362

albinism Genetic disorder characterized by a defect in pigment production. 71

albumin Plasma protein that helps the osmotic concentration of blood. 209

aldosterone Hormone secreted by the adrenal cortex that functions in regulating sodium and potassium excretion by the kidneys. 194, 332

alimentary canal Tubular portion of the digestive tract. 296

alkalosis Excessive accumulation of bases in body fluids. 23, 333

ALL See *acute lymphoblastic leukemia.* 213

allantois Extraembryonic membrane that serves as a source of blood vessels for the umbilical cord. 371

allele Different forms of a gene. 395

allergen Foreign substance capable of stimulating an allergic reaction. 268

allergy Immune response to substances that usually are not recognized as foreign. 268

all-or-none law Law that states that muscle fibers either contract maximally or not at all, and that neurons either conduct a nerve impulse completely or not at all. 122

alopecia Loss of hair. 72

alveolus Air sac of a lung (pl., *alveoli*). 279

Alzheimer's disease Brain disorder characterized by a general loss of mental abilities. 145

amino acid Unit of a protein that takes its name from the fact that it contains an amino group ($-NH_2$) and an acid group ($-COOH$). 28

amniocentesis Method of retrieving fetal cells for genetic testing in which a long needle is used to withdraw a sample of amniotic fluid. 390

amnion One of the extraembryonic membranes; a fluid-filled sac around the embryo. 371

ampulla Expansion at the end of each semicircular canal that contains receptors for rotational equilibrium. 181

anabolic steroid Synthetic steroid that mimics the effect of testosterone. 197

anaphase Stage in mitosis when replicated chromosomes separate and move to opposite poles of the cell. 50

anaphylactic shock A severe systemic form of anaphylaxis involving bronchiolar constriction, impaired breathing, vasodilation, and a rapid drop in blood pressure with a threat of circulatory failure. 268

anatomy Branch of science dealing with the form and structure of body parts. 2

androgen Male sex hormone. 197, 343

anemia Condition characterized by a deficiency of red blood cells or hemoglobin. See also *iron deficiency anemia, pernicious anemia.* 214

anencephaly Congenital absence of the cranial vault, with cerebral hemispheres completely missing or reduced to small masses attached to the base of the skull. 382

aneurysm Saclike expansion of a blood vessel wall. 239

angina pectoris Condition characterized by thoracic pain resulting from occluded coronary arteries; precedes a heart attack. 228

ankle-jerk reflex Automatic, involuntary response initiated by tapping the Achilles tendon just above its attachment to the calcaneus (heel bone). 155

anorexia nervosa Eating disorder caused by the fear of becoming obese; includes loss of appetite and inability to maintain a normal minimum body weight. 319

antagonist Muscle that acts in opposition to a prime mover. 124

anterior Pertaining to the front; the opposite of *posterior.* 3

anterior pituitary Front lobe of the pituitary gland. 188

antibody Protein produced in response to the presence of some foreign substance in the blood or tissues. 212

antibody-mediated immunity Resistance to disease-causing agents resulting from the production of specific antibodies by B lymphocytes; humoral immunity. 261

antibody titer Amount of antibody present in a sample of blood serum. 266

anticodon Three contiguous nucleotides of a transfer RNA molecule that are complementary to a specific mRNA codon. 48

antidiuretic hormone (ADH) Hormone released from the posterior lobe of the pituitary gland that enhances water conservation by the kidneys; sometimes called vasopressin. 188

antigen Foreign substance, usually a protein, that stimulates the immune system to produce antibodies. 212, 260

antigen-presenting cell (APC) The cell that displays the antigen to the cells of the immune system so they can defend the body against that particular antigen. 262

antigen receptor Receptor proteins in the plasma membrane of immune system cells whose shape allows them to combine with a specific antigen. 260

anus Outlet of the alimentary canal. 304

aorta Major systemic artery that receives blood from the left ventricle. 242

aortic body Receptor in the aortic arch sensitive to oxygen content, carbon dioxide content, and blood pH. 283

aplastic anemia Insufficient number of red blood cells brought on by damage to the red bone marrow due to radiation or chemicals. 213

apnea Temporary cessation of breathing. 284

apoptosis Programmed cell death. 46, 260

appendicitis Infected swelling of the appendix. 307

appendicular portion Pertaining to the upper limbs (arm) and lower limbs (legs). 4

appendicular skeleton Part of the skeleton forming the upper limbs, pectoral girdle, lower limbs, and pelvic girdle. 97

appendix Small, tubular appendage that extends outward from the cecum of the large intestine; see also *vermiform appendix*. 257

aqueous humor Watery fluid that fills the anterior cavity of the eye. 171

arachnoid membrane Weblike middle covering (one of the three meninges) of the central nervous system. 146

areola Dark, circular area surrounding the nipple of the breast. 356

arrector pili Smooth muscle in the skin associated with a hair follicle. 72

arrhythmia Abnormal heart rhythm. 20, 231

arterial duct See *ductus arteriosus*. 246

arteriole Branch from an artery that leads into a capillary. 234

arteriosclerosis Thickening and hardening of arterial walls. 234

artery Vessel that takes blood away from the heart; characteristically possesses thick elastic walls. 234

articular cartilage Hyaline cartilaginous covering over the articulating surface of the bones of synovial joints. 84

articulation Joining together of bones at a joint. 84

artificial insemination by donor (AID) Placement of donated sperm in the vagina so that fertilization followed by pregnancy might occur. 360

ascending colon Portion of the large intestine that travels superiorly as it extends from the entry of the small intestine to the transverse colon. 304

assisted reproductive technologies (ART) Medical techniques, sometimes performed in vitro, that are done to increase the chances of pregnancy. 360

association area Region of the cerebral cortex related to memory, reasoning, judgment, and emotional feelings. 149

aster Short, radiating fibers about the centrioles at the poles of a spindle. 49

asthma Condition in which bronchioles constrict and cause difficulty in breathing. 268, 288

astigmatism Visual defect due to errors in refraction caused by abnormal curvatures in the surface of the cornea or lens. 177

atherosclerosis Condition in which fatty substances accumulate abnormally beneath the inner linings of the arteries. 228

athlete's foot Skin disease caused by fungal infection, usually of the toes and soles of the foot. 74

atlas First cervical vertebra; it supports and balances the head. 95

atom Smallest unit of matter. 2, 18

ATP See *adenosine triphosphate*. 32

atria Referring to the chambers of the heart. 226

atrial natriuretic hormone (ANH) Substance secreted by the atria of the heart that accelerates sodium excretion so that blood volume decreases. 194, 238, 332

atrioventricular (AV) bundle Part of the cardiac conduction system that extends from the AV node to the bundle branches. 230

atrioventricular (AV) node Small region of neuromuscular tissue located near the septum of the heart that transmits impulses from the SA node to the ventricular walls. 230

atrioventricular (AV) valve Valve located between the atrium and the ventricle. 226

atrium Chamber; particularly an upper chamber of the heart that lies above the ventricles (pl., *atria*). 226

atrophy Wasting away or decrease in size of an organ or tissue. 123

auditory canal Tube in the outer ear that leads to the tympanic membrane. 178

auditory (eustachian) tube Air tube that connects the pharynx to the middle ear. 178

autoimmune disease Disease that results when the immune system mistakenly attacks the body's own tissues. 269

autonomic motor system Sympathetic and parasympathetic portions of the nervous system that function to control the actions of the visceral organs and skin. 156, 172

autosome Chromosome other than a sex chromosome. 390

AV bundle See *atrioventricular bundle*. 230

AV node See *atrioventricular node*. 230

AV valve See *atrioventricular valve*. 226

axial portion Pertaining to the body's axis. 4

axial skeleton Portion of the skeleton that supports and protects the organs of the head, neck, and trunk. 89

axis Second cervical vertebra upon which the atlas rotates, allowing the head to turn. 95

axon Process of a neuron that conducts nerve impulses away from the cell body. 142

B

ball-and-socket joint The most freely movable type of joint (for example, the shoulder or hip joint). 105

balloon angioplasty Procedure for treating a blocked coronary artery: A flexible guide wire is pushed into the coronary artery, and a miniature balloon catheter is pushed down the wire to the blockage; repeated inflations of the balloon decrease or relieve the blockage. 229

basal body Cytoplasmic structure that is located at the base of and may organize cilia or flagella. 42

basal cell carcinoma Form of skin cancer that begins in the epidermis and rarely metastasizes but has the capacity to invade local tissues. 74

basal nuclei Mass of gray matter located deep within a cerebral hemisphere of the brain. 151

base Solution in which pH is more than 7; a substance that contributes or liberates hydroxide ions in a solution; alkaline; opposite of *acid*. 23

basophil Leukocyte with a granular cytoplasm and that is able to be stained with a basic dye. 212

benign prostatic hyperplasia (BPH) Enlargement of the prostate gland. 338

benign tumor Mass of cells derived from a single mutated cell that has repeatedly undergone cell division but remained at the site of origin. 80

bicarbonate ion The form in which carbon dioxide is carried in the blood; HCO_3^-. 285

bicuspid valve Atrioventricular valve between the left atrium and the left ventricle; also known as the mitral valve. 226

bile Secretion of the liver that is temporarily stored in the gallbladder before being released into the small intestine, where it emulsifies fat. 302

biopsy Removal of sample tissue by plungerlike devices to diagnose a disease. 66

birth control pill Oral contraception containing estrogen and progesterone. 358

blastocyst Early stage of embryonic development that consists of a hollow ball of cells. 373

blind spot Area where the optic nerve passes through the retina and where vision is not possible due to the lack of rod cells and cone cells. 173

blood Connective tissue composed of cells separated by plasma. 61

blood pressure Force of blood against a blood vessel wall. 236

blood transfusion Introduction of whole blood or a blood component directly into the bloodstream. 218

B lymphocyte Type of lymphocyte that is responsible for antibody-mediated immunity. 212, 260

bolus Small lump of food that has been chewed and swallowed. 298

bone Connective tissue having a hard matrix of calcium salts deposited around protein fibers. 61

brachiocephalic Pertaining to the arm and head, as in the brachiocephalic artery and vein. 242

bradycardia Slow heart rate, characterized by fewer than 60 heartbeats per minute. 231

brain stem Portion of the brain that includes the midbrain, pons, and medulla oblongata. 151

Braxton-Hicks contractions Strong, late-term uterine contractions prior to cervical dilation; also called false labor. 384

breech birth Birth in which the baby is positioned rump first. 379

bronchiole Smaller air passages in the lungs. 279

bronchus One of the two major divisions of the trachea; leads to the lungs (pl., *bronchi*). 279

buffer Substance or compound that prevents large changes in the pH of a solution. 23, 333

bulbourethral gland Gland located in the pelvic cavity that adds secretions to seminal fluid within the urethra. 343

bulimia nervosa Eating disorder characterized by binge eating followed by purging. 318

bursa Saclike, fluid-filled structure, lined with synovial membrane, that occurs near a joint (pl., *bursae*). 104

bursitis Inflammation of any of the friction-easing sacs called bursae within the knee joint. 104

C

calcaneus Heel bone. 103

calcitonin Hormone secreted by the thyroid gland that helps regulate the level of blood calcium level. 192

capillary Microscopic vessel located in the tissues connecting arterioles to venules; molecules either exit or enter the blood through the thin walls of capillaries. 235

carbaminohemoglobin Hemoglobin carrying carbon dioxide. 285

carbohydrate Organic compounds with the general formula $(CH_2O)_n$, including sugars and glycogen. 24

carcinogen Any agent that causes cancer. 80

carcinoma Cancer arising from epithelial tissue. 66

cardiac cycle Series of myocardial contractions that constitutes a complete heartbeat. 232

cardiac muscle Heart muscle (myocardium) consisting of striated muscle cells that interlock. 63, 114

cardiac vein Blood vessel that returns blood from the venules of the myocardium to the coronary sinus. 228

cardioregulatory center Portion of the medulla oblongata that regulates the heartbeat rate. 233

caries Destruction of tooth enamel by oral bacteria. 297

carotid artery Either of two arteries branching off the aortic arch and supplying blood to the head and neck. 242

carotid body Structure located at the branching of the carotid arteries that contains chemoreceptors. 283

carpals Bones of the wrist. 100

carrier Molecule that combines with a substance and actively transports it through the plasma membrane. 396

cartilage Type of avascular connective tissue containing a firm, jellylike matrix embedded with protein fibers that provides support & protection. 66

cartilaginous joint Two or more bones joined by cartilage. 104

cataract Opaqueness of the lens of the eye, making the lens incapable of transmitting light. 171

CCK See *cholecystokinin*. 304

cecum Blind pouch, such as the one below where the small intestine enters the large intestine. 304

cell Structural and functional unit of an organism; smallest structure capable of performing all the functions necessary for life. 2

cell body Portion of a nerve cell that includes a cytoplasmic mass and a nucleus, and from which the nerve fibers extend. 142

cell cycle Life cycle of a cell consisting of G_1 (growth), S (DNA synthesis), G_2 (growth), and mitosis (division). 46

cell-mediated immunity Immunological defense provided by killer T cells, which destroy virus-infected cells, foreign cells, and cancer cells. 263

cellular respiration Process that releases energy from organic compounds in cells. 41

cellulose Polysaccharide very abundant in plant tissues that human enzymes cannot break down. 25

central Situated at the center of the body or an organ. 3

central canal Tube within the spinal cord that is continuous with the ventricle of the brain and contains cerebrospinal fluid. 146

central nervous system (CNS) Brain and spinal cord. 141

centriole Short, cylindrical organelle that contains microtubules in a 9 + 0 pattern and is associated with the formation of the spindle during cell division. 42

cerebellum Part of the brain that controls muscular coordination. 151

cerebral cortex Outer layer of the cerebrum. 149

cerebral hemisphere One of the large, paired structures that together constitute the cerebrum of the brain. 149

cerebral palsy Spastic weakness of the arms and legs due to damage to the motor areas of the cerebral cortex. 149

cerebrospinal fluid (CSF) Fluid found within the ventricles of the brain and surrounding the CNS in association with the meninges. 146

cerebrovascular accident (CVA) Condition resulting when an arteriole in the brain bursts or becomes blocked by an embolism; stroke. 239

cerebrum Main portion of the vertebrate brain that is responsible for consciousness. 148

cervix Narrow end of the uterus that projects into the vagina. 349

Cesarean section Birth by surgical incision of the abdomen and uterus. 379

chemotherapy Use of drugs to kill cancer cells. 80

Cheyne-Stokes respiration Type of respiration characterized by alternate periods of deep, labored breathing and no breathing at all. 284

chlamydia Sexually transmitted disease caused by the bacterium *Chlamydia trachomatis*; often causes painful urination and swelling of the testes in men; is usually symptomless in women but can cause inflammation of the cervix or uterine tubes. 363

chloride shift Movement of chloride ions from the blood plasma into red blood cells as bicarbonate ions diffuse out of the blood cells into the plasma. 285

cholecystokinin (CCK) Hormone secreted by the small intestine that stimulates the release of pancreatic juice from the pancreas and bile from the gallbladder. 304

chordae tendineae Tough bands of connective tissue that attach the papillary muscles to the atrioventricular valves within the heart. 226

chorion Extraembryonic membrane that forms an outer covering around the embryo and contributes to the formation of the placenta. 371

chorionic villi Projections from the chorion that appear during implantation and that in one area contribute to the development of the placenta. 377

chorionic villi sampling (CVS) Method of retrieving fetal cells for genetic testing in which a long, thin tube is passed through the vagina into the uterus, and suction is used to obtain a sample of chorionic villi cells. 390

choroid Vascular, pigmented middle layer of the wall of the eye. 170

chromatids Two identical parts of a chromosome following replication of DNA. 47

chromatin Threadlike network in the nucleus that condenses to become the chromosomes just before cell division. 39

chromosome Rod-shaped body in the nucleus, particularly during cell division, that contains the hereditary units, or genes. 39

chronic bronchitis Obstructive pulmonary disorder that tends to recur, marked by inflamed airways filled with mucus, and degenerative changes in the bronchi, including loss of cilia. 287

chronic disease Long and continued but not acute. 12

chronic obstructive pulmonary disease (COPD) Continued interference with airflow in the lungs due to chronic bronchitis or emphysema. 287

chyme Semifluid food mass leaving the stomach. 301

cilia Membrane-bounded microtubular structures that project from a cell, and in multicellular animals facilitate the flow of materials over the cell surface. 42

ciliary body Structure associated with the choroid layer of the eye that secretes aqueous humor and contains the ciliary muscle. 170

ciliary muscle Muscle that controls the curvature of the lens of the eye. 170

cilium Short, hairlike projection from the plasma membrane, occurring usually in large numbers. 42

circadian rhythm Pattern of repeated behavior associated with the cycles of night and day. 200

circle of Willis Arterial ring located on the ventral surface of the brain. 246

circular fold Permanent transverse folds of the luminal surface of the small intestine, involving the mucosa and the submucosa. 302

circulation Movement of the blood through the heart and blood vessels. 236

circumcision Removal of the prepuce (foreskin) of the penis. 347

circumduction Conelike movement of a body part, such that the distal end moves in a circle, while the proximal portion remains relatively stable. 106

cirrhosis Chronic, irreversible injury to liver tissue; commonly caused by frequent alcohol consumption. 310

clavicle Bone extending from the sternum to the scapula. 97

cleavage Early, successive divisions of the blastocyst cells into smaller and smaller cells. 371

cleavage furrow Site of cell division of the fertilized egg that is unaccompanied by growth. Numerous small cells result. 50

clitoris Small, erectile, female organ located in the vulva; homologous to the penis. 353

clonal selection theory States that the antigen selects which lymphocyte will undergo clonal expansion and produce more lymphocytes bearing the same type of receptor. 260

CNS See *central nervous system*. 141

coagulation Blood clotting. 214

coccyx Caudal end of the vertebral column formed by the fusion of four vertebrae; tailbone. 95

cochlea Portion of the inner ear that contains the receptors for hearing. 178

cochlear canal Canal within the cochlea that bears small hair cells that function as hearing receptors. 179

cochlear implant Prosthetic device used to help persons with severe hearing impairment; the device converts sound to an electrical impulse that directly stimulates the auditory nerve. 182

cochlear nerve Either of two cranial nerves that carry nerve impulses from the spiral organ to the brain; auditory nerve. 179

codon Set of three nucleotides of a messenger RNA molecule corresponding to a particular amino acid. 48

collecting duct Tube that receives urine from several distal convoluted tubules. 324

colon Large intestine. 304, 327

color vision Ability to detect the color of an object, dependent on three kinds of cone cells. 172

colostomy Attachment of a shortened colon to a surgical opening in the abdominal wall. 306

colostrum First secretion of a woman's mammary glands after she gives birth. 356

compact bone Hard bone consisting of osteons cemented together. 61, 84

complement system Group of proteins in plasma that aid the general defense of the body by destroying bacteria; often called complement. 259

compound Chemical substance having two or more different elements in fixed ratio. 20

concha Shell-shaped structure, such as that seen in the bones of the nasal cavity (pl., *conchae*). 277

condom For males, a latex sheath used to cover the penis during sexual intercourse; for females, a large polyurethane tube with a flexible ring that fits onto the cervix. Both male and female condoms function as contraceptives and help minimize the risk of transmitting infection. 359

conduction deafness Hearing impairment due to fusion of the ossicles or other damage to the middle ear, thereby restricting the ability to transmit and magnify sound. 182

conduction system of the heart Neuromuscular tissue and fibers that control the cardiac cycle; includes the SA node, the AV node, the AV bundle and its branches, and the Purkinje fibers. 230

condyle Large, rounded surface at the end of a bone. 87

condyloid joint Bone with an oval-shaped projection at one end joined with a bone processing a complementary elliptical cavity. 105

cone cell Color receptor located in the retina of the eye. 172

congenital defect Body abnormality arising from birth and due to hereditary factors. 382

congestive heart failure Inability of the heart to maintain adequate circulation, especially of the venous blood returned to it. 239

connective tissue Type of tissue characterized by cells separated by a matrix; often contains fibers. 58

constipation Infrequent, difficult defecation caused by insufficient water in the feces. 306

COPD See *chronic obstructive pulmonary disease.* 287

cornea Transparent, anterior portion of the outer layer of the eyeball. 170

coronal suture Line of junction of the frontal bone with the two parietal bones. 104

coronary artery Artery that supplies blood to the wall of the heart (myocardium). 228

coronary bypass operation Therapy for blocked coronary arteries in which part of a blood vessel from another part of the body is grafted around the obstructed artery. 229

coronary sinus A large vessel on the posterior surface of the heart into which the cardiac veins drain. 228

corpus albicans White, fibrous tissue that replaces the regressing corpus luteum in the ovary in the latter half of pregnancy. 351

corpus callosum Mass of white matter within the brain, composed of nerve fibers connecting the right and left cerebral hemispheres. 149

corpus luteum Structure that forms from the tissues of a ruptured ovarian follicle and functions to secrete female hormones. 351

cortisol Glucocorticoid secreted by the adrenal cortex. 194

covalent bond Chemical bond created by the sharing of electrons between atoms. 21

coxal bone Bone of the pelvic girdle. 100

cranial cavity Hollow space in the cranium containing the brain. 6

cranial nerve Nerve that arises from the brain. 152

creatine phosphate Muscle biochemical that stores energy. 120

creatinine Nitrogenous waste, the end product of creatine phosphate metabolism. 324

crenation Shrinking of red blood cells often caused by osmotic conditions. 43

cretinism Condition resulting from a lack of thyroid hormone in an infant. 191

CSF See *cerebrospinal fluid.* 146

Cushing syndrome Condition characterized by thin arms and legs and a "moon face;" accompanied by high blood glucose and sodium levels due to hypersecretion of cortical hormones. 195

cutaneous membrane Pertaining to the skin. 66, 70

cyanosis Bluish cast to the skin due to an increased amount of deoxyhemoglobin in the blood; sometimes due to a defective atrial septum, which incompletely closes the foramen ovale after birth. 247

cyclic AMP Derivative of ATP that responds to messages entering a cell and triggers the cell's response; also known as cAMP. 201

cystic fibrosis (CF) Generalized, autosomal recessive disorder of infants and children, in which there is widespread dysfunction of the exocrine glands. 396

cystitis Inflammation of the urinary bladder. 334

cytokine Type of protein secreted by a T lymphocyte that attacks viruses, virally infected cells, and cancer cells. 262

cytokinesis Division of the cytoplasm following mitosis and meiosis. 46

cytoplasm Ground substance of cells located between the nucleus and the plasma membrane. 36

cytoskeleton System of microtubules and filaments that reinforces a cell's three-dimensional form; maintains cell shape and allows movement of cell and its contents. 36

cytotoxic T cell T lymphocyte that attacks and kills antigen-bearing cells. 263

D

dandruff Skin disorder characterized by flaking, itchy scalp; caused by accelerated keratinization of the scalp. 74

daughter cell Cell that arises from a parent cell by mitosis or meiosis. 49

decubitus ulcer Skin sore due to restricted blood flow to the area in bedridden patients; also called a bedsore. 71

deep Located away from the surface of the body or an organ. 3

defecation Discharge of feces from the rectum through the anus. 304

dehydration reaction Anabolic process that joins small molecules; synthesis. 24

delayed allergic response Allergic response initiated at the site of the allergen by sensitized T cells, involving macrophages and regulated by cytokines. 268

delirium tremens Alcohol withdrawal. 382

denaturation Loss of normal shape by an enzyme so that it no longer functions; caused by a less than optimal pH or temperature. 28

dendrite Process of a neuron, typically branched, that conducts nerve impulses toward the cell body. 142

dense connective tissue Type of tissue containing many collagen fibers packed together; found in tendons and ligaments, for example. 59

deoxyhemoglobin Hemoglobin not carrying oxygen. 211

deoxyribonucleic acid (DNA) Nucleic acid; the genetic material found in the nucleus of a cell. 31

depolarization Loss in polarization, as when a nerve impulse occurs. 143

depression Movement of a synovial joint that lowers a body part. 106

dermis Thick skin layer that lies beneath the epidermis. 71

descending colon That portion of the large intestine that travels inferiorly as it extends from the transverse colon to the sigmoid colon. 304

diabetes insipidus Condition characterized by an abnormally large production of urine, due to a deficiency of antidiuretic hormone. 188

diabetes mellitus Condition characterized by a high blood glucose level and the appearance of glucose in the urine, due to a deficiency of insulin. 25, 197

diagnosis Decision based on an examination to determine the nature of a diseased condition. 66

dialysate Material that passes through the membrane in dialysis. 335

diaphragm Sheet of muscle that separates the thoracic cavity from the abdominopelvic cavity; also, a birth control device inserted in front of the cervix in females. 6, 128, 358

diaphysis Shaft of a long bone. 84

diarrhea Frequent, watery defecation, often caused by digestive infection or stress. 306

diastole Relaxation of heart chambers. 232

diastolic pressure Arterial blood pressure during the diastolic phase of the cardiac cycle. 239

diencephalon Portion of the brain in the region of the third ventricle that includes the thalamus and hypothalamus. 151

differential white blood cell count Microscopic examination of a blood sample in which each type of white blood cell is counted. 213

differentiation Process by which a cell becomes specialized for a particular function. 371

diffusion Passive movement of molecules from an area of greater concentration to an area of lesser concentration. 43

disaccharide Sugar that contains two units of a monosaccharide; for example, maltose. 25

disease Any abnormal condition considered harmful to the body; an illness or disorder. 12, 255

distal Further from the midline or origin; opposite of *proximal.* 3

distal convoluted tubule Highly coiled region of a nephron that is distant from the glomerular capsule. 327

diuretic Drug used to counteract hypertension by inhibiting Na$^+$ reabsorption so that less water is reabsorbed in the nephron. 332

diverticulosis Presence of diverticula, or saclike pouches, in the colon. 307

DNA See *deoxyribonucleic acid*. 31

dominant allele Hereditary factor that expresses itself even when there is only one copy in the genotype. 395

dorsal Pertaining to the back or posterior portion of a body part; opposite of *ventral*. 3

Down syndrome Human congenital disorder associated with an extra chromosome 21. 392

ductus arteriosus Fetal connection between the pulmonary artery and the aorta; venous artery. 246

ductus venosus Fetal connection between the umbilical vein and the inferior vena cava; also called venous duct. 246

duodenum First portion of the small intestine into which ducts from the gallbladder and pancreas enter. 302

duplicated chromosome Chromosome having two sister chromatids held together by a centromere. 47

dura mater Tough outer layer of the meninges; membranes that protect the brain and spinal cord. 146

E

ECG See *electrocardiogram*. 231

ectopic pregnancy Implantation of the embryo in a location other than the uterus, most often in a uterine tube. 352, 375

eczema Form of noncontagious dermatitis that begins with itchy red patches that thicken and crust over. 74

edema Swelling due to tissue fluid accumulation in the intercellular spaces. 217, 254

ejaculation Ejection of seminal fluid. 343

ejaculatory duct Tube, formed by the joining of the vas deferens and the tube from the seminal vesicle, that transports sperm to the urethra. 343

EKG See *electrocardiogram*. 231

elastic cartilage Cartilage composed of elastic fibers, allowing greater flexibility. 61

electrocardiogram (ECG) Recording of the electrical activity that accompanies the cardiac cycle. 231

electroencephalogram (EEG) Graphic recording of the brain's electrical activity. 151

electrolyte Any substance that ionizes and conducts electricity; electrolytes are present in the body fluids and tissues. 23, 331

electron Small, negatively charged particle that revolves around the nucleus of an atom. 18

element The simplest of substances, consisting of only one type of atom (for example, carbon, hydrogen, oxygen). 18

elephantiasis Swelling of the arms, legs, or external genitalia due to failure of the lymphatic system to remove excess fluid. 257

elevation Movement of a synovial joint that raises a body part. 106

embolus Moving blood clot that is carried through the bloodstream. 214

embryo Organism in its early stages of development; in humans, the organism in its second week to two months of development. 375

embryonic development Period of development from the second through eighth weeks. 375

embryonic disk Flattened area between the amniotic cavity and the yolk sac from which the embryo arises. 375

emphysema Lung impairment caused by deterioration of the bronchioles, which traps air in alveoli. 287

emulsification Breaking up of fat globules into smaller droplets by the action of bile salts. 26

endocardium Inner layer of the heart wall. 226

endochondral ossification Ossification that begins as hyaline cartilage that is subsequently replaced by bone tissue. 86

endocrine gland Gland that secretes hormones directly into the bloodstream or body fluids. 186

endocytosis Process in which a vesicle is formed at the plasma membrane to bring a substance into the cell. 44

endomembrane system Collection of membranous structures involved in transport within the cell. 40

endometriosis Implantation of uterine endometrial tissue in the abdominal cavity, possibly as a result of irregular menstrual flow. 360

endometrium Lining of the uterus that becomes thickened and vascular during the menstrual cycle. 349

endoplasmic reticulum (ER) Complex system of tubules, vesicles, and sacs in cells; sometimes has attached ribosomes. 40

enzyme Protein catalyst that speeds up a specific reaction or a specific type of reaction. 29

eosinophil Granular leukocyte capable of being stained with the dye eosin. 212

epicardium Visceral portion of the pericardium on the surface of the heart. 226

epidermis Organism's outer layer of cells. 70

epididymis Coiled tubules next to the testes where sperm mature and may be stored for a short time. 343

epidural hematoma Bleeding between the dura mater and the bone, as a result of a head injury. 146

epiglottis Structure that covers the glottis during the process of swallowing. 278, 298

epinephrine Hormone produced by the adrenal medulla that stimulates "fight-or-flight" reactions; also called adrenaline. 193

epiphyseal plate Cartilaginous layer within the epiphysis of a long bone that functions as a growing region. 86

epiphysis End segment of a long bone, separated from the diaphysis early in life by an epiphyseal plate, but later becoming part of the larger bone. 84

episiotomy Surgical procedure performed during childbirth in which the opening of the vagina is enlarged to avoid tearing. 385

epithelial tissue Type of tissue that lines the body's internal cavities and covers the body's external surface. 55

erectile dysfunction Failure of the penis to achieve erection. 347

erythrocyte Nonnucleated, hemoglobin-containing blood cell capable of carrying oxygen; red blood cell. 211

erythropoietin (EPO) Kidney hormone that promotes red blood cell formation. 211, 324

esophagus Tube that transports food from the mouth to the stomach. 299

essential amino acid Amino acid that is necessary in the diet because the body is unable to manufacture it. 314

essential fatty acid Fatty acid that is necessary in the diet because the body is unable to manufacture it. 314

estrogen Female sex hormone secreted by the ovaries that, along with progesterone, promotes the development and maintenance of the primary and secondary female sex characteristics. 197

eupnea Easy or normal respiration. 284

eversion Movement of the foot in which the sole is turned outward. 106

excretion Elimination of metabolic wastes. 324

exocrine gland Particular glands with ducts, such as salivary glands, whose secretions are deposited into cavities. 186

exocytosis Process in which an intracellular vesicle fuses with the plasma membrane so that the vesicle's contents are released outside the cell. 44

exophthalmic goiter Enlargement of the thyroid gland, accompanied by an abnormal protrusion of the eyes. 191

expiration Process of expelling air from the lungs; exhalation. 276

expiratory reserve volume Volume of air that can be forcibly exhaled after normal exhalation. 281

extension Movement that increases the angle between parts at a joint. 106

external auditory meatus Opening through the temporal bone that connects with the tympanum and the middle ear chamber and through which sound vibrations pass. 92

external genitals Sex organs that occur outside the body. 347

external respiration Exchange of oxygen and carbon dioxide between alveoli and blood. 285

extraembryonic membranes Membranes that are not a part of the embryo but that are necessary to the embryo's continued existence and health. 371

extra cellular fluid Fluid found around tissue cells that contains molecules that enter from or exit to the capillaries. 216

extrinsic muscle Muscle that anchors and moves the eye. 169

F

facilitated diffusion Use of a plasma membrane carrier protein to move a substance into or out of a cell from higher or lower concentration; no energy required. 44

fascia Tough sheet of fibrous tissue that binds the skin to underlying muscles, also supports and separates muscles. 115

fascicle Small bundle of muscle fibers. 114

fat Organic molecule that the body uses for long-term energy storage. 26, 71

fatigue Failure of a muscle fiber to continue to contract, due to exhaustion of ATP. 122

fatty acid Molecule that contains a hydrocarbon chain and ends with an acid group. 26

feces Indigestible wastes expelled from the digestive tract; excrement. 307

femur Thighbone located in the upper leg. 102

fertilization Union of a sperm nucleus and an egg nucleus, which creates a zygote with the diploid number of chromosomes. 370

fetal alcohol syndrome (FAS) Babies born with decreased weight, height, and head size and with malformation of the head and face due to the mothers' consumption of alcohol during pregnancy. 382

fetal development Period of human development from the ninth week through birth. 379

fetus Human in its later developmental stages (from three months to term), following the embryonic stage. 379

fiber Dendrites and axons of neurons. 304

fibrillation Rapid but uncoordinated heartbeat. 231

fibrin Insoluble protein threads formed from fibrinogen during blood clotting. 214

fibrinogen Plasma protein that is converted into fibrin threads during blood coagulation. 214

fibroblast Cell that produces fibers and other intercellular materials in connective tissues. 59

fibrocartilage Cartilage with a matrix of strong collagenous fibers. 61

fibrous connective tissue Tissue composed mainly of closely packed collagenous fibers and found in tendons and ligaments. 59

fibrous joint Two or more bones joined by connective tissue containing many fibers. 104

fibrous pericardium External layer of the pericardium, consisting of fibrous tissue. 226

fibula Long, slender bone located on the lateral side of the tibia. 102

filament Protein molecule that makes up part of a myofibril. 116

filtration Passage of fluid through a membrane because of mechanical pressure, as when blood pressure forces water out of a capillary. 43

fimbria Fingerlike extension from the uterine tube near the ovary (pl., *fimbriae*). 352

flagellum Slender, long process used for locomotion—for example, by sperm (pl., *flagella*). 42

flexion Bending at a joint so that the angle between bones is decreased. 106

floating kidney Kidney that has been dislodged from its normal position. 325

focus Bending of light rays by the cornea, lens, and humors so that they converge and create an image on the retina. 171

follicle Structure in the ovary that produces the egg and particularly the female sex hormone estrogen. 351

follicle-stimulating hormone (FSH) Hormone secreted by the anterior pituitary gland that stimulates the development of an ovarian follicle in a female or the production of sperm cells in a male. 348

fontanel Membranous region located between certain cranial bones in the skull of a fetus or infant. 90, 379

foramen Opening, usually in a bone or membrane (pl., *foramina*). 87

foramen ovale Oval-shaped opening between the atria in the fetal heart. 246

foreskin Skin covering the glans penis in uncircumcised males. 347

formed element Cellular constituent of blood. 209

fovea centralis Region of the retina that consists of densely packed cones and is responsible for the greatest visual acuity. 171

fracture A break in a bone. 87

free radicals Atoms or molecules with an unpaired electron in their outermost shell; linked to several diseases and play a role in the aging process. 315

frontal lobe Area of the cerebrum responsible for voluntary movements and higher intellectual processes. 149

frontal plane Plane or section that divides a structure lengthwise into anterior and posterior portions; pertaining to the region of the forehead. 5

FSH See *follicle-stimulating hormone*. 348

G

gallbladder Saclike organ associated with the liver that stores and concentrates bile. 310

ganglion Collection of neuron cell bodies outside the central nervous system (pl., *ganglia*). 152

gap junction Junction between cells formed by the joining of two adjacent plasma membranes; it lends strength and allows ions, sugars, and small molecules to pass between cells. 65

gastric gland Gland within the stomach wall that secretes gastric juice. 301

gastrulation Formation of a gastrula from a blastula; characterized by an invagination of the cell layers to form a caplike structure. 377

gene Unit of heredity located on a chromosome. 395

gene therapy Method of replacing a defective gene with a healthy gene. 399

genital Pertaining to the genitalia (internal and external organs of reproduction). 342

genital herpes Sexually transmitted disease caused by herpes simplex virus and sometimes accompanied by painful ulcers on the genitals. 363

genital wart Raised growth on the genitals due to a sexually transmitted disease caused by human papillomavirus. 363

genome All of the DNA in a cell of an organism. 399

genotype Combination of genes present within a zygote or within the cells of an individual. 395

gestation Period of development, from the start of the last menstrual cycle until birth; in humans, typically 280 days. 373

GH See *growth hormone*. 188

gingivitis Inflammation of the gums. 297

gland Epithelial cell or group of epithelial cells that are specialized to secrete a substance. 65

glaucoma Increasing loss of field of vision, caused by blockage of the ducts that drain the aqueous humor, creating pressure buildup and nerve damage. 171

gliding joint Two bones with nearly flat surfaces joined together. 105

globulin Type of protein in blood plasma. There are alpha, beta, and gamma globulins. 209

glomerular capsule Double-walled cup that surrounds the glomerulus at the beginning of the kidney tubule; also known as Bowman's capsule. 326

glomerular filtrate Liquid that passes out of the glomerular capillaries in the kidney into the glomerular capsules. 329

glomerular filtration Process whereby blood pressure forces liquid through the glomerular capillaries in the kidney into the glomerular capsule. 329

glomerulus Cluster of capillaries surrounded by the glomerular capsule in a kidney nephron. 326

glottis Slitlike opening between the vocal cords. 278

glucagon Hormone secreted by the pancreatic islets that causes the release of glucose from glycogen. 196

glucocorticoid Any one of a group of hormones secreted by the adrenal cortex that influences carbohydrate, fat, and protein metabolism. 193

glucose Blood sugar that is broken down in cells to acquire energy for ATP production. 24

glycerol Three-carbon molecule that joins with fatty acids to form fat. 26

glycogen Polysaccharide that is the principal storage compound for sugar in animals. 25

glycosuria Presence of glucose in the urine, typically indicative of a kidney disease, diabetes mellitus, or other endocrine disorder. 197

Golgi apparatus Organelle that consists of concentrically folded membranes and that functions in the packaging and secretion of cellular products. 40

gonad Organ that produces sex cells: the ovary, which produces eggs, and the testis, which produces sperm. 197

gonadotropic hormone Type of hormone that regulates the activity of the ovaries and testes; principally, follicle-stimulating hormone and luteinizing hormone. 188

gonorrhea Sexually transmitted disease caused by the bacterium *Neisseria gonorrhoeae* that causes painful urination and swollen testes in men and is usually symptomless in women, but can cause inflammation of the cervix and uterine tubes. 363

gout Joint inflammation caused by accumulation of uric acid. 324

granular leukocyte White blood cell with prominent granules in the cytoplasm. 212

Graves disease Autoimmune disease with swollen throat due to an enlarged, hyperactive thyroid gland; patients often have protruding eyes and are underweight, hyperactive, and irritable. 191

gravitational equilibrium Maintenance of balance when the head and body are motionless. 181

gray matter Nonmyelinated nerve fibers in the central nervous system. 146

greater sciatic notch Indentation in the posterior coxal bone through which pass the blood vessels and the large sciatic nerve to the lower leg. 100

growth Increase in the number of cells and/or the size of these cells. 371

growth factor Chemical signal that regulates mitosis and differentiation of cells that have receptors for it; important in such processes as fetal development, tissue maintenance and repair, and hematopoiesis; sometimes a contributing factor in cancer. 200

growth hormone (GH) Hormone released by the anterior lobe of the pituitary gland that promotes the growth of the organism; also known as somatotropin. 188

gyrus Convoluted elevation or ridge (pl., *gyri*). 149

H

hair Consists of a cylindrical shaft and a root, which is contained in a flasklike depression (hair follicle) in the dermis and subcutaneous tissue. The base of the root is expanded into the hair bulb, which rests upon and encloses the hair papilla. 72

hair cell Mechanoreceptor in the inner ear that lies between the basilar membrane and the tectorial membrane and triggers action potentials in fibers of the auditory nerve. 178

hair follicle Tubelike depression in the skin in which a hair develops. 72

hard palate Anterior portion of the roof of the mouth that contains several bones. 296

hay fever Seasonal variety of allergic reaction to a specific allergen. Characterized by sudden attacks of sneezing, swelling of nasal mucosa, and often asthmatic symptoms. 268

hCG See *human chorionic gonadotropin.* 354, 375

head Pertaining to the skeleton, an enlargement on the end of a bone. 102

heart Muscular organ located in the thoracic cavity that is responsible for maintenance of blood circulation. 225

heart attack See *myocardial infarction.* 228

heart block Impairment of conduction of an impulse in heart excitation. 231

heartburn Burning pain in the chest occurring when part of the stomach contents escapes into the esophagus. 299

heart murmur Clicking or swishing sounds often due to leaky valves. 232

heart valve Valve found between the chambers of the heart or between a chamber and a vessel leaving the heart. 227

helper T cell Secretes lymphokines, which stimulate all kinds of immune cells. 263

hematopoiesis Production of blood cells. 84, 210

heme Iron-containing portion of a hemoglobin molecule. 211

hemodialysis Mechanical way to remove nitrogenous wastes and to regulate blood pH when the kidneys are unable to perform these functions. 45, 335

hemoglobin Pigment of red blood cells responsible for oxygen transport. 211, 285

hemolysis Bursting of red blood cells with the release of hemoglobin; can be caused by osmotic conditions. 211

hemolytic anemia Insufficient number of red blood cells caused by an increased rate of red blood cell destruction. 213

hemolytic disease of the newborn Destruction of a fetus's red blood cells by the mother's immune system, caused by differing Rh factors between mother and fetus. 213

hemophilia Most common of the severe clotting disorders caused by the absence of a blood clotting factor. 214

hemorrhagic bleeding Escape of blood from blood vessels. 214

hemorrhoids Abnormally dilated blood vessels of the rectum. 235

hemostasis Stoppage of bleeding. 214

hepatic portal system Portal system that begins at the villi of the small intestine and ends at the liver. 245

hepatic portal vein Vein leading to the liver and formed by the merging blood vessels of the small intestine. 245

hepatitis Inflammation of the liver; often due to a serious infection by any of a number of viruses. 310

hernia Protrusion of an organ through an abnormal opening, such as the intestine through the abdominal wall near the scrotum (inguinal hernia) or the stomach through the diaphragm (hiatal hernia). 346

herniated disk Fibrous ring of cartilage between two vertebrae that has ruptured. 94

heterozygous Different alleles in a gene pair. 395

hexose Six-carbon sugar. 24

hinge joint Type of joint characterized by a convex surface of one bone fitting into a concave surface of another so that movement is confined to one place, such as in the knee or interphalangeal joint. 105

hirsutism Excessive body and facial hair in women. 72

histamine Substance produced by basophil-derived mast cells in connective tissue that causes capillaries to dilate; causes many of the symptoms of allergy. 259

HLA (human leukocyte-associated) antigen Protein in a plasma membrane that identifies the cell as belonging to a particular individual and acts as an antigen in other organisms. 262

Hodgkin disease Cancer of the lymph glands that is normally localized in the neck region. 257

homeostasis Constancy of conditions, particularly the environment of body cells: constant temperature, blood pressure, pH, and other body conditions. 10

homozygous dominant Possessing two identical alleles, such as *AA,* for a particular trait. 395

homozygous recessive Possessing two identical alleles, such as *aa,* for a particular trait. 395

hormone Substance secreted by an endocrine gland that is transmitted in the blood or body fluids. 186, 304

human chorionic gonadotropin (HCG) Hormone produced by the placenta that helps maintain pregnancy and is the basis for the pregnancy test. 354, 375

human immunodeficiency virus (HIV) Virus responsible for AIDS. 264

humerus Heavy bone that extends from the scapula to the elbow. 98

Huntington disease Genetic disease marked by progressive deterioration of the nervous system due to deficiency of a neuro-transmitter. 145

hyaline cartilage Cartilage composed of very fine collagenous fibers and a matrix of a glassy, white, opaque appearance. 61

hydrocephalus Enlargement of the brain due to abnormal accumulation of cerebrospinal fluid. 146

hydrogen bond Weak attraction between a partially positive hydrogen and a partially negative oxygen or nitrogen some distance away; found in proteins and nucleic acids. 22

hydrolysis reaction Splitting of a bond by the addition of water. 24

hydrolytic enzyme Enzyme that catalyzes a reaction in which the substrate is broken down by the addition of water. 311

hydrophilic Type of molecule that interacts with water by dissolving in water and/or forming hydrogen bonds with water molecules. 22

hydrophobic Type of molecule that does not interact with water because it is nonpolar. 22

hyperglycemia Excessive glucose in the blood. 197

hyperopia Inability to see nearby objects. 177

hyperpnea Deep and labored breathing. 284

hypertension Elevated blood pressure, particularly the diastolic pressure. 20, 239

hyperthermia Abnormally high body temperature. 78

hypertonic solution Solution that has a higher concentration of solute and a lower concentration of water than the cell. 43

hypertrophy Increase in the size of an organ, usually by an increase in the size of its cells. 123

hypodermic needle Slender, hollow instrument for introducing material into or removing material from or below the skin. 71

hypodermis Mainly composed of fat, this loose layer is directly beneath the dermis; subcutaneous. 70

hypoglycemia Insufficient amount of glucose in the blood. 197

hypothalamic-inhibiting hormone One of many hormones produced by the hypothalamus that inhibits the secretion of an anterior pituitary hormone. 188

hypothalamic-releasing hormone One of many hormones produced by the hypothalamus that stimulates the secretion of an anterior pituitary hormone. 188

hypothalamus Region of the brain; the floor of the third ventricle that helps maintain homeostasis. 151, 188

hypothermia Abnormally low body temperature. 78

hypotonic solution Solution that has a lower concentration of solute and a higher concentration of water than the cell. 43

hysterectomy Surgical removal of the uterus. 352

I

ileum Lower portion of the small intestine. 302

ilium One of the bones of a coxal bone or hipbone. 100

immediate allergic response Allergic response that occurs within seconds of contact with an allergen; caused by the attachment of the allergen to IgE antibodies. 268

immune system All the cells in the body that protect the body against foreign organisms and substances and also against cancerous cells. 255

immunity Resistance to disease-causing organisms. 255

immunization Strategy for achieving artificial immunity to the effects of specific disease-causing agents. 266

immunoglobulin (Ig) Globular plasma proteins that function as antibodies. 261

immunosuppressive Inactivating the immune system to prevent organ rejection, usually via a drug. 269

impetigo Contagious skin disease caused by bacteria in which vesicles erupt and crust over. 74

implantation Attachment and penetration of the embryo to the lining (endometrium) of the uterus. 349, 375

incontinence Involuntary loss of urine. 336

incus The middle of three ossicles of the ear; serves with the malleus and the stapes to conduct vibrations from the tympanic membrane to the oval window of the inner ear. 178

infant respiratory distress syndrome Condition in newborns, especially premature ones, in which the lungs collapse because of a lack of surfactant lining the alveoli. 279

inferior Situated below something else; pertaining to the lower surface of a part. 3

inferior vena cava Large vein that enters the right atrium from below and carries blood from the trunk and lower extremities. 242

infertility Inability to have as many children as desired. 360

inflammatory reaction Tissue response to injury that is characterized by dilation of blood vessels and accumulation of fluid in the affected region. 259

inhibin Hormone secreted by seminiferous tubules that inhibits the release of follicle-stimulating hormone from the anterior pituitary. 348

inner cell mass An aggregation of cells at one pole of the blastocyte, which is destined to form the embryo proper. 373

inner ear Portion of the ear, consisting of a vestibule, semicircular canals, and the cochlea, where balance is maintained and sound is transmitted. 178

inorganic molecule Type of molecule that is not an organic molecule; not derived from a living organism. 22

insertion End of a muscle that is attached to a movable part. 124

inspiration The act of breathing in; inhalation. 276

inspiratory reserve volume Volume of air that can be forcibly inhaled after normal inhalation. 281

insulin Hormone produced by the pancreas that regulates glucose storage in the liver and glucose uptake by cells. 196

insulin-dependent diabetes mellitus (IDDM) Type 1 diabetes mellitus characterized by abrupt onset of symptoms, dependence on exogenous insulin, and a tendency to develop ketoacidosis. 197

integration Summing up of excitatory and inhibitory signals by a neuron or by some part of the brain. 145

integument Pertaining to the skin. 70

integumentary system Pertaining to the skin and accessory organs. 8, 70

interatrial septum Wall between the atria of the heart. 226

intercalated disk Membranous boundary between adjacent cardiac muscle cells. 63

interferon Protein formed by a cell infected with a virus that can increase the resistance of other cells to the virus. 259

interleukin Class of immune system chemicals (cytokines) having varied effects on the body. 263

internal respiration Exchange of oxygen and carbon dioxide between blood and tissue fluid. 285

interneuron Neuron found within the central nervous system that takes nerve impulses from one portion of the system to another. 142

interstitial cell Hormone-secreting cell located between the seminiferous tubules of the testes. 345

intervertebral disk Layer of cartilage located between adjacent vertebrae. 94

intramembranous ossification Bone that forms from membranelike layers of primitive connective tissue. 86

intrauterine device Solid object placed in the uterine cavity for purposes of contraception; IUD. 358

intrauterine insemination (IUI) Process of achieving pregnancy in which donated sperm are deposited in the uterus. 360

inversion Movement of the foot so that the sole is turned inward. 106

in vitro fertilization (IVF) Process of achieving pregnancy in which eggs retrieved from an ovary are fertilized in a laboratory; viable embryos are then placed into the woman's uterus. 360

ion A charged atom. 20

ionic bond Chemical attraction between a positive ion and a negative ion. 20

iris Muscular ring that surrounds the pupil and regulates the passage of light through this opening. 170

iron deficiency anemia Abnormally low amount of red blood cells or hemoglobin, due to a lack of iron in the diet. 213

ischemic heart disease Insufficient oxygen delivery to the heart, usually caused by partially blocked coronary arteries. 228

ischial spine Projection of the coxal bone into the pelvic cavity. 100

isotonic solution Solution that contains the same concentration of solutes and water as does the cell. 43

isotope One of two or more atoms with the same atomic number that differs in the number of neutrons and, therefore, in weight. 19

IVF See *in vitro fertilization*. 360

IVI See *intrauterine insemination*. 360

J

jaundice Yellowish tint to the skin caused by an abnormal amount of bilirubin in the blood, indicating liver malfunction. 310

jejunum Middle portion of the small intestine. 302

joint Union of two or more bones; an articulation. 104

jugular Any of four veins that drain blood from the head and neck. 245

juxtaglomerular apparatus Structure located in the walls of arterioles near the glomerulus that regulates renal blood flow. 332

K

karyotype Arrangement of all the chromosomes from a nucleus by pairs in a fixed order. 390

keratin Insoluble protein present in the epidermis and in epidermal derivatives, such as hair and nails. 71

ketonuria Abnormal presence of acidic molecules called ketones in the urine. 198

kidney Organ in the urinary system that forms, concentrates, and excretes urine. 325

Klinefelter syndrome Condition caused by the inheritance of XXY chromosomes. 393

knee-jerk reflex Automatic, involuntary response initiated by tapping the ligaments just below the patella (kneecap). 155

kyphosis Increased roundness in the thoracic curvature of the spine; also called "hunchback." 94

L

labia majora Two large, hairy folds of skin of the female external genitalia. 353

labia minora Two small folds of skin inside the labia majora and encircling the clitoris. 353

lacrimal apparatus Structures that provide tears to wash the eye, consisting of the lacrimal gland and the lacrimal sac with its ducts. 168

lactation Production and secretion of milk by the mammary glands. 356

lacteal Lymph vessel in a villus of the wall of the small intestine. 302

lacuna Small pit or hollow cavity, as in bone or cartilage, where a cell or cells are located (pl., *lacunae*). 61

lambdoidal suture Line of junction between the occipital and parietal bones. 104

Langerhans cells Specialized epidermal cells that assist the immune system. 71

lanugo Short, fine hair that is present during the later portion of fetal development. 379

large intestine Portion of the digestive tract that extends from the small intestine to the anus. 304

laryngitis Inflammation of the larynx. 286

laryngopharynx Lower portion of the pharynx near the opening to the larynx. 278

larynx Structure that contains the vocal cords; also known as the voice box. 278

lateral Pertaining to the side. 3

lateral malleolus Rounded protuberance on the lateral surface of the ankle joint. 103

lens Clear, membranelike structure that is found in the eye behind the iris and that brings objects into focus. 170

leptin Hormone produced by adipose tissue that acts on the hypothalamus to signal satiety. 200

leukemia Form of cancer characterized by uncontrolled production of leukocytes in red bone marrow. 213

leukocytes Several types of colorless, nucleated blood cells that, among other functions, resist infection; white blood cells. 212

leukocytosis Abnormally large increase in the number of white blood cells. 213

leukopenia Abnormally low number of leukocytes in the blood. 213

LH See *luteinizing hormone*. 348

ligament Strong connective tissue that joins bone to bone. 59, 104

limbic system System that involves many different centers of the brain and that is concerned with visceral functioning and emotional responses. 151

lipase Enzyme secreted by the pancreas that digests or breaks down fats. 308

lipid Group of organic compounds that are insoluble in water—notably, fats, oils, and steroids. 26

liver Largest organ in the body, located in the abdominal cavity below the diaphragm; performs many vital functions that maintain homeostasis of blood. 308

loop of Henle Portion of a nephron between the proximal and distal convoluted tubules; functions in water reabsorption. 327

loose connective tissue Tissue that is composed mainly of fibroblasts separated by collagenous and elastin fibers and that is found beneath epithelium. 59

lordosis Exaggerated lumbar curvature of the spine; also called "swayback." 94

lumen Space within a tubular structure such as a blood vessel or intestine. 299

lung Internal respiratory organ containing moist surfaces for gas exchange. 279

lunula Pale, half-moon–shaped area at the base of nails. 73

luteinizing hormone (LH) Hormone produced by the anterior pituitary that stimulates the development of the corpus luteum in females and the production of testosterone in males. 348

lymph Fluid having the same composition as tissue fluid; carried in lymph vessels. 217, 254

lymphadenitis Infection of the lymph nodes. 257

lymphangitis Infection of the lymphatic vessels. 257

lymphatic organ Organ other than a lymphatic vessel that is part of the lymphatic system; includes lymph nodes, tonsils, spleen, thymus gland, and bone marrow. 255

lymphatic system Vascular system that takes up excess tissue fluid and transports it to the bloodstream. 8, 254

lymphatic vessel Vessel that carries lymph. 217, 254

lymph node Mass of lymphatic tissue located along the course of a lymphatic vessel. 257

lymphocyte Type of white blood cell characterized by agranular cytoplasm; lymphocytes usually constitute 20–25% of the white cell count. 212

lymphoma Cancer of lymphatic tissue (reticular connective tissue). 257

lysosome Organelle involved in intracellular digestion; contains powerful digestive enzymes. 40

M

macromolecule Large molecule composed of smaller molecules. 2

macrophage Enlarged monocyte that ingests foreign material and cellular debris. 212, 259

macular degeneration Disruption of the macula lutea, a central part of the retina, causing blurred vision. 174

malignant The power to threaten life; cancerous. 80

malleolus Rounded projection from a bone. 103

malleus First of three ossicles of the ear; serves with the incus and stapes to conduct vibrations from the tympanic membrane to the oval window of the inner ear. 178

maltase Enzyme that catalyzes conversion of maltose into glucose. 311

mammary gland Milk-secreting gland that develops within the breast in pregnancy and lactation; only minimally developed in the breast of a nonpregnant or nonlactating woman. 73

mast cell Cell to which antibodies, formed in response to allergens, attach, bursting the cell and releasing allergy mediators, which cause symptoms. 259

mastoiditis Inflammation of the mastoid sinuses of the skull. 90

matrix Secreted basic material or medium of biological structures, such as the matrix of cartilage or bone. 58

medial Toward or near the midline. 3

mediastinum Tissue mass located between the lungs. 6

medulla oblongata Lowest portion of the brain; concerned with the control of internal organs. 151

medullary cavity Within the diaphysis of a long bone, cavity occupied by yellow marrow. 84

megakaryocyte Large bone marrow cell that gives rise to blood platelets. 214

meiosis Type of cell division in which the daughter cells have 23 chromosomes; occurs during spermatogenesis and oogenesis. 342

melanin Pigment found in the skin and hair of humans that is responsible for coloration. 71

melanocyte Melanin-producing cell. 71

melanocyte-stimulating hormone (MSH) Substance that causes melanocytes to secrete melanin in lower vertebrates. 188

melanoma Deadly form of skin cancer that begins in the melanocytes, pigment cells present in the epidermis. 74

melatonin Hormone, secreted by the pineal gland, that is involved in biorhythms. 200

memory B cell Cells derived from B lymphocytes that remain within the body for some time and account for active immunity. 260

meninges Protective membranous coverings around the brain and spinal cord (sing., *meninx*). 6, 66, 146

meningitis Inflammation of the meninges around the brain and spinal cord usually caused by bacterial infection. 6, 66

meniscus Piece of fibrocartilage that separates the surfaces of bones in the knee (pl., *menisci*). 104

menopause Termination of the menstrual cycle in older women. 355

menses (menstruation) Loss of blood and tissue from the uterus. 354

menstrual cycle Female reproductive cycle characterized by regularly occurring changes in the uterine lining. 354

mesentery Fold of peritoneal membrane that attaches an abdominal organ to the abdominal wall. 66

messenger RNA (mRNA) Nucleic acid (ribonucleic acid) complementary to genetic DNA; has codons that direct cell protein synthesis at the ribosomes. 39

metabolism All of the chemical changes that occur within cells. 29

metacarpal Bone of the hand between the wrist and the finger bones. 100

metaphase Stage in mitosis when chromosomes align in the center of the cell. 50

metastasis Mechanism of cancer spread in which cancer cells break off from the initial tumor, enter the blood vessels or lymphatic vessels, and start new tumors elsewhere in the body. 80

metatarsal bones Bones found in the foot between the ankle and the toes. 103

microtubule Hollow rod of the protein tubulin in the cytoplasm. 42

microvillus Cylindrical process that extends from some epithelial cell membranes and increases the membrane surface area (pl., *microvilli*). 302

micturition Emptying of the bladder; urination. 325

midbrain Small region of the brain stem located between the forebrain and the hindbrain; contains tracts that conduct impulses to and from the higher parts of the brain. 151

middle ear Portion of the ear consisting of the tympanic membrane, the oval and round windows, and the ossicles, where sound is amplified. 178

mineral Inorganic substance; certain minerals must be in the diet for normal metabolic functioning of cells. 315

mineralocorticoid Hormones the adrenal cortex secretes that influence the concentrations of electrolytes in body fluids. 193

mitochondrion Organelle in which cellular respiration produces the energy molecule ATP. 41

mitosis Type of cell division in which two daughter cells receive 46 chromosomes; occurs during growth and repair. 46

mixed nerve Nerve that contains both the long dendrites of sensory neurons and the long axons of motor neurons. 152

mole Raised growth on the skin due to an overgrowth of melanocytes. 74

molecule Smallest quantity of a substance that retains its chemical properties. 2, 20

monoclonal antibody Antibody of one type that is produced by cells derived from a lymphocyte that has fused with a cancer cell. 268

monocyte Type of white blood cell that functions as a phagocyte. 212

mononucleosis Viral disease characterized by an increase in atypical lymphocytes in the blood. 213

monosaccharide Simple sugar; a carbohydrate that cannot be decomposed by hydrolysis. 24

mons pubis The rounded fleshy prominence over the pubic symphysis. 353

morphogenesis Establishment of shape and structure in an organism. 371

morula Early stage in development in which the embryo consists of a mass of cells, often spherical. 373

motor neuron Neuron that takes nerve impulses from the central nervous system to an effector; also known as an efferent neuron. 142

motor unit Motor neuron and all the muscle fibers it innervates. 122

mouth Opening through which food enters the body. 296

MS See *multiple sclerosis*. 269

mucous membrane Membrane lining a cavity or tube that opens to the outside of the body; also called mucosa. 66

multiple sclerosis (MS) Disease in which the outer, myelin layer of nerve fiber insulation becomes scarred, interfering with normal conduction of nerve impulses. 269

muscle fiber Muscle cell. 114, 116

muscle twitch Contraction of a whole muscle in response to a single stimulus. 122

muscular dystrophy Progressive muscle weakness and atrophy caused by deficient dystrophin protein. 136

muscular tissue Major type of tissue that is adapted to contract; the three kinds of muscle are cardiac, smooth, and skeletal. 62

myalgia Pain in a muscle or muscles. 136

myasthenia gravis Muscle weakness due to an inability to respond to the neurotransmitter acetylcholine. 136, 168, 269

myelin sheath Fatty plasma membranes of Schwann cells that cover long neuron fibers and give them a white, glistening appearance. 64, 142

myocardial infarction Damage to the myocardium due to blocked circulation in the coronary arteries; a heart attack. 228

myocardium Heart (cardiac) muscle consisting of striated muscle cells that interlock. 226

myofibril Contractile portion of muscle fibers. 116

myoglobin Pigmented compound in muscle tissue that stores oxygen. 116

myopia Inability to see distant objects clearly. 177

myosin Thick myofilament in myofibrils that is made of protein and is capable of breaking down ATP; see also *actin*. 116

myxedema Condition resulting from a deficiency of thyroid hormone in an adult. 191

N

nasal cavity Space within the nose. 277

nasopharynx Portion of the pharynx associated with the nasal cavity. 298

natural killer cell (NK) Lymphocyte that causes an infected or cancerous cell to burst. 259

negative feedback Mechanism that is activated by a surplus imbalance and acts to correct it by stopping the process that brought about the surplus. 10

nephron Anatomical and functional unit of the kidney; kidney tubule. 326

nerve Bundle of long nerve fibers that run to and/or from the central nervous system. 64, 152

nerve deafness Hearing impairment that usually occurs when the cilia on the sensory receptors within the cochlea have worn away. 182

nerve fiber Thin process of a neuron (i.e., axon, dendrite). 142

nerve impulse Change in polarity that flows along the membrane of a nerve fiber. 143

nervous tissue Tissue of the nervous system, consisting significantly of neurons and neuroglia. 64

neuroglia Nonconducting nerve cells that are intimately associated with neurons and function in a supportive capacity. 64, 142

neurolemmocyte Type of neuroglial cell that forms a myelin sheath around axons; also called a Schwann cell. 142

neuromuscular junction Junction between a neuron and a muscle fiber. 118

neuron Nerve cell that characteristically has three parts: dendrite, cell body, and axon. 64, 142

neurotransmitter Chemical made at the ends of axons that is responsible for transmission across a synapse. 145

neutral fat A triglyceride. 26

neutron Electrically neutral particle in an atomic nucleus. 18

neutrophil Phagocytic white blood cell that normally constitutes 60–70% of the white blood cell count. 212

node of Ranvier Gap in the myelin sheath of a nerve fiber. 64

nondisjunction Failure of the chromosomes (or chromatids) to separate during meiosis. 392

noninsulin-dependent diabetes mellitus NIDDM Type 2 diabetes mellitus, usually characterized by gradual onset with minimal or no symptoms of metabolic disturbance and no required exogenous insulin to prevent ketonuria and ketoacidosis. 198

norepinephrine Hormone secreted by the adrenal medulla to help initiate the "fight-or-flight" reaction. 145, 193

nose Specialized structure on the face that serves as the sense organ of smell and as part of the respiratory system. 277

nostril One of the external orifices of the nose. 277

nuclear envelope Membrane surrounding the cell nucleus and separating it from the cytoplasm. 39

nuclear pore Opening in the nuclear envelope. 39

nucleic acid Large organic molecule found in the nucleus (DNA and RNA) and in the cytoplasm (RNA). 31

nucleolus Organelle found inside the nucleus and composed largely of RNA for ribosome formation (pl., *nucleoli*). 36

nucleotide Building block of a nucleic acid molecule, consisting of a sugar, a nitrogen-containing base, and a phosphate group. 31

nucleus Large organelle that contains the chromosomes and acts as a cell control center. 36

O

obesity Excess adipose tissue; exceeding desirable weight by more than 20%. 318

occipital condyle One of two processes on the lateral portions of the occipital bone; for articulation with the atlas. 90

occipital lobe Area of the cerebrum responsible for vision, visual images, and other sensory experiences. 149

occluded coronary arteries Blocked blood vessels that serve the needs of the heart. 229

oil Substance, usually of plant origin and liquid at room temperature, formed when a glycerol molecule reacts with three fatty acid molecules. 26

olfactory cell Cell located high in the nasal cavity that bears receptor sites on cilia for various chemicals and whose stimulation results in smell. 167

oocyte Developing female gamete. 343

oogenesis Production of eggs in females by the process of meiosis and maturation. 351

oophorectomy Surgical removal of one or both ovaries. 352

opportunistic infection Disease that arises in the presence of a severely impaired immune system. 265

optic nerve Nerve composed of the ganglion cell fibers that form the innermost layer of the retina. 171

organ Structure consisting of a group of tissues that perform a specialized function; a component of an organ system. 2

organelle Part of a cell that performs a specialized function. 2, 36

organic molecule Carbon-containing molecule. 24

organism Individual living thing. 2

organ of Corti See *spiral organ*. 179

organ system Group of related organs working together. 2

organ transplantation Replacement of a diseased or defective organ with a healthy one. 8

orgasm Physical and emotional climax during sexual intercourse; results in ejaculation in the male. 347

origin End of a muscle that is attached to a relatively immovable part. 124

oropharynx Portion of the pharynx in the posterior part of the mouth. 298

osmosis Movement of water from an area of greater concentration to an area of lesser concentration across the plasma membrane. 43

osmotic pressure The amount of pressure needed to stop osmosis; the potential pressure of a solution caused by nondiffusible solute particles in the solution. 209

ossicles Tiny bones in the middle ear; malleus (hammer), incus (anvil), and stapes (stirrup). 178

ossification Formation of bone. 86, 370

osteoarthritis Disintegration of the cartilage between bones at a synovial joint. 107

osteoblast Bone-forming cell. 86

osteoclast Cell that causes the erosion of bone. 86

osteocyte Mature bone cell. 86

osteoporosis Weakening of bones due to decreased bone mass. 107

osteoprogenitor cells Cells found on or near all of the free surfaces of bone, which undergo division and transform into osteoblasts. 86

otitis media Inflammation of the middle ear. 286

otolith Granule that lies above, and whose movement stimulates, ciliated cells in the utricle and saccule. 181

otosclerosis Overgrowth of bone that causes the stapes to adhere to the oval window, resulting in conductive deafness. 182

ototoxic Damaging to any of the elements of hearing or balance. 182

outer ear Portion of the ear consisting of the pinna and the auditory canal. 178

oval window Membrane-covered opening between the stapes and the inner ear. 178

ovarian cancer Cancer of an ovary. 352

ovariohysterectomy Surgical removal of the ovaries and uterus. 353

ovary Female gonad; the organ that produces eggs, estrogen, and progesterone. 197, 349

ovulation Discharge of a mature egg from the follicle within the ovary. 351

oxygen deficit Amount of oxygen needed to metabolize the lactic acid that accumulates during vigorous exercise. 120

oxyhemoglobin Hemoglobin bound to oxygen in a loose, reversible way. 285

oxytocin Hormone released by the posterior pituitary that causes contraction of uterus and milk letdown. 188

P

pacemaker Small region of neuromuscular tissue that initiates the heartbeat; also called the *SA node*. 230

palatine tonsil Either of two small, almond-shaped masses located on either side of the oropharynx, composed mainly of lymphatic tissue; believed to act as sources of bacteria-killing phagocytes. 298

pancreas Endocrine organ located near the stomach that secretes digestive enzymes into the duodenum and produces hormones, notably insulin. 196, 308

pancreatic amylase Enzyme that digests starch to maltose. 308

pancreatic islets (of Langerhans) Distinctive groups of cells within the pancreas that secrete insulin and glucagon. 196

papillary muscle Muscle that extends inward from the ventricular walls of the heart and to which the chordae tendineae attach. 226

Pap smear Sample of cells removed from the tip of the cervix and then stained and examined microscopically. 352

paranasal sinus One of several air-filled cavities in the maxillary, frontal, sphenoid, and ethmoid bones that is lined with mucous membrane and drains into the nasal cavity. 277

paraplegia Paralysis of the lower body and legs, due to injury to the spinal cord between vertebrae T1 and L2. 147

parasympathetic division Portion of the autonomic nervous system that usually promotes those activities associated with a normal state. 157

parathyroid gland One of four small endocrine glands embedded in the posterior portion of the thyroid gland. 192

parathyroid hormone (PTH) Hormone secreted by the parathyroid glands that raises the blood calcium level primarily by stimulating reabsorption of bone. 192

parental cell Cell that divides so as to form daughter cells. 59

parietal Pertaining to the wall of a cavity. 66

parietal lobe Area of the cerebrum responsible for sensations involving temperature, touch, pressure, pain, and speech. 149

parietal pericardium Outer layer of the two layers of the serous pericardium, lining the fibrous pericardium. 7, 226

parietal peritoneum Lines the abdominal and pelvic walls and the inferior surface of the thoracic diaphragm. 7

parietal pleura Membrane that lines the inner wall of the thoracic cavity. 7

Parkinson disease Progressive deterioration of the central nervous system due to a deficiency in the neurotransmitter dopamine; also called paralysis agitans. 145

parturition Processes that lead to and include the birth of a human and the expulsion of the extraembryonic membranes through the terminal portion of the female reproductive tract. 384

passive immunity Protection against infection acquired by transfer of antibodies to a susceptible individual. 266

patella Bone of the kneecap. 102

pathogen Disease-causing agents, such as bacteria and viruses. 209

pathologist Person trained in knowledge of diseases and their symptoms, allowing for diagnosis of disease. 66

pectoral girdle Portion of the skeleton that provides support and attachment for the upper limbs. 97

pedigree Chart showing the relationships of relatives and which ones have a particular trait. 397

pelvic cavity Hollow place within the ring formed by the sacrum and coxal bones. 7

pelvic girdle Portion of the skeleton to which the lower limbs are attached. 100

pelvic inflammatory disease (PID) Latent infection of gonorrhea or chlamydia in the vasa deferentia or uterine tubes. 363

penis Male excretory and copulatory organ. 347

pentose Five-carbon sugar; deoxyribose is the pentose sugar found in DNA; ribose is a pentose sugar found in RNA. 24

pepsin Protein-digesting enzyme produced by the stomach. 301

peptidase Enzyme that catalyzes the breakdown of polypeptides. 311

peptide bond Bond that joins two amino acids. 28

peptide hormone Type of hormone that is a protein, a peptide, or derived from an amino acid. 201

perforin Protein released by cytotoxic T cells that attaches to an antigen. 263

pericardial fluid Fluid found in small amounts in the potential space between the parietal and visceral laminae of the serous pericardium. 226

pericarditis Inflammation of the pericardium. 248

pericardium Protective serous membrane that surrounds the heart. 66, 226

periodontitis Inflammation of the periodontal membrane that lines tooth sockets, causing loss of bone and loosening of teeth. 297

periosteum Fibrous connective tissue covering the surface of bone. 84

peripheral Situated away from the center of the body or an organ. 3

peripheral nervous system (PNS) Nerves and ganglia of the nervous system that lie outside the brain and spinal cord. 141

peristalsis Rhythmical contraction that moves the contents along in tubular organs, such as the digestive tract. 299

peritoneum Serous membrane that lines the abdominopelvic cavity and encloses the abdominal viscera. 66

peritonitis Generalized infection of the lining of the abdominal cavity. 7, 304

peritubular capillary network Capillary network that surrounds a nephron and functions in reabsorption during urine formation. 326

pernicious anemia Insufficiency of mature red blood cells, due to poor absorption of vitamin B$_{12}$. 213

peroxisomes Small vessicles within a cell which contain enzymes for the break down of fatty acids and hydrogen peroxide. 45

Peyer patches Lymphatic organs located in small intestine. 257

phagocytosis Taking in of bacteria and/or debris by engulfing; also called cell eating. 212

phalanges Bones of the fingers and thumb in the hand and of the toes in the foot (sing., *phalanx*). 103

pharyngeal tonsil Diffuse lymphatic tissue and follicles in the roof and posterior wall of the nasopharynx. 298

pharynx Common passageway for both food intake and air movement; the throat. 277, 298

phenotype Physical manifestation of a trait that results from the action of a particular set of genes. 395

phlebitis Inflammation of a vein. 235

phospholipid Lipid that contains two fatty acid molecules and a phosphate group combined with a glycerol molecule. 27

pH scale Measure of the hydrogen ion concentration; any pH below 7 is acidic, and any pH above 7 is basic. 23

physiology Branch of science dealing with the study of body functions. 2

pia mater Innermost meningeal layer that is in direct contact with the brain and spinal cord. 146

pineal gland Small endocrine gland, located in the third ventricle of the brain, that secretes melatonin and is involved in biorhythms. 200

pinna Outer, funnel-like structure of the ear that picks up sound waves. 178

pituitary dwarfism Condition in which a person has normal proportions but small stature; caused by inadequate growth hormone. 190

pituitary gland (hypophysis) Endocrine gland attached to the base of the brain that consists of anterior and posterior lobes. 188

pivot joint The end of a bone moving within a ring formed by another bone and connective tissue. 105

placenta Structure formed from the chorion and uterine tissue, through which nutrient and waste exchange occurs for the embryo and later the fetus. 371

placental membrane Semipermeable membrane that separates the fetal from the maternal blood in the placenta. 378

plaque Accumulation of soft masses of fatty material, particularly cholesterol, beneath the inner linings of arteries. 228

plasma Liquid portion of blood. 61, 209

plasma cell Cell derived from a B lymphocyte that is specialized to mass-produce antibodies. 260

plasma membrane Membrane that surrounds the cytoplasm of cells and regulates the passage of molecules into and out of the cell. 36

platelet Cell-like disks formed from fragmentation of megakaryocytes that initiate blood clotting. 214

platelet plug platelets that stick and cling to each other in order to seal a break in a blood vessel wall. 214

pleura (pl., *Pleurae*) Serous membrane that covers the lungs and lines the walls of the chest and the diaphragm. 7, 66, 279

pleurisy Inflammation of the pleura. 8, 9

pneumonectomy Surgical removal of all or part of a lung. 288

pneumonia Infection of the lungs that causes alveoli to fill with mucus and pus. 286

PNS See *peripheral nervous system*. 141

polar body Small, nonfunctional cell that is a product of meiosis in the female. 351

polar molecule Combination of atoms in which the electrical charge is not distributed symmetrically. 22

polycythemia Abnormally high number of red blood cells in the blood. 213

polydipsia Chronic, excessive intake of water. 197

polyp Small, abnormal growth on any mucous membrane, such as in the large intestine. 306

polypeptide A compound formed by the union of many amino acid molecules. 28

polyphagia Excessive eating. 197

polyribosome String of ribosomes simultaneously translating regions of the same mRNA strand during protein synthesis. 39

polysaccharide Carbohydrate composed of many bonded glucose units—for example, glycogen. 25

polyuria Excessive output of urine. 197

poly-X female Female who has more than two X chromosomes. 394

pons Portion of the brain stem above the medulla oblongata and below the midbrain; assists the medulla oblongata in regulating the breathing rate. 151

portal triad Grouping of the tributaries of the hepatic artery, vein, and bile duct at the angles of the lobules of the liver. 308

positive feedback Process by which changes cause more changes of a similar type, producing unstable conditions. 11, 188

posterior Toward the back; opposite of *anterior*. 3

posterior pituitary (neurohypophysis) Portion of the pituitary gland connected by a stalk to the hypothalamus. 190

posterior (dorsal)-root ganglion Mass of sensory neuron cell bodies located in the dorsal root of a spinal nerve. 154

postganglionic fiber In the autonomic nervous system, the axon that leaves, rather than goes to, a ganglion. 157

prefrontal area Association area in the frontal lobe that receives information from other association areas and uses it to reason and plan actions. 150

preganglionic fiber In the autonomic nervous system, the axon that goes to, rather than leaves, a ganglion. 157

premature birth Child born before full term and weighing 5 pounds, 8 ounces, or less. 380

presbycusis Loss of hearing that accompanies old age. 181

primary germ layers Three layers (endoderm, mesoderm, and ectoderm) of embryonic cells that develop into specific tissues and organs. 377

primary motor area Area in the frontal lobe where voluntary commands begin; each section controls a part of the body. 149

primary respiratory center Group of neurons in the medulla oblongata that regulates respiration. 283

primary somatosensory area Area posterior to the central sulcus where sensory information arrives from the skin and skeletal muscles. 149

primary spermatocyte Cell dividing into two secondary spermatocytes. 345

prime mover Muscle most directly responsible for a particular movement. 124

PRL See *prolactin.* 188

progesterone Female sex hormone secreted by the ovaries that, along with estrogen, promotes the development and maintenance of the primary and secondary female sex characteristics. 197

prolactin (PRL) Hormone secreted by the anterior pituitary that stimulates milk production in the mammary glands; also known as lactogenic hormone. 188

pronation Rotation of the forearm so that the palm faces backward. 106

prophase Stage of mitosis when chromosomes become visible. 49

proprioceptor Sensory receptor that assists the brain in knowing the position of the limbs. 164

prostaglandins Hormones that have various and powerful effects, often within the cells that produce them. 200

prostate gland Gland in males that is located about the urethra at the base of the bladder; contributes to the seminal fluid. 343

protein Macromolecule composed of amino acids. 28

prothrombin Plasma protein made by the liver that must be present in blood before clotting can occur. 214

prothrombin activator Enzyme that catalyzes the transformation of the precursor prothrombin to the active enzyme thrombin. 214

proton Positively charged particle in an atomic nucleus. 18

proximal Closer to the midline or origin; opposite of *distal.* 3

proximal convoluted tubule Highly coiled region of a nephron near the glomerular capsule. 326

pseudostratified columnar Appearance of layering in some epithelial cells when, actually, each cell touches a baseline and true layers do not exist. 57

psoriasis Common chronic, inherited skin disease in which red patches are covered with scales; occurs most often on the elbows, knees, scalp, and trunk. 74

PTH See *parathyroid hormone.* 192

puberty Stage of development in which the reproductive organs become functional. 342

pubic symphysis Slightly movable cartilaginous joint between the anterior surfaces of the hip bones. 104

pulmonary artery Blood vessel that takes blood away from the heart to the lungs. 242

pulmonary circuit Path of blood through vessels that take O_2-poor blood to and O_2-rich blood away from the lungs. 242

pulmonary edema Excessive fluid in the lungs caused by congestive heart failure. 257

pulmonary embolism Blockage of a pulmonary artery by a blood clot that commonly originates in a vein of the lower legs. 235

pulmonary fibrosis Accumulation of fibrous connective tissue in the lungs; caused by inhaling irritating particles, such as silica, coal dust, or asbestos. 287

pulmonary tuberculosis Tuberculosis of the lungs, caused by the tubercle bacillus. 286

pulmonary vein Blood vessel that takes blood away from the lungs to the heart. 242

pulse Vibration felt in arterial walls due to expansion of the aorta following ventricular contraction. 238

pupil Opening in the center of the iris that controls the amount of light entering the eye. 170

Purkinje fiber Specialized muscle fiber that conducts the cardiac impulse from the AV bundle into the ventricular walls. 230

pus Thick, yellowish fluid composed of dead phagocytes, dead tissue, and bacteria. 259

pyelonephritis Inflammation of the kidney; due to bacterial infection. 334

Q

quadriplegia Paralysis of the entire body and all four limbs, due to injury to the spinal cord between vertebrae C4 and T1. 147

R

radioactive isotope Atom whose nucleus undergoes degeneration and in the process gives off radiation. 19

radius Elongated bone located on the thumb side of the lower arm. 99

recessive allele Hereditary factor that expresses itself only when two copies are present in the genotype. 395

recruitment Increase in the number of motor units activated as intensity of stimulation increases. 123

rectum Terminal portion of the intestine. 304

red blood cell See *erythrocyte.* 61, 211

red bone marrow Blood cell-forming tissue located in spaces within certain bones. 84, 225

reduced hemoglobin (HHb) Hemoglobin that is carrying hydrogen ions. 285

referred pain Pain perceived as having come from a site other than that of its actual origin. 165

reflex Automatic, involuntary response of an organism to a stimulus. 154, 298

renal artery Vessel that originates from the aorta and delivers blood to the kidney. 325

renal cortex Outer, primarily vascular portion of the kidney. 326

renal medulla Inner portion of the kidney, including the renal pyramids. 326

renal pelvis Inner cavity of the kidney formed by the expanded ureter and into which the collecting ducts open. 326

renal vein Vessel that takes blood from the kidney to the inferior vena cava. 325

renin Secretion from the kidney that activates angiotensinogen to angiotensin I. 194, 238, 332

replication Production of an exact copy of a DNA sequence. 47

repolarization Recovery of a neuron's polarity to the resting potential after the neuron ceases transmitting impulses. 143

residual volume Volume of air that remains in the lungs after normal exhalation. 281

respiration Transport and exchange of gases between the atmosphere and the cells via the lungs and blood vessels. 276

respiratory membrane Alveolar wall plus the capillary wall, across which gas exchange occurs. 280

retina Innermost layer of the eyeball that contains the rod cells and cones cells. 171

retinal A form of vitamin A. 173

rheumatoid arthritis Persistent inflamation of synovial joints, often causing cartilage destruction, bone erosion, and joint deformities. 107, 269

Rh factor Type of antigen on red blood cells. 219

rhodopsin Light-sensitive biochemical in the rod cells of the retina; visual purple. 172

ribonucleic acid (RNA) Nucleic acid that helps DNA in protein synthesis. 31

ribosomal RNA (rRNA) RNA (ribonucleic acid) occurring in ribosomes, structures involved in protein synthesis. 48

ribosome Minute particle, found attached to the endoplasmic reticulum or loose in the cytoplasm, that is the site of protein synthesis. 39

rickets Defective mineralization of the skeleton, usually due to inadequate vitamin D in the body. 20

RNA See *ribonucleic acid.* 31

rod cell Dim-light receptor in the retina of the eye that detects motion but not color. 172

rotation Movement of a bone around its own longitudinal axis. 106

rotational equilibrium Maintenance of balance when the head and body are suddenly moved or rotated. 181

rough ER Endoplasmic reticulum that is studded with ribosomes on the side of the membrane that faces the cytoplasm. See *smooth ER.* 40

round window Membrane-covered opening between the inner ear and the middle ear. 178

rubella An acute, infectious disease affecting the respiratory tract in children and nonimmune young adults; characterized by a slight cold, sore throat, and fever, and the appearance of a fine, pink rash. 382

rugae Deep folds, as in the wall of the stomach. 301

S

saccule Saclike cavity of the inner ear that contains receptors for gravitational equilibrium. 181

sacroiliac joint Connection between the coxal bone and the sacrum. 100

sacrum Bone consisting of five fused vertebrae that form the posterior wall of the pelvic girdle. 95

saddle joint Two bones joined, having convex and concave surfaces that are complementary. 105

sagittal plane Plane or section that divides a structure into right and left portions. 5

sagittal suture Line of junction between the two parietal bones in the cranium. 104

salivary amylase In humans, enzyme in saliva that digests starch to maltose. 296, 311

salivary gland Gland associated with the mouth, secretes saliva. 296

salpingectomy Surgical removal of the uterine tubes. 352

salt Compound produced by a reaction between an acid and a base. 20

SA node See *sinoatrial node*. 230

sarcoma Cancer that arises in striated muscle, cartilage, or bone. 66

sarcomere Structural and functional unit of a myofibril. 116

saturated fatty acid Organic molecule that includes a fatty acid molecule which lacks double bonds between the atoms of its carbon chain. 26

scapula Large bone in the posterior shoulder area (pl., *scapulae*). 97

Schwann cell Cell that surrounds a fiber of a peripheral nerve and forms the neurilemmal sheath and myelin. 64, 139

SCID See *severe combined immunodeficiency disease*. 270

sclera White, fibrous outer layer of the eyeball. 170

scoliosis Abnormal lateral (side-to-side) curvature of the vertebral column. 94

scrotum Pouch of skin that encloses the testes. 7, 343

sebaceous gland Gland of the skin that secretes sebum. 73

sebum Oily secretion of the sebaceous glands. 73

secondary sex characteristic Trait that is sometimes helpful but not absolutely necessary for reproduction and is maintained by the sex hormones in males and females. 348

secondary spermatocyte One of the two cells into which a primary spermatocyte divides, and which in turn gives rise to spermatids. 345

selectively permeable Membrane that allows some molecules through but not others. 43

sella turcica Saddle-shaped area of the sphenoid bone; houses the pituitary gland. 92

semen Sperm-containing secretion of males; seminal fluid plus sperm. 343

semicircular canal Tubular structure within the inner ear with ampullae that contain the receptors responsible for the sense of rotational equilibrium. 178

semilunar valve Valve resembling a half-moon located between the ventricles and their attached vessels. 226

seminal fluid Sperm-containing secretion of males; also called semen. 343

seminal vesicle Convoluted, saclike structure attached to vas deferens near the base of the bladder in males; contributes to seminal fluid. 343

seminiferous tubule Highly coiled duct within the male testes that produces and transports sperm. 344

sensory neuron Neuron that takes the nerve impulse to the central nervous system; also known as an afferent neuron. 142

sensory receptor Sensory structure specialized to receive information from the environment and to generate nerve impulses. 164

serous membrane Membrane that covers internal organs and lines cavities lacking an opening to the outside of the body; also called serosa. 6, 66

serum Light-yellow liquid left after clotting of the blood. 214

severe combined immunodeficiency disease (SCID) Congenital illness in which both antibody-and cell-mediated immunity are lacking or inadequate. 270

sex chromosome Chromosome responsible for the development of characteristics associated with maleness or femaleness; an X or Y chromosome. 390

sex-linked Allele that occurs on the sex chromosomes but may control a trait that has nothing to do with the sex characteristics of an individual. 396

sexually transmitted disease (STD) Illness communicated primarily or exclusively through sexual intercourse. 362

sickle-cell disease Hereditary disease in which red blood cells are narrow and curved so that they are unable to pass through capillaries and are destroyed, causing chronic anemia. 213

sigmoid colon Portion of the large intestine that is S-shaped and extends from the descending colon to the rectum. 304

simple columnar epithelium Covering of the internal and external surfaces of the body; composed of a single layer of tall, prismatic cells. 57

simple cubodial epithelium Covering of the internal and external surfaces of the body; composed of a single layer of cube-shaped cells. 55

simple (endemic) goiter Condition in which an enlarged thyroid produces low levels of thyroxine. 191

simple squamous epithelium Covering of the internal and external surfaces of the body; composed of a single layer of flattened, platelike cells. 55

sinoatrial (SA) node Small region of neuromuscular tissue that initiates the heartbeat; also called the pacemaker. 230

sinus Cavity; for example, the sinuses in the human skull. 90

sinusitis Inflammation of the mucous membrane lining a paranasal sinus. 286

sister chromatids Two chromatids of a chromosome, held together by a centromere. 47

skeletal muscle Contractile tissue that comprises the muscles attached to the skeleton; also called striated muscle. 62, 114

SLE See *systemic lupus erythematosus*. 269

sliding filament theory Muscles contract when the thin (actin) and thick (myosin) filaments move past each other, shortening the skeletal muscle cells. 116

small intestine Portion of the digestive tract that extends from the lower opening of the stomach to the large intestine. 302

smooth ER Synthesizes the phospholipids that occur in membranes, among other functions, depending on the particular cell. 40

smooth muscle Contractile tissue that comprises the muscles in the walls of internal organs; also called visceral muscle. 63, 114

soft palate Entirely muscular posterior portion of the roof of the mouth. 296

solute The substance dissolved in a solution. 43

somatic system Portion of the peripheral nervous system containing motor neurons that control skeletal muscles. 152

somatotropin See *growth hormone*. 188

spasm Sudden, violent, involuntary contraction of a muscle or a group of muscles. 136

sperm Male gamete having a haploid number of chromosomes and the ability to fertilize an egg, the female gamete. 345

spermatid Intermediate stage in the formation of sperm cells. 345

spermatogenesis Sperm production in males by the process of meiosis and maturation. 345

spermatozoa Developing male gametes. 345

sphincter Muscle that surrounds a tube and closes or opens the tube by contracting and relaxing. 299

spinal cord Portion of the central nervous system extending downward from the brain stem through the vertebral canal. 146

spinal meningitis Inflammation of the meninges of the spinal cord. 6

spinal nerve Nerve that arises from the spinal cord. 152

spindle Apparatus composed of microtubules to which the chromosomes are attached during cell division. 49

spiral organ Structure in the vertebrate inner ear that contains auditory receptors; also called organ of Corti. 179

spleen Large, glandular organ located in the upper left region of the abdomen that stores and purifies blood. 257

spongy bone Bone found at the ends of long bones; consists of bars and plates separated by irregular spaces. 61, 84

sprain Joint injury in which some of the fibers of a supporting ligament are ruptured, but the continuity of the ligament remains intact. 136

squamosal suture Type of suture formed by overlapping of the broad, beveled edges of the participating bones. 104

stapes The last of three ossicles of the ear; serves with the malleus and incus to conduct vibrations from the tympanic membrane to the oval window of the inner ear. 178

starch Polysaccharide that is common in foods of plant origin. 25

stem cell A precursor cell. 210

sternum Breastbone to which the ribs are ventrally attached. 96

steroid Lipid-soluble, biologically active molecules having four interlocking rings; examples are cholesterol, progesterone, and testosterone. 27

steroid hormone Type of hormone that has the same complex of four-carbon rings, but each one has different side chains. 201

stomach Saclike, expandable digestive organ located between the esophagus and the small intestine. 301

strain An overstretching or overexertion of some muscles. 136

stratified Layered, as in stratified epithelium, which contains several layers of cells. 55

stratified squamous epithelium Covering of the internal or external surfaces of the body composed of layered, flattened, platelike cells. 57

stratum basale Deepest layer of the epidermis, where cell division occurs. 70

stratum corneum Uppermost keratinized layer of the epidermis. 71

stroke See *cerebrovascular accident.* 149, 239

sty Inflammation of a sebaceous gland. 168

subclavian Either of two arteries branching off the aortic arch and supplying the arms. 245

subcutaneous injection Introduction of a substance beneath the skin, using a syringe. 71

subcutaneous tissue Tissue beneath the dermis that tends to contain fat cells. 71

subdural hematoma Accumulation of blood between the dura mater and the brain. 146

superficial Near the surface. 3

superior Toward the upper part of a structure or toward the head. 3

superior vena cava Large vein that enters the right atrium from above and carries blood from the head, thorax, and upper limbs to the heart. 242

supination Rotation of the forearm so that the palm faces forward when in the anatomical position. 106

surface tension Force that holds moist membranes together when water molecules attract. 279

surfactant Agent that reduces the surface tension of water; in the lungs, a surfactant prevents the alveoli from collapsing. 279

suture Type of immovable joint articulation found between bones of the skull. 90

sweat gland Skin gland that secretes a fluid substance for evaporative cooling; also called sudoriferous gland. 73

sympathetic division Part of the autonomic nervous system whose effects are generally associated with emergency situations. 157

synapse Region between two nerve cells where the nerve impulse is transmitted from one to the other, usually from axon to dendrite. 145

synaptic cleft Small gap between the synaptic knob on one neuron and the dendrite on another neuron. 145

syndrome A group of symptoms that characterize a disease condition. 195, 390

synergist Muscle that assists the action of the prime mover. 124

synovial fluid Fluid secreted by the synovial membrane. 104

synovial joint Freely movable joint. 104

synovial membrane Membrane that forms the inner lining of a capsule of a freely movable joint. 66, 104

synthesis reaction To build up, as in the combining of two small molecules to form a larger molecule. 29

syphilis Sexually transmitted disease caused by the bacterium *Treponema pallidum* characterized by a painless chancre on the penis or cervix; if untreated, can lead to cardiac and central nervous system disorders. 364

systemic circuit Part of the cardiovascular system that serves body parts other than the gas-exchanging surfaces in the lungs. 242

systemic disease Illness that involves the entire body or several body systems. 12

systemic lupus erythematosus (SLE) Syndrome involving the connective tissues and various organs, including kidney failure. 269

systole Contraction of the heart chambers, particularly the left ventricle. 232

systolic pressure Arterial blood pressure during the systolic phase of the cardiac cycle. 239

T

tachycardia Abnormally rapid heartbeat. 231

talus Ankle bone. 103

tarsal bones Bones of the ankle in humans. 103

taste bud Organ containing the receptors associated with the sense of taste. 166

Tay-Sachs Lethal genetic disease in which the newborn has a faulty lysosomal digestive enzyme. 40

tectorial membrane Membrane in the spiral organ (organ of Corti) that lies above and makes contact with the receptor cells for hearing. 179

telophase Stage in mitosis when newly formed cells separate. 50

template Pattern or guide used to make copies; parental strand of DNA serves as a guide for the production of daughter DNA strands, and DNA also serves as a guide for the production of messenger RNA. 47

temporal lobe Area of the cerebrum responsible for hearing and smelling and for the interpretation of sensory experience and memory. 149

tendinitis Inflammation of muscle tendons and their attachments. 136

tendon Tissue that connects muscle to bone. 59, 115

teratogen Any substance that produces abnormalities during human development. 382

testis Male gonad; the organ that produces sperm and testosterone (pl., *testes*). 197, 343

testosterone The most potent of the androgens, the male sex hormones. 197, 345

tetanic contraction Sustained muscle contraction without relaxation. 122

tetanus Acute infection caused by the toxin of the tetanus bacterium; results in a rigidly locked jaw. 136

tetany Severe twitching caused by involuntary contraction of the skeletal muscles due to a lack of calcium. 192

thalamus Mass of gray matter located at the base of the cerebrum in the wall of the third ventricle; receives sensory information and selectively passes it to the cerebrum. 151

thoracic cavity Hollow place within the chest. 6

thrombin Enzyme derived from prothrombin that converts fibrinogen to fibrin threads during blood clotting. 214

thrombocyte A blood platelet. 214

thrombocytopenia Insufficient number of platelets in the blood. 214

thromboembolism Obstruction of a blood vessel by a thrombus that has dislodged from the site of its formation. 214, 228

thrombus Blood clot that remains in the blood vessel where it formed. 214

thymosins Hormones secreted by the thymus. 200

thymus gland Lobular gland that lies in the neck and chest area and is necessary for the development of immunity. 200, 256

thyroid gland Endocrine gland, located just below the larynx and in front of the trachea, that secretes thyroid hormones. 191

thyroid-stimulating hormone (TSH) Hormone that causes the thyroid to produce thyroxine. 188

thyroxine (T$_4$) Hormone produced by the thyroid that speeds the metabolic rate. 191

tibia Shinbone found in the lower leg. 102

tidal volume Amount of air that enters the lungs during a normal, quiet inspiration. 281

tight junction Junction between cells when adjacent plasma membrane proteins join to form an impermeable barrier. 65

tissue Group of similar cells that performs a specialized function. 2, 55

tissue thromboplastin Clotting factor, released or derived from tissues, that interacts with platelets, calcium ions, and other clotting factors. 215

T lymphocyte One of two types of lymphocytes: a killer T cell that interacts directly with antigen-bearing cells and is responsible for cell-mediated immunity, or a helper T cell that stimulates other immune cells. 212, 260

tone Continuous, partial contraction of muscle. 123

tonicity Osmolarity of a solution compared to that of a cell. If the solution is isotonic to the cell, there is no net movement of water; if the solution is hypotonic, the cell gains water; and if the solution is hypertonic, the cell loses water. 43

tonsil Partly encapsulated lymph nodule located in the pharynx. 257, 286

tonsillectomy Surgical removal of the tonsils. 286

tonsillitis Inflammation of the tonsils. 286

trabeculae Branching bony plate that separates irregular spaces within spongy bone. 61

tracer Substance having an attached radioactive isotope that allows a researcher to track its whereabouts in a biological system. 19

trachea Windpipe; serves as a passageway for air. 278

tracheostomy Creation of an artificial airway by incision of the trachea and insertion of a tube. 278

tract Bundle of neurons forming a transmission pathway through the brain and spinal cord. 146

transcription Manufacturing RNA from DNA. 48

transfer RNA (tRNA) Molecule of RNA (ribonucleic acid) that carries an amino acid to a ribosome engaged in the process of protein synthesis. 48

translation Assembly of an amino acid chain according to the sequence of base triplets in a molecule of mRNA. 48

transverse colon Portion of the large intestine that travels transversely as it extends from the ascending colon to the descending colon. 304

transverse plane Plane or section that divides a structure horizontally to give a cross section. 5

tricuspid valve Atrioventricular valve between the right atrium and the right ventricle. 226

triglyceride Lipid composed of three fatty acids combined with a glyercol molecule. 26

triplet code Three-nucleotide base unit coding for a particular amino acid during protein synthesis. 48

trisomy State of having an extra chromosome—three instead of the normal two. 392

trophoblast Outer cells of a blastocyst that help form the placenta and other extra-embryonic membranes. 373

trypsin Protein-digesting enzyme produced by the pancreas. 308

TSH See *thyroid-stimulating hormone.* 188

T (transverse) tubule Membranous channel that extends inward from a muscle fiber membrane and passes through the fiber. 116

tubal ligation Method for preventing pregnancy in which the uterine tubes are cut and sealed. 359

tubular reabsorption Process that transports substances out of the renal tubule into the interstitial fluid from which the substances diffuse into peritubular capillaries. 329

tubular secretion Process of substances moving out of the peritubular capillaries into the renal tubule. 329

tumor Abnormal growth of tissue that serves no useful purpose. 80

Turner syndrome Condition caused by the inheritance of a single X chromosome. 393

tympanic membrane Membrane located between the external and middle ear; the eardrum. 178

U

ulcer Open sore in the lining of the stomach; frequently caused by bacterial infection. 301

ulna Elongated bone within the lower arm. 99

umbilical Pertaining to the umbilicus. 6

umbilical cord Cord through which blood vessels that connect the fetus to the placenta pass. 377

unsaturated fatty acid Organic compound that includes a fatty acid molecule having one or more double bonds between the atoms of its carbon chain. 26

urea Primary nitrogenous waste of mammals. 324

uremia High level of urea nitrogen in the blood. 334

ureters Tubes that take urine from the kidneys to the bladder. 325

urethra Tube that takes urine from the bladder to the outside of the body. 325, 343

urethritis Inflammation of the urethra. 334

uric acid Product of nucleic acid metabolism in the body. 324

urinalysis Examination of a urine sample to determine its chemical, physical, and microscopic aspects. 334

urinary bladder Organ where urine is stored before being discharged by way of the urethra. 325

urticaria Skin eruption characterized by the development of welts as a result of capillary dilation. 74

uterine tube Tube that extends from the uterus on each side toward an ovary and transports sex cells; also called fallopian tube or oviduct. 352

uterus Female organ in which the fetus develops. 349

utricle Saclike cavity of the inner ear that contains receptors for static equilibrium. 181

uvula A fleshy portion of the soft palate that hangs down above the root of the tongue. 296

V

vaccine Treated antigens that can promote active immunity when administered. 266

vagina Female copulatory organ and birth canal. 349

valve Structure that opens and closes, ensuring one-way flow; common to vessels, such as systemic veins, lymphatic veins, and veins to the heart. 235

varicose vein Irregular dilation of a superficial vein, seen particularly in the lower legs, due to weakened valves within the veins. 235

vascularization Process in which a tumor becomes supplied with blood vessels. 80

vas deferens Tube connecting the epididymis to the urethra; sperm duct (pl., *vasa deferentia*). 343

vasectomy Method for preventing pregnancy in which the vasa deferentia are cut and sealed. 359

vasomotor center Neurons in the brain stem that control the diameter of the arteries. 237

vein Blood vessel that takes blood to the heart. 235

venous duct See *ductus venosus*. 246

ventilation Breathing; the process of moving air into and out of the lungs. 282

ventral Toward the front or belly surface; the opposite of *dorsal*. 6

ventricle Cavity in an organ, such as the ventricles of the brain or the ventricles of the heart. 146, 226

venule Type of blood vessel that takes blood from capillaries to veins. 235

vermiform appendix Small, tubular appendage that extends outward from the cecum of the large intestine. 304

vernix caseosa Cheeselike substance covering the skin of the fetus. 379

vertebra Bone of the vertebral column. 94

vertebral canal Hollow place within the vertebrae containing the spinal cord. 6

vertebral column Backbone of vertebrates, composed of individual bones called vertebrae. 94

vertigo Dizziness and a sense of rotation. 181

vesicle Small, membranous sac that stores substances within a cell. 40

vesicular (Graafian) follicle Mature follicle within the ovaries that houses a developing egg. 351

vestibule Space or cavity at the entrance of a canal, such as the cavity that lies between the semicircular canals and the cochlea. 178, 296, 353

villi Fingerlike projections that line the small intestine and function in absorption (sing *villus*). 302

visceral Pertaining to the contents of a body cavity. 66

visceral pericardium The inner layer of the serous pericardium; it is in contact with the heart and the roots of the vessels of the heart. 7, 226

visceral peritoneum Membrane that covers the surfaces of organs within the abdominal cavity. 7

visceral pleura Membrane that covers the surfaces of the lungs. 7

visual accommodation Ability of the eye to focus at different distances by changing the curvature of the lens. 171

visual field Area of vision for each eye. 176

vital capacity Maximum amount of air a person can exhale after taking the deepest breath possible. 281

vitamins Organic molecules (usually coenzymes) that must be in the diet and are necessary in trace amounts for normal metabolic functioning of cells. 315

vitreous humor Substance that occupies the posterior cavity of the eye. 171

vocal cords Folds of tissue within the larynx that produce sounds when they vibrate. 278

vulva External genitalia of the female that lie near the opening of the vagina. 349

W

wart Raised growth on the skin due to a viral infection. 74

white blood cell See *leukocytes*. 61, 212

white matter Myelinated nerve fibers in the central nervous system. 146

X

X chromosome Female sex chromosome that carries genes involved in sex determination; see *Y chromosome*. 390

xenotransplantation Use of animal organs, instead of human organs, in human transplant patients. 269

X-linked Gene found on the X chromosome that controls traits other than sexual traits. 396

XYY male Male who has an X chromosome and two Y chromosomes in each nucleus. 394

Y

Y chromosome Male sex chromosome that carries genes involved in sex determination; see *X chromosome*. 390

yolk sac Extraembryonic membrane that serves as the first site of red blood cell formation. 371

Z

zygote Cell formed by the union of the sperm and egg; the product of fertilization. 342, 370

PHOTO CREDITS

Design Elements

(Background pattern): © BS7/Getty Images; (Stethoscope): © Vol. 59/Getty Images; (Fingerprint): © SS47/Getty Images; (Researcher): © Digital Vision

Chapter 1

Opener: © Corbis Royalty Free; Box 1B: © Scott Camazine; Box 1C-b: © Biomed Comm/Custom Medical Stock Photos; Box 1D-a: © Hank Morgan/Rainbow; Box 1D-b: © Mazzlota et al/Photo Researchers, Inc.

Chapter 2

Opener: © Vol. 72/Photo Disc/Getty Images; 2.2b: © Charles M. Falco/Photo Researchers, Inc.; 2.7: © Jeremy Burgess/SPL/Photo Researchers, Inc.; 2.8: © Don W. Fawcett/Photo Researchers, Inc.; 2.11(both): © Vol. 176/Corbis

Chapter 3

Opener: © The McGraw-Hill Companies, Inc./Dennis Strete, photographer; 3.3: Courtesy Stephen Wolfe; 3.4: © Barry F. King/Biological Photo Service; 3.5(top): © R. Rodewald/Biological Photo Service; 3.5(lower): Courtesy Tim Wakefield, John Brown University; 3.6a: Courtesy Dr. Keith Porter; 3.7a: © Oliver Meckes/Photo Researchers, Inc.; 3.7b: © Manfred Kage/Peter Arnold, Inc.; 3.8(basal body): © William L. Dentler/Biological Photo Service; 3.8(flagellum): © William L. Dentler/Biological Photo Service; 3.10a-c: © David M. Phillips/Visuals Unlimited; 3.17(all) :© Michael Abbey/Photo Researchers, Inc., Box 3B: Courtesy Dr. Michael Baird, Lifecodes Corporation

Chapter 4

Opener: © Cabisco/Visuals Unlimited; 4.1(right): © Ed Reschke; 4.2(right): © The McGraw-Hill Companies, Inc./Dr. Alvin Telser, photographer; 4.3(right): © Ed Reschke; 4.4(right): © Manfred Kage/Peter Arnold, Inc.; 4.5-4.8(right): © Ed Reschke; 4.9(right): © The McGraw-Hill Companies, Inc./Dr. Alvin Telser, photographer; 4.10, 4.11(right): © Ed Reschke; 4.12: © The McGraw-Hill Companies, Inc./Dr. Alvin Telser, photographer; 4.13-4.16(right): © Ed Reschke

Chapter 5

Opener: © The McGraw-Hill Companies, Inc./Dr. Alvin Telser, photographer; Box 5A: Courtesy of Kent Van De Graaff, photo by James M. Clayton; 5.2b: © CNRI/SPL/Photo Researchers, Inc.; 5.5a: © Ken Greer/Visuals Unlimited; 5.5b: © Dr. P. Marazzi/SPL/Photo Researchers, Inc.; 5.5c: © James Stevenson/SPL/Photo Researchers, Inc.; 5.8a: © Neil Borden/Photo Researchers, Inc.; 5.8b: Courtesy Dr. George Bogumill

Chapter 6

Opener: © Stockbyte/Punchstock; Box 6Aa: © Royce Blair/Unicorn Stock Photos; Box 6Ab,c: © Michael Klein/Peter Arnold, Inc.; 6.25b: © Lennart Nilsson, "Behold Man", Little Brown and Company, Boston; Box 6Ca: © Ralph T. Hutchings/Imagingbody.com; Box 6Cb: © Ralph T. Hutchings/Imagingbody.com

Chapter 7

Opener: © The McGraw-Hill Companies, Inc./Dr. Alvin Telser, photographer; 7.1(all): © Ed Reschke; 7.2c: © The McGraw-Hill Companies/Dennis Strete, photographer; 7.3: Courtesy of Hugh E. Huxley; 7.9: © G.W. Willis/Visuals Unlimited

Chapter 8

Opener 8: © Corbis Royalty Free; 8.5: Courtesy Dr. E.R. Lewis, University of California, Berkeley; 8.7: © Manfred Kage/Peter Arnold, Inc.; 8.8b: © Colin Chumbley/Science Source/Photo Researchers, Inc.; p.165: © Reuters/Corbis

Chapter 9

Opener: © Vol. 122/PhotoDisc/Getty Images; 9.3: © Omikron/SPL/Photo Researchers, Inc.; 9.9: © Lennard Nilsson, from "The Incredible Machine"; 9.10b: © The McGraw-Hill Companies, Inc./Dennis Strete, photographer; 9.13: © P. Motta/SPL/Photo Researchers, Inc.

Chapter 10

Opener: © Ed Reschke; 10.5: © Wellcome Library, London; 10.6: From Clinical Pathological Progression of Acromegaly Conference, "Acromegaly, Diabetes, Hyper metabolism, Proteniura and Heart Failure", American Journal of Medicine 20 (1956) 133, with permission from Excerpta Medica; 10.7: © Biophoto Associates/SPL/Photo Researchers, Inc.; 10.8: © John Paul Kay/Peter Arnold, Inc.; 10.12a,b: © NMSB/Custom Medical Stock Photo; 10.13a,b: "Atlas of Pediatric Physical Diagnosis," Second Edition by Zitelli & Davis, 1992. Mosby-Wolfe-Europe Limited, London, UK; Box 10A: Courtesy Robert P. Lanza; 10.15: © James Darell/Stone/Getty

Chapter 11

Opener: © National Cancer Institute/SPL/Photo Researchers, Inc.; 11.5: © Manfred Kage/Peter Arnold, Inc.; Box 11B: © Bill Longcore/Photo Researchers, Inc.; 11.9(top): © Jean Claude Revy/ISM/Phototake; 11.9(bottom): © Jean Claude Revy/ISM/Phototake

Chapter 12

Opener: © The McGraw-Hill Companies, Inc.; Box 12A: © Biophoto Associates/Photo Researchers, Inc.

Chapter 13

Opener: © Scott Camazine/CDC/Photo Researchers, Inc.; 13.2(thymus gland): © Ed Reschke/Peter Arnold, Inc.; 13.2(lymph node): © Fred E. Hossler/Visuals Unlimited; 13.2(red bone marrow): © R. Valentine/Visuals Unlimited; 13.2(spleen): © Ed Reschke/Peter Arnold, Inc.; 13.3a-e: © Ed Reschke/Peter Arnold, Inc.; 13.6b: Courtesy of

Dr. Arthur J. Olson, The Scripps Research Institute; 13.8b: © Boehringer Ingelheim International, photo by Lennart Nilsson; 13.10: © Estate of Ed Lettau/Peter Arnold, Inc.; p. 290: © Michael Newman/Photo Edit

Chapter 14

Opener: © Manfred Kage/Peter Arnold, Inc.; 14.7(right): © Yoav Levy/Phototake; 14.11a,b: © Martin Rotker/Martin Rotker Photography

Chapter 15

Opener: © Stockbyte/Punchstock; 15.3b: © Biophoto Associates/Photo Researchers, Inc.; 15.4b: © Ed Reschke/Peter Arnold, Inc.; 15.7c: © Manfred Kage/Peter Arnold, Inc.; 15.7d: Photo by Susumu Ito, from Charles Flickinger, "Medical Cellular Biology", W.B. Saunders, 1979; Box 15B: Courtesy GIVEN ® IMAGING; 15.9c: © Ed Reschke; 15.13(right): © Donna Day/Stone/Getty Images; 15.13(left): © Tony Freeman/PhotoEdit

Chapter 16

Opener: © SPL/Photo Researchers, Inc.; 16.4(top left): © R.G. Kessel and R.H. Kardon, "Tissues and Organs: A Text Atlas of Scanning Electron Microscopy", 1979; 16.4(top right, bottom left): © 1966 Academic Press, from A.B. Maunsbach, J. Ultrastruct. Res. 15:242-282; 16.10(left): © Hank Morgan/SPL/Photo Researchers, Inc.

Chapter 17

Opener: © Branx X/Corbis Royalty Free; 17.4b: © Biophoto Associates/Photo Researchers, Inc.; 17.10: © Ed Reschke/Peter Arnold, Inc.; 17.16a,c, d,f: © The McGraw-Hill Companies, Inc./Bob Coyle, photographer; 17.16b: © The McGraw-Hill Companies, Inc./Vincent Ho, photographer; 17.16e: © Gusto/Photo Researchers, Inc.; Box 17D(alligator): © Index Stock; Box 17D(fish): © Norbert WU; Box 17D(owl): © Art Wolf/Stone/Getty; Box 17D(woman): Steven Peters/Stone/Getty Images; Box 17E(left): © Vol. 161/Corbis; Box17E(right): © Vol. 178/Corbis; Box 17F: © Argus Fotoarchiv/Peter Arnold, Inc.

Chapter 18

Opener: © Dr. Yorgas Nikas/SPL/Photo Researchers, Inc.; 18.1a: © David M. Phillips/Visuals Unlimited; 18.6a: Lennart Nilsson "A Child is Born," Dell Publishing Company; 18.8: © James Stevenson/SPL/Photo Researchers, Inc.; Box18Bb-d: Streissguth, A.P., Landesman-Dwyer, S. Martin J.C., and Smith D.W. (1980). Teratogenic effects of alcohol in humans and laboratory animals. /it/Science/xit/:209(18):353-361

Chapter 19

Opener: © Meckes/Ottawa/Photo Researchers, Inc.; 19.1c-e: © CNRI/SPL/Photo Researchers, Inc.; 19.3: © Jill Cannefax/Coyote Crossing, Inc.; 19.4a,b: Photograph by Earl Plunkett. Courtesy of G. H. Valentine; Box19B: © Sygma/Corbis

INDEX

Crural region, 4
Cryptorchidism, 374
Cubital region, 4
Cubital vein, 266
Cuboid, 115
Cuboidal epithelium, 62–63, 64
Cuneiforms, 115
Cunnilingus, 395
Cupula, 197
Cushing syndrome, 214, 216
Cuspids, 325
Cutaneous membrane, 76. See also Skin
Cuticle, 83
Cyanosis, 81
  neonatal, 269
Cystic duct, 442
Cystic fibrosis, 50, 337, 431, 432
Cystitis, 362
Cytokines, 286, 287
Cytokinesis, 54, 56
Cytoplasm, 41
Cytoskeleton, 40, 46
Cytotoxic T cell, 287
Cytotoxins, 75

Dandruff, 85
Daughter cell, 54
Dead-space air, 308
Deafness, 200
Deamination, 338
Death, time of, 136
Decubitus ulcers, 165
Deep, definition of, 3
Deep brachial artery, 264
Deep-brain stimulation, in Parkinson disease, 177
Defecation, 333, 335
Degradation (decomposition) reaction, 33, 34
Dehydration, 51
Dehydration reaction, 26
Delirium tremens, 417
Deltoid, 140, 143–44, 145, 438, 439
Deltoid region, 4
Deltoid tuberosity, 110
Denaturation, 32
Dendrite, 72, 157
Dense bodies, 133
Dense connective tissue, 66
Dentin, 325
Dentition, 325
  of skeletal remains, 121
Deoxyhemoglobin, 230
Deoxyribonucleic acid (DNA), 34–35
  replication of, 52–53
  template, 52
Depolarization, of axon, 158, 159
Depo-Provera injection, 388
Depression, 119
Dermal papillae, 81
Dermis, 80, 81. See also Skin
DES (diethylstilbestrol), 417–18
Descending aorta, 263, 264, 265, 443, 444
Descending colon, 324, 333, 334, 441, 442, 443
Detrusor muscle, 353
Development, 404–15
  embryonic, 404, 405, 406, 408–10, 411
  extraembryonic membranes in, 404–5
  fetal, 405, 410, 412, 414–15
  pre-embryonic, 405–6
  of sex organs, 412–15
DEXA scan, 99
Diabetes insipidus, 208
Diabetes mellitus, 216–17
  islet cell transplantation in, 218
  in pregnancy, 419
Diagnosis, of cancer, 74
Dialysate, 363

Diaphragm (birth control), 387, 389
Diaphragm (respiratory muscle), 6, 142, 300, 324, 440, 441, 442, 443, 444
Diaphysis, 95, 96
Diarrhea, 333
Diarthrosis, 115
Diastole, 254, 255
Diastolic pressure, 262
Diencephalon, 168
Diet
  anemia and, 235
  in cardiovascular disease, 270–71
Diethylstilbestrol (DES), 417–18
Differential white blood cell count, 235
Differentiation, 404
Diffusion, 48, 49, 50, 238
Digestive enzymes, 336–37, 339, 340
Digestive system, 10, 15, 323–48. See also specific components
  accessory organs of, 336–39
  aging of, 340
  endocrine system interaction with, 222, 336–37
  in homeostasis, 13, 340–42
  skeletal system interaction with, 120
Digital region, 4
Dipeptide, 31
Disaccharases, 339
Disaccharides, 27
Disease, definition of, 13
Distal, definition of, 3
Distal convoluted tubule, 356, 357
Diuresis, 360
Diuretics, 361
Diverticulosis, 336
Dizziness, 198
DNA (deoxyribonucleic acid), 34–35
  replication of, 52–53
  template, 52
DNA fingerprinting, 57
DNA technology, 434–35
Dominant allele, 429
Dorsalis pedis artery, 264
Dorsal-root ganglion, 170
Dorsiflexion, 117
Dorsum, 4
Double bond, 23
Double helix, 34, 35
Douching, for birth control, 388
Down syndrome, 424, 426–27
Drugs
  abuse of, 270, 417
  ototoxicity of, 200, 201
  testing for, 363
Dual energy X-ray absorptiometry, 99
Duchenne muscular dystrophy, 150
Ductus arteriosus, 267, 269
Ductus venosus, 267, 269
Duodenum, 324, 330, 442, 443
Dural venous sinuses, 162
Dura mater, 162
Dwarfism, pituitary, 210

Ear, 196–201
  aging of, 200–201
  anatomy of, 196, 197
  functions of, 196–99
  infection of, 313
Eating disorders, 345–48
Eccrine glands, 84
Ectoderm, 408, 409
Ectopic pregnancy, 382, 406
Eczema, 85
Edema, 238–39, 263, 277
Efferent (motor) system, 156, 169

Ejaculation, 373, 377
Ejaculatory duct, 373, 376
Elastic cartilage, 67
Elastic fibers, 65, 81
Elastin, 65
Electrocardiogram, 254, 255
Electroencephalogram, 164
Electrolytes, 26, 358–61
  reabsorption of, 360–61
Electron, 21
Electronegative element, 23
Element, 21
  electronegative, 23
Elephantiasis, 239, 281
Elevation, 119
Embolism, 236, 258–59, 268
Embryo, 405, 406, 408–10, 411
Embryonic disk, 408
Emergency contraception, 388, 390–92
Emphysema, 310, 314–15
Emulsification, 29–30
Endocarditis, 248
Endocardium, 248
Endochondral ossification, 97
Endocrine system, 10, 14, 204–26
  aging of, 222
  contaminant effects on, 392–93
  glands of, 73, 204–26. See also specific glands
  in homeostasis, 13, 207, 222–24
  negative feedback in, 207
  positive feedback in, 207, 208
Endocytosis, 48, 50
Endoderm, 408, 409
Endomembrane system, 43–45
Endometriosis, 393
Endometrium, 379, 382
Endomysium, 127, 128
Endoplasmic reticulum, 40, 41, 43–44
Endoscopy, 332
Endothelium, 257, 258
Energy
  ATP for, 35–36
  for muscle contraction, 133–35
Enteroendocrine cell, 329
Enzymes, 32–34
  digestive, 336–37, 339, 340
  salivary, 324
Eosinophils, 229, 231, 232, 280
Ependymal cell, 72, 162
Epicardium, 248
Epicondyles, 113, 114
Epidermal growth factor, 221
Epidermis, 80–81. See also Skin
Epididymis, 373, 374, 375–76, 441
Epiglottis, 300, 302, 303, 304, 326
Epimysium, 128
Epinephrine, 206, 213–14
Epiphyseal plate, 97
Epiphysis, 95, 96
Epiretinal implant, 193
Episiotomy, 415
Epithelial tissue, 62–65
  columnar, 63–64, 65
  cuboidal, 62–63
  squamous, 62, 63
  transitional, 62, 64–65
Equilibrium, 197–99
Erectile dysfunction, 377
Erythrocytes. See Red blood cell(s)
Erythropoietin, 221, 230–31, 232, 352
Esophageal sphincter, 327, 328
Esophagus, 302, 324, 327, 443, 444
  in digestive process, 326
Essential amino acids, 343
Essential fatty acids, 342–43

Essure system, 389
Estrogen, 31, 205–6, 219, 384, 399
  menopause and, 385–86
  placental, 419
Estrogen therapy, 99
Ethmoid bone, 102, 103, 104
Ethnicity, of skeletal remains, 121
Eupnea, 311
Eustachian tube, 196, 199
Exercise
  benefits of, 149–50
  cardiovascular disease and, 271
  muscle size and, 137
  osteoporosis and, 99
  ventilation during, 309
Exocrine gland, 73, 74
Exocytosis, 48, 50
Expiration, 300, 306–7
Expiratory reserve volume, 308
Extension, 117–18, 119
Extensor carpi, 144, 145, 146
Extensor digitorum, 140, 144, 145, 146
Extensor digitorum longus, 146, 148
External auditory meatus, 102, 103, 104
External intercostal muscle, 142, 143, 439, 440, 444
External jugular vein, 440
External oblique muscle, 142, 143, 438, 439, 440
Extracellular junctions, 73
Extraembryonic membranes, 404–5
Extraocular muscles, 186–88
Eye, 186–96
  accessory organs of, 186–88
  aging of, 200
  anatomy of, 188–89
  bionic, 193
  corrective lenses for, 191
  muscles of, 186–88
  photoreceptors of, 190, 192
Eyebrows, 186, 187
Eyelashes, 186, 187
Eyelids, 186
Eyestrain, 190

Facial bones, 102, 103, 105
Facial expression, muscles of, 141
Facial nerve, 170, 171
Facial vein, 266
Facilitated transport, 48, 50
Falciform ligament, 440
Fallopian tube, 379, 380, 381–82, 442
False pelvis, 112
Farsightedness, 191
Fascia, 128
Fascicle, 127, 128
Fast-twitch muscle fibers, 137–38
Fat, 29, 81
  absorption of, 277
  on nutrition labels, 28–29
Fat tissue, 65–66, 67, 81, 221, 393
Fatty acids, 30
  essential, 342–43
Fatty streak, in coronary artery disease, 252, 253
Feces, 333
Feedback
  negative, 10–11, 12, 378
  positive, 11, 12, 384, 415
Fellatio, 395
Femoral artery, 264, 439, 440, 442
Femoral head, 113, 114
Femoral nerve, 171, 172, 439, 440, 441
Femoral region, 4
Femoral vein, 266, 438, 439, 440, 442

Femur, 101, 113–14, 444
Fermentation, 134, 135
Fertilization, 379, 382, 384–85, 403–4, 406
Fetal alcohol syndrome, 417
Fetus, 405, 410, 412, 414–15
  circulation of, 267–69
Fever, 89
Fiber, 333
Fibrillation, 255
Fibrin, 234, 236
Fibrinogen, 228, 230, 234, 236
Fibroblasts, 65–66, 67, 68
  in wound healing, 86–87
Fibrocartilage, 67
Fibromyalgia, 150
Fibrosis, pulmonary, 314
Fibrous connective tissue, 65–66
Fibrous joint, 115–16
Fibrous pericardium, 248
Fibula, 101, 114
Fibular artery, 264
Fibular head, 115
Fibularis, 140, 146, 148
Fight-or-flight response, 172–73, 224
Filtration, 49–50
Fimbriae, 382
Fingernails, 83, 84
Fingers, 101, 111, 112
Flagella, 40, 46, 47
Flexion, 117, 119
Flexor carpi, 144, 145, 146
Flexor digitorum, 144, 146
Flexor digitorum longus, 146, 148
Floating kidney, 352
Floating ribs, 108
Flu, 293
Fluid balance, 358–61
  lymphatic system in, 277
Fluid-mosaic model, of plasma membrane, 42
Fluoride, 345
Foam cell, 252
Focusing, visual, 189–90
Folic acid, 344
Follicle, ovarian, 380–81, 382
Follicle-stimulating hormone, 206, 208, 209, 378, 384, 385
Fontanels, 101, 412
Food pyramid, 342
Foot
  athlete's, 84
  bones of, 114
  muscles of, 146, 148
Foramen (foramina)
  obturator, 113
  sacral, 444
  vertebral, 106, 107
Foramen magnum, 102, 104, 162
Foramen ovale, 267, 269
Forced expiration, 306–7
Forearm
  bones of, 101, 111, 112
  muscles of, 144, 146
Forensic anthropologist, 121
Fossa ovalis, 269
Fourth ventricle, 164
Fovea capitis, 114
Fovea centralis, 188, 189
Fracture, 98
Free nerve endings, 80, 81, 183
Frontal bone, 101, 102, 103, 104
Frontalis, 140, 141
Frontal lobe, 166
Frontal (coronal) plane, 5
Frontal region, 4
Frontal sinus, 102, 302
Frostbite, 89
Fructose, 27
Functional electrical stimulation, 165
Fungal infection, 84

Galactose, 27
Galactosemia, 432